Planning Canada

A Case Study Approach

EDITED BY Ren Thomas

OXFORD
UNIVERSITY PRESS

Oxford University Press is a department of the University of Oxford. It furthers the University's objective of excellence in research, scholarship, and education by publishing worldwide. Oxford is a registered trade mark of Oxford University Press in the UK and in certain other countries.

Published in Canada by
Oxford University Press
8 Sampson Mews, Suite 204,
Don Mills, Ontario M3C 0H5 Canada

www.oupcanada.com

Library and Archives Canada Cataloguing in Publication
Planning Canada : a case study approach / edited by Ren Thomas.

Includes bibliographical references and index.
ISBN 978-0-19-900807-0 (bound)

1. City planning—Canada—Case studies. 2. Regional planning—Canada—Case studies.
3. Cities and towns—Canada—Case studies.
I. Thomas, Ren, editor

HT169.C3P63 2016 307.1'2160971 C2015-907278-6

Cover image: Richard Goerg/Getty Images

Oxford University Press is committed to our environment.
This book is printed on Forest Stewardship Council® certified paper and comes from responsible sources.

Printed and bound in the United States of America
1 2 3 4 — 19 18 17 16

Contents

Contributors vii

Preface xiii

PART A: FUNDAMENTALS OF PLANNING

1.0 An Introduction to Canadian Planning 2
Ren Thomas

PART B: CASE STUDIES

2.0 Community Development and Social Planning: Introduction **48**

2.1 Promising Practices in Social Plan Development and Implementation: Applying Lessons Learned from Canadian Urban Municipalities in the City of Richmond, British Columbia 51
John Foster, Olga Shcherbyna, and Leonora Angeles

2.2 Cultural Planning in Canada 64
Kari Huhtala

2.3 Cultivating Intercultural Understanding: Dialogues and Storytelling among First Nations, Urban Aboriginals, and Immigrants in Vancouver 75
Umbreen Ashraf, Kate Kittredge, and Magdalena Ugarte

2.4 Social Vulnerability to Climate Change: An Assessment for Yarmouth, Nova Scotia 84
Michaela Cochran

2.5 Reconceptualizing Social Planning: A Case Study of the Spatialized Impacts of Urban Poverty for Lone-Parent, Female-Headed Families in Vancouver 96
Silvia Vilches and Penny Gurstein

3.0 Urban Form and Public Health: Introduction 104

3.1 Design and Beyond: The Mobility and Accessibility of Community Gardens in the Region of Waterloo, Ontario 106
Luna Khirfan

3.2 A Comparison of Cross-Alberta and Cross-Canada Health Initiatives 116
Kyle Whitfield

3.3 Building up While Sprawling Out? Paradoxes of Urban Intensification in Ottawa 125
Donald Leffers

3.4 Planning for Diversity in a Suburban Retrofit Context: The Case of Ethnic Shopping Malls in the Toronto Area 134
Zhixi Cecilia Zhuang

4.0 Natural Resource Management: Introduction 143

4.1 Resource Development Proposals in Drybones Bay, Northwest Territories 145
Darha Phillpot and Todd Slack

4.2 Climate Adaptation Planning in British Columbia: The Elkford Approach 155
Timothy Shah

4.3 Government-to-Government Planning and the Recognition of Indigenous Rights and Title in the Central Coast Land and Resource Management Plan 168
Janice Barry

5.0 Housing: Introduction 176

5.1 Saskatchewan's Affordable Housing Challenge: Allocation of Public Funding in a Thriving Province 179
Jenna Mouck

5.2 We Call Regent Park Home: Tenant Perspectives on Redevelopment of Their Toronto Public Housing Community 187
Laura C. Johnson

5.3 Energy-Efficiency Retrofits and Planning Solutions for Sustainable Social Housing in Canada 196
Sasha Tsenkova

5.4 Meeting the Workforce Housing Needs of a Resort Municipality: The Whistler Example 208
Marla Zucht and Margaret Eberle

6.0 Participatory Processes: Introduction 217

6.1 People and Plans: Vancouver's CityPlan Process 220
Ann McAfee

6.2 Cultural Planning and Governance Innovation: The Case of Hamilton 231
Jeff Biggar

6.3 Taking It Online: How the City of Vancouver Became Comfortable with Engaging Residents in Their PJs: Vancouver's Use of Online Crowdsourcing to Engage Residents during the Development of the Greenest City Action Plan 241
Lisa Brideau and Amanda Mitchell

6.4 Reaching Youth: Tools for Participating in the Upgrading and Evaluation of Municipal Equipment and Services 251
Juan Torres and Natasha Blanchet-Cohen

6.5 I "Like" You, You Make My Heart Twitter, But . . . : Reflections for Urban Planners from an Early Assessment of Social Media Deployment by Canadian Local Governments 260
Pamela Robinson and Michael DeRuyter

7.0 Urban Design: Introduction 268

7.1 What's Public about Public Markets? Beyond Public and Private Space in Community-Building 271
Leslie Shieh

7.2 Planning as Placemaking? The Case of the East Bayfront Precinct Planning Process 282
James T. White

7.3 Developing the Community Plan and Urban Design Plan for Benalto, Alberta 290
Beverly Sandalack and Francisco Alaniz Uribe

7.4 Variations on Empire: Planning the US Embassy in Ottawa 304
Jason R. Burke

7.5 The Toronto Avenues and Mid-Rise Buildings Study 315
Cal Brook and Matt Reid

8.0 Urban Redevelopment: Introduction 325

8.1 Urban Regeneration in a Mid-sized City: A New Vision for Downtown Sudbury 328
Alison Bain and Ross Burnett

8.2 After "Ours": Creating a Sense of Ownership in Winnipeg's Downtown Plan 339
Gerald H. Couture

8.3 The Quartier des Spectacles, Montreal 348
Laurie Loison and Raphaël Fischler

8.4 Places to Grow: A Case Study in Regional Planning in a Rapidly Growing Urban Context 361
Jason Thorne

9.0 Transportation and Infrastructure: Introduction 370

9.1 The Death and Life of "Transit City": Searching for Sustainable Transportation in Toronto's Inner Suburbs 374
Anna Kramer and Christian Mettke

9.2 Declining Infrastructure and Its Opportunities: Gardiner East Environmental Assessment 384
Antonio Medeiros

9.3 Hamilton's Red Hill Valley Parkway: Fifty-Seven Years in the Making 396
Walter G. Peace

9.4 Travel Demand Management and GHG Emission Reductions: Meeting Multiple Objectives through Partnerships and Multi-Level Co-ordination 405
Ugo Lachapelle

Suggestions for Further Reading 415

Glossary 422

Index 433

Contributors

Leonora Angeles, PhD, is an associate professor at the School of Community and Regional Planning and the Women's and Gender Studies Undergraduate Program at the University of British Columbia.

Umbreen Ashraf is a PhD candidate at the Interdisciplinary Studies Graduate Program (ISGP) at the University of British Columbia. She has examined the pedagogical potential of the City of Vancouver's Dialogues Project since 2011 and participated in some of the project initiatives over the years.

Alison Bain, PhD, is an associate professor of Geography at York University and does research on cultural planning in Canadian cities and suburbs. *Creative Margins: Cultural Production in Canadian Suburbs* (2013) is her most recent book.

Janice Barry, PhD, is an assistant professor of City Planning at the University of Manitoba. Her ongoing research and teaching interests include Indigenous peoples' relationships to land-use planning, and the Central Coast Land and Resource Management Plan was the subject of her PhD dissertation. She also provided volunteer planning support to the Nanwakolas Council while conducting her doctoral research.

Jeff Biggar is a PhD candidate in planning in the Department of Geography at the University of Toronto. Jeff's research interests span a variety of topics: the relationships between planning policy, law, the built environment, urban design, the public realm, cultural and community planning. The big issue driving his research and practice is how to make more equitable and imaginative cities.

Natasha Blanchet-Cohen, PhD, is a professor in the Department of Applied Human Sciences at Concordia University. Her research centres on community youth development with a focus on rights-based approaches to programs and services, culture, and eco-citizenship.

Lisa Brideau was one of the city planners involved in designing the Greenest City public engagement process, including advocating for the use of an online forum. She has an MSc in planning from the University of British Columbia and is currently a sustainability specialist for the City of Vancouver, where one of her duties is to manage the Greenest City's ongoing social media outreach.

Cal Brook is an architect and planner and a principal of Brook McIlroy with more than 30 years of experience. Cal was the principal-in-charge and the primary author of the Toronto Avenues and Mid-Rise Buildings Project.

Jason R. Burke holds a PhD in planning from the University of Toronto and is currently a senior policy advisor at the Ministry of Municipal Affairs and Housing of the Government of Ontario. His work on the planning of the US Embassy in Ottawa complements his doctoral research on the history of security planning in Canadian cities.

Ross Burnett is a senior planner at Infrastructure Ontario, responsible for implementing development opportunities and maximizing provincial real estate assets. While working in the private sector as a planning associate, Ross was the project manager and lead planner for the Downtown Sudbury Master Plan.

Michaela Cochran is a planner in Edmonton. She has a masters of planning degree from Dalhousie University, where she conducted research to assess the social vulnerability of Yarmouth and Lunenburg through the Dalhousie Climate Change Working Group.

Gerald H. Couture is president of Urban Edge Consulting, a firm he established in 2007 following a 20-year career with the City of Winnipeg. Gerry was project manager for Winnipeg's CentrePlan, which was initiated in March 1993 and completed in January 1995.

Michael DeRuyter holds degrees in journalism and public administration from Ryerson University. He is a policy analyst for the Ministry of Transportation in Ontario.

Margaret Eberle, RPP, MCIP, is currently a housing planner with Metro Vancouver. She was previously a housing policy consultant working with a range of government and non-governmental organizations on affordable housing policy and planning initiatives in BC.

Raphaël Fischler, PhD, is director of the School of Urban Planning at McGill University. His research and teaching pertain to urban design and land development, land-use planning and regulation, and the history and theory of planning.

John Foster, RPP, MCIP, is manager of community social development in the City of Richmond Community Services Department.

Penny Gurstein, PhD, is a professor and director of the School of Community and Regional Planning at the University of British Columbia. Her current research focus is on affordable housing strategies. Previous research focused on the impact of government reduction of social services on single-parent families on income assistance, from which the Vancouver case study derived.

Kari Huhtala, MCIP, RPP, has more than 25 years' experience as a municipal planner with the cities of Vancouver and Richmond and the Corporation of Delta. Since 2006, he has been an arts and cultural consultant working with communities and the development industry in western Canada. Kari was the planning consultant involved in the development of the City of Kelowna "Thriving Engaging Inspiring" 2012–17 Cultural Plan.

Laura C. Johnson, PhD, RPP, is a professor at the School of Planning, University of Waterloo. Her teaching and research relate to housing, social planning, social research methods, and participatory planning. Since 2007, she and her students have been conducting a longitudinal qualitative research project tracking the first phase as residents of Toronto's Regent Park public housing go through the stages of displacement, relocation, and resettlement while the community is redeveloped.

Luna Khirfan is an associate professor at the School of Planning, University of Waterloo. She received her PhD in urban and regional planning from the University of Michigan in 2007. She investigates the cross-national transfer of planning knowledge, urban governance in Middle Eastern cities, and challenges that face historical cities as they adapt to meet the contemporary needs of their residents and to mitigate the pressures of tourism.

Kate Kittredge has a masters degree in Indigenous community planning from the School of Community and Regional Planning at the University of British Columbia and is working for a remote First Nation in northern British Columbia as the manager of their Health and Social Department. She has examined the pedagogical potential of the City of Vancouver's Dialogues Project since 2011 and participated in some of the project initiatives over the years.

Anna Kramer completed her PhD at the University of Waterloo with a dissertation on housing affordability in relation to public transit in North American cities. She currently works at Metrolinx in Toronto.

Ugo Lachapelle, PhD, is an assistant professor in the Urban Studies Department of the Université du Québec à Montréal and works on sustainable transport strategies, policies, and programs. He acts as a scientific advisor to the AQLPA and has completed a number of studies and contracts on vehicle-recycling and modal transfer with the organization.

Donald Leffers is a PhD candidate at York University in Toronto studying the power and politics of urban land-use planning and development. He was the primary researcher in the Ottawa case study.

Laurie Loison is a planner at EPA ORSA, France. She is a graduate of Science Po in Paris and of the School of Urban Planning at McGill University. Her research on the Quartier des spectacles earned her the Cornelia and Peter Oberlander Prize in Urban Design.

Ann McAfee, PhD, FCIP, RPP, was responsible for Vancouver's housing programs from 1974 to 1988 and co-director of planning from 1994 to 2006, including guiding the extensive public process leading to the adoption of CityPlan, Vancouver's first city-wide plan since 1930. She established City Choices Consulting, a firm specializing in strategic planning and public processes, in 2006. In 2007, she received the Kevin Lynch Award for Distinguished Planning Practice from the Massachusetts Institute of Technology.

Antonio Medeiros leads urban development, transportation, and infrastructure projects from inception to delivery with experience in Canada, the United States, and Asia. While at Waterfront Toronto he managed the Gardiner Expressway East Environmental Assessment through the spring of 2015. Antonio holds a master's degree in urban planning from Harvard University where he focused on real estate development, affordable housing delivery and public/private partnerships.

Christian Mettke is a PhD candidate in the area of spatial and infrastructure planning at the Technische Universität Darmstadt in Germany. His thesis compares the socio-technological trajectories of Toronto's and Frankfurt's public transit systems in the context of post-suburbanization. He spent a year at York University's City Institute in Toronto and interviewed many of the key decision-makers involved in public transit provision.

Amanda Mitchell is the public engagement specialist at the City of Vancouver, where she acts as a best practice advisor to staff throughout the City. During the Greenest City public engagement process, Amanda moderated the online forum, managed the social media accounts, and produced the larger public events. Amanda has a MASA from the University of British Columbia, is one of the founding directors of the Vancouver Public Space Network, and is a co-founder of *Re:place Magazine* (now *Spacing Vancouver*).

Jenna Mouck is the principal project director at SaskBuilds Corporation. She has previously worked in private-sector development as well as in other capacities within the Saskatchewan government, including with the Ministries of Environment and Health and the Housing Division's program and policy development branch. Her work in the Housing Division involved planning for housing at the provincial level as well as developing and administering the Summit Action Fund.

Walter G. Peace, PhD, recently retired from his position as associate professor in the School of Geography and Earth Sciences at McMaster University, where for 35 years he taught courses in human geography, the history of cartography, the regional geography of Canada, and urban planning. He has written numerous articles and book contributions about Hamilton's history; his edited book, *From Mountain to Lake: The Red Hill Creek Valley*, was published in 1998.

Darha Phillpot is the manager of the Land Use Planning Unit in the newly created Department of Lands with the Government of the Northwest Territories. She is a planner and a long-term Northerner. She worked for the Mackenzie Valley Environmental Impact Review Board on one of the environmental assessments in the Drybones Bay area.

Matt Reid is a planner, urban designer, and project manager at Brook McIlroy in Toronto. Matt was involved in the background review and site analysis for the Toronto Avenues and Mid-Rise Buildings Project.

Pamela Robinson, PhD, is an associate professor and graduate program director at Ryerson's School of Urban and Regional Planning. She began her planning career near Kingston, Ontario, working on an Official Plan review. This project sparked a career-long interest in exploring new ways for planners to engage the public.

Beverly Sandalack, PhD, FCSLA, MCIP, RPP, is associate dean (academic) in the Faculty of Environmental Design at the University of Calgary. She is research leader of the Urban Lab, which completed the Benalto Community Plan and Urban Design Plan.

Timothy Shah is a planner with Stantec Consulting in the area of environmental and socio-economic planning. In his masters degree in community and regional planning, he interviewed staff and other stakeholders involved in the District of Elkford's climate change adaptation planning process. The results of these interviews helped him to build a framework for local climate change adaptation planning, which he turned into his masters project.

Olga Shcherbyna has a MSc in planning from the University of British Columbia. She is a senior consultant at Diversity Clues Consulting Inc., a Vancouver-based company specializing in social policy development, project management, and community/ethnic outreach.

Leslie Shieh, PhD, is the co-founder of Take Root, a development firm specializing in adaptive re-use. She holds a doctorate in urban planning from the University of British Columbia and was the project planner in the redevelopment of the Westminster Quay public market.

Todd Slack is a long-serving regulatory specialist with the Yellowknives Dene First Nation, the sole staff member for a First Nation with a traditional territory larger than New Brunswick containing four diamond mines, two metal mines in final permitting, and associated exploration activities—all of which is occurring during a very large and very rapid decline in caribou populations, changing land-use patterns, and an absence of land-use planning. He has worked on three of the seven environmental assessments in Drybones Bay.

Ren Thomas, PhD, is a research and planning consultant in Toronto. Her recent work includes policy analysis for the Ontario Growth Secretariat, housing research for the Ministry of Municipal Affairs and Housing, research on transit-oriented development for the University of Amsterdam, and program evaluation for the British Columbia Non-Profit Housing Association and the Elizabeth Fry Society of Greater Vancouver.

Jason Thorne worked at the Ontario Growth Secretariat from 2004 to 2008, where he was one of the primary architects of the Growth Plan for the Greater Golden Horseshoe. He then worked as a principal with the Toronto-based urban planning and design firm planningAlliance and currently leads the Department of City Planning and Economic Development of the City of Hamilton.

Juan Torres, PhD, is a certified planner and professor in the École d'urbanisme et d'architecture de paysage (School of Urban Planning and Landscape Architecture) at l'Université de Montréal. He was part of the multidisciplinary team collaborating with Sainte-Julie's municipal staff in 2011, supported by Carrefour action municipal et famille.

Sasha Tsenkova, PhD, is a professor of planning and international development at the University of Calgary. Her research focuses on sustainable community planning and comparative housing policy. She carried out the first evaluation of the energy retrofit program in Canada presented in the chapter.

Magdalena Ugarte is a PhD candidate at the School of Community and Regional Planning (SCARP) at the University of British Columbia. She has examined the pedagogical potential of the City of Vancouver's Dialogues Project since 2011 and participated in some of the project initiatives over the years.

Francisco Alaniz Uribe is a PhD candidate and research associate in the in the Faculty of Environmental Design at the University of Calgary. He co-manages The Urban Lab.

Silvia Vilches, PhD, is a postdoctoral researcher at the University of Victoria. Her PhD research, involving a three-year research process with female lone parents in Vancouver, formed the basis of her case study.

James T. White, PhD, is a lecturer in urban design at the University of Glasgow and specializes in the tools that planners can use to make more design-sensitive decisions. He examined the case of the East Bayfront precinct planning process on Toronto's waterfront.

Kyle Whitfield, PhD, MCIP, RPP, is an associate professor in the Faculty of Extension at the University of Alberta. She teaches courses and conducts research related to community planning and health service planning.

Zhixi Cecilia Zhuang, PhD, is an assistant professor at the School of Urban and Regional Planning at Ryerson University. She has been conducting research in the Greater Toronto Area on suburban ethnic retailing and its implications for municipal planning.

Marla Zucht is the general manager of the Whistler Housing Authority, a subsidiary of the Resort Municipality of Whistler. She holds a masters degree in environmental studies and community planning from York University and has long-standing interests as a planner in housing policies, growth management strategies, and sustainable planning.

Preface

Community and regional planning is a discipline rooted in concerns about cities. From their physical forms and environmental conditions to their economic development and socio-cultural networks, there are few aspects of living in human settlements that planners have not attempted to improve. Policies and programs designed to improve quality of life, regulate growth, and minimize the negative consequences of development are commonplace in our municipalities and provinces. Yet many Canadians do not know how their neighbourhood parks are designed, how their transportation networks are funded, or which responsibilities rest with their local government. This book introduces the reader to the discipline of community and regional planning through real cases, which illustrate the multidisciplinary and participatory approach that planners use in the development of policies, plans, and programs.

Case studies, with their in-depth exploration of planning decisions, stakeholder relationships, and political realities, are often used to help develop policy solutions to the complex problems planners face in their own jurisdictions. They have been used to understand how plans were developed, the challenges encountered, and different implementation outcomes from one municipality to another—often telling a compelling story in the process. Case study as a research approach is also well used in planning, because it allows incorporation of multiple research methods to develop a deep understanding of a topic. Completed case studies can be also used in meta-analysis and cross-case analysis to develop a broader understanding of a key issue.

Because case studies are so established in the planning discipline, *Planning Canada: A Case Study Approach* uses them to introduce readers to the diversity of community and regional planning research and practice across Canada. I hope that the book reaches planners in both practice and research and that it will be used to introduce students to the fascinating discipline of planning. The cases are compelling enough to interest readers from any field in the complex issues facing our cities and regions, including climate change, equity, urban sprawl, and redevelopment.

A Case Study Approach for Planning Students

One application of *Planning Canada: A Case Study Approach* would be as a textbook in introductory planning courses. Canadian students in urban planning are required to take an introduction to planning history and theory course as part of their degree programs. Students in urban studies, geography, real estate development, landscape architecture, and sociology often take an introductory course in planning. The reading lists for these courses are usually a mix of book chapters, online articles, and government publications—rarely do instructors use a single textbook. Reading lists typically draw upon American authors, but major differences in planning law, history, and governance mean that these theories and analyses can have limited applications for those intending to pursue planning careers in Canada. My intent was to present cases developed within students' own political, cultural, and governance frameworks to allow them to easily see how the plans or policies could be implemented in their own cities and regions.

The format of the book, divided into nine sections, is intended to introduce students to the planning "sub-disciplines" by presenting three to five cases of policy development, plan implementation, or in-depth research in each area. Each of the cases works as a standalone reading, which allows them to be used in elective courses such as transportation planning or community development: for example, Jason Thorne's chapter on the Province of Ontario Growth Plan for the Greater Golden Horseshoe could be integrated into in a course on regional planning. Many of the chapters cross sub-disciplines, illustrating the inter-disciplinarity of planning, so they can be used in multiple ways. Gerry Couture's case study of Winnipeg's CentrePlan addresses public participation, urban regeneration, and community development; Tim Shah's case of climate change adaptation in Elkford, BC, could be used in courses on ecological planning, planning methods, or disaster and risk management. The length of the cases, a maximum of 4500 words each, allows an individual case to be combined with two or three in-depth articles on a key topic in an elective course.

The authors have each illustrated the ways in which planners were involved in the development of plans and policies, how planners dealt with challenges, and their reflections on the outcomes of the planning processes. In this way, students learn how typical challenges encountered in practice could be addressed. Planning theories highlighted within each case study help students to understand the intersection of theory and practice. The list of further readings at the end of the book will be useful in developing these specialized topic reading lists.

A Text for Planning Researchers

Another way to use the case studies would be in planning research. The authors have outlined the methods they used in the development of each case study, including interviews, focus groups, design charrettes, GIS, meta-analysis, policy analysis, and participant observation. Some, such as Laura Johnson in her case on Regent Park, go into considerable detail on their research methods and could be used in a masters research methods class or PhD colloquium on research design. Advanced discussion of the case study method can be found in the introduction and in Kyle Whitfield's case on health initiatives in Alberta, which illustrates the use of case studies in cross-case comparison to aid in analytic generalization. The case studies could also be used as qualitative data: the four cases on housing presented in Section 5 could be used in a policy analysis of the development of affordable housing in Canada; the five cases on participatory processes could be used in a meta-analysis of public participation in planning. Many of the cases discuss decision-making and public participation processes, which are often of interest to planning researchers.

Presenting Compelling Stories and Developing a Shared Vision

Although planners have their own set of skills and expertise, many decisions on planning policies, programs, and plans rest with elected officials, members of the public, developers, non-profit organizations, Aboriginal governments, and other stakeholders. It is good practice to exchange knowledge of planning goals, aspirations, and values with those involved in planning processes. Discussions of increased density in communities, implementation of cycling lanes, or redevelopment of former industrial spaces can then occur with common goals and understandings. Some municipalities, regional planning authorities, and non-profit organizations have been instrumental in distributing information on planning processes, developing new ways for getting involved in shaping the city, and developing online participation tools to decrease distance in rural areas. Increased knowledge of planning issues contributes to an understanding of the "common good" or the development of a future vision for a town or region. Those who participate in a plan's development share its ownership and assist in its implementation.

With this goal in mind, I hope to introduce readers from all age groups, all disciplines, and all regions to planners' unique skills and expertise, the challenges that they face in implementation, and the fundamental governance frameworks that shape planning decisions in Canada. These critical components of theory and practice ground the case studies in the daily realities of decision-making in municipalities, regions, and provinces.

Acknowledgments

This book first developed through a chance meeting between myself, Caroline Starr, acquisitions editor at Oxford University Press Canada, and Silvia Vilches, postdoctoral researcher at the University of Victoria. It was astonishing to see how quickly the idea of "a case study book in Canadian planning" attracted support from academic and practising planners across the country. Because most of the work took place during my two-year research posting at the University of Amsterdam, the authors became somewhat of a virtual community in the process of developing the book.

Special thanks must go not only to Caroline at OUP Canada but also to Jodi Lewchuk, Peter Chambers, and Dorothy Turnbull, who took the book through its developmental stages to publication. Gordon Price and Shawn Micallef spread the word on the call for papers and were undoubtedly the main reasons that I received so many high-quality case study proposals. Thanks to Penny Gurstein, Luca Bertolini, and Leonie Janssen-Jansen, models of excellent researchers and teachers. To Nick Doniere, everlasting gratitude for all your support. Finally, this book is dedicated to the staff, faculty, students, and alumni of the University of British Columbia School of Community and Regional Planning—without you, there would be no *Planning Canada*.

To all of those who have supported the authors in the development of their cases, thank you for helping us to develop a volume that we feel will introduce students, practitioners, urban enthusiasts, community activists, and many others to community and regional planning policies, plans, and programs in Canada.

Ren Thomas

PART A

Fundamentals of Planning

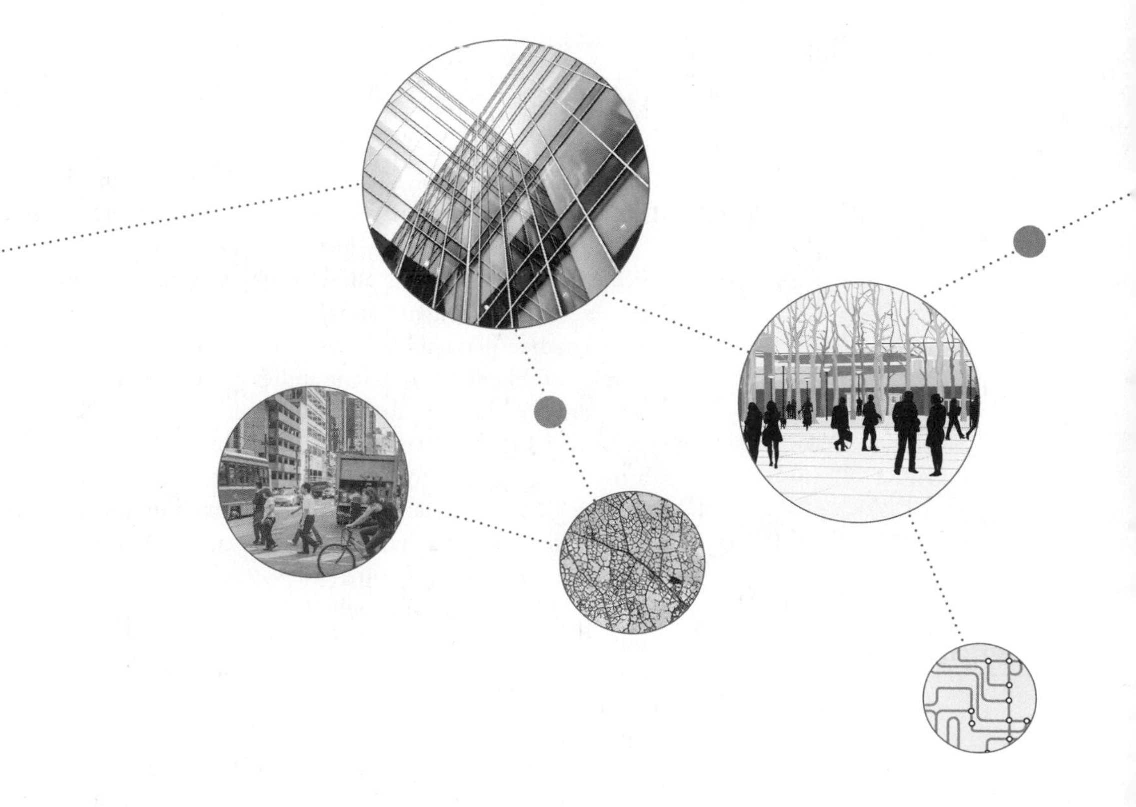

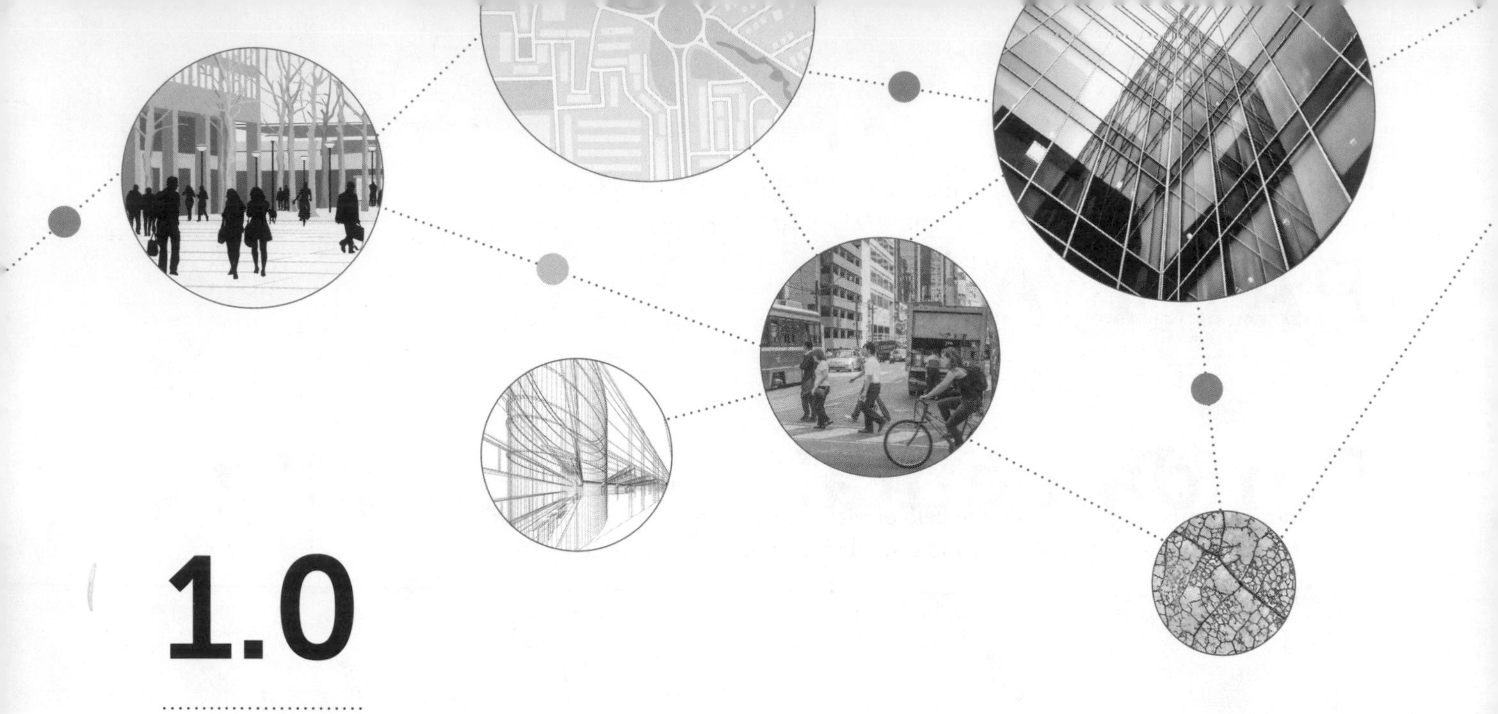

1.0

AN INTRODUCTION TO Canadian Planning

Why Plan?

Community and regional planning is practised in human settlements around the world. In Canada, with our sparsely populated, vast geography and a governance structure rooted in our rural history, planning has its own challenges. These challenges have inspired some fascinating and inspiring solutions, shaping the country's built environments and protecting its natural heritage.

Canadian planners have been working to improve the physical, social, and economic conditions in our communities since our earliest settlements were established. The roots of planning theory and practice in Canada can be traced back to human settlements in the seventeenth century, but the formal discipline originated less than a century ago. Since 1919, when the Town Planning Institute of Canada was founded, planning has integrated professionals trained in different disciplines such as architecture, engineering, public health, and social work. Even after the establishment of formal planning programs in the postwar period, many enter the discipline through other fields of study. The planner might be considered "a jack of all trades but master of none," but in fact planners possess a set of skills that places them at the centre of complex urban problem-solving and enables them to make professional judgments on matters of the public interest.

Most planners enter the field with the aim of improving towns, cities, and regions. In this sense, planning can be considered a normative discipline. Planners believe that they can to help build better, more compact, more sustainable, and even more equitable urban settlements and that they have been given the tools to do so. They use this knowledge to develop plans, policies, and programs. While "trying to change the world" may seem a lofty goal, planners work at different levels, from engaging community members in the design of a neighbourhood park to drafting provincial policies on climate change. They can be involved in everything from the physical design of streets to the development of partnerships between non-profit, private, and public organizations. However, planning is not a scientific discipline with strict principles that must be applied to cities and regions: there is no "perfect" city. Although there have been times when planning could be characterized as positivist, today the role of the planner can vary from facilitator of community dialogue to policy developer, from political advisor to advocate for under-represented groups. Within the field of planning, the roles of orator, analyst, designer, and project manager all exist, and a single individual could play all of these roles over time.

Planners do not have carte blanche to build cities; they work within complex governance frameworks including municipal, regional, provincial, federal, and Aboriginal governments. While planning can be said to be a local matter, Canada's legislative structure means that it is also a provincial responsibility. Planning issues can have implications far beyond the municipality (Cullingworth 1987). In addition to governments, community associations, non-profit organizations, business interest groups, advisory boards, school boards, and individuals assist in the development and implementation of plans, programs, and policies that affect our daily lives. Each of these players has interests and goals related to the social and spatial fabric of the city that may conflict with each other. Planners can work for, or with, any of these players. So as many planning theorists have noted, aspects of power and authority influence the actions of planners, from the political party whose members are local councillors to the business owner who sits on the regional development board. Many different types of information, such as statistical data, the opinions of residents, housing forecasts, and environmental concerns, are used in decision-making.

Sometimes planners fail. Sometimes higher aspirations for their policy, plan, program, or **demonstration (pilot) project** remain unrealized. Monitoring of the program over time might show that it did not meet its intended goals: the demolition of low-income urban neighbourhoods in the 1950s to build new public housing projects is a classic example. Like other departments within municipalities, planning departments can be severely constrained in their budgets, particularly in rural areas, in emerging policy areas such as food policy, or during economic downturns. This can seriously hamper their efforts. Powerful interests may collaborate to prevent key planning legislation from passing or hamper implementation efforts, even at the neighbourhood level.

Planning does not exist in spite of these challenges—it exists because of them. These conflicts between the state and the individual, between the public good and private concerns, are as old as civilization itself. It was because of rapidly changing cities, their air pollution, the effects of unsafe and unsanitary housing on their working classes, and their social inequalities that the discipline of planning came into being. Planners must continually face new challenges, develop new solutions, and work in new ways to "change the world." The aim of this volume is to highlight the ordinary and extraordinary ways in which Canadian planners have helped communities across the country achieve greater **resilience**, create robust strategies, and develop a sense of ownership in their own futures.

Why Use Case Studies?

The idea for a book of case studies in Canadian planning originated at the University of British Columbia, where many of the PhD candidates are involved in in-depth studies of particular policies, programs, or relationships between key actors in a planning process. Case studies have developed a niche for themselves in the discipline, whether in academia or in practice. Planning is, after all, a complex process involving residents, municipal and regional planners, Aboriginal communities, business associations, non-profit organizations, and environmental groups, to name but a few. Delving into the intricacies of these players and their roles in developing a program or helping to shape a vision for the future is infinitely interesting to those who seek to understand and improve upon the profession.

For practising planners, sharing their progress on a new **bylaw** or regional governance model occurs in day-to-day practice; exchanging success stories and challenges is common in any profession. Canadian planners share their work more formally through meetings with local Canadian Institute of Planners (CIP) affiliates and publications in the professional journal *Plan Canada* or informally in local newspapers, online publications, and websites. Exchanges also take place between planning departments that visit each other to learn, for example, about the latest in food security policies or sustainable transportation initiatives. Implicit in this type of exchange is the desire for policy transfer, a process in which policy ideas, administrative arrangements, institutions, and ideas can be transferred from one context to another (Dolowitz and Marsh 2000). Although there is a hunger for "best practices," planners are also interested in the challenges to developing an **Official Community Plan (OCP)** or the barriers to the adoption of a new policy. For example, planners might be interested in how a municipality engaged **immigrants** or **youth** in a planning process and which particular techniques worked better than others.

In academic planning, case study is a well-established research methodology. A number of authors have gone into considerable depth on the methods, research design, and typology of case study research, e.g., single-case, multiple-case, and cross-case research. Robert K. Yin, in his influential volumes *Case Study Research: Design and Methods* (1984) and *Applications of Case Study Research* (1991), outlined in detail the approaches researchers could take, the variety of methods that could be used to gather qualitative and quantitative data within a case study framework, and the ability to generalize from case study results. Central to the planning discipline is Yin's argument that single-case studies can be used in analytic generalization in which a theory can be tested in another similar setting to further define its explanatory power. This explains the attraction of books such as John Punter's *The Vancouver Achievement* (2004), whose detailed interviews and document analysis tell the story of how some of the city's urban design and planning achievements, such as discretionary control of major developments and **Development Cost Levies**, came into existence. In *Making Social Science Matter: Why Social Inquiry Fails and How It Can Succeed Again* (2001), planning theorist Bent Flyvbjerg writes that the richness of case studies, including interviews with practising planners and politicians, analysis of values and power, and integration of data from different sources, is essential in the development of theory in the social sciences. Flyvbjerg argues that the social sciences are not, and never have been, cumulative and predictive in nature, making the aim of social science research quite different from that of the natural sciences. Cross-case analysis, which compares a number of similar cases using a variety of techniques (Miles and Huberman 1994), can advance the development of theory in planning by enhancing understanding and explanation of the specific conditions under which a finding can occur. Case comparison can also help practising planners understand similarities and differences between policies, actors, or governance models that might be useful in policy implementation (Thomas and Bertolini 2014).

There are a number of reasons that a volume that teaches planning through a case study approach is relevant today. First, cities face new challenges such as climate change, rising income inequality, and an unstable economy; public servants and policy-makers may be dissatisfied with the ability of their current policies to respond to these problems (Marsden and Stead 2011). Conditions of uncertainty, such as the absence of a scientific consensus, lack of information, new problems, policy disasters, crisis, and political conflict, may contribute to the need for new solutions; uncertainty acts as a powerful force for imitation in policy transfer (DiMaggio and Powell 1983). Second, we live in an era of heightened competition among cities, regions, and countries. Every city wants to attract new residents to its territory, to become a magnet for creative professionals (Florida 2002), and to sustain its economy. When a city is successful, whether socially, economically, or environmentally, other cities notice and want to achieve the same outcomes. Finally, in today's digital world, policy-makers, politicians, and public servants have more opportunities to network with each other and share policy ideas (Marsden and Stead 2011). We, the editor and authors involved in the creation of this book, hope that the cases serve as inspiration and learning (Spaans and Louw 2009) for students studying planning and as lessons for planners across the country, who may be able to adapt these policies to fit their own local contexts, developing hybrid policies or ideas in the process (Stone 2004).

The book is arranged into nine sections:

1. An Introduction to Canadian Planning
2. Community Development and Social Planning
3. Urban Form and Public Health
4. Natural Resource Management
5. Housing
6. Participatory Processes
7. Urban Design
8. Urban Regeneration
9. Transportation and Infrastructure

The eight sections following this introduction reflect prominent "sub-disciplines" of planning. In this way, students of planning programs can learn about the different activities that planning entails. For those new to planning, this breakdown of subject areas provides a brief introduction to the main concepts, theories, and authors in each area, grounding these with practical examples rooted in Canadian governance frameworks. But as will become clear from reading the chapters, many of them cross multiple subject areas, which illustrates the complexity and multidisciplinarity of the profession. So don't limit yourself to the examination of cases in one or two areas; it is very likely that cases in other sections will offer insights into your area(s) of interest. For example, Juan Torres and Natasha Blanchet-Cohen discuss two cases of youth participation in planning processes in Sainte-Julie, Quebec, which may also interest readers with sustainable transportation and urban design backgrounds. Michaela Cochran's examination of the social vulnerability to climate change in Yarmouth, Nova Scotia, would appeal to readers interested in natural resource management or urban form and public health. The eight sections roughly represent the range of planning sub-disciplines in which a planning student or practising planner could specialize. But many individuals change their area(s) of focus over time or work in several areas simultaneously. A planner working at a small municipality is involved in a variety of projects, from parks planning to development bylaws, and a planner in private practice may specialize in transportation and infrastructure projects ranging from a new LRT line to district energy solutions. New areas of practice, such as food policy or climate change adaptation, contribute to the evolution of the sub-disciplines.

A Brief History of Planning in Canada

How did planning come into existence in Canada? How has the discipline evolved over time? This brief introduction to the history of planning is not intended to familiarize readers with all the details, names, and dates necessary for a thorough understanding of the topic. Indeed, other volumes, such as Jill Grant's *Reader in Canadian Planning: Linking Theory and Practice* (2008) or *Shaping the Urban Landscape: Aspects of the Canadian City-Building Process* (1982) by Alan Artibise and Gilbert Stelter, would be more suitable for that purpose. Since most students of planning will take an entire course on the history and theory of planning, this section merely introduces the main stages in the development of the profession and key innovations in Canadian planning for those who are unfamiliar with the field. Canadian planning practice has been largely shaped by developments in the United Kingdom and the United States, but Canadian governance arrangements, geography, and demographics have contributed to a unique planning culture (Hodge 1985).

While the "official" history of the planning profession begins with the establishment of the Town Planning Institute of Canada in 1919, the broader notion of planning dates back much further. A rich history of Aboriginal (**First Nations**, Inuit, and Métis) settlements in strategic locations across the country existed for thousands of years, but European colonization wiped out much of the evidence of Aboriginal settlement patterns (Wolfe 2004). Colonial powers or powerful corporations, such as the Hudson's Bay Company, were instrumental in developing all but a few of the places that later became cities (Hodge 1985).

As in other countries, many towns in Canada were laid out with consideration of the transportation networks used for shipping, physical structures used for military defence, and a consistent pattern for roads and streets. The earliest planned communities were Quebec City (1608), Trois-Rivières (1634), and Montreal (1642), all of which had the street layout and character of medieval towns. British settlements were typically laid out relative to a survey baseline along the harbour (Hodge 1985), such as Charlottetown (1768), Saint John (1783), and Lunenberg (1753). In the Prairies, settlements at this time were mostly scattered posts established by the Hudson's Bay Company or missions, such as Portage la Prairie, Prince Albert Mission, and Fort Albert. In 1870, Winnipeg had only 100 inhabitants and a few frame structures, hotels, and retail buildings (Artibise 1981). Montreal became Canada's major industrial city, growing rapidly from 1840 onwards. In 1880, the Canadian Pacific Railway established its head office there, and much of Western Canada's grain was shipped from its port (Artibise 1981).

Confederation and Industrialization: 1867–1900

Two catalysts were instrumental in the establishment of non-Aboriginal settlements across Canada: Confederation in 1867 and the construction of the railway network, completed in 1885. The first railway reached Winnipeg in 1878; between 1871 and 1901, the population of the Prairies increased from 70,000 to more than 400,000, including three incorporated cities, 25 towns, and 57 villages (Artibise 1981). New settlements established by the Canadian Pacific Railway were typically gridiron in layout regardless of the site's topography, such as Brandon, Manitoba, and New Westminster, British Columbia. Even when the initial townsites did not coincide with the area railway terminus, as in Saskatoon, Regina, Calgary, and Edmonton, the station areas became the dominant commercial centres. Dramatic growth occurred along the main CPR line where Moose Jaw, Swift Current, Medicine Hat, and Calgary were all located (Artibise 1981). Settlements that had existed before the railway, such as Edmonton, Prince Albert, and Saskatoon, struggled during this time. Toronto, as the nexus of Ontario's railway network, became the centre of regional trade, transportation, and finance.

Until the late 1800s, Canada was a rural nation. Urbanization at a rapid rate had the same social and environmental effects in this country that it had in other industrialized countries of the era. Air pollution, water contamination, and contagious diseases became common problems as large numbers of people were drawn to urban centres for the first time in the nation's history. The economy transitioned from agriculture to the manufacturing and processing of raw materials. As urban centres developed rapidly,

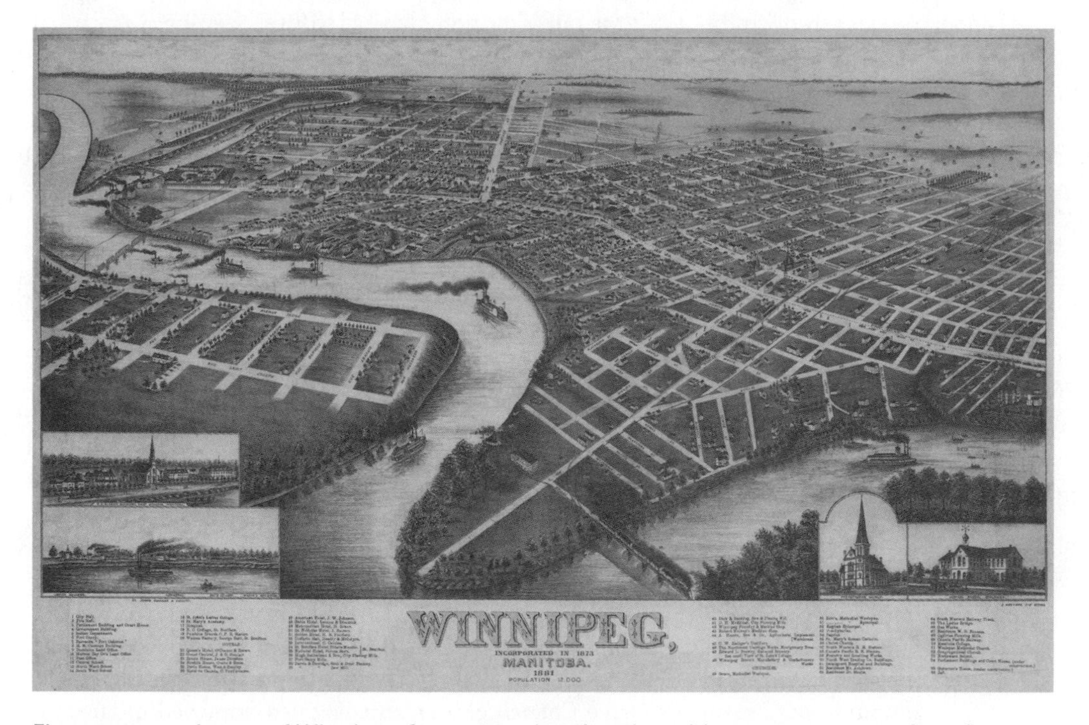

Figure 1.0.1 • A map of Winnipeg from 1881 showing the grid street pattern against the river. Its population at this time was 12,000.

Source: City of Winnipeg.

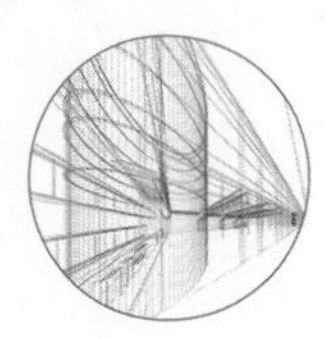

PLANNING THEORY

Governance and Canadian Municipalities

Canada's governance structure places the balance of responsibility on the federal and provincial governments, with less for municipalities. With the signing of the British North America Act, 1867, now known as the Constitution Act, 1867, the responsibilities of the federal and provincial governments were defined and remain today:

- Areas of federal concern include: defence, foreign policy, interprovincial trade and communication, currency, navigation, criminal law, citizenship, agriculture (shared), immigration (shared).
- Areas of provincial concern include: municipal government, education, health, natural resource development, environmental management, property and civil rights, land use, highways, agriculture (shared), immigration (shared).
- Under Section 92 of the Act, the responsibility for creating municipal institutions lies with the provinces. Municipalities are thus granted responsibilities by the provinces, including social and community health, recycling programs, transportation and utilities, snow removal, policing, firefighting, and emergency services. In some large metropolitan areas, provincial governments have created and granted responsibility to regional governments or organizations to co-ordinate region-wide policies, plans, or services. Section 92 reflects the predominantly rural character of the country at the time of Confederation. It is also important to note the absence of Aboriginal governments, which were not recognized at this time.

Federalism allows different provinces to adopt policies tailored to their own populations and gives provinces the flexibility to experiment with new ideas and policies. Some areas of planning, such as health, are constitutionally concerns of the provincial government. However, public health effects were seen in Canada's burgeoning cities as the population began to concentrate, forcing local players to take action.

Because the provinces have the responsibility for property rights and land use, they have each developed Planning Acts that describe the activities and procedures that municipalities must follow, such as the development and regular review of **Official Community Plans**, the rules for public consultation, and the co-ordination of municipal planning with provincial policies. They describe how land uses are regulated and the tools that can be used. Municipalities regulate land use within their jurisdiction by preparing Official Community Plans and **zoning bylaws**, ensuring that they conform to the Planning Acts and any other relevant provincial legislation.

middle- and upper-class reformers became concerned about the living conditions of the working class, which they saw as related to increases in urban poverty, crime, and child neglect. J.S. Woodsworth, a social worker, Methodist minister, and politician, was one such reformer who worked with immigrant "slum" dwellers in Winnipeg (Perks and Simmins 2013); members of the Council of Women advocated improved urban conditions for children in Canadian cities (Wolfe 2004). The implications of the physical conditions of housing on the family structure, in particular, were discussed in newspapers and urban affairs journals (Purdy 1997). While reformers were concerned with preventing the contagious diseases (typhus, cholera, and tuberculosis) that seemed to spread within slum areas, they also sought to eliminate what they considered the moral failings of slum dwellers. Spurred by declining birth rates, industrialization, and women's increasing participation in the workforce at the turn of the nineteenth century, reformers focused on threats to the family structure.

Geographers, surveyors, and economists were instrumental in documenting many of the conditions in early Canadian cities (Wolfe 2004), including Sir Herbert Brown Ames in his *City below the Hill* (1897), a statistical examination of poverty, overcrowding, and unhealthy living conditions in Montreal's west end. Once it was understood that waterborne microbes could spread diseases in a community's water supply, efforts to secure sources of pure water and provide piped water to all buildings began. Ottawa initiated its public water system in 1872 and Winnipeg in 1882, but it was decades before either had a pure source of water (Hodge 1985). Clean water supplies were among the first services provided to urban residents.

There was a growing emphasis on the health aspects of public parks, ideas that were shared among American, British, and Canadian planners beginning in the 1880s (Hodge 1985). The Parks and Playgrounds movement, spurred by branches of the Council of Women, lobbied for open space, playgrounds, and child welfare in Canada's largest cities (Wolfe 2004). After his pioneering designs for Central Park in New York (1857) and Prospect Park in Brooklyn (1867), Frederick Law Olmsted brought his social concerns and democratic ideals to Canada. Olmsted oversaw the design of Mount Royal in Montreal (1876). Calvert Vaux, Olmsted's co-designer on Central Park, designed Rockwood Park in Saint John. The Halifax Public Garden, Assiniboine Park in Winnipeg, and the linked parks along Stratford's Avon River were also designed at this time. Canada's first National Park, Banff, was designated in 1885.

Closely linked to the desire for open spaces was innovation in transportation: rail networks developed rapidly to allow working people greater access to clean air and natural surroundings. Electric streetcar lines were established in many Canadian cities: for the first time, working-class people could live outside of the dense industrial inner-city areas where they worked. These streetcar suburbs were built through railway entrepreneurs who sold plots of land adjacent to their streetcar lines, contributing to linear grids of streets and roads within walking distance of the streetcar stops. Shaughnessy, in Vancouver, is one such example (see Figure 1.0.2). Private interests were also responsible for other services at this time, such as waterworks, electric power, and telephone systems (Hodge 1985).

At this time, then, planning as a profession did not exist. Concerned individuals and groups, private developers, and railway companies were responsible for the layout of streets and other urban infrastructure and for raising issues of social concern (e.g., overcrowded housing). Delivery of urban services such as fresh water and electric power was often through private companies, but the design of parks and open spaces usually involved public funds and lands.

The First Stage of the Profession: 1900–30

Rapid urban growth continued in the first decades of the twentieth century. European immigrants settled in the Prairies, drawn to cheap agricultural land and the rapid expansion of the railways. The Northwest Territories established its government in 1897, the provinces of Alberta and Saskatchewan were created in 1905, and Manitoba's boundaries were extended in 1912. Between 1901 and 1916, 14 prairie cities, 125 towns, and 366 villages were incorporated (Artibise 1981). Winnipeg was a large manufacturing centre, while Calgary, Edmonton, and Saskatoon also developed political, legal, and

Figure 1.0.2 • The First Shaughnessy neighbourhood in Vancouver, built by the Canadian Pacific Railway in 1907. At the time, CPR was the largest property developer in Canada.

Source: City of Vancouver.

educational institutions. Other cities across the country expanded rapidly, but in 1901 less than 38 per cent of Canadians lived in urban areas (Harris 2000).

The first fledgling steps toward a planning profession occurred during this time. The Commission of Conservation, established in 1909 to advise Canadian governments on the conservation of human and natural resources, directed research in seven fields: agricultural land, water and water power, fisheries, game and fur-bearing animals, forests, minerals, and public health. The commission published hundreds of books, reports, and scientific papers (Smith 2013). Among its key achievements were the organization of an international planning conference in Toronto in 1914 and the appointment of Thomas Adams, a prominent British planner of the era, as planning advisor to the commission. In the years before World War I, Adams disseminated his ideals of efficiency and functionality to the fledgling members of the profession and used them to draft provincial planning legislation and advise governments on planning programs. Three utilitarian principles embraced by Adams included the notion of social progress, the emphasis on reason to determine the solution to social problems, and the acceptance of government intervention to achieve the public good, if the weight of objective evidence suggests that course of action (Hodge 1985). This emphasis on order and efficiency became embedded in the definition of town planning and was included in the masthead of the *Journal of Town Planning Institute of Canada* in the 1920s:

> Town planning may be defined as the scientific and orderly disposition of land and buildings in use and development with a view to obviating congestion and securing economic and social efficiency, health, and well-being in urban and rural communities.

The presence of private interests in planning (civic improvement associations, boards of trade, and professional associations) clearly served the interests of property-owners: they advocated visual order, efficient transportation, and municipal reform to protect property values and promote land development (Hodge 1985). These groups assumed a right to membership on planning boards, municipal councils, and other local boards and commissions:

> that such boards and commissions could be more effective by using the skills of people from the business community is an assumption that has had enduring effects on local government and its planning activities (Hodge 1985, 18).

Urban development slowed during an economic lull preceding World War I, and population growth in cities slowed during the wartime years. Growth rebounded in the 1920s: Toronto, in particular, boomed in the 1920s as it became the financial and head-office centre for Ontario (Artibise 1981). New resource-based towns began to be laid out according to prevailing theories of the era, including the Garden City ideals, separation of land uses, and masterplanning. Témiskaming, Quebec (1917), Kapuskasing, Ontario (1922), and Corner Brook, Newfoundland (1923), are all examples of towns solely based on the existence of pulp and paper companies, whose main concern was worker housing (Perks and Simmins 2013). Témiskaming was designed by Thomas Adams and showed the influence of Garden City principles (Hodge 1985).

For several decades, the Town Planning Institute comprised a few hundred members requiring no professional training: engineers, surveyors, landscape architects, and architects were welcome to become members (CIP 2014) if they were members of their own professional associations (Wolfe 2004). Many worked across what are now seen as firm professional boundaries (Hess 2009): engineers wrote texts on city planning, and planners contributed to early developments in traffic engineering. In these early years, there was considerable consensus about the common good (Grant 2005).

During World War I, planning activities in Canada were curtailed, but there were important legislative achievements during this time period. Influenced by the first British Planning Act (1909), all but one province in Canada passed provincial Planning Acts by 1925, in part to combat the chaotic **subdivision** of land in suburban areas as private companies promoted speculative growth (Hodge 1985), e.g., urban railway/streetcar companies. However, the legislation did not require provincial governments take an active part in planning or municipalities to prepare plans. This fact can be traced to the distribution of power and responsibility in Canada and its rural character at the time of Confederation. Planning at first was limited to suburban or fringe areas to be developed; built-up lands were excluded (Hodge 1985). All planning was subject to close scrutiny by central government authorities. One innovation in Alberta's 1913 Act allowed municipalities to acquire up to 5 per cent of any new suburban development's area for park purposes at no cost to itself (Hodge 1985). Both zoning and subdivision control reflected the role of the provincial government in Canadian cities at the time. In some provinces, the subdivision control process was administered entirely by the provincial ministry, and zoning laws had to be approved by the province (Hodge 1985). The federal government, on the other hand, provided funding and/or advice, such as that offered by the Commission of Conservation until it was discontinued in 1921. The City of Kitchener adopted Canada's first zoning bylaw in 1924, and zoning became formally recognized in planning law in 1925.

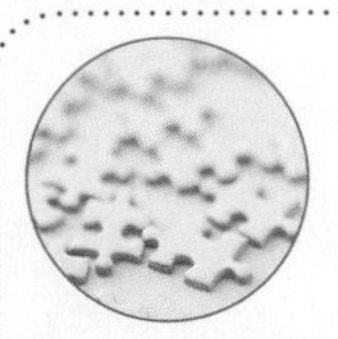

KEY ISSUE

Property Rights and the Public Interest

From the very beginning, planning law in Canada limited individual property-owners' rights in order to secure benefits for the whole community, a key difference from its counterpart in the United States, which prioritizes individual rights. English common law is the foundation on which land-use issues in Canada are based, particularly those dealing with property rights and the "law of nuisance" (Hodge 1985, 17); here, private landowners are considered to be "very privileged tenants of the Sovereign." That is, the nation can demand reasonable land use on the part of its "tenants." So although zoning developed along the same general lines in Canada and the US, the Canadian Constitution does not spell out personal property rights and the limits of state intervention; under the Constitution Act (1867) municipalities were deemed to have the statutory power to regulate land use. This gives **zoning bylaws** in Canada more scope, as do municipal bylaws in Britain; the only issue is whether the bylaw is discriminatory in pursuing the **public interest** (Hodge 1985). In Canada, restrictive **covenants** predating zoning bylaws protected many residential areas from noxious uses such as tanneries but also barred settlement of various ethnic groups (Wolfe 2004).

About half of all urban households in Canada owned their housing by 1921, with lower levels in the Atlantic provinces and Quebec (Harris 2000). Montreal, in particular, was dominated by triplexes and apartment buildings and had a higher proportion of collective accommodation, especially for women. Many families took in boarders to help finance homeownership or turned to private lenders rather than to banks. Economic forces were also changing: by the 1920s Vancouver had replaced Winnipeg as the leading city for trade in western Canada (Artibise 1988).

With these early steps toward formalizing the profession of planning, regulations on land use and development began to shape the rural and urban settlements that we recognize today. Although formal education or training was not required for membership in the planning association, members shared some common visions for orderly, well-designed cities.

Planning during the Great Depression and the Second World War: 1930–44

During the Great Depression, regional differences across the country emerged. Accordingly, the federal government made major efforts to assist the hardest-hit areas: the Prairie Farm Rehabilitation Act (1935) and a system of equalization grants between the federal and provincial governments were established (Robinson and Webster 1985). The Maritime Marshland Rehabilitation Administration attempted to reverse saltwater intrusion into coastal agricultural lands in the Maritimes (Hodge 2004). A major area of concern during the 1930s was social welfare, including the improvement of housing conditions: mirroring the Chicago School of social ecology, citizen's groups and social welfare agencies in Canadian cities undertook surveys of living conditions and mapped overcrowding, juvenile delinquency, accidents, and substandard housing (Wolfe 2004). The League for Social Reconstruction was established and recommended old age pensions, unemployment insurance, mothers' allowances, health care, town planning, and **social housing**. The first national housing policy, the Dominion Housing Act (1935) (later called the National Housing Act), aimed at improving housing conditions and creating jobs in the housing industry by granting loans for the construction of new housing units with an approved mortgage. However, the need for federal intervention was not as great as in the United States; virtually none of the Canadian banks, trusts, or loan companies went bankrupt, and fewer owners had mortgages on which they might default (Harris 2000). Development of a National Building Code also proceeded at this time and was finalized by 1941 (Nicholls 1938).

Metropolitan and regional issues became important during the wartime years. Winnipeg's Metropolitan Planning Commission, the first metropolitan planning agency in Canada, was established in 1943 as an advisory body that aimed to foster cooperation between the municipalities in the region (Hodge 1985). There were also major developments in housing: from 1941 to 1945, the Crown corporation Wartime Housing Ltd built 19,000 rental houses (Wolfe 2004). Sweeping changes to the National Housing Act in 1944 allowed the construction of new houses, facilitated repair of older housing, offered community planning, and provided employment. To implement these programs, **Canada Mortgage and Housing Corporation (CMHC)** was established in 1946, and for the first time the country had a national planning agency with strong regulatory and financial power. Because CMHC insured residential mortgages and provided joint funding for new housing along with the banks, the Crown corporation has had a great deal of control over the design of new housing developments.

Like its counterpart in the United States, the National Housing Authority, CMHC published guidelines and drawings illustrating the ideal street layouts for optimum health and efficiency. These guidelines were based on the ideas of land-use separation, which was by then well established in zoning bylaws across the country, and the provision of as much light and air as possible, along the lines of late-nineteenth-century ideas about public health and the spread of diseases. They also contained elements of class and income segregation: suburban housing ideals were increasingly marketed to nuclear families and the native-born rather than to single-person, non-family, or immigrant households. The resulting suburbs emphasized low-density, single-family homes and street layouts that were much more closed than the grid street layout that had dominated most urban areas for decades.

A number of other planning standards that would have major consequences on the built form of postwar cities were also developed during this time period. Provincial subdivision and zoning rules were established, and the Official Plan was introduced in the Province of Ontario in 1946 (Wolfe 2004). The Transportation Association of Canada developed traffic control ordinances and design standards, which were adopted by highway departments and municipal traffic departments (Hess 2009). They would have a lasting effect on the layout and classification of roads in Canadian cities. Ontario passed a Conservation Authorities Act (1946) to establish regional conservation authorities to plan river basins in the province, resulting in flood control measures, wetlands conservation, and water-based recreation principles (Hodge 2004).

Chief developments in this period were the establishment of zoning bylaws, provincial planning, metropolitan planning, and the beginnings of regional planning perspectives. A national housing policy and agency and the development of design standards and traffic control ordinances would also have a major influence on the spatial and social fabric of Canadian cities. At the same time, during the Depression and wartime years, the official planning institute was disbanded.

The Dawn and Dusk of the Planning Expert: 1944–70

The years immediately following World War II were a time of immense growth and formalization in the profession of planning: programs in planning were established at several universities, aided by the federal government and CMHC through Part V of the 1944 National Planning Act. The first planning schools were founded at McGill University (1947), the University of Manitoba (1949), the University of British Columbia (1950), and the University of Toronto (1951). The Canadian Institute of Planners was revived in 1952. Planning rapidly became a rational comprehensive process, with the planner seen as the expert who used scientific methods to solve problems. There was a general agreement that planning had a major role to play in promoting economic and societal advancement (Filion 1999); the courses taught in planning programs were designed to give planners the tools to do this.

During the two decades following World War II, Canada became an urban nation. Immigration increased rapidly: changes to the Immigration Act in 1951 brought an influx of Italian, Greek, and other European immigrants, and further amendments in 1967 allowed citizens of non-European countries to settle in Canada. By the 1960s, one-third of Canadians had origins that were neither French nor English (Minister of Public Works and Government Services Canada 2009). Many immigrants settled in inner-city neighbourhoods, starting their own businesses and repairing aging housing, replacing families who were moving to new suburban communities. This trend enabled many Canadian cities to remain vibrant in the neighbourhoods adjacent to their central business districts—a marked contrast with American inner cities, which began to show the stresses and disinvestment associated with rapid suburban growth and residential segregation related to ethnicity.

But Canada did share America's preference for modernist public housing over "slums" at this time:

> In the 1950s, a broad popular consensus saw the public interest as growth and progress. Planners helped to redesign cities to accommodate rising affluence. We advised governments to tear down blighted neighbourhoods, rebuild civic centres and accommodate the poor in upgraded public housing. We separated pedestrian and vehicular traffic. Our profession praised the resulting projects (Grant 2005, 48).

CMHC shared the cost of urban renewal with the provincial and federal governments, with the first "slum clearance" project begun in 1948: Toronto's Regent Park. From 1948 to 1964, more than 50 redevelopment studies were undertaken across Canada and 22 large projects completed. CMHC granted loans for sewage treatment and assisted in the construction of 250,000 public housing units from 1947 to 1986 (Wolfe 2004). In Montreal, the city's urban renewal program included the widening of Dorchester Boulevard in 1955 (now known as René Lévesque Boulevard) and the construction of Place Ville-Marie (see Figure 1.0.3), which in turn stimulated the construction of the city's extensive underground pedestrian network (Artibise 1988).

Suburbs and highways developed rapidly to meet the increases in population growth, facilitated by CMHC's accepted design standards. Don Mills, Canada's first corporate suburb, was constructed between 1952 and 1962 by a single developer. With its community shopping centre, high school, and four neighbourhood units, the project became a model for the subsequent pattern of urban development (Wolfe 2004). Edmonton suburb Mill Woods is another early example, with three neighbourhoods and a town centre.

Figure 1.0.3 • Place Ville-Marie in Montreal, (top) under construction (1960) and (bottom) after completion (1971).

Source: City of Montreal.

Another important planning approach from British planning, development control, was initially introduced in the 1950s as an interim regulatory device: permits would be required while community plans and zoning bylaws were being prepared (Hodge 1985). Later, development control regulations began to be used to establish conditions for development on a property-by-property basis, a system in which planners would need to decide whether certain types of developments would be allowed.

Metropolitan and regional planning came to be seen as a means of coping with rapid postwar population growth: cooperative, advisory metropolitan planning was established in Vancouver (1949), Edmonton (1950), and Calgary (1950). Regulatory power for regional authorities began when Metro Toronto was created in 1953 and the Lower Mainland Planning Board in BC in 1954, along with Alberta's regional planning districts (Wolfe 2004). In rural areas, regional planning commissions (Alberta and New Brunswick), county planning departments (Nova Scotia and Ontario), and regional district planning departments (British Columbia) were established (Hodge 2004). By 1961, just over half of Canadians lived in the 22 census metropolitan areas (those with more than 100,000 residents), which created pressure on local agricultural and urban areas. Regional planning agencies were created as a response to the perceived threat to provinces that were mainly agricultural at the time, such as Alberta (Robinson and Webster 1985).

Provincially created regional planning agencies had to prepare and adopt regional plans to guide and/or regulate development, provide assistance to member municipalities in preparing municipal plans and bylaws, offer advice and assistance to member municipalities on planning-related matters, review and approve subdivision applications, solve inter-municipal problems, deliver rural services to surrounding unincorporated areas, and encourage public interest in the planning process. Planning responsibilities were sometimes split: for example, in Metro Toronto responsibilities for roads and development control were split between the metropolitan and local municipalities. Metro developed expressways and a grid of arterials, while municipalities were responsible for detailed land-use planning, including residential streets, to meet broad density targets defined by Metro, partly based on the service capacity of the arterial roads and infrastructure (Hess 2009). From 1945 to 1959, car use and ownership increased fourfold in Toronto (Filion 1999), but Metro was an earlier innovator in public transit: the Bloor-Danforth and University subway lines were built in the 1950s.

In rural areas, the Agricultural Rehabilitation and Development Act (1961) began to have an effect on enlarging farms, establishing community pastures, and establishing market roads. Regional rural projects included a grassroots redevelopment program in the Gaspé, the Bureau d'aménagement de l'est du Québec (BAEQ), involving all sectors of the community in what would now be called "self-management" of the region's future, and the Interlake Region program in Manitoba. To balance economic development in poorer regions of the country, the Agricultural Rehabilitation and Development Act (1961) was renamed the Agricultural and Rural Development Act in 1966 with a broader mandate, and the Atlantic Development Board was established (1962). A fund for Regional Economic Development was established in 1965 (Robinson and Webster 1985). A federal Department of Regional Economic Expansion was established in 1969 to deliver programs with the provinces.

As in the United States, there was a backlash against heavy-handed planning decisions, particularly in cities. Citizen movements against urban renewal and highway projects, such as the Stop Spadina movement against the Spadina Expressway in Toronto, began in the mid-1960s. Neighbourhood residents across the country began to fight for the preservation of historic buildings rather than accepting their demolition and replacement. There was considerable opposition to apartment buildings, whether public or private (Filion 1999). By 1964, amendments to the National Housing Act included the redevelopment of non-residential areas and renovation, rather than demolition, of substandard housing. Municipal governments became preoccupied with urban renewal: from 1964 to 1972, more than 300 renewal studies were conducted with research assistance from CMHC. Many provincial governments set up housing agencies, which used CMHC funding to develop housing in municipalities (Wolfe 2004). Montreal's urban development continued with the opening of the metro system, each station designed by a different architect (Artibise 1981), which had international exposure during Expo '67.

The federal government played a role in the development of remote areas, such as First Nations and Inuit settlements, including providing very basic housing from 1959 to 1966 and more developed low-rent housing from 1966. However, Aboriginal governments were not recognized at this time, nor were residents allowed to participate in the development of these programs until the 1970s (Chabot and Duhaime 1998).

In the United States, Canada, and the United Kingdom, planners and communities began to realize that planning was not an objective activity and that the rational comprehensive approach to planning had major social and spatial consequences.

> The prospect for future planning is that of a practice which openly invites political and social values to be examined and debated. Acceptance of this position means rejection of prescriptions for planning which would have the planner act solely as a technician (Davidoff 1965, 331).

Conflicting views on planning solutions emerged (Filion 1999): should cities make bold infrastructure investments in mass transit, in highways, in slum clearance, in housing rehabilitation? Planners began to disagree with federal agencies and programs (Davidoff 1965). Citizens began to voice their opinions at planning board and city council meetings, and ratepayer/homeowner organizations were consulted in the preparation of official plans. New ideas about planning generated considerable debate and ushered in the postmodern era, where planners would once again need more than technical skills to solve urban problems. No longer could planners consider themselves experts—they would have to acknowledge multiple points of view and different types of knowledge. New possibilities for employment also emerged: planners could now work as advocates for citizen, environmental, or other interest groups as opposed to within public agencies (Davidoff 1965). Nine new university planning programs were developed from 1960 to 1971 (Wolfe 2004), and a federal Ministry of Urban Affairs was established in 1970.

This era ushered in major changes in an increasingly urban Canada: rapid expansion of suburban areas, regional planning to encourage economic growth, and more efficient, scientific approaches to housing and transportation problems. The planning profession became much more formal with the creation of university planning programs and the development of a curriculum that emphasized tools and problem-solving skills. Planners became experts, with a positivist, rational approach to decision-making. However, within two brief decades community activists began to challenge planners' visions, plunging the new profession into a profound crisis.

The Postmodern Era: 1970–2000

A key concept during the postmodern era is the splintering of formerly homogenous theories, such as **modernism**, into multiple viewpoints. No longer was there a consensus on the public interest, good urban form, or what elements contributed to strong communities (Ploger 2004; Lloyd 2006; Grant 2005). Similarly, planning returned to addressing social, community, and environmental issues, not just the physical planning that had become predominant during the 1950s and 1960s. In academic planning, research approaches began to include community views through focus groups, interviews, life histories, and other methods that allowed in-depth understanding of the complex issues of particular groups in contrast to large-scale quantitative studies that used massive datasets and tended to target aggregated populations or areas.

As the role of the expert planner faded, planners refocused on concerns outside of the built form: social equity, environmental degradation, and energy conservation became pressing issues. One viewpoint acknowledged in planning's postmodern era was that of environmental protection. The first environmental impact legislation was introduced at the federal and provincial levels. The Canada Water Act (1970), which provided a framework for federal–provincial study and action through joint water quality agencies, was a major advance in legislative provision designed to help solve regional water quality management issues (Cullingworth 1987). Ten national parks were established between 1968 and 1972 in response to major increases in park use and recreational demand (Cullingworth 1987) and maintained in cooperation with federal and provincial governments and public and voluntary agencies. The vast hydroelectric project at James Bay, Quebec, was built in 1971 (Chabot and Duhaime 1998). The 1973 oil crisis, which identified Canada as having the highest per capita energy consumption in the world (Wolfe 2004), also ushered in change.

A number of innovative federal programs to address housing quality and affordability were introduced in the 1970s. In 1973, CMHC introduced the Residential Rehabilitation Assistance Program to help homeowners renovate housing in key urban areas and the Neighbourhood Improvement Program, which offered grants to municipalities to repair and upgrade infrastructure and amenities. Cooperative housing secured housing for low- and middle-income families; in Quebec, technical resource groups became extremely skilled in helping interested groups to establish this type of housing. In Montreal, a 1970 health report showed that half of the province's poor residents lived in the city, most concentrated in its east end; thousands of homes and working-class neighbourhoods had been destroyed in the rapid downtown redevelopment of the 1960s (Artibise 1988).

Municipalities tried to develop architecture and rehabilitate neighbourhoods in more sensitive ways than they had in the previous two decades. After Toronto experienced a growth in modernist skyscrapers, including Mies van der Rohe's Toronto-Dominion Centre, I.M. Pei's Commerce Court, and Edward Durell Stone's First Canadian Place (the tallest building in the British Commonwealth in 1983), measures were taken to limit the height of buildings to less than 45 feet. Inner-city neighbourhoods adjacent to downtown, such as Cabbagetown with its sturdy brick houses and small front gardens, were restored; former warehouse areas such as St Lawrence were redeveloped into residential neighbourhoods (Artibise 1988). In Vancouver, the redevelopment of False Creek and Granville Island redefined the former industrial areas, drawing residents to neighbourhoods adjacent to the downtown. The downtrodden Gastown and Chinatown were also redeveloped in the 1970s; Chinatown residents who had resisted redevelopment in

Figure 1.0.4 • One of CMHC's Neighbourhood Improvement Program (NIP) areas in St John's, Newfoundland, 1978.

Source: Canada Mortgage and Housing Corporation (CMHC).

the 1960s saw buildings rehabilitated through CMHC's Neighbourhood Improvement Program. However, rental housing became less profitable to build because of changes to the Income Tax Act (1971), and many provinces introduced legislation permitting condominiums during this decade.

Polycentric growth across metropolitan areas, contrasted with the predominantly monocentric idea of the city, was an important growth management tool as cities tried to determine what their "growth centres" could contribute to their urban systems (Hodge 1968). The Greater Vancouver Regional District, for example, designated regional town centres in Port Coquitlam, Surrey, New Westminster, and Burnaby in 1976. Entirely new lines of research were developed in related fields such as regional science, which attempted to determine the most rational methods to control growth and prevent urban decline: was it the age of a community, the education level, ethnic/religious segregation, or the density that mattered most? During this decade, new areas of expertise developed for planners: neighbourhood planning, public participation, historic conservation, energy efficiency, regional and resource planning, environmental impact assessment, and non-profit housing. These areas demanded an understanding of the physical, social, and environmental processes and characteristics that would contribute to community and regional planning solutions.

By the 1970s, the primary planning role of the provinces was to monitor the quality and consistency of local planning and adherence to provincial planning statutes (Hodge 1985). Regional development programs in the Atlantic provinces focused on building infrastructure and creating employment, while in Ontario controlling, managing, and decentralizing growth in the Toronto region was more important (Robinson and Webster 1985). Key growth areas in Toronto included financial and legal services, insurance, real estate, engineering, architecture, accounting, and advertising (Artibise 1981); many corporations moved their headquarters from Montreal to Toronto in the 1970s. Thriving theatres, opera houses, and art galleries also kept Toronto vibrant. In Alberta, the provincial government designated areas in the southeast and north for future development to balance the booming industry in Edmonton and Calgary. Often, regional planning agencies acted to prevent local agencies from exerting control over major proposals for resource development, highways, large hydroelectric developments, or developments on provincial Crown lands; thus, they were often reactionary rather than initiatory (Robinson and Webster 1985). In both Alberta and British Columbia, political tensions between the provinces and their provincially created regional planning agencies began to limit the latter's planning functions: the British Columbia Municipal Act (1983) eliminated the requirements for regional districts to prepare official regional plans, reducing them to advisory boards.

Canada's Multiculturalism Act (1971) and changes to the Immigration Act that allowed Canadian citizens to sponsor extended family members contributed to more diverse cities in the 1970s. Toronto, Montreal, and Vancouver were particular immigrant destinations during this time period, becoming "testing ground[s] for Canada's commitment to multiculturalism" (Artibise 1988, 258). These changes, coupled with the locations of many ethnic neighbourhoods adjacent to central business districts, meant that many Canadian inner cities remained vibrant while persistent residential segregation and disinvestment continued to affect American cities.

Canadian planners continued to transform inner cities in the 1980s, increasingly through **public–private partnerships** as governments began to decrease public expenditures. This meant that more focus went into the efficiency and cost-effectiveness of projects and their potential to raise revenues, e.g., through property taxes or tourism. Major reurbanization efforts occurred as former industrial lands, rail yards, and harbours in cities were rezoned and redesigned for residential and multi-use purposes, e.g., Queen's Quay in Toronto, the Lachine Canal in Montreal. Economic and community development became established as new areas of expertise for planners. In Quebec, the James Bay and Northern Quebec Agreement led to major structural changes. Between 1981 and 1992, the Société d'habitation du Québec built 1465 social housing units in its northern communities, partly financed by CMHC.

Concern about existing zoning and restrictive bylaws started to emerge as residents began to appreciate retail or commercial uses close to home: the work of new urbanists Andres Duany and Elizabeth Plater-Zyberg in the United States was influential in rethinking urban form. The Healthy Cities Movement, placing emphasis on a well-designed physical environment and supportive community, was introduced in 1986 and adopted by the CIP and the Canadian Public Health Association (Wolfe 2004). Both of these movements emphasized the dense, pedestrian-oriented, mixed land-use patterns common to historic urban areas. By 1986, more than 75 per cent of Canadians lived in cities; 61 per cent lived in the 28 metropolitan areas with populations over 100,000.

The Brundtland Report (World Commission on Environment and Development 1987) emphasized environmental protection and sustainable development, heightening planners' concerns for future development. Planning documents began to reflect environmental values: recommendations for additional green space, restoration of environmentally significant areas, and energy conservation in the built form and transportation were common (Filion 1999). Planners, politicians, and advocacy groups coalesced around higher-density development to increase public transit ridership, reduce car use, protect agricultural land, and decrease public infrastructure costs

Figure 1.0.5 • The Lachine Canal, used for shipping into Montreal from 1825 until 1970, earning it a spot on the Canadian Register of Historic Places for its role in the industrial development of the country. After it closed in 1970, areas around the canal were extensively redeveloped through the 1980s and 1990s; the canal reopened for recreational boating in 2002.

Source: City of Montreal.

(Filion 1999). Pre-war urban forms, main streets with upper-storey apartments, and traditional street layouts trended away from the one-storey, car-oriented suburbs and retail strips of the 1950s and 1960s. However, the predominant growth in residential areas still followed postwar suburban layouts and favoured detached, single-family housing.

Important structural changes also occurred during the 1980s as the federal government began to slow financing and construction of public housing and all levels of government were forced to decrease spending during the economic downturn at the end of the decade. By the mid-1980s, regional planning was poorly integrated institutionally and in the planning process; there was no co-ordinated approach to land use, environmental quality, or socio-economic development (Robinson and Webster 1985).

Economically, Canada further transitioned from a resource-based to a service-based economy during a neo-liberal economic agenda, contributing to major recessions in

1981–2 and 1989–94. Environmental, social, and democratic visions of the city clashed with the neo-liberal agenda to create "world-class" cities, as seen in Toronto, Vancouver, and Calgary. The 1986 Expo in Vancouver triggered a wave of redevelopment of former industrial lands through private investment. American-owned companies and American branch plants were increasingly drawn to Toronto in the 1980s (Artibise 1988). Drastic deregulation of the development process was proposed to combat the economic slowdown and housing affordability crisis in Toronto. Nearly three-quarters of housing starts in the Toronto region from 1986 to 1991 were outside of Metro Toronto, contributing to a fall in transit usage of nearly 20 per cent (Filion 1999). Vancouver began to combat sprawl and traffic congestion by attracting residential development downtown to allow inner-city workers to live near their places of work.

Many cities, such as Halifax and Vancouver, moved toward development of neighbourhood or area plans during the 1980s and 1990s (Grant 1994). These plans allowed deeper participation of area residents and more detailed visions for future development, designating key streets for redevelopment or increased density. Public participation in planning had become standard by this time, although planners had already begun to deal with **nimbyism**: often, residents seemed to oppose projects that seemed to be in the best public interest. Rather than opposing the massive social upheavals and destruction of historic streetscapes characterized by urban renewal in the mid-1960s, citizens in the postmodern era began to fight high-rise apartment towers, projects that might damage heritage buildings, and projects believed to increase traffic levels in their neighbourhoods.

Inequalities between Canadian cities continued to develop in the 1980s and 1990s: in Calgary and Edmonton, average household incomes began to increase, while almost all metropolitan areas in Quebec had declining average household incomes (Walks 2013). Some metropolitan areas also developed considerable income inequality within their urban areas, such as Vancouver, Montreal, Halifax, Hamilton, and London (Walks 2013). Some regions, such as the Maritimes, suffered persistent economic decline, which drove municipalities such as Halifax to pursue urban development at any cost (Grant 1994).

The federal government withdrew funding for new public housing and ended its cooperative housing program in 1993, beginning an era of housing affordability concerns in Canada's largest cities. In 1996, the federal government transferred administration of federal housing programs to provinces and territories without providing appropriate funding. Municipalities have struggled to develop affordable housing without the cooperation of the federal government, although some provinces such as Saskatchewan have developed comprehensive **Housing Plans** to deal with increased competition for affordable housing. Many planning innovations have been introduced to address affordable housing: secondary suites, laneway housing in older neighbourhoods, restrictions on condominium conversions to preserve rental housing, **Community Amenity Contributions** from developers, and affordable housing trusts. In 2009, the United Nations declared that Canada had an affordable housing crisis.

Important innovations in the postmodern era included the integration of public participation into planning processes and urban redevelopment practices that respected neighbourhoods' social and historic resources. Growth management focused on polycentric and regional growth patterns as the nation transitioned to a post-industrial economy; spatially, industrial lands gave way to mixed-use development in inner-city areas. There was a weakening in regional authorities, and the state began to withdraw from large-scale planning and infrastructure projects, a pattern that continued until the 1990s. Municipalities faced increased responsibility for issues such as affordable housing provision, and urban inequalities began to increase. Finally, the seeds of neo-traditional built form were sown with the new urbanist, healthy cities, and sustainable development movements. Planning perspectives became very diverse; planner's interests expanded to include community economic development, **advocacy planning**, and environmental and **social planning**. Planners' roles returned to co-ordinating, sharing, and collecting different points of view, paired with more normative actions, e.g., determining land uses and community service needs. Planners also began to specialize in one or more sub-disciplines during the postmodern era, developing a specific set of skills and expertise.

Current Planning Challenges in Canadian Cities and Regions

The vast majority of Canadians live in cities: 82 per cent, according to the 2011 census. Canadian cities are constantly changing from one kind of settlement to another, reinventing themselves (Artibise 1988). Many consider our cities to be examples of strong urban centres, liveable neighbourhoods, and diverse communities.

> Surprisingly, Canada has developed as an urban nation without the benefit of a national urban policy, yet something in the Canadian psyche has molded from a cold, northern landscape dozens of modern, efficient, and healthy cities. It is no mean achievement (Artibise 1988, 260).

However, planners will always face challenges in Canadian cities. Many of the problems that planners deal with today are similar to those in other industrialized nations: urban sprawl, growth management, inequality, climate change, and urban governance. More fundamental is that planning, as a postmodern practice, must accept diverging points of view; there is no longer a consensus on the public interest (Grant 2005). Instead, our understanding of the common good is fluid and context dependent, and planners must use their professional judgment to determine the course of action that balances multiple public interests.

Because Canada grew so rapidly during the decades immediately following the Second World War, the urban forms predominant in that era define our towns and cities. Although redevelopment, mixed-use development, and higher densities have increased since the late 1980s, the vast majority of development has taken place in suburban and exurban areas.

> It is difficult to reverse car dependency when land-use patterns have been tailored to this form of transportation by decades of infrastructure development and land-use control. It is equally arduous to promote mixed use when functional and social segregation has been imprinted in the minds of a generation ignorant of other forms of urbanization (Filion 1999, 438).

Many cities have tried to combat sprawl through growth management policies, such as permitting secondary suites in low-density residential areas, encouraging higher densities on main streets, and protecting agricultural land. But others feel that they cannot lose important revenue raised through new development and place fewer restrictions on developers to locate in strategic areas. Because municipalities have fewer revenue-generating mechanisms than the provincial or federal governments, property taxes are their most important source of revenue. This means that there is a conflict between requiring developers to locate in urban growth centres or corridors and providing a climate that favours development, such as low property taxes in suburban and exurban areas. Competition among municipalities within metropolitan areas can often drive development outwards. Municipalities with constant growth, such as Metro Vancouver, have been successful in obtaining developer contributions for community amenities, while those with slower growth are often grateful for any new developments. As Grant (1994, 62) wrote about Halifax, "In Halifax, it is unpatriotic to oppose development (with the promise of prosperity it offers).... Haligonians do not argue about whether they want development, but rather about whether a particular project is appropriate or good."

It is acknowledged today that planning is not only a technical process but also a highly political one. Planners often struggle to balance the public interest with the views of stakeholders, the needs of different socio-economic groups, and pressures to work within tight operational budgets. In the late 1960s when public participation began to be integrated into the planning process, there was already concern that the public did not understand enough about planning to be able to provide input:

> In a world in which the march of events and flow of related information are accelerating rapidly and becoming increasingly difficult to digest, very few of the body politic have more than a cursory and often misleading partial knowledge of city planning. Since most people are absorbed in their own personal problems and priority affairs, they think about city planning only when one of its concerns touches them personally.... People must be aware that the methods and extent of city planning are rooted in such fundamental products of the democratic process as existing equities, acceptable powers of control, the respective roles of government and private enterprise, and tax policies (Branch and Robinson 1968, 268).

Many Canadian cities now attempt to involve the public through online methods, which supplement the town hall meetings, focus groups, and **design charrettes** introduced in the 1960s. In addition to this, efforts at public education can be seen in cities such as Calgary, Vancouver, and Saskatoon where **participatory planning**, marketing of planning programs, and neighbourhood-based plans have become commonplace. Consultation with Aboriginal governments and communities has increased, particularly in areas where **co-management** of natural resources is possible and in urban areas, e.g., the Tsawwassen First Nation and the Musqueam in Vancouver.

Canadian cities today are mostly growing because of immigration; some provinces such as Saskatchewan are also growing through migration from other provinces. Suburban cities, such as Brampton, Ontario, and Surrey, British Columbia, are among the fastest-growing areas of the country (Mahoney 2006). But today's suburbs are shadows of their postwar single-use, single-family pasts: increasingly, high-rise apartments and mixed-use developments have made their way into the suburban fabric. In Port Moody and White Rock, British Columbia, more than 90 per cent of new housing is apartments or townhouses (Bula 2011). Residents, too, are much more diverse, with immigrants making up about half of the residents in the suburbs around Toronto and Vancouver: in 2006, more immigrants lived in Markham, Ontario; Burnaby, British Columbia; and Richmond, British Columbia, than in the large cities they bordered. Housing affordability is a pervasive problem across the country, even in small cities such as Kelowna, British Columbia; this drives many non-traditional families, such as seniors, singles, and single parents, to settle in the suburbs (Mahoney 2006). While suburbs are still predominantly car-oriented, there are signs of change: more suburban residents, driven outwards by affordable housing prices (Hulchanski 2005), are now part of the "new breed of suburbanites" (Bula 2011) who fight for local shops and services and for better public transit, cycling, and walking options.

Many cities have a post-industrial economic base, and inter-municipal competition for development increases during economic downturns: in metropolitan areas, the member municipalities often cannot agree on growth management targets or locations. Commercial space in urban centres is supplemented by rapid growth in suburban areas: office parks and big-box retailers seek to pay lower property taxes to municipalities, and municipalities often cannot turn down such an important source of revenue. While industrial land uses have dwindled, some regions, such as Metro Vancouver, have attempted to preserve some industrial land in urban areas for the future.

Canadian cities must also deal with urban inequalities: with the shift to a post-industrial economy, high-income service-sector jobs in finance, business, insurance, and real estate are balanced by a rapid growth in low-income service-sector jobs in retail and food services (Hutton 2004; Walks 2011). Income inequality in Canada has increased as steady, middle-class jobs have been replaced by high-income service-sector jobs or those paying low wages and requiring few skills (Walks 2013). Immigrants are more concentrated in low-income industries and occupations such as manufacturing, childcare, and home support; many immigrants face persistent barriers to working in the industries and occupations in which they were trained (Kelly, Astorga-Garcia, and Esguerra 2009; Thomas 2013). While Canada does not have ghettos (Walks and Bourne 2006), neighbourhoods with the highest concentrations of single visible minority groups typically had higher incomes and lower levels of poverty than areas of greater diversity and, in some cases, than the metropolitan averages in the early 2000s.

Today, planning continues to develop as a multidisciplinary profession: the Canadian Institute of Planners has more than 7500 members, and thousands of other planners who have not sought membership work in the non-profit, public, and private sectors. Planning is a well-established program of study: nine undergraduate programs and 19 graduate programs provide training at Canadian universities (CIP 2014). In graduate school, planners continue to enter the profession through allied disciplines such as geography, landscape architecture, and environmental studies, just as they did 100 years ago. But one of the consequences of the modern and postmodern approaches to planning is that disciplinary "silos" exist: while separating tasks and areas of specialization seemed logical in the decades immediately following World War II, planning problems have grown more complex. Planners, many of whom are trained in multiple disciplines, are uniquely placed to understand and bring together different points of view. For example, in the area of transportation, planners have been instrumental in integrating larger goals such as sustainability and the needs of pedestrians and cyclists into a discipline whose primary concern was moving motorized traffic as quickly as possible. As planners work to develop more sustainable urban growth, an understanding of urban design principles, economic development, and land-use policies helps in collaborations between architects, local business associations, and property developers. In rural areas, regional planning authorities often balance the needs of local residents, agricultural landowners, Aboriginal governments, and conservation agencies.

An Introduction to Planning Theory

Planning theory can be either broadly or narrowly defined. Theories about urban life, whether they are normative ideas about society and **culture** or reflections on our own role in communities, link planning to related fields such as sociology, political science, urban design, and economic geography. Because planning is such a complex and multidisciplinary field, planners often situate their work within other theoretical frameworks; e.g., in an article exploring the ideal built form, Emily Talen and Cliff Ellis (2002) integrated ideas from epistemology, philosophy, and political theory. Trends in land-use and physical planning also reflect societal norms and aspirations, such as the desire for progress during the years immediately after the Second World War or the desire for social equity during the 1960s civil rights and feminist movements. More narrowly, theorists examine planning processes and the interactions between planners, the public, and other groups. The following subsections describe theories related to rationality and positivism; power, communication, and justice; economic and structural changes; and the built form.

Just as the previous section sketched a brief introduction to Canadian planning history, this section introduces readers to the main threads of theory that have influenced planning practice. The limited length and scope of this section means that it cannot offer a thorough grounding on this subject. Excellent texts exploring planning theory, such as John Friedmann's *Planning in the Public Domain* (1987) and Bent Flyvbjerg's *Making Social Science Matter: Why Social Inquiry Fails and How It Can Succeed Again* (2001), offer more thorough discussions of key areas of interest for academic and practising planners. Planning theorists have produced many monographs on the subjects of public participation, power relationships, and dialogue and will be cited in the following subsections. Key theories from the disciplines of geography and urban studies relevant to the planning discipline will also be mentioned briefly.

As a further note, most planning theory seems to have originated outside of Canada, mainly in the United States and the United Kingdom; in contrast to the previous introduction to planning history, it was not possible to focus on the work of Canadian authors. Several, such as Jill Grant and Pierre Filion, have written extensively from theoretical perspectives in planning; Grant's *Reader in Canadian Planning: Linking Theory and Practice* (2008) relates planning theory to the Canadian reality. It should also be noted that in the list of required readings for professional certification exams in planning, only a handful were written by Canadian authors. In this summary of planning theory, readers should be aware that the application of American or British theories to Canadian cities, rural areas, and regional contexts could be problematic. The application and use of theories in Canadian settings is discussed whenever possible.

Rationality and Positivism

Rationality is a strong organizing force in planning, beginning with the "orderly" layout of streets in the earliest Canadian settlements and continuing with the development of zoning regulations and **master plans**. Sound, rational judgment links the planner to the lawyer and regulator; planning is firmly established in the legal traditions of the federal, provincial, and municipal governments. The **rational comprehensive model** of planning, in which scientific analysis is used to solve planning problems and logic is highly valued, often clashes with the view that planning should value different types of knowledge. In planning practice, scientific approaches are used in a number of ways, e.g., calculating population forecasts, measuring environmental impacts, and program evaluation. Numbers appeal to the public and to politicians in an era of cost-effectiveness and efficiency; scientific, generalizable findings often seem to be easier for non-planners to understand and apply. However, community realities, individual experiences, and other non-quantifiable data are just as valid. It is important for actors in planning processes to remember that every source of data used in decision-making is inherently biased. For example, in Canada the views and concerns of groups such as Aboriginal peoples, immigrants and refugees, women, youth, and the low-income population have often been ignored in positivist traditions, which undervalue non-scientific facts such as oral histories or cultural preference.

Rationality and positivism were at their zenith in the postwar decades, when planners turned to the new ideas of modernism and technical optimism to redesign neighbourhoods; it seemed logical to replace "slum" housing with new, functional buildings in the latest architectural style. Large-scale aggregate datasets were used to make sweeping changes to transportation systems; smaller subgroups of travel demand, such as pedestrians' travel patterns, were not considered in the development of **arterials** and freeways, because the car was perceived as the travel mode of the future.

> In the early postwar period, politicians and planners saw people in aggregates, as "the masses," or different classes and income groups, or people in different age-groups (Healey 1996, 208).

By ignoring the consequences of physical planning decisions, planners of the 1950s and 1960s tended to replicate social and economic inequities, e.g., making major public investments in highway infrastructure, which the lower-income population was still unable to use (Davidoff 1965). Oren Yiftachel argues that women, ethnic minorities, and other groups have been marginalized and excluded through planning and that

> Public policies, particularly planning policies, should thus be viewed as the products of a dialectical process that is shaped and reshaped by an ongoing tension between oppression and reform (Oren Yiftachel 1998, 400).

Yiftachel (1998) and Sandercock (1998) maintain that there is a "dark side" to planning and that it has historically been used as a tool of oppression and means of societal control. As planning became a more formalized profession from the 1950s to the 1970s, most planners worked for the state; institutions with a tremendous amount of social and administrative power developed policies affecting those with very little power. As citizens began to fight back and critics began to ask who benefitted from these planning efforts, many planners realigned themselves with environmental and citizen groups to put an end to the displacement of residents and to improve living conditions for racial and ethnic groups (Fainstein 2000). A profound shift in the perception of non-objective and disaggregate types of information—notably the views of long-time residents of the affected communities—took place. This era marked the greatest crisis point in the profession of planning.

The emergence of dissenting voices, citizen groups protesting urban expressways and urban renewal, and advocacy organizations promoting environmental issues and historic preservation occurred alongside the transition to postmodernism. Between the late 1960s and the early 1970s, postmodernism was triggered by a rejection of modernist values and a fragmentation of values and lifestyles (Filion 1999). Important social and structural changes related to feminism, post-Fordism, and civil rights radically transformed ways of living and working, social hierarchies, and economic processes. Rather than one predominant, normative, and technocratic approach to planning, postmodernism acknowledged some of the adverse effects of modernist projects and rejected some of its confidence in progress (Filion 1999).

Planning research began to integrate these different voices through the development of qualitative and participatory research methods: focus groups, interviews, participatory action research, discourse analysis, and case study, to name a few. These developments spurred major changes in the way practical and academic planning worked on a daily basis, increasing citizen involvement and oversight in planning processes, decreasing overreliance on large-scale quantitative datasets and quantitative modelling, and introducing new methods of inquiry. The development of many different views on planning prevented the dominance of one theory or approach in the discipline. In fact, by 1981 Kevin Lynch admonished planners for failing to develop a normative basis for planning (Talen and Ellis 2002).

However, planning is still a normative activity: planners still believe that they possess the skills and expertise to plan land uses and develop master plans, and their values underlie the development and implementation of policies that affect communities. Planners often believe that they are fighting for the common good, even though it can be difficult to define (Ploger 2004; Lloyd 2006; Campbell and Marshall 2002). Planning is considered a profession, which means that it is a self-identified community of people doing similar work, with a unique body of knowledge, performing a beneficial role in society (Vigar 2012). Planning practice goes beyond mere technical skills, requiring professional judgment on a daily basis, e.g., when evaluating development proposals or balancing competing needs. Planners' expertise has been somewhat tempered by the inclusion of participatory processes and acknowledgment of professional biases, but as long as planning exists, there will be conflicts between planners' visions and those of other actors.

While the positivist tradition is not as strong as it was some decades ago, Canada, like the United States and the United Kingdom, has been in the throes of a neo-liberal political ideology since the 1970s. This means that ideas such as cost-effectiveness, efficiency, and effectiveness remain very important measures in planning. Susan Fainstein (2000) argues that efficiency is the most important criterion used to evaluate public policy: hence the widespread use of cost-benefit analysis and program evaluation methods in planning and public policy. But the difficulty in translating planning outcomes, such as increased use of a public park by low-income residents, into economic terms makes cost-benefit analysis a difficult method by which to assess efficiency or a positive economic result (Beukers, Bertolini, and Te Brömmelstroet 2012). For citizens, quality of life measures, such as an improved feeling of safety, a sense of belonging, or an appreciation of aesthetic qualities, might be important in their consideration of successful planning outcomes.

While much of the criticism arising in the postmodern era stemmed from a profound crisis in planning, a new optimism can be seen in the development of theories such as the just city, communicative and **collaborative planning**, and the renewed focus on the built form (**new urbanism, Smart Growth**, and sustainable development).

> The progressives of the previous period spent much of their energy condemning traditional planning for authoritarianism, sexism, the stifling of diversity, and class bias. More recent theorizing has advanced from mere critique to focusing

> instead on offering a more appealing prospect of the future. . . . At the millennium's end, then, planning theorists have returned to many of the past century's preoccupations. Like their nineteenth-century predecessors, they are seeking to interpose the planning process between urban development and the market to produce a more democratic and just society (Fainstein 2000, 472).

Power, Communication, and Justice

Closely related to rational and positivist approaches to planning are the concepts of power, communication, and justice. As a profession, planning has a claim to expertise, and planners often believe that they are acting in the public interest. For example, Grant (1994) writes that in Halifax, politicians, most planners, and citizens with an interest in keeping taxes low generally believe that elected representatives have the skills and wisdom to decide important matters and that widespread involvement of the citizenry could lead to anarchy and expressions of self-interest. However, each actor in the planning process has their own claim to legitimacy:

> celebrities have fame and notoriety; citizens claim roots in the community and rights as residents; consultants hold degrees and professional affiliations; developers have wealth; planners hold official positions in the local bureaucracy (Grant 1994, 68).

As Grant shows, actors in planning processes frequently use logical reasoning and scientific evidence in their speeches, and planners occasionally accept the legitimacy of other types of knowledge. Patsy Healey (2003) maintains that all social relations have a power dimension, expressed in the dynamics of interaction between actors, the process through which some actors dominate the ways others work, and cultural assumptions and practices.

These differences in reasoning, evidence, and power contribute to difficulties among the various actors in communicating with each other, developing common understandings, and contributing to long-term goals. In local planning matters in Canada, provincial planning laws require that public participation is early and ongoing during the planning process. The city council has the right to decide whether a particular project or municipal plan will be approved: council members listen to community residents, staff, and other concerned parties and then make a decision that represents the best option for the municipality. However, it is often unclear how the council uses public input in their decision-making: do they eliminate controversial aspects of a plan or modify bylaws? In non-urban areas such as the Northwest Territories, federal interests and Aboriginal governments may be involved. Decision-making in provinces such as British Columbia, where Aboriginal groups do not have recognized/formalized treaty or land rights, can be particularly contentious. What is the role of these long-time residents in the planning process?

Theoretically, there are a number of ways to address conflicts between the various actors. Susan Fainstein (2000) describes communicative planning theory as emphasizing the role of the planner as a mediator among stakeholders in the planning process. It arose as a response to expert-driven, rationality-based, top-town planning. Based on the approaches of American pragmatism (John Dewey) and communicative rationality (Jürgen Habermas), communicative theory suggests that the main function of the planner is to listen to people's stories, assist in developing a consensus among different viewpoints, and prevent the interests of one group from dominating (Fainstein 2000). The planner provides information to participants in the planning process rather than technocratic leadership. Healey (1997, 29) emphasizes that communicative theory acknowledges that all forms of knowledge are socially constructed, that knowledge and reasoning can take different forms, that individuals develop their views through social interaction, that people have diverse interests and expectations, and that public policy needs to draw upon and use these different types of knowledge and reasoning.

Paul Davidoff (1965) suggested that debate on policy development should be aided by the production of multiple plans; producing a single plan, in his view, discouraged the full participation of citizens in the planning process. If the different political, social, and economic interests in a city produced their own plans, planners would act as advocates of government, groups, organizations, or individuals. This would help balance the needs of the general population as well as those of special interest groups such as ethnic minorities, allowing citizens to propose their own goals and future actions rather than reacting to public agency proposals. John Friedmann (1973) describes planning as a form of mutual learning in which the process is as important as the outcome: knowledge or theory influences practice and vice versa. John Forester (1989) advocates dialogue as the means to meaningful public participation and balancing the power that planners often hold. Healey (1997) suggests that collaborative planning can allow participants to arrive at an agreement on an action that expresses their mutual interests. Her concept of collaborative planning recognizes that there are different social worlds, viewpoints, and practices that coexist in urban contexts with complex power relationships within them, yet collaborative processes could

> have the potential to be transformative, to change the practices, cultures, and outcomes of "place governance," and, in particular, to explore how, through attention to process design, such processes could be made more socially just, and, in the context of the multiplicity of urban social worlds, more socially inclusive (Healey 2003, 107).

However, in practice these theories are based upon ideas of rationality: Davidoff believed that "lively political dispute aided by plural plans could do much to improve the level of rationality in the process of preparing the public plan" (1965, 332). The successful application of these theories requires that all the players in the process understand their roles, have full knowledge of the subject, understand the rules of debate and rhetoric, and are able to express their views. As Fainstein (2000, 455) writes, "There is the assumption that if only people were reasonable, deep structural conflict would melt away." But individuals and groups do not have equal access to information, abilities to participate in planning processes, or understandings of political and social processes. Early in the development of the communicative and collaborative concepts, theorists raised the concern that members of the public needed to "become well informed about the underlying reasons for planning proposals, and be able to respond to them in the technical language of professional planners" (Davidoff 1965, 332). Davidoff considered this education component, which would help citizens become less reactionary, as falling within the role of the advocate planner. Groups that do not traditionally participate in planning or governance processes may need to be encouraged to participate and given the tools to do so in a meaningful way. Planners themselves cannot be completely disinterested; although communicative theory emphasizes the mediator role that planners should have, it can be difficult for planners to avoid the expression of their own biases and values in planning processes.

There is therefore no guarantee that communicative or collaborative processes will result in better planning outcomes. Fainstein (2000) observes that communicative theorists avoid discussing what to do when open processes produce unjust results: individuals or groups that historically have more power may continue to dominate. Flyvbjerg (1998, 234) acknowledged, "When we understand power we see that we cannot rely solely on democracy based on rationality to solve our problems." Fainstein (2014) argues that a situation in which all social classes are represented proportionately will rarely occur, that representatives of the poor may be co-opted or manipulated, and that planners have little independent power so they may only succeed as advocates with the support of grassroots movements or progressive politicians. Communicative processes are also time-consuming and can contribute to burnout and disillusionment among citizens (Fainstein 2000). However, people can become aware of their actions through collaborative processes, including the development of a "fragile, incomplete, and contestable" consensus (Healey 2003, 114).

The planning profession is simultaneously rooted in concerns for social equity and social control: Victorian-era reformers cared as much about the poor housing conditions of urban workers as about their non-traditional family and social structures. Postmodern concerns with social justice can be seen in David Harvey's *Social Justice and the City* (1973), which focused on the relationship between urban space and the social situation of residents; Harvey (1985) also discussed the key problems within the capitalist economic system, whose crises of accumulation lead to rapid disinvestment and impoverishment. Sherry Arnstein (1969) laid out her now-famous Ladder of Citizen Participation in the midst of the planning crisis: manipulation and therapy comprise the bottom section; informing, **consultation**, and placation are in the middle; and partnership, delegation, and citizen control are at the top. The model continues to serve as a touchstone for planners, politicians, and citizen groups and has been used extensively to avoid **tokenistic** inclusion of the public in planning processes.

A new focus on ethics in planning began the 1990s (Fainstein 2000), including the development of theories on just cities. Iris Marion Young, in *Justice and the Politics of Indifference* (1990), considered the development and differentiation of social groups both inevitable and desirable, since they are defined by shared identities and interests; the challenge, then, is how to achieve social differentiation without exclusion. Just city theorists believe that "progressive social change results only from the exercise of power by those who previously had been excluded from power" (Fainstein 2000, 467), that the redistribution of social benefits can happen through mobilization of a public. Unlike Marxism, the just city concept does not imply that accumulation of capital is problematic; rather, increased wealth for the majority goes hand-in-hand with providing welfare. Participation by **stakeholders** in this context cannot be assumed to be transformative or superior to decisions undertaken by the state; both the state and its citizens can do good and bad. Social rights are just as critical as political rights in the just city.

> If the aim is justice, the purpose of inclusion in decision-making is to have interests fairly represented, not to value participation in and of itself (Fainstein 2014, 12).

Edward Soja (2010) wrote about environmental and spatial justice issues, such as the location of waste treatment and other noxious land uses in low-income neighbourhoods.

This concept may be particularly apt in the United States, where there is much higher residential segregation by income and ethno-cultural background than in Canada. However, the Africville case in Halifax illustrates the practice in a black Canadian community, including the location of a fertilizer plant, slaughterhouses, a prison, a waste facility, and an infectious diseases hospital during the second half of the nineteenth century (Tattrie 2014). Soja describes coalition-building between groups in Los Angeles as a way of achieving greater equity for marginalized populations.

In Canada, one ongoing social justice concern is that of Aboriginal land rights; this can be illustrated by the growing interest of First Nations peoples in managing their own natural resources as governments alongside the country's federal and provincial governments. Another issue is inclusion under an official regime of multiculturalism, although the experiences of ethno-cultural groups in Canada cannot compare to those in the US, where the legacy of slavery has contributed to long-standing patterns of residential and occupational segregation. Immigrants in Canada face persistent labour market and housing market barriers (Kelly, Astorga-Garcia, and Esguerra 2009; Pendakur and Pendakur 2011; Haan 2005), which may pose a barrier to their integration and ability to participate in civic activities. Immigrants may also come from countries in which participation in planning activities is not accepted or expected. Likewise, while incomes in Canada are not as sharply polarized as in the US (Walks 2013), income inequality could affect the ability of citizens to participate in planning processes.

Economic and Structural Changes in Cities

Conceptions of rural and urban areas are strongly related to their contemporary economic systems. Economic systems contribute significantly to the urban form and social geography of our cities, and structural changes in the economy have contributed to major changes in Canadian settlements over time.

Early models of urban growth reflect the reality of the industrial city: Charles Booth's (1902) concept of the British city was concentric, while E.W. Burgess's concept of concentric growth (1925) acknowledged that the industrial city's manufacturing base was physically located in the city centre. Homer Hoyt's (1939) extensive studies of American cities showed that variations in topography, transportation infrastructure, and low- and high-rent districts created different patterns in each city. Burgess theorized that the rising socio-economic status of a group led to their trajectory outwards to more desirable neighbourhoods away from industrial areas and their associated health problems. The development of pre-war suburbs in Canada, as in other countries, was enabled by innovations in transportation that gave the working class access to areas with better air quality and more natural surroundings. Manufacturing and transportation of natural resources by sea or rail contributed to largely monocentric towns and cities with dominant port areas, such as Halifax, Montreal, and Winnipeg. Countering Burgess's observations on monocentric cities, Hoyt's research showed growth at the edges of the city where manufacturing was beginning to relocate along the earliest highways and railway junctions. Both Burgess and Hoyt described different patterns of residential segregation based on their theories: Burgess wrote about progressive spatial assimilation while Hoyt described entrenched segregation in low-income areas. These theories have been applied in understanding growth and change in Canadian cities, such as the development of ethnic enclaves and spatial integration of immigrants (e.g., Hiebert and Ley 2003; Walks and Bourne 2006).

The 1950s **Fordist** era could be characterized by the development of a consumer culture, stable wages, and steady work schedules securing markets for mass production (Filion 1999). Increased productivity through economies of scale, massive investments in public infrastructure, automation, and the progress of trade unions in achieving middle-class wages for blue-collar and public-sector workers were hallmarks of the Fordist era (Filion 1999). As discussed in the previous subsections, the postwar belief in scientific and social progress was strong. In planning, architecture, engineering, and many other disciplines, there was a strong desire to break with tradition and solve problems in new and unique ways. This transformation from the pre-war built form took shape in low-density, spatially and socially segregated suburbs, and the city crossed with freeways, which spoke to the idea of progress (Filion 1999). Another typical spatial form was the industrial park, with its wealth of parking and single-storey assembling plants, which would construct the goods suburban residents would need. Because land-use separation was strictly enforced, industrial land uses were segregated from residential areas so that workers needed to commute daily. Planners were technical optimists at a time when the profession espoused technical expertise: for example, many believed that technical solutions to air and water pollution were just around the corner (Filion 1999). Therefore, the widespread growth in automobile use was not seen as problematic, nor was the transition from rail-based to road-based transport of goods that enabled manufacturers to locate near highways or in dispersed areas rather than city centres.

With the earliest inklings of postmodernism, the transition from Fordist to **post-Fordist** economies first became visible. As globalization dawned in the early 1970s, initial job losses took place in the manufacturing sector, later spreading throughout other sectors, and workers' bargaining power weakened as companies shifted their operations offshore. Manufacturing shifted to suburbs, exurbs, and other countries. Replacing manufacturing jobs in the inner city were service-sector jobs such as finance, insurance, and real estate (Hutton 2004). Full-time, well-paid, permanent jobs began to give way to part-time, low-paid, temporary, and contract jobs, a trend that continues today. Within the new neo-liberal economic framework, market processes were considered as producing the most efficient allocation of resources, providing incentives for innovation and economic growth, and rewarding merit, but only without the interference of state welfare policies to reduce inequality (Fainstein 2014). Redeveloped downtowns with high-rises and plazas, and the metropolitan region with new towns and green belts, were still on the agenda.

Another structural change could be seen in the changing character of institutions such as governments, planning authorities, and non-profit organizations, which carry their own values, biases, and organizational styles. When planning was a state-led activity in the immediate postwar era, strong conformity to the state's views was necessary for planners. A shift away from the state began when planners started to advocate for the needs of under-represented groups.

> It is not within the power of municipal governments to achieve transformational change. Only the nation state has this kind of leverage, and the more egalitarian cities of Europe are in fact underpinned by strong national welfare programmes. At the same time, local policies make life better or worse for people. There are many decisions, especially involving housing, transport, and recreation, made at the local level that differentially affect people's quality of life (Fainstein 2014, 14).

In policy and governance frameworks, Lowi (1972) described the period 1880–1930 in the United States as the regulatory era (centralization of welfare, regulatory commissions), 1930–70 the redistributive era (expansion of welfare-state protections), and post-1970 as the re-regulatory era (reduction in public-sector expenditure through privatization, contracting out, enhancement of private-sector competition, downsizing). During the regulatory era, policy implementation was mostly through the legal system (legislation, rules, and regulation); during the redistributive era, implementation was through the state system (plans and macro-level bargaining); during the re-regulatory era, it occurred through the market system (auctions, contracts, subsidies, tax incentives, penalties) (Considine 2001; Howlett 2009). These general trends can be seen in Canada, although it may be argued that the redistributive era began earlier.

At the smaller scale, Healey is particularly concerned with the daily routines and practices of governance—that is, the characteristics of institutions such as municipalities and neighbourhood organizations. For example, she challenges

> the well-established "sectoral" way of organizing the articulation and delivery of government policies through functions, such as health, education, employment, housing, crime management, etc. While this produced enormous benefits in improving the quality of people's lives in many respects, it led to an organization of government that was often ignorant of the way different policies interacted as experienced by specific people and firms in specific places (Healey 2003, 116).

Healey's intent is not to produce a "formula" for governance processes but to build capacities that may lead to changes in the broader culture of governance. As governance has shifted from traditional public-sector agencies and hierarchies to diverse private, non-profit, and community agencies, planners have also had to adjust to a much more diverse public. The expansion of the service sector has changed traditional family structures, and the balance of work opportunities has shifted from favouring well-paid full-time opportunities to temporary and part-time opportunities. People choose to live in neighbourhoods with others with similar lifestyles, and residents can become defenders of the quality of their neighbourhoods. In Healey's view, these changes have decreased the opportunities for addressing issues of collective concern. Building relationships between different groups increases trust, allowing them to share ideas and opportunities: this can be done through the design of working groups that allow all stakeholders to communicate and collaborate on an issue (e.g., engaging in managing a project or developing strategies), finding ways to allow those closest to the consequences of strategic choices to become involved in decision-making, and fostering a discussion style that allows different viewpoints to be explored (Healey 1996).

Another critical perspective on structural changes is that of the transitions perspective in which a transition is described as a structured social change that is the result of alterations in intertwined systems that support each other (Grin 2010, 1). The multi-level perspective (Geels and Schot 2007) claims that "transitions result from the interaction between innovative practices, novelties, incremental change induced by actors who operate at what we call the regime level and quasi-autonomous macro-dynamics, or the 'landscape' level" (Grin, Schot, and Rotmans 2011). Switzer, Bertolini, and Grin (2013) describe the "landscape" as made up of long-term exogenous trends, such as macroeconomic developments, cultural trends, and demographic developments; the "regime" as the dominant practices, rules, and artefacts in the dominant socio-technical system; and the "novelties" as alternatives with unstable practices, rules, and artefacts that can become stable. This theory is useful in understanding how significant change happens in planning practice and in the broader community, such as the shift from the car-dominant transportation system to one focused on sustainable mobility.

Major structural changes in the economies of Canadian cities can be seen in spatial forms such as office parks and retail strips, the decline of the monocentric city centre, and the development of luxury condominiums in city centres to provide nearby housing for high-end service-sector employees. The rise of polycentric cities, where economic development and growth is concentrated in a few nodes across a metropolitan area rather than just one, could be seen as representative of the transition to the service-sector economy. Institutionally, there has been a continual **downloading** of responsibility of the federal and provincial governments to the municipalities in Canada and the development of partnerships with non-profit organizations, business groups, and others to deliver important planning programs such as social housing and immigrant services. These economic and institutional changes have influenced the form of our towns and cities as much as our efforts to control the built form.

Built Form

From the earliest foundations of Canadian cities, there have been ideas about the best way to lay out streets and buildings in our settlements. Several key theories influenced the earliest cities. As discussed in the introduction to history, the perceived need for light and air at the dawn of the industrialized nation in Europe and the US arose from the theory that poor air quality caused disease. Ebenezer Howard's Garden City (1898) was a powerful visual illustration of these ideas, with its emphasis on urban nodes connected by railways and buffered by green zones. The winding pathways and thoughtfully designed natural landscapes of Frederic Law Olmsted's public parks showed how the built environment could promote health and encourage socializing among different groups of people. Both were strongly influenced by English landscape gardening traditions from the eighteenth and nineteenth centuries and the growing industrialization of English cities. The Garden City and landscaped park concepts shaped the development of residential neighbourhoods, parks, and some of the earliest suburbs in Canada; to this day, most public parks are designed in the English landscape style popularized by Olmsted, and municipal bylaws enforce building separation and building setbacks to encourage light to penetrate to street level.

The concentric model of urban growth (Burgess 1925) held that central cities were less desirable than zones farther out, since urban workers would not want to live in close proximity to manufacturing land uses. Burgess identified lower-class neighbourhoods and areas with concentrated poverty in inner cities, even as Homer Hoyt (1939) showed that patterns of high and low income varied considerably from city to city. Ideas about the physical and social problems in inner cities influenced the development of streetcar suburbs as much as theories about the spread of diseases; working-class people, immigrants, and others with low incomes could now afford to live farther away from the dense, overcrowded cities where they worked.

Clarence Perry (1929) held that neighbourhoods should be strongly defined and protected from external threats. Although this idea emerged from his desire to protect pedestrians from the growing prevalence of the car, Perry's Neighbourhood Unit, surrounded by arterial roads with an internal street hierarchy and community buildings within, became popular as a way to segregate communities from the larger urban forces around them. This idea influenced many suburbs in Canada, including Don Mills,

Ontario, and Mill Woods, Alberta. The theory could also be seen in the layout and design of public housing projects in later decades, which were often removed from the existing street networks and featured interior spaces intended solely for the use of residents.

The rational, positivist approach to planning in the 1940s and 1950s emphasized land-use separation between residential and industrial uses but also the separation of low-income from high-income housing. As Friedan (1963) discussed, gender stereotypes also influenced the development of suburbs so that women were more isolated in the home; Sennett (1971) made a powerful case that conformity was reinforced in modern suburbs. The heavily zoned postwar metropolis spurred the widespread construction of new building types and configurations: high-rise apartments clustered around a park, bungalows along "loops and lollipops," retail strip development along highways. Influential architects of the time included Le Corbusier, whose *Radiant City* (*La ville radieuse*, 1935) inspired futuristic high-rises and functionally separated transportation infrastructure, and Frank Lloyd Wright, whose Broadacre City concept (1932) introduced the suburban bungalow prototype.

The housing career model, widely used in planning, economics, and geography, strongly influenced residential location and the supply of housing units. The model is based upon the idealized human lifecycle, which includes pre-child, childbearing, childrearing and launching, post-child, and later life stages. The idea of rational choice is implicit: families choose the most appropriate type of housing for their lifecycle stage. Families are assumed to move from rental apartments in the pre-child stage toward single-family homeownership in the childrearing and launching stage, then to downsize in the post-child stage. In both Canada and the US, homeownership was redefined during the postwar years as more stable and socially acceptable than renting (Lands 2008). Through CMHC and the National Housing Act, homeownership has been promoted in Canada, and housing programs have supported ownership since the 1930s (Thomas 2013; Oberlander and Fallick 1992). Mortgage insurance policies and guidance from CMHC encouraged single-family homes in suburban locations on quiet, curving streets rather than in city centres, meaning that families tended to move outward as they had children.

Since the postmodern era, our understanding of the social, economic, psychological, and other effects of the physical form of cities on the population has increased. One of the earliest critics of modernist architecture and planning was Jane Jacobs, whose book *The Death and Life of Great American Cities* (1961) described the devastating effects that large-scale architecture had upon the fine-grained, **pedestrian-oriented** urban form of historic cities. The perception of masterplanning, in which one designer (often an architect) designs an entire neighbourhood or area of a city, had been that it was a rational, scientific method of achieving the perfect built form free of historical constraints; this perception rapidly evolved to a damaging process that tore communities apart. A Canadian example of destructive redevelopment is Africville in Halifax, whose residents were forced to relocate between 1964 and 1967 after the city proposed the A. Murray MacKay Bridge highway construction and port facilities in the area. This Black Canadian community, while neglected and underserved by the city, had existed since the early 1800s. Davidoff (1965) wrote that physical relationships and conditions had no meaning or quality apart from the way they served their users: the profession's fixation with the physical environment, particularly the urban renewal projects of the 1950s and 1960s, demonstrated the high cost of such biases. Even if planners acknowledged the existence of social, economic, and other consequences of physical planning decisions, "they have not sought or trained themselves to understand socioeconomic problems, their causes or solutions" (Davidoff 1965, 336).

As a result of this crisis in the large-scale physical design and layout of cities, a greater part of planning theory became concerned with the process of planning: dialogue, debate, negotiation, and power dynamics.

> As conventional conceptions of the public interest floundered, planning theory turned increasingly to examining the planning process (Grant 2005, 49).

However, as planners shifted away from the physical design of communities, architects and landscape architects continued to search for the elusive "perfect built form." *A Pattern Language* (Alexander et al. 1977) went into great detail on the specific dimensions of public squares, streets, sidewalks, and other physical forms from individual buildings all the way up to the regional scale. Their detailed drawings and descriptions were based on historical precedents across a variety of countries. In *Good City Form* (1981), Kevin Lynch presented a set of "dimensions of performance": vitality (the support of biological requirements), sense (mental perception and differentiation of a settlement), fit (matching pattern to behaviour), access to resources, control of the use of the settlement, cost-efficiency, and equity.

In the 1990s, Andres Duany, Elizabeth Plater-Zyberk, and members of the Congress for New Urbanism advanced principles of good urban design based on historical precedents. Just like the proponents of earlier theories of built form, **new urbanists** take a normative stance: they believe that the physical design of communities can have social effects. They advocate complete communities, **mixed-use** town centres, and traditional building types such as rowhouses. **Transit-oriented development (TOD)**, first promoted by Peter Calthorpe (1993) and Doug Kelbaugh (1989) as a means of organizing higher-density land uses around public transit stations and corridors, can be considered another component of new urbanism. Another significant development in Canada followed the publication of the Brundtland Report (World Commission on Environment and Development 1987): sustainability, in the form of compact development, increased transportation options, healthy communities, affordable housing, and environmental responsibility, quickly became a planning buzzword. Grant (2005) writes that by the mid-1990s, the idea of sustainable development fused with the concept of new urbanism in cities such as Calgary, leading to policies to create sustainable suburbs. As Grant (1994) observes, two dominant "camps" within local residents in planning processes are often those who cite Jane Jacobs and HRH Prince Charles, both of whom wrote extensively on traditional neighbourhood forms (although from quite different perspectives) and those who envision a bustling, prosperous urban environment with high-rise towers in the urban core, the primary concern being growth and progress. Both groups believe that they are defending the public interest.

Smart Growth is a concept arising from the American political movement promoting planning at a regional scale to manage land use and transportation systems, achieve a better jobs–housing balance, and intensify urban development. By 2000, Smart Growth principles had begun to influence provincial policies in British Columbia and Ontario (Grant 2009). The Ontario government disseminated reports and manuals, appointed Smart Growth panels, and adopted the Places to Grow Act (2006). But evaluations of Canadian practice have pointed out the inability of governments to actually meet Smart Growth objectives (e.g., Filion and McSpurren 2007). McKenzie Towne, one of the first new urbanism projects in Canada, began in 1995 near Calgary. Cornell, near Markham, Ontario, is another new urbanist development, which also began in the mid-1990s. East Clayton, near Surrey, BC, is an example of a sustainable or Smart Growth suburb.

Through the 1990s and 2000s, a convergence in opinion about good city form began to develop based on historical precedents: the concept that walkable, diverse, mixed-use landscapes produce the highest quality of urban life (Talen and Ellis 2002; Grant 2009). New urbanism, for example, takes a normative stance on planning by advocating the types of urban forms that exist in cities around the world as an antidote to the suburban sprawl, environmental degradation, and social isolation that many see as inherent to suburbia. A parallel development from a transportation planning focus has arisen amid concerns about public health, with walking and cycling seen as antidotes to the high rates of obesity and diabetes in the United States, Canada, and other industrialized nations (Frumkin, Frank, and Jackson 2004). In order to support these non-motorized modes of transportation, the physical environment should include a permeable street pattern to allow better pedestrian connections than suburban street layouts, short blocks rather than long impenetrable blocks, improved layout of streets emphasizing pedestrian and cycling infrastructure, mixed land uses to contribute to shorter trip distances to work and other destinations, and higher densities to support the use of public transit.

Some criticisms of new urbanist communities include the lack of participation in planning processes during their development and their social conformity. While design charrettes are one way that citizens can become involved in the planning of these communities, this method suffers from the same problems that many other participatory methods do: it assumes a certain level of knowledge and ability to engage (e.g., ability to speak the dominant language, understanding of basic design concepts, ability to express ideas about urban form). Many proponents of the new urbanism are trained in architecture, urban design, or landscape architecture and are firm believers in the principles outlined by the Congress for New Urbanism; they are, in fact, experts, and the high design quality of their plans attests to the fact that there may be a limited role for the public in the design process. Many see the architectural style and form of new urbanism as conformist, with its lists of acceptable paint colours and housing types and its lack of attention to racial, ethnic, and class diversity (Harvey 1997). In addition, the principle of developing small shops near the residential units has been difficult to implement; although theoretically there are enough residents to support them, many have not been successful (Grant 2009). Transit-oriented development and Smart Growth also face significant barriers as many communities fight increased densities and public transit infrastructure. There are also important issues such as **gentrification** and displacement of lower-income people during redevelopment; TOD and Smart Growth must include provisions to protect affordable housing (Centre for Neighbourhood Technology 2006).

Talen and Ellis argue that the planning profession needs a renewed focus on substance rather than process; they write that formless urban sprawl "is in a profound way linked to the lack of a solid theory of good city form" (2002, 38). They believe that planning theories focusing on process leave planners without guidance when making critical decisions on a daily basis; make the profession appear weak, uncertain, and divided; result in little training in physical planning and urban design; and reinforce the perception that planning is bureaucratic and administrative in nature. But Grant (2009, 14) writes that although most planners are well versed in mixed-use development, intensification, compact form, **walkability**, and transit-oriented development, they practise in contexts where suburbs are "highly valued as desirable landscapes full of prized real-estate commodities."

Grant (2009) writes that municipalities across Canada have adopted principles of Smart Growth and new urbanism, encouraging urban intensification, mixed use, transit-oriented development, integrated housing, and connected street grids in their plans. Although Canadian suburbs have higher residential densities and greater use of public transportation than American suburbs (Filion 2003), they generally fall short of Smart Growth objectives. Research has shown that institutional, political, economic, and socio-cultural barriers in Canada prevent the implementation of new urbanism and Smart Growth ideas (Grant 2009):

- Institutional: city councillors and planners often have conflicting goals for growth; other municipal departments (such as engineering or traffic) may not support planning principles such as narrow streets or laneways; council may favour growth at any cost during poor economic times.
- Political: councillors often have worked in the development industry and may not always think beyond their political terms; municipalities must get the development community to support the cost of infrastructure in the absence of funding from the provincial and federal governments.
- Economic: developers may not always deliver some of the benefits promised because of real or perceived market resistance; the architectural details may be implemented without the broader design principles.
- Socio-cultural barriers: there may be local resistance to planning principles.

Much of the implementation of new urbanist developments has been in suburban rather than urban settings because of the ease of assembling land and dealing with site constraints on greenfield sites and policies restricting these types of built forms in inner-city areas. This means that most of the buyers are interested in suburban lifestyles and will drive rather than demand the transit that could work because of the higher densities. Smart Growth and sustainable development principles, on the other hand, can be used in intensification or small-site redevelopment projects.

Planning theory addresses many aspects of the built form, planning processes, power, institutions, and social justice. Theoretical perspectives allow planners to test new ideas that can be applied to practical settings and vice versa—Bent Flyvbjerg's powerful work on power in the planning process was spurred by his planning work in the city of Aalborg, Denmark. Theories about public health, the vices of cities, and industry still influence the design and layout of our streets, parks, and residential communities. Ideas, then, can be quite powerful. Planners are not immune to error; many planning theories have now been disproved, and many are contentious. Planning theorists continue to debate ideas and to propose new ways of understanding and undertaking practice, either in person during conferences or in writing in academic journals. New ideas about cities or communities continually arise from other disciplines to influence planning: witness Richard Florida's *The Rise of the Creative Class* (2002), which spurred the whole creative cities movement, or Robert Putnam's *Bowling Alone* (2000), which vividly illustrates the importance of social capital in communities. Also important is the use of multiple theories in the work of academic planners, since planning is such a multidisciplinary profession.

Conclusions

Planning in Canada is complex. The country's unique legislation, history, and demographics have contributed to a highly urbanized, but largely low-density, pattern of settlement and a governance system that makes planning transportation, housing, social services a shared endeavour. Our settlements are remarkably diverse, from the rural towns of the Prairies, the small Aboriginal settlements of northern British

Columbia and Yukon, to the multicultural metropolises of Ontario. Immigration continues to contribute strongly to the country's growth and shape its social fabric, particularly in its largest cities. These considerations pose several challenges for planners, who seek to plan orderly, well-designed cities that will be resilient to growth and change in the future. However, with their multidisciplinary training, planners are uniquely poised to become today's problem-solvers; they possess the communication skills and technical and subject-matter knowledge to work with professionals such as architects, urban designers, developers, politicians, and the public. Planners are often called upon to make professional judgments and exercise considerable skill in balancing their own interests with those of the organization for which they work and the multiple publics affected by any planning decision (Bolan 1983).

Most Canadian planners work within an overall governance context that values efficiency so that policies and programs that protect economic interests and are more affordable to implement are often the chosen alternatives. Generally, the state has retreated from the large-scale urban planning and redevelopment initiatives common in the postwar decades and has granted more responsibilities to the lower levels of government. Regular economic recessions have contributed to an atmosphere of constraint in some areas of the country, such as the Maritimes and Ontario, while provinces like British Columbia and Saskatchewan currently have more freedom to implement broader-scale initiatives. Public–private partnerships have made efficiency and cost-effectiveness important considerations. Demonstration or pilot projects, which allow municipalities to test new ideas temporarily, are increasingly used to justify longer-term investments. For example, the City of Vancouver piloted a separated bicycle lane on the Burrard Bridge in 2009; one year later, a 24 per cent increase in cycling over the bridge was recorded while car traffic remained stable and the accident rate decreased (Mayor of Vancouver 2010). By 2015, 100,000 trips were made by bike each day, and the city reached its goal of increasing the combined mode share of walking, cycling, and public transit to 50 per cent of all trips.

Values and ideals from the postwar era, including preferences for single-family housing on large lots, travelling exclusively by car, and land-use zoning that separates residential uses from commercial and industrial areas, continue to shape our cities. It has been difficult for planners to move forward with sustainable development, Smart Growth, or new urbanist built forms, or to plan communities that encourage a range of income, ethnic, and age groups, when ideals from the postwar decades seem entrenched in the minds of the public, developers, and politicians. However, some progress has been made, including sustainable transportation alternatives, multi-family and smaller housing types, and a percentage of affordable housing in new developments (inclusionary housing). Suburban neighbourhoods are now integrating more compact housing types, rental housing, and sustainable features such as stormwater management designs. Demographic shifts, such as the continued preference on the part of young people for urban living locations, have also begun to affect the predominant postwar paradigm.

Although Canadian planning law places less emphasis on private property rights than American planning law does, residents are often concerned about the impacts of a new policy or development on their own property values. There have been times throughout history when there was a consensus on the public interest, or the common good, but there no longer is such agreement. Planners work to reconcile differing points of view and must always take into consideration a variety of different perspectives on an issue. This can be very challenging; in Ontario, where the Ontario Municipal Board often reviews Official Plans during appeals from landowners or powerful business interests, it can be discouraging at times. Planning is political, and the implementation of plans, policies, and programs relies upon the interest and willpower of city councillors, mayors, and other public servants. A vast and diverse public, who may not always know the theory or reasoning behind planning decision-making, are also critical in the planning process. However, since the profound crisis in planning of the late 1960s, planning processes have increasingly included focus groups, town hall meetings, and, more recently, online surveys and interactive methods to involve the public in decision-making. In some towns and cities, this has contributed to a public that is quite knowledgeable on planning issues and concepts and that participates quite broadly in municipal initiatives; in others, there is still considerable work to be done. In many places across the country, planning is just as much about the process as the outcome.

Planning is a long-term endeavour; in the postwar decades, growth seemed to take place overnight, but since the 1970s economic crisis, governments and the private sector have been more cautious about investing in public infrastructure such as public transit, affordable housing, and community services. It can take a great deal of time before planners themselves understand the impacts of their decisions on communities and learn from and adjust their practices. Planning is an iterative process, perhaps appropriate to the modern search for resilience. If the history of Canadian planning is any indicator, a combination of public and private, individual and community efforts will be needed to build and sustain our communities in the long run. Just as the cities of the streetcar suburb era relied upon a combination of privately constructed infrastructure, publicly funded open spaces, and community-driven social services, so can our future cities develop their own partnerships.

The role of planners has changed over the years, just as their education and training has changed to include communication and negotiation skills, urban design concepts, Geographic Information Systems (GIS), and cost-benefit analysis. They are still concerned with the effects of the urban form on behaviour, public health considerations, regulating land use, and growth management. New areas of planning, such as food security, cultural planning, and **climate change adaptation**, are always emerging. Some focus areas, like social planning and public health, have recently re-emerged among the planning community. Because a good deal of planning is initiated by public authorities, which can be risk-averse, it takes some time before new concepts make their way into policies and plans. Quicker movement may be seen through the efforts of non-profit agencies and community organizations, who have always been the first to address social, environmental, and local economic issues. As transition theory suggests, what are now novelties will someday become embedded ideas in the planning regime. So we see in Canadian planning a mixture of top-down and bottom-up approaches to solving complex societal problems, the necessary collaboration of energetic individuals at the community level working alongside the bureaucratic administration of the state.

This introduction to urban planning, in all its complexity, offers a counterpoint to the insightful and inspiring case studies to follow. Considering the Canadian planning legacy, with its British and American theoretical underpinnings, helps us to understand the remarkable achievements of our municipalities, towns, and Aboriginal communities. Understanding the theories and values behind planning ideals is essential to daily practice and academic research. Finally, acknowledging the many challenges that planners face situates the modern planning student in precisely the same place as the Victorian-era urban reformer: both facing rapidly changing, congested, and unhealthy cities and both feeling so strongly about the conditions in their communities that they seek to improve them. This is the best and highest aspiration a planner can have: to improve the world in which we live in a myriad of ways both large and small.

KEY TERMS

Advocacy planning
Arterials
Bylaw (or by-law)
Canada Mortgage and Housing Corporation (CMHC)
Climate change adaptation planning
Collaborative planning
Co-management
Community Amenity Contributions
Consultation
Covenant
Culture
Demonstration (or pilot) project
Design charrette
Development Cost Levies
Downloading
First Nations
Fordist
Gentrification
Housing plan
Immigrants
Master plan
Mixed-use building
Modernism
New urbanism
nimbyism
Official Community Plan (OCP)
Participatory planning
Pedestrian-oriented
Post-Fordist
Public interest
Public–private partnerships (PPPs)
Rational comprehensive model (RCM)
Resilience
Smart Growth
Social housing
Social planning
Stakeholder
Subdivision
Tokenism
Transit-oriented development (TOD)
Walkability
Youth
Zoning bylaws/schedules

QUESTIONS FOR CONSIDERATION

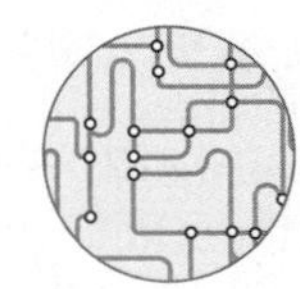

1. In what ways do practising planners use case studies? How does this differ from the ways in which academic planners use case studies?
2. Why are some planning activities, such as planning for public health or public transportation, difficult for municipal and regional planning authorities to accomplish on their own?
3. What is a key difference between Canadian and American planning law? How does this difference affect planning practice in Canada?
4. What particular skills and expertise does a planner have? Should a planner be considered an expert in the development of plans and policies?
5. What was the extent of urban renewal projects in Canada during the postwar era? Are there any examples in your city or region?
6. Regional planning has a considerable history in many Canadian provinces. What are some of the responsibilities of regional planning authorities?
7. Attend a public meeting held by your municipal planning department to observe the ways in which planners and members of the public share information. How were differing points of view exchanged? How were comments from members of the public to be used in the planning process?

(continued)

8. Planners can work for a number of different public, private, and non-private organizations. How might the planner's role change while working in these different sectors?
9. What are some challenges that planners face in their efforts to decrease urban sprawl through increased densities and neo-traditional built forms?
10. Planners try to include a range of stakeholders in planning processes. Which ways would be most suitable for small towns or rural areas?
11. Discuss an ongoing social justice issue affecting the planning of communities in Canada.
12. How have structural changes in the economy affected the spatial form and layout of Canadian cities?
13. Historically, Canadian governance has gone through a regulatory era, a redistributive era, and a re-regulatory era. Give examples of the planning policies or programs implemented during each of these eras.
14. Many inner-city neighbourhoods and former industrial areas in Canada were redeveloped in the 1970s. Choose a major urban development project located in or near your city. How was the area redeveloped, and what challenges were encountered?
15. Several areas of planning theory seem to have converged upon the ideas of walkable communities with a medium-density, mixed-use built form and a range of transportation choices. What are these strands of research and their main characteristics?

REFERENCES

Alexander, C., S. Ishikawa, M. Silverstein, M. Jacobson, I. Fiksdahl-King, and S. Angel. 1977. *A Pattern Language*. New York: Oxford University Press.

Ames, H. 1897. *The City below the Hill: A Sociological Study of a Portion of the City of Montreal, Canada*. Montreal: Bishop Engraving and Printing Company.

Arnstein, S. 1969. "A Ladder of Citizen Participation." *Journal of the American Planning Association* 35(4): 216–24.

Artibise, A. 1981. *Prairie Urban Development*. Canadian Historical Association Booklet no. 34. Ottawa: Canadian Historical Association.

———. 1988. "Canada as an urban nation." *Daedalus* 117(4): 237–64.

———. and G. Stelter. 1982. *Shaping the Urban Landscape: Aspects of the Canadian City-Building Process*. Ottawa: Carleton University Press.

Beukers, E., L. Bertolini, and M. Te Brömmelstroet. 2012. "Why cost benefit analysis is perceived as a problematic tool for assessment of transport plans: A process perspective." *Transport Research Part A* 46: 68–78.

Booth, C. 1902. *Life and Labour of the People in London*. 3rd edn. London: Macmillan.

Branch, M.C., and I.M. Robinson. I.M. 1968. "Goals and objectives in civil comprehensive planning." *Town Planning Review* 38(4): 261.

Bula, F. 2011. "Home in the suburbs, heart in the city." *Globe and Mail* 8 April.

Burgess, E.W. 1925. "The growth of the city: An introduction to a research project." In R.E. Park, E.W. Burgess, and R.D. McKenzie, eds, *The City*, 47–62. Chicago: University of Chicago Press.

Calthorpe, P. 1993. *The Next American Metropolis*. New York: Princeton Architectural Press.

Campbell, H., and R. Marshall. 2002. "Utilitarianism's bad breath? A re-evaluation of the public interest justification for planning." *Planning Theory* 5(1): 163–87.

Centre for Neighborhood Technology. 2006. *Preserving and Promoting Diverse Transit-Oriented Neighborhoods*. Chicago: Centre for Transit-Oriented Development.

Chabot, M., and G. Duhaime. 1998. "Land-use planning and participation: The case of Inuit public housing (Nunavik, Canada)." *Habitat International* 22(4): 429–47.

CIP (Canadian Institute of Planners). 2014. "History of CIP." http://www.cip-icu.ca/web/la/en/pa/9AB7D1E80D2A-48BD915C86479FBFCB92/template.asp.

Considine, M. 2001. *Enterprising States: The Public Management of Welfare-to-Work*. Cambridge: Cambridge University Press.

Cullingworth, J.B., ed. 1987. *Urban and Regional Planning in Canada*. New Brunswick, NJ: Transaction Books.

Davidoff, P. 1965. "Advocacy and pluralism in planning." *Journal of the American Institute of Planners* 31(4): 331–8.

DiMaggio, P., and W. Powell. 1983. "The iron cage revisited: Institutional isomorphism and collective rationality in organisational fields." *American Sociological Review* 48: 147–60.

Dolowitz, D., and D. Marsh. 2000. "Learning from abroad: The role of policy transfer in contemporary policy making." *Governance* 13(1): 5–24.

Fainstein, S.S. 2000. "New directions in planning theory." *Urban Affairs Review* 35(4): 451–78.

——. 2014. "The just city." *International Journal of Urban Sciences* 18(1): 1–18.

Filion, P. 1999. "Rupture or continuity? Modern and postmodern planning in Toronto." *International Journal of Urban and Regional Research* 23(3): 421–44.

——. 2003. "Towards Smart Growth: The difficult implementation of alternatives to urban dispersion." *Canadian Journal of Urban Research: Canadian Planning and Policy* 12(1): 48–70.

——. and K. McSpurren. 2007. "Smart Growth and development reality: The difficult coordination of land use and transportation objectives." *Urban Studies* 44(3): 501–23.

Florida, R. 2002. *The Rise of the Creative Class and How It's Transforming Work, Leisure and Everyday Life*. New York: Basic Books.

Flyvbjerg, B. 1998. *Rationality and Power*. Chicago: University of Chicago Press.

——. 2001. *Making Social Science Matter: Why Social Inquiry Fails and How It Can Succeed Again*. Cambridge: Cambridge University Press.

Forester, J. 1989. *Planning in the Face of Power*. Berkeley: University of California Press.

Fowler, T.M. 1881. *Birds-Eye View of Winnipeg*. Ottawa: Mortimer, A. Lith.

Friedan, B. 1963. *The Feminine Mystique*. New York: Dell.

Friedmann, J. 1973. *Retracking America: A Theory of Transactive Planning*. Garden City, NJ: Anchor Press/Doubleday.

——. 1987. *Planning in the Public Domain: From Knowledge to Action*. Princeton, NJ: Princeton University Press.

Frumkin, H., L.D. Frank, and R. Jackson. 2004. *Urban Sprawl and Public Health: Designing, Planning, and Building for Healthy Communities*. Washington: Island Press.

Geels, F.W., and J.W. Schot. 2007. "Typology of socio-technical transition pathways." *Research Policy* 36(3): 399–417.

Grant, J. 1994. "On some public uses of planning 'theory': Rhetoric and expertise in community planning disputes." *Town Planning Review* 65(1): 59.

——. 2005. "Rethinking the public interest as a planning concept." *Plan Canada* 45(2): 48–50.

——. ed. 2008. *Reader in Canadian Planning: Linking Theory and Practice*. Scarborough, ON: Thomson Nelson Canada.

——. 2009. "Theory and practice in planning the suburbs: Challenges to implementing new urbanism, smart growth, and sustainability principles." *Planning Theory and Practice* 10(1): 11–33.

Grin, J. 2010. "Understanding transitions from a governance perspective." In J. Grin, J. Rotmans, and J. Schot, eds, *Transitions to Sustainable Development: New Directions in the Study of Long Term Transformative Change*, 47–80. London: Routledge.

——. J. Schot, and J. Rotmans. 2011. "On patterns and agency in transition dynamics: Some key insights from the KSI programme." *Environmental Innovations Societal Transitions* 1(1): 76–81.

Haan, M. 2005. "The decline of immigrant home-ownership advantage: Life-cycle, declining fortunes and changing housing careers in Montreal, Toronto, and Vancouver." *Urban Studies* 42(12): 2191–212.

Harris, R. 2000. "More American than the United States: Housing in urban Canada in the twentieth century." *Journal of Urban History* 26(4): 456–78.

Harvey, D. 1973. *Social Justice and the City*. Baltimore: Johns Hopkins University Press.

——. 1985. *The Urbanization of Capital*. Baltimore: Johns Hopkins University Press.

——. 1997. "The new urbanism and the communitarian trap." *Harvard Design Magazine* winter/spring. www.gsd.harvard.edu/hdm/harvey.htm.

Healey, P. 1996a. "Planning through debate: The communicative turn in planning theory." In S. Campbell and S.S. Fainstein, eds, *Readings in Planning Theory*, 234–57. Oxford: Blackwell.

——. 1996b. "Consensus-building across difficult divisions: New approaches to collaborative strategy making." *Planning Practice and Research* 11(2): 207–16.

——. 1997. *Collaborative Planning*. Hampshire, UK: Macmillan.

——. 2003. "Collaborative planning in perspective." *Planning Theory* 2(2): 101–23.

Hess, P.M. 2009. "Avenues or arterials: The struggle to change street building practices in Toronto, Canada." *Journal of Urban Design* 14(1): 1–28.

Hiebert, D., and D. Ley. 2003. "Assimilation, cultural pluralism, and social exclusion among ethnocultural groups in Vancouver." *Urban Geography* 24(1): 16–44.

Hodge, G. 1968. "Urban structure and regional development." *Papers of the Regional Science Association* 21(1): 101–23.

——. 1971. "Comparison of urban structure in Canada, the United States, and Great Britain." *Geographical Analysis* 3(1): 83–90.

——. 1985. "The roots of Canadian planning." *Journal of the American Planning Association* 51(1): 8–22.

——. 2004. "Regional planning: The Cinderella discipline." *Plan Canada 75th Anniversary Special Edition*.

Howard, E. 1898. *To-Morrow: A Peaceful Path to Real Reform*. London: Swan Sonnenschein.

Howlett, M. 2009. "Governance modes, policy regimes and operational plans: A multi-level nested model of policy instrument choice and policy design." *Policy Sciences* 42: 73–89.

Hoyt, H. 1939. *The Structure and Growth of Residential Patterns in American Cities*. Washington: Federal Housing Administration.

Hulchanski, J.D. 2007. "The three cities within Toronto: Income polarization among Toronto's neighbourhoods, 1970–2000." Research Bulletin 41. Toronto: University of Toronto Centre for Urban and Community Studies.

Hutton, T.A. 2004. "The new economy of the inner city." *Cities* 21(2): 89–108.

Jacobs, J. 1961. *The Death and Life of Great American Cities*. New York: Random House.

Kelbaugh, D. 1989. *The Pedestrian Pocket Book: A New Suburban Design Strategy*. New York: Princeton Architectural Press.

Kelly, P., M. Astorga-Garcia, and E. Esguerra. 2009. "Explaining the deprofessionalized Filipino: Why Filipino immigrants get low-paying jobs in Toronto." Working Paper no. 75. Toronto: Joint Centre of Excellence for Research on Immigration and Settlement (CERIS).

Lands, L. 2008. "Be a patriot, buy a home: Re-imagining home owners and home ownership in early 20th century Atlanta." *Journal of Social History* 41(1): 943–65.

Le Corbusier. 1935. *Radiant City* (*La ville radieuse*). Paris: Éditions de l'architecture d'aujourd'hui.

Lloyd, G. 2006. "Planning and the public interest in the modern world." Sir Patrick Geddes Commemorative Lecture, Royal Town Planning Institute in Scotland.

Lowi, T.J. 1972. "Four systems of policy, politics, and choice." *Public Administration Review* 32(4): 298–310.

Mahoney, J. 2006. "Suburban myths demolished." *Globe and Mail* 31 July.

Marsden, G., and D. Stead. 2011. "Policy transfer and learning in the field of transport: A review of concepts and evidence." *Transport Policy* 19: 492–500.

Mayor of Vancouver. 2010. "Burrard Bridge bike lane reaches one million cyclists." 8 July. http://www.mayorofvancouver.ca/burrard-bridge-bike-lanes-reach-one-million-cyclists.

Miles, M.B., and A.M. Huberman. 1994. *Qualitative Data Analysis: An Expanded Sourcebook*. Thousand Oaks, CA: Sage.

Minister of Public Works and Government Services Canada. 2009. *Discover Canada: The Rights and Responsibilities of Citizenship*. Ottawa: Minister of Public Works and Government Services Canada.

Nicholls, R.W. 1938. *Housing in Canada, 1938. Housing Yearbook 1938: National Association of Housing Officials, Chicago, 1938*. Toronto: Urban Policy Archive, Centre for Urban and Community Studies, University of Toronto.

Oberlander, P.H., and A.L. Fallick. 1992. *Housing a Nation: The Evolution of Canadian Housing Policy*. Ottawa: Centre for Human Settlements, University of British Columbia, and Canada Mortgage and Housing Corporation.

Pendakur, K., and R. Pendakur. 2011. "Colour by numbers: Minority earnings in Canada 1996–2006." Working Paper 11-05. Vancouver: Metropolis British Columbia.

Perks, W.T., and G. Simmins. 2013. "Urban and regional planning." *The Canadian Encyclopedia*. http://www.thecanadianencyclopedia.com/en/article/urban-and-regional-planning.

Perry, C. 1929. *The Neighbourhood Unit*. London: Routledge/Thoemmes.

Ploger, J. 2004. "Ethics in Norwegian planning: Legitimacy, ambivalence, rhetoric." *Planning, Practice and Research* 19(1): 49–66.

Punter, J. 2004. *The Vancouver Achievement*. Vancouver: University of British Columbia Press.

Purdy, S. 1997. "Industrial efficiency, social order and moral purity: Housing reform thought in English Canada, 1900–1950." *Urban History Review* 25(2): 30–40.

Putnam, R. 2000. *Bowling Alone: The Collapse and Revival of American Community*. New York: Simon and Schuster.

Robinson, I.M., and D.R. Webster. 1985. "Regional planning in Canada: History, practice, issues, and prospects." *Journal of the American Planning Association* 51(1): 23–33.

Sandercock, L., ed. 1998. *Making the Invisible Visible: A Multicultural Planning History*. Berkeley and Los Angeles: University of California Press.

Sennett, R. 1971. *The Uses of Disorder*. New York: W.W. Norton.

Smith, P.J. 2013. "Commission of Conservation." http://www.thecanadianencyclopedia.com/en/article/commission-of-conservation.

Soja, E.W. 2010. *Seeking Spatial Justice*. Minneapolis: University of Minnesota Press.

Spaans, M., and E. Louw. 2009. *Crossing Borders with Planners and Developers and the Limits of Lesson-Drawing*. City Futures in a Globalising World. Madrid: University Rey Juan Carlos of Madrid.

Stone, D. 2004. "Transfer agents and global networks in the 'transnationalization' of policy." *Journal of European Public Policy* 11(3): 545–66.

Switzer, A., L. Bertolini, and J. Grin. 2013. "Transitions of mobility systems in urban regions: A heuristic framework." *Journal of Environmental Policy and Planning* 5(2): 141–60.

Talen, E., and C. Ellis. 2002. "Beyond relativism: Reclaiming the search for good city form." *Journal of Planning Education and Research* 22: 36–49.

Tattrie, J. 2014. "Africville." *The Canadian Encyclopedia*. http://www.thecanadianencyclopedia.ca/en/article/africville.

Thomas, R. 2013. "Resilience and housing choices among Filipino immigrants in Toronto." *International Journal of Housing Policy* 13(4): 408–32.

——. and L. Bertolini. 2014. "Beyond the case study dilemma in urban planning: Using a meta-matrix to distil critical success factors in transit-oriented development." *Urban Policy and Research* DOI 10.1080/08111146.2014.882256.

Vigar, G. 2012. "Planning and professionalism: Knowledge, judgement and expertise in English planning." *Planning Theory* 11(4): 361–78.

Walks, R.A. 2011. "Economic restructuring and trajectories of socio-spatial polarization in the twenty-first-century Canadian city." In L. Bourne, T. Hutton, R. Shearmur, and J. Simmons, eds, *Canadian Urban Regions: Trajectories of Growth and Change*, 125–59. London and Toronto: Oxford University Press.

——. 2013. "Income inequality and polarization in Canada's cities: An examination and new form of measurement." Research Paper 227. Toronto: University of Toronto, Cities Centre.

——. and L.S. Bourne. 2006. "Ghettos in Canadian cities? Racial segregation, ethnic enclaves and poverty concentration in Canadian urban areas." *The Canadian Geographer* 5(3): 273–97.

Wolfe, J. 2004. "Our common past: An interpretation of Canadian planning history." *Plan Canada 75th Anniversary Special Edition*.

World Commission on Environment and Development (WCED). 1987. *Our Common Future*. London: Oxford University Press.

Yiftachel, O. 1998. "Planning and social control: Exploring the dark side." *Journal of Planning Literature* 12(4): 395–406.

Yin, R.K. 1984. *Case Study Research: Design and Methods*. Thousand Oaks, CA: Sage.

——. 1991. *Applications of Case Study Research*. Washington: Cosmos.

Young, I.M. 1990. *Justice and the Politics of Difference*. Princeton, NJ: Princeton University Press.

PART B

Case Studies

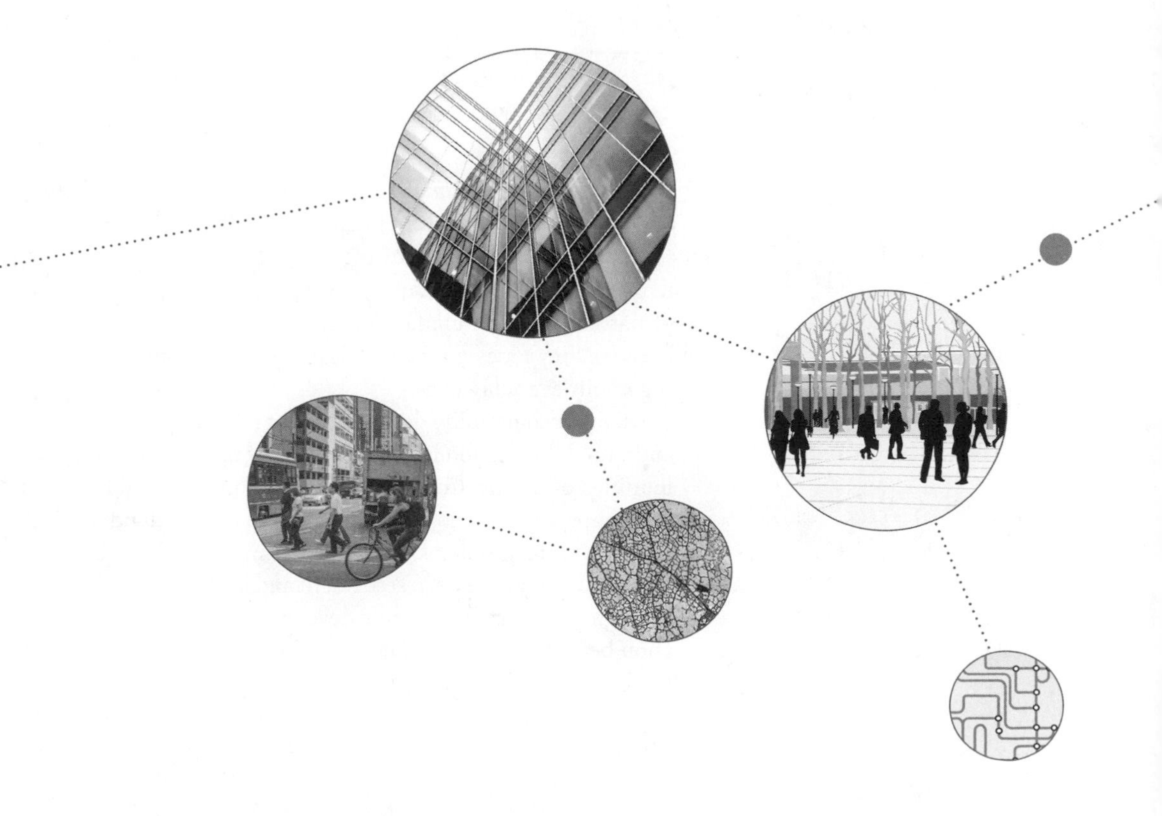

2.0

INTRODUCTION

Community Development and Social Planning

This first section of *Planning Canada* examines the areas of community development and social planning. Like others in this volume, these case studies span several subtopics or sub-disciplines of planning: Michaela Cochran's exploration of social vulnerability to climate change, for example, could also fit under the broad category of natural resource management. However, these cases have been grouped together because they focus on the physical and social communities that have so much meaning in our everyday lives.

Neither community development nor social planning is easily defined, but all of the authors in this section have explored how planners can improve the well-being of communities over time. Community associations, charitable organizations, business interest groups, advisory boards, school boards, and individuals have historically addressed pressing social problems, from advocating better housing for workers in the late 1800s to assisting immigrants in settlement and integration today. Supporting local businesses through community economic development, encouraging communication and cooperation between groups, and empowering community members to achieve change are all key ideas in social planning.

Many planners in this sub-discipline are concerned with the rights of women, Aboriginal peoples, immigrants, and the low-income population and act as community advocates themselves. Although planning law in Canada has limited individual property-owners' rights in order to achieve community-wide benefits (Hodge 1985), in reality there are always groups with more power or influence in our society. Planners and advocates often empower or educate local residents to achieve change, such as establishing community organizations or providing their own local knowledge to inform planning processes. Planners' roles include being experts in community engagement, acting as facilitators, and conducting community-based research to improve public service delivery.

Community development and social planning, like other sub-disciplines in the field, involve the input of municipal, provincial, and federal governments in providing social and community services and funding programs and plans that impact neighbourhoods and cities. However, governance roles have changed over time. Early in Canadian planning history, informal and community organizations were particularly influential in addressing urban issues—for example, the Council of Women advocated for better conditions for children in urban areas during the reform era (Wolfe 2004). The federal and provincial governments have responsibility for social infrastructure, as outlined in the Constitution Act, but municipal governments also influence local residents' quality of life (FCM 2010). As the planning profession became more established and Canada's welfare state strengthened by the 1930s, policies involving unemployment benefits, assistance to working families, and old age pensions were established.

Strong state involvement in community services began to recede during the 1970s as privatization, contracting out, and downsizing aimed to decrease public-sector expenditures; this process continued until the mid-1990s. It was at this time that community-based and non-profit organizations and new types of businesses emphasizing community benefits (e.g., social enterprises) began to expand their activities in social planning and community development to help fill the gap left by the upper levels of government. Municipalities have also taken a more active role in this area. Some municipalities now have dedicated staff for social planning, such as Vancouver and Richmond, while others work in partnership with the non-profit sector to achieve community benefits.

In the first case study, John Foster, Olga Shcherbyna, and Leonora Angeles explore the development of the Social Development Strategy in Richmond, British Columbia. With an extensive community engagement process and a co-learning process for municipal staff, the City of Richmond was able to develop a strategy that met the needs of a population that is very diverse in age, immigration status, and income level. The case shows how the City managed to engage multiple publics using traditional and digital tools and also provides an excellent description of the roles of the different levels of government in social planning.

In the second case, Kari Huhtala offers an introduction to cultural planning using three case studies: Kelowna, British Columbia; Kingston, Ontario; and Moncton, New Brunswick. Huhtala defines cultural planning and the main actors in developing community goals and programs. He describes the process of developing and implementing cultural plans in these three cities and describes the main challenge as engaging community stakeholders, many of whom have no background in culture or the arts. Cultural planning is a relatively new and rapidly developing area of municipal and regional planning, and these three cases offer many lessons for smaller cities with limited financial and staff resources.

Authors Umbreen Ashraf, Kate Kittredge, and Magdalena Ugarte explain how the City of Vancouver attempted a community-building exercise aimed at increasing intercultural understanding between First Nations, urban Aboriginals, and immigrants. Over a three-year period, the Dialogues Project adopted an appreciative inquiry approach under which participants shared their stories with each other, completed cultural exchange visits, and engaged in creative collaborative projects. Several outcomes, including the Newcomers Guide to First Nations and Aboriginal Communities and the development of stronger relationships between governments and community organizations, are relevant for other municipalities.

Michaela Cochran presents the case of Yarmouth, Nova Scotia. Undertaken as one of a series of studies for the Atlantic Climate Adaptation Solutions Association (ACAS), the case study explores how the Town and Municipality of Yarmouth could reduce its social vulnerability to climate change. Using methods that could easily be transferred to other small municipalities, this case shows how careful consideration of the needs of vulnerable groups can reduce the long-term impacts in the case of storm outages, flooding, or other effects of climate change.

In their exploration of the impacts of urban poverty on lone-parent female-headed households, Silvia Vilches and Penny Gurstein depict social planning as a process that is currently being transformed by neo-liberal policy shifts in Canada. By examining the impact of policies originating at the provincial level of government on lone-parent, low-income women in East Vancouver, the case study clearly illustrates the gaps created by changing governance roles and responsibilities in the provision of social services. It highlights the role of high-level policy on local lives but also the agency of individuals.

Clearly, all of these cases emphasize the planning process and the role of dialogue and engagement in understanding the needs of communities. Building relationships between the various actors, institutions, and demographic groups is increasingly difficult in Canadian cities as they become more diverse. Not only are many neighbourhoods becoming more multicultural, they also include a diversity of age and income groups with aging-in-place options and limited affordable housing options. Listening to and understanding the needs of these different groups, rather than one undifferentiated "public," can be difficult.

Partnerships between governments and local actors are common in implementing cultural and social plans and delivering services. In many cases, governance has become more complex as senior governments have shifted responsibilities, without necessary funding, to municipalities. On the other hand, municipalities have shown a great deal of capacity to develop plans in areas that were previously unknown to them, such as cultural planning, with the help of grants from provincial and federal governments. Small municipalities have also begun to build partnerships with non-profits and community organizations to help them implement plans. There is a lot of potential for local people to have a say in the social and cultural well-being of their neighbourhoods, particularly on pressing issues such as climate change. Planners need to balance these needs with plans and policies that are flexible and can be monitored and adjusted over time to ensure their responsiveness to local social and economic conditions.

REFERENCES

FCM (Federation of Canadian Municipalities). 2010. *Mending Canada's Frayed Social Safety Net: The Role of Municipal Governments*. Theme report no. 6. Quality of Life in Canadian Communities Series. Ottawa: FCM.

Hodge, G. 1985. "The roots of Canadian planning." *Journal of the American Planning Association* 51(1): 8–22.

Wolfe, J. 2004. "Our common past: An interpretation of Canadian planning history." *Plan Canada 75th Anniversary Special Edition*.

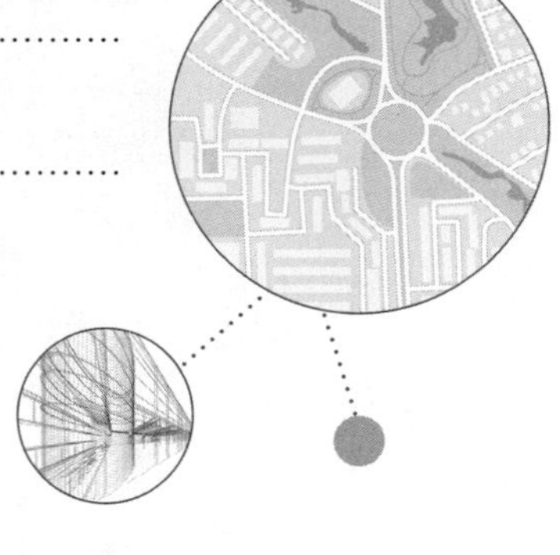

CASE STUDY 2.1

Promising Practices in Social Plan Development and Implementation: Applying Lessons Learned from Canadian Urban Municipalities in the City of Richmond, British Columbia

John Foster, Olga Shcherbyna, and Leonora Angeles

SUMMARY

The City of Richmond, British Columbia, is used as a case study to examine some of the challenges and solutions in developing municipal social policies in a fast-changing multicultural municipality facing multiple social issues, such as housing affordability, childcare, immigrant integration, opportunities for public engagement, and planning for an aging population. The case study demonstrates some promising practices in social planning and social plan and strategy development based on lessons learned from other Canadian municipalities. Examining the governance framework, tools used, and process for developing the Richmond Social Development Strategy, the case study distills some anticipated outcomes of the strategy implementation that Richmond and other cities need to consider when developing policies and programs on social issues.

OUTLINE

- The Canadian governance context, along with an understanding of global trends, was critical in Richmond's approach to social planning.
- Several social challenges in the City of Richmond, including demographics, affordability, and migration, influenced the development of Richmond's Social Development Strategy. Extensive public consultation contributed to a strong strategy.
- The implementation strategies and lessons learned are instructive for other communities undertaking social plans.

Introduction

Despite the lack of mandate and resources to address social issues on their own, many Canadian cities have developed comprehensive long-term Social Development Plans aimed at improving their communities' well-being.

The City of Richmond, as the fourth-largest municipality in British Columbia, presents an interesting case study in Social Development Strategy preparation. The city has the highest immigrant population in Canada (59 per cent) and among the highest proportions of visible minorities (65 per cent) of any census subdivision (Statistics Canada 2006a). While the city is generally regarded as an affluent community, with average household incomes above the national average, it also has BC's second-highest rates of households/families living in poverty or below Statistics Canada's **Low Income Cut-offs** (20.9 per cent and 18.9 per cent, respectively) (Statistics Canada 2006a). The city faces many social challenges, including planning for aging and immigrant population groups, insufficient affordable housing and childcare spaces, emerging ethnic-specific recreational needs, and growing social inequality.

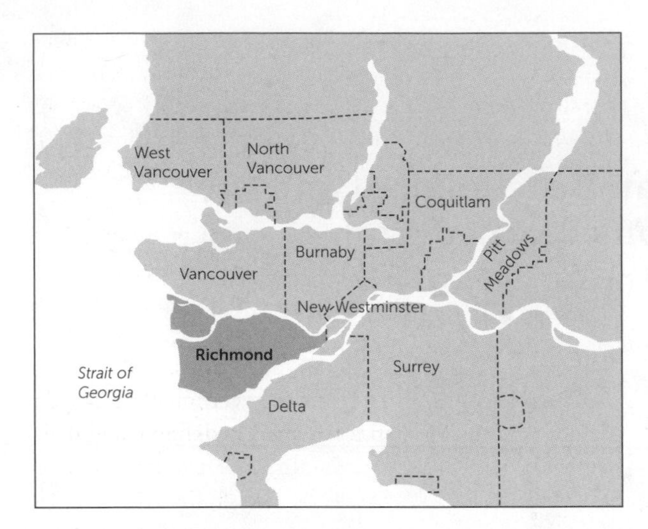

Figure 2.1.1 • Location map.

This case study presents the process and outcomes of the City of Richmond's journey in preparing its Social Development Strategy—a journey that was informed by "best practices" research, considering lessons learned from other Canadian municipalities.

Over the years, the City had developed a number of social policies to address social issues, including a Child Care Plan, Child Care Needs Assessment and Strategy, Homelessness Needs Assessment and Strategy, and Community Wellness and Affordable Housing Strategies. Notwithstanding these policies, Richmond Council recognized that with municipal social issues becoming more diverse and complex, a broader, more comprehensive, co-ordinated, and strategic approach was required. In 2009, the council commissioned preparation of the Ten-Year Richmond Social Development Strategy. The draft strategy was completed in late 2012 and adopted by City Council in September 2013. Its preparation involved an extensive community engagement process, municipal staff co-learning processes, and "best practices" research on existing Canadian municipal social plans (Shcherbyna 2011).

This case study begins by clarifying the federal, provincial, and municipal governance and policy frameworks that informed Richmond's social planning efforts. It covers the role of planners and the public in the process and the alternatives considered by the City based on the "best practices" research undertaken by a graduate student

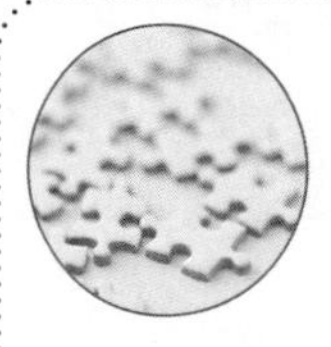

KEY ISSUE

Lessons from Three Canadian Municipalities

In 2011, social planners from the City of Burnaby, City of Surrey, and City of Edmonton were interviewed as a part of the research conducted by a social policy intern for the City of Richmond. Among other findings, a number of lessons in developing and implementing Social Strategies/ Plans were identified. These lessons included:

- Identify clear objectives for the Social Strategy.
- Secure the leadership of the council and mayor.
- Allocate sufficient time/human resources to all phases of planning.
- Use innovative techniques to engage different community groups in public consultations.
- Spend time on developing background information, including an inventory of programs and services and demographic profiles.
- Ensure that all materials related to the Social Strategy are easily accessible in terms of both language and distribution.
- Ensure that the public understands the limited role of municipal government in addressing social needs.
- Identify and engage potential partners in the early stages of planning.
- Emphasize the actions, and identify the first steps for implementation.
- Develop a solid measurement and reporting framework.
- Think bigger than basic needs.

In addition, the following were considered important for the preparation of a Social Strategy policy document:

- using plain and consistent language;
- using simple graphics to explain complex processes and comprehensive concepts;
- using images to convey the messages.

intern. It also analyzes the assumptions of risks and challenges expected to be encountered by Richmond in its Social Development Strategy implementation. Finally, it examines the tools the City considered to address the challenges that often emerge from siloed or compartmentalized planning approaches.

Federal, Provincial, and Municipal Governance and Policy Frameworks Informing Social Planning in Richmond

Limited Social Role of Canadian Municipalities

Social planning involves a variety of processes aimed at improving the well-being and quality of life in local communities. Traditionally, Canadian municipalities had limited jurisdiction over social issues. As articulated in the Canadian Constitution, the federal and provincial governments have exclusive legislative powers in the areas of social infrastructure. In reality, however, all levels of government influence local residents' quality of life (FCM 2010). While federal, provincial, and municipal governments play different roles in the social well-being of Canadian communities, there are other players as well, such as school boards, not-for-profit community agencies, grassroots informal networks and movements, local foundations, and other organizations.

The roles of the three levels of government in Canada are not particularly distinct or sharply defined, and their responsibilities in the social realm often overlap. The situation is further complicated because local governments vary significantly in terms of their form and function (e.g., different levels of delegated responsibilities from province to province). Notwithstanding these differences, Canadian municipalities share a common set of duties, as summarized in Table 2.1.1.

Table 2.1.1 • Canadian Government Roles and Responsibilities

Level of Government	Roles and Responsibilities
Federal government responsibility	Employment insurance Fiscal equalization National standards to ensure comparable levels of public service across the country
Joint federal/provincial responsibilities	Pension and income support Housing Post-secondary education, training, and research Public health
Provincial responsibility	Primary and secondary education Health care Municipal government Social assistance and societal services
Municipal responsibility (delegated through provincial enabling legislation)	Water Sewage Parks and recreation Fire services Police Waste management Public transit Land-use planning Animal control Economic development Libraries Maintenance (repairing municipal roads, clearing snow, cleaning public areas)

Note: A more detailed breakdown of government responsibilities in Canada can be found in FCM 2006, 33.

Downloading Social Services onto Municipalities

From 1970 to 1995, Canada experienced increased budget deficits similar to those in the US, the UK, France, and many other countries (Broadbent 2010). This prompted the Government of Canada to introduce measures commonly referred to as **downloading**. Municipalities were affected by this fiscal squeeze.

As result of downloading, the Canadian federal government balanced its budget in the 1990s. Many other Western countries, in contrast, took advantage of growing economies and managed to bring their budgets in order without such significant program cuts (Broadbent 2010).

Canadian municipalities experienced decades of increasing demand for social and other services with limited sources of revenue available to meet this demand. Canadian municipalities have no access to income and sales taxes, the two largest revenue producers. They largely relied on property tax, a vehicle that does not adequately prepare municipalities for the demands of economic growth (FCM 2012; 2013b).

Municipal Social Roles in British Columbia

In July 1994, the Legislative Assembly of the Province of British Columbia passed revisions to the Municipal Act, which provided the legal basis for BC's municipalities to address social concerns. The revisions explicitly enabled municipalities to undertake social planning and incorporate social planning components into the **Official Community Plan (OCP)** (Government of British Columbia 1994).

While the legislation enabled municipalities to undertake social planning, it did not elaborate on the scope of the municipal role in this realm—nor did it address the issue of municipal capacity to respond to increasing demand in social services. These limitations set the context for the social challenges faced by BC cities like Richmond.

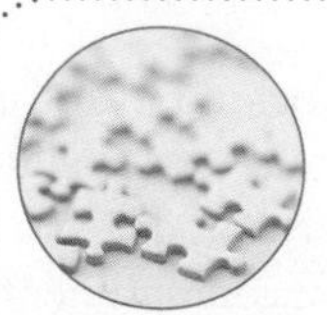

KEY ISSUE

Global Trends Affecting Canadian Municipalities

Urbanization According to UNFPA (United Nations Population Fund), in 2008, for the first time in history more than 50 per cent of the world's population lived in towns and cities. It is estimated that by 2030, the number of urban dwellers will grow to 5 billion people. In Canada between 1976 and 2006, the urban population increased by 30 per cent, and by 2006 more than 80 per cent of the Canadian population was urban (UNFPA website).

Migration In recent times, the number of people migrating in search of a better life has increased. In 2005, it was estimated that there were 185 to 192 million migrants worldwide, which represents 2.9 per cent of the global population. Nearly 20 per cent of Canada's population, or 6,186,950 people, are foreign-born, making it second only to Australia in terms of how many people make up the non-native portion of its total population (Statistics Canada 2006b; International Organization of Migration 2005; Statistics Canada 2006c).

Globalization Economic restructuring—as a result of globalization—has brought significant changes to the economy. Canada is in the "global race for talent," and in order to attract investments and global workforce, it will need cities with a high quality of life, developed public transportation systems, and affordable housing (FCM 2011).

Climate Change Canadian municipalities, along with other cities in the world, are faced with the effects of climate change. Global warming, increased extreme weather, sea-level rise, and altered rainfall patterns have created increased risk to municipal assets and infrastructures. According to the Conference Board of Canada, Canada is one of the world's largest per capita greenhouse gas emitters (Conference Board of Canada 2010).

Setting the Context: Richmond's Social Development Challenges

The City of Richmond is one of Metro Vancouver's 22 municipalities, and it is the fourth-most populous city in British Columbia with 199,302 residents

(City of Richmond, n.d.). It is located on Canada's Pacific coast in close proximity to downtown Vancouver and the US border. It is also home to the Vancouver International Airport and the Olympic Oval venue, which hosted track speed skating competitions for the Vancouver 2010 Olympic and Paralympic Winter Games.

Some of the key challenges specifically affecting Richmond are discussed below.

Aging Population

Richmond is known for its long life expectancy—four years longer than the national average of 81.2 years (City of Richmond 2011a) and greater than that of Japan (81.4 years), considered as one of the world's healthiest nations (Beun-Chown 2007).

In 2011, roughly 28 per cent of Richmond residents were 55 years old or older (Statistics Canada 2011). Geographically, the city is well suited for older residents because of its flat landscape and accessible streetscapes. The City is still facing challenges in responding to expanding and diverse social and related infrastructure needs. A "one size fits all" approach is not appropriate, given the large variation among the older adult population (e.g., baby boomers who remain in the workforce, the recently retired, immigrant seniors, and those with declining physical and mental abilities). In planning for housing, it is thus essential that consideration be given to the mobility and financial challenges faced by many older adults. For recreational and civic engagement strategies, ethno-cultural differences must be reflected.

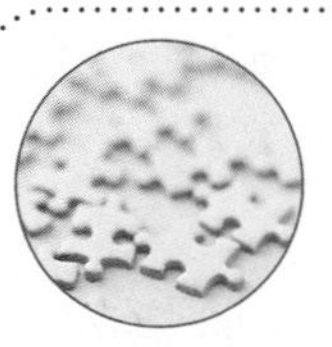

KEY ISSUE

Characteristics of the Age-Friendly City

According to the Federation of Canadian Municipalities, a number of age-friendly planning processes and services can keep seniors engaged and active members of our society, such as low-floor buses, safe sidewalks in winter, developed public transit systems, and access to housing within walking distance to amenities (FCM 2013a, 2).

Diverse Richmond

Known for its dynamic immigrant and Asian communities, Richmond went through significant population growth in the early 1990s when there was a large influx of immigrants from Hong Kong. With one of the highest percentages of immigrants in Canadian cities, Richmond saw nearly two-thirds of its immigrants arriving during the past 20 years. Close to 80 per cent of the newcomers came from China, Hong Kong, East Asia, the Philippines, India, and the United Kingdom and contributed significantly to local economic growth, particularly in the small-business and retail sectors. However, many newcomers face economic, social, and cultural integration challenges, such as lower employment and income rates than their Canadian-born counterparts (Aydemir and Skuterud 2005; Picot 2008; Shcherbyna 2012), limited language and/or communication skills, and lower rates of access to municipal services and participation in civic life. Further, immigrants have their own ethno-cultural norms and previous home-country experiences that shape their needs and preferences on such matters as housing, public spaces, and recreational activities (Qadeer 2009). Richmond, like other Canadian municipalities, is under pressure to develop new strategies and services to meet the growing demand for services to address the needs of newcomer residents and ensure their successful integration (FCM 2011). These demands include expanding and/or modifying existing services, such as access to affordable housing, recreational activities, libraries, childcare, and public transit, but also accommodating new realities, such as supporting the efforts of local settlement agencies or introducing programs to enhance intercultural harmony and strengthen intercultural cooperation. Balancing the values and needs of diverse ethnic groups and the non-immigrant population and avoiding an "us–them" dynamic are among the biggest challenges facing the community.

Other Challenges

While Richmond is home to a rather affluent population, it also has a high number of families with children who spend a disproportionate amount of their income on basic necessities such as food, clothing, and shelter. A growing number of individuals and families either who spend more than 50 per cent of their income on housing and/or whose living spaces do not meet minimum health and safety standards are at risk of homelessness. From 2007 to 2012, Richmond had a higher-than-average increase in apartment prices (21 per cent), the highest in Metro Vancouver

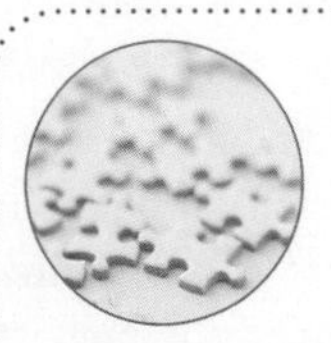

KEY ISSUE

Immigration and Diversity

Richmond has a larger proportion of foreign-born residents than the City of Toronto or the City of Vancouver. While traditionally, immigrants chose larger urban centres like Vancouver, Toronto, or Montreal as their arrival destinations, in the past decade the settlement patterns of newcomers in Canada have shifted in favour of suburbs. The high cost of living is pushing immigrants to move away from the major metropolitan areas. For example, "in pursuit of the North American dream," newcomer homeowners often look for bigger houses on larger lots, which are more affordable in suburbs (FCM 2011, 6, 7).

One of the best examples of the challenges the City faces with respect to diversity is the recent controversy over Chinese language signs in Richmond. A small group of English-speaking concerned citizens managed to collect more than 1000 signatures on a petition requesting that 70 per cent of signs in Richmond (businesses, shops, etc.) be in English and/or French. While the City decided not to "enforce" the English-language sign bylaw, it is constantly looking for innovative ways to create public spaces for community discussions to encourage cross-ethnic collaboration and collective efforts to break down barriers and mistrust in the community.

(City of Richmond 2012); hence, the issue of affordable and special-needs housing is an important concern.

Another key social challenge is limited municipal funding and resources for children, youth, and families. Healthy families and children are a cornerstone of a socially sustainable community. Considerable progress has been made in ensuring that Richmond becomes increasingly child- and family-friendly, such as providing recreational subsidies for children from low-income families and securing City-owned childcare facilities. However, the lack of provincial and national childcare frameworks poses significant challenges for Richmond and other Canadian municipalities in their efforts to support families with young children. Although a number of non-profit organizations provide essential support to children, youth, and families, their ability to meet growing service demand and expand their facilities is hindered by inadequate funding from senior levels of government and the increasing cost of securing office and programming space.

In addition to the challenges cited above, local residents do not always fully understand the limited role of municipalities in the social realm. Thus, one of the City's emerging responsibilities is to educate its residents on these constraints, its role, and how it collaborates with senior levels of government and other stakeholders to address social issues.

Process for Preparation of Richmond Social Development Strategy

In November 2009, Richmond City Council endorsed the goals, principles, and process for preparation of a Social Development Strategy for the city. The purpose of the Richmond Social Development Strategy was to:

- identify social development priorities for City attention over the next 10 years;
- clarify the roles of the City (and other stakeholders) with respect to addressing particular social development topics;
- provide a foundation for a more integrated, co-ordinated, and sustainable approach for social development in Richmond for the future.

In launching the strategy, council was aware of four key factors:

1. The City had a strong, proud, and effective legacy regarding social development. For example, it hired its first social planner in the early 1970s and, over time, had adopted numerous progressive social policies and pursued a diversity of social programs and initiatives. In planning for the future, the City was not starting from scratch. It had a strong foundation upon which to build.
2. The City was devoting considerable resources to social development concerns. The question

in moving forward was whether the resources would be deployed effectively and whether they would be targeted to the highest-priority needs.

3. While the City faced numerous social development challenges, such as downloading from senior governments, rising expectations from the community, and the increasing complexity of social issues, it also had a host of opportunities to build on. These opportunities included a strong tradition of community engagement, a robust network of community agencies, and a strong history of collaboration and consultation among government and non-government partners.
4. Finally, like other Canadian cities, the City of Richmond lacked the resources and mandate to address social issues on its own. According to FCM, Canadian municipalities receive only eight cents of every tax dollar, the remainder going to their federal and provincial counterparts (FCM 2006). Richmond Council thus recognized that to effectively address future social development issues, the City must be increasingly strategic and follow a multi-partnership approach.

The City's Corporate Sustainability Policy, adopted in 2010, and Richmond's Official Community Plan, adopted in 2012, have been linked with the Social Development Strategy. Sustainability principles have been reflected in the preparation of the Social Development Strategy:

- Social, economic, and environmental factors were considered in the strategy actions.
- Extensive and varied community consultation efforts were undertaken.
- A broad interdepartmental engagement was established.
- Preparation focused on the implications of today's decisions for future generations.

To help co-ordinate preparation of the strategy and to provide political oversight to the process, council directed that a Council/Staff Liaison Committee be established. The committee consisted of two members of council and four core staff members working on the strategy. An interdepartmental staff team was also struck to provide information and advice as the strategy preparation proceeded. The manager of the Community Social Development Division assumed the leadership role in developing the strategy and was primarily assisted by one social planning staff member from his department. The process outlined for preparation of the strategy involved four phases, described in Table 2.1.2.

During its Phase One activities, the City made a concerted effort to engage the community in preparation of the strategy. A variety of traditional and innovative approaches were used to ensure a broad engagement of local residents. These approaches included presentations, questionnaires, public forums, social media and online discussion forums, and small study circles with immigrants.

The presentations and discussions were held with the City's socially oriented advisory committees, community stakeholder groups, and others. The purpose was to provide information on and receive input regarding the Social Development Strategy.

A public forum on the strategy was widely advertised in print and online newspapers and the City website, yet only 24 local residents attended. While the attendance was disappointing, having an intimate group of interested residents allowed for a rich discussion and informed the development of the policy.

Table 2.1.2 • Creating Richmond's Social Development Strategy

Phases	Timelines
Phase One: Initial community engagement	November 2009–December 2010
Phase Two: Analysis and draft strategy preparation	January 2011–November 2012
Phase Three: Consultation, revision, and strategy adoption	January 2012–August 2013
Phase Four: Implementation, monitoring, and reporting on the strategy	September 2013 and onwards

Further, with consultant assistance, the City facilitated three workshops with key stakeholders (e.g., City advisory committee members, community agency representatives). The aim of these workshops was to identify Richmond's social development strengths, weaknesses, opportunities, and threats (SWOT analysis) and develop a preliminary 10-year Social Development Vision Statement for the City. As well, strategy options and priorities for consideration in the strategy and appropriate City roles in addressing the priority options were identified.

Concerted efforts were made during both public engagement phases to engage *non-traditional public engagement groups*, i.e., those who do not attend public forums and workshops. For example, considering the high percentage of Chinese-speaking residents, online and paper questionnaires were produced in English and Chinese.

The City also worked in conjunction with the Richmond Civic Engagement Network, a small non-profit organization, to plan and host four study circles: one for Mandarin-speaking residents, one for Cantonese-speaking residents, one for recent immigrants from various countries, and one for the general public. Each study circle group met on three occasions to identify issues and provide suggestions for the strategy. A wrap-up session involving all study circle groups was held at the conclusion of the individual group sessions. Although only 48 people participated in the sessions, this innovative method proved to be a valuable vehicle for engaging segments of the community that would not normally participate in city planning processes. New immigrants face a number of hurdles to their participation in planning and public engagement processes, from limited language proficiency to unfamiliarity with planning processes in Canada (Palermo 2012; Halifax Regional Municipality 2011). Holding small group discussions has been recognized as an effective tool for engaging newcomers with language barriers in planning processes (Schachter and Liu 2005).

In its efforts to reach out to a broader audience, including youth and young professionals, the City utilized various social media tools in its public consultations. The questionnaires were posted online, and all the documents and reports related to the development of the strategy were available on the City's website. In conjunction with work being done for the update of the City's Official Community Plan, the City used an online discussion forum (Let's Talk Richmond) to provide information and yield comments on the Social Development Strategy.

After the Phase One efforts ended, Phase Two work began, focused on the analysis and draft strategy preparation. From the analysis, it was clear that Richmond residents and other stakeholders appreciated the quality of life and breadth of services available in the city. They also cared passionately about the future social development of the community. A plethora of issues and suggestions were identified; however, no consensus existed on the priorities deserving City attention. In preparing the draft Social Development Strategy, staff considered the information generated through the consultations, balancing it with knowledge of existing commitments, current demographics and future growth projections, and the political and administrative context for moving forward. The essence of the strategy is captured in the Framework graphic (Figure 2.1.2 below).

Figure 2.1.2 • Richmond Social Development Strategy Framework.

Source: City of Richmond

The Framework consists of the Vision, three Goals, and nine Strategic Directions. For each Strategic Direction, recommended actions and associated timelines are specified, along with information on City roles and proposed partners.

- **Timelines** The actions are categorized according to four time frames for initiation: short term (0–3 years), medium term (4–6 years), long term (7–10 years), or ongoing.
- **Roles** The range of potential City roles are identified for each action, including undertaking planning, research, and policy development; delivering programs and services; engaging and empowering communities; collaborating and establishing partnerships; establishing infrastructure; providing land, space, or funding, advocating; and securing external contributions.
- **Proposed partners** A key assumption of the Social Development Strategy is that the City cannot address all social issues on its own. Therefore, for each action a range of proposed partners is identified, including senior governments, government agencies, non-profit agencies and community groups, Vancouver Coastal Health, Richmond School District, post-secondary institutions, faith and ethno-cultural groups, developers, businesses, community members, advisory committees, and other community partners.

The draft Social Development Strategy was presented to City Council in January 2013. Council in turn authorized staff to circulate the strategy for public comment, thus launching Phase Three of the strategy preparation process.

As with the Phase One consultation, a concerted effort was made to involve the community and secure comments on the draft strategy during Phase Three. Vehicles used included presentations to City advisory committees and other stakeholder groups, an open house, the City's website, and the Let's Talk Richmond online discussion forum. Additionally, a local immigrant-serving agency, one of the city's long-term community partners, launched an initiative to seek comments from their visitors/clients and received 36 comments both in English and in Chinese.

Responses to the draft were generally favourable. Those commenting were pleased with the thorough consultation that had occurred in the earlier phases of the strategy preparation. It was suggested that while the draft did not capture all their priorities, it was a solid, balanced document for guiding the City's future social development activities. As a result of Phase Three consultations, minor changes and revisions to the strategy were introduced; however, a major overhaul or rethinking was not required.

City Council adopted the final version of the strategy in September 2013. Council was highly complimentary to all involved and was particularly gratified by the extensive consultation processes undertaken and the strong community support shown for the document.

Reflecting on the Consultation and Social Policy Development Processes

Using "Best Practice" Lessons

The process of developing a city-wide comprehensive Social Development Strategy was a challenging and multifaceted process. Looking back, it became evident that the City managed to incorporate many lessons from the previous "best practice" research.

Like other municipalities, the City recognized the importance of allocating sufficient time to pursue broad and meaningful public engagement, ensuring that residents' and community partners' views were adequately reflected in the final draft of the strategy.

Given that no clear priorities were identified during the first phase of consultations, the staff had to evaluate a number of alternative frameworks in formulating major goals and strategic directions. Assessments of Richmond social and demographic data, existing City social policies and programs, and related materials were helpful in identifying the emerging trends while clarifying the priorities. Other useful sources included information from previous public engagement processes, comments from other departments, and the knowledge and expertise of the City's social planning staff.

Securing the leadership and support of the council was critical for the advancement of Richmond's Social Development Strategy. Council identified development of the strategy as a key priority. Strong political and administrative support thus existed, and sufficient resources were allocated for completion of the project. Moreover, two

Table 2.1.3 • Richmond Social Development Strategy: Public Consultations in Numbers

Engagement Tool	Phase One: Initial Community Engagement	Phase Three: Consultation, Revision and Strategy Adoption
Presentations and discussions	• 12 presentations • 8 written submissions	• 11 PowerPoint presentations by staff • 77 submissions from community partners and advisory committees
Stakeholder workshops	• 3 stakeholder workshops	

Questionnaires in Phase One	• 278 survey responses in total:
	  • 208 online surveys
	  • 70 printed surveys	

Public forum—Phase One	• 24 attendees of public forum	
Open house—Phase Three		• 52 attendees of open house • 11 comment sheets submitted by open house participants • 36 submissions from visitors of local immigrant-serving agency
Social media:		
Website	• 1003 distinct viewers • 2964 visits	• 507 distinct viewers • 965 visits
Let's Talk Richmond	• 7562 page views • 139 document downloads	• 1692 page views • 415 document downloads
Immigrant engagement: study circles—Phase One	• 4 study circle groups • 13 meetings of study circle groups • 48 attendees	
Submissions from a local immigrant-serving agency—Phase Three		• 36 comment sheets (29 Chinese, 7 English)

Note: For more details, please refer to City of Richmond 2013; 2011b; 2009.

Source: City of Richmond.

council liaisons provided their support, leadership, and a "sounding board" to guide the planning process.

Concerted efforts were made to engage a variety of potential partners in the early stages of strategy development. Presentations were made to and comments were solicited from various community organizations, the health authority, the school board, and others. As a result, the City enhanced partnerships, raised awareness about priority social issues, and helped to build commitment for a shared response.

As mentioned, deliberate efforts were made to include "unusual suspects" in planning processes using non-traditional public engagement strategies. A number of newcomers and long-term immigrant residents were introduced to planning processes and contributed to the strategy development. Additionally, a variety of social media tools were used to enable residents to share their views. The public expressed its satisfaction with the many consultation opportunities available. However, it became evident from the comments and written submissions that more education was needed to raise public awareness on municipalities' limited role in addressing social issues.

The strategy, as a document, was well received by the public and City partners. People complimented its style, breadth, and content. The 85-page draft document deliberately used plain and consistent language accessible to a broad public. It introduced the local context information (the city's demographic profile), used easy-to-read graphics, incorporated a variety of interesting facts and statistics, and clearly identified the actions and the City's roles.

It also recognized the financial commitment that the City has to make in the initial phases of the strategy's implementation. The strategy and related documents are accessible online, and summaries were developed and translated into Chinese. Finally, the document went beyond identification of challenges, using text and images to highlight the diversity of places, people, and programs in the city. It adopted an asset-based approach, with proposed directions acknowledging and building upon the City's existing strong policies, programs, people, and resources.

Anticipated Outcomes of Richmond's Social Development Strategy Implementation

In September 2014, Richmond Council received the first of what will be periodic reports regarding implementation of the Social Development Strategy. The purpose of the report was to provide an update on strategy implementation and identify key areas of focus over the near term. While it is too soon to determine the ultimate effectiveness or impacts of the strategy, several anticipated outcomes are as follows.

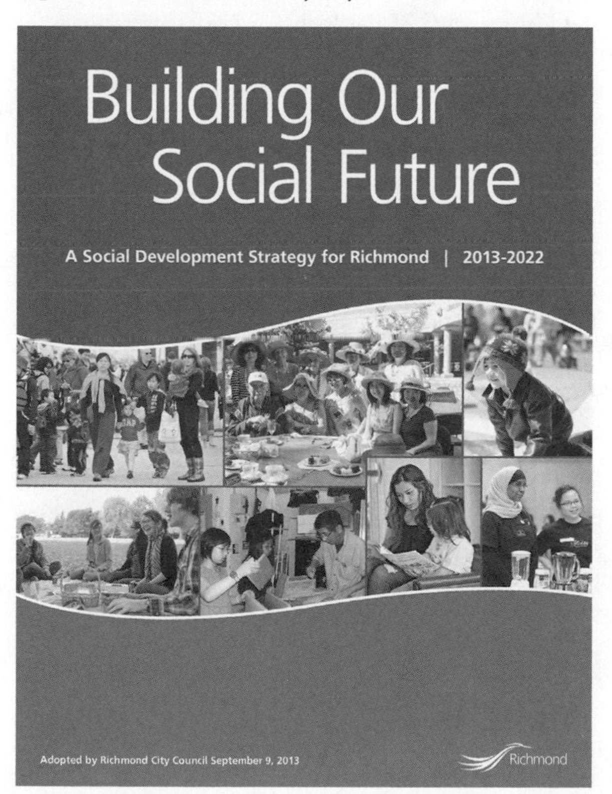

Figure 2.1.3 • Richmond Social Development Strategy.

Source: City of Richmond.

A more proactive and focused approach to social development

The Social Development Strategy has generated considerable discussion, as well as a strong sense of shared ownership and pride, among City Council, staff, and the community. It has raised overall awareness of social issues in Richmond and the importance of having a strategy in place to address them. While the City cannot address all social issues on its own, the strategy should be a valuable resource, helping to legitimize and focus the City's future social development efforts.

Before preparing the Social Development Strategy, the City had been very active in social development. Activities evolved over time, however, often in response to a particular concern or issue. They were not founded on a cohesive, comprehensive social development plan but rather reflected the City's ability to address the most urgent issues. The strategy enables Richmond to take a more proactive and strategic approach to social development, identifying emerging issues, setting priorities, and clarifying appropriate roles for itself and other partners (e.g., external organizations, community members).

A stronger and more connected community of partners

Community engagement was a key element in strategy preparation—and will remain critical as implementation proceeds. Such engagement should be a catalyst for a stronger, more connected community. Particular attention will be needed to ensure that Richmond's large recent immigrant population and other hard-to-reach groups can be effectively engaged.

Additionally, one of Richmond's key assets is its strong network of advisory committees, non-profit agencies, community groups, and other organizations. The Social Development Strategy should help to strengthen these networks, engaging and coalescing them as work proceeds on strategy implementation.

Finally, while not solely responsible, the business community and developers will play a key role in several of the actions included in the strategy. As implementation proceeds on the strategy, it is expected that enhanced partnerships and collaboration should emerge among businesses, developers, and the broader community.

Enhanced effectiveness and efficiency of service delivery

While the City has a strong social development history (e.g., policies, programs, services, funding), its activities have not been guided by a comprehensive vision or plan. The Social Development Strategy could help the City focus on priority service delivery areas that appropriately fit within the municipal realm. It should also be a useful resource for external partners as they plan their own service delivery.

With the benefit of an adopted strategy, the City should be more effective and efficient in addressing future social development concerns. Thus, quality of life in the community should be enhanced, and customer service for local residents should improve.

Sharing knowledge and building capacity for addressing social issues

The preparation and implementation of the strategy has created a shared understanding of Richmond's social needs, issues, and initiatives, which has been and will continue to be critical. By committing to share data, research, and other pertinent information with the community and external organizations, the City should be able to build broader buy-in and capacity for addressing Richmond's social development concerns.

Richmond has a myriad of assets, including its strong community networks, culturally diverse population, large and committed volunteer base, and an array of progressive social policies and programs. In moving forward on strategy implementation, a key thrust will be to build on existing community assets—not just to focus on problems or "reinvent the wheel" in developing solutions.

In moving from strategy adoption to implementation, several challenges can be expected. Based on Richmond's experience, other cities considering preparation of a similar strategy need to be prepared for the following:

- Managing expectations from elected officials, external organizations, and the community. This includes setting realistic work plans with appropriate timelines and resource allocations.
- Expecting the unexpected—using the strategy to guide future planning and decision-making but maintaining flexibility in light of new issues or challenges that will inevitably emerge.
- Continued and increasing downloading from senior governments.
- Broad community engagement (including targeted outreach to immigrants, youth, single parents, low-income families, and other hard-to-reach groups) to ensure that the strategy is embraced by and targeted to various segments of the population, not just a select a few.

Notwithstanding these challenges, and recognizing that plans must be backed up by specific, measurable, attainable, relevant, and time-bound (SMART) actions, the City believes that the strategy will yield positive outcomes and provide a valuable resource to guide future social development in Richmond.

Conclusion

The key message of the Richmond strategy is that "the City cannot do it alone" and needs to establish strategic and sustainable partnerships with senior governments and other stakeholders, particularly community partners and other government agency representatives.

This analysis of one particular municipal case and how it went about formulating its social development plan could prove useful to social planners working in multicultural cities facing similar challenges posed by urbanization, demographic shifts, gentrification, resource constraints, and transition to an economy more dependent on the service and financial sectors. Richmond's social planning initiative, aimed at addressing the fast-growing city's unique contextual social issues, has broader inter-municipal, regional, and provincial implications from which other metropolitan regions can learn.

The process of developing a Social Development Strategy was a challenging but important exercise for the City. It allowed the City to identify social development priorities, clarify its role, foster existing partnerships, and build new collaborations. Most important, the consultation processes brought together a variety of groups and individuals and provided opportunities for mutual learning and collaboration that can strengthen community capacity and resilience to address future social challenges.

KEY TERMS

Downloading
Low Income Cut-off
Official Community Plan (OCP)

REFERENCES

Aydemir, A., and M. Skuterud. 2005. "Explaining the deteriorating entry earnings of Canada's immigrant cohorts, 1966–2000." *Canadian Journal of Economics* 38(2): 641–72.

Beun-Chown, J. 2007. "Canada's healthiest city. Get the secrets to longevity from the residents of Richmond, B.C." *Canadian Living* August. http://www.canadianliving.com/health/prevention/canadas_healthiest_city.php.

Broadbent, A. 2010. "Governments claim they are getting their fiscal house in order. In reality, they are passing the buck." *Mark* 10 May. http://www.themarknews.com/articles/1103-beware-government-downloading.

City of Richmond. n.d. "Population and Hot Facts." http://www.richmond.ca/__shared/assets/Population_Hot_Facts6248.pdfhmond.ca/shared/assets/Population_Hot_Facts6248.pdf.

——. 2011a. *Richmond Health Profile*. February. Richmond: City of Richmond.

——. 2011b. *Planning Committee Report*. February. Richmond: City of Richmond.

——. 2012. *Richmond Social Development Strategy Draft*. December. Richmond: City of Richmond.

——. 2013. *Planning Committee Report*. January. Richmond: City of Richmond.

Conference Board of Canada. 2010. "Greenhouse gas (GHG) emissions." http://www.conferenceboard.ca/hcp/details/environment/greenhouse-gas-emissions.aspx.

FCM (Federation of Canadian Municipalities). 2006. *Building Prosperity from the Ground Up: Restoring Municipal Fiscal Balance*. Ottawa: FCM.

——. 2010. *Mending Canada's Frayed Social Safety Net: The Role of Municipal Governments*. Theme Report no. 6. Quality of Life in Canadian Communities Series. Ottawa: FCM.

——. 2011. *Starting on Solid Ground: The Municipal Role in Immigrant Settlement*. Ottawa: FCM.

——. 2012. *The State of Canada's Cities and Communities 2012*. Ottawa: FCM.

——. 2013a. *Canada's Aging Population: The Municipal Role in Canada's Demographic Shift*. Ottawa: FCM.

——. 2013b. *The State of Canada's Cities and Communities 2013: Opening a New Chapter*. Ottawa: FCM.

Government of British Columbia. 1994. Bill 25, 1994 Municipal Affairs Statutes Amendment Act, 1994. http://www.leg.bc.ca/35th3rd/3rd_read/gov25-3.htm.

Halifax Regional Municipality. 2011. *Newcomer Engagement Manual: A Guide for HRM*. Halifax: Cities and Environment Unit.

International Organization of Migration. 2005. *World Migration Report 2005: Costs and Benefits of International Migration*. Geneva: International Organization of Migration.

McMillan, M.L. 2003. "Municipal relations with the federal and provincial government: A fiscal perspective." Draft. April. http://queensu.ca/iigr/conf/Arch/03/03-2/McMillan.pdf.

Palermo, F. 2012. "Towards attracting and retaining newcomers in Halifax Regional Municipality." Working Paper no. 45. Halifax: Atlantic Metropolis Centre.

Picot, G. 2008. *Immigrant Economic and Social Outcomes in Canada: Research and Data Development at Statistics Canada*. Statistics Canada Catalogue no. 11F0019M. Analytical Studies Branch Research Paper Series no. 319. Ottawa: Statistics Canada.

Qadeer, M. 2009. "What is this thing called multicultural planning? Welcoming communities: Planning for diverse populations." *Plan Canada* 49(4): 11.

Schachter, H., and R. Liu. 2005. "Policy development and new immigrant communities: A case study of citizen input in defining transit problems." *Public Administration Review* 65(5): 614–23.

Shcherbyna, O. 2011. "Identifying promising practices in the development and implementation of social plans: Learning lessons from the Canadian urban municipalities." (Internship Report, prepared for the City of Richmond).

——. 2012. "Do policies of the lowest common denominator bring about system-level change? Examining the success factors of the Toronto Regional Immigrant Employment Council." Master's thesis, University of British Columbia. https://circle.ubc.ca/handle/2429/43046.

Statistics Canada. 2006a. *Census Community Profiles: Richmond BC*. Ottawa: Statistics Canada.

——. 2006b. *Immigration in Canada: A Portrait of the Foreign-Born Population*. Ottawa: Statistics Canada.

——. 2006c. Population, Urban and Rural, by Province and Territory. 2006 Census: Driver of Population Growth. Ottawa: Statistics Canada.

UNFPA (United Nations Population Fund). "Linking population, poverty and development. Urbanization: A majority in cities." http://www.unfpa.org/pds/urbanization.htm.

CASE STUDY 2.2

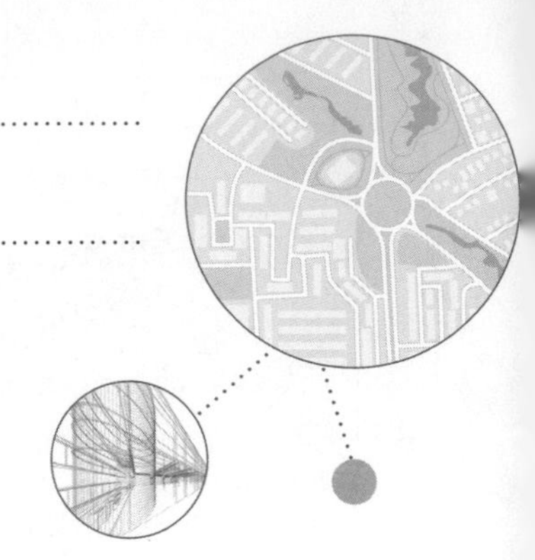

Cultural Planning in Canada

Kari Huhtala

SUMMARY

This chapter discusses Canadian cultural planning as a public process that identifies local cultural resources and shows how these resources can help a community achieve its objectives and aspirations. It presents three Canadian communities that have recently completed community cultural plans: Kelowna, British Columbia; Kingston, Ontario; and Moncton, New Brunswick.

Writers, community leaders, and scholars interested in the creative economy have also examined the role of artists in culture-led economic development. They have noted that communities can be "havens" for artists because of factors leading to the attraction of creative talent, such as natural amenities, arts infrastructure, and the cost of living. Developing a comprehensive understanding of the unique local resources that can support a creative economy is a critical early step in this process.

Culture contributes to the overall attractiveness and vitality of a community, which in turn increases its economic viability.

OUTLINE

- As cities' economies shift from manufacturing to the knowledge-based industry, cultural planning has become more integral to economic development and quality of life.
- Cultural planning is a recent addition to planning practice, involving stakeholder engagement and the development of community and business partnerships to achieve implementation.
- Case studies of the cultural planning processes in Kelowna, British Columbia; Kingston, Ontario; and Moncton, New Brunswick, outline the purposes of the plans; the roles of planners, the public, and others; plan implementation and success; and lessons learned from all three cases.

Rationale for Cultural Planning

> *If creative cities are the end, cultural planning is the means.*
>
> Municipal Cultural Planning Incorporated, Ontario

This is an opportune time for cultural planning in Canada. Municipalities are undergoing a profound shift in their traditional economic base. For many communities, it is a shift away from reliance on traditional industries stressing natural resources and large manufacturing companies as the foundation of local economies and toward a creative economy driven by ideas, innovation, knowledge, and collaboration.

This shift is related to the growing number of small and medium-sized creative businesses responding to new consumer demands for original and place-based products and services in the fields of information technology, graphic design, food and beverage products, and hospitality, to name a few.

Awareness of the importance of culture in Canadian cities has been promoted by many individuals. Jane Jacobs, Glen Murray, and Richard Florida are among those who have provided notable coverage and reaction.

Jane Jacobs (1916–2006), author, in her book *Dark Age Ahead* (2004), identifies five pillars of our culture that we depend on but which are in serious decline: community and family; higher education; the effective practice of science; taxation and government; and self-policing by learned professions. The decay of these pillars, Jacobs contends, is behind such ills as environmental crisis, racism, and the growing gulf between rich and poor. The continued degradation could lead us into a new Dark Age, a period of cultural collapse in which all that keeps a society alive and vibrant is forgotten.

Glen Murray, politician and urban issues advocate, has championed the role of culture and how it can play in creating a prosperous city. He believes that cities, rather than countries, will compete for human capital. Cities are centres for innovation and creative thinking, and, consequently, arts and culture are just as necessary to cities as bridges and airports.

Richard Florida, professor and author, says that Canada is shifting from the Industrial to the Creative Age in which creativity, "the ability to create meaningful new forms" (Webster's Dictionary 2013), has become the decisive source of competitive advantage. The Creative Class, as he termed it (2002), at the turn of the twentieth century comprised less than five per cent of the workforce, growing to 31 per cent by mid-century, and surging in the 1980s. Today, 5 million Canadian workers, or roughly 30 per cent of the workforce, are members of the Creative Class, which Florida (2012) describes as workers in science and technology, the arts, culture, and entertainment, as well as professional knowledge workers in health care, law, and management.

These and other cultural advocates note that successful municipalities will be those that:

- offer appealing and attractive communities that are diverse and welcoming;
- have interesting public spaces;
- celebrate history, art and culture, and urban design; and
- understand that linking these elements creates the magnet to attract people and talent.

Understanding Culture

A broad definition of culture not only takes into account the traditional art forms (e.g., visual art, drama, and painting) but also includes the lifestyle and activities of residents and visitors that are both uplifting and energizing. When looked at through this broad lens, it is evident that culture contributes significantly to the quality of life in communities. Access to all forms of culture, nature, outdoor activities, and the benefits growing from that culture is the primary reason that people want to live, visit, and celebrate in these communities. This strong sense of culture attracts new residents and retains talent, fosters economic endeavours, provides a setting for strong, collaborative partnerships, and cultivates a shared community identity.

Given this definition of culture, the following describes a range of ways in which culture contributes to economic and broader community development agendas (Creative City Network of Canada 2010).

- *Culture as an economic driver:* Growth is inevitable when culture is used as a tool. It brands cities and creates job growth, spin-off businesses, and economic competition.
- *Urban renewal and revitalization:* Culture-based initiatives have been essential to urban revitalization and urban renewal programs in Canada. The arts ensure that a community's environment reflects who residents are and how they live.
- *Building community identity and pride:* The arts have been instrumental in facilitating social cohesion, bringing tourism to unlikely places, fostering a sense of belonging, and preserving collective memory.
- *Arts and positive change in communities:* Arts and culture make considerable and necessary contributions to the well-being of communities. Arts and culture are powerful tools with which to engage communities in various levels of change. They are a means to public dialogue, contribute to the development of a community's creative learning, create healthy communities capable of action, provide a powerful tool for community mobilization and activism, and help to build community capacity and leadership.
- *Quality of life and place:* Culture, long overlooked as a tool for garnering quality of life, is recognized as a means of building community; encouraging outdoor activity, healthy lifestyles, and life-long learning; increasing accessibility to programs for all levels of society; and celebrating diversity and cultural differences.

- *Personal and social development:* The ability to motivate and engage the community (particularly youth) from all socio-economic levels in education and community life is a respected strength of arts and culture.

Planning and Culture

The planning profession is a recent player in the development of culture in communities. The profession has an important role to play in assisting community members and organizations as they inventory and explore their community context. It provides a grounding mechanism to ensure that the wealth of forgotten or unknown community information is revealed, enhanced, and maintained over time. It assists other stakeholders in anchoring the narrative about the authentic elements of a community and its resources and helps its talent to emerge and thrive.

Cultural planning promotes a partnership among the government, community organizations, and citizens to explore, measure, and assess the values, resources, and assets of the community. It relies heavily on the interaction

Figure 2.2.1 • Case study cities in Canada.

Source: Odette Hildalgo.

and participation of a broad range of stakeholders (from youth and adult residents to community-based organizations, academics, public officials, and policy-makers) to achieve its goals, objectives, and results.

The primary challenge encountered in cultural planning is how to engage community stakeholders (government, business, community organizations, and residents), many of whom have little or no background in culture or the arts. A number of different tools are available to facilitate and aid the dialogue: the world café model, focus groups/roundtable format, open houses, community visioning sessions, Web-based community engagement, and other techniques. Through discussion and consideration, the involved partners select the participatory tools to ensure that all participants' ideas and thoughts are included in the plan-making process.

Three case studies of cities where **cultural plans** have recently been completed present transferable key cultural planning characteristics and values that can offer direction to other community-based cultural endeavours.

The discussion of each case study will assist the reader in understanding the planning processes used, problems being addressed, stakeholders involved, and successes.

Kelowna Cultural Plan: "Thriving Engaging Inspiring"

Kelowna, with a population of 117,312 (2011 census) is located in south-central British Columbia, about two hours north of the Washington, US, border. Kelowna is the seat of the Regional District of the Central Okanagan and the third-largest metropolitan area in British Columbia after Vancouver and Victoria. With scenic lake vistas and a dry, mild climate, Kelowna has become one of the fastest-growing cities in North America.

Figure 2.2.2 • The "Bear" artwork by Brower Hatcher (2010), looking north along Lake Okanagan.

Source: City of Kelowna.

The service industry is the biggest employer in Kelowna, the largest city in the tourist-oriented Okanagan Valley. Kelowna produces wines that have received international recognition. Okanagan College and the University of British Columbia (Okanagan Campus) are the predominant centres for post-secondary education.

Kelowna's Cultural District is located in downtown Kelowna and was once the centre of the Okanagan fruit-packing industry. Situated within the eclectic six-block Cultural District are the Kelowna Community Theatre, Kelowna Art Gallery, Rotary Centre for the Arts, the main branch of the Kelowna Library (Okanagan Branch of the Okanagan Library System), four Kelowna museums, and other exciting facilities such as the Prospera Place Arena and the Actors Studio Dinner Theatre. The area also houses private galleries, restaurants, and unique shops as well as 18 pieces of public art created by local, regional, national, and international artists.

Kelowna holds a number of festivals, such as the Canadian Culinary Championship, the Fat Cat Children's Festival, the Okanagan Wine Festival, and the Keloha Music Festival, to name a few.

Cultural Plan Purpose

The Kelowna Cultural Plan "Thriving Engaging Inspiring" (Kari Huhtala + Associates 2011), adopted in 2011, plays an important role in support of Kelowna's evolution into a sustainable city with a unique identity and a prosperous and diverse economy, all toward the goal of Kelowna becoming the most liveable mid-sized city in North America.

The plan recognizes that culture makes many contributions to Kelowna's quality of life. Culture creates a place where people want to live, celebrates diversity, attracts and retains talent, fosters entrepreneurship and innovative businesses, cultivates collaboration and partnerships, and builds an authentic, shared identity (Kari Huhtala + Associates 2011). Culture is an important economic driver in the community, creating jobs and generating $143.8 million annually (Kari Huhtala + Associates 2011).

Cultural development in Kelowna had reached a stage in which most of the initiatives set out in a cultural plan had been achieved: cultural infrastructure, funding support for cultural facilities, municipal staff dedicated to culture, grant programs for cultural organizations, and a public art program, for example. The City then proactively considered what the next phase of cultural development should be, given the considerable changes in demographics and community perspective since the period 1995 to 2005 when most of the cultural development occurred.

Role of Municipal Staff, the Public, and Other Stakeholders

The City's Cultural Services staff led the planning process with assistance from staff from the Communications and the Policy and Planning Departments. An eight-member internal liaison team representing several other departments was also engaged periodically to ensure that strategies were consistent with broader objectives and practices. A contracted consultant was invaluable during the public consultation process. Based on community consultations, he established the framework of the plan in terms of vision and goal statements. The consultant also assisted with early drafts of the plan as a whole, writing some of its sections, and was a resource for review of the plan as it moved through the final drafting stages.

The consultation process included six public events, three community surveys, and the City's first use of Facebook as an engagement tool. Each of the six events was designed to attract a particular audience, ranging from the business community to young adults, stakeholders, the general public, and cultural organizations. Public comments were captured whenever possible and included in an appendix to the final plan document. Direct quotations from the public are also scattered throughout the plan. In some instances, strategies in the plan reflect specific feedback on a particular issue. For the most part, public comments and ideas were filtered throughout the plan as themes, values, and broadly expressed goal statements.

Each goal statement and strategy went through several iterations and levels of review before being finalized. Early drafts of the goals and strategies were shared with the public during an open house and feedback event that was very influential in determining the final content of the plan. Some of the ideas were tested informally during an Arts and Culture Summit in 2011, which provided perspective on which items could be considered priorities for early implementation. Extending the timelines for completion of the plan to allow these opportunities for public feedback was very important and beneficial. The internal staff team was also helpful in articulating the content for certain strategies and providing advice on implementation.

The challenges encountered concerned the general public interest in matters such as new infrastructure, which was not an immediate priority in terms of the plan. The general public had less vested interest in cultural planning and development than did cultural organizations, which tended to be the major contributors of public input. This fact reinforced the importance of ensuring that cultural organizations engage with the public in their own strategic planning rather than developing plans in isolation from public consultation.

Although the City was very pleased with the level of public engagement in the plan (more than 1350 interactions), the reality is that cultural planning will never elicit the same level of interest in the community as, for example, a controversial new development or roadway plan. The important question is "Are the right people here?" rather than "Are enough people here?"

Plan Implementation and Successes

Currently, implementation is proceeding throughout the five-year span of the plan (2012–17). Although no new resources were provided for implementation at the time of the plan's approval in 2011, the Cultural Services Department has obtained modest annual budget increases to facilitate plan implementation. Annual work plans reflect plan strategies that can reasonably be fulfilled in any given year with reliance on partnerships and collaborations as they emerge.

Cultural organizations seeking access to City of Kelowna grant-funding are asked to ensure that their operations and projects are linked to the Cultural Plan and to demonstrate that they are actively working to fulfill the community priorities as reflected in the plan. Organizations receiving funding from the City are learning more about what public-sector funders expect in terms of value and impact in the community. However, this is a long-term conversation, and it is too early to gauge success.

The plan states clearly that the work of implementation involves collaborative leadership and energy from multiple sources and that the creation of culture is, ultimately, led by the community with City support. Consistent and steady communication about the plan is key to building awareness and generating ideas in the community.

Strategies in the Cultural Plan have been successfully implemented in both small and significant ways over the past two years, ranging from the provision of infrastructure for posting handbills in the newly revitalized downtown to providing more operating support for professional arts organizations and a major investment in the theatre's amplified and acoustic sound systems. Community participation in national Culture Days, which provide grassroots, free cultural experiences, has grown significantly. Progress in 2015 will be the subject of the first "Report Card" on implementation of the plan, an opportunity to document change and assess Kelowna's progress in a wider context. At this point, one main challenge is that other City departments are largely unaware of the plan and unaware that they might have a role in implementing it.

Kingston Culture Plan

Kingston is located in eastern Ontario where the St Lawrence River meets Lake Ontario. Growing European exploration in the seventeenth century and the desire to establish a presence close to First Peoples for trade led to the founding of a French trading post there, known as Fort Frontenac, in 1673. The fort became a focus for settlement.

Today, Kingston's population is 123,363 (2011 census). Its economy relies heavily on public-sector institutions. The most important sectors are health care, education (Queen's University, the Royal Military College of Canada, and St Lawrence College), government (including the military and correctional services), tourism, and culture. Manufacturing and research and development play a smaller role than they did in the past. Because of the city's central location between Toronto, Ottawa, Montreal, and Syracuse, NY, a trucking and logistics warehousing industry has developed.

Figure 2.2.3 • Jazz Festival, Kingston: Porch Jazz.

Source: City of Kingston.

Kingston is known for its historic properties, as reflected in the city's motto: "Where History and Innovation thrive." Some 700 of the properties in the municipality are listed in the heritage register it maintains pursuant to the Ontario Heritage Act.

Kingston hosts several festivals during the year, including the Kingston WritersFest, the Limestone City Blues Festival, the Kingston Canadian Film Festival, Artfest, the Kingston Buskers' Rendezvous, the Reelout Film Festival, Feb Fest, the Skeleton Park Arts Festival, and the Wolfe Island Music Festival.

Plan Purpose

Until 2008, the City of Kingston did not have a stand-alone department devoted to culture. Prior to that time, the City had grouped culture either with recreation (Recreation and Culture Department) or with heritage (Culture and Heritage Department). In the former structure, culture was managed as part of a larger Community Services portfolio but later became part of Sustainability and Growth, now Corporate and Strategic Initiatives. In 2008, Culture and Heritage were separated, and the City established a Cultural Services Department for the first time.

During this period, the City of Kingston was also pursuing the development of an Integrated Community Sustainability Plan, known as the Sustainable Kingston Plan, which meant that culture was being recognized for the first time as one of four pillars of sustainability.

The Kingston Culture Plan (KCP) (Canadian Urban Institute 2010), adopted in 2010, articulates a sustainable long-term vision for cultural vitality in Kingston. The plan identifies possibilities for connections between cultural organizations and other stakeholders as well as opportunities for collaboration among City departments to achieve municipal strategic objectives, and it contains strategic directions, initiatives and recommendations for action, and an implementation timeline.

Role of Municipal Staff, Public, and Other Stakeholders

In 2006, a group called Imagine Kingston was established by a local businessman, and Glen Murray was brought in to help the group develop a vision. Culture was recognized as an important aspect of Kingston's long-term prosperity. Richard Florida was also invited to Kingston to discuss his theories and how they might be applied within a more localized context. The CAO of the City of Kingston was very much a part of these discussions, and much of this thinking influenced the creation of a municipal cultural plan, supported by the community.

The Culture Plan was developed in three stages over the course of a year, each stage building on the knowledge and insights gained during the previous stage. As part of the process of consultation and engagement, many discussions, meetings, focus groups, and public workshops were facilitated at each stage of the project to ensure that the perspectives and priorities of the participants were captured in order to inform and guide the work of City staff and the consulting team led by the Canadian Urban Institute.

At the outset, there was scepticism within the community that the plan would do anything other than sit on a shelf. Initially, some people were loath to get involved because of their past experiences working with the City. A clear commitment to community consultation, however, fostered dialogue and engagement and turned sceptics into advocates once the KCP went before council for consideration.

A "big move" identified early on in the process was the need to leverage Kingston's many heritage assets as an economic development opportunity though cultural tourism. A lot of criticism emerged in response to this suggestion, because many people wanted the plan to focus on the community first before shifting attention to attracting visitors. This represented an interesting moment in the process, because a decision was made to shift the focus based on the input received, even though the evidence showed that cultural tourism represented a significant area of growth for the community.

Plan Implementation and Successes

The Culture Plan as approved by council included an Implementation Plan with 60 recommendations, mapped out according to the following time periods: first year; one to three years; three to five years; and five years and beyond.

The Implementation Plan provided council and City staff a framework within which to prioritize the recommendations and the specific initiatives, actions, and resources needed to make the KCP a reality. Council committed to investing an additional $2.25 million into the Cultural Services budget over four years (an investment that continues to be honoured but over a longer time period as part of a larger corporate effort to manage annual tax increases). Within Cultural Services, the KCP was used to devise a multi-year strategic plan and to clarify who should take the lead for each of the recommendations. Cultural Services maintains a tracking chart to monitor the progress being made and has also committed to publishing an annual update report to provide council with tangible illustrations of the progress being made.

To date, the implementation of the KCP has been extremely successful given the support it has received within the community and from council. Progress is being made on all 60 recommendations, and Cultural Services is making plans for a five-year review in 2015. Cultural Services also had the opportunity to collaborate on the creation of a guidebook, devised by the Canadian Urban Institute and published in 2011, called *Municipal Cultural Planning: Indicators and Performance Measures*. The publication of this guidebook was timely in that the City of Kingston is dedicated to developing a corporate-wide strategy involving indicators and performance measures. As part of this process, the guidebook has proved to be helpful as a way of quantifying the impacts and benefits of culture and cultural investment at the municipal level.

Since the adoption of the Kingston Culture Plan, the City has completed work on an Integrated Cultural Heritage and Cultural Tourism Strategy. The strategy explores the issues that were set aside during the early stages of developing the KCP to make the case for leveraging Kingston's cultural heritage assets in relation to cultural tourism for the benefit of the city as a whole. An initial pilot project was launched in 2014 with the intention of bringing together a diverse group of stakeholders to develop the cultural tourism market by establishing a set of standards and guidelines unique to Kingston and to identify and create more market-ready products and experiences.

Moncton Cultural Plan: "Telling Our Story"

Metro Moncton's population is 138,644 (2011 census). The city is located in Westmorland County in southeastern New Brunswick. Situated in the Petitcodiac River Valley, it lies at the geographic centre of the Maritime Provinces. The city has earned the nickname "Hub City" because of its central location and because it is the railway and road transportation hub for the Maritimes.

Figure 2.2.4 • Capitol Theatre, Moncton.

Source: City of Moncton.

The local economy is based on Moncton's heritage as a commercial, distribution, transportation, and retailing centre. This is due to Moncton's central location in the Maritimes and its large catchment area, with 1.4 million people living within a three-hour drive of the city. Today, insurance, information technology, and the educational and health care sectors are major factors in the local economy, with the city's two hospitals alone employing more than 5000 people.

Tourism is an important industry in Moncton and historically owes its origins to the presence of two natural attractions, the tidal bore of the Petitcodiac River and the optical illusion of Magnetic Hill. Today, Magnetic Hill is the city's most famous attraction. In addition to the phenomenon itself, a golf course, a major water park, a zoo, and an outdoor concert facility have been developed in this area. A $90-million casino/hotel/entertainment complex opened at Magnetic Hill in 2010.

Moncton boasts a varied and vibrant cultural scene. The city hosts a number of arts and cultural experiences, such as Robinson Fridays Summer Concert Series, Acoustica (Moncton's free outdoor summer concert series), the Mosaïq Multicultural Festival, the Acadie Rock Festival, St George Neighbourhood Street Party, the Festival international du cinéma francophone en Acadie, the Frye Festival, the Moncton Highland Games, and many more. In addition to the numerous festivals and events, Moncton is home to a wide-ranging collection of fine art, much of which is part of the City of Moncton's fine arts collection and the Université de Moncton's extensive art collection.

The city also owes much of its artistic and cultural vitality to numerous artists practising a variety of art disciplines who call Moncton home, as well as to several institutions that have helped to establish the city as an important cultural centre in New Brunswick and in Atlantic Canada. They include Université de Moncton, the Aberdeen Cultural Centre, the Galerie d'art Louise-et-Reuben-Cohen, Théâtre l'Escaouette, the Atlantic Ballet-Theatre, Botsford Street Station, Galerie sans nom, Galerie 12, Éditions Perce-Neige, FilmZone, DansEncorps, and several others.

Plan Purpose

The City of Moncton has put in place important cultural services and programs over the years. Initiation of the position of cultural officer in the early 1990s, creation of a Cultural Task Force in 1997 to develop Moncton's first Cultural Policy (adopted in 2000), and recognition of arts, culture, and heritage as priorities in several strategic City documents over the past decade illustrate the City's desire to support the community's cultural aspirations.

Other City of Moncton policies that relate to arts, culture, and heritage—the Percent for Public Art Policy, the Film Policy, the Art Acquisition Policy, and the Heritage Preservation Bylaw—have also helped to shape the City's cultural services offering.

The Moncton Cultural Plan, "Telling Our Story," adopted in 2010, is the city's first cultural plan. The creation of the plan was a natural next step in the city's cultural evolution. It establishes a clear, collective vision for culture in Moncton; outlines priorities and actions that strengthen the arts, culture, and heritage sectors; and promotes collaborations in the community and identifies key partners that will help move the priorities forward.

The plan sets the City's cultural priorities for the next five years, gathers the cultural priorities identified in previously adopted guiding documents and policies, and combines them with priorities identified in the Cultural Plan consultation process to form one overall strategy for cultural development in Moncton.

Role of Municipal Staff, Public, and Other Stakeholders

During the development of the Cultural Plan, staff provided information on current arts, cultural, and heritage programs and assisted in hosting community forums and workshops during which the opinions of numerous stakeholder groups were received. In the initial stages of the plan's process, a consultant was retained to synthesize the information gathered through these sessions and to identify priority areas for development.

City staff consulted with key demographic groups when developing the Cultural Plan, including the Mayor's Youth Advisory Group and the Seniors Advisory Group. Efforts were made to reach immigrant and multicultural communities through partners such as the Multicultural Association of the Greater Moncton Area and immigration support groups. Because Moncton has a large francophone population, the community engagement process included participation by Acadian and francophone members through bilingual community forums where simultaneous translation was provided.

No alternatives to the plan were discussed or considered, except that the plan needed a much broader, overarching city-wide arts and culture policy.

A number of different visions emerged from these public consultation sessions.

Often, the end goals in the various sessions were similar (e.g., a more vibrant downtown, a larger number of cultural activities, better support systems for organizations), but some of the actions for achieving the goals could vary. The suggestions received from the community also needed to be evaluated in terms of the necessary investments. Some projects might have been left out of the plan if they were deemed too cost-prohibitive.

Plan Implementation and Successes

Staff oversee the implementation of the Cultural Plan with the help of the Cultural Board, which was established upon adoption of the plan. The main implementation responsibility lies with the cultural development officer, but certain areas of the plan involve other City staff and departments, such as the Moncton Museum and the Economic Development and Events Department. The Moncton Cultural Board oversees the progress of the plan and helps to guide the priorities from year to year.

The City is currently more than halfway through the plan's five-year implementation program. A few of the initiatives have been completed, such as developing the Heritage iTour and creating a new cultural grants program. Several are close to completion—for example, expansion of the Moncton Museum to create a Transportation Discovery Centre. And a few have been pushed back for the time being, such as creation of a cultural district, which will be considered once the municipal cultural mapping project is further along. A number of initiatives require ongoing work and will never be "done," including building Moncton's reputation as an entertainment hub and exploring partnerships and collaboration opportunities. A public forum was held in 2012 to give stakeholders an update and assess their satisfaction with the progress to date. In general, they were pleased with the progress and felt that things were moving in the right direction. A few factors did create some delay in plan implementation: some departments were reorganized, and the cultural development officer position was vacant for several months. But otherwise, many of the initiatives outlined in the plan will be completed or undertaken by the time it is renewed.

Lessons Learned

The lessons learned in reviewing the development of cultural plans in Kelowna, Kingston, and Moncton can be summarized as follows.

Municipal Community Partnership

The cultural plan is a useful tool to guide municipal priorities and policies when it comes to arts, culture, and heritage. If meaningful public consultation has occurred and the public's priorities are reflected in the plan, the community and local government have a role to play in implementing the plan. Ongoing sharing of the plan, talking about it, linking it to ongoing initiatives, and holding funding recipients accountable to it are starting to succeed in that language from the plan is becoming commonly understood and meaningful in the cultural community. The plan has built-in measuring mechanisms, such as biannual cultural forums that allow a regular check-in with stakeholders on progress.

Data Collection

Data collection in the cultural sector is an emerging field, and each of these communities has come to realize how little data there is and how much more needs to be collected and analyzed in order to understand the City's return on investment and the level of value and impact culture has within the community at large. Although cultural data indicators are included in the plans, there is a large educational task ahead for the community's cultural sector in order to encourage artists and cultural organizations to gather good data and share it.

Common Issues

Common primary issues to be addressed in the plan are:

- the lack and affordability of production space for artists;
- building more awareness of the value of culture in the community at large and, with it, building more cultural citizenship and creative vitality;
- nurturing sustainability and organizational health for non-profit groups, which are an important part of the cultural service delivery network. This includes funding, of course, but also skills development for good governance and accountability for the use of public-sector funds;
- exploring other funding options (e.g., senior governments, private and non-profit foundations) to reduce the strain on taxpayer resources and expand the reach of the resources that are available.

Community and Corporate Cultural Development

Municipal cultural planning is about community cultural development as well as corporate cultural development. The focus is not so much about service delivery as it is about how to work differently in order to leverage different kinds of assets to benefit the corporation as well as the community. Given the emphasis on creativity, social cohesion, quality of place, economic development, and organizational development, the process of municipal cultural planning is entirely transferable. However, it does require an openness to internal collaboration, external partnerships, and new ways of working that necessitate a certain level of risk-taking over the long term, all of which is inherently challenging within the context of municipal government.

Implementation

The plans include an implementation framework (e.g., five or 10 years) that provides direction with regard to actualizing the vision, goals, resources (facilities, programs, and funding), and strategies in the cultural plan. The progress of each strategy is evaluated annually and the findings presented to all stakeholders for further discussion, consideration, and action.

Concluding Comments

Cultural planning is a way of looking at all aspects of a community's cultural life as community assets. Cultural planning considers the increased and diversified benefits these assets could bring to the community in the future, if planned strategically. Understanding local culture and cultural activity as resources for human and community development promotes our understanding of the broad view of culture, resulting in more assets with which we can address civic goals (Creative City Network of Canada 2010).

Although the development of cultural planning in Canada is a recent phenomenon, the four areas where communities are showing encouraging outcomes include:

1. *Creative economy:* Community cultural activities strengthen and diversify the economy.
2. *Creative places:* Cultural landscapes (e.g., buildings, places, and activities) are being integrated into land-use planning, infrastructure development, and programming.
3. *Creative people:* Individuals, organizations, and industries are developing creative capacity, enterprises, and activities.
4. *Creative identity:* Shared cultural understanding is making the community a place where people want to live, work, learn, play, visit, and invest.

The key characteristics of successful cultural planning are its ability to be wide-ranging, fluid, and situational.

KEY TERMS

Cultural plan

Culture

CULTURAL PLANS FROM OTHER JURISDICTIONS

Canadian Urban Institute. 2010. "City of Kingston Cultural Plan." http://www.cityofkingston.ca/documents/10180/14469/Plan_KingstonCulturePlan_2010.pdf/e5a2c4ec-3de7-4186-b533-32b8b1061ce4.

City of Kitchener. 2004. *Culture Plan II*.

City of London. Creative City Task Force. 2004. *The Creative City Report*.

City of Richmond. Parks, Recreation and Cultural Services. 2006. *Live. Connect. Grow. A Master Plan for 2005–2015*.

City of Toronto. Culture Division. 2003. *Culture Plan for the Creative City*.

City of Vancouver. Creative City Task Force. 2008. *Cultural Plan for Vancouver 2008–2018*.

Corporate Research Group and Euclid Canada. 2005. *Leveraging Growth and Managing Change: Prince Edward County Strategic Cultural Plan*. Report prepared for the County of Prince Edward, ON, Economic Development Department.

Creative Capital Advisory Council. 2011. *Creative Capital Gains: An Action Plan for Toronto*. Report prepared for the City of Toronto Economic Development Committee and Toronto City Council.

Edmonton Arts Council. 2008. *The Art of Living 2008–2018: A Plan for Securing the Future of Arts and Heritage in the City of Edmonton*.

Legacy Heritage Consultants and Professional Environmental Recreation Consultants. 2001. *Cultural Policy and Plan*. Report prepared for the City of Port Coquitlam.

Kari Huhtala + Associates. 2011. "City of Kelowna Cultural Plan." http://www.kelowna.ca/CityPage/Docs/PDFs//Cultural%20District/CulturalPlan-WEB.pdf.

Kari Huhtala + Associates. 2011. *Pemberton and Area Cultural Plan*.

McMeekin Leffler. 2001. *City of Ottawa Arts Plan*.

Miller, Dickinson, Blais. 2012. *Town of Stony Plain Cultural Plan*.

REFERENCES

Canadian Urban Institute. 2011. *Municipal Cultural Planning: Indicators and Performance Measures*. Toronto: Canadian Urban Institute.

Creative City Network of Canada. 2010. *Cultural Planning Toolkit*. Vancouver: 2010 Legacies Now and Creative City Network of Canada.

Florida, R. 2002. *The Rise of the Creative Class and How It's Transforming Work, Leisure, Community and Everyday Life*. Cambridge, MA: Basic Books.

———. 2012. *The Rise of the Creative class: Revisited: Revised and Expanded*. New York: Basic Books.

Jacobs, J. 2004. *Dark Age Ahead*. New York: Random House.

Kari Huhtala + Associates. 2011. "City of Kelowna Cultural Plan." http://www.kelowna.ca/CityPage/Docs/PDFs//Cultural%20District/CulturalPlan-WEB.pdf.

Webster's Dictionary. 2013. Random House Reference. New York: Random House.

QUESTIONNAIRE RESPONSES

Sandra Kochan, manager of Culture Services, City of Kelowna.

Roxanne Richard, cultural development officer, City of Moncton.

Colin Wigginton, acting cultural director, Cultural Services, City of Kingston.

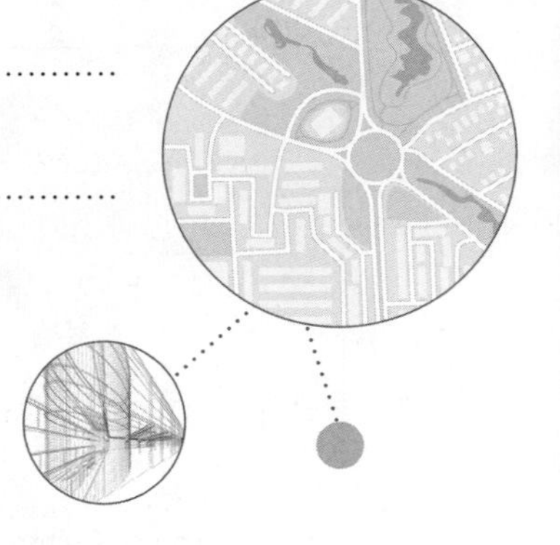

CASE STUDY 2.3

Cultivating Intercultural Understanding: Dialogues and Storytelling among First Nations, Urban Aboriginals, and Immigrants in Vancouver

Umbreen Ashraf, Kate Kittredge, and Magdalena Ugarte

SUMMARY

"Dialogues between First Nations, urban Aboriginal, and immigrant communities in Vancouver" (hereafter, the Dialogues Project) was a community-building **demonstration project** convened by the City of Vancouver (CoV) in collaboration with diverse community partners, which ran between January 2010 and March 2013. The goal of the project was to build increased intercultural understanding and strengthened relations among the fast-growing populations of **First Nations**, **urban Aboriginals**, and **immigrants** in the city. It was the first municipal project in Canada to explicitly address the issue of Indigenous/immigrant relationships in urban settings.

OUTLINE

- We explain context, origins, and goals of the Dialogues Project.
- We present how the community shared stories, cultivating intercultural understanding and building community.
  - Storytelling and listening: The transformative power of conversation.
  - Dialogic and experiential learning.
  - Appreciative inquiry approach: Building on community assets.
- We present the challenges, lessons, and transferability of this case, including time and resources, engagement, and evaluation.

Introduction

Drawing on in-depth interviews, participant observation, and official document reviews, in this chapter we present an analysis of the Dialogues Project. We explore its successes and challenges from a planning perspective, focusing on the project's three strategic pillars: storytelling and listening, dialogic and experiential learning, and appreciative inquiry. We examine some of the main challenges and lessons learned and offer guiding questions to support other planners who may see this model's potential to be adapted in other municipalities facing similar challenges. We pay particular attention to how the case offers insights into the use of a dialogic approach to community-building in highly diverse communities with complex histories.

Context, Origins, and Goals of the Dialogues Project

First Nations, urban Aboriginals, and immigrants are among the fastest-growing populations in Vancouver, a city that is located on the traditional territory of the Coast Salish peoples, including the Musqueam, Squamish, and Tsleil-Waututh Nations. The eighth-largest urban centre in Canada, Vancouver is also home to many First Nations, Inuit, and Métis people from other communities. With a population of 603,502, the city has the third-largest urban Aboriginal population in the country, and 44 per cent of its residents are foreign-born immigrants (Statistics Canada 2011). Together with Toronto, Vancouver ranks as one of the two most culturally heterogeneous metropoles in Canada.

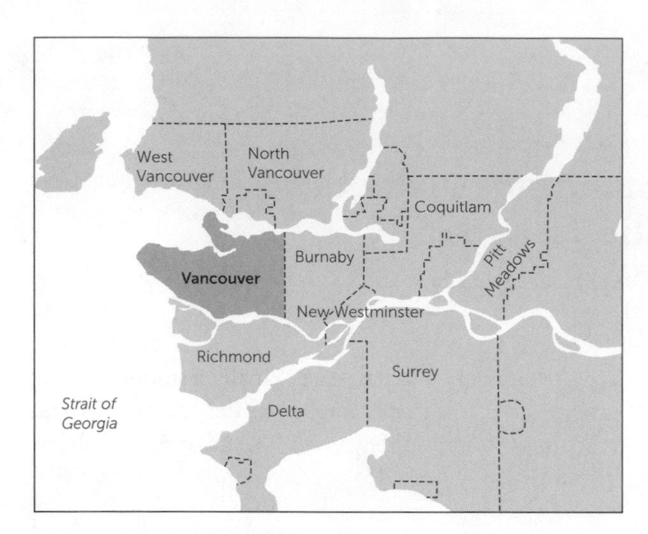

Figure 2.3.1 • Location map.

Source: Umbreen Ashraf, Kate Kittredge, and Magdalena Ugarte.

Although these groups share the place in which they live, have rich cultural histories, and often have common experiences of displacement, marginalization, and racism, there is limited interaction among them (Gray 2011; Yu 2011). Moreover, new immigrants have little exposure to the colonial history of Canada and the complexities of Indigenous/**settler** relations (Little Bear 2009; Regan 2010; Yu 2011). Municipal officials at CoV identified this lack of intercultural awareness and a prevalence of negative stereotypes. In response, the Dialogues Project emerged and became a pioneer social planning initiative.

The goal of the Dialogues Project was to foster "increased understanding and stronger relationships" among these groups through the development of diverse community-building initiatives. Social planners at CoV explored innovative ways to bring these communities together and help bridge this communication and information gap, with a vision to "create cohesive communities by using dialogue to explore the stories of Vancouver's First Nations, urban Aboriginal and immigrant communities" (City of Vancouver 2011).

The Dialogues Project was conceived as a demonstration project and ran for 38 months, from January 2010 until March 2013. It was led by the Social Policy, Community Services Group of CoV and had the support of the City Council, the Mayor's Working Group on Immigration, and a Steering Committee of 27 community partners, including members of the three local First Nations, urban Aboriginal leaders, and immigrants. The Government of Canada and the Province of British Columbia provided key funding through the BC Welcoming and Inclusive Communities and Workplaces Program (WICWP), an initiative that funds projects that aim to build inclusive and vibrant communities in BC. The total funding for the Dialogues Project was $600,000, which was complemented by in-kind contributions from City Council and community partners. The Dialogues Project aligned with the capacity-building goals of the WICWP to create cohesive communities "where immigrants can realize their full potential, racism is eliminated and cultural diversity is valued and celebrated" (Welcome BC 2013).

Development and Working Structure of the Dialogues Project

The development of the Dialogues Project was built on the recognition that municipal governments have an important role to play in promoting inclusive planning practices and helping to build strong communities in multicultural settings. While CoV had significant experience working with the Indigenous and immigrant communities individually, there had never been a concerted effort to bring them together.

Based on the demographic realities described above, the Dialogues Project was conceived as an engagement and public education process to foster deeper intercultural knowledge and understanding through dialogue among these groups. To achieve this goal, the Dialogues Project was grounded in three strategic pillars that guided the design of all of its key initiatives: focusing on storytelling and listening, promoting dialogic and experiential learning, and adopting an **appreciative inquiry** approach. As will be detailed in the next section, the project recognized the transformative potential of dialogue and storytelling in facilitating productive face-to-face interactions among groups traditionally positioned apart and seemingly aloof to each other's realities (see Planning Theory Box).

The planning and execution of the Dialogues Project were fairly centralized. The Social Policy Division at CoV was responsible for the initial formulation of the project, which included defining the goal and key initiatives, writing the funding proposal for WICWP, gauging interest and support among community groups, and seeking preliminary support from City Council. The Steering Group and other community partners provided feedback on the project proposal and endorsed it before it was formally brought before City Council for approval. Implementation of all the initiatives was led by the Social Policy Division, with the direct input, support, and participation of 49 community organizations that work with the Indigenous and immigrant communities.

Given the demonstration nature of the Dialogues Project, the planners mostly innovated and adapted their strategies at various programmatic and execution levels based on feedback received from participants, the Steering Group, and third-party evaluators. Initially envisioned as an 18-month project, the Dialogues Project extended to more than three years because of the positive reception among the target communities. This perceived success led to the continuation of funding from WICWP, CoV, and community partners. Overall, 9000 people participated in the project.

In Phase I (2010–11), participants engaged in five key initiatives, all of them dialogue-centred. More than 100 people participated in nine facilitated dialogue circles, which focused on story-sharing among members of the three communities. They aimed to offer an arena for people to describe their experiences as the original inhabitants of or as newcomers to this land. Fifteen community cultural exchange visits allowed more than 750 participants to engage with other cultures. Hosts included the Musqueam and Tsleil-Waututh Nations and the Chinese, Afghan, Mayan, and Ismaili communities. A collaborative photo-voice project brought together youth and elders to explore intergenerational issues. Finally, a story-gathering project gave voice to members of the Indigenous and immigrant communities in the Woodlands-Grandview neighbourhood, resulting in the publication of a book. In addition, CoV conducted surveys, interviews, and a literature review to gather people's perceptions about intercultural relations and the availability of information about Indigenous issues to newcomers. This research confirmed the lack of knowledge and the need to strengthen relations.

Figure 2.3.2 • Dialogues Project book cover.

Source: City of Vancouver.

Phase II (2011–13) built on the feedback gathered from the initial activities. The dialogue circles were discontinued, and more focus was placed on youth engagement, particularly through a youth summit and youth dialogue sessions. More than 120 youths participated in the weekend-long summit during which they engaged in workshops and discussed some of the challenges of cultural plurality in Vancouver. The cultural exchange visits were a key component of the second phase, especially since several community groups took the lead and organized their own visits. Toward the last year of the Dialogues Project, the social planners at CoV adopted more arts and cultural approaches to dialogues and storytelling, engaging numerous community artists and activists through art-based programs. Particularly noteworthy were creative programs like Food Dialogue, Sharing Traditional Healing Practices, and Kayachtn, a community gathering that invited 40 First Nation, urban Aboriginal, and immigrant artists; community leaders; and residents to participate in an integrated culinary, art-making, and dialogue lunch event. In order to encourage and mobilize more community organizations to launch similar initiatives, a how-to handbook was compiled, offering guidelines for organizing cultural exchange visits.

The Dialogues Project has inspired several spin-off initiatives among community partners, including seniors' multicultural activities and cultural exchange visits hosted by local First Nations. From CoV's perspective, the Dialogues Project also:

Figure 2.3.3 • Dialogues Project overview diagram.

Source: Umbreen Ashraf, Kate Kittredge, and Magdalena Ugarte.

- contributed to the formulation of the council's First Nations and urban Aboriginal engagement strategy;
- facilitated joint programs between the Vancouver School Board and local First Nations;
- developed the first Newcomers Guide to First Nations and Aboriginal Communities, to be launched in 2014;
- paved the way for the first citizenship ceremonies held on a BC First Nation reserve; and
- helped to build momentum for the proclamation of Vancouver's Year of Reconciliation, 2013–14.

Sharing Stories, Cultivating Intercultural Understanding, Building Community

> We have already heard the dominant story, over and over and over again. But what we have not heard is this community's (First Nations) story.
>
> Social planner, City of Vancouver

The Dialogues Project is a rich planning case to study for a number of reasons. To better understand this unique initiative, we conducted in-depth interviews with Dialogues Project organizers, facilitators, and participants; attended some of the key events of the Dialogues Project and community-led spin-off initiatives; and carried out an extensive review of official government documents, articles, and book chapters. Here we highlight some of our main findings: notably, the centrality given to dialogue, storytelling, experiential learning, and appreciative inquiry as community-building tools in highly diverse communities with complex histories (see Planning Theory Box). We argue that the direction taken by CoV's Social Policy Division reflects an awareness of this complexity. The City's direction also highlights the need to address the root causes of weak intercultural interaction and understanding in order to build strong and sustainable communities in urban contexts. The three pillars that guided the Dialogues Project echo this understanding.

Storytelling and Listening: The Transformative Power of Conversation

Sharing stories was the cornerstone of the Dialogues Project. This emphasis was especially clear in the dialogue circles, youth and elders program, story-gathering projects, and youth summit. Focusing on storytelling was a way to honour the oral traditions of the original inhabitants of the land. Equally important, it was an

Figure 2.3.4 • Cultural exchange visit, Tsleil-Waututh Nation.

Source: City of Vancouver.

avenue for the recognition of untold stories (Sandercock 1998; Yu 2011). The planners saw the potential of storytelling in facilitating productive exchanges among these groups, which coexist in Vancouver without having strong interactions and knowledge of each other's challenges, cultures, and histories. All participants were encouraged to share their experiences about the past, present, and future but also to listen to others' stories with attention and respect (Forester 1989). These close encounters were aimed at awakening people's genuine interest in other cultures and perspectives, exploring common ground, and laying the foundation for future relations (Gray 2011).

The Dialogues Project was conceived as a relationship-building and public education exercise. The planners understood that profound transformations in the perceptions and behaviours of people do not happen overnight. They also recognized the challenges and risks of fostering face-to-face cross-cultural interactions among communities marked by traumatic histories. Therefore, experienced facilitators were invited to help guide the dialogue circles. By developing ground rules and protocols, and encouraging people to listen with open hearts and minds, the facilitators helped to create safe and deliberative spaces where these sometimes unsettling conversations could take place (Forester 2009).

Stories about racism, exclusion, traumatic displacement, and loss emerged in these dialogues. Although they were often difficult—but in the view of some participants did not get to the core of the issues—those initial conversations provided a space to envision strategies for building relationships and inspiring change.

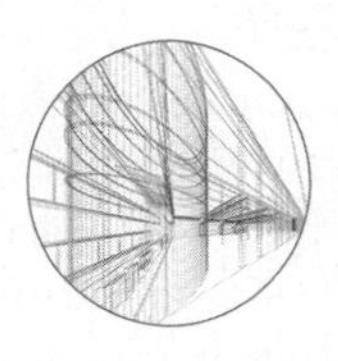

PLANNING THEORY

Impacts of Colonization for Planning

Planning theorists and practitioners recognize that traumatic experiences such as colonization, racism, displacement, and exclusion—and the subsequent social cleavages they generate—pose important challenges to planning processes (Forester 2009; Jojola 2008; Sandercock 2003). These challenges become evident in many planning efforts, particularly community-building, and are even more difficult to address when deep cultural differences exist in communities.

Planning Theory: Potential of Dialogue and Storytelling

Dialogue and storytelling have the potential to enable difficult conversations and inspire social change, making them particularly useful planning tools in communities marked by histories of violence, neglect, racism, and social exclusion (Dale 1999; Forester 1989; Sandercock 2003). Facilitating productive exchanges based on dialogic interaction among groups, which coexist without having knowledge of each other's past and present realities, can result in previously untapped understanding and even new and lasting relationships.

Dialogic and Experiential Learning

The second pillar of the Dialogues Project was creating spaces for direct dialogic engagement and experiential learning among participants. The project valued different ways of knowing, in particular the efficacy of learning from non-verbal evidence (Sandercock 2003). The project also involved deliberate efforts "to tap into people's tacit knowing" (Sandercock 2003, 77–8) to help explore those experiences that people struggle to express in conversations. The cultural exchange visits illustrate the potential the social planners at CoV saw in allowing participants to *experience* other cultures as opposed to reading about them or relying on stereotypes.

Different groups of First Nations, urban Aboriginals, and immigrants received members of other communities in places of strong cultural significance. There, the hosts shared their histories, traditions, beliefs, and practices with their guests through guided tours, food-sharing, and conversation. These encounters enabled factual knowledge exchange, which was critical given the lack of information available to newcomers, as the initial research stages of the Dialogues Project had revealed. Most important, they allowed the visitors to put themselves in other people's shoes, contemplate, and have an embodied approach to different cultures. In this way, the Dialogues Project went beyond theoretical and abstract knowledge—such as socio-economic statistics, indicators, and dominant historical narratives—and highlighted local knowledge instead. Moreover, these experiential initiatives sought to unsettle and empower the participants and inspire them to generate change in their own communities.

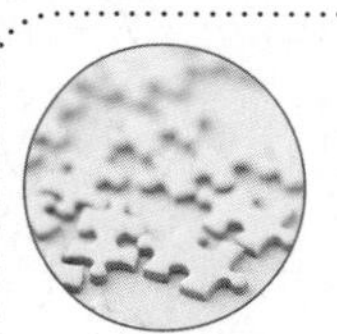

KEY ISSUE

Strengths-Based Approach to Planning

COV approached the Dialogues Project using a strengths-based lens. They identified and celebrated the assets that are already abundant in the communities and then worked to build on these assets through dialogic processes, storytelling, and appreciative inquiry.

Appreciative Inquiry Approach: Building on Community Assets

The third pillar of the Dialogues Project was adopting an appreciative inquiry approach. In other words, all the initiatives aimed at exploring, emphasizing, and building on the strengths, resources, resilience, and shared experiences of the First Nations, urban Aboriginal, and immigrant communities as opposed to focusing on problems and deficiencies. Given the goals of the Dialogues Project, the approach seems pertinent and empowering, in line with assets-based community development (Kretzmann and McKnight 1996). It highlighted the point that the potential to increase intercultural understanding, build stronger relationships, and lessen negative stereotypes lies with the communities themselves. While there was space to explore existing differences and challenges, the ultimate goal was to acknowledge them in order to envision ways of moving forward together.

Challenges, Lessons, and Transferability

Any demonstration project is sure to pose noteworthy challenges. This is particularly so when the project involves something as crucial—yet intangible—as relationship-building between cultural groups that have been previously distanced and may have faced traumatic experiences of exclusion, racism, and marginalization. While it is early to assess the long-term effects of the Dialogues Project, it is fair to say that COV not only successfully explored innovative ways of bringing these communities together and triggered several community-led spin-off initiatives but also posed a challenge to the "framework of rules and patterns of resource allocation" that tends to prevail in typical governmental processes (Healey 1992). Here we present some of the main challenges and lessons learned and offer guiding questions to support other planning practitioners who may see the Dialogues Project as a model to be adapted in other municipalities.

Time and Resources

From a municipal perspective, limited time and a lack of resources posed one of the greatest challenges for the Dialogues Project. Because of its scope, tight time frame, and budget, the small team of co-ordinators was stretched as it worked to achieve ambitious goals. Given the dialogic spirit of the project, allocating more time and more sessions for some of the initiatives—such as the dialogue circles—could have also encouraged greater depth of discussion and cross-cultural understanding.

It is challenging for municipalities to secure funding for projects that focus on Indigenous peoples and immigrants together. Similarly, it is difficult to garner political "buy-in" and secure funding for planning processes—such as relationship- and trust-building—that take time and added resources to meaningfully develop and sustain over time. Their results are also difficult to tangibly measure. As some of the organizers recognized, with limited human resources to explore and articulate new initiatives and spin-off ideas or even to seek out other funding opportunities, the potential to expand the project beyond the demonstration phase was limited. However, like ripples in a pond, the Dialogues Project sparked momentum at the community level, inspiring great potential for future projects.

How do we emphasize the process over the product or outcome? How do we encourage funding bodies to do the same? What is lost or gained when a planning initiative depends to a great extent on political agendas?

Engagement

> I think people are afraid of conflict, especially in this society; everybody tries to be so nice. Things are there, but they don't talk about them. I am more passionate about things. I don't fight, but it's [about] expressing my feelings and thoughts . . . for me it's part of my culture. How are we going to come and talk if we are not going to be honest saying what we really think? There is no point, really, [in] touching just the superficial things and being nice.
>
> Participant, Dialogue Circles

The essence of social transformation depends on who is engaged in the initiative. Expanding the invitation to participate beyond representatives from select organizations to the broader community would have allowed for the coming together of a more diverse group of people from different backgrounds and experiences and could have also inspired change at more grassroots levels. What is lost when participants are "cherry-picked"? On the other hand, what is at risk when you open up the process to the broader community?

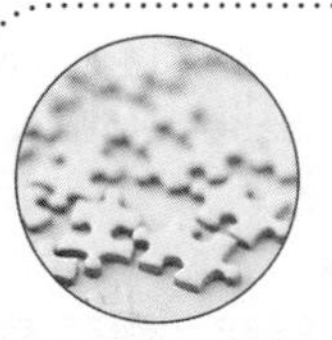

KEY ISSUE

Capacity-Building at the Municipal Level

By embarking on the Dialogues Project, COV and City Council have the additional confidence, greater capacity, and deeper understanding to take next steps forward and embark on other innovative initiatives with the Indigenous and immigrant communities in the city.

It is imperative to consider stakeholder inclusion from the outset of the project design process. Some interviewees felt that COV did not collaborate enough with community groups in the initial planning stages to set the program's agenda. Was this a missed opportunity to inspire deeper engagement with, and ownership of, the process? How can a municipality continue to move away from top-down approaches to planning when faced with the reality of tight timelines and resource constraints?

Although COV planners define all non-Indigenous Canadians as immigrants, the Dialogues Project was conceived and funded for building relationships between Indigenous people and new immigrants only and did not include the settler community. A number of the participants felt that the dialogue circles, in particular, should have been opened up to this group and that it was a missed opportunity to connect and educate cross-culturally at a deeper level.

The Dialogues Project was intentionally designed to mitigate the potential risks of conflict. The profile of the participants, the choice of highly skilled facilitators, and the creation of communication ground rules in some of the initiatives are all markers of this careful control. In line with this, COV cautions never to assume that people arrive ready to engage but to ensure that they are prepared and willing to engage beforehand. Part of this preparation includes providing potential participants with background materials. In this way, they can make an informed decision whether to be involved or not and—if they choose to be—arrive ready to participate with a basic understanding of the issues.

Controlling the variables in these ways increased the probability of a positive experience for those involved, likely contributed to the new relationships that were built, and increased capacity at the municipal level. Positive outcomes notwithstanding, can dialogue be truly transformative, addressing the root causes of conflict, when the "messiness" and risks are so carefully mitigated? How can municipal governments and city councils be encouraged and supported to meaningfully undertake such complex dialogic processes?

Evaluation

A co-ordinated and in-depth evaluation process—supported by adequate resources—at all stages and levels of the project is a necessary and valuable undertaking.

In this case, CoV was required to follow the funder's (WICWP) evaluative format and contracted a third-party evaluator to carry out a formative evaluation of the project, which included both a qualitative and quantitative assessment of the initiatives. In general terms, it "measured" the perception of participants and community partners immediately after their involvement in a specific initiative through the administration of short feedback surveys. In addition, CoV gathered informal feedback from the Steering Committee and community partners, who offered their comments and suggestions. Despite these efforts at evaluation through both phases of the Dialogues Project, there are fundamental challenges with evaluating a demonstration project such as this. How do we trace social change over time? How does one measure unintended outcomes and their "ripple effects" at the individual and societal levels? How does one create an evaluative framework attuned to such complexity, nonlinearity, and ultimate uncertainty (Quinn Patton 2011)?

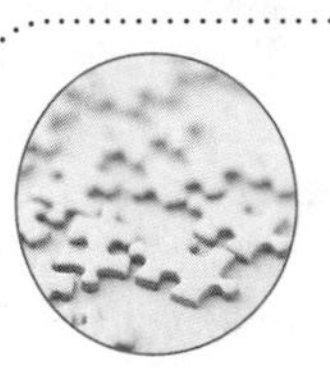

KEY ISSUE

Importance and Challenges of Evaluation

The Dialogues Project experience highlights the need to plan community-building initiatives with short- and long-term evaluation in mind and draws attention to the complexities of assessing changes in people's perceptions, values, and behaviour.

Transferability

Other communities across Canada are experiencing similar challenges related to cultural plurality. Consequently, the Dialogues Project has attracted the attention of other city governments, as well as human rights groups and Indigenous councils. There is immense potential for the model to be adapted in other municipalities across Canada, perhaps even internationally, provided all parties involved—particularly at the governmental level—not only acknowledge the need for such interactions and share a common vision but also collaboratively commit financial, personnel, and infrastructural resources from the project's inception through to its conclusion. Perhaps more important, it is central that the various communities be involved in the planning process from the beginning and able to shape the agenda, as a number of participants made clear.

Conclusions

In cities as culturally diverse as Vancouver, building strong and sustainable communities is not without its challenges, and current demographic trends can make these challenges even more difficult. Problematically, newcomers have little exposure to the country's colonial roots and the complexities of Indigenous/settler relations, and there is limited interaction among First Nations, urban Aboriginals, and immigrants in the city. The Dialogues Project aimed to explore innovative ways of bringing these communities together and fostering increased understanding and stronger relationships among them by developing community-building initiatives focused on dialogue, storytelling, experiential learning, and appreciative inquiry. While it is early to assess the project's long-term impacts, the experience to date offers inspiring evidence of the potential of transferability of this approach in similar contexts. The approach could be used to strengthen relations and build trust between municipal governments and community organizations working with these communities, trigger community-led spin-off initiatives and expand the message to the broader public, and enhance awareness among groups that historically have been apart. The main challenges for municipal planners remain: garnering sufficient resources and time for plan implementation, engaging the relevant communities from the outset, building political momentum and support to sustain the project, and developing strategies to assess the long-term effects of such pivotal initiatives.

KEY TERMS

Appreciative inquiry
Demonstration (or pilot) project
First Nations
Immigrants
Settler
Urban Aboriginals

REFERENCES

City of Vancouver. 2011. "Dialogues Project." http://vancouver.ca/commsvcs/socialplanning/dialoguesproject/index.htm.

Dale, N. 1999. "Cross-cultural community-based planning: Negotiating the future of Haida Gwaii (BC)." In L. Susskind, S. McKearnan, and J. Thomas-Larmer, eds, *The Consensus Building Handbook*, 923–50. Thousand Oaks, CA: Sage.

Forester, J. 1989. *Planning in the Face of Power*. Berkeley: University of California Press.

——. 2009. *Dealing with Differences: Dramas of Mediating Public Disputes*. New York: Oxford University Press.

Gray, L. 2011. *First Nations 101: Tons of Stuff You Need to Know about First Nations People*. Vancouver: Adaawx Publishing.

Healey, P. 1992. "A planner's day: Knowledge and action in communicative practice." *Journal of the American Planning Association* 58(1): 9–20.

Jojola, T. 2008. "Indigenous planning: An emerging context." *Canadian Journal of Urban Research* 17(1): 37–47.

Kretzmann, J., and J.P. McKnight. 1996. "Assets-based community development." *National Civic Review* 85(4): 23–9.

Little Bear, L. 2009. "Jagged worldviews colliding." In M. Battiste, ed., *Reclaiming Indigenous Voice and Vision*. Vancouver: University of British Columbia Press.

Quinn Patton, M. 2011. *Developmental Evaluation: Applying Complexity Concepts to Enhance Innovation and Use*. New York: Guilford Press.

Regan, P. 2010. *Unsettling the Settler Within: Indian Residential Schools, Truth Telling, and Reconciliation in Canada*. Vancouver: University of British Columbia Press.

Sandercock, L., ed. 1998. *Making the Invisible Visible: A Multicultural Planning History*. Berkeley: University of California Press.

——. 2003. *Cosmopolis 2: Mongrel Cities of the 21st Century*. New York: Continuum.

Statistics Canada. 2011. *NHS Focus on Geography Series—Vancouver*. http://www12.statcan.gc.ca/nhs-enm/2011/as-sa/fogs-spg/Pages/FOG.cfm?lang=E&level=4&GeoCode=5915022.

Welcome BC. 2013. "BC Welcoming and Inclusive Communities and Workplaces Program (WICWP)." http://www.welcomebc.ca/Communities-and-Service-Providers/Service-Providers/funded-services/inclusive-communities/wicwp.aspx.

Yu, H. 2011. "Nurturing dialogues between First Nations, urban Aboriginal and immigrant communities in Vancouver." In A. Mathur, J. Dewar, and M. DeGagne, eds, *Cultivating Canada: Reconciliation through the Lens of Cultural Diversity*, 299–308. Aboriginal Healing Foundation. http://speakingmytruth.ca/downloads/AHFvol3/20_Yu.pdf.

CASE STUDY 2.4

Social Vulnerability to Climate Change: An Assessment for Yarmouth, Nova Scotia

Michaela Cochran

SUMMARY

Climate change is predicted to increase the frequency and severity of natural hazards. Climate change researchers have observed that areas with the highest **social vulnerability**, rather than those with the greatest biophysical risk, are the most heavily affected by natural hazards. This is because socially vulnerable populations have limited access to the social, political, and economic resources necessary to prepare for, withstand, and recover from these events.

The best way to reduce overall impacts from these events is by working to decrease social vulnerability and limit negative impacts on socially vulnerable populations. While reducing social vulnerability itself requires initiatives from all levels of government, incorporating considerations of social vulnerability into emergency management planning can enable communities to minimize climate change impacts.

The Town and Municipality of the District of Yarmouth are located on the southern tip of Nova Scotia. This area is subject to severe weather such as tropical storms and hurricanes; physical impacts may include wind damage and storm surge flooding leading to property damage and loss of electrical power or road access. Populations in Yarmouth that are most vulnerable to these impacts include those with low socio-economic status, who may have limited resources to prepare, withstand, and recover, and seniors, who are more likely to face physical challenges such as limited mobility. Isolation from essential services is a concern for rural residents.

OUTLINE

- Socially vulnerable populations experience the most severe consequences of natural hazards because they have less access to social and economic resources to help them prepare, withstand, and recover.
- Knowledge of the geographic distribution of demographic characteristics that contribute to social vulnerability can enable the direction of appropriate mitigation and emergency management measures to the populations most in need.
- This case study evaluates the geographic distribution of social vulnerability to climate change impacts in Yarmouth, Nova Scotia.

Introduction

Scientists predict that climate change will cause an increase in the frequency and severity of natural hazards such as severe storms and droughts. To help minimize negative outcomes from these and other climate change effects, planners have been called on to incorporate **climate change adaptation** into planning practice.

Throughout both the developing and developed world, researchers have observed that the impacts of natural hazards are unequal: the most severe consequences are experienced by underprivileged people, who may lack access to resources that could allow them to better prepare for, withstand, and recover from the impacts of a natural hazard. The most vulnerable populations vary somewhat depending on location but have in common the experience of marginalization or lack of social and economic capital and almost always include the poor, minorities, disabled persons, young children, and the elderly. These populations are described as socially vulnerable (Adger et al. 2004; Cutter et al. 2009; Tapsell et al. 2010). The observation that socially vulnerable people are hardest hit by natural hazards caused by climate change has led to the study of **social vulnerability to climate change and natural hazards.**

An important component of research on social vulnerability to climate change and natural hazards is the development of methods for measuring the geographic distribution of social vulnerability so that climate change adaptation, emergency management, and disaster relief efforts may be directed to those areas. By incorporating an assessment of social vulnerability into climate change adaptation and **emergency management planning**, planners can help to reduce negative consequences from climate change and natural hazards.

This case study describes the research approach and findings of an evaluation of the geographic distribution of social vulnerability to climate change impacts in Yarmouth, Nova Scotia. The planning approach incorporates aspects of climate change adaptation, emergency management, and **social planning**.

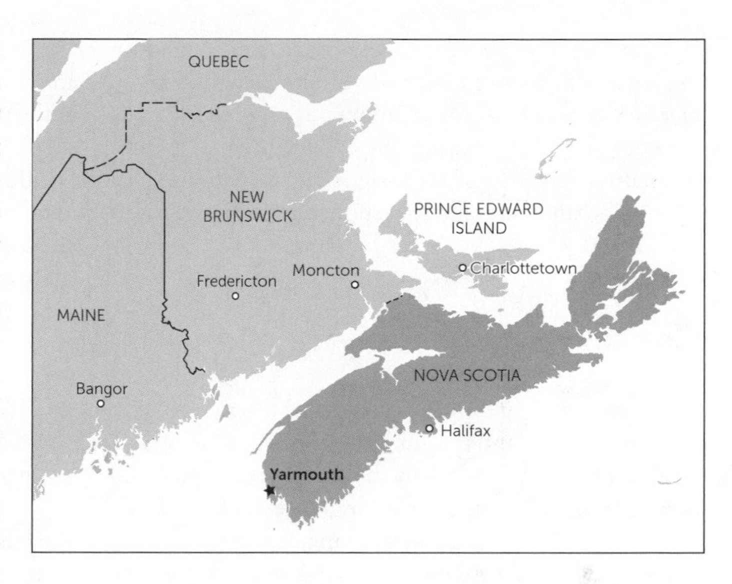

Figure 2.4.1 • Location map.

Social Vulnerability to Climate Change

Social, cultural, and political forces accord different levels of access to resources to different groups in society. Socially vulnerable people are those who have less access to social, economic, and political resources and may experience marginalization. Social vulnerability is created in a social, cultural, political, economic, and historical context through relations, structures, and processes in wider society. Therefore, people have little to no control over the forces that render them socially vulnerable. Because social vulnerability is socially determined, not the result of inherent traits, it takes a different shape in different places (Enarson and Walsh 2007; Mikkonen and Raphael 2010).

It is important to understand the scale on which social vulnerability is generated: entire social groups are affected by the social forces that limit access to resources, so members of those social groups are more likely than other members of society to be socially vulnerable. However, this does not mean that every member of those social groups is socially vulnerable. Because social vulnerability is not determined by inherent traits, it is also not static: people can become more or less vulnerable if their situation changes. For example, as a person enters old age or if a disability becomes more severe, they may become more socially vulnerable; if a person's employment changes and they earn a higher income that allows them access to resources previously unavailable to them, they may become less socially vulnerable (Tapsell et al. 2010).

Social vulnerability exists independently of and prior to the event of a natural hazard. Social vulnerability is significant when a natural hazard occurs, because socially vulnerable people have limited access to social, economic, and political resources and these resources are necessary in order to prepare for, withstand, and recover from natural hazard impacts. Additionally, the effects of a natural hazard can worsen existing social vulnerability. Social vulnerability to natural hazards can be conceptualized as the interaction of pre-existing social vulnerability and natural hazard impacts.

Researchers studying social vulnerability to climate change have worked to develop methods for measuring its geographic distribution. The purpose is to identify the most socially vulnerable areas so that adaptation, emergency management, and disaster relief efforts may be directed to those locations because they will experience the most severe consequences from a natural hazard, even if other less socially vulnerable areas experience the same physical exposure. Understanding the particular nature of social vulnerability in a given location is necessary for an appreciation of the various types of problems that may be experienced as a result of a natural hazard and for insight into the most helpful adaptation and emergency response measures (Tapsell et al. 2010).

Context

This project was undertaken as part of a series of studies for the Atlantic Climate Adaptation Solutions Association (ACAS), a collaboration of government and non-government partners working to enhance Atlantic Canada's resilience to climate change impacts. ACAS is one of six Regional Adaptation Collaboratives operating under the auspices of Natural Resources Canada.

The purpose of the ACAS projects was to assist participating municipalities in meeting the requirement to prepare a Climate Change Action Plan in order to receive infrastructure support funding from higher levels of government, as well as serving as an example for other municipalities.

ACAS projects for the Yarmouth, Nova Scotia, study area were performed by the Dalhousie University Climate Change Working Group. Studies included local sea-level rise and storm surge scenarios; assessments of climate change impacts on infrastructure, social assets, and socially vulnerable populations; and an evaluation of municipal capacity to respond to climate change effects. Project supervisors were Patricia Manuel and Eric Rapaport of the Dalhousie University School of Planning.

Study Area

The study area includes the Town of Yarmouth and the Municipality of the District of Yarmouth, both of which volunteered to participate in ACAS. The Town of Yarmouth (the Town) has a population of approximately 7000; the Municipality of the District of Yarmouth (the District) has a population of approximately 10,000. Throughout this case study, the two municipalities together will be referred to as Yarmouth. Yarmouth is located at the southwestern tip of Nova Scotia on the Gulf of Maine. The area is subject to severe weather, including tropical storms and hurricanes, which may produce a storm surge. In and around Yarmouth, storm surge has caused serious damage in the past and poses the most significant climate change risk for this area.

The Yarmouth area is experiencing significant economic challenges, resulting in long-term effects such as unemployment, low income, and population loss, especially among younger age cohorts. These factors contribute to social vulnerability in Yarmouth.

Purpose and Objectives

Purpose

The purpose of the study was to evaluate social vulnerability to climate change impacts in Yarmouth, Nova Scotia. The study assessed the prevalence of characteristics that contribute to the social vulnerability of individuals and households, using data aggregated at the **dissemination area** level.

The Town and District of Yarmouth may use the results of this study to minimize storm surge impacts by focusing adaptation and emergency management planning on high-risk populations and areas. It will also serve as an example for other Nova Scotia municipalities.

Objectives

- To generate an index of social vulnerability for application in Nova Scotia.
- To identify residential areas and populations in Yarmouth with high levels of social vulnerability and community services that serve socially vulnerable people.
- To suggest the possible impacts of climate change, specifically storm surge flooding, on socially vulnerable populations and areas in Yarmouth.
- To suggest opportunities for reducing adverse impacts on socially vulnerable populations in Yarmouth.

Methodology

Because there is no established method of evaluating social vulnerability in Canada, it was necessary to design one. To inform the methodology, an extensive

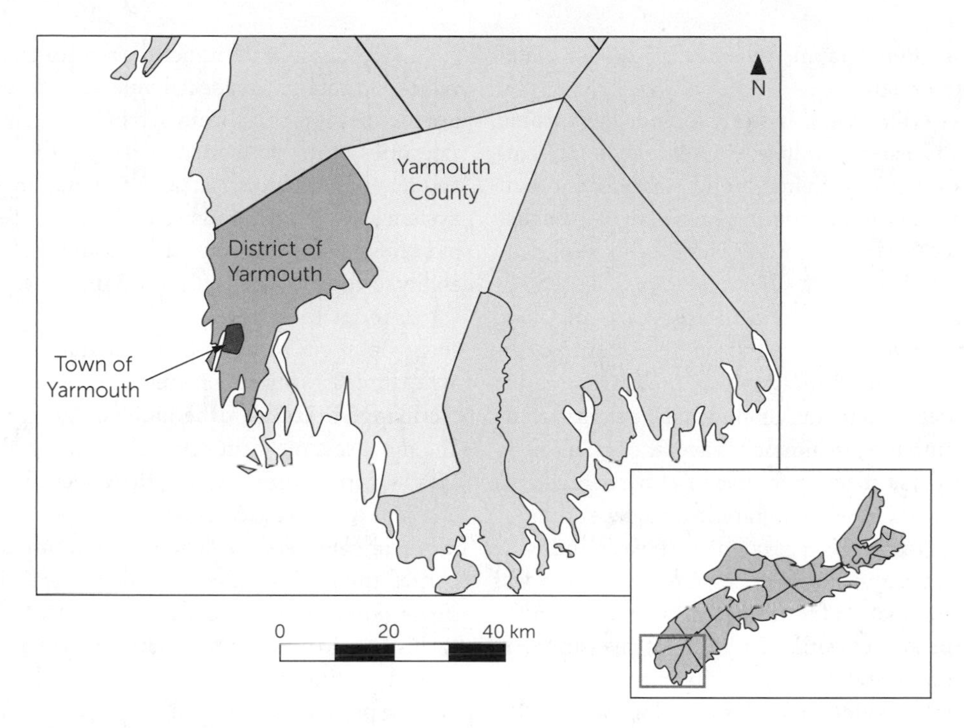

Figure 2.4.2 • Study area.

review of literature was conducted to address various interpretations of social vulnerability and the strengths and weaknesses of different methodologies for assessing it. Based on this research, a **mixed methods** approach was selected: first, an index of social vulnerability measures data from statistical sources was developed; second, consultations with persons possessing local knowledge relevant to issues of social vulnerability provided context for the index results and additional local detail.

The purpose of the ACAS project series was both to assist participating municipalities in their climate change planning process and to provide an example for other Nova Scotia municipalities. Therefore, the study methodology was designed so that it could be applied in municipalities throughout Nova Scotia, using readily available data and minimal technological requirements.

Index

An index of social vulnerability uses data from statistical sources to measure the prevalence of various indicators of social vulnerability in different geographic areas. In order to create the index, it was necessary to identify which factors contribute to social vulnerability in the study area.

Because social vulnerability is not an inherent trait but rather determined by social, cultural, and political forces, the factors that contribute to social vulnerability are different in different places. The indicators selected for use in the index of social vulnerability in this study are based primarily on the Canadian **Social Determinants of Health**. This choice is consistent with research performed by Drs Lindsay, Enarson, and Walsh of the Brandon University Department of Applied Disaster Studies and the Canadian Red Cross (Enarson and Walsh 2007; Lindsay 2003).

The Social Determinants of Health are a set of conditions that contribute to marginalization and exclusion, which in turn cause negative health outcomes. While this evaluation of social vulnerability does not address the health impacts of the Social Determinants of Health, it benefits from this established body of research regarding demographic groups with limited access to social and economic resources in Canada.

Where possible, the index directly measures the prevalence of characteristics that are Social Determinants of Health in Canada. Where a Social Determinant of Health could not be directly measured, this study calculates the prevalence of demographic characteristics associated with that determinant.

Indicators utilized in this study include low income, government transfer payments, unemployment, children, seniors, seniors living alone, lone-parent families, no secondary education, no knowledge of English or French, recent immigrants, visible minorities, and Aboriginal identity.

The prevalence of each indicator was measured using statistical data sourced from the 2006 Census of Canada. The unit of analysis was the dissemination area—the smallest scale at which census data are available. Dissemination areas each have a population of approximately 400 to 700 people or 250 households but may vary significantly in geographic area depending on population density. The prevalence of each indicator of social vulnerability in each dissemination area was measured in comparison to the average for all dissemination areas in Nova Scotia. The unit of measurement was the **standard deviation** of results in Nova Scotia for that indicator. This demonstrates not only *whether* the prevalence of a characteristic that contributes to social vulnerability is higher or lower than the mean for Nova Scotia but *how much* higher or lower than the mean it is in comparison to the average of how much all Nova Scotia dissemination areas differ from the mean. This demonstrates the prevalence of social vulnerability in the dissemination area relative to the average prevalence of social vulnerability throughout Nova Scotia.

In addition to identifying how each dissemination area compares to the average for Nova Scotia for each separate indicator of social vulnerability, the scores for each indicator were combined to produce an overall score of social vulnerability. All indicators were weighted equally, with the exception of the "low income" category, which was accorded double weighting. This choice was made to reflect widespread evidence that low income has the most profound impact among the Social Determinants of Health in Canada because of its influence on living conditions, psychological functioning, health-related behaviours, and access to health services (Public Health Agency of Canada 2003). Perhaps most important in relation to natural hazards, low income predisposes people to material and social deprivation, limiting their ability to obtain potentially life-saving necessities such as food, clothing, and housing (Mikkonen and Raphael 2010).

The results for each indicator were scaled so that higher values indicate greater social vulnerability and lower values indicate less social vulnerability. Scores were assigned categories of low to high vulnerability, then colour-coded and mapped using Esri ArcMap **Geographic Information System** software. This index method closely resembles that of Dr Susan Cutter, a prominent researcher of social vulnerability to climate change (Cutter, Boruff, and Shirley 2003).

In order to estimate the possible impacts of storm surge on socially vulnerable populations in the study area, three storm surge scenarios were overlaid with the mapped results of the index of social vulnerability. The most severe storm surge scenario was based on the area's worst recorded storm, the Groundhog Day Storm of 1976, which caused serious damage. Potential storm surge impacts were calculated by counting the number of properties either inundated or isolated (areas surrounded by water or where road access was lost) in each scenario. Community services were also mapped in order to estimate impacts on services that socially vulnerable populations rely on.

A benefit of the index of social vulnerability is that it can be performed relatively quickly, using readily available census data. By providing a quantifiable measure, it also allows the comparison of levels of social vulnerability geographically and over time. This creates a justifiable basis for directing adaptation and emergency management initiatives to those areas in greatest need.

Consultations

The next step in the study's methodology was to carry out consultations with sources of local knowledge. This process helped to ensure the inclusion of relevant factors in the index of social vulnerability as well as to compensate for limitations in the index and provide further local context. Additionally, consultees provided insight into municipal and emergency management policy and practice. This type of community-based approach is consistent with the work of a number of researchers studying the impacts of climate change on Canadian communities (Dolan and Walker 2003; Wall and Marzall 2006). While most assessments of social vulnerability use only either a statistical or a community-based approach, the use of both approaches in a single study means it will benefit from the strengths of each method while counterbalancing their limitations.

Persons selected for consultation included municipal councillors, who have knowledge of local decision-making processes and issues of concern in the community; municipal planners, who are aware of community demographics and development patterns; community services, public health and community health representatives, who are knowledgeable about health and socio-economic concerns in the community and have direct contact with socially vulnerable persons; representatives of social service organizations, who serve socially vulnerable persons; and the local Emergency Management Organization co-ordinator, who understands local emergency management priorities and capacity. The consultees provided insight into the strengths and weaknesses in community capacity to meet the needs of socially vulnerable persons in an extreme weather event.

This method was particularly important in the Municipality of the District of Yarmouth: dissemination areas in rural areas are quite large, so the index of social vulnerability was less able to illustrate patterns of social vulnerability in the District than in the Town. The consultations with knowledgeable local persons provided detailed information about how social vulnerability might manifest in rural areas and the types of needs that socially vulnerable persons might experience in an extreme weather event.

The strengths of this community-based research approach are that it does not require expert knowledge or technological capacity and that it benefits from local knowledge and experience.

Review of Municipal Planning Policy and Emergency Management

This component of the research involved a document review of the Municipal Planning Strategy and Integrated Community Sustainability Plan for each of the Town and District of Yarmouth, as well as consultation with municipal planners and emergency management personnel. The purpose was to determine the extent to which considerations of social vulnerability are integrated into municipal planning and emergency management.

Integration and Conclusions

In the final phase of the study, information obtained from all of the methods described above was integrated to produce a picture of social vulnerability to climate change impacts in Yarmouth. These observations formed the basis of recommendations for reducing vulnerability.

Limitations

The methodology used in this study has a number of limitations, such as subjectivity in methodological choices, data limitations, and simplifying assumptions.

Methodological choices in the study included the indicators selected, weighting of indicators, methods of comparison, aggregation techniques, and the scale of analysis. While extensive research was undertaken to inform these choices, a change in methodology could have a considerable impact on the results obtained.

The indicators selected in this study were limited to those for which accurate demographic data were available. A number of factors relevant to social vulnerability, such as disability, medical dependence, and homelessness, could not be included because of a lack of available data. Additionally, it may not be possible to replicate this study in the future because of changes in the Canadian census. For the 2011 collection, the long-form census return was not mandatory, resulting in a significantly reduced response rate in comparison to previous censuses (Rennie 2012). This is particularly problematic in regard to the availability of accurate data for an index of social vulnerability, since socially vulnerable persons are less likely to fill out a voluntary census form; prominent Canadian epidemiologists and population researchers have stated that data from the long-form census was not valid for research purposes, particularly for studies pertaining to the Social Determinants of Health (Clark 2010; Groome 2010). The mandatory long-form census was reinstated in 2015 in time for the 2016 collection.

A final limitation is that an index of social vulnerability requires simplifying assumptions: clearly, not all persons in a subpopulation that tends to experience social vulnerability are socially vulnerable, and not all persons in a geographic area with high social vulnerability are socially vulnerable. However, the fact that assessing social vulnerability is challenging does not mean that the task should be abandoned. Rather, the limitations in indices of social vulnerability must be acknowledged and further research undertaken to improve methodology.

Findings

The Town of Yarmouth has above-average overall social vulnerability for Nova Scotia. Social vulnerability is concentrated in the southern half of the Town, which has a high proportion of residents with low socio-economic status. People with low socio-economic status may lack the resources to prepare for or recover from an extreme weather event. For example, they may not have access to a personal vehicle for evacuation purposes, a stock of emergency supplies, or materials and equipment necessary to repair damage.

However, the northern half of the Town has the highest proportion of seniors. While many people over the age of 65 are healthy and able-bodied, this age cohort is more likely than younger cohorts to experience physical limitations and require assistance in the event of a natural hazard. Seniors might have hearing or vision loss that prevents them from hearing or understanding warnings about a natural hazard or might be unable to evacuate because of reduced mobility.

Few residential buildings are affected by storm surge in the scenarios studied. However, damage to waterfront businesses, such as restaurants, fisheries infrastructure, and industrial operations, could result in loss of employment and thereby increase social vulnerability.

The District of Yarmouth has average social vulnerability for Nova Scotia. The geographic distribution of social vulnerability varies widely among different factors. The District of Yarmouth was heavily affected

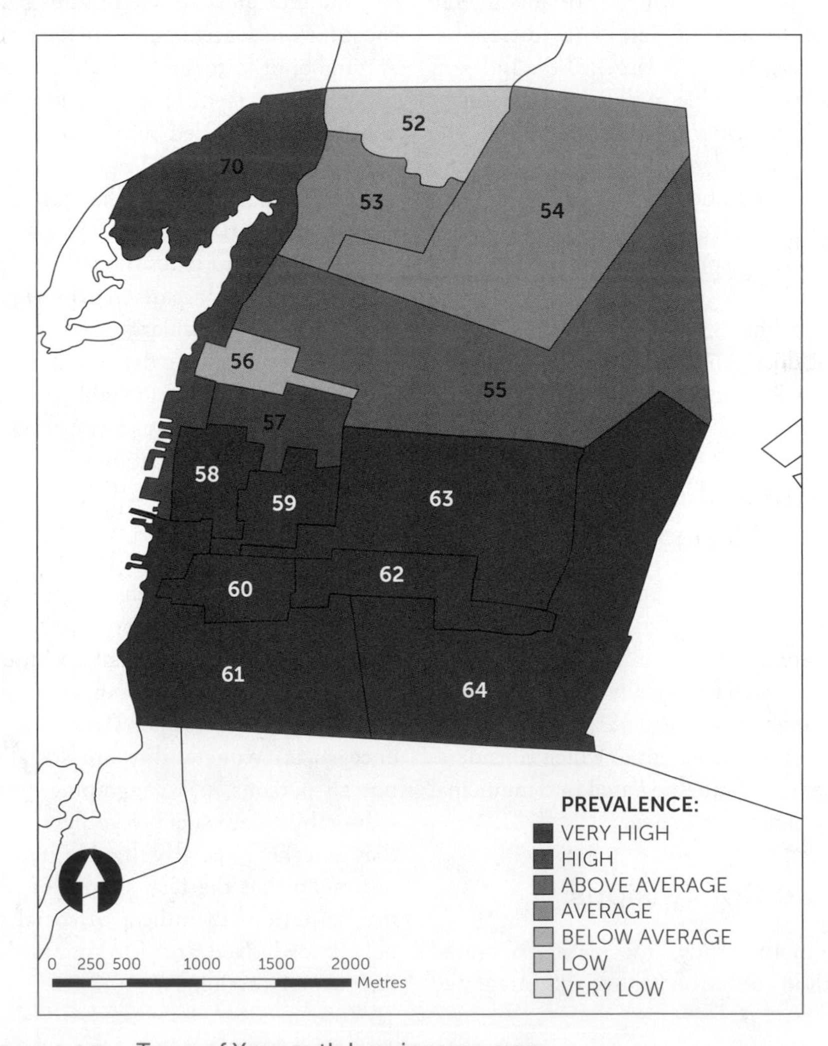

Figure 2.4.3 • Town of Yarmouth low-income map.

under the storm surge scenarios, with dozens of residences inundated and hundreds isolated. Loss of road access and electrical power are significant concerns in the District because of the many low-lying coastal roads very close to the open ocean or in marshy areas at or near sea level. Damage to fisheries infrastructure, boats, and related gear could result in loss of livelihood for many fishers in the area.

None of the services that socially vulnerable people rely on were inundated in the storm surge scenarios; however, access could be lost to one seniors' care facility. The most significant concern is the potential loss of bridge access from the Town of Yarmouth to the Yarmouth Regional Hospital.

Next Steps

The assessment of social vulnerability provided the Town and District of Yarmouth with the knowledge necessary to incorporate considerations of social vulnerability into mitigation measures and emergency management planning. Understanding the strengths and weaknesses of various geographic areas will allow local authorities and service providers to target each area with the most appropriate type of assistance in order to reduce overall vulnerability.

Voluntary organizations that serve socially vulnerable groups and representatives of socially vulnerable groups themselves can provide valuable input to emergency management planning (Enarson and Walsh

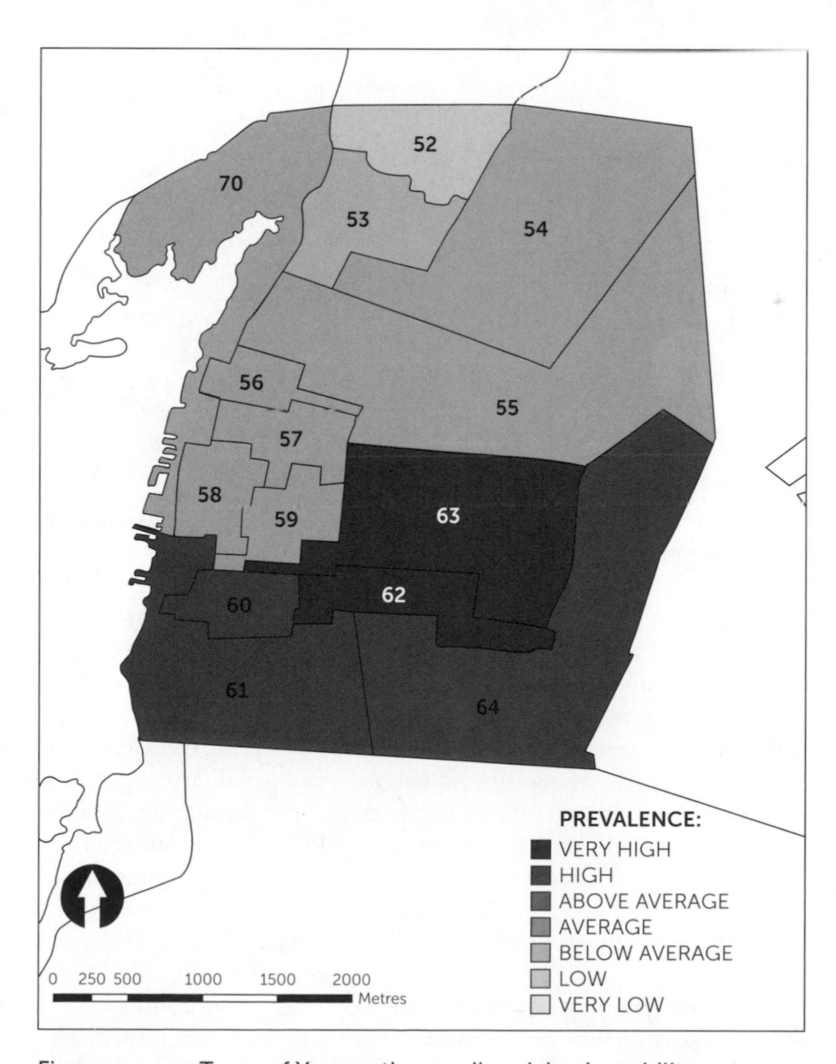

Figure 2.4.4 • Town of Yarmouth overall social vulnerability map.

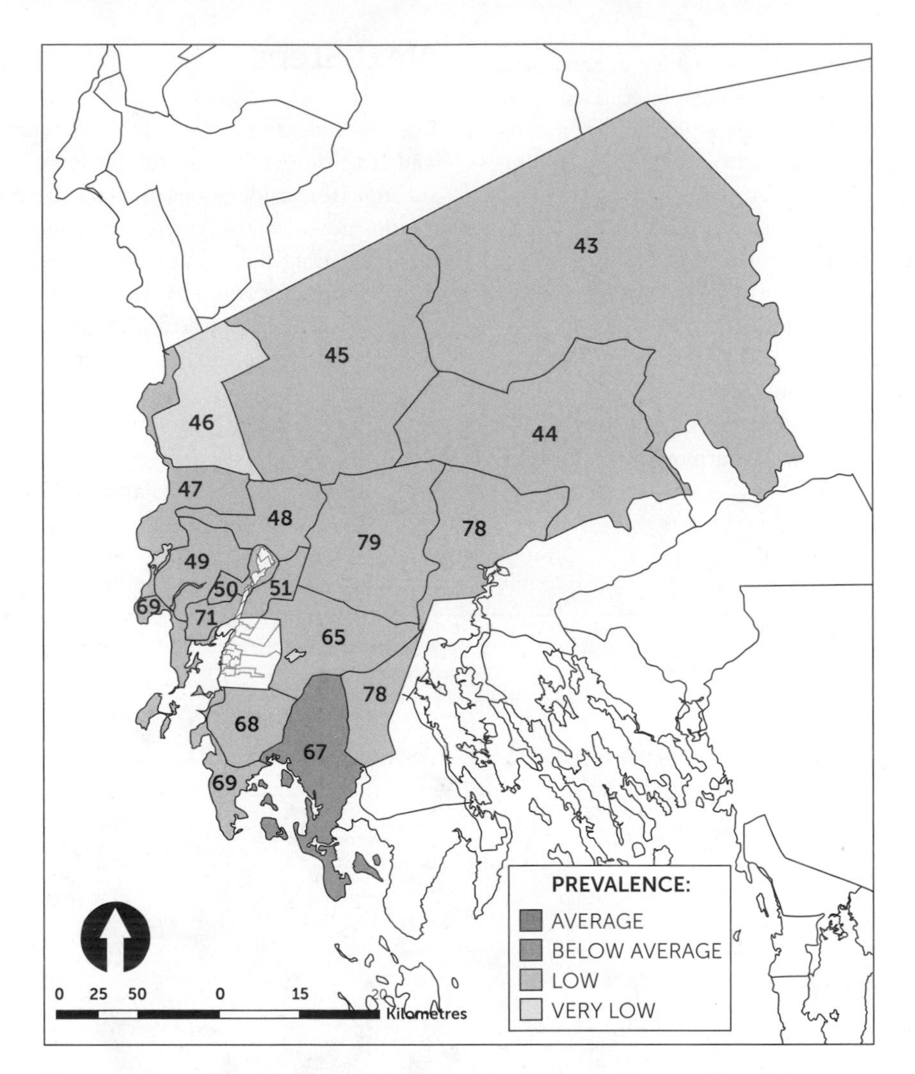

Figure 2.4.5 • Municipality of the District of Yarmouth overall social vulnerability map.

2007). Consultation regarding process would be important in determining the preferred form of involvement. Including these groups and individuals in planning processes can help to avoid misconceptions and stereotypes among planners and emergency management staff about the needs and capacities of socially vulnerable groups; this would work to counteract the social forces that generate social vulnerability in the first place, as well as ensuring that the most effective strategies are targeted toward the areas of greatest need, thus avoiding any wasted effort on misdirected or inappropriate approaches.

Some members of society may be truly unable to help themselves; examples include the infirm elderly, the severely disabled, and medically dependent persons. While many residents are aware of community members who might require assistance, individuals who are less well known, such as new residents or persons who are confined to their homes, could go unnoticed. Therefore, the involvement of government and voluntary organizations that are knowledgeable about the number, location, and specific needs of these individuals is essential in ensuring their well-being in the event of a natural hazard.

The Community Geomatics Centre and Accessibility Sault Ste Marie have developed the concept of a Vulnerable Persons Registry (Sault Ste Marie Innovation Centre 2011). This free, voluntary service provides first responders with key information about those choosing to register and helps to protect their

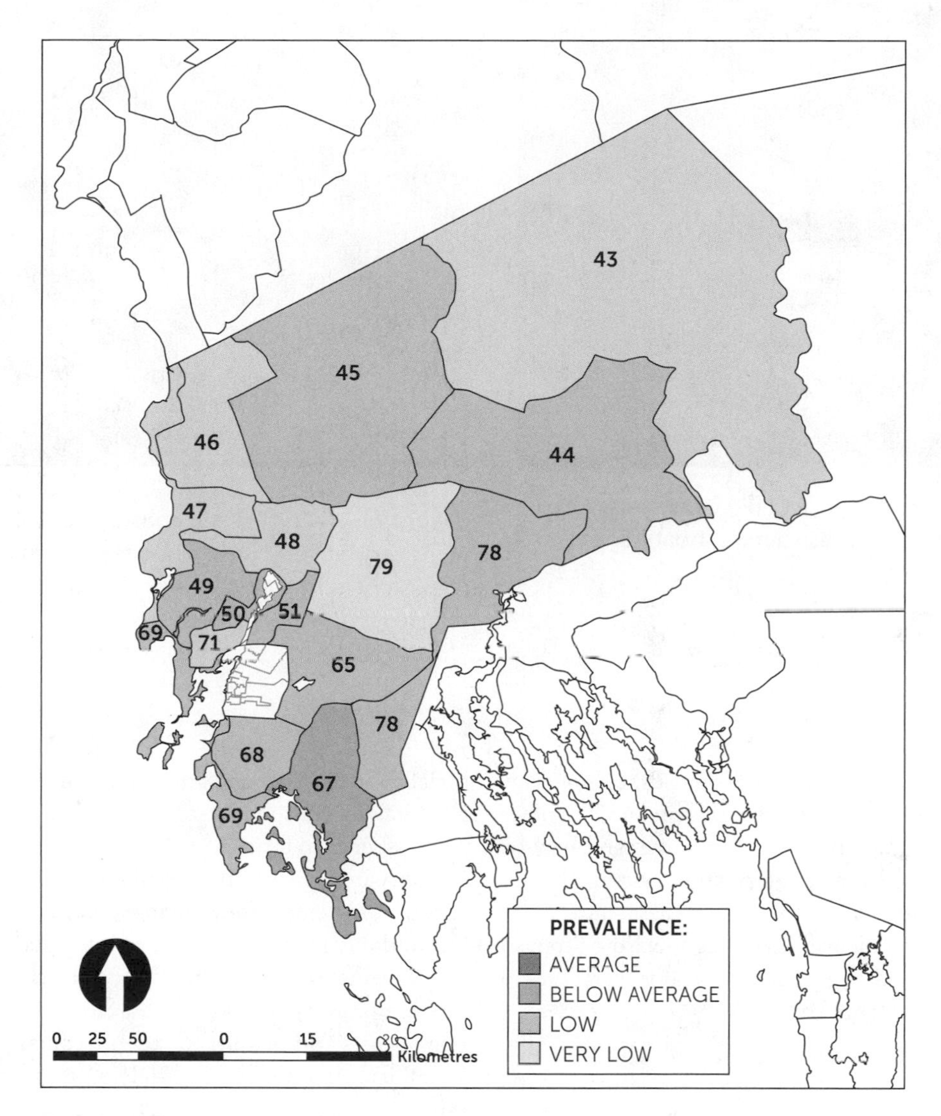

Figure 2.4.6 • Municipality of the District of Yarmouth low-income map.

safety in the event of an emergency. Home support groups in Yarmouth, such as the Victorian Order of Nurses, already have a protocol in place to check on clients and could be involved in informing residents about the option to register in such a database. The usefulness of this type of database is determined by its accuracy. While maintaining a database containing names, addresses, and information about an individual's needs does not demand extensive resources, it would require organization and regular updating. This type of proactive approach would offer Yarmouth a means of helping to prevent avoidable tragedies.

The purpose of the ACAS project series was both to assist participating municipalities in their climate change planning process and to provide an example for other Nova Scotia municipalities. Therefore, the study methodology was designed to be transferable. The factors considered in the index of social vulnerability are valid throughout Nova Scotia, while consultations with local knowledge sources ensure that locally specific conditions are considered. Many municipalities in Nova Scotia are small and have limited capacity and resources; therefore, the methodology for this study uses readily available census data and requires minimal technological capacity or expert knowledge. Index

Figure 2.4.7 • Storm surge could cause extensive damage to fisheries infrastructure, boats, and fishing gear.

Source: Zoë Wollenberg.

Figure 2.4.8 • Low-lying coastal roads and power lines very close to the open ocean, or in marshy areas, could be inundated or damaged, causing loss of access and electrical power.

Source: Zoë Wollenberg.

calculations can be performed with spreadsheet software. Sea-level rise and storm surge scenarios produced in other ACAS projects are available to municipalities and may be processed using Geographic Information System software or estimated using elevation maps. These methodical choices will allow municipalities throughout Nova Scotia to replicate this method of evaluating social vulnerability to climate change and natural hazards.

Conclusion

An assessment of the spatial distribution of social vulnerability provides the necessary knowledge to target adaptation and emergency planning efforts to those areas most in need. Additionally, it suggests the types of needs that socially vulnerable populations may have in the event of a natural hazard so that emergency planners can provide the most needed forms of assistance.

Decisions about priorities for climate change adaptation and emergency management planning will unavoidably privilege one set of interests over another. Considering social vulnerability in climate change planning moves beyond mere accounting for the economic costs and benefits of adaptation and introduces an ethical component to decision-making. However, this approach is also a practical one: since socially vulnerable populations are the hardest hit by natural hazards, the best way to reduce overall impacts from these events is by working to decrease social vulnerability and limit negative impacts on socially vulnerable populations.

KEY TERMS

Climate change adaptation planning
Dissemination area
Emergency management planning
Geographic Information System
Mixed methods research
Social Determinants of Health
Social planning
Social vulnerability
Social vulnerability to climate change and natural hazards
Standard deviation

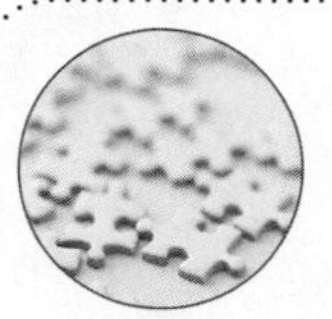

KEY ISSUE

Canadian Social Determinants of Health

The Social Determinants of Health are a set of conditions that contribute to marginalization and exclusion, which in turn cause negative health outcomes. A strong body of Canadian health and social sciences research demonstrates that these social determinants, rather than lifestyle choices or medical treatments, have the most profound impact on health. An important observation of this research is that certain demographic groups are more likely than others to experience a range of adverse living conditions that result in negative health outcomes (Mikkonen and Raphael 2010). Canadian research in this field emerged in the 1970s, and the Social Determinants of Health are recognized by major health organizations such as the Public Health Agency of Canada, the World Health Organization, and the Centres for Disease Control and Prevention.

The Canadian Social Determinants of Health

- Social support networks
- Education and literacy
- Employment/working conditions
- Social environments
- Physical environments
- Personal health practices and coping skills
- Healthy child development
- Biology and genetic endowment
- Health services
- Gender
- Culture (Public Health Agency of Canada 2003)

REFERENCES

Adger, W.N., N. Brooks, G. Bentham, M. Agnew, and S. Eriksen. 2004. *New Indicators of Vulnerability and Adaptive Capacity*. Norwich, UK: Tyndall Centre for Climate Change Research.

Clark, C. 2010. "Scrapped mandatory census cuts even deeper for disability advocacy group." *Globe and Mail* 24 July.

Cutter, S., C. Emrich, J. Webb, and D. Morath. 2009. *Social Vulnerability to Climate Variability Hazards: A Review of the Literature*. Oxfam America. Columbia, SC: Hazards and Vulnerability Research Institute, University of South Carolina Department of Geography.

Cutter, S., B. Boruff, and W.L. Shirley. 2003. "Social vulnerability to environmental hazards." *Social Science Quarterly* 84(2): 242–61.

Dolan, A H., and I.J. Walker. 2003. "Understanding vulnerability of coastal communities to climate change related risks." *Journal of Coastal Research* 39: 1316–23.

Enarson, E., and S. Walsh. 2007. *Integrating Emergency Management and High-Risk Populations: Survey Report and Action Recommendations*. Canadian Red Cross. Brandon, MB: Brandon University.

Groome, P. 2010. "Census changes harm Canadians' health: Health professionals warn of health impact of the loss of the mandatory long form census." Toronto: Social Planning Toronto. http://www.savethecensus.ca/savethecensus.ca/Resources.html.

Lindsay, J. 2003. "The determinants of disaster vulnerability: Achieving sustainable mitigation through population health." *Natural Hazards* 28: 291–304.

Mikkonen, J., and D. Raphael. 2010. *Social Determinants of Health: The Canadian Facts*. Toronto: York University, School of Health Policy and Management.

Public Health Agency of Canada. 2003. "What makes Canadians healthy or unhealthy?" http://www.phac-aspc.gc.ca/ph-sp/determinants/determinants-eng.php.

Rennie, S. 2012. "Lower response rates threaten census data in some places." Canadian Press. http://www.cbc.ca/news/politics/lower-response-rates-threaten-census-data-in-some-places-1.1217470.

Sault Ste Marie Innovation Centre. 2011. "Vulnerable Persons Registry." http://www.ssmic.com/index.cfm?fuseaction=content&menuid=15&pageid=1169.

Statistics Canada. 2012. "Census dictionary." http://www12.statcan.gc.ca/census-recensement/2011/ref/dict/geo021-eng.cfm.

Tapsell, S., S. McCarthy, H. Faulkner, and M. Alexander. 2010. *Social Vulnerability to Natural Hazards*. CapHaz-Net WP4 Report. London: Flood Hazard Research Centre, Middlesex University.

Wall, E., and K. Marzall. 2006. "Adaptive capacity for climate change in Canadian rural communities." *Local Environment* 11(4): 373–97.

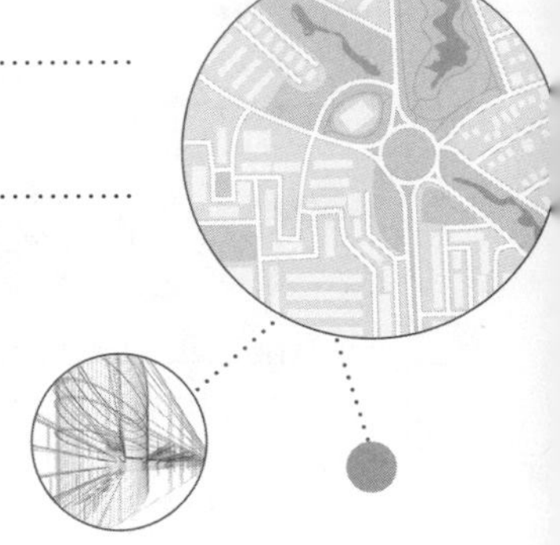

CASE STUDY 2.5

Reconceptualizing Social Planning: A Case Study of the Spatialized Impacts of Urban Poverty for Lone-Parent, Female-Headed Families in Vancouver

Silvia Vilches and Penny Gurstein

SUMMARY

Social planning, as a process that addresses societal conditions in communities, is being transformed by neo-liberal policy shifts in Canada from a service delivery model driven almost entirely by the public sector to a model that is increasingly oriented toward strategic partnerships between the public, private, and non-profit sectors. This shift challenges our conceptualization of what social planning is and how to achieve social justice. We explore these challenges and opportunities through a case study of lone parents living in extreme poverty in Vancouver, BC. We identify the impact that neo-liberal policies are having on social supports for impoverished populations and make the argument that while accountability is threatened, changes offer new opportunities to enact social justice.

Data is drawn from a qualitative longitudinal study of Aboriginal and non-Aboriginal lone-parent, female-headed families conducted from 2003 to 2006 (Gurstein et al. 2008; Gurstein and Vilches 2010). Spatialized impacts of reforms were apparent in the way women struggled within their neighbourhoods to sustain their lives and their families. Spatial planning tools could improve the quality of life for this segment of the population, but we argue that the greater challenge for social justice lies in addressing these women's perceptions of being "second-class citizens" with no effective rights to citizen participation and engagement. This is a particularly important issue for the increasingly urban Aboriginal lone parents, as the intergenerational effects of social and political exclusion of Aboriginal people suggest a special responsibility for planning engagement.

OUTLINE

- The delivery of social planning through a social welfare model is changing as neo-liberal ideals affect policy development.
- In order to examine the effects of these changes over time, lone-parent families headed by women in Vancouver were studied over a three-year period.
- Although a number of innovations have been introduced recently to respond to social needs in Canadian municipalities, low-income lone mothers still have unmet needs and are not recognized as active agents in their communities.

Introduction

Social planning, as a process that addresses societal conditions in communities, is being challenged by neo-liberal policy shifts in Canada. The social welfare delivery model that developed after World War II in Canada was a system designed to meet a wide range of human needs, driven almost entirely by the public sector. Attention has been moving to models that are based on partnerships between the public, private, and non-profit sectors. As this happens, the government role is changing from one of provider to that of facilitator. To understand the dynamics that are occurring and implications for social planning, this chapter focuses on the most vulnerable in society—those in extreme poverty—and problematizes social planning responses for this population using a social justice lens.

Neo-liberal welfare reforms, while unique from country to country, reorient social policies away from meeting the basic needs of impoverished citizens to engagement in the market economy (Fox Piven 2002; Hudson, Hwang, and Kuhner 2008). Welfare reforms have been driven by an interest in rationalizing government through reducing costs and shrinking the overall size of government (Morgen and Maskovsky 2003). In Canada, welfare reforms are complicated by the constitutional distribution of responsibilities, which allocates health, welfare, and education to the provinces but leaves municipalities to respond to the community-based needs of citizens (Cameron 2006). As a result, local governments find themselves responding to provincial-level policy decisions about income supports (FCM 2010). As neo-liberal reforms are implemented and senior government roles change from that of provider to facilitator, communities are challenged to find new models for meeting social needs. However, we argue that local social planning responses are more important than ever for addressing social justice.

One approach to social planning focuses on the value of social planning as a process in addition to its outcomes (Community Tool Box website), such as when a group or community determines its social goals and strategies with the assistance of a social planner. The focus of this social planning approach is to address the equitable distribution of social benefits and the externalities that inhibit equitable redistribution. A variety of models are available, which engage an increasingly diverse array of partners, including, often, the non-profit and private sectors and Indigenous organizations (Success by Six in BC website; Community Action Initiative website). While these approaches address the lack of participation that may occur in top-down models, participatory models also have challenges. Community-based approaches rely on the capacity within a community to advocate for and implement needed changes—a capacity that can be severely limited in impoverished communities where day-to-day survival depletes resources that residents rely on, such as time (McGrath et al. 1998). In addition, if a move to consultative social planning is accompanied by budget constraints, increased accountability targets, and decentralization of services, the very populations services are intended to reach may be inadvertently disadvantaged (Aimers 2011). The Canadian Federation of Municipalities recognizes this challenge and is working to identify new approaches that can enhance the quality of life for all of a community's citizens (FCM 2010).

Using a case study of the experiences of lone-parent families in the East Side of Vancouver, we explore ways in which social planning can and must play a role in attending to the effects of the increasing inequalities in our society. In the historical context of a strong social planning tradition in Canada, this case offers the opportunity

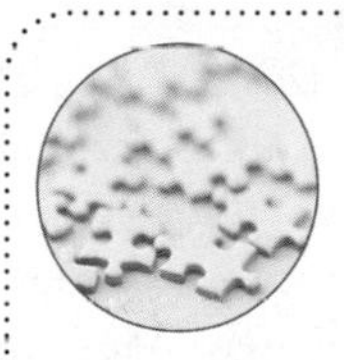

KEY ISSUE

Evolution of Social Planning

Contemporary social planning may be traced back to urban sanitation, or city hygiene, movements in the mid-nineteenth century, as well as to the growth of social services after World War II. However, the models in use have fluctuated dramatically over the years. Post–World War II social planning evolved into a top-down model in which policy-makers provided universal benefits but also decided what was good for a community or a population (Graham, Swift, and Delaney 2003). Policy models were implemented through laws, regulations, and education campaigns. This worked well in large-scale initiatives that benefit many people, such as the smoking ban that is being imposed in public places in a number of jurisdictions. However, state-led top-down initiatives have often failed in local community contexts where, without engagement and buy-in from community residents, the policies may be seen as heavy-handed and deleterious to the community spirit. For example, large-scale urban renewal in the 1950s and 1960s of what was seen as blighted "slum" neighbourhoods in North American cities was intended to create safer, more economically healthy communities, but as Jacobs (1961), Gans (1962), and others have chronicled, the result was an opposite effect in which strong social networks within the existing communities were severed.

to problematize new strategies that social planners may use to support social justice.

Lone-Parent Families in BC

Lone-parent, female-headed families have been an object of moral concern because of their perceived chronic public dependency on income assistance (Mead 2004). Epidemiological research, which shows that low-income families are vulnerable to poorer health, lower career achievement, lower education levels, and poorer childhood outcomes, has been used to create a discourse of risk that rationalizes neo-liberal welfare reforms by linking dependency to individual and societal risk (Morgen and Maskovsky 2003). The discourse of risk means that it is important to know the facts about lone parents. Approximately one-fifth of Canadian families with dependent children are lone-parent families, and 80 per cent are female-headed (Statistics Canada 2011). Of these, 36 per cent are living with low income (Statistics Canada 2011), but the proportion of lone-parent families has been stable at about one-fifth of all families since the late 1990s, with a bimodal distribution of parents in their early 20s and early 30s (Ambert 2006). Aboriginal families, though, are particularly at risk of low income, conflating racialization and poverty. A review of urban Aboriginal lone parents with federal Indian status in Winnipeg found that more than two-thirds lived below the **Low Income Cut-off** (LICO) (Hallett 2006). Approximately one-third of all Aboriginal children and one-fifth of non-Aboriginal children live in lone-parent families (Statistics Canada 2013). It is not surprising, therefore, that 38 per cent of food bank clients in Canada are youth and children (Food Banks Canada 2011). However, Canada has one of the highest rates of working lone parents, with approximately 68 per cent with children under six in the workforce (Statistics Canada 2006). The picture is thus one in which the lone-parent family form is vulnerable to poverty but is also a stable, enduring part of the makeup of Canadian households.

In 2001, a newly elected BC provincial government implemented a broad range of income assistance reforms, reducing eligibility for income assistance supports, lowering rates, and increasing barriers for disability pensions. These changes were accompanied by reductions in government department staffing and social support services (Gurstein and Vilches 2009). The changes emphasized a workforce engagement approach and shifted the social service model from meeting needs to motivating individuals to participate in the workforce. This is known as activation, or "workfare" (Skevik 2005). It has been observed to negatively impact female lone-parent families by increasing hardship without addressing structural barriers to women's earning opportunities (Lewis and Giullari 2005). As a result, it was particularly significant that in BC, childcare was simultaneously restructured by lowering parent subsidy thresholds, eliminating a childcare wage supplement, lowering operating grants, and levelling differences between fully licensed and home-based childcare, resulting in the closure of training and childcare centres (Kershaw, Forer, and Goelman 2005). As the social service provision shifted radically at the provincial level, we looked to see how social justice was affected at the neighbourhood level.

Living a Pedestrian Life: The Neighbourhood Experiences of Welfare Reform of Lone Parents

To study the effect of these shifts in benefits in BC, 17 diverse female-headed, lone-parent families living in the low-income neighbourhoods of East Vancouver were followed for three years (Gurstein et al. 2008). The study sample was selected to include families with young children, because under the new regulations, parents would be required to seek employment when the youngest child turned three, down from seven years of age previously. The study sample was selected to include an even number of lone-parent families with young children under the age of three and those with children between three and six. It was expected that the regulatory changes would mean families would require childcare. To ensure representation of other dimensions of interest, half of the women had been receiving benefits for less than two years, half had only one child, and half self-identified as Aboriginal.

East Vancouver was chosen because of its concentration of lone-parent families, social services, and low-income households (City of Vancouver 2010). Poor neighbourhoods may be defined in many ways but are often seen as places with fewer opportunities. Dear and

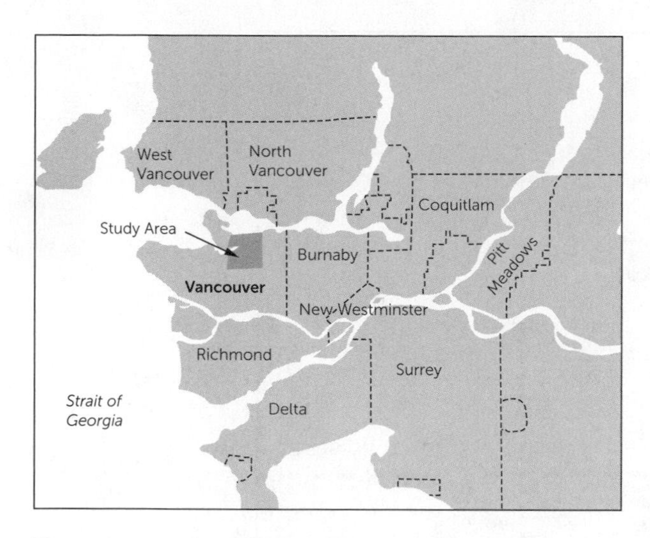

Figure 2.5.1 • Location map.

Wolch (1987) describe the collusion of support needs with high-intensity service areas as service-dependent neighbourhoods and theorized that they may generate further concentrations of populations in need. The women in this study viewed the affordable East Vancouver neighbourhoods as a place that represented an opportunity to survive or make changes in their lives. The lower cost of older rental accommodation and a dense network of social services gave them opportunities to obtain food, clothing, shelter, and other supports within their neighbourhood. However, in spite of finding the best conditions they could afford, their lives were difficult.

Women's urban lives were dominated by pedestrian experiences, as Phillipa demonstrated when her welfare cheque was withheld:

> Every time I go anywhere I have to get everybody dressed and there's a lot to that. . . . I have all the kids and [I don't] even have a stroller that would fit all the kids. Day in, day out [the] experiences are quite time-consuming (Phillipa).

Women in the study not only walked to their income assistance offices but also to food banks, food distribution centres, school early-breakfast programs, and support programs. They also walked to participate in mandatory volunteering or social programs. The pedestrian experience of the urban environment highlights how policy changes at the provincial level affect community-level living. Income assistance does not provide transit passes unless a person is eligible for disability benefits, and although children under six ride for free in Vancouver, adults do not. The result was that the women's needs demanded long days of walking to gain access to supports that are unplanned and awkwardly located. The lack of pedestrian-oriented amenities such as public washrooms or even curb cuts added to the rigours of commuting by foot and stroller in the city.

While housing in East Vancouver was somewhat affordable, women often had to accept low quality and inadequate space. Poor housing conditions represented an additional interplay between policy and their lives, as mediated by market housing conditions.

> It is really hard to find a place with children and being on welfare. Nobody wants to rent to you. I think we're one of the best people because you're guaranteed that you're going to get your money, and we have kids, so generally we're going to be responsible, but landlords don't usually see it that way. So I moved into this place and there's no bathtub, the ceilings are just above my head, and my landlord lives upstairs . . . but he has no children. . . . His ideas are totally unrealistic. Like, there's no privacy between the floors (Kate).

The small, crowded, and inadequate spaces meant that women used public spaces for their domestic life. Sometimes this meant using corporate spaces like malls and fast-food restaurants as places where for the price of a cheap cup of coffee, they could let their children play on play equipment while they visited with friends out of the rain and cold. However, they also used parks, public beaches, and streets as extensions of living space. One woman, who sublet one of her two bedrooms to help with the rent and slept in the living room, remarked that the only place her five-year-old daughter could get privacy was in the bathroom. As one mother remarked, this kind of need means that privacy is a middle-class privilege.

> Your family is way more open—if you're a [middle-class] family you have a lot of barriers, you have the van, and you've got a lot of privacy. But the main difference too from being poor to being rich is the

Figure 2.5.2 • City of Vancouver, Northeast Quadrant, showing approximate locations of study participants.

Source: Adapted from City of Vancouver, VanMap, used with permission.

> more money you have the more barriers you can have, and the more poor you are the more that you're going to be out in public, you're going to be around people, you're going to be interacting (Phillipa).

Inadequate housing meant that public space was more than just aesthetic green space or optional athletic and leisure space; it provided these women with release from the stress of overcrowding and helped them to manage parenting responsibilities. At the same time, public space was also a complicated space where lone mothers were exposed to more public observation of childrearing practices and lifestyle habits.

While women were required to accelerate their work retraining and job re-entry processes, the restructuring of social supports across government resulted in reductions of funding to smaller local agencies that provided more specialized support programs. As a result, supports for the difficult task of parenting under stress increased (Gurstein and Vilches 2009). In spite of the pressure, women in the study refused, and often failed at, short-term, part-time, or shift work because childcare was difficult to obtain. The three women in the study who did go on to sustainable employment disobeyed regulations and opted to forgo benefits in order to achieve their goals (in the information technology sector, as a personal support worker in a job with benefits, and as a unionized worker in the construction industry). However, women who were struggling with inadequate housing, such as Ali, who made 16 moves in one-and-a-half years trying to find affordable adequate space, had great difficulty and sometimes risked their family cohesion. Ali dropped out of the study early, likely because of increasing instability. Stable housing, often in social or subsidized situations, gave women adequate space for their own and their children's needs.

As a result, they did not have to spend their food money on rent, and we observed that they were more stable (Gurstein and Vilches 2009; Vilches 2011).

The inability of citizens to care for themselves or their families affects communities. This is especially true for women addressing very difficult circumstances, such as one participant who was involved in street-level sex work. When she joined the study, she was focused on maintaining sobriety and leaving the streets so that she could retain custody of her youngest child. She was not successful before the study ended and lost custody of her two children. This woman's need for extra supports was directly impacted by provincial policy decisions, which in turn impacted the community environment—for example, through a greater need for municipal policing, child protection, and mental wellness supports. While only some of the women in the study required such intensive supports, the income support reductions led to an increased need for emergency food services and childcare among all of the women in the study.

Reconceptualizing Social Planning for Lone Mothers in Poverty

The everyday experiences of the lone mothers demonstrate how the distribution of responsibilities between the provincial government and the local government create place-based dilemmas. The women gain access to very localized resources to accommodate the mismatch between their needs and the benefits they receive by walking everywhere, sleeping on couches or subletting their spaces, or enduring poor housing (Gurstein et al. 2008; Vilches 2011). This intensifies their dependency on local space and resources for survival, including green space, while increasing public surveillance of their families. These localized experiences suggest opportunities for social interventions that are spatial, such as better designs for pedestrian neighbourhoods, improved or new social housing initiatives, and a better geographic distribution of resources such as food banks, parks and public spaces, and social programs, including childcare. As Fainstein (2010) argues, planners have an opportunity to enact social justice through spatial planning. BC's reforms challenge local governments to think about supports beyond income and suggest that local governments may be the next site of innovation in responding to structural poverty.

The challenges the women in this study faced demonstrate how neo-liberal reforms at the federal and provincial levels are impacting the neighbourhood context. Cities are being forced to design new approaches to addressing housing and homelessness, community development, food security, and health issues. They are doing so while non-profit organizations that provide social and community support initiatives are also impacted (City of Vancouver website). However, the non-profit sector is also a significant and knowledgeable partner, and women and others who live in low-income circumstances may also be active partners in adapting their environment to their needs. In this context, new emerging models of community-based facilitation offer an opportunity to rethink what social planning is.

Recent Innovations Addressing Low-Income Residents

One key for the lone mothers in this study was the provision of adequate housing that would give them the stability they need to move forward in attaining their goals and providing for their children. Municipalities, such as the City of Vancouver, have developed innovative initiatives such as rent banks that partner with donor agencies to give one-time loans to prevent evictions or loss of essential utilities (City of Vancouver website). The Streetohomefoundation, modelled after other foundations in Toronto and elsewhere, brings together business, government, and community leaders to work together on **Housing First** solutions (Streetohomefoundation website). These initiatives do not develop housing themselves but provide the seed funding for other non-profit housing organizations, which helps to address the fact that Vancouver is one of the least affordable cities in the world (Demographia 2013) and that the stock of affordable housing has not kept up with demand.

Innovations around housing demonstrate how local governments and their partners can leverage specific knowledge about needs in their areas to generate specific solutions. The role of social planners shifts toward advocacy at the provincial level but also toward engaging with localized knowledge. As facilitators, they may engage the expertise of local service providers, not only for service provision but also for planning to address the causes of social inequality. Social planners also become researchers; for example, this study suggests that addressing specific issues such as public transit subsidies, localized events, and adapted cityscapes may lead to opportunities to improve the quality of life for all urban residents.

Conclusion

While municipalities are trying to respond to social needs, they are not addressing the structural impediments facing women that result from the dynamics of their gendered experiences as low-income lone mothers or, for Aboriginal women, their racialized experiences (Baspaly 2003; Benoit, Carroll, and Chaudhry 2003). Nor do municipalities recognize that lone parents are active agents in their communities (Gurstein and Vilches 2010). However, actively engaging women as authors of their own experiences and recognizing them as citizens shifts the social planning process to one of co-creation. As Moffat et al. (1999) suggest, the concept of social planning needs to be expanded to include practices that focus on advancing citizenship, local democracy, and bearing witness to inequality. Shifting from service delivery to one of co-creation means that social planners must leverage the specific knowledge of municipalities and local service providers.

To achieve more effective results, social planners need to recognize their shifting roles and how to use their tools in the active pursuit of accountability for all citizens. Shifting from service delivery to one of co-creation with women means that social planners must leverage the specific knowledge of municipalities and local service providers. This demands a more active, advocacy role for local governments that engage in co-achieving outcomes.

KEY TERMS

Housing First
Low Income Cut-off
Social planning

REFERENCES

Aimers, J. 2011. "The impact of New Zealand 'Third Way' style government on women in community development." *Community Development Journal* 46(3): 302–14.

Ambert, A.M. 2006. *One Parent Families: Characteristics, Causes, Consequences and Issues*. Ottawa: Vanier Institute of the Family.

Baspaly, D. 2003. *GVRD Aboriginal Homelessness Study 2003 (Abridged)*. Vancouver: Lu'ma Native Housing and dbappleton consulting.

Benoit, C., D. Carroll, and M. Chaudhry. 2003. "In search of a Healing Place: Aboriginal women in Vancouver's Downtown Eastside." *Social Science and Medicine* 56: 821–33.

Cameron, B. 2006. "Social reproduction and Canadian federalism." In M. Luxton and K. Bezanson, eds, *Social Reproduction: Feminist Political Economy Challenges Neo-Liberalism*, 44–74. Montreal and Kingston: McGill-Queen's University Press.

City of Vancouver. 2010. *Neighbourhood Profiles*. Vancouver: City of Vancouver, Community Services.

———. website. http://vancouver.ca/people-programs.aspx.

———. Rent Bank Program. http://vancouver.ca/people-programs/financial-aid.aspx.

Community Action Initiative website. http://www.communityactioninitiative.ca.

Community Tool Box website. http://ctb.ku.edu/en/tablecontents/sub_section_main_1055.aspx.

Dear, M., and J.R. Wolch. 1987. *Landscapes of Despair: From Deinstitutionalization to Homelessness*. Princeton, NJ: Princeton University Press.

Demographia. 2013. "9th Annual Demographia international housing affordability survey: 2013 Ratings for metropolitan markets." http://www.demographia.com/dhi.pdf.

Fainstein, S.S. 2010. *The Just City*. Ithaca, NY: Cornell University Press.

FCM (Federation of Canadian Municipalities). 2010. *Mending Canada's Frayed Social Safety Net: The Role of Municipal Governments*. Ottawa: FCM.

Food Banks Canada. 2011. *Hunger Facts 2010*. Toronto: Food Banks Canada.

Fox Piven, F. 2002. "Welfare policy and American politics." In F. Fox Piven, J. Acker, M. Hallock, and S. Morgen, eds, *Work, Welfare, and Politics: Confronting Poverty in the Wake of Welfare Reform*, 19–33. Eugene: University of Oregon Press.

Gans, H. 1962. *The Urban Villagers: Group and Class in the Life of Italian-Americans*. New York: Free Press.

Graham, J., K. Swift, and R. Delaney. 2003. *Canadian Social Policy: An Introduction*. Toronto: Prentice Hall Canada.

Gurstein, P., M. Goldberg, S. Fuller, P. Kershaw, J. Pulkingham, and S. Vilches. 2008. *Precarious and Vulnerable: Lone Mothers on Income Assistance*. Burnaby, BC: Social Planning and Research Council of BC.

Gurstein, P., and S. Vilches. 2009. "Re-visioning the environment of support for lone mothers in extreme poverty." In M. Cohen and J. Pulkingham, eds, *Public Policy for Women: The State, Income Security, and Labour*, 226–47. Toronto: University of Toronto Press.

———. 2010. "The just city for whom? Re-conceiving active citizenship for lone mothers in Canada." *Gender, Place and Culture* 17(4): 421–36.

Hallett, B. 2006. *Aboriginal People in Manitoba 2006*. Catalogue no. SG5-2/2006E. Winnipeg: Aboriginal Single Window of Service Canada and Aboriginal Affairs Secretariat of the Manitoba Department of Northern Affairs and the Manitoba Department of Family Services and Housing.

Hudson, J., G.-J. Hwang, and S. Kuhner. 2008. "Between ideas, institutions and interests: Analysing third way welfare reform programmes in Germany and the United Kingdom." *Journal of Social Policy* 37(2): 207–30.

Jacobs, J. 1961. *The Death and Life of Great American Cities*. New York: Random House.

Kershaw, P., B. Forer, and H. Goelman. 2005. "Hidden fragility: Closure among child care services in BC." *Early Childhood Research Quarterly* 20: 417–32.

Lewis, J., and S. Giullari. 2005. "The adult worker model family, gender equality and care: The search for new policy principles and the possibilities and problems of a capabilities approach." *Economy and Society* 34(1): 76–104.

McGrath, S., K. Moffat, U. George, and B. Lee. 1999. "Community capacity: The emperor's new clothes." *Canadian Review of Social Policy* 44: 9–23.

Mead, L.M. 2004. *Government Matters: Welfare Reform in Wisconsin*. Princeton, NJ: Princeton University Press.

Moffat, K., U. George, L. Lee, and S. McGrath. 1999. "Advancing citizenship: A study of social planning." *Community Development Journal* 34(4): 308–17.

Morgen, S., and J. Maskovsky. 2003. "The anthropology of welfare 'reform': New perspectives on U.S. urban poverty in the post-welfare era." *Annual Review of Anthropology* 32: 315–38.

Skevik, A. 2005. Women's citizenship in the time of activation: The case of lone mothers in 'needs-based' welfare states." *Social Politics: International Studies in Gender, State and Society* 12(1): 42–66.

Statistics Canada. 2006. "Census families in private households by family structure and presence of children, by province and territory, 2006 Census." Ottawa: Statistics Canada. http://www40.statcan.ca/l01/cst01/famil121g-eng.html.

———. 2011. *Census Families in Private Households by Family Structure and Presence of Children, by Province and Territory for Manitoba, Saskatchewan, Alberta, British Columbia, 2011 Census*. Catalogue no. 98-312-XCB. Ottawa: Statistics Canada.

———. 2013. *Aboriginal Peoples in Canada: First Nations People, Métis and Inuit*. Ottawa: Minister of Industry

Streetohomefoundation website. http://www.streetohome.org/about-streetohome/streetohome-plan.

Success by Six in BC website. http://www.successby6bc.ca.

Vilches, S. 2011. *Dreaming a Way Out: Social Planning Responses to the Agency of Lone Mothers Experiencing Neo-liberal Welfare Reform in Canada*. Unpublished dissertation, University of British Columbia, Vancouver.

3.0

INTRODUCTION

Urban Form and Public Health

Urban planning had its origins in public health concerns in rapidly industrializing cities, whether it was the spread of contagious diseases, the quality of worker housing, or the provision of open spaces (e.g., Frumkin, Frank, and Jackson 2004). We can trace some of our most widespread practices, such as zoning bylaws regulating building height in residential areas, to public health concerns. As our cities continue to grow and change over time, public health has remained a significant area of interest for urban and regional planners: new areas of concern include active transportation options for schoolchildren, urban agriculture, and service provision in rural areas. Our ideas about public health can often be seen in the built form, including the density and configuration of high-rise buildings, the distribution of industrial land uses, and the provision of spaces for social interaction.

Since the earliest notions of urban planning, members of the profession have firmly believed that the orderly, rational distribution of land was critical to the development of healthy, safe, and socially cohesive cities. As many authors have argued, planners have not always wielded this power wisely; for example, Leonie Sandercock (1998) argues that there has always been a "dark side" to planning that sought to control and dominate those who fell outside of society's norms. Planners have always faced opposition from community residents on issues affecting the physical layout of their cities; residents may feel that their needs should not be sacrificed for the public good. Women, youth, immigrants, and single-person households have often been ignored in planning processes spanning the past six decades, even when, as Betty Friedan (1963) and Richard Sennett (1971) observed, physical planning decisions significantly affected their lives. The critical crisis in urban planning emerged in the 1960s over key concepts in urban form, such as ideas of progress, modernity, and efficiency. The demolition of inner-city neighbourhoods to make space for interstate highway infrastructure and the concentration of low-income populations into public housing projects are examples of a planning paradigm that focused on urban form without realizing its social or health impacts.

From the 1970s onward, with the advocacy of community groups and local residents, planners have made more efforts to involve a variety of groups and individuals in the planning process; planning charrettes have allowed communities to develop designs for their parks, schools, and streetscapes. Projects and plans of a certain size and scale are required to include early and ongoing public participation so that social ties, community spaces, and natural areas can be identified and preserved. Planning for healthy cities has returned to prominence as health professionals in Canada, the US, and other countries have found increasing obesity and diabetes rates even among schoolchildren (e.g., Canning, Courage, and Frizzell 2004), and food security issues have begun to emerge (e.g., Feagan 2007) as agricultural land becomes scarcer.

The cases in this section share links to other planning sub-disciplines: Kyle Whitfield's exploration of health service delivery in rural areas could easily be included in community development and social planning, while Donald Leffers's case on intensification policies in Ottawa could easily fit in to the urban design section. If this volume aims to illustrate ways in which planning plans, policies, and processes integrate knowledge from different disciplines and streams of thought, these cases are excellent examples.

Community gardens perfectly illustrate the intersection between urban form and public health: they contribute to food security and provide health, social, and economic benefits to local people. Luna Khirfan's case describes the process of designing four community gardens in the Region of Waterloo, a partnership between local organizations, the regional planning agency, and academic planners and students. Using a design charrette, students and community members together designed the Trinity Village Garden and the Chandler Mowat Community Garden in Kitchener, the Good Earth Garden in Waterloo, and the Preston Community Garden in Cambridge.

Kyle Whitfield is perhaps the only author in this volume to discuss the case study research method in detail. She explains different approaches to case study: single-case, multiple-case, and cross-case comparisons. Using local, provincial, and national settings, she explores planning for rural health services in Canada using a community-based research approach. The single-case study focused on citizen engagement in hospice care services, the multiple-case study focused on social and community care for seniors, and the cross-case comparison examined the evolution of hospice palliative care across seven provinces. Whitfield illustrates the different ways case studies can assist in our understanding of trends and how we can use generalizable patterns from case study in planning.

Donald Leffers's case describes the conflict between urban infill and sprawl. In his analysis of urban intensification attempts in an Ottawa neighbourhood, he highlights the City's policies on intensification and the interpretation of the policies through specific projects in Old Ottawa South. As he found during this study, intensification policies relying on the efficient use of space and infrastructure (e.g., through density requirements) ignore other related issues such as demolition and construction processes, building code requirements, and as-of-right development. This complicates intensification efforts and makes it difficult to determine whether the policies have had the intended effects.

Zhixi Cecelia Zhuang focuses on the social aspects of urban form at the urban boundary between the City of Toronto and the City of Markham. Through a study on ethnic shopping malls in this area, she portrays a complex story of suburban retrofits spanning two decades as recent waves of immigration contributed to a higher percentage of Chinese businesses and the need for community gathering places. Planning challenges such as mixed-use development, the lack of large "anchor" stores, parking, and traffic concerns have engaged municipal planners. The case illustrates how local planners have developed more flexible responses to an urban form that was at first unfamiliar to them but contained many of the mixed-use components that planners advocate.

All of these chapters show the importance of the built form to the health and well-being of our communities, whether they are urban, suburban, or rural. Several cases illustrate the complex governance involved in solving planning challenges and partnerships that can develop as a result. Innovative urban forms such as ethic shopping malls may take some time to be accepted into Canadian planning practice, but they offer a number of opportunities to increase community interaction, a goal that Frederick Law Olmsted had in mind when he designed New York's Central Park at a time when the planning discipline was in its early stages. Thus, public health concerns in planning have come full circle, and in many cases they have intersected with ideas about the built form of neighbourhoods and cities.

REFERENCES

Canning, P.M., M.L. Courage, and L.M. Frizzell. 2004. "Prevalence of overweight and obesity in a provincial population of Canadian preschool children." *Canadian Medical Association Journal* 171(3): 240–2.

Feagan, R. 2007. "The place of food: Mapping out the 'local' in local food systems." *Progress in Human Geography* 31(1): 23–42.

Friedan, B. 1963. *The Feminine Mystique*. New York: Dell.

Frumkin, H., L.D. Frank, and R. Jackson. 2004. *Urban Sprawl and Public Health: Designing, Planning, and Building for Healthy Communities*. Washington: Island Press.

Sandercock, L., ed. 1998. *Making the Invisible Visible: A Multicultural Planning History*. Berkeley: University of California Press.

Sennett, R. 1971. *The Uses of Disorder*. New York: W.W. Norton.

CASE STUDY 3.1

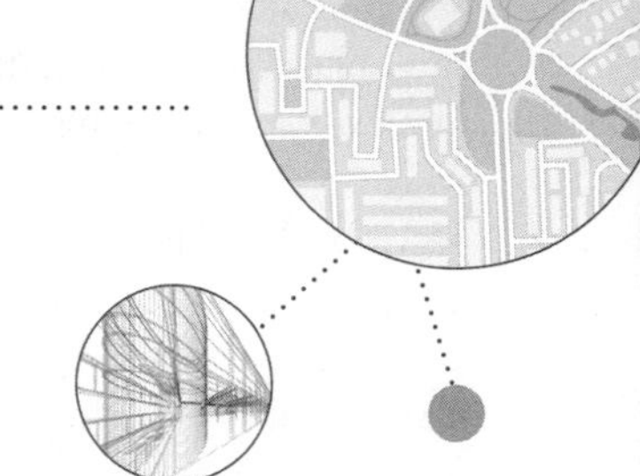

Design and Beyond: The Mobility and Accessibility of Community Gardens in the Region of Waterloo, Ontario

Luna Khirfan

SUMMARY

This chapter discusses an initiative that involved local and regional community organizations (the Community Garden Council of Waterloo and Opportunities Waterloo), a regional planning agency (the Region of Waterloo Public Health), and academic planners and students (the School of Planning at the University of Waterloo). The initiative included a day-long design charrette during which the local community gardeners shared their visions and aspirations for inclusive and accessible community gardens in the Region of Waterloo. The urban design students at the University of Waterloo articulated these visions and aspirations into design proposals for four different community gardens in the Region of Waterloo. For two weeks after the design charrette, the academic planner and the students conducted additional research into accessible and inclusive designs and accordingly further refined the gardeners' visions into final design proposals. This chapter therefore addresses the links between accessibility and inclusiveness in the design of community gardens. It also offers an example of planning academics and students positively contributing their knowledge to a regional planning initiative through research and design of the four local community gardens. The emerging blueprints from these designs also have the potential to influence regional, and eventually provincial, policies on the accessibility and inclusiveness of community gardens through proposing new design guidelines and building codes.

OUTLINE

- This case presents the design process for a group of inclusive community gardens in the Region of Waterloo, Ontario. Community gardens have become important contributors to the urban quality of life as concerns over food security and health issues have grown.
- Student involvement in a participatory design activity, integrating the needs and concerns of gardeners, has been a critical element of this project.
- The garden designs have been accepted by the Region and the community and are now in the process of being implemented.

Introduction

This chapter relays an initiative that began in October 2008 when the Region of Waterloo Public Health (RWPH) in southwestern Ontario enlisted the academic expertise of the University of Waterloo's School of Planning. The RWPH is one of three partners in the Diggable Communities Collaborative (DCC), an initiative that was "seeking to strengthen existing and expand the number of **community gardens** within the Region." The RWPH's message continued to emphasize that the DCC was "interested in creating accessible gardens as part of this work" (Khirfan 2008). The other two partners are the Community Garden Council of Waterloo and Opportunities Waterloo Region.

In discussions on how best to achieve these objectives through a mutual collaborative initiative, many options were proposed, including the supervision of a graduate student with funding from the DCC. Eventually, during a meeting on 26 February 2009, the idea of a **design charrette** that would involve the students at the School of Planning was adopted. Design charrettes are public events that bring together members of the community, design professionals, and volunteers and are thus considered a powerful planning tool because they combine a diversity of interests and disciplinary expertise in order to explore different design options and, accordingly, generate visual solutions that reflect

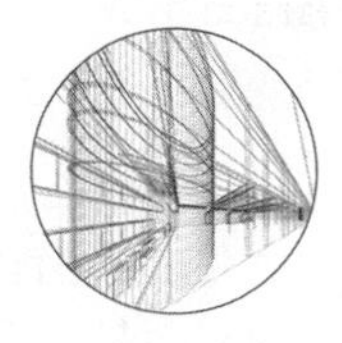

PLANNING THEORY

Participatory Planning

Participatory planning and design take place through the collaborative involvement of youth, students, new immigrants, elderly citizens, and citizens with mobility challenges. The particular issues to be addressed in this case include sustainable food production through accessible and inclusive means and strengthening food security and community engagement.

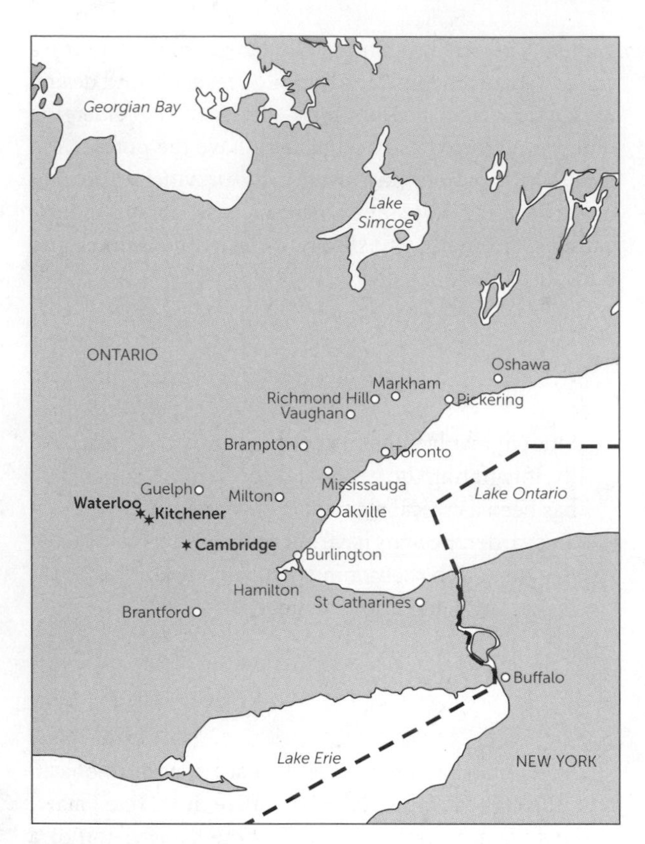

Figure 3.1.1 • Location map.

Source: Luna Khirfan.

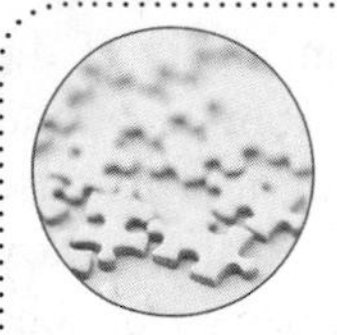

Key Issue

Community Gardens

Community gardening empowers the marginalized members of the community to assume leadership roles and to mobilize toward collective goals and thus provides them with an inclusive platform (Glover, Shinew, and Parry 2005; Holland 2004). Culturally, community gardens facilitate access to the ethnic fruits and vegetables that may otherwise be unavailable for minorities and new immigrants (Wakefield et al. 2007).

these options. Because of their enhanced potential to stimulate discussion beyond other, more conventional methods, design charrettes have been widely adopted in Canada as elsewhere (Canada Mortgage and Housing Corporation 2002, 6). Project samples from students in urban design studio courses at the School of Planning were shared with the members of the DCC, resulting in increased enthusiasm among members regarding the charrette proposal. Thus, by mid-March 2009 an agreement was reached whereby the DCC would provide the necessary funding to hold the charrette, while the University of Waterloo's School of Planning faculty partner would deliver all other recruitment, administrative, and logistical support.

The initiative began with a review of the relevant literature on community gardens in general and their design and accessibility in particular. The review revealed that while there is widespread recognition that community gardens represent an effective grassroots response to a wide range of issues, including food security, health improvement, and community development, there is still a need for empirical research that addresses the accessibility and inclusiveness of community gardens and the links between them.

This chapter therefore asks: How can a community garden achieve both accessibility and inclusiveness? How are accessibility and inclusiveness linked in the design of community gardens? And most important, how can urban planning academics and students contribute their knowledge in a participatory design process? This chapter specifically discusses an approach that considered the perspective of local gardeners in identifying the design elements that influence the accessibility and inclusiveness of community gardens. Simultaneously, the same approach also involved graduate and undergraduate students from the University of Waterloo's School of Planning, who investigated the viability of these design elements and went on to propose design solutions for four local community gardens in the Region of Waterloo. The gardens are the Trinity Village and Chandler Mowat Community Gardens in Kitchener, the Good Earth Garden in Waterloo, and the Preston Community Garden in Cambridge.

The Benefits of Community Gardens

Community gardens contribute to food security. They ensure the safety, stability, and consistency of local food production and hence liberate the food supply system from the insecurity of fluctuating global and national economies (Feagan 2007). Evidence abounds on the health, social, and economic benefits of locally

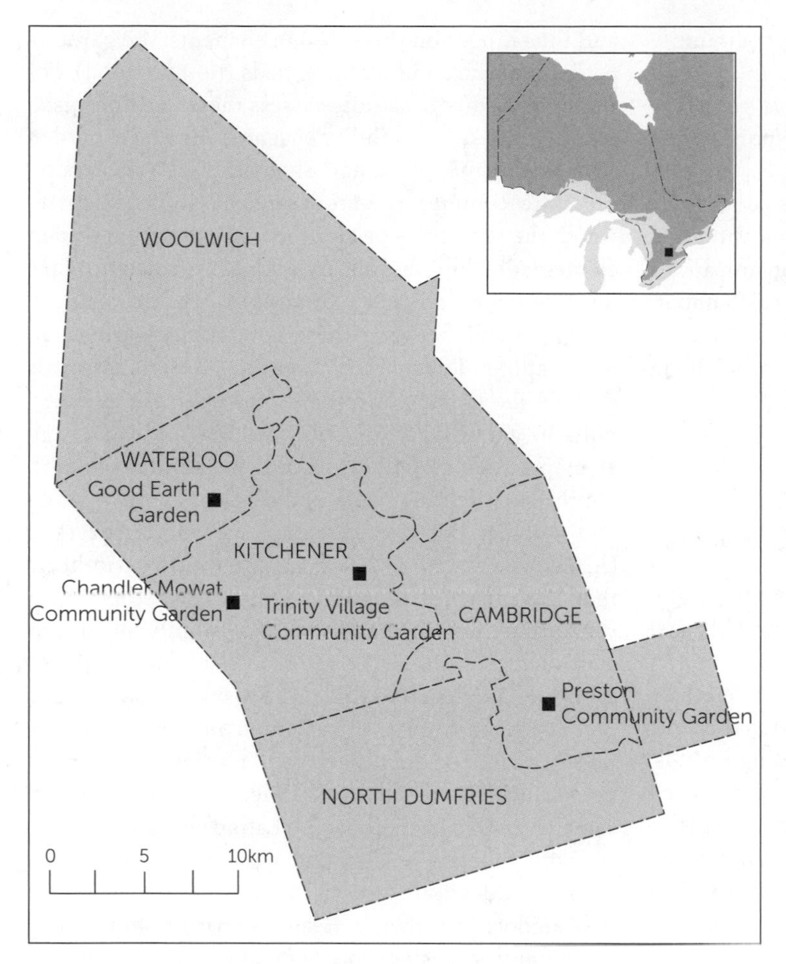

Figure 3.1.2 • Community gardens case study context.

Source: Luna Khirfan.

produced food from community gardens. For example, empirical studies demonstrate that community gardens facilitate access to a more nutritious diet and that the households of community gardeners consume significantly more fruits and vegetables than households that do not partake in community gardening activities (Allen et al. 2008). Gardening is also considered a moderate physical activity that triggers health benefits (Park and Shoemaker 2009).

In Canada, where the rapidly growing urban populations parallel a concentration of low-income households with the elderly, minorities, and/or new immigrants, community gardens offer many benefits, including economic benefits (Houseman 1983; Wakefield et al. 2007). Indeed, the popularity of community gardens in Canadian urban centres stems from city dwellers' perceptions of urban agriculture as an effective tool for local and sustainable food and flower production (Sumner, Mair, and Nelson 2010). A Public Health Study conducted in 2003 in the Region of Waterloo offers justification for increasing the number of community gardens. The study found that 58 per cent of the Region's residents consumed less than the recommended daily serving of fruits and vegetables and that nearly 50 per cent were overweight or obese (Desjardins, Lubczynski, and Xuereb 2011).

Beyond food security and health, empirical research indicates that community gardens also offer socio-cultural and economic benefits derived from developing human and community social capital and by triggering neighbourhood improvement, which subsequently enhance the sense of community (Hanna and Oh 2000). The community garden is also a relatively inexpensive and effective tool to boost a sense of place by facilitating more constructive relationships and interactions between residents and underutilized spaces and lots within their neighbourhood (Harris 2010). The process of community gardening also promotes social integration and inclusiveness by linking people through a positive communal activity, regardless of their abilities, ages, ethnicities, racial groups, and cultural backgrounds (Moore and Cosco 2007). For example, the Global Roots Garden at The Stop! Community Food Centre in Toronto is the outcome of cooperation with CultureLink, a local settlement agency for recent immigrants. The Global Roots Garden offers culturally themed communal plots to seniors and youth from several of Toronto's diverse cultural communities, thus helping new immigrants to combat social isolation while simultaneously bringing together multiple generations to foster the continual exchange of local food-based knowledge (The Stop! Community Food Centre 2011).

Empirical research has also established the economic benefits of community gardens in terms of poverty alleviation through financial savings (Mougeot 2005; Redwood 2009; Schischka, Dalziel, and Saunders

2008). Indeed, a significant increase in community and domestic gardening was observed in the United States as a result of the recent economic recession (Butterfield 2009). Similarly, the 2010 report of the P-Patch Community Gardens initiative in Seattle estimates that around 30 community gardens collectively donate more than 11,340 kilograms of fruits and vegetables annually to nearly two dozen local organizations (City of Seattle 2007–9). Research findings also underscore the positive economic impacts of community gardens on the values of nearby property, especially in poor neighbourhoods (Voicu and Been 2008).

Design: Functionality and Accessibility of Community Gardens

Existing research into the design of community gardens may be classified into two groups. The first focuses on the effective functionality and operation of community gardens, such as whether it is more effective to place them within existing conventional public parks or on vacant lots (Linn 1999). Building on this, some studies address the adaptation of existing raised beds, patios, rooftop containers, and ground planting beds at school campuses, parks, and community centres for community gardening activities (Twiss et al. 2003). Other research focuses on whether locating community gardens near underground utilities and hydroelectric lines could have negative health impacts stemming from long-term exposure to electromagnetic fields (McKeown 2008). Another avenue of research delves further into functional design details such as the appropriate orientation to optimize access to sunlight and by association to establish the relative ratios of open space to the heights of the surrounding buildings (DeKay 1997). Such aspects of layout and context have also been linked to the size, the number of planned plots, and the anticipated number of future users in the context of residential densities in the surrounding areas (DeKay 1997).

The second group of research underscores issues of accessibility and distinguishes between the mobility within and gaining access to community gardens. Mobility within refers to the accessibility of the garden's facilities for people with mobility challenges and thus hinges on three design elements: the garden's paths, contours, and planting beds (Rothert 1994). For example, numerous studies assess the impact of raised planting beds on the ability of senior citizens to garden (Houseman 1983; Park and Shoemaker 2009). Gaining access to community gardens refers to the extent to which the presence—or lack thereof—of certain design elements influences an individual's choice to take up gardening activities (Kurtz 2001). For example, in a comparison between three community gardens in Minneapolis, Kurtz (2001) identifies design elements such as the fencing and the use of padlocks as deterrents to gaining access. Other studies discuss design elements such as bulletin boards for announcements, appropriate seating, and paths that not only facilitate mobility but also encourage walking for exercise (Armstrong 2000). Other empirical findings indicate that the presence of ethnic symbols, whether structures or plants, increases the accessibility of groups who identify with such symbols (Rishbeth 2004). For example, Salvidar-Tanaka and Krasny's (2004) study of the Latino Community Garden in New York City reveals that the community garden's "casitas" or small houses, which are traditional Mexican wooden structures, attract the neighbouring Latino community.

Both mobility within and gaining access to community gardens are especially relevant to the Region of Waterloo. Together 4 Health, a partnership among community agencies and individuals that promotes healthy lifestyle choices and chronic disease prevention, noted that in 2001, nearly 16 per cent of the Region's inhabitants suffered from some sort of a physical disability—a proportion that is expected to increase by 5.3 per cent annually until 2025 (Together 4 Health 2010). Furthermore, the Region of Waterloo was, until 2006, home to 105,375 new immigrants to Canada—a significant statistic given its total population of 473,260 (Statistics Canada 2007). These data underscore the significance of addressing how community gardens can provide better mobility and accessibility so that they can enhance their potential as catalysts for positive social, economic, and health-related change in the Region. Certainly, this data resonated with the concerns of the three DCC partners. Consequently, plans to organize a design charrette proceeded, and the event was scheduled for Saturday, 3 October 2009.

The Design Charrette: Encouraging Students' Participation

Typically, there are four categories of design charrettes depending on the participants' backgrounds (Condon 2008; Lennertz, Lutzenhiser, and Failor; Walters 2007):

- the educational charrette for architecture and/or urban design students;
- the leadership forum for citizen activists, elected officials, and representatives of non-profit organizations;
- the traditional problem-solving charrette for ordinary citizens;
- the interdisciplinary team charrette for practising professionals.

These categories are not mutually exclusive, and in fact, many argue that a combination of expertise and backgrounds within each team of any charrette contributes to a stronger participatory approach (Condon 2008; Sutton and Kemp 2006). Indeed, this was the case when from the outset, student involvement was viewed as an asset, and thus a mix of graduate and undergraduate students from the School of Planning was sought by means of an announcement soliciting their participation. Beyond the student body, volunteer facilitators were also recruited, including University of Waterloo professors and involved citizen activists with local NGOs such as the Zonta Club for Kitchener-Waterloo. Lastly, the DCC recruited community gardeners and garden co-ordinators, who came from a diversity of backgrounds (age, gender, ethnicity, and physical ability). The gardeners were distributed equally among the four participating gardens, namely: Chandler-Moat, Good Earth, Preston, and Trinity.

The student contributions covered many aspects of the design charrette. Initially, three graduate students were involved: one wrote a major research paper on the accessibility of community gardens; another offered one of three workshops on the day of the design charrette; and a third served as a facilitator. An undergraduate student worked as a research assistant and handled the logistical preparations, including obtaining the approval of the research ethics board, compiling and printing the **orthomaps**, and preparing the stationery (e.g., flip charts, coloured markers, trace rolls, and pencils).

Close to 10 undergraduate students volunteered to help with the logistics on the day of the charrette, spreading out over the University of Waterloo campus to guide the gardeners, especially the physically challenged, as they navigated their way to the location of the design charrette. They also handled the registration of participants at the door and served refreshments and lunch. Most important, 36 undergraduate students responded to the call for participation as designers in the charrette. The solicitation had prioritized third- and fourth-year full-time students during the academic year 2009–10, ensuring that the participants would possess the necessary urban design and planning skills to effectively contribute to and benefit from the design charrette. Interestingly, the selected students, after inquiring about the participating community gardens, took the initiative to visit the gardens before the design charrette. They also founded a Facebook page for the event.

Several measures were taken to enhance the students' participation in the design charrette, including holding three workshops. They were scheduled on the day of the charrette and introduced various aspects related to the design of community gardens. A planner from RWPH led the first workshop, which underscored the health and economic benefits of community gardens, sharing current design policies and associated policy gaps with regard to accessibility. A PhD student offered a second workshop, which discussed the adaptation of plant beds at high schools for community gardening activities. And the School of Planning faculty partner offered a workshop on accessibility and design, presenting as an example the Guelph Enabling Garden in southwestern Ontario, where design caters to various types of mobility challenges while simultaneously considering cultural accessibility.

Additional incentives to encourage student involvement in the design charrette included participation certificates signed by the dean of the Faculty of Environment and coverage by the RWPH of all the expenses of producing, printing, and mounting the final design panels.

The Design Charrette: Bringing Communities Together

A refreshment break followed the three workshops during which all the participants—students, community gardeners, and facilitators—socialized in an effective icebreaker session. During the break, the student volunteers rearranged the room into four meeting areas, each dedicated to one of the four participating community gardens. The gardeners were seated in their appropriate areas, and the 36 student participants were allocated equally among the four groups, ensuring that note-takers would be available at each table to document the conversation for reference at the later stages of the design process. Students were assigned to their respective teams in accordance with the norms established for design charrettes in which teams range between eight and 12 participants (Condon 2008; Sanoff 2000). Each team included one facilitator, who helped to bridge any communication gaps between the undergraduate students and the community gardeners (Condon 2008).

The actual charrette activity began with an audio recording of a physically challenged gardener (her voice distorted to ensure anonymity) who relayed her personal challenges and needs in the pursuit of community gardening—an activity she loves. Then each team tackled the specific challenges of their particular community garden. Interactive dialogue around the table ensued during which the participants used flip charts, coloured markers, and even collective online searches. Typical of design charrettes, the teams worked collaboratively to solve the problems, producing drawings and diagrams as a final outcome (Condon 2008; Sanoff 2000). As the day progressed, the participants debated the best design options that catered to the accessibility needs of the gardeners until, by the end of the day, the student designers—with the direct involvement of the community gardeners—gradually produced a series of sketches of different design alternatives for each garden (Figure 3.1.3).

While contemporary design charrettes may range from one day to two weeks (Walker and Seymour 2008), in this instance the event lasted for a day. Over the next two weeks, however, the students continued, under supervision in the University of Waterloo School of Planning, to refine the outcomes from the design charrette into more detailed design proposals.

Outcomes

New Accessibility Parameters

The design charrette yielded new insights that otherwise would not have been attainable. The detailed notes taken throughout the charrette captured the needs of the gardeners, who underscored the importance of the paths and the planting plots (i.e., raised beds), which overlap with two of Rothert's (1994) conditions for the design of an enabling garden. The participants also provided novel insights based on their critical needs regarding accessibility to gardening facilities such as composting, watering, and storage of tools. They also offered insights on the details of designing raised beds, such as a ledge to sit on while performing tasks. Most important, transcript analysis from the design charrette revealed new accessibility parameters that the gardeners identified—namely, the larger context of the garden and its location within the urban setting. The gardeners indicated that the perception of a culturally accessible community garden results from a combination of the availability of public transit, the cultural diversity of its surrounding neighbourhood, and the presence of signage that is visual and/or multilingual.

The data analysis also revealed that for the gardeners, the accessibility of a community garden means more than addressing mobility challenges. The gardeners specifically referred to visual and hearing impairment and that colour coding could be used to

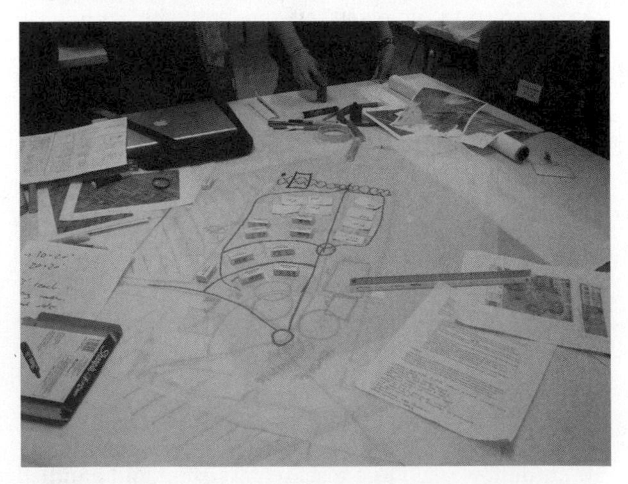

Figure 3.1.3 • An example of a sketch showing the progress of the design process during the charrette.

Source: Luna Khirfan.

distinguish different types of paths and work areas as well as to indicate the start and end of slopes. They also indicated that signage should depend on large visual symbols (as opposed to text) to accommodate the visually impaired as well as new immigrants with limited language skills, who constitute an important segment of the Region's residents (Statistics Canada 2007). Indeed, one of the gardeners shared her own insights on inclusiveness: "Coming from an immigrant family, I know there is a challenge when colloquial or slang terms are used. Pictures and very basic simple language should be used."

Thus, as the themes unfolded during the data analysis, the links between physical accessibility and cultural inclusiveness became more apparent. A design that caters to accessibility seems to inherently cater to the needs of a diversity of other users as well. For example, visual rather than textual signs simultaneously accommodate the visually impaired, children, and new immigrants. Similarly, distinguishing the work areas by using different surfacing colours enhances their accessibility for visually impaired gardeners and at the same time demarcates these work areas for activities by other users such as children. As simple as they can be, such design elements have the potential to transform a community garden into an inviting and welcoming place for various socio-cultural groups in the community, enhancing the community garden's accessibility and inclusiveness. The designs that emerged from the design charrettes reflected all these insights (see Figure 3.1.4).

Policy Implications

This design charrette and its outcomes represent an example of a research project that is tied both to the Region's planning initiatives and to food policies while also serving as a participatory planning activity. The efforts were rewarded when in August 2011, nearly two years after the design charrette, the Ontario Trillium Foundation, a major funding agency of the Government of Ontario, granted the members of the DCC $140,400 to implement adapted designs from the design charrette (Khirfan 2010) based on feedback from the Grand River Accessibility Committee and other partners. This funding provided the financial support needed to improve the accessibility and inclusiveness of the four community gardens. Most important, it paved the way to developing concrete regional, and eventually provincial, policies to overcome the barriers that hinder the accessibility and inclusiveness of community gardens through design guidelines and building codes. The funding enabled the establishment of three of the designed gardens in 2012: the Chandler Mowat, Trinity Village, and Good Earth Community Gardens (Region of Waterloo Public Health 2013). The fourth site, at Preston Community Centre in Cambridge, was slated for development in spring 2014 and opened in time for the June–October 2015 season.

In addition, a number of official and research documents have since emerged to aid in the accessible and inclusive design of other community gardens. These documents include a guide to barrier-free community gardening developed by the Region of Waterloo (Ross and Popovic 2012) as well as various University of Waterloo student research studies that include recommendations for design and policy (Jung, Keys, and McCarthy 2011; Shabbir 2010). While no accessibility-specific policies are currently in place in the Region of Waterloo, the use of the charrette approach is a useful visioning and creative tool for collecting feedback on community requirements. As these gardens attract visitors, their experiences will be an important indicator of the need for progressive policy development in the area of accessibility of community garden facilities.

KEY TERMS

Community gardens
Design charrette
Orthomaps

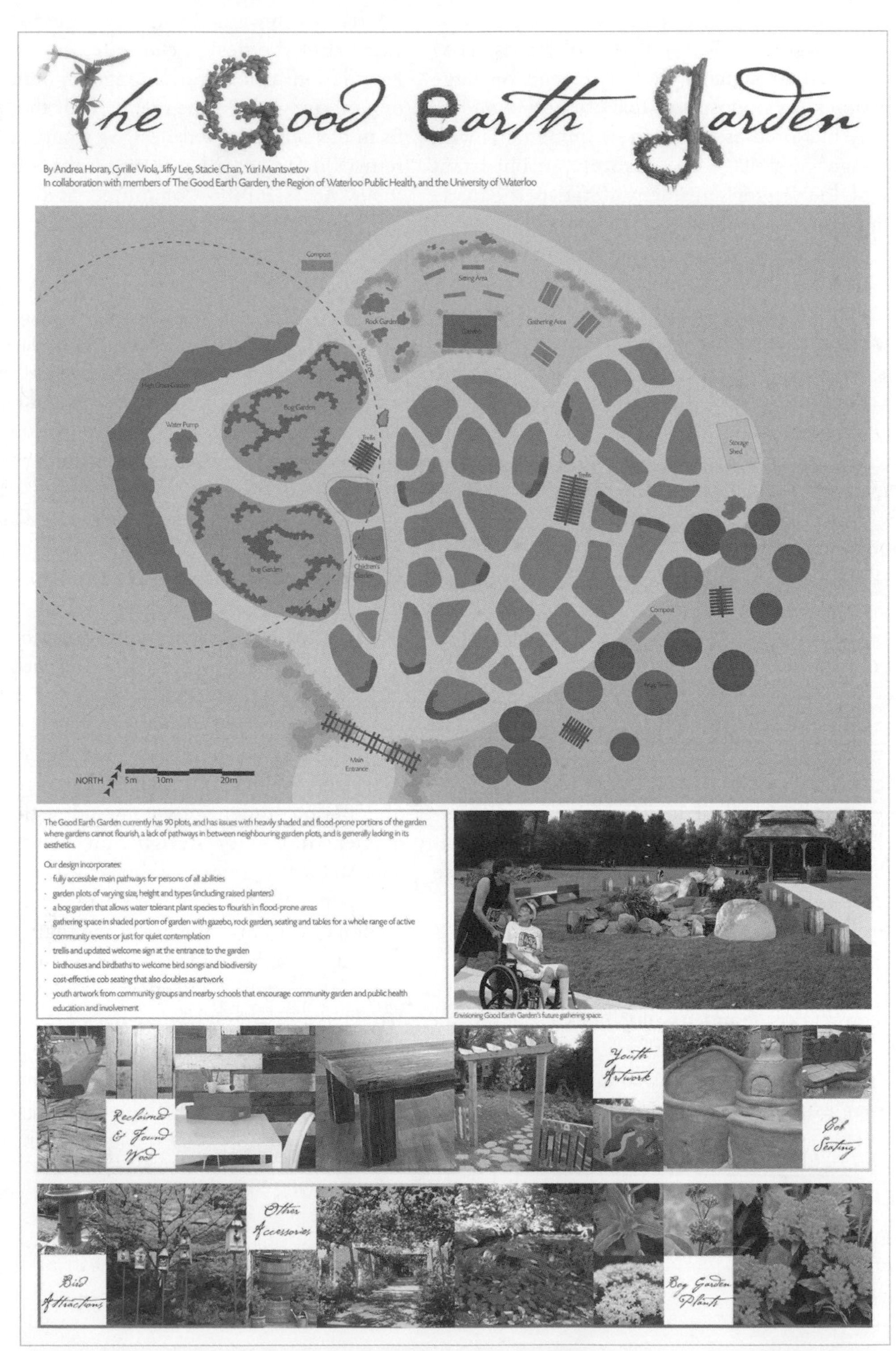

Figure 3.1.4 • The final design for the Good Earth Garden as proposed by the students.

Source: Luna Khirfan.

REFERENCES

Allen, J.O., K. Alaimo, D. Elam, and E. Perry. 2008. "Growing vegetables and values: Benefits of neighborhood-based community gardens for youth development and nutrition." *Journal of Hunger and Environmental Nutrition* 3(4): 418–39. DOI: 10.1080/19320240802529169

Armstrong, D. 2000. "A survey of community gardens in upstate New York: Implications for health promotion and community development." *Health and Place* 6(4): 319–27.

Butterfield, B. 2009. *The Impact of Home and Community Gardening In America*. South Burlington, VT: National Gardening Association.

Canada Mortgage and Housing Corporation. 2002. *Sustainable Community Planning and Development: Design Charrette Planning Guide*. Research Highlights Socioeconomic Series no. 103. Ottawa: Canada Mortgage and Housing Corporation.

City of Seattle. 2007–9. "P-Patch Community Gardens." http://www.solid-ground.org/PROGRAMS/NUTRITION/P-PATCH/Pages/default.aspx.

Condon, P.M. 2008. *Design Charrettes for Sustainable Communities*. Washington: Island Press.

DeKay, Mark. 1997. "The implications of community gardening for land use and density." *Journal of Architectural and Planning Research* 14(2): 126–49.

Desjardins, E., J. Lubczynski, and M. Xuereb. 2011. "Incorporating policies for a healthy food system into land use planning: The case of Waterloo Region, Canada." *Journal of Agriculture, Food Systems, and Community Development* 2(1): 127–40.

Feagan, R. 2007. "The place of food: Mapping out the 'local' in local food systems." *Progress in Human Geography* 31(1): 23–42.

Glover, T.D., K.J. Shinew, and D.C. Parry. 2005. "Association, sociability, and civic culture: The democratic effect of community gardening." *Leisure Sciences* 27: 75–92. DOI: 10.1080/01490400590886060

Hanna, A.K., and P. Oh. 2000. "Rethinking urban poverty: A look at community gardens." *Bulletin of Science, Technology and Society* 20(3): 207–16.

Harris, E.M. 2010. "Eat local? Constructions of place in alternative food politics." *Geography Compass* 4(4): 355–69. DOI: 10.1111/j.1749-8198.2009.00298.x

Holland, L. 2004. "Diversity and connections in community gardens: A contribution to local sustainability." *Local Environment* 9(3): 285–305. DOI: 10.1080/1354983042000219388

Houseman, D.H. 1983. "Food cooperatives and community gardens save money for the elderly." *Aging* 33(7): 19–25.

Jung, J., C. Keys, and K. McCarthy. 2011. *The Diggable Communities Collaborative: The Power of Partnership in Strengthening Community Gardens in the Region of Waterloo*. Course Project, Environment and Resource Management. Waterloo, ON: University of Waterloo. http://www.wrfoodsystem.ca/files/www/Diggable_Communitities_Collaborative.pdf.

Khirfan, L. 2008. Personal email correspondence with the Region of Waterloo Public Health, 15 October.

———. 2010. Personal email correspondence with the Region of Waterloo Public Health and the Ontario Trillium Foundation.

Kurtz, H. 2001. "Differentiating multiple meanings of garden and community." *Urban Geography* 22(7): 656–70.

Lennertz, B., A. Lutzenhiser, and T. Failor. 2008. "An introduction to charrettes." *Planning Commissioners Journal* 71: 1–3.

Linn, K. 1999. "Reclaiming the sacred commons." *New Village* 1: 42–9.

McKeown, D. 2008. *Reducing Electromagnetic Field Exposure from Hydro Corridors*. Toronto: Toronto Public Health.

Moore, R.C., and N.G. Cosco. 2007 "What makes a park inclusive and universally designed?" In T.C. Ward and P. Travlou, eds, *Open Space People Space*, 85–110. London: Taylor and Francis.

Mougeot, L., ed. 2005. *AGROPOLIS: The Social, Political, and Environmental Dimensions of Urban Agriculture*. Ottawa: International Development Research Centre.

Ontario Trillium Foundation. 2010. "Grants awarded: Waterloo, Wellington & Dufferin 2011–2012." http://grant.otf.ca/grantlistings.aspx?year=2011&catchment=Waterloo, Wellington %26 Dufferin&lang=en&id=19.

Park, S.-A., and C.A. Shoemaker. 2009. "Observing body position of older adults while gardening for health benefits and risks." *Activities, Adaptation and Aging* 33: 31–38. DOI: 10.1080/01924780902718582.

Redwood, M., ed. 2009. *Agriculture in Urban Planning: Generating Livelihoods and Food Security*. London and Ottawa: Earthscan and International Development Research Centre.

Region of Waterloo Public Health. 2013. "The health of Waterloo Region's food system: An update (May)." http://chd.region.waterloo.on.ca/en/researchResourcesPublications/resources/WRFoodSystemHealth_Update.pdf.

Rishbeth, C. 2004. "Ethno-cultural representation in the urban landscape." *Journal of Urban Design* 9(3): 311–33. DOI: 10.1080/1357480042000283878

Ross, K., and C. Popovic. 2012. *Barrier-Free Community Gardening in Waterloo Region*. Waterloo, ON: Region of Waterloo Public Health.

Rothert, G. 1994. *The Enabling Garden: Creating Barrier-Free Gardens*. Dallas, TX: Taylor Publishing.

Salvidar-Tanaka, L., and M.E. Krasny. 2004. "Culturing community development, neighbourhood open space, and civic agriculture: The case of Latino community gardens in New York City." *Agriculture and Human Values* 21(4): 399–412.

Sanoff, H. 2000. *Community Participation Methods in Design and Planning*. New York: John Wiley and Sons.

Schischka, J., P. Dalziel, and C. Saunders. 2008. "Applying Sen's capability approach to poverty alleviation programs: Two case studies." *Journal of Human Development* 9(2): 229–46.

Shabbir, M. 2010. "Finding the relationship between designing of community gardens and issues of local economic development." Waterloo, ON: University of Waterloo. http://www.wrfoodsystem.ca/files/www/shabbir_accessible_community_gardens.pdf.

Statistics Canada. 2007. Waterloo, Ontario (Code3530) (table). "2006 Community profiles. 2006 Census." Statistics Canada Catalogue no. 92-591-XWE. 13 March. http://www12.statcan.ca/census-recensement/2006/dp-pd/prof/92-591/index.cfm?Lang=E.

Sumner, J., H. Mair, and E. Nelson. 2010. "Putting the culture back into agriculture: Civic engagement, community and the celebration of local food." *International Journal of Agricultural Sustainability* 8(1/2): 54–61. DOI: 10.3763/ijas.2009.0454

Sutton, S.E., and S.P. Kemp. 2006. "Integrating social science and design inquiry through interdisciplinary design charrettes: An approach to participatory community problem solving." *American Journal of Community Psychology* 38: 125–39. DOI: 10.1007/s10464-006-9065-0

The Stop! Community Food Centre. 2011. "Global Roots Garden." http://www.thestop.org/global-roots-gardens.

Together4Health. 2010. "Making your garden accessible." http://www.together4health.ca/workgroups/community-gardens-waterloo-region/accessible-gardens.

Twiss, J., J. Dickinson, S. Duma, T. Kleinman, H. Paulsen, and L. Rilveria. 2003. "Community gardens: Lessons learned from California healthy cities and communities." *American Journal of Public Health* 93(9): 1435–8.

Voicu, I., and V. Been. 2008. "The effect of community gardens on neighboring property values." *Real Estate Economics* 36(2): 241–83.

Wakefield, S., F. Yeudall, C. Taron, J. Reynolds, and A. Skinner. 2007. "Growing urban health: Community gardening in south-east Toronto." *Health Promotion International* 22(2): 92–101.

Walker, J.B., and M.W. Seymour. 2008. "Utilizing the design charrette for teaching sustainability." *International Journal of Sustainability in Higher Education* 9(4): 157–69. DOI: 10.1108/14676370810856305

Walters, D. 2007. *Designing Community: Charrettes, Masterplans and Form-Based Codes*. Burlington, MA: Elsevier.

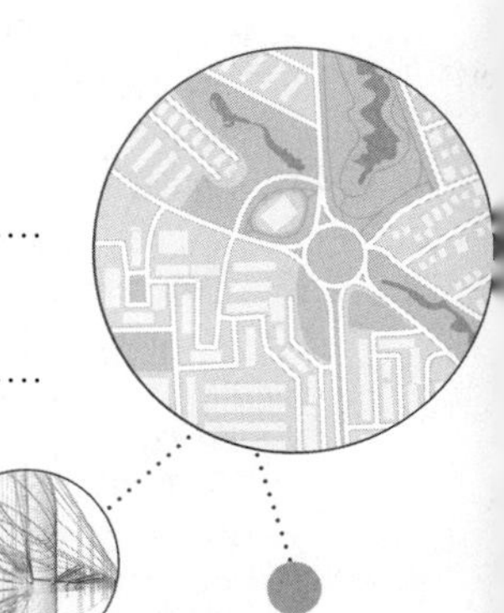

CASE STUDY 3.2

A Comparison of Cross-Alberta and Cross-Canada Health Initiatives

Kyle Whitfield

SUMMARY

This chapter describes how the case study method was used in three planning studies that all examine a certain health service. Because the studies were conducted in three different setting types (local, provincial, and national), a comparison highlights key findings after the case study method is used. Compared are: a single-case study used in a local context; a multiple-case study used in a provincial situation; and a cross-case comparison used at a national level. The questions guiding this chapter are: Why does the field of planning use "cases" in community-based planning research, and what are the associated benefits and dilemmas? The case study method is used extensively in planning scholarship; however, little exploration or critical analysis has gone into exploring implications. After investigating key characteristics of the case study method, this chapter presents relevant characteristics of community-based planning research as an investigative approach that can be used in planning scholarship to enhance the case study method.

OUTLINE

- Case study as a methodology is well established in planning, including single-case studies, multiple-case studies, and cross-case comparisons.
- An example of each of the three types, all involving planning for rural health services, shows how the approach can be used to understand key trends in service delivery at different scales (community, regional, national).
- The community-based planning research approach in all the cases allowed the research to be grounded in issues pertinent to local experts.

Introduction

Scholars studying various aspects of the planning field use the case study quite often as a research method. It is used both as a method of teaching planning theories, concepts, and skills in planning schools and as a methodology in planning research. Although using cases to understand particular situations or events is common in planning scholarship, little critical analysis or exploration has gone into asking why the case study approach is used and what its benefits and downfalls are (Mukhija 2010). Even less scholarly critique has gone into exploring the implications of using different types of case studies in planning research. Presented here are three types of case examples, all used to examine health initiatives in health planning studies conducted by the author. The three types examined here are: a single-case study; a multiple-case study comparing health and social service initiatives in one province; and a cross-case comparison study of similar cases from across Canada. The use of these three case study examples helps to form the underlying questions guiding this chapter:

- Why does the field of planning use "cases" in its research?
- What are the associated benefits and dilemmas?

After investigating key characteristics of the case study method, this chapter presents relevant characteristics of **community-based planning research** as an investigative approach that can be used in planning scholarship because it aligns well with the underlying purpose of the case study research method.

The Case Study Method in Planning Scholarship

Planning is loosely defined as a scholarly discipline that takes on or integrates several other disciplines. It is influenced by urban design and the methods and background of architecture and by many social science methods and process-oriented theories and practices (Zanon 2012). The field of planning has a long history of using "a case" to study the field's many facets, such as urban planning, community planning, transportation planning, land-use planning, health service planning, and environmental planning. But there are few published works that critically evaluate why or how the use of a certain "case" may be effective for successful research that explores planning initiatives (Mukhija 2010) and health planning initiatives in particular. As shown in Table 3.2.1, planning is replete with examples from a wide variety of studies that use cases to examine certain aspects of planning.

A very quick and informal selection of articles from three key academic planning journals examined between 2012 and 2013 showed that by using key search terms ("case," "case study," "case example"), an abundance of articles can be found, describing a range of topics. Clearly, planning research uses the case study quite often in the published planning literature.

Table 3.2.1 • Examples of Titles of Planning Cases from Three Academic Planning Journals*

- "Collaborative planning in a complex local context: **The case of an Islamic school** in Sydney, Australia," by Laura Beth Bugg. *Journal of Planning Education and Research* June 2013 33: 204–14.
- "Immigrant organizations in pursuit of inclusive planning: Lessons from a **municipal annexation case**," by Michelle C. Kondo. *Journal of Planning Education and Research* September 2012 32: 319–30.
- "Connecting growth and wealth through visionary planning: **The case of Abu Dhabi** 2030," by Michael Murray. *Planning Theory and Practice* June 2013 14(2): 278–82.
- "At the crossroads between urban planning and urban design: Critical lessons from **three Italian case studies**," by Pier Carlo Palermo and Davide Ponzini. *Planning Theory and Practice* September 2012 13(3): 445–60.
- "Spatial inequality between and within urban areas: **The case of Israeli cities**," by Daniel Shefer and Malka Antonio. *European Planning Studies* March 2013 21(3): 373–87.
- "Balancing competitiveness and cohesion in regional innovation policy: **The case of Finland**," by Mika Kautonen. *European Planning Studies* December 2012 20(12): 1925–43.

*Key planning journals from which the above articles are pulled after using key words "case," "case study," and "case example": (1) *Journal of Planning Education and Research*; (2) *Planning Theory and Practice*; and (3) *European Planning Studies*.

Defining the Case

Research that explores the primary stages of most planning practice processes begins by formulating issues and problems. The first stage, problem identification, sets the planning process in motion toward next phases, i.e., identifying key influencing factors, etc. (Fishler 2000). Problem identification, the first stage, is usually grounded in stories about situations, events, and scenarios that are complex, ambiguous, and even uncertain. This is where planning begins with a case (Stone and Redmer 2006). The study of, and interest in, a particular case is referred to as a case study. A case is an event or instance of a class of events (George and Bennett 2004). Interest in a case means that there is a curiosity about a certain situation or occurrence that has limits and boundaries around it. To Stake (1995), a case study is a method or a unit of analysis, whereas to Yin (2009), it is more of a process of inquiry. Cases can be typical of a certain phenomenon or chosen because they are atypical or extraordinary for some reason, but all cases aim to examine a specific phenomenon latent within an event, person, process, program, situation, or social group (White, Drew, and Hay 2009). Cases help us to more fully understand people's day-to-day, lived experience (Mills, Durepos, and Wiebe 2010).

In the field of planning, the case study is also used a great deal in teaching and learning situations. In such formal contexts, cases are used to take apart planning theory or actions, allowing for the examination of various levels of success in the planning intervention or process. Using cases coincides well with practical experience and projects and is used effectively in the planning field because, as Zanon (2012, 103) describes, "the discipline is engaged in providing operational methods and tools and it accompanies political actions," making the case study a useful research tool in the planning field. Because such a significant part of planning is about action, the phenomenon of social action creates another good association between the use and study of a case and the planning field (Olsson and Hysing 2012). Fischler (2000) suggests that planning departments—i.e., in municipalities—do a lot of evaluation of projects, again making the case study an excellent tool for investigating relevant planning concepts, ideas, and practices.

As discussed, case examples are used significantly in the field of planning practice and research for a number of reasons (Mukhija 2010). They help to simplify and explain complex and often misunderstood phenomenon, helping to convey what planners do (Beauregard 2012). Planning uses cases to describe the work of planning theorists because "theorists [help to] situate planners in actual settings," says Beauregard (2012, 183). Using a particular case to gain planning knowledge shows how planning occurs in a certain place and space and at a certain time and often with certain consequences that are outcomes of planning interventions (Beauregard 2012). But because planning draws from so many disciplines and is mostly an operational discipline "whose goal is the development of practical outputs" (Zanon 2012, 103), more methodological reflection and critique is needed.

Contextualization

"Context is everything," and this is true in the evaluation of the use of the case study method in planning-related research. The power of the case study, says Stake (2006), is that its attention is always on the local context. And since things are socially constructed, we have to study the problem or issue at hand in its situated place, in the real world context of practicality, in its day-to-day experiences (Smith 1987). Planning studies need to move away from the highly theoretical, abstract level to a place where the point of entry is a particular person; in other words, the case itself. To Smith (1987), the everyday world is problematic, and everyday experiences, or cases, therefore must always be examined and considered in their contextual state.

Contextualization is a significant feature of case-based planning research. In planning research, a case is context-specific because it "seek[s] to understand specific social processes in a contextualized way, as parts of a wider configuration of social relations" (Mills, Durepos, and Wiebe 2010, 231). It is hard to know where context begins and ends. If a case uses principles of "internal contextualization," then contextualization places the case being studied in the social context that it operates within. For example, a case could examine specific features of events and their associated narratives. Alternatively, "external contextualization" principles would place more focus on the wider setting within which the case is grounded, such as provincial or national policies that affect a school, where the school is the case.

Dilemmas of Case Study Use

There are dilemmas associated with using the case study method in planning-focused research. Darke, Shanks, and Broadbent (1998) admit that it is difficult to scope case study research and ensure that it adequately answers the research question. Not only is data collection time-consuming, but large amounts of data are produced when using this research approach. The findings of planning studies using the case study are said to be limited because the background of the researcher (i.e., cultural, educational, socio-economic) contributes quite significantly to the interpretation of the data (Stake 2006). But the use of this research strategy in planning scholarship is highly relevant, especially when a certain phenomenon is underdeveloped and is in its formative stages.

Using Community-Based Planning Research to Examine Health Initiatives

Within planning research, there is also a tendency to use a community-based research approach within the case study method (Minkler and Wallerstein 2008). This is because case study research, especially in the field of planning, needs to be highly relevant to the community where the research is occurring (Darke, Shanks, and Broadbent 1998). Community-based planning research is community-situated and begins with a research question that is carried out in and pertinent to the community at that time (Centre for Community-Based Research website). Studies that use community-based planning research also need to be collaborative whereby community members and researchers share in creating the research design, the implementation of the research, and the interpretation of the data as well as in its dissemination. Finally, community-based planning research that uses the case study method must be oriented toward making some positive social change or actions. Community-based planning research is well aligned to planning scholarship because it is driven by the same principles of community planning (Grant 2008), such as empowerment, the need to build supportive relationships, ongoing learning as part of the research process, and respect for differing views (Centre for Community-Based Research website).

Community-Based Planning Research Case Examples: Comparing Similarities in Cross-Alberta and Cross-Canada Health Initiatives

The three case examples described on page 120 each convey how a certain type of case study was used in a community-based planning research study about a common health care service. Throughout these examples, the main questions being critically examined are:

- Is there value in using the case study method in community-based planning research to examine health planning processes? What is it, and, therefore, what can be concluded about using a variety of case study types to do so?

- What are the implications for planning action when community stakeholders and government decision-makers not only are involved in but, more important, have significant influence on the outcomes?

Case example: Single-case study

This study (Whitfield 2012) used a single case to examine the social phenomenon of citizen engagement in a **rural health service planning** context. The research question asked how a certain group of community members of one geographical town engaged in advancing hospice care and the implications of that engagement. It examined the everyday experience of the individual citizen in the context of planning for certain health needs (i.e., hospice care needs); it also examined the factors influencing that local community engagement (i.e., factors at a regional level).

Case example: Multiple cases across several cases

This study (Whitfield and Daniels 2014) compared multiple cases across cases within one province to examine the phenomenon of social support for seniors. The main research question asked how seniors centres currently meet the needs of older adults in Alberta and what changes were required to meet their future needs. The case study used principles of external contextualization to examine how factors at the local, community, regional, and provincial levels impacted the phenomenon of social and community care for older adults. This community-based planning research study explored the benefits and challenges of taking single cases and comparing them with one another. Out of approximately 400 seniors centres in Alberta, eight were compared for their similarities and differences, their representativeness, and their ability to promote aging well in Alberta.

Case example: Cross-case study

The third community-based planning research study (Williams et al. 2010) compared health initiatives across Alberta and across Canada using a cross-case analysis to compare multiple cases that spanned the nation. This study aimed to understand the phenomenon of the evolution of one health service over time to generate findings that were then placed in an external context of regional to provincial, then more broadly to a national, view. The study tracked the policy development of hospice care in seven provinces across Canada from 1930 to 2005. The question guiding the study asked how hospice care services and policies evolved in province X and then, when compared to six other provinces, what could be learned about factors affecting the policy development of this particular health initiative.

Reflecting on the Strength of Generalizability of the Three Examples

In the single-case study, internal contextualization was significant because of the relative effect of regional decision-making on the local level in this province. In this single case, similar and different themes evolved after many accounts from key individuals within the case community were compared, which offered some explanatory strength to the findings (Ayres, Kavanaugh, and Knafl 2003). A particular group of citizens engaging in discussions, making plans, and implementing support for their community hospice care needs was used to investigate **citizen engagement** in rural health service planning in one geographical rural community. In the example that used multiple cases across several cases in one province, examining commonalities and differences from across multiple cases, explanatory strength is enhanced, moving findings more toward generalizability because they can be applied beyond one case study to others. In the final example, a cross-case study, because of the breadth of the case study comparison—i.e., regional to provincial to national—the findings offer a high level of generalizability. In particular, the study found that the health service (i.e., hospice care), across seven provinces, remains at "the margins" of the overall health care system and needs to be integrated into existing primary care systems to ensure that the service can best respond to ongoing and significantly increasing hospice care needs. Table 3.2.2 offers a summary of the three community-based planning research studies.

Table 3.2.2 • Summary of Three Community-Based Planning Research Studies Using Single-Case, Multiple-Case, and Cross-Case Research in Health Service Planning

	Single-Case Study (in one community)	Multiple-Case Study (across several cases within one province)	Cross-Case Comparison (across a nation)
Contexts viewed in the study	Individual–community–regional	Community–regional–provincial	Regional–provincial–national
Research question	How are (a certain group of) community members engaged in advancing hospice palliative care in this community, and what are the effects of that engagement?	How do seniors centres meet the needs of older adults in Alberta currently, and what changes are required to enhance quality care in the future?	How have palliative and end-of-life care policies and services evolved in province X, and what can be learned about the policy development of health initiatives?
Phenomenon explored	Citizen engagement in rural health service planning	Social and community care for seniors	The evolution of health services across a country (differences/similarities)
Data convey . . .	Similar themes in many individual accounts (of one case) (Ayres, Kavanaugh, and Knafl 2003)	Commonality and some differences of themes from across cases	That hospice palliative care in seven provinces, remains at "the margins" of the overall health care system and should be integrated into existing primary care systems to ensure the service can best respond to needs
Generalizability of findings	Results in explanatory strength	Results in explanatory strength moving toward generalizability*	Results in high level of generalizability
Contextualization	Internal contextualization	External contextualization	External contextualization
Reference	(Whitfield 2012)	(Whitfield and Daniels 2014)	(Williams et al. 2010)

* generalizability = applicability of findings beyond one case study (Ayres, Kavanaugh, and Knafl 2003).

Discussion

With researchers in the field of planning using cases to study health planning processes in a community-based planning research framework, critical analysis needs to go into asking about any associated benefits and dilemmas, especially for those delving into the implications of using differing case study types for better planning scholarship and research effectiveness (Zanon 2012). In other words, does using cases to study certain community health planning phenomena help us to assess and describe planning scholarship in any evolved way? Does it further our current state of knowledge?

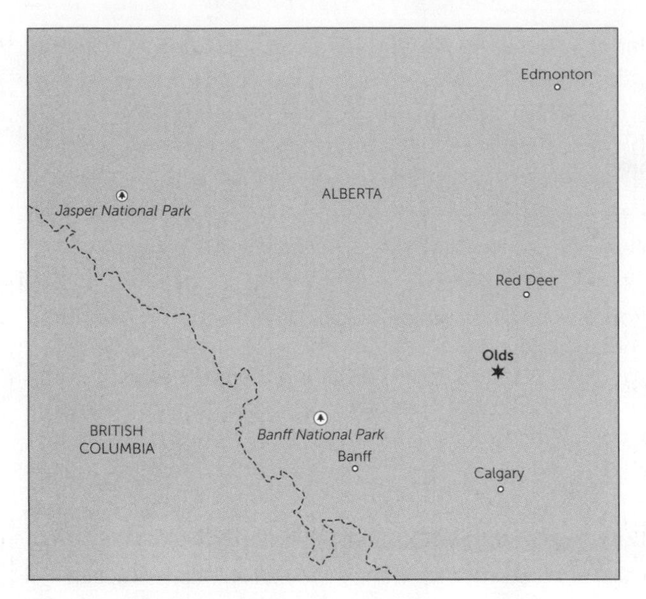

Figure 3.2.1 • Location map.

Case Example: Single-Case Study

A single-case study was used to explore one community's experience of some of its members engaging in the development of hospice care services and their day-to-day experience of that engagement. Using an ongoing local community activity as a "case" emphasized that context is everything. An important piece of the context in this health initiative began with two community members sharing stories. One person shared with the other that her mother had recently passed away but that it was such a positive experience because of the excellent care provided by a nearby hospice. As this story was shared more broadly, this group of two became a group of four, then six, and then eight, and so on. They are now the Olds and District Hospice Society, which exists to further the cause (Olds and District Hospice Society website). A single case can be chosen to examine a certain phenomenon because it is unique in some way or because it is representative of the situation in many other communities. When the study of this case began, it was thought to be a rare activity, but over time it became clear that, in fact, other communities were conducting similar activities throughout the province. The research findings from this single-case study determined that there is a social movement of sorts occurring in other regions of the province necessitating further exploration, and possibly future study, to examine issues of common community and regional-level concern.

The research question here explored how a specific group of community members engaged in advancing hospice palliative care in their community. The effects of that engagement dictated the methodological use of an ecological model, since there was a strong association between the individuals, their community, and the region. In this community-based planning research study, the individual citizen was understood to be responding to growing hospice care needs at the community level, but regional decision-making structures and supports influenced that relationship significantly. Multiple units of analysis, such as individual interviews with community group members, one focus group interview with the community group, a community scan, document analysis, and participant observation, were used in order to see citizen engagement from a holistic view. The findings supported this view. Individuals acted in relation to their community's needs, which were also highly influenced by regional-level factors such as decisions supporting community actions.

Case Example: Multiple Cases across Several Cases

Use of a multiple-case study approach to investigate how seniors centres are meeting current and future needs of older adults in Alberta (Whitfield and Daniels 2014) required the use of several data-gathering actions: a survey of a large number of seniors centres in the province; a survey of individual seniors attending an annual aging-focused conference; site visits to eight representative seniors centres where individual interviews with the director, a focus group interview of users of seniors centres, and participant observation took place.

When using a multiple-case study approach, it was challenging to be thorough because of the large amount of data generated and the difficulty in maintaining the scope of the research. But when comparing similar services from many regions, Varro and Lagendijk (2013) suggest that although it is necessary to take a view that sees relationships, connectivity, and association among regions, it is important to come back to understanding the local, and specifically, the local democratic politics. We also found this to be true. Although our findings from this community-based planning research offered a very broad view of issues facing seniors centres from three system levels (town/city, regional, and provincial), what maintained a high degree of relevance was the local context, both political and social. Although having the provincial government as a funding partner strengthened the need to see seniors centres in a geographical community context, a regional context, and a provincial context, the research question itself suggested this interconnection from the beginning.

Naturally, there are regional influences on seniors centres that are scattered throughout the province. In Alberta, as elsewhere in Canada, the provincial government has increasingly devolved responsibility for health and social services to regional decision-making bodies (Alberta Health 2012). But if seniors centres are to be hubs of support to enhance the quality of life of our aging population, then regional policies need to surround, facilitate, and support these hubs while, at the same time, maintaining a local focus. As well, seniors centres must not become immersed in only their local context; they must also contribute to regional-level policies in terms of regional planning and decision-making. Both the region and the local community have a responsibility to improve the quality of their place, which requires good collaborative planning (Healey 1997). Directives set out in provincial government reports, such as *Elevate* (Community Sustainability Task Force 2012) and the *Seniors Programs and Services Information Guide 2013–2014* (Alberta Health 2012), influenced the findings in this community-based planning research. In good case-based health planning research, provincial-level health service directives are viewed as one of the limits or boundaries of studies. Although they play a significant role in assessing context, provincial guidelines and policies can often act as an endpoint in defining the scope of the study. Equally, multi-case research provides findings that can be generalized so that policy and thus practice can be facilitated (Stake 2006). Ayres, Kavanaugh, and Knafl (2003) define generalizability as the ability to apply findings beyond one case study; this study questioned how seniors centres are meeting current and future needs of older adults in Alberta and will be generalizable to many other communities.

Case Example: Cross-Case Study

In a cross-case comparison study of how hospice care policies and services evolved in seven Canadian provinces (Williams et al. 2010), findings provided a broader view than that offered by the community-based planning research multi-case and single-case studies. Since a cross-case analysis produces findings that are aggregated, many cases are compared using a common issue or research question. And in this study, the final interpretation was a synthesis of all seven cases (provinces) in which themes, issues, and phenomena were compared for similarities. For example, a few common conclusions were made about all cases: that hospice care is not a priority for the health care system because of gaps in foundational health policies, such as the Canada Health Act, and the fact that planning is urban- and not rural-focused (Williams et al. 2010). A cross-comparison study also highlights the unique qualities of each case. This study found, for example, that in western provinces, Pallium educational initiatives funded through a federal fund had a very positive impact on hospice service. This was especially true in rural communities. The eastern provinces, on the other hand, did not offer such grassroots education, resulting in a slower community-focused approach to hospice care services in that part of the country (Williams et al. 2010; Pallium Project Resource Development Model website). Stake (2006) reminds us that when using a case-comparison approach, sometimes the details of each case can get lost. Use of a cross-case comparison in this study offered results that covered a regional-to-provincial-to-national focus, making findings more strongly generalizable than those of the single- or multiple-case study examples.

Conclusion

Research conducted in the field of planning is replete with studies that use the case study method. This is because the first stage of most planning initiatives begins with the identification of a problem, and a problem is always grounded in a story or case (Stone and Redmer 2006). Using different types of case study approaches helps to advance research that explores planning associated with health initiatives. In this chapter, several case types are used—the single case, multiple cases, and a cross-case analysis—to investigate community planning approaches to improve aging-related services. Studying such research techniques as the case study method will help to determine the value of its contribution to the planning field (Zanon 2012).

To corroborate the work of Stake (2006), the case study always brings the focus back to the local context, and involving local communities and responding to their specific needs facilitates good community-based planning research. These studies also concluded that when using a community-based planning research approach, both depth and breadth are necessary in health service planning research. The case may bring the focus back to the local level (Stake 2006), but as found in these three case examples, context is not only of a local nature but of a community, regional, and national nature as well. Thinking ecologically when using the case study method is necessarily important, because all context levels are intimately interconnected.

Using a community-based planning research approach to health planning in these three studies expanded the findings of each study. They were carried out in a community and were also highly pertinent to the community at that time. Data collected, therefore, involved the insights and knowledge of community experts (by way of key informant interviews). Flyvbjerg (2006, 222) defines experts as "operat[ing] on the basis of intimate knowledge of several thousand concrete cases in their areas of expertise. Context-dependent knowledge and experience are at the heart of expert activity [which] lie at the center of the case study as a research . . . method." Greater use of a community-based planning research focus in health planning research could help to extend our knowledge of the importance of context in case studies and the role of community experts.

Reflection, along with action, in the field of planning is therefore essential. If we continue to study research measures like the case study and its use in planning research, it can only help to determine its contribution to the productivity of the planning field (Zanon 2012).

KEY TERMS

Citizen engagement
Community-based planning research
Contextualization principles
Rural health service planning

REFERENCES

Alberta Health. 2012. *Seniors Programs and Services Information Guide 2013–2014*. Edmonton: Alberta Health.

Ayres, L., K. Kavanaugh, and K. Knafl. 2010. "Within case and across case approaches to qualitative data analysis." *Qualitative Health Research* 13(6): 871–83.

Beaugard, R. 2012. "Planning with things." *Journal of Planning Education and Research* 32(2): 182–90.

Centre for Community-Based Research website. http://www.communitybasedresearch.ca.

Community Sustainability Task Force. 2012. *Elevate*. Edmonton: City of Edmonton.

Darke, P., G. Shanks, and M. Broadbent. 1998. "Successfully completing case study research: Combining rigour, relevance and pragmatism." *Information Systems Journal* 8: 273–89.

Fishler, R. 2000. "Linking planning theory and history: The case of development control." *Journal of Planning Education and Research* 19: 233–41.

Flyvbjerg, B. 2006. "Five misunderstandings about case study research." *Qualitative Inquiry* 12(2): 219–45.

George, A., and A. Bennett. 2004. *Case Studies and Theory Development*. Cambridge, MA: MIT Press.

Grant, J., ed. 2008. "The nature of Canadian planning." In *A Reader in Canadian Planning: Linking Theory and Practice*, 3–20. Toronto: Thomson Nelson.

Healey, P. 1997. *Collaborative Planning: Shaping Places in Fragmented Societies*. Vancouver: University of British Columbia Press.

Mills, A., G. Durepos, and E. Wiebe, eds. 2010. *Encyclopedia of Case Study Research: Volumes 1 and 2*. Los Angeles: Sage.

Minkler, M., and N. Wallerstein. 2008. *Community-Based Participatory Research for Health: From Process to Outcomes*. 2nd edn. San Francisco: Jossey-Bass.

Mukhija, V. 2010. "N of one plus some: An alternative strategy for conducting single case research." *Journal of Planning Education and Research* 29(4): 416–26.

Olds and District Hospice Society website. http://oldshospice.com/about-us.

Olsson, J., and E. Hysing. 2012. "Theorizing inside activism: Understanding policymaking and policy change from below." *Planning Theory and Practice* 13(2): 257–73.

Pallium Project Resource Development Model website. www.pallium.ca.

Smith, D. 1987. *The Everyday World as Problematic: A Feminist Sociology*. Hanover, NH: University Press of New England.

Stake, R. 1995. *The Art of Case Study Research*. London: Sage.

——. 2006. *Multiple Case Study Analysis*. New York: Guilford Press.

Stone, G., and T. Redmer. 2006. "The case study approach to scenario planning." *Journal of Practical Consulting* 1(1): 7–18.

Varro, K., and A. Lagendijk. 2013. "Conceptualizing the region: In what sense relational?" *Regional Studies* 47(1): 18–28.

White, J., S. Drew, and T. Hay. 2009. "Ethnography versus case study: Positioning research and researchers." *Qualitative Research Journal* 9(1): 18–27.

Whitfield, K. 2012. "The Story of Olds, Alberta: One community's experience in attending to their community's hospice care needs." Presentation of preliminary research findings to the board members of the Olds and District Hospice Society. September.

——. and J. Daniels. 2014. *Examining Seniors' Centres in Alberta as Centres of Excellence: Identifying Their Needs and Capacities*. Report to the Government of Alberta, Ministry of Health.

Williams, A., V.A. Crooks, K. Whitfield, M. Kelley, J. Richards, L. DeMiglio, and S. Dykeman. 2010. "Tracking the evolution of hospice palliative care in Canada: A comparative case study of seven provinces." *BMC Health Services Research* 10:147.

Yin, R. 2009. *Case Study Research: Design and Methods*. 4th edn. London: Sage.

Zanon, B. 2012. "Research quality assessment and planning journals: The Italian perspective." *Italian Journal of Planning Practice* 2(2): 96–123.

CASE STUDY 3.3

Building up While Sprawling Out? Paradoxes of Urban Intensification in Ottawa

Donald Leffers

SUMMARY

This chapter analyzes urban intensification, focusing on a case study of a **first-ring suburb** in Ottawa. This suburb, Old Ottawa South, is fairly typical of an early twentieth-century middle-class suburb: it was connected to the downtown by a streetcar, which was removed in the 1960s. In recent years, planning policies attempting to curb outward urban expansion through intensification of existing urban areas have led to significant changes in Old Ottawa South. This chapter focuses on some of the place-based effects of urban intensification in Ottawa, noting that while intensification policies have normative goals based on notions of urban sustainability, they also face challenges in practice that potentially undermine these "progressive" origins.

OUTLINE

- Intensification efforts have become common among municipalities, including developing denser residential areas and limiting development to areas with existing servicing infrastructure.
- The City of Ottawa has used zoning, bylaws, and other planning tools to implement intensification, but the case of Old Ottawa South illustrates that this has been a conflicted process.

- A number of challenges in legislating urban sustainability remain, including controls on demolition and construction, existing building codes, as-of-right redevelopment, and the consideration of factors other than density.

Introduction

The city of Ottawa, like many other North American cities, currently grows primarily through low-density greenfield suburban subdivision development. With a population of approximately 0.9 million (2011), the City of Ottawa has recently tried to change the form and function of the city through policies that attempt to foster a more sustainable, land-intensive form of urbanization. It has done so within a planning and governance system in which municipalities operate constitutionally within the jurisdiction of provincial governments (Rogers and Butler 2005), albeit in an economic and political climate where provinces have **downloaded** many responsibilities to municipal governments while cutting funding (Graham, Phillips, and Maslove 1998). In Ontario, therefore, the provincial level of government retains the constitutional authority to greatly influence local planning decisions (Rogers and Butler 2005), but an underlying provincial neo-liberal agenda burdens local planning in Ontario cities (Keil 2009).

This chapter shows the various challenges and contradictions that characterize the task of moving toward more sustainable urbanism through policies of residential urban intensification. Williams (2007, 5) concludes that

> it is blatantly clear that simply increasing densities and mixing uses will not lead to sustainable outcomes. High quality infrastructure needs to be provided, public transport needs to be well managed, affordable and reliable, noise and air pollution have to be maintained at acceptable standards, basic services such as water, drainage and electricity need to be provided, and levels of public facilities such as health care and education have to be appropriate for the high numbers of city dwellers.

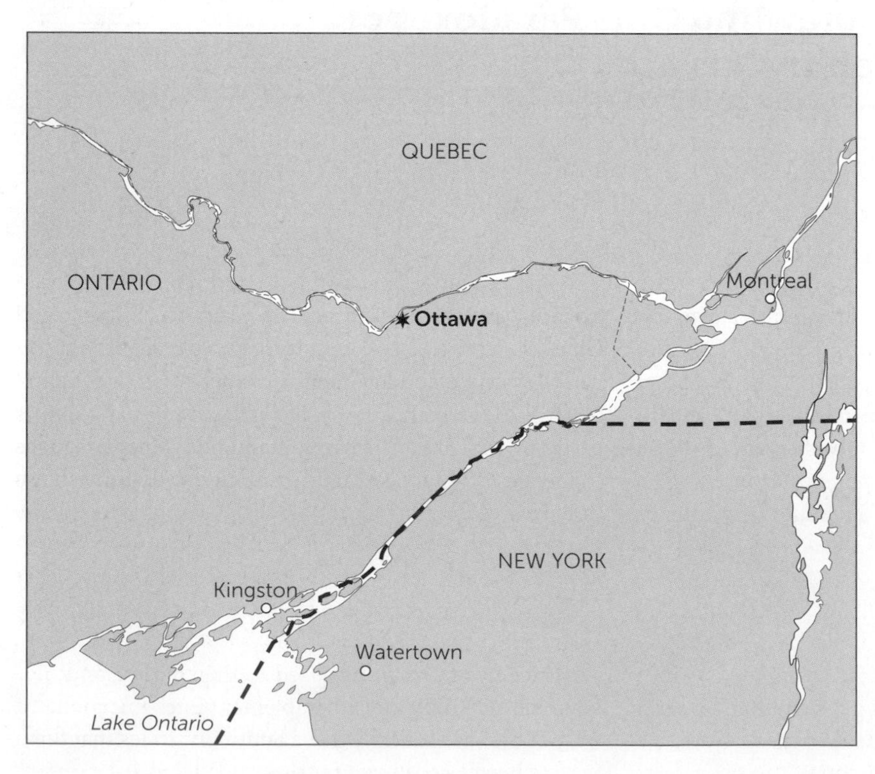

Figure 3.3.1 • Location map.

Similarly, Westerink et al. (2013) show that there are both positive and negative aspects of urban intensification, influenced by the historical, economic, and biophysical contexts of particular places. The key message is that while intensification may indeed be an important part of urban sustainability, it is only one part of a complex process: a part that, if deployed as a singular solution, is likely to run into problems.

Although intensification can be deployed in many different ways (Westerink et al. 2013), it is often linked to notions of urban sustainability and draws on compact city and **Smart Growth** discourses (Bunce 2004; Williams 1999). Little research has explored the place-based effects of these sustainability initiatives in

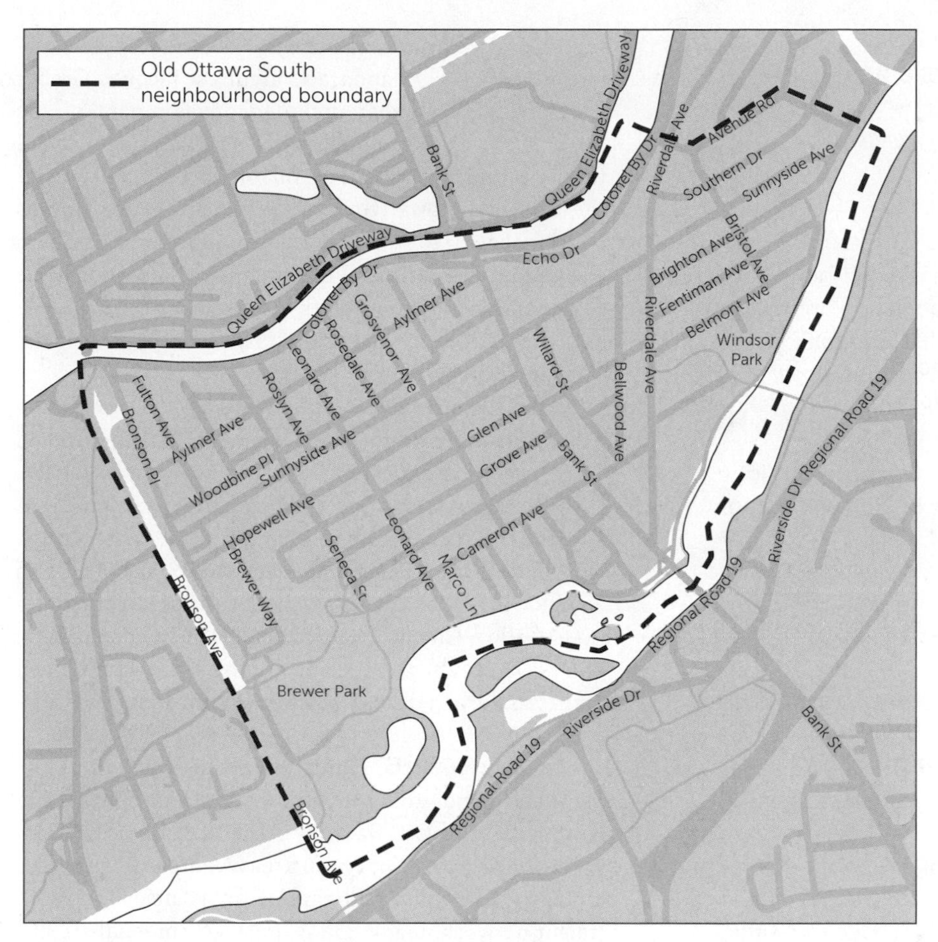

Figure 3.3.2 • Study Area: Old Ottawa South.

Canadian cities (Bunce 2004; Searle and Filion 2011). This chapter draws on empirical research conducted in Old Ottawa South over a one-year period, including participant observation, interviews, and document analysis. The case shows that despite dominant claims that intensification leads to a more sustainable future, the sustainability benefits are often lacking in practice.

Conceptualizing Urban Intensification

In some ways, urban intensification is a fairly straightforward concept, simply describing a process of development that is more concentrated than is the norm for a particular place.

In many intensification discourses, the environment, or "nature," is a key focus of protection. A consensus has emerged within planning and academic research that urban sprawl has many negative elements, particularly for the environment (Blais 2010; Filion 2003; Winfield 2003).

A key aspect of intensification is the notion of increased economic activity in a given area (Bunce 2004). Campsie (1995) also notes this dual meaning of intensification: it generally refers to increases in building or population density but can also refer to increases in investment in a given area with or without increases in building or population density. Bunce (2004) argues that Smart Growth intensification combines these two senses of the term "intensification," creating denser and more sustainable cities through increased economic activity.

Two complications are the scale and jurisdiction of growth regulation. In the Ottawa region, despite increased efforts at intensification within the City of Ottawa, greenfield development continues in adjacent municipalities. Even within the boundaries of the City of Ottawa, urban intensification does not preclude further greenfield development, which continues at approximately 60 to 70 per cent of total development (City of Ottawa 2012a). Only 20 per cent of the city's 2795 square kilometres is "urban" and is designated as such by an urban boundary, which allows municipal planners to control urban expansion (Finlay 2010). Rural villages near Ottawa have their own growth objectives and boundaries (City of Ottawa 2012b).

Prior to 2001, Ottawa was part of a regional entity called the Regional Municipality of Ottawa-Carleton. In 2001, restructuring resulted in the formation of a single new City of Ottawa from 11 former municipalities; this new municipal area is large and approximately 80 per cent rural, comprising 23 wards. There is no longer a regional governance body. Provincial legislation on regional

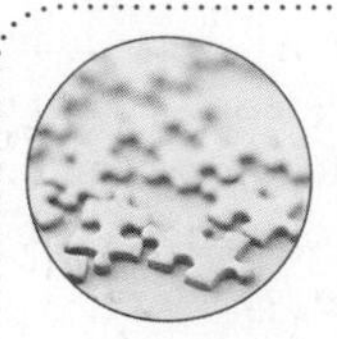

KEY ISSUE

Urban Intensification

In Ontario, urban intensification or densification describes processes of increasing the density of the urban population and intensifying economic activity (Bunce 2004; Campsie 1995). In the City of Ottawa, this is measured in a number of ways, including:

- increased number of people and jobs per hectare (City of Ottawa 2009; 2012b)
- increased number of building units per hectare (City of Ottawa 2009)
- increased number of high-density buildings in a given area, often measured in number of storeys or building height (City of Ottawa 2012b)
- greenfield development at a higher density than is the historical norm (City of Ottawa 2009)
- increased amount of economic activity in a given space (City of Ottawa 2008a)
- any development that occurs within the existing urban boundary where services such as pipes and roads already exist (City of Ottawa 2012a)

The benefits of adopting one or all of these elements are more efficient use of land, protection of farmland and "natural" areas, reduced air pollution, reduced infrastructure costs, and revitalization of city centres (Hayek, Arku, and Gilliland 2010; Searle and Filion 2011).

growth, the Places to Grow Act (2005), fills this void for the Toronto region (Ontario Ministry of Infrastructure 2012); the Ottawa region lacks a similar framework.

The Origins of Urban Intensification in Ottawa

Although urban intensification policies are relatively recent in Ottawa, concern with urban form has preoccupied planners and architects for decades. In the late 1940s, French landscape architect Jacques Gréber developed a plan that included development of the National Capital Greenbelt, a ring of park and agricultural land surrounding the imagined future extent of the City of Ottawa (Gréber 1950). The greenbelt was never intended to stop growth beyond its limits; growth was to follow a "nuclear" or nodal pattern of "complete self-contained communities" of about 20,000 people each, accessible to the central city through transportation corridors (Gréber 1950, 191).

With the greenbelt established by the 1960s, the first **Official Plans** for the Region of Ottawa-Carleton (RMOC 1974) and for the City of Ottawa (City of Ottawa 1970) were primarily concerned with the management of growth. The sustainable development discourse became evident in Official Plans in the late 1990s (City of Ottawa 1991; RMOC 1988), and intensification appears much later (RMOC 1997; City of Ottawa 2007). Density targets (City of Ottawa 2012b) and specific **zoning bylaws** developed to increase urban density (City of Ottawa 2008b) are very recent.

The latest Official Plan attempts to achieve employment and residential growth targets primarily through zoning (City of Ottawa 2007). The plan defines intensification as development in areas where infrastructure already exists and also targets areas near the rapid transit network, major roads, busy commercial streets, areas containing large tracts of so-called "vacant" or "underdeveloped" land, and smaller spaces in the general urban area.

Planning Context in Ontario and Ottawa

The Planning Act (1990) gives municipalities the authority to develop Official Plans and zoning regulations under specific guidelines, subject to approval by the Ministry of Municipal Affairs and Housing (MMAH), which can be appealed to the Ontario Municipal Board (OMB). The OMB is a provincially appointed "quasi-judicial" appeals body that has the authority to interpret municipal Official Plans, municipal bylaws, and other land-use planning laws and to adjudicate disputes based on these laws (Rogers and Butler 2005). OMB decisions tend to favour those with the best "planning evidence" (Moore 2013); developers and municipalities generally do well because they have experts to testify at OMB cases, while community groups and individuals tend to fare poorly at OMB hearings.

The City of Ottawa enacted a new comprehensive zoning bylaw in 2008 (City of Ottawa 2008b) that specifies permitted land uses and building parameters for each zone: every parcel of land in the City of Ottawa has a well-documented and detailed zoning designation. In Ottawa, zoning can be modified through either minor zoning variances or more significant zoning amendments (City of Ottawa 2013b). Minor variances are small changes to the zoning whereby proposed uses do not represent a significant zoning change (City of Ottawa 2013b). Minor variance applications trigger a specific planning process, including notification of neighbours and the ward councillor and a public meeting at the Committee of Adjustment, which decides on the case (City of Ottawa 2013b). Similar to the OMB, the Committee of Adjustment in Ontario is a quasi-judicial tribunal empowered through the Planning Act (1990), whose decisions can be appealed to the OMB (City of Ottawa 2013b).

Zoning amendment applications constitute more significant changes to zoning and trigger a more complex process that often includes pre-consultation with city staff, notification of neighbours, the ward councillor, community associations, and other city departments, and the posting of a "development application" sign on the subject property (City of Ottawa 2013b). Municipal planning staff advise council through reports and City Council Planning Committee meetings. City Council either supports applications by proposing new bylaws or rejects them. The public has 20 days to respond to proposed bylaws, after which they are either accepted or rejected by council. Final council decisions can be appealed to the OMB.

It is in this planning context that urban intensification policies have been developed and disputed in the City of Ottawa and many other Ontario municipalities. Intensification is in some ways a high-level policy but in other ways very local and incremental.

Contesting Urban Intensification

In Old Ottawa South, intensification has produced opposition to a variety of redevelopment projects. Three examples are discussed in the following sections.

Example 1: Infill on a Single Lot

The first development involved the replacement of a small house on a relatively large lot with a large, semi-detached infill development that filled the lot. The original three-bedroom house was 1058 square feet (MPAC 2008) and was purchased by the developer for $540,000. The new infill development consisted of two semi-detached two-bedroom houses of 3376 square feet each, which were each listed for more than $1 million (Cook 2008b). This development conformed to new zoning intended to promote increased residential density (City of Ottawa 2008b). The project adds one more residential unit, and perhaps three more people, to the existing urban area; technically, it could be considered intensification.

This form of intensification is characterized as "small-scale intensification [that] can and does occur **as-of-right**" (City of Ottawa 2008a, 56). However, in keeping with the Ontario Building Code Act (1992), the developers had to apply for a "building permit to demolish" and a "building permit to construct" and, as with all construction, had to comply with local building codes. No public notice was required other than the posting of demolition and building permits on the construction site, no public meetings were required, and planning and political actors were not directly involved (Rogers and Butler 2005).

Local community members opposed the infill development, referring to the City of Ottawa's **design guidelines** (City of Ottawa 2005a) and to a white paper on intensification (City of Ottawa 2008a). They interpreted this development as a result of intensification, which is plausible since it could not have happened as-of-right prior to 2008. Yet as they found out when they met with city planners, design guidelines only apply when zoning changes (either minor variances or amendments) are being sought (Ostling 2008b).

Although many of the opponents' objections were based on "community character," which is often interpreted as **nimbyism** (Burningham 2000; Freudenberg and Pastor 1992), they also focused on the size of the houses, the dominance of garages and parking pads, and the loss of green space (Ostling 2008a). They were concerned that the Smart Growth principles of walkable communities, preservation of "natural" space, and compact building design—found in the City's publications on intensification—were being undermined.

This example shows the shift from a regulatory to a political case. The opponents met with the developers, the local councillor, and planning staff (Ostling 2008b).

The development was politicized when local newspaper the *Ottawa Citizen* published a series of articles on the development (Cook 2008a, 2008b, 2009), emphasizing the size of the infill and its energy consumption. The ward councillor suggested that the main cause of this form of intensification is the general nature of Official Plans and the absence of specific community design plans (Cook 2008b). A senior planner for the City of Ottawa suggested that since the infill required no zoning changes, no further scrutiny of the project was available to the planning department apart from what was covered in the building code (Cook 2009). Nevertheless, he stated that "infill is indeed an illustration of what we don't want more of" and suggested that the City was working on revising the tools available to improve the process (Cook 2009).

While these types of small-scale infill development can indeed be referred to as intensification because they increase the density of the urban area, broader notions of urban sustainability are not as clear. Land-use elements of residential infill (e.g., setbacks, residential density, building height) are all regulated by zoning. But Smart Growth principles such as walkability, compact housing design, and preservation of green space are not easily addressed through zoning.

Although previous Official Plans more fully incorporated urban design as a legitimate planning tool (Kumar 2002; Lanktree 1994), design guidelines have become strictly voluntary in the City of Ottawa. Developers are expected to voluntarily conform to design guidelines.

Figure 3.3.3 • Infill development has become common in Old Ottawa South.

Source: Donald Leffers.

However, infill developers in Old Ottawa South have built mainly higher-end projects, usually single-family with the frontage dominated by garages.

> Design Considerations are not meant to be prescriptive and will not constitute a checklist. None of the Design Considerations will be expressed as policy, but rather are expected to act as a stimulus to development proponents to demonstrate how individual proposals will further the City's Design Objectives. Proponents are free to respond in creative ways to the Design Objectives and Principles and are not limited only to those suggested by the Design Considerations (City of Ottawa 2007, 2–34).

> The current approach to negotiating design through the development process is based on guidance, rather than regulation. The Council approves "design guidelines" for developments, but they do not have the statutory weight of the Official Plan policy or regulation (City of Ottawa 2008a, 51).

The City of Ottawa is currently proposing to integrate design guidelines more fully into the Official Plan, which would permit planning decisions to be defensible at OMB hearings (City of Ottawa 2013a).

Example 2: Commericial Infill

Another contested redevelopment saw the building of a large-format two-storey retail drugstore on a site previously occupied by a gasoline station and parking lot. Despite the fact that the City of Ottawa and the local community had recently completed a zoning study for the "**traditional main street**" in Old Ottawa South, the first zoning amendment proposed for this street was passed with very little debate. Some opponents claimed that the proposed single-use, two-storey commercial building with surface parking was not intensive enough (Tansey 2008), since the site is zoned for mixed-use four-storey buildings with underground parking, retail on the first floor, and two or three storeys of residential or office space.

Some local residents and small-business owners noted that the "big-box" format undermined the small-business context of the neighbourhood (Cairns 2008). Others were largely supportive of the development, pointing to its status as a brownfield redevelopment that would result in a "higher" use of the land (Doucet 2008); the development

Figure 3.3.4 • Mixed-use four-storey building along Bank Street, the "traditional main street" in Old Ottawa South.

Source: Donald Leffers.

required soil to be removed and remediated, which is a costly procedure and had thus far resulted in "underdevelopment" of the site (Pinchin Environmental 2007). Still others saw the building as only the first stage in the long-term development of the site (Harper 2008).

Relying on zoning bylaws and site-specific amendments to achieve intensification as a sustainability initiative is not ideal. Whether this particular development will contribute to urban sustainability is not really the point; relying on spot rezoning and incremental decision-making indicates the absence of an overall plan. Current planning also fails to move beyond simple zoning parameters and building uses to issues such as pedestrian-oriented, mixed-use urban environments.

Example 3: Redevelopment of a School

Competing narratives around appropriate use of space and protection of "nature" can be found in another contested development within the case study area: the redevelopment of an abandoned neighbourhood school into luxury townhouses. Of the three examples within the case study, this one was framed most explicitly in terms of Smart Growth and intensification as a sustainability initiative. In approving this development, Ottawa City Council drew on the Official Plan to legitimate this type of development:

> rezoning the subject land for residential development will advance the intent of the residential intensification policies in the new Official Plan by providing a residential infill development at a suitable urban location where services already exist (City of Ottawa 2005b, 3).

This redevelopment was spurred by years of declining enrolment in a small neighbourhood school and the restructuring of the school board so that small schools no longer fit the new model of education delivery. Although the community petitioned to retain the school, or at least the property, in "public" hands as institutional space, opposition to redevelopment included notions of improving a derelict space and providing additional housing in the central urban area. When framed in terms of sustainability, however, most interview participants wondered what the effects of residential intensification would be when evaluated against the loss of trees and green space and the privatization of a public space.

> [F]or the city, intensification is . . . all about getting more tax revenue out of the same parcel of land . . . when the City says "intensification," that's what they mean. They don't mean more schools. They don't mean more green space. . . . They mean: are they getting more money, more tax revenue to service the same amount of land? (C. Doucet, city councillor, personal communication, 7 July 2009)

Redevelopment rationalized by intensification is but the endpoint of a long history of municipal amalgamations, school board amalgamations, and restructuring processes that began under the leadership of Mike Harris, premier of Ontario from 1995 to 2002 (Jeffrey 1999). Municipal councillors are also at fault: they could have voted to retain institutional zoning rather than rezoning it to "general urban area" under the rationale of urban intensification.

However intensification is defined and assessed, these examples illustrate that it is not as coherent as municipal policies claim. It does not simply refer to processes that are inherently more sustainable. Intensification policies are not immune to misuse and co-optation when implemented locally; Bunce (2004, 188) asserts that intensification can be used to mask "straight-ahead plan[s] for economic revitalization," using the discourse of sustainability to legitimate an underlying growth agenda.

Conclusions: Challenges to Legislating Urban Sustainability

In some ways, urban sustainability is a relatively simple and intuitively grasped concept. According to the Brundtland Report (WCED 1987), sustainable development consists of three interrelated pillars—the social, economic, and environmental—that must be maintained and nurtured into the future. In some ways, it seems only logical that urban intensification would reduce the amount of land and infrastructure required and provide opportunities for more pedestrian- and transit-oriented living through increased density.

Yet what sustainability actually means for cities is neither clear nor homogenous, and it cannot be addressed with singular initiatives like increased density. In the City of Ottawa, intensification relies primarily on the efficient use of space and infrastructure but says little about green space, local pollution, the ecological footprints of buildings, or social elements such as affordable housing. Intensification policies in Ottawa rely on incremental and local land-use regulation but pay little attention to regional environmental processes. While provincial government policies, such as the Provincial Policy Statement (Ontario Ministry of Municipal Affairs and Housing 2005), attempt to protect wetlands, agricultural land, and natural resources, these high-level policies are often lost through incremental planning and development at the local level.

Intensification could only be useful in fostering more sustainable land use in the City of Ottawa if significant changes were made. First, by requiring small projects to go through more planning approval, by placing more elaborate controls on demolition and construction, by reforming the building code itself, or by revisiting the zoning bylaw that results in these forms of redevelopment being permitted as-of-right. Second, by inserting important elements of sustainability, such as pedestrian-oriented development, compact design, and maximization of green space into the Official Plan (City of Ottawa 2013a). Third, by including factors other than density, such as affordable housing provision: "demolition replacements," when new, larger buildings replace existing housing strictly for profit, fail to increase density at the regional scale. Finally, a regional planning process and a regional scale of governance are needed in order to fulfill the economic, social, and environmental elements of sustainable urbanism.

In summary, the City of Ottawa must control development more diligently if it expects urban intensification to remain a credible sustainability strategy. It has promised to do this by formulating a "strategic directions" policy framework with the goal of directing 90 per cent of its growth to urban areas where services already exist (City of Ottawa 2012b). But it must also move beyond a strictly land-use calculus of urban sustainability. If intensification is to become a genuine component of urban sustainability in Ottawa, local governments and planners need to accept this complexity rather than attempt to bypass it.

KEY TERMS

As-of-right
Design guidelines
Downloading
First-ring suburb
Nimbyism
Official Plan
Smart Growth
Traditional mainstreet zoning
Zoning bylaws/schedules

REFERENCES

Blais, P. 2010. *Perverse Cities: Hidden Subsidies, Wonky Politics, and Urban Sprawl*. Vancouver: University of British Columbia Press.

Building Code Act Consolidation [1992] S.O. 1992, c. 23.

Bunce, S. 2004. "The emergence of 'Smart Growth' intensification in Toronto: Environment and economy in the new official plan." *Local Environment* 9(2): 177–91.

Burningham, K. 2000. "Using the language of NIMBY: A topic for research, not an activity for researchers." *Local Environment* 5(1): 55–67.

Cairns, H. 2008. "This is my business." *The Oscar* 38(11): 12.

Campsie, P. 1995. *Social Consequences of Planning Talk: A Case Study in Urban Intensification*. Toronto: Centre for Urban and Community Studies, University of Toronto.

City of Ottawa. 1970. *Official Plan of the Ottawa Planning Area*. Ottawa: City of Ottawa Planning Board.

———. 1991. *City of Ottawa Official Plan (Volume 1): The Primary Plan*. Ottawa: City of Ottawa.

———. 2005a. *Urban Design Guidelines for Low-Medium Density Infill Housing*. Ottawa: City of Ottawa Planning and Growth Management Department.

——. 2005b. "Zoning—88 Bellwood: Report to Planning and Environment Committee and Council." 4 January. http://www.ottawa.ca/calendar/ottawa/citycouncil/occ/2005/02-23/pec/ACS2005-DEV-APR-0024.htm#_msocom_6.

——. 2006. *Urban Design Guidelines for Development along Traditional Mainstreets*. Ottawa: City of Ottawa Planning and Growth Management Department.

——. 2007. *City of Ottawa Official Plan: Consolidation (Vol. 1)*. Ottawa: City of Ottawa.

——. 2008a. *Residential Intensification: Building More Vibrant Communities [White Paper]*. Ottawa: City of Ottawa Planning, Transit and Environment Department.

——. 2008b "Zoning By-law 2008-250 Consolidation." http://www.ottawa.ca/residents/bylaw/a z/zoning/parts/index_en.html.

——. 2009. *Residential Land Strategy for Ottawa 2006–2031*. Ottawa: Ottawa Department of Infrastructure Services and Community Sustainability Planning Branch.

——. 2012a. *Annual Development Report, 2011*. Ottawa: City of Ottawa Planning and Growth Management Research and Forecasting Unit.

——. 2012b. "City of Ottawa Official Plan." http://www.ottawa.ca/en/city_hall/planningprojectsreports/ottawa2020/official_plan/vol_1/02_strategic_directions/growth/index.htm.

——. 2013a. "Official Plan and Master Plan review: Urban design and compatibility." http://ottawa.ca/en/preliminary-policy-proposals/9-urban-design-and-compatibility.

——. 2013b. "Zoning By-law Amendment." http://ottawa.ca/en/development-application-review-process-0/zoning-law-amendment.

Cook, M. 2008a. "Intensification versus 'uglification.'" http://www.canada.com/ottawacitizen/news/city/story.html?id=57f7a703-2a12-4de0- 95e9-d452b6f15100.

——. 2008b. "Small victory in Old Ottawa South." http://www.canada.com/ottawacitizen/news/city/story.html?id=fd58bceb-e2c2-4851-87dc-a5fd-89b8ab46&k=14005&p=1.

——. 2009. "Big infill: Why Brighton St. allowed by QED will be 'test case.'" *Ottawa Citizen* 23 September. http://communities.canada.com/OTTAWACITIZEN/blogs/designingottawa/archive/2009/09/23/big-infill-why-brighton-st-was-allowed-but-qed-will-be-quot-test-case-quot.aspx.

Doucet, C. 2008. "City Councillor's report." *The Oscar* 36(10): 7.

Filion, P. 2003. "Towards Smart Growth? The difficult implementation of alternatives to urban dispersion." *Canadian Journal of Urban Research* 12(1): 48–70.

Finlay, B. 2010. Witness Statement, OMB File 100206.

Graham, K.A., S.D. Phillips, and A.M. Maslove. 1998. *Urban Governance in Canada: Representation, Resources, and Restructuring*. Toronto: Harcourt Brace Canada.

Gréber, J. 1950. *Plan for the National Capital: General Report Submitted to the National Capital Planning Committee. Federal District Commission*. Ottawa: National Capital Planning Service.

Harper, K. 2008. "The realities of the Shopper's Drug Mart proposal." *The Oscar* 36(11): 11.

Hayek, M., G. Arku, and J. Gilliland. 2010. "Assessing London, Ontario's brownfield redevelopment effort to promote urban intensification." *Local Environment* 15(4): 389–402.

Jeffrey, B. 1999. *Hard Right Turn: The New Face of Neo-Conservatism in Canada*. Toronto: HarperCollins.

Keil, R. 2009. "The urban politics of roll-with-it neoliberalization." *City* 13(2/3): 230–45.

Kumar, S. 2002. "Canadian urban design practice: A review of urban design regulations." *Canadian Journal of Urban Research* 11(2): 239–63.

Lanktree, C. 1994. "A vision for Ottawa: Urban design policy in Ottawa's Official Plan. *Plan Canada* 34(5): 17–20.

Moore, A.A. 2013. *Planning Politics in Toronto: The Ontario Municipal Board and Urban Development*. Toronto: University of Toronto Press.

MPAC (Municipal Property Assessment Corporation). 2008. "AboutMyProperty." Toronto: MPAC. https://portal.mpac.ca/wps/portal.

Ontario Ministry of Infrastructure. 2012. *Growth Plan for the Greater Golden Horseshoe 2006: Office Consolidation January 2012*. Toronto: Queen's Printer for Ontario.

Ontario Ministry of Municipal Affairs and Housing (MMAH). 2005. *Provincial Policy Statement*. Toronto: MMAH.

Ostling, K. 2008a. "Development at 35 Brighton Avenue." *The Oscar* 36(7): 8.

——. 2008b. "Update on 35 Brighton Avenue." *The Oscar* 36(10): 17.

Pinchin Environmental. 2007. *Phase II Environmental Site Assessment, 1080 Bank Street and 297–305 Sunnyside Avenue, Ottawa, Ontario*. Ottawa: Pinchin Environmental.

Places to Grow Act [2005] S.O. 2005, c. 13.

Planning Act [1990] R.S.O 1990, c. P.13.

RMOC (Regional Municipality of Ottawa-Carleton). 1974. *Official Plan, Ottawa-Carleton Planning Area*. Ottawa: Office of the Regional Clerk.

——. 1988. *Official Plan of the Regional Municipality of Ottawa-Carleton (Volume 1)*. Ottawa: RMOC.

——. 1997. *Draft Official Plan of the Regional Municipality of Ottawa-Carleton*. Ottawa: RMOC.

Rogers, I.M., and A.S. Butler. 2005. *Canadian Law of Planning and Zoning*. 2nd edn. Toronto: Thomson Carswell.

Searle, G., and P. Filion. 2011. "Planning context and urban intensification outcomes." *Urban Studies* 48(7): 1419–38.

Tansey, B. 2008. "What's wrong with the Shopper's application." *The Oscar* 36(11): 10.

WCED (World Commission on Environment and Development). 1987. "Towards sustainable development." In G.H. Brundtland, ed., *Our Common Future*, 43–66. Oxford: Oxford University Press.

Westerink, J., D. Haase, A. Bauer, J. Ravetz, F. Jarrige, and C.B.E.M. Aalbers. 2013. "Dealing with sustainability trade-offs of the compact city in peri-urban planning across European city regions." *European Planning Studies* 21(4): 473–97.

Williams, K. 1999. "Urban intensification policies in England: Problems and contradictions. *Land Use Policy* 16(3): 167–78.

——. 2007. "Can urban intensification contribute to sustainable cities? An international perspective." Oxford: Oxford Brookes University Oxford Centre for Sustainable Development. http://unpan1.un.org/intradoc/groups/public/documents/APCITY/UNPAN026009.pdf.

Winfield, M. 2003. *Building Sustainable Urban Communities in Ontario: Overcoming the Barriers*. Toronto: Pembina Institute.

INTERVIEWS

C. Doucet, city councillor, 7 July 2009. In-person interview.

CASE STUDY 3.4

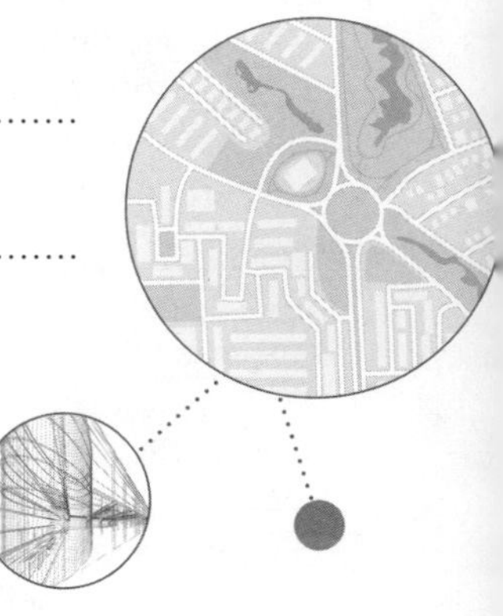

Planning for Diversity in a Suburban Retrofit Context: The Case of Ethnic Shopping Malls in the Toronto Area

Zhixi Cecilia Zhuang

SUMMARY

Recent waves of immigration have had a dramatic impact on urban economies and the landscapes of Canadian metropolitan regions. With increasing suburbanization of immigrant settlement, ethnic shopping malls have emerged as a noticeable phenomenon in suburban regions of the Greater Toronto Area (GTA). The dynamics of ethnic retailing generate significant community changes and raise questions for planners in terms of land use, built form, parking capacity, economic development, and community-building.

This chapter investigates the development and retrofit processes of several ethnic shopping malls located at the major intersection of Steeles Avenue East and Kennedy Road bordering the City of Toronto and the City of Markham. Specifically, it examines how these ethnic malls were developed since the 1990s in response to the growing Chinese population in the area and the booming Asian-oriented businesses and how they successfully regenerated the area once affected by business decline but also presented unprecedented challenges to the planning system. The chapter also examines the evolving role of city planners in facilitating the new retail form and addressing the challenges it poses, followed by a summary of lessons learned.

The findings presented in this chapter reveal that the suburban ethnic mall is not a standalone phenomenon; instead, it should be treated as an important part of the community. Planners must think beyond technicalities and exert more control on the free market in order to help nurture and sustain the emerging ethnic market that can, in turn, be a lucrative tool for the larger economy and contribute to community-building.

OUTLINE

- New trends of immigration, urban dynamics, and new suburban retail forms are occurring at the suburban boundary of Markham and Toronto.

- The transformation of the study site from a country theme to an Asian theme and the planning challenges of Asian-theme mall development brought up many concerns among residents and planners alike.
- The evolving role of planners toward a new holistic approach demonstrates how ethnically significant urban forms can be valuable additions to Canadian communities.

Introduction

Immigration has long been a key factor in Canadian population and economic growth. The recent waves of immigration have had a dramatic impact on urban economies and the landscapes of Canadian metropolitan regions through immigrants' social, economic, and spatial settlement (Hoernig and Zhuang 2010). With increasing suburbanization of immigrant settlement, ethnic shopping malls have emerged as a noticeable phenomenon in suburban regions of the Greater Toronto Area. The dynamics of ethnic retailing produce significant community changes and raise questions for planners in terms of the functions and land uses that will serve the community as a whole in the future: how should planners respond to the needs of ethno-culturally diverse communities, and how can a new ethnic retail form be integrated with the existing neighbourhood and facilitate the suburban retrofit process? Typical issues are related to land use, built form, parking capacity, economic development, and community-building (Preston and Lo 2000; Zhuang 2013). Planners do provide for variety in community life, and they often deal with change. They should not be absent in ethnic retail developments, which offer an opportunity to help stabilize and sustain these areas along with city-wide developments.

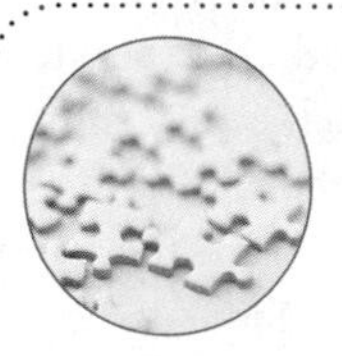

KEY ISSUE

Immigration

The recent waves of immigration to Canada have generated significant community changes and challenged planners on how to respond to the needs of ethno-culturally diverse communities. This case study specifically uses ethnic-mall development as a lens to discuss how to integrate a new ethnic retail form with existing suburban neighbourhoods and planners' roles in facilitating the long-term sustainability of these ethnic retail areas in a suburban retrofit context.

This chapter begins with an examination of the new trends in immigration, the associated urban dynamics, and the emerging suburban ethnic retail form. Next, it investigates the development and retrofit processes of several ethnic shopping malls located at the major intersection of Steeles Avenue East and Kennedy Road bordering the City of Toronto and the City of Markham (Figure 3.4.2). It examines how these ethnic malls were developed since the 1990s in response to the growing Chinese population in the area and the booming Asian-oriented businesses and how they successfully regenerated an area once affected by business decline but at the same time presented unprecedented challenges to the planning system. Third, the chapter examines the evolving role of city planners in facilitating the new retail form and addressing the challenges it presents, followed by a summary of lessons learned. The findings presented in this chapter are based on two research projects conducted from 2005 to 2011. Primary research data were collected through interviews with key informants (e.g., merchants, developers, planners, city councillors, economic development officers), business surveys of merchants, and intercept surveys of shoppers at the existing ethnic malls.

New Trends in Immigration, Urban Dynamics, and New Suburban Retail Forms

Today's new immigrants represent more diverse cultural backgrounds and possess more human and economic capital than their predecessors. According to the 2011 Canadian census, foreign-born individuals represented 20.6 per cent of the total population. Of the immigrants who arrived between 2006 and 2011, 56.9 per cent came from Asia. Within Canada's population, 19.1 per cent self-identified as visible minority people, with the three largest groups—South

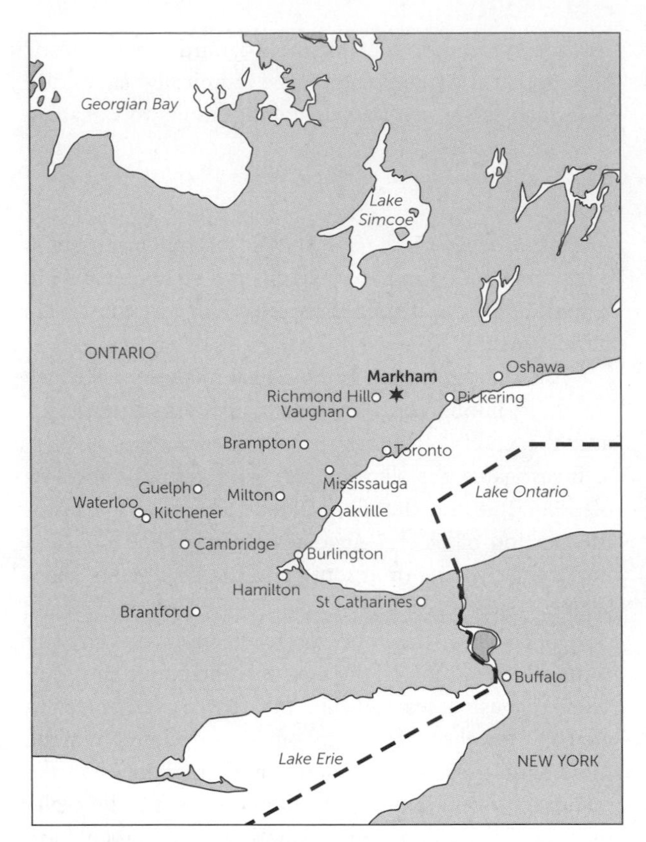

Figure 3.4.1 • Location map.

Asians, Chinese, and Blacks—accounting for 61.3 per cent of the visible minority population (Statistics Canada 2013).

The arrival of the new type of immigrants presents unprecedented settlement experiences and integration issues in Canada's immigration history (Hoernig and Zhuang 2010). From the perspective of geographical settlement distribution, immigration has been unevenly distributed across the country. According to the 2011 census, the country's three "gateway cities"—Toronto, Vancouver, and Montreal—attracted 63.4 per cent of the immigrant population. In contrast, just over one-third (35.2 per cent) of Canada's total population lived in these three areas (Statistics Canada 2013). These data show that contemporary immigrants tend to settle in large urban regions rather than in smaller urban centres or rural areas as their predecessors did.

It is difficult to depict today's immigrants' settlement experiences accurately. Immigrants may follow multiple paths and undergo unpredictable transformations in their social and economic lives. Traditionally, the inner city played an important role in immigrant settlement as a "port of entry" and reception area for immigrants. Today in Canadian cities, the inner city is no longer a typical trajectory for immigrants. They tend to bypass the inner city and directly settle in suburban areas; they either settle in a dispersed pattern or remain concentrated in ethnic clusters, or **ethnoburbs** (Li 1998). Many immigrants settle in suburbs because of housing affordability (Hulchanski 2010).

Retail developers are finding opportunities in this settlement pattern, with suburban ethnic retail locations like strip malls and shopping centres an emerging phenomenon in areas of major immigrant settlement such as the Greater Toronto Area (Zhuang, Hernandez, and Wang 2015). Notably, a rapidly growing new retail form, the **Asian-theme mall** (or Chinese shopping mall) generates significant impacts on suburban landscapes.

Transformation of the Study Site: From Country Theme to Asian Theme

The study site at the intersection of Steeles Avenue East and Kennedy Road is one of the most famous Asian-theme shopping destinations in the GTA, with three indoor shopping malls, the Pacific Mall, Market Village Markham ("Market Village"), and Kennedy Corners Country Shoppes ("Kennedy Corners") built on a large tract of land (Figure 3.4.3). The three retail complexes have developed different specialty businesses: the Pacific Mall features apparel and electronic products, especially for trend-setting youth; Market Village provides more restaurants and grocery stores for family-oriented activities; and Kennedy Corners focuses on banking and other services. In a way, the complexes complement each other, and the commercial triangle works as an integrated site.

The transformation of the study site since the 1990s reflects demographic changes in the then–Town of Markham that have transformed a rural Anglo-Saxon small town into a fast-growing ethnoburb. The first retail complex on the site, the Cullen Country Barns ("The Barns"), opened in 1983. The Barns was a popular Markham tourist spot featuring a country market theme of bygone times. This theme was maintained in the later developments of Market Village and Kennedy Corners; they both extensively used the same architectural language to achieve an old-town atmosphere. The names of the three retail complexes also reinforce the traditional country theme. However, "the market for a 'country theme' commercial development weakened and Market Village Markham, which opened during the 1989–1990 recession, was unable to attract more than a few tenants, most of which quickly failed" (Heaslip 2011, 3). Then, with the influx of immigrants and

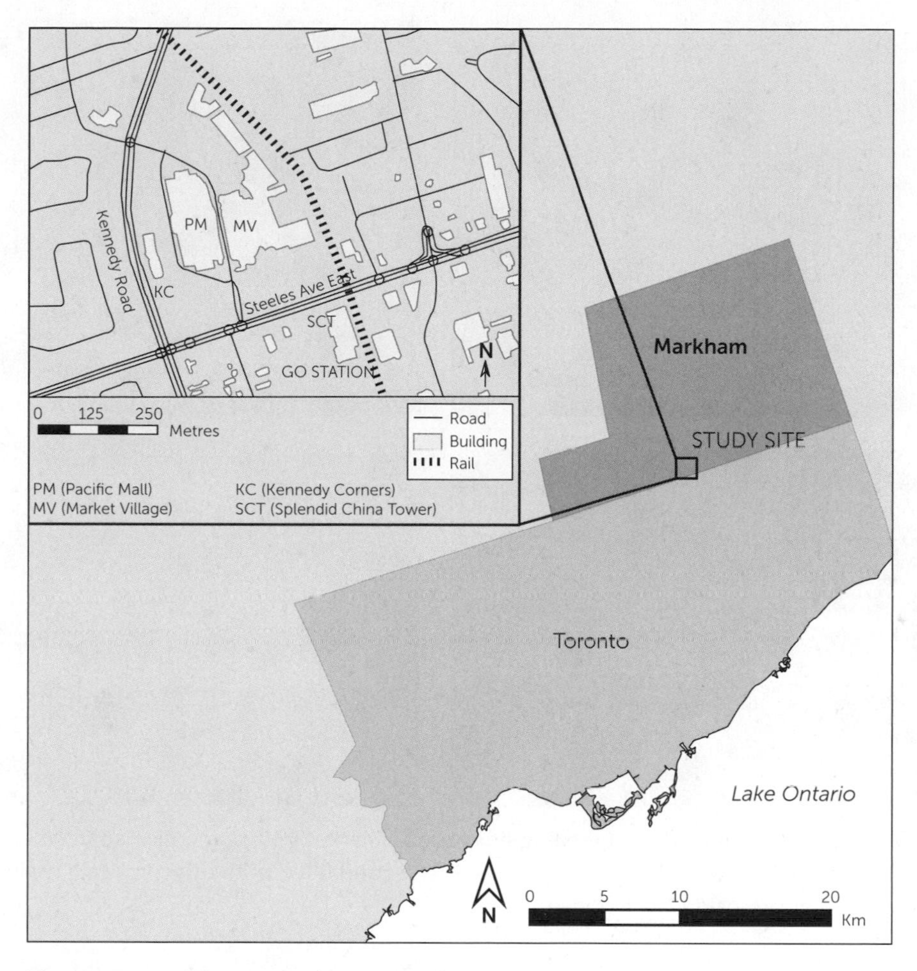

Figure 3.4.2 • The study site.

Source: Cornel Campbell.

capital into Markham, Asian-theme businesses began to be featured inside Market Village and Kennedy Corners to meet the changing demographics of the surrounding areas.

Because of a decline in business, the Barns closed in 1994. In response to the growing Chinese population in the area and the booming Asian-oriented businesses, a retail replacement, later known as the Pacific Mall, opened in 1997. It features innovative condominium ownership, which mainly targets the Chinese community, who prefer to purchase or own properties for residential or business purposes. The suburban shopping mall configuration with condominium ownership seems to be a successful new trend, attracting ethnic entrepreneurs and investors, and the Pacific Mall and other Asian-theme malls offer alternatives for combining both the retail and social functions in a highly concentrated and desirable manner that caters to suburbanized immigrants.

Planning Challenges of Asian-Theme Mall Development

The development of condominium retailing is an adaptive approach in that it provides more flexibility and authorizes more ownership power to individual business owners and investors.

However, the tenure structure distinctions from conventional leasehold retailing have caused a series of conflicts regarding land use in the planning system and public concern. Conventional retail planning policies tend to use a hierarchical system to classify shopping centres at the neighbourhood, community, and regional levels based on several criteria, such as floor area, store numbers, market size, and sales (Jones 2005). However, when planning ethnic malls, this hierarchical system is irrelevant

Figure 3.4.3 • Exteriors and interiors of the Pacific Mall and Market Village (clockwise from upper left corner: main entrance of the Pacific Mall; Heritage Town on the second floor of the Pacific Mall; main entrance of Market Village; food court inside Market Village).

in determining parking standards, retail sizes and uses, retail facilities (e.g., anchor stores), and retail forms.

Because of the large number of small store units and the resulting intensity of commercial activity, as well as its regional draw of customers, the Pacific Mall development has inevitably generated higher demand for parking and increased traffic. The Barns' parking standard (1095 spaces), based on use breakdown, and the Town's "Shopping Centre" standard (718 spaces), based on gross leasable area, could not accommodate Pacific Mall's increased parking demands. The Torgan Group, a long-established Ontario-based developer, the Planning Department, and the Town Council each hired external consultants to conduct parking studies. In the end, the developer provided the most adequate number of parking spaces (1505), exceeding the then–governing bylaw requirement and the other consultants' recommendations in order to maximize lot capacity and gain site plan approval. Since there was no need to rezone the site to build the mall, only a site plan approval was required, without public meetings. However, years after the mall was built and opened in 1997, parking and traffic still remain ongoing concerns for the business area and surrounding neighbourhoods, and the mall was thus described as "the victim of its own success" (Zhuang 2013, 107).

In addition to land-use or technical issues, the ever-emerging Asian-theme mall phenomenon has triggered racial tensions centred around this new retail format. The most publicized social and political dispute regarding the Asian-theme mall was the controversial "Carole Bell incident." In 1995, the deputy mayor, Carole Bell, made inflammatory comments on the concentration of Asian-theme malls, calling it "racial monopoly" and the cause of social conflicts that made long-time residents move away (*Toronto Star* 1995). Bell's comments deeply offended many people in the community, who later accused her of racism. From a planning perspective, as the chair of the Town's planning committee noted, restricting developers from building something that caters to a specific group is unfair and would restrain trade. It is also beyond the function of planning to determine whether the market can bear the development or not (Krivel 1995).

A group of 12 mayors from across the GTA signed a statement condemning Bell's offensive remarks. A Mayoral Advisory Committee comprised of Markham Council and members of the public was created in order to heal the growing rift in the community.

Nearly two decades after the racial tensions, today's Markham is a highly diverse ethno-cultural community, and the political scene has changed along with the social transformations. The Asian-theme mall is now widely recognized as a multi-faceted phenomenon that responds to the emerging demands of an ethnic-oriented market. The existing malls on the study site serve as important social hubs and one-stop shopping destinations for immigrants and contribute to the local economy and the retrofit of suburban spaces. Their developments reflect the interplay of social and cultural values and economic activities. Decision-makers must consider these factors and think beyond technicalities.

Following the success of the Pacific Mall, another condominium Asian-theme mall, the Splendid China Tower, was opened in 2007, right across from the Pacific Mall on the Toronto side. A former Canadian Tire store on the site was converted into an ethnic shopping centre. Several years later, the Splendid China Tower proposed a Phase II addition. Meanwhile, the Market Village is to be demolished to make way for a mega–Asian-theme mall called the Remington Centre. The Pacific Mall is also proposing an expansion plan. Table 3.4.1 provides details of existing and proposed retail spaces. All these proposals are mega-complex developments with larger massing, higher density, multiple functions, and distinct designs. None of the developers is of Chinese background. It seems that developers, who are generally profit driven, have an instinctive sense of the rapidly changing ethnic market, and they have seized the opportunity for profit.

The new expansion and redevelopment proposals present new planning challenges: other than planning approval, what is the role of municipalities in planning ethnic retail in the long run, especially in the context of suburban retrofit and community-building? How can these mega-malls be recreated as community focal points that can help sustain existing neighbourhoods and treated as an avenue for immigrant integration and community-building?

The Evolving Roles of Planners: A New Holistic Approach

In the 1990s, the development of Asian-theme malls was largely market driven, and its potential impact on local communities remained unknown to planners. In practice, planners must work within legislative boundaries; they abide by the Ontario Planning Act and follow Official Plans and zoning bylaws that regulate and guide land developments with a city-wide perspective regardless of people's ethno-cultural differences. In the case of the Pacific Mall, planners played only a reactive and regulatory role to deal with issues related to parking and traffic (Zhuang 2009; 2013). The council's final approval was based largely on projected economic impacts recommended by the Economic Development Department, which outweighed social, cultural, or technical aspects. The Town Council and staff preferred to follow market forces in search of a signal on whether to proceed with an ethnic-oriented development or not. Thus, municipal intervention was minimal.

Table 3.4.1 • Retail Spaces of Existing Ethnic Malls and New Development Proposals

Ethnic Mall	Current GFA (sq. ft.)	Proposed GFA (sq. ft.)	Total GFA (sq. ft.)	Projected Floor Area Ratio (FAR)	Tenure
Pacific Mall	276,000	375,000	651,000	1.1	Condominium
Market Village/ Remington Centre	352,000	856,000	856,000	0.85	Leasehold; to be changed to condominium
Splendid China Tower	90,000	247,000	337,000	1.3	Condominium
Kennedy Corners	45,000	n/a	45,000	0.38	Leasehold

Note: GFA = gross floor area

Sources: City of Toronto 2007; Heaslip 2011.

Nearly 20 years later, when the three existing malls have proposed mega-scale mixed-use redevelopment to rejuvenate the business area, more holistic planning interventions can be seen. First of all, planners and professionals involved in the redevelopment not only address the long-standing parking and traffic issues but also take into consideration the demographic changes and diverse needs of the local community. This results in the acknowledgment of the contributions of the Asian-theme malls to community life and the retrofit of the area and a stronger need to connect them with other uses, such as housing, transit, and public space. One architect comments on the community functions the existing malls serve:

> [What] really intrigues me is that this kind of mall more so than the conventional type of North American suburban mall, it acts more like a community centre in a lot of ways. . . . People would come for the day, meet their friends, they go for lunch, their children go there to meet their friends, there're events happening on the stage in the middle of the mall . . . so there's really a whole day of activities. . . . So we have to deal with that sort of phenomenon (architect, private firm, personal communication, 2011).

In a staff report submitted to the Scarborough Community Council with regard to the redevelopment of the Splendid China Tower, it is well acknowledged that the studied retail area

> has become a tourist attraction and destination not only to local shoppers but also shoppers from the Pacific Rim. In the past, "Chinatown" as it is known in downtown Toronto carried that label, today this section of the City and others rival Chinatown because of their location, product selection, competitive pricing and proximity to large consumer markets (City of Toronto 2007, 6).

Markham planners are open to the discussion about including the residential component in the future phase of redevelopment, which reflects a retail format commonly found in many Asian cities, as one planner interviewee indicates:

> It makes sense. I think we are understanding that residential is a natural fit with retail. Again, there is an issue of understanding culture [that is] evolved. I mean, if someone here told the planner 15 years ago that somebody would live in apartment on top of a shopping centre, he'd be "you gotta be kidding, it has to be some place where it's got schools and parks and trees and like that." And now we realize that there is a different market up there, and the market really wants to be [like that] (planner, private firm, personal communication, 2011).

Second, recognizing that pedestrian and cyclist activities are already frequent and transit uses are high in the retail area that serves as a community hub, the City of Markham emphasizes urban design issues and improved transit uses to mitigate traffic impacts on the area. Planners and architects work closely with the applicants of the proposals to recreate a new retail place that is pedestrian friendly, transit accessible, and architecturally identifiable with a strong urban relationship to the street and surrounding area. The guiding principle of the redevelopment is to create a "people place" (architect, private firm, personal communication, 2011). Other outstanding issues, such as year-round useable open space, quality streetscapes, and public art, are also addressed by City staff in recommendations to refine the redevelopment plans (Heaslip 2011). Specifically, a large central plaza space is proposed for the Remington Centre as "a centre for the community," featuring a reflecting pond that will be turned into an artificial skating rink in the winter (Figure 3.4.4). Also proposed are graded separate entrances for automobiles to go below grade and into a multi-level parking structure and maximization of surface space for pedestrians and urban design details. The separation of automobiles from ground level will ensure that pedestrians, cyclists, and transit users do not conflict dramatically with automobiles. An on-site transit hub connecting local and regional bus lines, as well as the regional GO train service, is also incorporated in the development plans.

Third, planners facilitated fruitful discussions among various stakeholders so as to plan holistically for the future of the community. Since the study area is located at the intersection bordering Toronto and Markham, it is a complex site involving local, regional, and provincial governments as well as their affiliated agencies. A forward-thinking transportation planner retained by the developers has been instrumental in creating the synergy among stakeholders and a strong advocate for

Figure 3.4.4 • Architectural rendering of the public plaza at the proposed Remington Centre.

Source: Cornel Campbell/Kohn Partnership Architects Inc.

an urban public realm in a suburban retrofit context. Thanks to the planner's continuous efforts, a steering committee has been formed with representation from the City of Toronto, the City of Markham, the Regional Municipality of York, the Toronto Transit Commission, Metrolinx, GO Transit, and York Region Transit to tackle traffic and parking issues throughout the study site and surrounding neighbourhoods. The parties involved have been in active discussion about finding solutions to the site access and traffic problems and working together to move the redevelopment along. It is not easy to get various stakeholders on board, yet all parties seem to have a good understanding of the Asian-theme mall phenomenon and see the advantage of the redevelopment. As one planner remarked about the different attitudes between then and now:

> I think we are becoming much more mature, as a community. You know, we've learned a lot. So I think we were more challenged than we are now . . . If [what's proposed now] had been proposed in the 1990s . . . [we] wouldn't have known what to do. Now [we] look at that kind of density and scale, and very modern and very urban stuff, no problem! Conceptually [we] are on board, and conceptually the community is on board as well. We're not getting the fundamental "stop stop stop." In the '90s, that's what we would get from the community which would just be like "WHOA STOP let me off." We don't get that anymore. So I think we are a lot less challenged than we were then (planner, private firm, personal communication, 2011).

The case shows a major shift in the role of planners over some 20 years. In the past, planners were unfamiliar with the new retail format, and the Pacific Mall case indicated a lack of experience, among both planning professionals and city officials, in dealing with ethno-cultural diversity and related developments. Today's planners demonstrate a better understanding of the ethnic-mall phenomenon and are more confident about finding holistic and creative planning solutions to tackle challenges facing the community as a whole.

Conclusions: Learning from the Case Study

This chapter examines the evolving roles of planners in response to the challenges posed by the (re)development of Asian-theme malls. Although it is still too early to evaluate the success of the retrofit of the retail area, there are lessons to be learned from the case study that will enrich our knowledge of the nature of suburban ethnic-mall developments and inform public policy on immigrant settlement, retail development, suburban retrofit, and community-building.

First of all, immigrants are inevitably involved in the various aspects of urban life in Canadian cities—socially, economically, and culturally—and cities are rebuilt as a result. Policy-makers and professionals at the forefront of urban development such as city planners, who are key players in the facilitation and promotion of the physical growth of cities, should acknowledge the complexity of today's immigration and incorporate this knowledge into their decision-making and daily practices.

Second, ethnic malls act as community focal points for ethnic groups and play multiple social and economic roles.

The ethnic mall is not a standalone phenomenon; instead, it should be treated as an important part of the community. Its development is closely associated with various planning functions, such as urban design, housing, transportation, neighbourhood design, and community services. Planners should see the bigger picture and adopt a holistic approach to connect ethnic malls with other aspects of community life as well as the surrounding neighbourhood. Simply leaving the long-term sustainability of these malls and surrounding neighbourhoods to the market will not work. Planning intervention is still required, such as housing intensification, improved transit use, and pedestrian design elements to promote local shopping and recreate community focal points. Planners must think beyond technicalities and exert more control on the free market in order to help nurture and sustain the emerging ethnic market that can, in turn, be a lucrative tool for the larger economy and contribute to community-building.

Third, planning for diversity is a work in progress, and planners cannot work alone. Planners must be more collaborative with external stakeholders. No single discipline can work alone to tackle complex urban issues such as ethnic-mall development. Ethno-cultural diversity adds another layer of complexity to urban issues.

KEY TERMS

Asian-theme mall

Ethnoburbs

REFERENCES

City of Toronto. 2007. "Staff report: 4675 Steeles Avenue East (Splendid China Square Inc.) Zoning application—Status report." http://www.toronto.ca/legdocs/mmis/2007/sc/bgrd/backgroundfile-4717.pdf.

Heaslip, S. 2011. *Preliminary Recommendation Report: The Remington Group (Market Village Markham) and Pacific Mall North East Kennedy Road and Steeles Avenue East Applications for Site Plan Approval for Two Proposed Commercial Developments*. Report to Development Services Committee, Town of Markham. 21 June.

Hoernig, H., and Z.C. Zhuang. 2010. "New diversity: Social change as immigration." In T. Bunting, P. Filion, and R. Walker, eds, *Canadian Cities in Transition: New Directions in the 21st Century*, 150–69. Toronto: Oxford University Press.

Hulchanski, D. 2010. *The Three Cities within Toronto: Income Polarization among Toronto's Neighbourhoods, 1970–2005*. Toronto: University of Toronto, Cities Centre.

Jones, K. 2005. "Retail sector planning in the Greater Toronto Area." In B.M. Massam and S.S. Han, eds, *Urban Planning Overseas 2005: Special Issues on Urban Planning in Toronto*, 66–75. Toronto: York University, Urban Studies Program.

Krivel, P. 1995. "Will bylaws or market determine growth of Chinese theme malls?" *Toronto Star* 10 November: NY1.

Li, W. 1998. "Los Angeles's Chinese ethnoburb: From ethnic service center to global economy outpost." *Urban Geography* 19(6): 502–17.

Preston, V., and L. Lo. 2000. "Asian theme malls in suburban Toronto: Land use conflict in Richmond Hill." *The Canadian Geographer* 44(2): 182–90.

Statistics Canada. 2013. "Immigration and ethnocultural diversity in Canada." http://www12.statcan.gc.ca/nhs-enm/2011/as-sa/99-010-x/99-010-x2011001-eng.cfm#a2.

Toronto Star. 1995. Editorial: "Grumbling about each new set of immigrants is as old as Canada itself." 24 August: A26.

Zhuang, Z.C. 2009. "Ethnic retailing and implications for planning multicultural communities." *Plan Canada* special edition: 79–82.

——. 2013. "Rethinking multicultural planning: An empirical study of ethnic retailing." *Canadian Journal of Urban Research* 22(2): 90–116.

——. T. Hernandez, and S. Wang. 2015. "Ethnic Retailing." In H. Bauder and J. Shields, eds, *Immigrant Experiences in North America*, 223–47. Toronto: Canadian Scholars Press.

INTERVIEWS

Architect, private firm, 1 December 2011. In-person interview.

Planner, private firm, 5 July 2011. In-person interview.

4.0

INTRODUCTION
Natural Resource Management

As issues such as climate change and habitat preservation become increasingly important, planners are involved in different aspects of managing natural resources: developing policy in provincial or federal governments, building partnerships with non-profit and community groups, and advising local communities in the identification of natural areas in need of protection. We have all witnessed extreme weather events in recent years in our own communities; we also all know of cases in which the protection of a body of water or forested landscape has had a major impact on the region. Although many of the cases in this volume address issues of sustainability, the cases in this section focus on the planning challenges associated with our land, water, and mineral resources. They illustrate how, from the local to the federal level, legislation protecting our natural resources often comes into conflict with the existence and development of our towns and cities.

The role of planners in managing natural resources integrates facilitation and policy development with state actors, such as Parks Canada and the Ministry of Natural Resources, as well as local residents, Aboriginal governments, and environmental organizations. Because Canada is a country rich in natural resources such as forests, minerals, and fresh water, managing these precious reserves has long been a concern of the federal and provincial governments. Under the Constitution Act [1867], provincial governments have responsibility for natural resource development and environmental management, and in provinces like British Columbia, much of the land is owned by the Crown. However, a key distinction in landownership involves First Nations and Aboriginal governments, who have recognized traditional claims on lands even though, in some cases, these claims were never spelled out in formal treaties. This fundamental legal ambiguity has created long-standing conflicts between the governments, as two cases in this section illustrate, although innovative partnerships have begun to develop to co-manage resources. In many cases, community-led efforts to preserve natural resources have also been instrumental, and environmental organizations help to protect our land, water, and air from overdevelopment and pollution. Environmental assessments (EAs) are critical in forecasting potential impacts on local communities and natural resources, although the implementation of EA recommendations is not always transparent or complete.

Natural resource management crosses several disciplines, including planning, public policy, forestry, geology, and meteorology. The cases in this section could be cross-listed with other sections of this volume. Tim Shah's case on climate change planning in Elkford, for example, could easily fit into Section 8 (Participatory Processes) and offers insights for other small communities for developing knowledge about resilience and climate change through a community-building process. Planning for climate change and increasing adaptivity (Moser et al. 2008) have recently become a policy focus in Canada (e.g., British Columbia 2012). This sub-discipline of planning affects rural and urban areas equally, small and large municipalities alike.

Darha Phillpot and Todd Slack narrate a fascinating case conflict between the allocation of mineral rights and pre-existing Aboriginal rights in their chapter on the Yellowknives Dene First Nation. In an area without a land-use plan to guide decisions, the Yellowknives Dene have been engaged in an epic struggle to exert their rights to co-management of the important cultural landscape around Drybones Bay. The chapter outlines a series of environmental assessments and the resulting judicial review that is currently underway.

In his study of the District of Elkford, British Columbia, Tim Shah describes how the rural community of 3000 residents developed a climate change adaptation planning strategy while updating their Official Community Plan. In partnership with the Columbia Basin Trust, the process helped residents identify potential climate impacts, assess local risks and vulnerabilities, and develop adaptation actions. The process itself became one of community-building, since a collaborative planning approach helped residents move from scepticism to acceptance. The case can be instructive for other communities considering integrating climate change adaptation strategies into their official plans or frameworks.

Janice Barry presents a complex environmental planning process that involved the Nanwakolas Council: the Central Coast Land and Resource Management Plan. First Nation representatives were included in all of the major governance bodies, and the case is seen as one of the first in "government-to-government" planning that recognizes Aboriginal rights. In that the case goes beyond the duty to consult with First Nations, it illustrates the creation of a workable long-term governance structure to make critical environmental planning decisions.

These cases illustrate the complexity of protecting natural resources, particularly the variety of actors involved in the planning process and the governance structures, which increasingly involve First Nations and Aboriginal communities as equal partners (e.g., *Delgamuukw v. British Columbia* 1997). Canada is a sparsely populated country; many of our natural resources are located in areas with very small or scattered human settlements. However, the impact of development, especially in resource-extractive industries such as mining, can be quite significant both for the ecosystems and for residents of these communities. Understanding legal and land rights can be complex and involve a collaborative approach to problem-solving (e.g., Picketts et al. 2012). This can be difficult, considering the very established institutions and governance responsible for managing natural resources (e.g., Parks Canada, Ministry of Natural Resources). However, there are many cases across Canada involving co-management. As the repercussions of climate change are as yet unknown, protecting these valuable resources and developing resilient responses has become critical.

REFERENCES

British Columbia. 2012. "Climate Action Charter." http://www.env.gov.bc.ca/cas/mitigation/charter.html.

Delgamuukw v. British Columbia [1997] 3 SCR 1010.

Moser, S.C., R.E. Kasperson, G. Yohe, and J. Agyeman. 2008. "Adaptation to climate change in the northeast United States: Opportunities, processes, constraints." *Mitigation and Adaptation Strategies for Global Change* 13: 643–59.

Picketts, I.M., A.T. Werner, T.Q. Murdock, J. Curry, S.J. Déry, and D. Dyer. 2012. "Planning for climate change adaptation: Lessons learned from a community-based workshop." *Environmental Science and Policy* 17: 82–93.

CASE STUDY 4.1

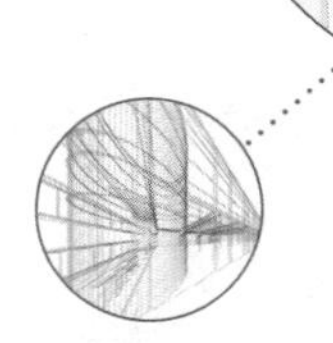

Resource Development Proposals in Drybones Bay, Northwest Territories

Darha Phillpot and Todd Slack

Summary

The Drybones Bay case study examines a series of proposed small-scale resource development projects in an area of high cultural significance for the Yellowknives Dene First Nation (YKDFN). The area is without a land-use plan to guide land-use decisions. The case describes a series of seven **environmental impact assessments** that took place between 2003 and 2011, resulting in a judicial review.

The case explores key planning issues such as the conflict between the **free entry system** of allocating mineral rights and pre-existing **Aboriginal rights**, the challenge of addressing public concern about the impact on Aboriginal rights, and the potential for land-use planning to positively contribute to the goal of reconciliation between Aboriginal rights and interests and broader public interests.

OUTLINE

- Drybones Bay, in the Northwest Territories, is an area with considerable importance to the Yellowknives Dene. The establishment of mineral rights in the Drybones Bay Area and the land and resource management regime in the Mackenzie Valley are the subject of this case.
- The concept of land-use planning as a path to reconciliation is explored through a chronology of development proposals in Drybones Bay, including development proposals and environmental assessments, further consideration of the environmental assessment measures, the application for judicial review, and consideration of public concern in one particular environmental assessment.

Methodology

This case study was written based on the observations and experiences of the authors, who were directly involved in the Debogorski Diamond Environmental Assessment. Todd Slack worked for the Yellowknives Dene First Nation as a research and regulatory specialist, and Darha Phillpot worked for the Mackenzie Valley Environmental Impact Review Board as an environmental assessment officer.

Background

Importance of Drybones Bay to the Yellowknives Dene

Drybones Bay is located on the shore of Great Slave Lake, approximately 50 kilometres southeast of Yellowknife, the capital city of the Northwest Territories. Adjacent to Yellowknife are the two communities of the Yellowknives Dene—Ndilo and Dettah. The Yellowknives Dene First Nation comprises the Indigenous people who have always used and occupied the lands and waters around *Weledeh-Cheh* (Yellowknife River and Bay), north to the Barrenlands. The Yellowknives Dene First Nation has approximately 1400 members, primarily residing in the communities of Ndilo, Dettah, and Yellowknife.

Drybones Bay is within the Yellowknives Dene's traditional territory, referred to as the Chief Drygeese Territory, and is considered a **cultural landscape** for the First Nation. There are several known historical villages, hundreds of known archaeological sites, including graves, birth sites, and other cultural places, as well as many family camps scattered throughout the area. Drybones Bay is of ongoing cultural importance and continues to be used today by the Yellowknives Dene for hunting, trapping, fishing, and providing youth with cultural exposure to traditional activities and the land. The Yellowknives Dene state that the Drybones Bay area is vital to their identity and is irreplaceable in terms of cultural heritage. A letter from seven former and current chiefs of the Yellowknives Dene First Nation explains:

> The Drybones Bay area is a special place to the YKDFN. Culturally, this area is without parallel and the highest level of protection is needed. The people's use of this area has been significantly impacted by the level of development and the subsequent effects that arise out of those impacts and we have seen our Treaty Rights considerably degraded over the last decades (Yellowknives Dene 2011).

Until recently, the Drybones Bay area has had relatively low resource development or recreational-use pressures from City of Yellowknife residents. However, the diamond staking rush of the 1990s and increased access to the area by Yellowknife residents for recreational purposes (Ski-Doos and boats) have changed this situation.

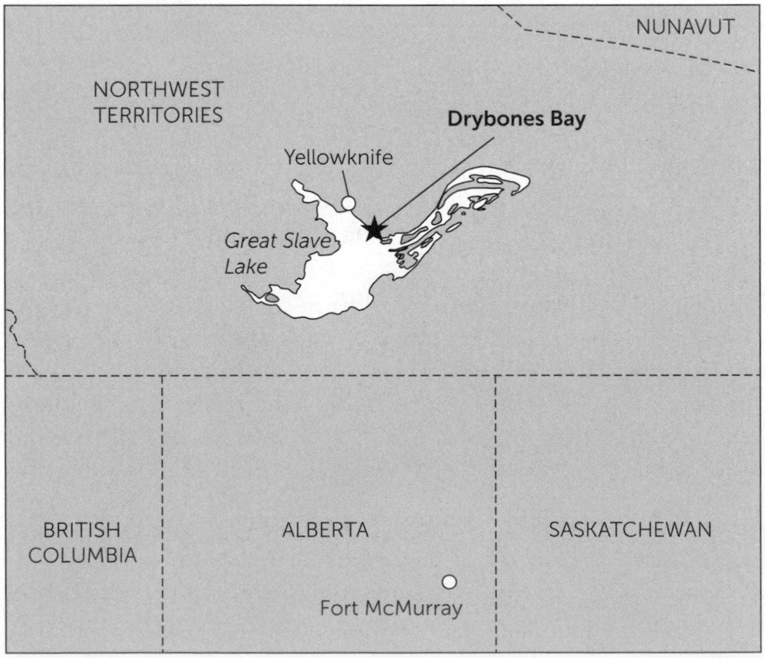

Figure 4.1.1 • Location map.

Establishment of Mineral Rights in the Drybones Bay Area

In the NWT, like much of Canada, mineral rights to Crown land are allocated

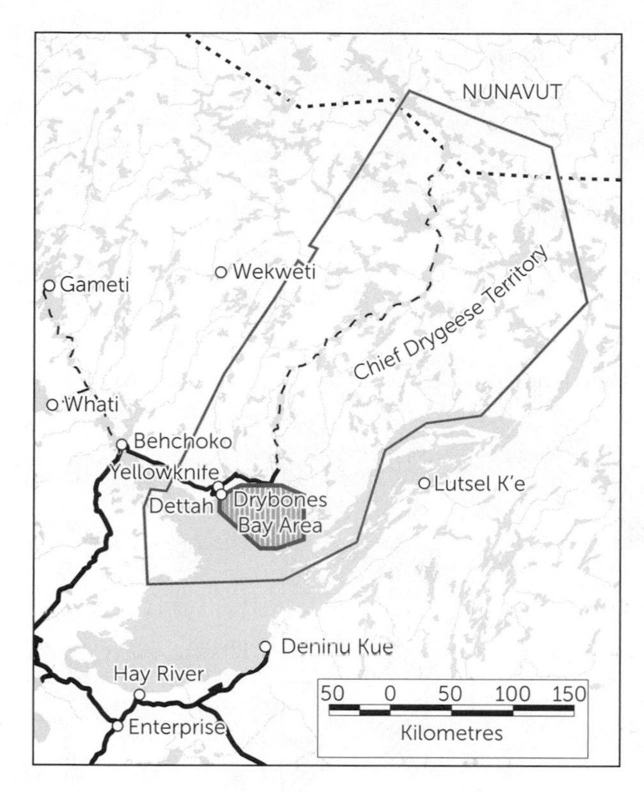

Figure 4.1.2 • Yellowknives traditional territory: Chief Drygeese Territory.

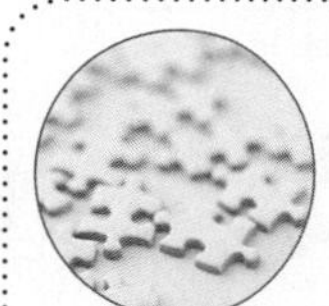

KEY ISSUE

Conflict between the Free Entry System and Aboriginal Rights

Consistent with the free entry system, the NWT and Nunavut mining regulations do not include discretionary powers in the registration of mineral claims. Once claims are registered, mineral rights are established. The purpose of the system is to encourage mineral development by requiring mineral-claim holders to do a certain amount of work each year to develop their claim or the claim will lapse. The free entry system presumes that mineral development is the highest and best value for land. This system of allocating mineral rights originated prior to case law about Aboriginal rights and, as such, comes into conflict with the Crown's duty to consult.

according to the free entry system. Under this system, land can be freely staked and mineral claims can be registered, thereby establishing legal rights to develop mineral resources. The discovery of diamonds in the NWT triggered a staking rush in the late 1990s. Much of the Chief Drygeese Territory was staked during that time, including parts of the Drybones Bay.

Some of the original diamond-rush claims in the Chief Drygeese area have now lapsed. However, there are still a number of companies that maintain their claims and seek to further develop them. From the developers' perspective, mineral rights in the Drybones area were secured through appropriate legal means. It is reasonable, from their perspective, that they be free to exercise their rights and pursue their interests within the Drybones Bay area and thus unlock the mineral potential that would benefit all the citizens of the Northwest Territories, including the Yellowknives Dene.

The land-use conflicts in the Drybones Bay area between the mineral-rights holders, Aboriginal rights holders, and the institutions of public government who manage Crown land is exacerbated by the establishment of mineral rights prior to a planning process that considers competing values in the area. The specific events that resulted from the conflict were the product of the legislative and legal land and resource management regime.

The Land and Resource Management Regime in the Mackenzie Valley

Concurrent with the diamond boom, the legislative and legal landscapes in the NWT was changing. In 1998, the Mackenzie Valley Resource Management Act (MVRMA) came into force, resulting in significant changes to the development-permitting process in the NWT and providing new opportunities for community involvement and participation in the resource management processes and decisions, including regional land-use planning.

The MVRMA introduced an integrated system of land and water management in the Mackenzie Valley of the NWT, which comprises all of the NWT except the Inuvialuit Settlement Region. The Act

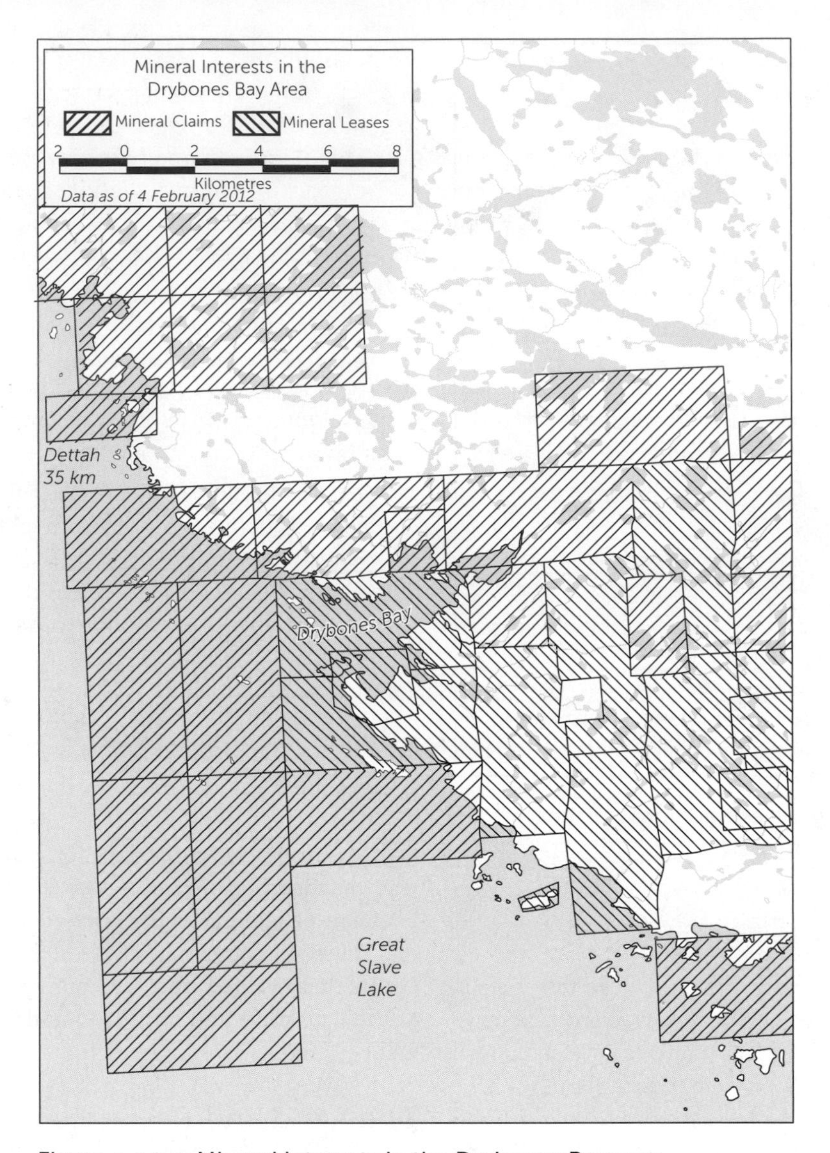

Figure 4.1.3 • Mineral interests in the Drybones Bay area.

established institutions of public government, including regional land-use planning boards, land and water boards responsible for land-use permitting, and an Environmental Impact Review Board for the Mackenzie Valley. The MVRMA is a unique **co-management** regime that was negotiated as a component of comprehensive **land claim agreements** in the NWT. The federal, territorial, and Aboriginal governments nominate individuals to sit on regional land-use planning and land and water boards and the Mackenzie Valley Environmental Impact Review Board.

In the region where Drybones Bay is located, comprehensive land claim agreements are still being negotiated between Canada, the Government of the NWT, and Aboriginal groups. As a result, the integrated system as envisioned through the MVRMA remains incomplete. There is no land-use planning board and no work underway to develop a land-use plan in the near future. There is nothing in place to guide decisions about appropriate land use. As part of the land claim negotiation process, in 2007 an Interim Land Withdrawal was negotiated between the Crown and

Akaitcho, the regional Aboriginal group to which the Yellowknives Dene belong. The purpose of the Interim Land Withdrawal was to ensure no further mineral rights were issued until the negotiations were complete; however, it had a number of limitations. It applied only to certain areas of the YKDFN's traditional territory and had no effect on existing mineral rights that had already been granted, including those in the Drybones Bay area.

Adding to the dynamic, pivotal Supreme Court decisions about Aboriginal rights and **Crown consultation** radically changed government's requirement to consider potential impacts on Aboriginal and treaty rights in decisions about resource development.

Chronology of Development Proposals in Drybones Bay

First Wave of Development Proposals and Environmental Assessments

In 2003, developers applied to the Mackenzie Valley Land and Water Board for land-use permits to conduct four separate mineral exploration drilling projects across the Drybones Bay area. The Mackenzie Valley Land and Water Board (MVLWB) conducted **preliminary screenings** of the project applications and concluded that they might cause significant public concern and referred to the projects to the Mackenzie Valley Environmental Impact Review Board (the Review Board) to conduct an environmental impact assessment (EIA).

The MVRMA sets out a tiered system of EIA. The first tier is an environmental assessment (EA) that considers "if" a project is likely to have significant adverse impacts on the environment or be the cause of significant public concern. At the end of the EA, the Review Board can: (1) recommend that the project move to permitting, with or without legally binding recommendations attached; (2) order a more in-depth environmental impact review; or (3) when the development is likely to cause an adverse impact on the environment so significant that it cannot be justified, recommend that the proposal be rejected without an environmental impact review. The Review Board's report of EA is distributed to the responsible ministers, who have the options of accepting the report, referring the project for environmental impact review, referring the report back to the board for further considerations, or consulting with the board to adopt recommendations with proposed modifications. If the report is accepted, any recommendations become legally binding.

The Review Board conducted assessments of the proposed projects. The following quote from the report of environmental assessment sums up the central issue:

> The Review Board finds that cultural impacts are being caused by the increasing number of developments, including the proposed project, in this important area, and that these cultural impacts are at a critical threshold. Unless certain actions are taken, this would result in a diminished cultural value of this particular area, which would be an unacceptable cumulative impact on Aboriginal land users (MVEIRB 2004d).

The Review Board recommended the approval of three of the four projects, but the reports of EA included recommendations intended to mitigate specific potential impacts of the project. These recommendations included drill site surveys, selection of access routes, and archaeological buffers. The reports also included a number of non–legally binding suggestions intended to address broader planning issues beyond project-specific impacts (MVEIRB 2004a; 2004b; 2004d). They included:

- development of a model for Crown consultation;
- development of an approach to provide Aboriginal communities an opportunity to provide input into staking areas;
- a thorough archaeological, burial, and cultural site survey of the area;
- no new land-use permits to be issued for proposed developments in the area until a plan has been developed to identify the vision, objectives, and management goals based on the resource and cultural values for the area (MVEIRB, 2004d).

The Review Board recommended that one of the four projects be rejected outright (MVEIRB 2004c). The federal minister approved the reports of EA, including the recommendation to reject one of the projects.

Second Wave of Development Proposals and Reviews

In 2005, two new proposed development projects in the Drybones Bay area were referred to the Review Board for EA. The board's findings were consistent with those of earlier EAs for proposed projects in the area in that recommendations were made to address potential project-specific impacts. In addition, noting that responsible agencies had done nothing to implement suggestions from earlier reports, this time the Review Board recommended that "A Plan of Action must be produced by government working with Aboriginal groups to make clear recommendations about future developments, considering the cultural values of Aboriginal peoples" (2007, v). Additional recommendations included the development of a monitoring program and archaeological and heritage studies. The reports of EA from the 2005 projects were sent to the federal minister, where they sat until 2010.

Further Consideration of the Environmental Assessment Measures

In April 2010, almost four years after the reports of EA had been sent to the minister's office, the minister invoked the provision in the MVRMA that allows for "further consideration." The report of EA for Consolidated Goldwin Ventures (CGV) was perhaps the most contentious, because it contained measures that, if approved, would require the federal government to engage in a planning process for the Drybones Bay area. From the perspective of the government, there are a number of reasons that the idea of entering into a planning process with the YKDFN should not be supported. Foremost among these reasons is that land-use planning in the NWT is typically done pursuant to a *settled* land claim agreement. While there is a land-use planning process underway in the Dehcho region of the NWT pursuant to the Interim Measures Agreement (IMA) of the land claim process, this approach has not been used in the Akaitcho region. During the judicial review, the Crown expressed its opposition to a land-use plan, calling it a discretionary public policy response and noting that it might delay development for years.

The minister asked the board to set aside four of its recommendations: three were associated with planning, impact monitoring, and a heritage resource inventory. The minister wrote:

> The proposed "Plan of Action" and long-term monitoring program are considered excessive for a proposed small-scale exploration project" and... the Responsible Ministers do not consider the proposed development to be of a scale that warrants a "thorough heritage resources assessment of the Shoreline Zone" (Strahl 2010).

Figure 4.1.4 • Yellowknives Dene gravesites in the Drybones Bay Area.

The ball was back in the Review Board's court. Between the time the report was issued, then returned for further consideration, a number of changes had been made to board membership. This had important procedural fairness implications because of the administrative law doctrine "*who hears must decide*." The procedural fairness issue was resolved when the parties collectively agreed that a new hearing focusing on the reconsideration was an appropriate step forward. This hearing was held in September 2011.

Figure 4.1.5 • The Yellowknives Dene present at the 2011 CGV hearing. From left to right: Councillor Peter D. Sangris, Chief Ted Tsetta, Fred Sangris, Greg Empson, Todd Slack, Chief Edward Sangris, Councillor Alfred Baillargeon.

Source: Mackenzie Valley Environmental Impact Review Board.

Figure 4.1.6 • Respected Yellowknives Dene Elder Michel Paper listens at one of the hearings. Michel Paper passed away in the fall of 2013.

Source: Mackenzie Valley Environmental Impact Review Board.

On 16 November 2011, the Review Board issued a new decision in the CGV environmental assessment. The decision restated the key measures from the earlier report of EA, particularly relating to environmental monitoring and planning.

The Debogorski Diamond Exploration Project

Around the same time that the Consolidated Goldwin report of EA was returned to the board for further consideration, local prospector, long-time Yellowknife resident, and "ice road trucker" Alex Debogorski applied for a permit to work his mineral claim in the Drybones Bay area.

Mr Debogorski proposed a diamond exploration project that included up to 10 drill holes over a five-year period within the claim. The developer initially proposed to drill two holes on or directly beside already-disturbed areas. The location of the remaining eight sites was unknown and would depend on the results of the initial two.

Consistent with previous preliminary screenings, the project was referred to EA based on the Review Board's statement that the "cumulative cultural impacts [in the Drybones and Wool Bay areas] are at a critical threshold" and that there was "significant public concern regarding the integrity of the cultural and spiritual values associated with the Drybones Bay area with continued development" (MVEIRB 2012, 4).

At the public hearing for the environmental assessment, Alex Debogorski explained the circumstances that had led to his application for a land-use permit and the eventual environmental assessment. He said:

> I had a couple of Section 81s, which means that because of mitigating circumstances, I can't do work on the claim, that we could set the work off for a couple of years. This [EA] process basically started August of 2010 when INAC [Indian and Northern Affairs Canada] gave me a Section 81, but at the same time told me that I had to make an effort to develop the property (MVEIRB 2012, 28).

Mr Debogorski was referring to a provision in the NWT and Nunavut mining regulations that gives the government discretion to grant "relief" from the regulatory requirement to work a claim when the parties involved are prevented from doing so through circumstances beyond their control. Between 2003 and 2011, the NWT Mining Recorder Office granted 48 orders for "Section 81 relief" involving 89 mineral claims held by 14 individuals or companies. Although it is not known how many were granted directly as a result of the unresolved land-use conflicts in the Drybones Bay area, it appears that Section 81 relief was being used to alleviate the land-use conflict in the area. It is also not known why Mr Debogorski's application for relief was treated differently from the others. However, it did result in conflict between the parties. The Yellowknives Dene mobilized to participate once again in an EA process, bringing forward the same concerns.

On 6 January 2012, the Review Board released their report of EA and the reasons for their decision, indicating that the project could proceed without mitigations. The Review Board found that concern was not significant and that the proposed project would not contribute to cumulative cultural impacts. The report did not directly address any of the landscape-level planning requirements from earlier EAs but relied on the as-yet unapproved measures.

Application for Judicial Review

On 7 February 2012, the Yellowknives Dene applied to the federal court for judicial review. The application focused on two issues: first, that the Crown had failed in its constitutional duty to consult with the YKDFN and that the Crown was aware of the importance of the area and the existing level of impacts but chose to rely on the Review Board's process rather than launch a separate process of deep consultation and accommodation; second, that the Review Board's decision was unreasonable and failed to live up to the guiding principles in the MVRMA.

The Yellowknives Dene sought relief in three ways: first, to have the decision quashed and set aside; second, a declaration that the Crown's duty to consult and accommodate had not been met; and third, that a land-use planning process represented a necessary component toward meeting this duty. The Yellowknives Dene also asked the court to issue an order prohibiting the issuance of further licences and permits in the area.

On 4 November 2013, the federal court dismissed the Yellowknives Dene request. On the issue of the Yellowknives argument that the board's conclusion that "the proposed project is not likely to significantly contribute to the previously identified cumulative adverse impacts on land use and culture," the court found the decision reasonable, noting that the board is entitled to considerable deference in making decisions fully within its mandated expertise.

On 12 December 2013, the Yellowknives Dene filed an appeal with the Federal Court of Appeals, and the land-use conflict in the region continues.

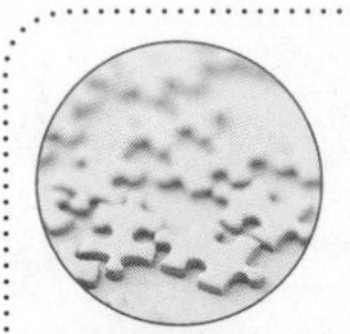

KEY ISSUE

Aboriginal Rights and Title

Contemporary land-use conflicts between the Crown and Aboriginal people can be traced to the Crown's assertion of sovereignty in the face of prior Aboriginal occupation. According to international law at the time that Canada was settled, sovereignty could be established in an inhabited land through conquest or treaty. Across much of Canada, historical treaties were signed to pave the way for settlement and resource development. Ancestors of the Yellowknives Dene signed an adhesion to Treaty 8 in 1900 (Fumeleau 2004). The conflicting interpretation of historical treaties is an ongoing issue today. Put very simply, the conflicting perspectives are as follows: the Crown's view is that whatever Aboriginal rights may have existed were extinguished and replaced by treaty rights; Aboriginal groups argue that the treaties were about peace and friendship, that lands were never ceded, and Aboriginal rights, including title, prevail today. Aboriginal groups across Canada have successfully argued through the courts to have their Aboriginal rights recognized (*Delgamuukw v. British Columbia* 1997 and *Paulette et al. v. The Queen* 1977). These decisions have led to the negotiations of land claim agreements, also referred to as modern treaties.

Consideration of Public Concern in the Debogorski EAs

Rational and rational comprehensive theories of planning place the public interest as the central objective guiding planning processes and decisions (Altshuler 1973). In this case, the Review Board had a legislated responsibility to make a determination about whether the proposed development would cause significant public concern, which is an important component of the public interest but must also be considered within the specific legislative context of the MVRMA. This determination of public concern turned on the question of whether, for the Review Board, the concerns raised primarily by the Yellowknives Dene were significant.

During the Debogorski environmental assessment, the Yellowknives Dene expressed concern that active participation in seven EAs dealing with development in the Drybones Bay area, over a period of eight years, had led to no concrete action to protect the values in the area. Chief Tsetta said,

> Within the present regulatory environment, it is not possible for the Board to create the mechanisms to institute appropriate mitigations or accommodations for future operations, and this project must be rejected (MVEIRB 2012, 35).

Evidence was presented in writing and orally at the hearings, with a sizable percentage of the adult Yellowknives Dene population speaking at hearings. A letter supporting the Yellowknives Dene and echoing their concern was submitted to the record by the Dene Nation and other regional First Nations groups. Some concern was also raised by non Aboriginal groups and individuals during the first wave of development proposals.

The Review Board considered:

1. concern with development proceeding in a sensitive area without a land-management framework or land-use plan in place;
2. concern with the federal agencies' lack of action to approve and implement other Review Board recommendations intended to mitigate cumulative cultural impacts in the Drybones Bay area;
3. concern with competing interest in land uses in the Drybones Bay area;
4. concern that adequate Crown consultation and accommodation had not occurred for the proposed project.

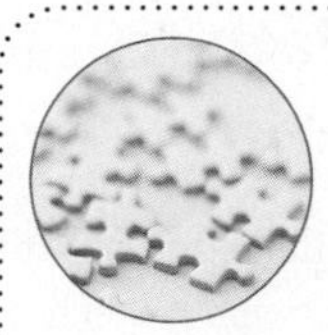

KEY ISSUE

How Values Influence Decisions of Significance

Much has been written about the complexity of the notion of a single public interest and to what degree the diversity of political and social *values* should be explicitly front and centre in decision-making (Davidoff 1973). There is an unavoidable value-based judgment inherent in decisions about whether concern is significant. What is significant to one may not be to another. How values are brought to bear on decisions in an EA process are in part through institutional design, through the nomination of individuals to the board to act on behalf of Canadians and, in particular, for the residents of an area like the Mackenzie Valley. Once appointed, members make decisions in the public interest; they do not represent the group that appointed them. However, the fact that they bring their experiences, world views, and values to bear on decisions, including decisions about the significance of public concern, is an important part of the institutional design negotiated through the land claim agreement. From YKDFN's perspective, it is important that they have not been in a position to nominate anyone to the Review Board because the land claim agreement has not been concluded.

The board concluded that "the underlying concern over ongoing competing land uses within the Drybones Bay area is regional in nature, and that the Debogorski project will not contribute to them significantly," and that "the effects of the Debogorski project [would] not be material and that they do not provide a basis for significant public concern." The Review Board's conclusions focused on the specific physical impacts of the proposed project, was silent on previous board findings that cumulative cultural impacts were at a critical threshold, and positioned the issue of competing interests and issues to a regional issue, not relevant to the specific project review decision.

How the concerns of one group are considered and weighed against the broader public interest is a complex one; it becomes more challenging when Aboriginal groups raise concerns that constitutionally protected rights are being infringed upon. In such cases, like the proposed Debogorski project, Aboriginal groups can turn to the courts to request a review of the decision. However, neither the EA process nor the judicial review process explicitly deal with issues of value.

Land Use Planning as a Path to Reconciliation

The Yellowknives Dene used the tools available to them to ensure that their rights and interests in land were protected: through active participation in seven project-specific EAs, appeals for consultation and accommodation, and ultimately court action. To date, these processes have been ineffective in addressing the underlying conflicts, are costly to all involved, and, perhaps most important, have resulted in a breakdown of relationships and trust. The Yellowknives Dene have strongly advocated for the opportunity to engage in a land-use planning process for the Drybones Bay area.

Within natural resource management and planning literature, land-use planning is presented as a potential tool for resolving complex land issues, including issues of reconciliation between Aboriginal people and the state (Stevenson and Natcher 2010; Berger, Kennett, and King 2010; Kennett 2010). On the one hand, state-led planning processes are criticized for falling short of this ideal "and further entrenching the state's authority and control over Aboriginal people, lands and resources" (Stevenson and Natcher 2010, 4). The failure has been attributed to a general lack of capacity to engage in state-led planning processes and the tendency for the rational comprehensive paradigm to marginalize the cultural perspectives of Aboriginal groups (King 2010). Others argue that the opportunity to participate in land-use planning is viewed as a significant social and political achievement (Lane 2006). The Yellowknives Dene's preferred approach aligns with the latter perspective. In the absence of a land-use planning process and seeking a genuine role in decision-making, the Yellowknives Dene continue to leverage other processes to advance their interests and protect their rights.

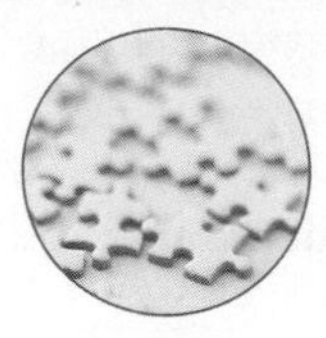

KEY ISSUE

Reconciliation

The fundamental principle behind the negotiation of land claim agreements and the Crown's duty to consult is the ongoing process of reconciliation. In the *Mikisew* decision about consultation over a land-use issue, Justice Binnie stated, "The fundamental objective of the modern law of aboriginal and treaty rights is the reconciliation of aboriginal peoples and non-aboriginal peoples and their respective claims, interests and ambitions" (para 1). Put simply, reconciliation seeks to find a way for the original inhabitants and the new inhabitants to make decisions that ideally provide for both groups to continue their way of life and respective pursuits. It is decisions over land use where these two conflicting interests intersect most intensely; it is also where the opportunities for reconciliation are greatest.

KEY TERMS

Aboriginal rights
Co-management
Crown consultation
Cultural landscapes
Environmental impact assessment
Free entry system
Land claim agreements
Preliminary screening

REFERENCES

Altshuler, A. 1973. "The goals of comprehensive planning." In A. Faludi, ed., *A Reader in Planning Theory*, 193–210. Oxford: Pergamon Press.

Berger, T., S. Kennett, and H. King. 2010. *Canada's North: What's the Plan?* Ottawa: Conference Board of Canada.

Davidoff, P. 1973. "Advocacy and pluralism in planning." In A. Faludi, ed., *A Reader in Planning Theory*, 277–96. Oxford: Pergamon Press.

Delgamuukw v. British Columbia [1997] 3 SCR 1010.

Fumeleau, R. 2004. *As Long as This Land Shall Last: A History of Treaty 8 and Treaty 11, 1870–1939*. Calgary: University of Calgary Press.

Kennett, S. 2010. "Integrated landscape management in Canada: Getting from here to there." Canadian Institute of Resource Law Occasional Paper #17. http://dspace.ucalgary.ca/bitstream/1880/47192/1/OP17Landscape.pdf.

King, H. 2010. "Give it up: Land and resource management in the Canadian North: Illusions of Indigenous power and inclusion." In T. Berger, S. Kennett, and H. King, *Canada's North: What's the Plan?* 75–107. Ottawa: Conference Board of Canada.

Lane, M. 2006. "The role of planning in achieving Indigenous land, justice and community goals." *Land Use Policy* 23: 385–94.

MVEIRB (Mackenzie Valley Environmental Impact Review Board). 2004a. "Report of environmental assessment and reasons for decision on the Consolidated Goldwin Ventures preliminary diamond exploration in Drybones Bay." http://www.reviewboard.ca/upload/project_document/EA03_002/Report_of_EA/EA03-002_REA_040211_CGV.pdf.

———. 2004b. "Report of environmental assessment and reasons for decision on the New Shoshoni Ventures preliminary diamond exploration in Drybones Bay." http://www.reviewboard.ca/upload/project_document/EA03-004_REA_040211_NSV_1144786911.pdf.

———. 2004c. "Report of environmental assessment and reasons for decision on the North American General Resources Corporation preliminary diamond exploration in Wool Bay." http://www.reviewboard.ca/upload/project_document/EA03_003/Report_of_EA/EA03-003_REA_040211_NAGR.pdf.

———. 2004d. "Report of environmental assessment and reasons for decision on the Snowfield Development Corporation's diamond exploration program." http://www.reviewboard.ca/upload/project_document/EA03-006_Report_of_Environmental_Assessment_and_Reasons_for_Decision_1305582664.pdf.

———. 2007. "Report of environmental assessment on Consolidated Goldwin Ventures mineral exploration program." http://www.reviewboard.ca/upload/project_document/EA0506-005_CGV_Report_of_Environmental_Assessment_1305583396.pdf.

———. 2012. "Report of environmental assessment and reasons for decision in EA1112-001: Debogorski Diamond exploration project, Drybones Bay." http://www.reviewboard.ca/upload/project_document/EA1112-001_Report_of_Environmental_Assessment_and_Reasons_for_Decision__1325887535.pdf.

Parks Canada. 1994. *Guiding Principles and Operational Policies*. Ottawa: Department of Canadian Heritage.

Stevenson, G.M., and D.C. Natcher, eds. 2010. *Planning Co-existence: Aboriginal Issues in Forest and Land Use Planning*. Edmonton: CCI Press.

Strahl, Chuck. 2010. Letter to the Mackenzie Valley Review Board. 13 April. http://www.reviewboard.ca/upload/project_document/EA0506-005_Letter_from_Minister_Strahl_1296500229.pdf.

Yellowknives Dene First Nation. 2011. Letter to the Mackenzie Valley Review Board. 1 September.

Yellowknives Dene First Nation v. The Minister of Aboriginal Affairs and Northern Development Canada [2013] Federal Court of Canada.

CASE STUDY 4.2

Climate Adaptation Planning in British Columbia: The Elkford Approach

Timothy Shah

SUMMARY

The following case study explores how the planning profession is acting on climate change adaptation in the province of British Columbia. The case study specifically examines how the District of Elkford, a remote community in eastern BC, is responding to climate change by using its **Official Community Plan (OCP)** to guide strategy and planning decisions. Climate change is an emerging phenomenon plaguing planning departments province-wide. While planners are uniquely positioned to address climate change adaptation, they face several challenges, since it is hard to predict, respond to, and mitigate, given competing planning priorities, constrained financial resources, or limited staff time (Hallegatte 2009; Füssel 2007; Picketts et al. 2012).

The District of Elkford successfully manoeuvred through a planning process to identify both short-term and long-term climate impacts and risks. A one-year planning engagement process resulted in a strategy that outlined several adaptation actions that could be implemented by the District over a 20-year period. The case study explores how the District of Elkford navigated this planning process and outlines major take-away lessons for planning practice. The overall purpose of this case study is to argue why and how the planning profession must have a more proactive role in responding to the climate change challenge by embracing adaptation efforts.

OUTLINE

- This case illustrates that communicating the risks associated with climate change is of utmost importance when conducting an adaptation planning process in smaller communities.
- Developing a climate change adaptation planning strategy while simultaneously revising an Official Community Plan can be highly effective, because it allows both documents to be fully integrated.
- The District of Elkford is a rural community in the eastern part of the province and the focus of this case study. It was the first community in BC to incorporate a Climate Change Adaptation Strategy into its Official Community Plan.

Introduction

Over the past 30 years, the world has seen a drastic increase in the number of extreme weather events—from temperature fluctuations, wildfires, and droughts to floods and intense storms; the overall loss trend from these extreme events is beginning to exceed $150 billion per year (Chu and Majumdar 2012). In North America, one of the most destructive extreme weather events in recent memory was Hurricane Sandy, which devastated New York City in October 2012 (Klinenberg 2013). In Canada, extreme weather events have been manifested as heatwaves in cities such as Toronto, severe flooding in the Prairies such as the devastating flood in Calgary in June 2013, and landslides in western Canada (Pengelly et al. 2007; CBC 2013; Owen 2012; Pike et al. 2010). According to Jakob and Church (2011), Canadians spent $2 billion on damages from flooding in the twentieth century.

The province of British Columbia has seen a rise in the number of extreme weather events, including the Lower Mainland heatwave of July 2009 and the Johnson's Landing landslide in 2012 (Procter 2010; CBC 2012). According to the BC Ministry of the Environment, coastal communities in the province already face flood risks related to precipitation and river flows. Climate change is projected to present new risks such as sea-level rise and storm surges. Further, approximately 3000 to 12,000 BC homes near the coast could be "at risk" of flooding by 2050 (British Columbia 2013b).

With the frequency of these events, policy-makers and academics have turned their attention to the need for climate change adaptation in coping with the impacts. Internationally, the 2011 Climate Change Conference in Durban, South Africa, produced the Green Climate Fund—a commitment from governments to proceed with plans to finance global adaptation efforts (Jones, Hole, and Zavaleta 2012). This was the first fund of its kind to focus on climate adaptation.

Much of the funding and discussion in the policy community, hitherto, has focused on climate change mitigation—actions that limit or reduce the magnitude of long-term climate change. To date, British Columbia has been proactive with climate change mitigation—namely, through implementation of the Climate Action Charter (2007) and the Carbon Tax Act (2008) (British Columbia 2012; 2013a). Climate change adaptation has been less studied, although it is starting to gain some salience in the planning community, e.g., the Planning Institute of British Columbia (PIBC) conference in 2008 was titled "Planning for Change," focusing on the role planners have in mitigating and adapting to climate change. This is true not only in the BC context but in the larger context of academic research on climate change (Berrang-Ford, Ford, and Paterson 2011).

This chapter is structured as follows: First, a short discussion is offered on the academic and professional connection between planning and climate change adaptation. Next, the District of Elkford is introduced as a case study. Elkford is reviewed with discussion and analysis of how

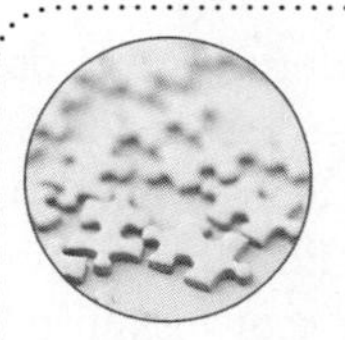

KEY ISSUE

Planning and Climate Change

The planning profession has an important role to play in helping communities adapt to climate change. Because planners can help influence policy decisions—namely, how land-use decisions are made—their role in climate adaptation planning will only become more critical amid more uncertain climate change. Climate change adaptation is a local and regional context-dependent phenomenon, unlike mitigation, which might require more international cooperation and consensus on which measures to take and how (Picketts et al. 2012).

Planners work at the local and regional scales, exploring issues over a long term. They work under uncertainty, evaluate trends, and make changes to policy accordingly. Climate change is one such issue, embedded with high uncertainty (see Sandford 2011) and with variable impacts that will occur over the long term, making it an appropriate subject for the planning profession to address.

adaptation planning occurred, including the governance framework, the key stakeholders who participated, how climate risks were communicated to the public, and Elkford's use of a risk management framework that led to the climate change adaptation strategy.

The author conducted an extensive literature review on the subject of climate change adaptation, reviewing academic articles, policy reports and briefing notes, and technical documents. Four semi-structured interviews were conducted with participants involved in Elkford's climate change adaptation process. The author then built a prescriptive framework of climate change adaptation planning and used this framework to assess the Elkford case study.

Climate Change Adaptation and Planning

One of the most central concepts underlying climate adaptation is **adaptive capacity**. Moser et al. (2008, 645) describe this as the "ability of a system to anticipate and adapt to the potential or experienced impacts of climate change." The capacity is determined by how well the system's institutions and networks learn from each other, interact, and store new information to cope with new threats.

While planners face a number of constraints that could limit a community's adaptive capacity, adaptive capacity is much higher when the public and local stakeholders understand *why* adaptation is a concern and how they can be engaged in the process. Continuous community engagement can encourage future buy-in and support for implementation and promotes a transparent process (Picketts et al. 2012).

Climatic uncertainty is another concept in climate change adaptation planning. Planners have long embraced the notion of uncertainty in their practices; issues are not always well understood at the beginning, and therefore acknowledging uncertainty can allow for more flexibility in decision-making (Lempert and Schlesinger 2000), a greater menu of choices in adaptation strategies, and a higher safety margin.

For example, water managers in Copenhagen account for uncertainty in how they calibrate local drainage infrastructure by using runoff figures that are 70 per cent greater than their current level. The 70 per cent figure is not a precise calibration, because current climate models cannot provide accurate forecasts of climatic change in the future (Hallegatte 2009). However, the increased rate after accounting for uncertainty is designed to enable the city to withstand heavier rains in the future should they arise from a changing climate. Designing a drainage system meant to cope with increased precipitation is relatively inexpensive compared to the costs of modifying the infrastructure after it has been constructed (Hallegatte 2009).

Planners recognize the importance of such contingency planning, working with the information they currently have to strive toward policy changes that improve societal well-being. It is clear that climate is changing, but it is unclear how quickly and to what extent (Silva, Dijkman, and Loucks 2004).

A final concept in climate change adaptation that pertains to planning is **risk communication**. The planning profession was born out of work relating to public health in response to epidemics such as cholera that posed several risks to the masses and killed millions. Planners have adopted a number of tools, such as mapping, to help the public better understand the impacts of a health, environmental, or social issue through visual illustration. Planners work in multidisciplinary capacities and environments

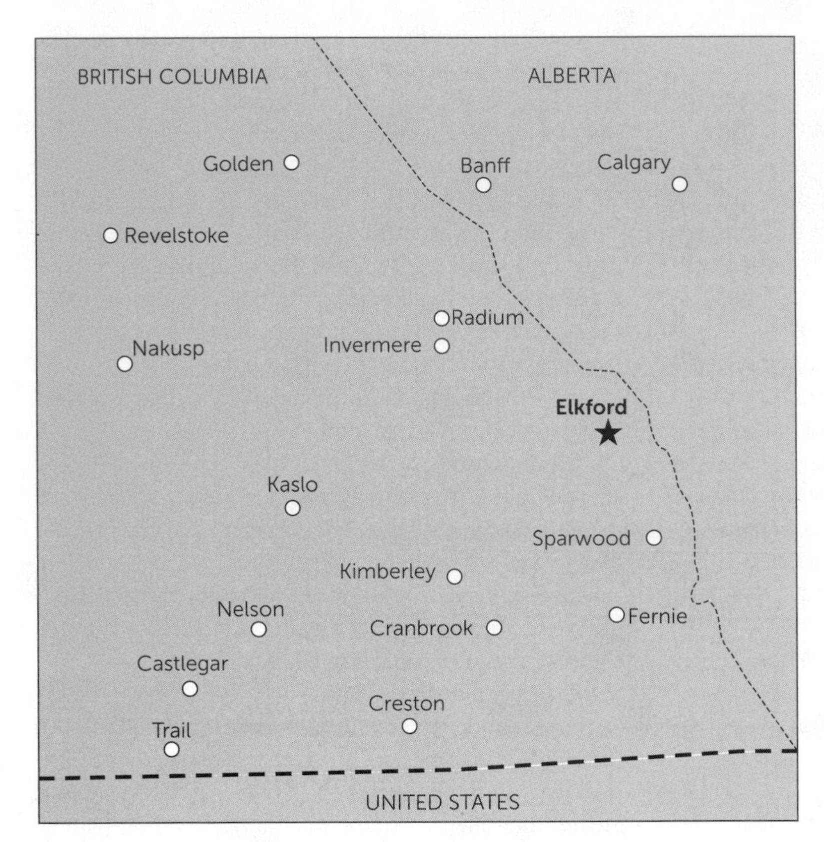

Figure 4.2.1 • Location map.

and are trained to take information from other experts (such as climatologists) and communicate in ways that are accessible for citizens.

Adaptive capacity, climatic uncertainty, and risk communication are all central concepts in planning practice for climate change. This case study illustrates how, and makes the case for why, planners can use these concepts to strengthen their communities in preparing for and adapting to the unknown impacts of climate change.

Context of the Case Study: The District of Elkford, British Columbia

The District of Elkford is located in southeastern British Columbia in the Upper Elk Valley between Boivin Creek and the Elk River. According to the 2011 Canadian census, Elkford had a population of 2523 residents (BC Stats). More than 50 per cent of the employment in the District is in the resource sector (District of Elkford 2010). The District has seen changing climate conditions since the 1980s: between 1986 and 2000, there was an 8.6 per cent loss of glacial cover in the Elk River drainage basin and a 16 per cent loss in the Columbia Basin as a whole (Gorecki, Walsh, and Zukiwsky 2010).

Projections of future climate impacts point to an increase in average temperatures and greater amounts of precipitation. Based on a global climate model (GCM) from the **Pacific Climate Impacts Consortium (PCIC)**, Elkford and surrounding area will see an increase from about 2.82 mm of precipitation per year to about 3.31 mm per day by 2050 (UNFCCC 2011). Historically, flooding, droughts, and wildfires have been the District's major natural hazards and are projected to be exacerbated by climate change (Gorecki, Walsh, and Zukiwsky 2010). As shown in Figure 4.2.2, the mean average temperature for Elkford and surrounding area is projected to be 2°C to 3°C warmer by the 2050s. Given the District's history of climate hazards and projected exacerbation risks, there was an impetus to take action (Shah 2012).

In the spring of 2008, the combination of increasing interest in climate change and growing leadership at the District level led to a year-long initiative that engaged residents in a planning process on adapting to local climatic change (Gorecki, Walsh, and Zukiwsky 2010). The District of Elkford was selected as one of two communities to partner with the **Columbia Basin Trust** to participate in this initiative. Its purpose was to help communities identify the range of potential climate impacts to be expected; assess local risks, vulnerabilities, and sensitivities; and develop adaptation actions to respond to these climate impacts (Welk 2010). When the process was first introduced to the community, some members of the public expressed scepticism about climate change; debates ensued about whether climate change was human-induced or a natural phenomenon.

As the process evolved, the conversation moved away from debating about the "causes" of climate change to a discussion that explored the "impacts" that residents were noticing in their community (Shah 2012). Once the community began to see the projected localized impacts of climate change, there was greater buy-in and a willingness

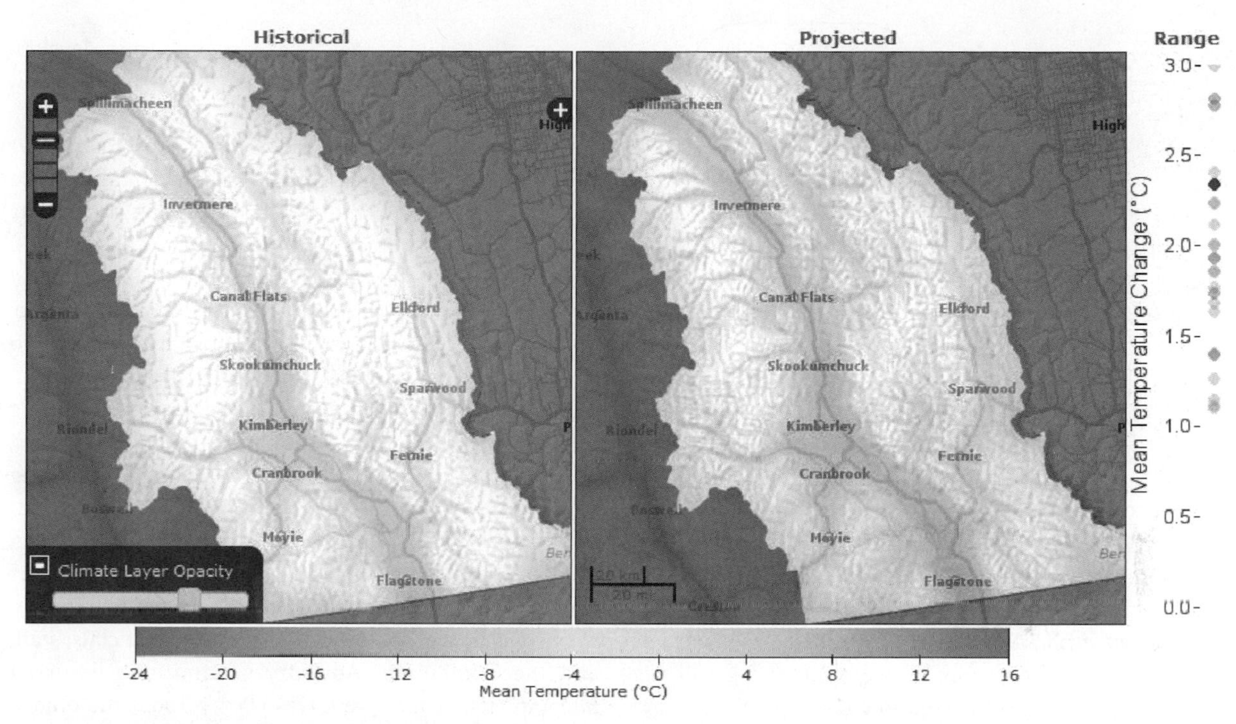

Figure 4.2.2 • Maps of Elkford's historical (1961–1990) and projected (2050) temperature and bar plot of range of projected changes.

Source: Pacific Climate Impacts Consortium.

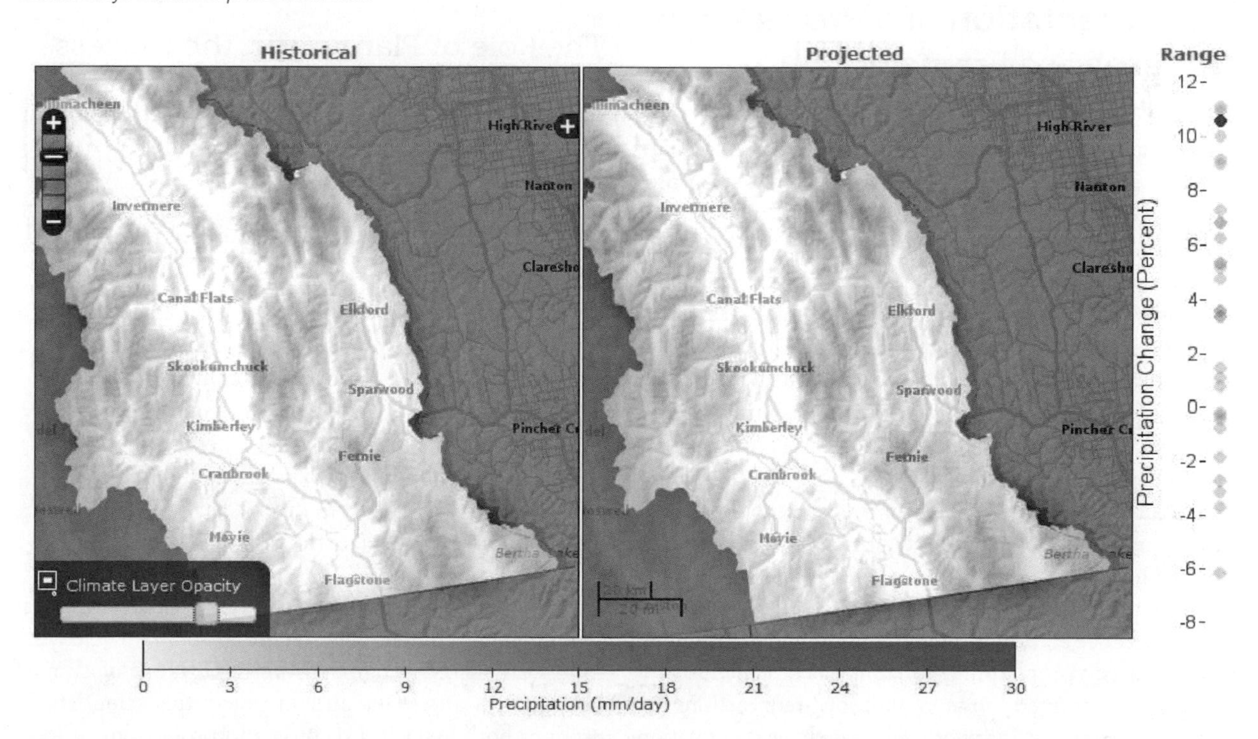

Figure 4.2.3 • Maps of Elkford's historical (1961–1990) and projected (2050) precipitation and bar plot of range of projected changes.

Source: Pacific Climate Impacts Consortium.

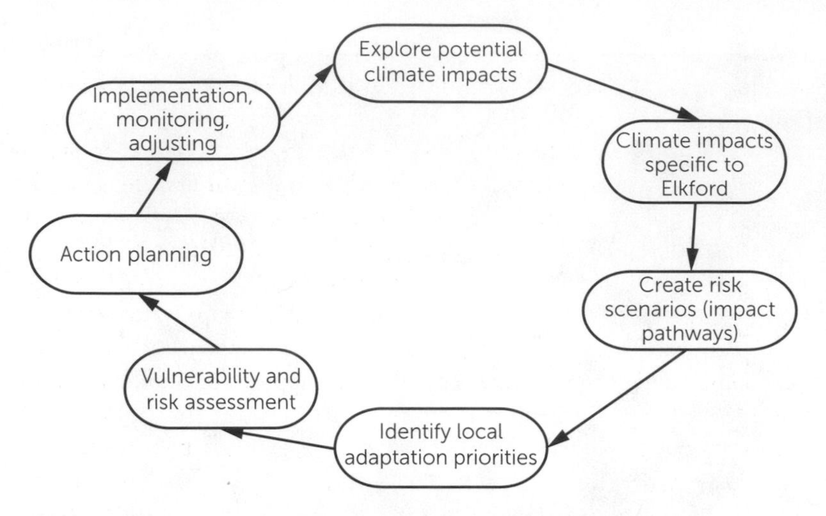

Figure 4.2.4 • Seven-step model: Elkford adaptation planning process.

to act. At the same time, the community was beginning to express concerns about the frequency of climate impacts in the region. In particular, it was experiencing a higher frequency of floods and wildfires (Gorecki, Walsh, and Zukiwsky 2010).

Implementation of a Process and Governance Framework

As a response to concerns surrounding future climate impacts, Elkford engaged in an extensive planning process that was designed to better inform the community about local climate risks and how to adopt actions to address them. The adaptation planning process in Elkford was guided by a local planning policy framework called the Official Community Plan (OCP). The District was in the midst of revising its OCP and intended to use the results of the climate change adaptation planning process to create a Climate Adaptation Strategy as an integral part of its new OCP.

According to primary research with stakeholders (Shah 2012), there was more credibility about and acceptance of the adaptation planning process once it was made clear how the adaptation strategy and OCP would be fully integrated. Because an OCP is a long-range planning document, typically a 20-year planning horizon, it is designed to outline immediate, medium-term, and long-term actions for implementation. Furthermore, climate change adaptation planning can be most effective when it is co-ordinated with local land-use objectives or fed into an already existing management activity and development plan (Füssel 2007; McDaniels, Longstaff, and Dowlatabadi 2006). Picketts and Curry (2011) found that planners in BC prefer adaptation to be incorporated into existing documents rather than separate standalone plans. In 2008, Elkford had the foresight to use its OCP as its planning policy framework to justify and give credence to adaptation actions.

The climate adaptation planning process was conducted in seven steps, led by a local planning consulting firm, Zumundo Consultants. The consultants worked in collaboration with another firm that specialized in Smart Growth strategy and community engagement. Together, the two firms worked closely with District officials, staff, and the Community Advisory Committee (discussed later) to ensure that there was clarity about the objectives of the process and how an adaptation strategy could be created that complemented the District's vision.

The Role of Planners in the Process

The stakeholders in the climate change adaptation planning process included but were not limited to:

- planning consultants (Zumundo Consulting);
- scientists (University of British Columbia and PCIC);
- Community Advisory Committee (representing several facets of the community, including seniors, youth, the chamber of commerce, the local forest industry, and First Nations, as well as a developer, a farmer, and a teacher);
- members of the District staff, including the chief administrative officer (CAO);
- members of the District Council.

The planning consultants worked with several stakeholders and professionals, including climate scientists from PCIC. They recognized the importance of climate information and data and engaged the scientists to produce both historical records and projections of local climate changes to the year 2050. The climate impacts predicted included the potential for more intense wildfires, droughts, and flooding. Once these impacts were established, the planning consultants worked with the District to create **impact pathways** to show risk scenarios.

As shown in Figure 4.2.5, impact pathways illustrated the various climate impacts on the community and ecosystems. The planning consultants used these models as risk communication tools to help build literacy around climate impacts and to foster more social acceptability around the need to take action. The impact pathways were a powerful tool to show risks such as flooding and its potential consequences on the community, including economic disruption, increased risk of property damage, and impacts on public health and safety. The planning consultants had a significant role in creating these impact pathways, which synthesized information on climate impacts in a way that was accessible for the lay public.

Previous research on this topic has shown why it is crucial to involve the local community in adaptation planning to help foster social acceptability. Jacques (2006) and Picketts et al. (2012) explain how community members possess important local knowledge of the unique social, environmental, and economic conditions of the area, which can be a valuable input to the adaptation planning process. In addition, engaging with local stakeholders promotes greater understanding and awareness of climate change and its impacts and encourages future buy-in and support for implementation (Picketts et al. 2012).

The impact pathways exercise is one of the most instructive lessons from the Elkford experience. The planning consultants were able to harness the expertise of other professionals (in this case, climate scientists) and translate their projections into information that was accessible for the lay public. Impact pathways

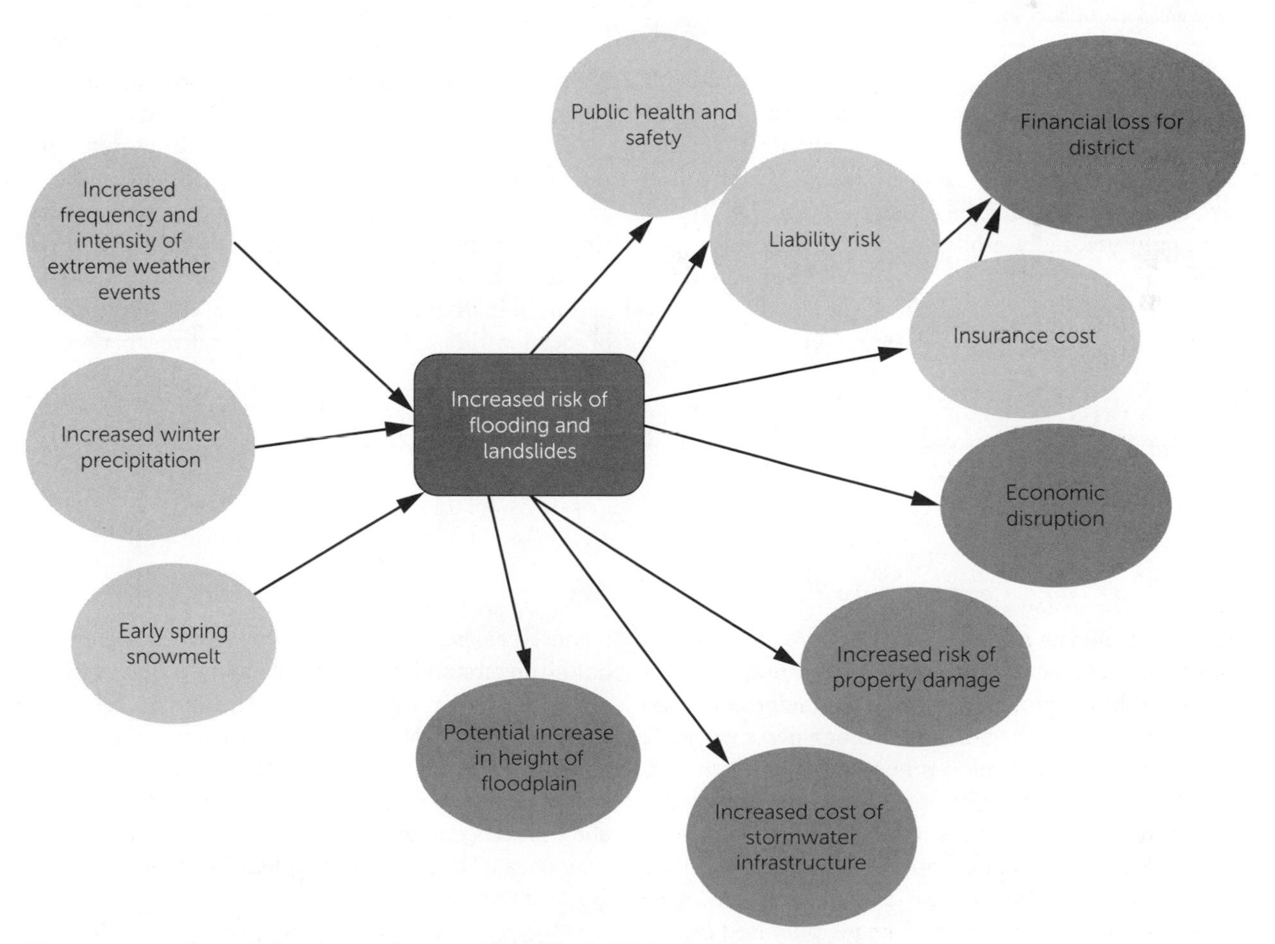

Figure 4.2.5 • Sample impact pathway of flooding in Elkford.

Source: Adapted from Gorecki, Walsh, and Zukiwsky 2010.

Figure 4.2.6 • Going through the impact pathways exercise to develop risk scenarios.

Source: Michelle Laurie.

not only demonstrated how risks like wildfires could compound flooding, they also demonstrated what the "consequences" of these risks would be on the social and economic fabric of the district. For instance, flooding could gradually erode road infrastructure, which would negatively affect the resource sector—the dominant economic sector of the district, employing more than 50 per cent of the labour force in 2006 (District of Elkford 2010). The impact pathways also showed linkages between planning priorities and triggered a conversation about the trade-offs that would have to be made.

Implications of Planning Theory

While the climate scientists and planning consultants were the "experts" in the adaptation planning process—through conveying relevant and local climate risks to the community—they did not fundamentally drive the decision-making process. In addition, while the seven-step process might include some elements similar to those in the **rational comprehensive model (RCM)**—a planning theory premised on the planner defining the problem and determining the scientific methods to solve it—the model does not fully characterize the Elkford experience. The Elkford adaptation planning process draws more on communicative planning theory.

Healey's (2003) work on collaborative planning is also instructive here. This theory sees planning as an iterative process and a governance activity occurring in complex and dynamic institutional environments. Healey (2003, 107) explains that the "rational" approach is ill-equipped to deal with the "multiplicity of social worlds, rationalities, and practices that coexist

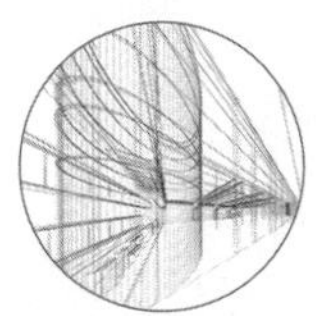

PLANNING THEORY

Communicative Planning

Communicative planning is an iterative planning paradigm that balances power among multiple stakeholders to help inform, develop, and produce a series of alternatives to achieve a planning outcome. Participation is important to this model; communication is used as a tool to help different interests in the process to understand each other. Unlike the rational comprehensive model, which searches for the optimal outcome or the "holy grail," communicative planning is premised on the planner seeking to convey to decision-makers a better understanding of the planning problem by engaging other stakeholders (Hostovsky 2006). Healey (1996, 239) argues that "planners need a new planning paradigm, a communicative conception of reality, to replace that of the self-conscious autonomous subject using principles of logic and scientifically formulated empirical knowledge to guide actions."

in urban contexts and the complexity of the power relations within and between them, resulting in typically dispersed or diffused power contexts." In other words, collaborative planning was the foundation of the Elkford planning process largely because power was distributed among the consultants, District officials, and the Community Advisory Committee (CAC).

The fourth step in the process, "identifying local adaptation priorities," allowed for community engagement and public comment. This step involved several outreach tactics, including surveys and **kitchen table meetings**, to help determine public sentiment around climate change. These outreach methods produced results identifying the top concerns the community had about climate change.

Subsequently, an open house was held, which integrated general information on climate impacts, the specific projected climate impacts on Elkford, and the six planning area impact pathways that the consultants had identified. At this time, a number of community members offered their local observations on potential climate changes and offered input to the adaptation priorities and actions that could be feasible. To ensure that the community's inputs would be valued and considered throughout the planning process, a Community Advisory Committee was formed.

The CAC was a group of people representing diverse interests in the community. Their mandate was to advise council, District staff, and the consultants on the OCP and Climate Change Adaptation Strategy process (District of Elkford 2008). Reviewing drafts, discussing policy options, and organizing meetings, the group was consulted throughout the process to ensure that the community's values were upheld and embedded in the OCP. While the committee was strictly advisory—it had no delegated authority from the District of Elkford—it remains emblematic of the principles of collaborative planning, since the consultants were not permitted to solely drive the process.

Step five of the process, "vulnerability and risk assessment," was a continuation of the community engagement process. The planning team led a one-day workshop with stakeholders, including Elkford staff, council members, and CAC members to determine the vulnerabilities of the priority planning areas (e.g., flooding). The overall purpose of the workshop was to identify the "high-frequency" and "high-consequence" climate-related impacts and the actions that could be undertaken to address these impacts (Gorecki, Walsh, and Zukiwsky 2010). Climate risks were identified, then judged for their sensitivity, adaptive capacity, and vulnerability (for a more detailed explanation of this method, see Gorecki, Walsh, and Zukiwsky 2010).

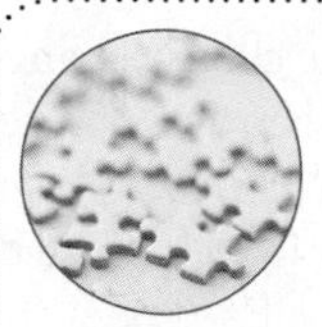

KEY ISSUE

Vulnerability and Risk Assessment

Planners working on climate issues use vulnerability and risk assessments to prioritize planning actions. In the Elkford case study, the assessment included several steps, beginning with an identification of the potential impact, followed by determining the vulnerability of the community to the impact, assigning the probability of occurrence for each impact, and then asking the community to determine what level of risk they were willing to accept, which is called risk tolerance.

Alternatives That Were Considered

After the completion of a comprehensive risk and vulnerability assessment process, the various stakeholders convened to produce a strategy prioritization table. This part of the process, called "action planning" in the seven-step model, enabled the planning team to produce four specific adaptation goals:

1. Elkford is a resilient FireSmart community.
2. Elkford prepares for and mitigates flood risk.
3. Elkford understands the status of water supply and manages the resource effectively.
4. Climate change adaptation is considered in future planning decisions.

As part of good planning, each goal was accompanied by a set of objectives and strategies. As an example, goal two, "Elkford prepares for and mitigates flood risk," included an objective to reduce the vulnerability of infrastructure to flooding. One action to satisfy this objective was the redesignation of the floodplain to incorporate new climate science and impact projections, in part to acknowledge the uncertainty surrounding climate change and to account for new information when it emerged. Through engaging the stakeholders, each of the goals outlined above was finalized and included in the District's OCP, which was published in 2010 (see Welk 2010).

To conclude the process, the stakeholders moved toward step seven, the implementation of the action items. Each action item was ranked by its urgency, including immediate implementation (0–2 years), in the near future (3–10 years), and in the more distant future (more than 10 years).

Lessons for Planning Practice

Climate change adaptation planning is context specific (Hallegatte 2009). What works in one jurisdiction may not work in another. However, what is clear from the Elkford experience is the importance of risk communication, both as a means of engaging the public and other stakeholders and as a means of prioritizing actions for adaptation. Pidgeon and Fischhoff (2011) explain how the usefulness of climate science depends on how effectively the analytical results can be communicated and how relevant they are to decision-makers. The Elkford experience began with the question "What changes are you seeing in your community?" This was complemented with information communicated by scientists about the potential localized impacts of climate change.

The question above builds on previous climate adaptation research in BC. Shepherd, Tansey, and Dowlatabadi (2006) examined a climate change adaptation process in the Okanagan and describe an adaptation process that began with the question "Are climate change impacts significant relative to other societal developments?" and concluded with "Are lessons learned from the evaluation incorporated in or influencing future decisions?"

It is clear from the Elkford experience that climate adaptation was seen as a significant issue of concern relative to other societal issues. Importantly, the projected climatic impacts were conveyed to local stakeholders and to the public. The results from the Elkford case study and a similar adaptation planning process in Prince George, BC, (see Picketts et al. 2012) found that the sharing of historical and projected climate change is indispensable for fostering adaptive capacity and for reaching the implementation stage. Elkford went one step further by using impact pathways to illustrate how climate hazards could impact the economic, social, and environmental fabric of the district. Impacts pathways grounded the issue and offered context.

What is more, Elkford used a Community Advisory Committee made up of 14 people from the community to deliberate, share ideas, and communicate directly with the larger public. In the spirit of collaborative planning, this type of participation is analogous to the "control" step on Sherry Arnstein's ladder of citizen participation (Arnstein 1969). In this case, members of the community who might not have been engaged otherwise had a chance to participate directly in planning and policy development. Their input was directly incorporated into the District's Climate Change Adaptation Strategy.

Other communities could benefit from integrating a climate change adaptation strategy into their respective OCPs or major planning frameworks. However, getting to this step requires a meticulous and comprehensive planning process such as the one employed by Elkford. For planners who work in contexts where conversations about climate change adaptation are beginning to take place, it is recommended that they consider the risk communication model as shown in Figure 4.2.7. As the model illustrates, the use of a risk communication team takes the Elkford approach one step further, recognizing the complexity of climate change—the difficulties with prediction, the uncertainty about costs associated with policy options—and thus the importance of drawing on multiple disciplines and professions to communicate the risks of climate change to a community.

Conclusion

There is consensus that climate change is a real phenomenon and that temperatures will continue to rise in the future (Pengelly et al. 2007). As shown by Chu and Majumdar (2012), these climate impacts are becoming more frequent and resulting in more severe and costly damage. Experts agree that to alleviate the risks of larger and more profound climate impacts, the earth's temperature must not exceed 2°C above pre-industrial levels (Meinshausen et al. 2009). A number of climate mitigation initiatives could help the world to meet this target by significantly reducing greenhouse gas emissions (e.g., Blok et al. 2012). However, historical evidence shows that even with international accords, there are practical difficulties associated with finding a consensus on how to go about lowering emissions and by when.

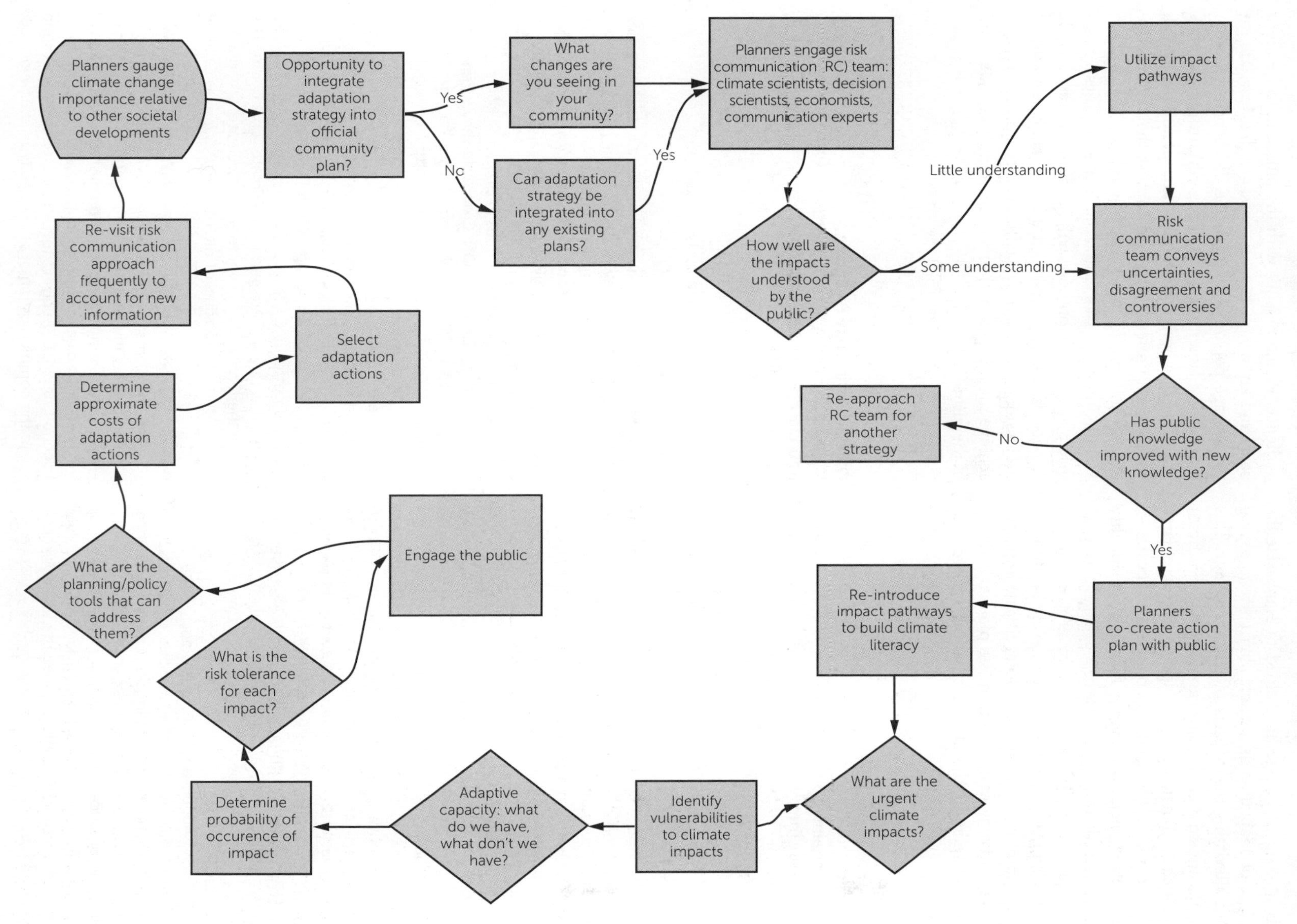

Figure 4.2.7 • Risk communication model.

Source: Timothy Shah.

This inherent challenge has helped to emphasize the importance of adaptation as a viable alternative to coping with and alleviating the impacts of climate change. This chapter has argued that planners can make two significant contributions to climate change adaptation practice. First, planners can use their community's planning policy framework (called an Official Community Plan in BC) to identify the climate hazards, determine how best to engage the public, prioritize the climate hazards/risks in terms of when to respond to them, and use existing and new policy tools to assist the community in adapting to local climate impacts.

The second major contribution is a planner's ability to synthesize relevant climate information for a general audience. Planners work with the public on a frequent basis and know how to craft messages that resonate for the average person who might be ill-informed, unconvinced, and unwilling to act on a certain policy issue. The Elkford experience revealed how planners can work with several professionals and harness their knowledge to help the public better understand a complex issue and the inherent risks embedded within it. These cross-disciplinary collaborations will be imperative as climatic impacts intensify. Planners are in the most appropriate professional position to facilitate these processes and achieve meaningful action for communities today and in the future.

KEY TERMS

Adaptive capacity
Climatic uncertainty
Columbia Basin Trust
Impact pathways
Kitchen table meetings
Official Community Plan (OCP)
Pacific Climate Impacts Consortium (PCIC)
Rational comprehensive model (RCM)
Risk communication

REFERENCES

Arnstein, S.R. 1969. "A ladder of citizen participation." *Journal of the American Institute of Planners* 35(4): 216–24.

BC Stats. "2011 Census total population results: Census subdivisions." http://www.bcstats.gov.bc.ca/StatisticsBySubject/Census/2011Census/PopulationHousing/CensusSubdivisions.aspx.

Berrang-Ford, L., J.D. Ford, and J. Paterson. 2011. "Are we adapting to climate change?" *Global Environmental Change* 21: 25–33.

Blok, K., N. Hohne, K.V. Leun, and N. Harrison. 2012. "Bridging the greenhouse-gas emissions gap." *Nature Climate Change* 2: 471–4.

British Columbia. 2012. "Climate Action Charter." http://www.env.gov.bc.ca/cas/mitigation/charter.html.

——. 2013a. "Carbon tax review, and carbon tax overview." http://www.fin.gov.bc.ca/tbs/tp/climate/carbon_tax.htm.

——. 2013b. "Climate change impacts." http://www.env.gov.bc.ca/cas/impacts.

Brooks, N., W.N. Adger, and P.M. Kelly. 2005. "The determinants of vulnerability and adaptive capacity at the national level and the implications for adaptation." *Global Environmental Change* 15: 151–63.

CBC (Canadian Broadcasting Corporation). 2012. "Third body recovered from Johnson's Landing landslide." 26 July. 2012 http://www.cbc.ca/news/canada/british-columbia/story/2012/07/26/bc-johnsons-landing-body-recovered.html.

——. 2013. "Calgary flood damage to cost city $256M." 2 July. http://www.cbc.ca/news/canada/calgary/calgary-flood-damage-to-cost-city-256m-1.1318371.

Chu, S., and A. Majumdar 2012. "Opportunities and challenges for a sustainable energy future." *Nature* 488: 294–303.

Cope, M. 2005. "Coding qualitative data." In I. Hay, ed., *Qualitative Research Methods in Human Geography*, 2nd edn, 223–33. Don Mills, ON: Oxford University Press.

Corner, A. 2012. "Communicating climate change: Where next?" *The Guardian* 9 July. http://www.guardian.co.uk/sustainable-business/blog/communicating-climate-change-where-next?intcmp=122&utm_source=buffer&buffer_share=a3997.

de Bruin, K, R.R. Dellink, A. Ruijs, L. Bolwidt, A. van Buuren, J. Graveland, R.D. de Groot, P.J. Kuikman, S. Reinhard, R.P. Roetter, V.C. Tassone, A. Verhagen, and E.C. van Ierland. 2009. "Adapting to climate change in The Netherlands: An inventory of climate adaptation options and ranking of alternatives." *Climatic Change* 95: 23–45.

District of Elkford. 2008. *District of Elkford Official Community Plan and Climate Change Adaptation Strategy*. Terms of Reference. Citizens Advisory Committee.

——. 2010. "District of Elkford community profile." http://www.elkford.ca/include/get.php?nodeid=132.

Fischhoff, B. 2011. "Applying the science of communication to the communication of science." *Climatic Change* 108: 701–5.

Frederick, K.D. 1997. "Adapting to climate change impacts on the supply and demand for water. *Climate Change* 37:141–56.

Füssel, H.-M. 2007. "Adaptation planning for climate change: Concepts, assessment approaches, and key lessons." *Sustainability Science* 2: 265–75.

Gibson, R.B. 2006. "Sustainability assessment: Basic components of a practical approach." *Impact Assessment and Project Appraisal* 24(3): 170–82.

Gorecki, K., M. Walsh, and J. Zukiwsky. 2010. "District of Elkford: Climate Change Adaptation Strategy." Zumundo Consultants. http://elkford.ca/include/get.php?nodeid=93.

Gregory, R., T. McDaniels, and D. Fields. 2001. "Decision aiding, not dispute resolution: Creating insights through structured environmental decisions." *Journal of Policy Analysis and Management* 20(3): 415–32.

Groves, D.G., and R.J. Lempert. 2007. "A new analytic method for finding policy-relevant scenarios." *Global Environmental Change* 17: 73–85.

Hallegatte, S. 2009. "Strategies to adapt to an uncertain climate change." *Global Environmental Change* 19: 240–7.

Healey, P. 2003. "Collaborative planning in perspective. *Planning Theory* 2(2): 101–23.

Hoekstra, G. 2012. "Climate change could bring about more Johnson's Landing disasters, experts say." *Vancouver Sun* 18 July. http://www.vancouversun.com/news/Climate+change+could+bring+about+more+natural+disasters/6955199/story.html.

Hostovsky, C. 2006. "The paradox of the rational comprehensive model of planning." *Journal of Planning Education and Research* 25: 382–95.

Jacques, P. 2006. "Downscaling climate models and environmental policy: From global to regional politics." *Journal of Environmental Planning and Management* 29(2): 301–7.

Jakob, M., and M. Church 2011. "The trouble with floods." *Canadian Water Resources Journal* 36(4): 287–92.

Jones, H.P., D.G. Hole, and E.S. Zavaleta. 2012. "Harnessing nature to help people adapt to climate change." *Nature Climate Change* 2: 504–9.

Klinenberg, E. 2013. "Adaptation: How can cities be 'climate proofed'?" *The New Yorker*. http://www.newyorker.com/reporting/2013/01/07/130107fa_fact_klinenberg.

Lempert, R.J., and M.E. Schlesinger. 2000. "Robust strategies for abating climate change. *Climate Change* 45: 387–401.

McDaniels, T.L., H. Longstaff, and H. Dowlatabadi. 2006. "A value-based framework for risk management decisions involving multiple scales: A salmon aquaculture example." *Environmental Science and Policy* 9: 423–38.

Meinshausen, M., N. Meinshausen, W. Hare, S.C.B. Raper, K. Frieler, R. Knutti, D.J. Frame, and M.R. Allen. 2009. "Greenhouse gas emission targets for limiting global warming to 2°C." *Nature* 458: 1158–62.

Moser, S.C., R.E. Kasperson, G. Yohe, and J. Agyeman. 2008. "Adaptation to climate change in the northeast United States: Opportunities, processes, constraints." *Mitigation and Adaptation Strategies for Global Change* 13: 643–59.

Owen, B. 2012. "From floods to drought fears." *Winnipeg Free Press* 25 January. http://www.winnipegfreepress.com/local/from-floods-to-drought-fears-138026608.html.

Pengelly, L.D., M.E. Campbell, C.S. Cheng, C. Fu, S.E. Gingrich, and R. Macfarlane. 2007. "Anatomy of heat waves and mortality in Toronto: Lessons for public health protection." *Canadian Journal of Public Health* 98(5): 364–8.

Picketts, I.M., and J. Curry. 2011. "Planning for climate change adaptation in British Columbia communities: Lessons for planners." *International Journal for Sustainable Society* 3: 397–413.

Picketts, I.M., A.T. Werner, T.Q. Murdock, J. Curry, S.J. Déry, and D. Dyer. 2012. "Planning for climate change adaptation: Lessons learned from a community-based workshop." *Environmental Science and Policy* 17: 82–93.

Pidgeon, N., and B. Fischhoff. 2011. "The role of social and decision sciences in communicating uncertain climate risks." *Nature Climate Change* 1: 35–41.

Pike, R.G., K. Bennett, T.E. Redding, A. Werner, D. Spittlehouse, R.D. Moore, T. Murdock, J. Beckers, B. Smerdon, K. Bladon, V. Foord, D. Campbell, and P. Tschaplinski. "Climate change effects on watershed processes in British Columbia (ch. 19)." In R.G. Pike et al., eds, *Compendium of Forest Hydrology and Geomorphology in British Columbia*. British Columbia Ministry of Forests and Range Forest Science Program, Victoria, and FORREX Forum for Research and Extension in Natural Resources, Kamloops. Land Management Handbook no. 66. www.for.gov.bc.ca/hfd/pubs/Docs/Lmh/Lmh66.htm.

Procter, A. 2011. "Adaptation-mitigation conflicts in municipal planning: The case heat wave preparedness in Vancouver, Canada." Unpublished masters project, University of British Columbia.

Sandford, R.W. 2011. "Background report: Climate change adaptation and water governance." Adaptation to Climate Change Team, Simon Fraser University.

Shah, T.M. 2012. "An evaluation of climate adaptation planning in practice: A case study of the District of Elkford, British Columbia." Unpublished masters project, University of British Columbia.

Shepherd, P., J. Tansey, and H. Dowlatabadi. 2006. "Context matters: What shapes adaptation to water stress in the Okanagan? *Climatic Change* 78: 31–62.

Silva, W., J.P.M. Dijkman, and D.P. Loucks. 2004. "Flood management options for The Netherlands." *International Journal of River Basin Management* 2(2): 101–12.

Slovic, P. 1992. "Perceptions of risk: Reflections on the psychometric paradigm." In S. Krimsky and D. Golding, eds, *Social Theories of Risk*, 117–52. New York: Praeger.

Smit, B., I. Burton, R.J.T. Klein, and J. Wandel. 2000. "An anatomy of adaptation to climate change and variability." *Climatic Change* 45: 223–51.

Socolow, R.H. 2011. "High-consequence outcomes and internal disagreements: Tell us more please." *Climatic Change* 108: 775–90.

State Government of Victoria. 2013. "Kitchen table discussion." http://www.dse.vic.gov.au/effective-engagement/toolkit/tool-kitchen-table-discussion.

UNFCCC (United Nations Framework Convention on Climate Change). 2011. "Assessing the costs and benefits of adaptation options: An overview of approaches." http://unfccc.int/files/adaptation/nairobi_work_programme/knowledge_resources_and_publications/application/pdf/2011_nwp_costs_benefits_adaptation.pdf.

Welk, E. 2010. "Official Community Plan: District of Elkford 2010." http://www.elkford.ca/official_community_plan.

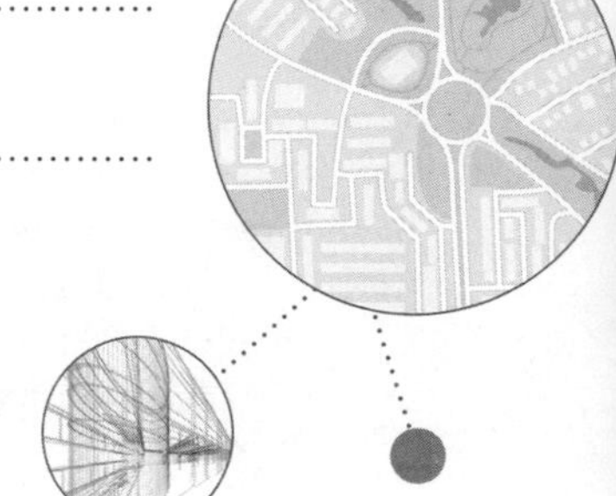

CASE STUDY 4.3

Government-to-Government Planning and the Recognition of Indigenous Rights and Title in the Central Coast Land and Resource Management Plan

Janice Barry

SUMMARY

This chapter explores how the recognition of Indigenous peoples' rights and title has the potential to alter both the process and outcomes of planning. These issues are approached through a detailed case study of the development of British Columbia's Central Coast Land and Resource Management Plan (CCLRMP), a complex environmental planning process that became a key site for the development of a new "government-to-government" approach to working with First Nations. Drawing on the experiences of the Nanwakolas Council, it examines how this approach evolved out of both court decisions and the day-to-day work of planners. The ability to translate complex legal principles into concrete ways of working was found to be a key planning skill. At the same time, this study also underscores the potential political and institutional limits of individual planners. Finding a meaningful and workable way of responding to Indigenous rights and title required the political commitment of senior officials as well as supportive policy changes and a well-designed governance structure. Government-to-government planning was most effective when it clearly differentiated between (1) the political discussions over the nature and extent of the Indigenous rights that need to be accommodated and (2) the professional discussions over the practical planning tools and approaches through which this accommodation could be achieved.

OUTLINE

- The development of a new "government-to-government" approach to engaging Indigenous peoples in strategic natural resource planning is explored in the case of British Columbia's Central Coast Land and Resource Management Plan.
- There is a relationship between new precedent-setting legal decisions with regard to Indigenous rights and title and the behaviour of planners.
- The case highlights the importance of cultivating planning professionals who are able to recognize opportunity within a rapidly changing legal and political environment and who are able to translate these changes into concrete ways of working.

Introduction

In the face of increased global interest in sustainability and public participation, a number of Canadian provinces have adopted a **collaborative planning** approach to the management of public lands and resources. In British Columbia, these collaborative initiatives have resulted in 30 **strategic natural resource plans**, which are often positioned as a way of securing increased biodiversity conservation, providing certainty for resource developers, and resolving conflicts (BC Integrated Land Management Bureau 2006). Although the overseeing agencies and names of the primary planning documents have changed numerous times over this ambitious program's approximately 20-year history, the overall vision and purpose remained the same: to apply a shared decision-making approach to the creation of "comprehensive, broadly accepted" management frameworks that would "guide resource development and more detailed planning" at the regional scale (BC Integrated Resource Planning Committee 1993). Yet in the face of precedent-setting high-court decisions, these processes became key sites for responding to another contemporary planning problem: the recognition of **Indigenous rights and title**.

This problem will be explored through a qualitative case study of the Central Coast Land and Resource Management Plan, one of BC's longest-running and most complex strategic natural resource planning processes. While perhaps better known for its innovative approach to ecosystem-based management and efforts to support sustainable approaches to regional economic development (see Clapp and Mortenson 2011; Saarikoski, Raitio, and Barry 2012), the CCLRMP also resulted in a unique "government-to-government" arrangement between the BC government and affected First Nations. At its core, government-to-government planning is about attending to First Nations' inherent and legally recognized right to be involved in decisions about their **traditional territory**. It required building the structures that would allow First Nations to participate in these collaborations as a unique form of government—not as a **stakeholder**. Although the possibility of government-to-government planning had been alluded to in previous provincial documents, this chapter explores how the concept came to fruition during the CCLRMP. Drawing on the experiences of the Nanwakolas Council (the only First Nation planning coalition to be involved since the inception of the CCLRMP), it highlights how the new approach to working with First Nations was shaped by a series of legal decisions as well as the creative and strategic choices made by planners.

Numerous aspects of this government-to-government approach have since been exported to other planning processes across the province. Yet the CCLRMP has implications beyond BC. Planners across the country face the challenge of developing new ways to recognize the rights of Indigenous people, and several provinces are also using the language of a government-to-government relationship (e.g., Manitoba 2007; Ontario Native Affairs Secretariat 2005). These challenges are particularly prevalent in resource planning, but there are potential lessons for urban planning contexts as well. For although Indigenous rights and title have been rendered somewhat invisible in the built environment (Porter 2013), these places are also traditional Indigenous territory, and those concerned might therefore learn from the innovative governance arrangements that are emerging out of processes such as the CCLRMP.

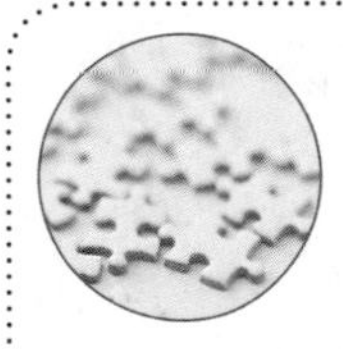

KEY ISSUE

Planning with Indigenous Governments

1. Developing a new approach to planning with Indigenous governments that:
   - responds to new legal direction on Indigenous rights and title;
   - builds professional and political rapport among Indigenous and non-Indigenous representatives;
   - maintains the spirit of the initial planning process and continues to involve various stakeholders.
2. Expanding and then formalizing the new approach by:
   - finding and making strategic choices around "windows of opportunity" in the broader political and policy landscape;
   - understanding the limits of planning professionals.

Indigenous Recognition and the CCLRMP

Straddling one of the most significant decades in terms of defining Canadian governments' responsibilities toward Indigenous rights and title, the CCLRMP is an important touchstone in terms of understanding how the planning profession is beginning to re-examine its relationship with Indigenous peoples. This 10-year process, which ran from 1996 to 2006, focused on a 4.6-million-hectare region of great ecological and economic importance (Figure 4.3.1). The area is perhaps best known for its large concentration of unlogged watersheds and the genetically distinct white "spirit bear." But it also supports a diverse, albeit declining, resource-based economy, including outdoor tourism, mining, commercial fisheries, and aquaculture. The majority of the approximately 5000 area residents are First Nation,

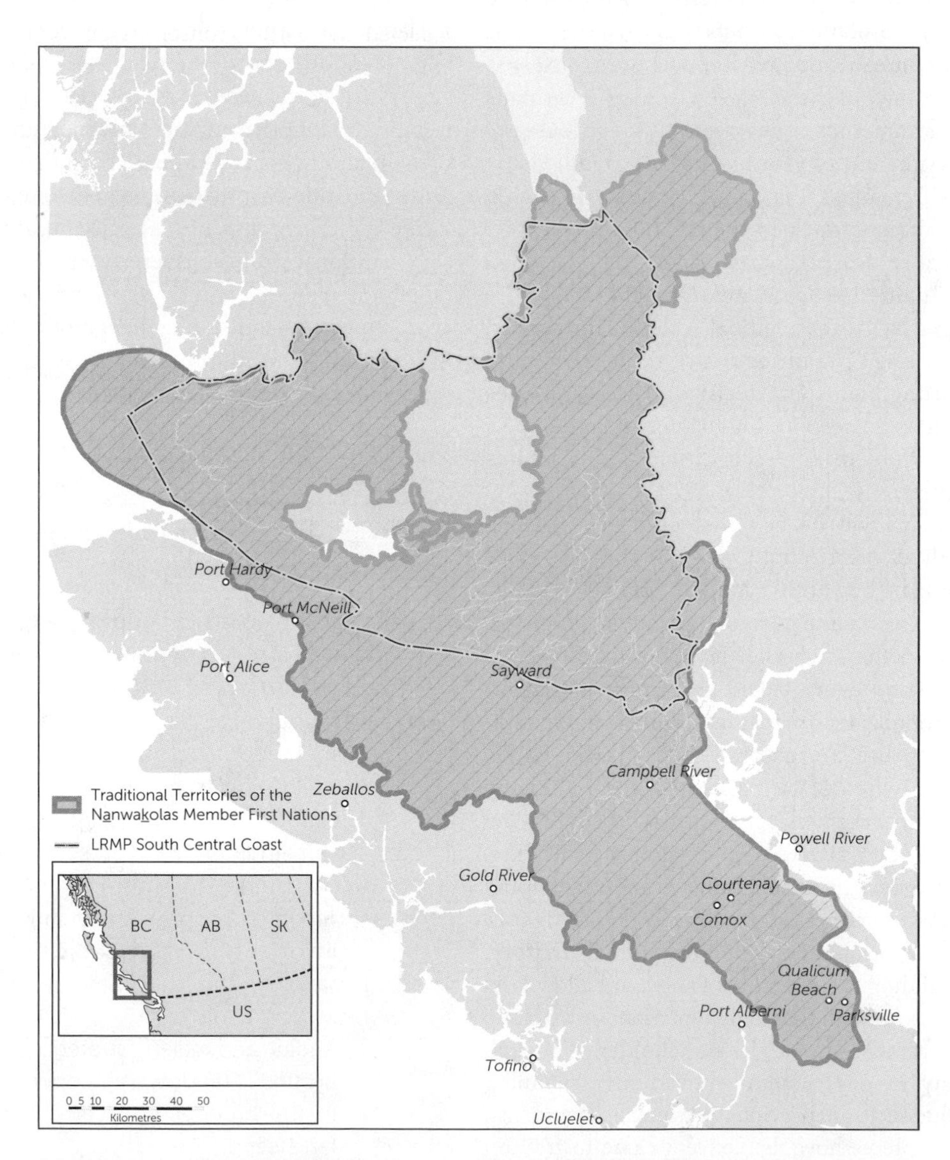

Figure 4.3.1 • The traditional territories of the Nanwakolas Council Member First Nations in relation to the southern portion of the CCLRMP.

Source: Nanwakolas Council 2013, used with permission.

many of whom experience various forms of socio-economic marginalization (e.g., unemployment, poorer health, lower levels of education). These factors not only created a strong potential for conflict between the various interest groups, but they also demanded a careful balance between socio-economic and environmental concerns.

Like previous planning processes in other parts of BC, provincial staff would provide the bulk of the technical, policy, and organizational expertise, but there would also be an extensive **multi-stakeholder planning** forum to consider the various land-use zoning scenarios and seek consensus on the final CCLRMP. First Nations were invited to participate in this planning forum, as well as the Province's own inter-agency caucuses and planning teams. The initial terms of reference even made some reference to the need for a "government-to-government" relationship, although there was little discussion of what the term meant in practical terms or what would be the markers of a successful government-to-government process. As one provincial employee involved in setting up the CCLRMP recalls, the idea of a government-to-government relationship was not a "foundational piece" of the initial process: "First Nations certainly talked about it, but government itself hadn't resolved or reconciled what government-to-government meant yet" (personal interview, 17 April 2009).

By the end of the process, the government-to-government relationship had become a defining feature of both the particular planning outcomes and the broader governance framework. The CCLRMP process did not result in an actual land-use plan but in a series of bilateral agreements with affected First Nations. These agreements reflected the parties' consensus on land-use zones and other resource management directives, while also establishing the framework for ongoing implementation and adaptation (Figure 4.3.2). First Nation representatives are included in all of the major governance bodies, from the scientifically driven Ecosystem-Based Management Working Group to the strategically oriented Land and Resource Forum. First Nations maintain links with the stakeholder groups that were involved in the original planning process by sitting on the Plan Implementation and Monitoring Committee. These agreements brought certainty not only to the socio-economic and ecological future of the

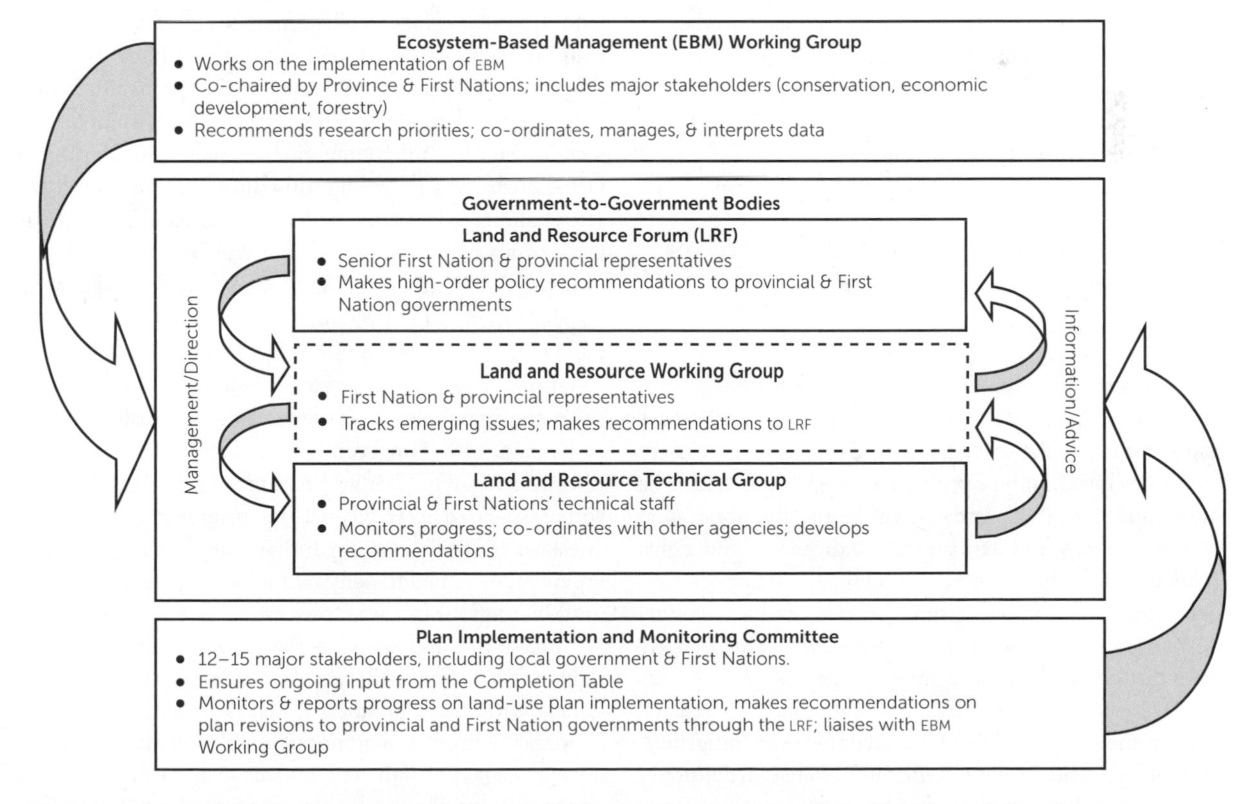

Figure 4.3.2 • Government-to-government decision-making framework for the Central Coast.

region but to the government-to-government relationship that had evolved over the previous 10 years. Many of the ambiguities around the nature of government-to-government planning had effectively been removed in that First Nations were no longer seen as a mere participant or as a provider of information on the nature and extent of their Indigenous rights and title. They were now at the heart of the long-term governance framework, acting both as a receiver of information and advice and as a provider of policy recommendations and management direction.

Although numerous area First Nations played a significant role in the CCLRMP, each contributing to the development of the long-term government-to-government framework, this chapter focuses on the experiences of a coalition of Kwakwaka'wakw Nations whose traditional territories are in the southern portion of the planning area. Now formally incorporated as the Nanwakolas Council, this organization represents the interests and priorities of almost half of the First Nations with traditional territories in the Central Coast region and was the only First Nation coalition to be involved since the very outset of the CCLRMP process. As a result, the Nanwakolas representatives were both key witnesses and active agents in the development of government-to-government planning. The chapter relies on detailed analysis of various planning documents (e.g., meeting minutes, terms of reference, memoranda), as well as nine semi-structured interviews with provincial and Nanwakolas representatives directly involved in the CCLRMP.

The Legal Context and Implications for BC's Approach to Strategic Natural Resource Planning

To understand the evolution of this government-to-government relationship, it is necessary to examine the broader legal context. Unlike other parts of Canada, there are very few treaties in BC, and the province was developed under the assumption that European settlement and the application of provincial law had extinguished Indigenous land rights (Tennant 1990). These destructive colonial assumptions have been questioned through a series of court cases, including *Sparrow* (1990) and *Delgamuukw* (1997), which respectively paved the way for the establishment of the BC Treaty Process (BC Treaty Commission 2010) and subsequent legal decisions on the duty to consult and accommodate Indigenous people (e.g., *Haida Nation v. British Columbia (Minister of Forests)* 2004). Both of these developments have significant implications for natural resource planning.

Officially established in September 1992, the BC Treaty Process was designed to facilitate modern treaty-making. Indigenous governance, lands, resources, and financial compensation were identified as some of the relevant issues (BC Claims Task Force 1991). Given the complexity of these issues, the BC Treaty Process has long asserted that the parties must develop mechanisms to "balance their conflicting interests until these negotiations are concluded" and has often promoted the development of interim measures, such as an increased role in decision-making, as well as restrictions or moratoriums on land and resource use (BC Claims Task Force 1991, 23). Indigenous participation in strategic natural resource planning was originally construed as an interim measure in that its zoning provisions provided a potential mechanism to protect Indigenous interests until a treaty was in place. BC's collaborative planning approach was also seen as a potential avenue for the advancement of joint management (BC Land Use Coordination Office 1998). In reality, Indigenous participation in strategic natural resource planning has historically been marginal at best (Barry 2011), and many First Nations continued to turn to the courts.

One such case, *Delgamuukw v. British Columbia* (1997), represents a watershed moment in provincial–First Nations relations, because it confirmed that Indigenous title does still exist in BC and took significant strides toward articulating the Crown's duty and approach to recognizing this Indigenous right. As Chief Justice Lamer wrote in the official ruling, this duty "will vary with the circumstances. In occasional cases, when the breach is less serious or relatively minor, it will be no more than a duty to discuss important decisions that will be taken with respect to lands held pursuant to aboriginal title. . . . In most cases, it will be significantly deeper than mere consultation. Some cases may even require the full consent of an aboriginal nation" (*Delgamuukw v. British Columbia* 1997, para. 168). However, it was some years before this direction began to filter into provincial policy, particularly with respect to strategic natural resource planning. As one lawyer specializing in Indigenous law suggests, the Province continued to deny that it had any duty to consult until Indigenous title was "proven" (Mandell 2004).

The source and scope of the duty toward Indigenous rights and title was further refined in *Haida Nation v. British Columbia (Minister of Forests)*. Released in late 2004, this Supreme Court of Canada decision clearly stated that the duty to engage in fair and meaningful consultation processes with Indigenous people does not arise out of the formal acknowledgment of an Indigenous claim or the signing of a treaty but rather out of the more general "honour of the Crown" (*Haida Nation v. British Columbia (Minister of Forests)* 2004, para. 16). In addition, the level of consultation must be in proportion to the strength of the Indigenous claim, regardless of whether or not that claim has been formally proven. Chief Justice McLachlin also drew specific attention to the importance of consulting on strategic planning decisions, since these processes tend to be characterized by multi-year decision-making and often establish the general parameters for all other resource management activities. Clearly, the Province could no longer afford to take a "business as usual" (Mandell 2004, 2) approach to strategic natural resource planning. It had to develop new methods for ensuring that Indigenous involvement upheld the principles being developed by the courts.

The Role of "Planners": Interpreting and Strategically Responding to Changing Conditions

Yet as significant as these court decisions were in terms of demanding a new approach to working with Indigenous peoples, they alone did not result in the formation of a government-to-government relationship. Government-to-government planning is about attending to the duty to consult and accommodate Indigenous people, but it is equally about creating a workable long-term governance structure. It is also about Indigenous and non-Indigenous land-use professionals learning to trust one another and finding ways to bridge organizational, political, and cultural divides. The legal changes often provided the impetus for change, but it was the creative and strategic behaviour of the planners that gave shape to the government-to-government relationship and that provided the rationale for its continuation. To be sure, neither of the Nanwakolas representatives would self-identity as planners, but both played a foundational role in the CCLRMP and, over the length of the process, were able to start building the technical, policy, and communicative competencies exhibited by provincial planners.

Released approximately 18 months after the CCLRMP process was formally initiated, the *Delgamuukw* decision had a significant impact on the formative years of the process—not in terms of changing the formal design of the planning process (that would come later) but in terms of shifting the perception of Indigenous peoples. As one Nanwakolas representative recalls, "you almost saw some lights turn on in some people's heads about 'Oh, that's why they're so adamant about that!' And that's when the relationship building was able to start. It wasn't so much about issues against issues; it was people dealing with certain variables in the area that have to be managed" (Nanwakolas representative, personal communication, 25 March 2009). Increased acceptance of Indigenous peoples' distinct role in the CCLRMP process not only opened up the space for a different kind of discussion, but it also resulted in a change in strategy. The Nanwakolas representatives continued to be very clear with the other table members that they needed to be treated as another form of government—not stakeholders. But the feeling of no longer needing to fiercely demand recognition of their rights and title allowed them to start thinking about building bridges and "learning the businesses of the stakeholders that were around the table" (Nanwakolas representative, personal communication, 17 April 2009).

Although they refused to engage in actual negotiations with non-government actors and did not vote on any of the table motions, the Nanwakolas representatives were starting to use the multi-stakeholder forum to raise contentious issues and to make their specific land-use interests known—a strategic input of information that would "usually dial the table back into coming up with some alternate [recommendation]" (Nanwakolas representative, personal communication, 31 March 2009). They were also using their increased understanding of the various stakeholder perspectives and growing competency with the language of natural resource planning to come up with new and potentially more politically viable proposals. Perhaps more important, they were beginning to differentiate between the political and professional functions of the government-to-government relationship. High-level discussions about the nature and extent of Indigenous rights and title tended to be delegated to higher levels of political and administrative authority (e.g., Nanwakolas's president and First Nation chiefs and councils on the Indigenous side; ministers and deputy ministers on the provincial side). As a result, Nanwakolas's executive director and the provincial planning representatives were able to focus on what the recognition of those rights meant in practical planning terms. For the provincial planners, this gradual differentiation of roles provided them with a First Nation "professional body that we could relate to" (Nanwakolas representative, personal communication, 31 March 2009).

After the 2001 provincial election, this differentiation became even more prominent. Discussions over the accommodation of each First Nation's unique rights, title, and long-term land-use interests (which could involve culturally sensitive information about spiritual sites and/or politically sensitive information about funding and treaty negotiations) increasingly occurred away from the multi-stakeholder forum. In fact, First Nation agreement on the entire CCLRMP was achieved through approximately two years of focused bilateral negotiations between the provincial government and each individual First Nation. At the same time, the Province also required a common framework for plan implementation, necessitating joint negotiations over long-term governance issues. Thus, the success of the government-to-government relationship was at least partially driven by the parties' ability to create a complex governance framework that incorporated multiple roles and forms of expertise and that struck an appropriate balance between the competing needs for separate and common discussion forums.

While the provincial and Nanwakolas representatives had informally crafted a structure for enacting the government-to-government relationship, they lacked the political will and concrete policy direction that would allow them to formalize this new way of working. The influential *Haida* decision had been making its way through the lower courts during the CCLRMP process and was beginning to influence how some senior BC officials conceived of a government-to-government relationship, but there had not yet been any formal changes in provincial policy or planning procedure. This situation would dramatically change in the months following the final Supreme Court of Canada ruling. Released not long after the bilateral negotiations that drew the CCLRMP to a close, the court decision triggered a series of high-level discussions between the premier and BC's three major Indigenous organizations. These discussions resulted in a five-page vision and statement of the principles. The *New Relationship Policy Statement* promised a new approach to working with First Nations: a government-to-government relationship based on the respect, recognition, and accommodation of Indigenous rights and title. Natural resource planning was explicitly identified as an area where these kinds of shared decision-making arrangements were to be pursued.

Despite its brevity and general lack of concrete direction, the *New Relationship Policy Statement* provided both the provincial and First Nation negotiators with the political support needed to better address the design of a new governance framework. Provincial staff inside these negotiations started to view the process as a place to explore the pragmatic and administrative aspects of the new relationship and to establish some benchmarks to guide future processes. It provided them with some clear political support and made them feel as though they had a firm leg to stand on but still left ample room to explore and to adapt to the current context. The new relationship provided the Nanwakolas representatives with additional bargaining power in the government-to-government negotiations, especially when they felt that the Province was trying to retreat to established policies and procedures. The *New Relationship Policy Statement*, therefore, began to function as a normative yardstick. It inspired the provincial negotiator to think outside the confines of existing governance structures and provided the Nanwakolas First Nations with a new mechanism to hold the Province to account when it was falling short. Thus, the arrival of the *New Relationship Policy Statement* provides another example of the essential role planners often play in responding to changes in the broader political landscape. It was this interplay between external forces and the creative and strategic actions of individual "planners" that led to the development of a government-to-government relationship, as well as the creation of a lasting governance framework for its enactment and further development (Figure 4.3.2).

Lessons for Other Planning Initiatives

Given that the legal decisions that triggered the CCLRMP process's experimentation with new ways of working with Indigenous people apply to land-use decision-making across Canada, the lessons drawn from this case study are potentially of broad application. Perhaps most significantly, the Nanwakolas Council's experience of government-to-government planning speaks to the importance of designing an appropriate regional governance framework. The multi-level framework that was developed for the Central Coast (which ensures that planning professionals collaborate with other planning professionals and politicians with politicians) represents a relatively simple yet quite innovative arrangement that can potentially be altered to suit other planning contexts. Beyond the particularities of Indigenous recognition, this case study provides valuable lessons for planning more broadly. It illustrates some of the legal and political factors that come to bear on the day-to-day practice of planning. In doing so, it highlights the importance of cultivating planning professionals who are able to recognize opportunity within a rapidly changing legal and political environment and who are able to translate these changes into concrete ways of working.

KEY TERMS

Collaborative planning
Indigenous rights
Indigenous title
Multi-stakeholder planning
Stakeholder
Strategic natural resource planning
Traditional territory

REFERENCES

Barry, J. 2011. "Building collaborative institutions for government-to-government planning: The Nanwakolas Council's involvement in Central Coast land and resource management planning." Unpublished doctoral dissertation, School of Community and Regional Planning, University of British Columbia.

BC Claims Task Force. 1991. *The Report of the British Columbia Claims Task Force*. Vancouver: British Columbia Land Claims Task Force.

BC Integrated Land Management Bureau. 2006. "A new direction for strategic land use planning in BC: Synopsis." Victoria: BC Ministry of Agriculture and Lands. http://archive.ilmb.gov.bc.ca/slrp/lrmp/policiesguidelinesandassessements/lrmp_policy/stmt.htmhtm.

BC Integrated Resource Planning Committee. 1993. "Land and resource management planning: A statement of principles and process." Victoria: Province of British Columbia. http://archive.ilmb.gov.bc.ca/slrp/lrmp/policiesguidelinesandassessements/lrmp_policy/stmt.htmhtm.

BC Land Use Coordination Office. 1998. "Inter-Agency Management Committee orientation to First Nations: Instructors guide—Lesson plans." LRMP Training Modules prepared by Trelawny Consulting Group. http://ilmbwww.gov.bc.ca/lup/training/First-Nations/FN0-MO-2.html.

BC Treaty Commission. 2010. News release: "Sparrow case moved BC government to negotiate treaties." 31 May. http://www.bctreaty.net/files/pdf_documents/news-release_sparrow-case-moved-bc-gov-to-negotiate.pdf.

Clapp, R.A., and Mortenson, C. 2011. "Adversarial science: Conflict resolution and scientific review in British Columbia's Central Coast." *Society and Natural Resources* 24(9): 902–16.

Delgamuukw v. British Columbia [1997] 3 SCR 1010.

Haida Nation v. British Columbia (Minister of Forests) [2004] 3 S.C.R. 511, 2004 SCC 73.

Mandell, M.L. 2004. *"Haida Nation v. B.C. (Minister of Forests)* ('Haida') and *Taku River Tlingit First Nation v. B.C. (Project Assessment Director)* 'Taku': Summary of Supreme Court of Canada Rulings of 18 November 2004." *Impact of the Haida and Taku River Decisions: Consultation & Accommodation with First Nations*. Vancouver: Pacific Business and Law Institute.

Manitoba. 2007. News release: "Landmark government-to-government accord on east side of Lake Winnipeg signed between Province of Manitoba and First Nations." 3 April. www.gov.mb.ca/chc/press/top/2007/04/2007-04-03-155047-1407.html.

Ontario Native Affairs Secretariat. 2005. *Ontario's New Approach to Aboriginal Affairs*. Toronto: Queen's Printer for Ontario.

Porter, L. 2013. "Coexistence in cities: The challenge of Indigenous urban planning in the 21st century." In R. Walker, T. Jojola, T. Kingi, and D. Natcher, eds, *Reclaiming Indigenous Planning*. Montreal: McGill-Queen's University Press.

Saarikoski, H., K. Raitio, and J. Barry. 2012. "Understanding 'successful' conflict resolution: Policy regime changes and new interactive arenas in the Great Bear Rainforest." *Land Use Policy* 32: 271–80.

Tennant, P. 1990. *Aboriginal Peoples and Politics: The Indian Land Question in British Columbia, 1849–1989*. Vancouver: University of British Columbia Press.

INTERVIEWS

Nanwakalas representative, 25 March 2009. In-person interview.

Nanwakalas representative, 31 March 2009. In-person interview.

Nanwakalas representative, 17 April 2009. In-person interview.

5.0

INTRODUCTION
Housing

Housing is such a basic need among the Canadian population that it has preoccupied the planning discipline for more than a century, yet the topic of housing is value laden. Some of the earliest housing reformers were concerned with the living conditions of the working class—not only the physical conditions of workers' dwellings but also the implications of these conditions for the family structure (Purdy 1997). While there was concern over preventing the contagious diseases that could spread from "slum" areas to more affluent neighbourhoods, reformers were also interested in eliminating what they considered the moral failings of slum dwellers. Spurred by declining birth rates, industrialization, and women's increasing participation in the workforce at the turn of the twentieth century, reformers focused on threats to the family structure. Characteristic building reforms of this time period included proper sanitary facilities, more access to light and air, and more space per occupant; programs in industrial hygiene and housing began to teach women how to use innovations such as electric lighting, water, and cooking appliances to maintain efficient and sanitary conditions in their homes.

It was not until the Great Depression and Second World War that hundreds of government and academic reports, commissions, and surveys on the severe housing shortage facing Canadians catalyzed the development of a national housing program. Through the first national housing policy (Dominion Housing Act [1935]) and the establishment of the Central Mortgage and Housing Corporation (CMHC) (now Canada Mortgage and Housing Corporation) in 1946, homeownership became possible for many residents and presented opportunities for builders and developers.

Some of the first public housing in the country was enabled through amendments to the National Housing Act in 1949. CMHC was also instrumental in the demolition of "slum" housing in areas such as Regent Park in Toronto in order to build new housing in the modernist style, with units based on a "strict notion of nuclear family life, delimiting individual aspirations, especially those of women" (Purdy 1997, 37). From the 1930s to the 1950s, planners sought more efficient, rational, and orderly cities, expressed through zoning bylaws controlling the location of housing, industrial, and commercial uses. Single-family residential areas were placed in the best locations, which at the time were perceived as being far from the industrial and moral pollution of the central city.

A suburban lifestyle went hand-in-hand with postwar consumerism, since it was assumed that the acquisition of property was now the measure of personal achievement (McCann 1999). Suburbs of this time were often the rural-facing edges of the built-up inner city, such as Leaside in Toronto or Point Grey in Vancouver. Major landowners would sometimes group together and petition the provincial governments to incorporate their land, which would introduce municipal bylaws controlling housing type and size, lot size, building setback, construction values, and even minimum house values. Guidelines based on single-family homes on curving roads with a hierarchy of streets quickly replaced the more compact, linear pattern of streetcar suburbs. Suburban living emphasized the nuclear family, with multiple, extended, and single-person families often excluded by means of the housing type (single-family), tenure (ownership), and size (four-person family). Between 1941 and 1952, the rate of homeownership in urban Canada jumped from 41 per cent to 56 per cent (McCann 1999, 130). Increasing access to homeownership has been a goal of the federal government since the 1930s (Oberlander and Fallick 1992); the Dominion Housing Act (1935) allowed individuals to borrow up to 60 per cent of the cost of new owner-occupied or rental housing, and the National Housing Act (1938) increased the threshold to 80 per cent (Nicholls 1938).

Persistent affordability concerns began in the 1970s because of changes to the Income Tax Act, changes in federal and provincial housing policies, and the introduction of provincial Condominium Acts. Fewer purpose-built rental buildings were constructed because developers could no longer profit from building rental housing. Increased competition for central business districts throughout the 1990s and early 2000s also contributed to the construction of trendy luxury condominiums, which are beyond the means of the average middle-class household. Many municipalities have taken steps to increase affordable housing, particularly as the federal government withdrew from the construction of social housing in 1993. Some steps include charging developers fees, asking developers to contribute to affordable housing funds, and placing restrictions on condominium conversions to protect rental housing. Affordability concerns are particularly severe among low-income, renter, and immigrant households (e.g., Thomas 2013; Hulchanski 2010). In 2009, the United Nations declared that Canada had an affordable housing crisis.

Although infill housing, new urbanism, and Smart Growth have made inroads into Canadian neighbourhood development (Grant 2009), the trend still skews toward postwar suburbs built on greenfield sites (Filion and McSpurren 2007). The "Canadian dream" of the single detached house has remained remarkably resilient, but affordability concerns are pushing some developers toward townhouses, apartments, and condominiums (Bula 2011). The cases in this section illustrate many of these trends.

They show the complex governance affecting housing plans and programs, from the municipal, provincial, and federal governments to developers and public–private partnerships.

Jenna Mouck presents the development of a new approach to housing in Saskatchewan. With increased immigration, interprovincial migration, and natural growth during the past decade making increased demands on housing, the Province has had to develop a new growth plan and housing strategy. Using the housing continuum model, the Province has encouraged municipalities to plan for different housing types and tenures by funding the development of housing plans and specific actions to achieve their housing goals. The Province also funds new technology through an innovation fund and will invest in new multi-family units through the sale of some of its single-family units.

Laura Johnson's study of Regent Park examines the redevelopment of Canada's first and largest public housing project. Built as a "slum clearance" project in the 1940s and 1950s, the buildings fell into disrepair within a couple of decades. Critics blamed its modernist design and isolation from the Toronto street network. The 15- to 20-year redevelopment of the project through a public–private partnership is now at the halfway mark. The new development features a mix of market-rate condominiums and social housing and introduced mixed uses such as a pharmacy, coffee shop, restaurant, bank, indoor aquatic centre, and arts and culture centre; 1100 new jobs were created for residents. However, relocation and resettlement has not been easy for the residents who have been rehoused within, or adjacent to, Regent Park.

Sasha Tsenkova addresses a new challenge for Canada's public housing projects: skyrocketing energy costs. Through the federal Renewable Energy Initiative (REI), introduced in 2009, and Canada's Economic Action Plan (CEAP), funding for energy-efficiency upgrades in existing social housing projects became available in order to decrease costs for tenants and governments alike. Tsenkova examined the allocation and use of these funds in three provinces: Ontario, British Columbia, and Alberta. While the investment of funds was necessary to pay for capital repairs and system upgrades in aging social housing stock, substantial energy savings were also realized.

Marla Zucht and Margaret Eberle present the housing innovations of the Whistler Housing Authority, a unique response to the shortage of employee housing. Enabled through its status as a resort municipality, Whistler has developed a solution to persistent affordability problems for employees of the local tourism industry. By requiring developers of commercial, tourist, and industrial land to build resident housing or contribute cash in-lieu to a designated housing fund, more than 300 restricted beds for the local workforce were created by 1996. The Whistler Housing Authority now manages almost 1900 units of resident-restricted housing, with 57 per cent affordable ownership units and 43 per cent resident-restricted rental units. Covenants registered on land titles prevent resale at far above market prices, maintaining the long-term affordability of the housing. As a result, 80 per cent of Whistler's workforce resides locally.

These cases offer but a glimpse into the complex area of housing in Canadian cities. Housing is a basic human necessity for an increasingly diverse population: single-person and single-parent households, the elderly who would like to continue living in suburban neighbourhoods, and young people who desire urban lifestyles and connections to public transit. Canadian municipalities have been able to innovate in a number of areas to ensure more housing choices, but more innovative partnerships and programs must be implemented to combat high housing costs, displacement, and urban sprawl.

REFERENCES

Bula, F. 2011. "Home in the suburbs, heart in the city." *Globe and Mail* 8 April.

Filion, P., and K. McSpurren. 2007. "Smart Growth and development reality: The difficult coordination of land use and transportation objectives." *Urban Studies* 44(3): 501–23.

Grant, J.L. 2009. "Theory and practice in planning the suburbs: Challenges to implementing new urbanism, Smart Growth, and sustainability principles." *Planning Theory and Practice* 10(1): 11–33.

Hulchanski, J.D. 2010. *The Three Cities within Toronto: Income Polarization among Toronto's Neighbourhoods, 1970–2005*. Update to Research Bulletin 41. Toronto: University of Toronto Cities Centre.

McCann, L. 1999. "Suburbs of desire: The suburban landscape of Canadian cities, c. 1900–1950." In Richard Harris and Peter Larkham, eds, *Changing Suburbs: Foundation, Form and Function*, 111–43. Toronto: Routledge.

Nicholls, R.W. 1938. *Housing in Canada, 1938. Housing Yearbook 1938: National Association of Housing Officials, Chicago, 1938*. Toronto: Urban Policy Archive, Centre for Urban and Community Studies, University of Toronto.

Oberlander, P.H., and A.L. Fallick. 1992. *Housing a Nation: The Evolution of Canadian Housing Policy*. Ottawa: Centre for Human Settlements, University of British Columbia, and Canada Mortgage and Housing Corporation.

Purdy, S. 1997. "Industrial efficiency, social order and moral purity: Housing reform thought in English Canada, 1900–1950." *Urban History Review* 25(2): 30–40.

Thomas, R. 2013. "Resilience and housing choices among Filipino immigrants in Toronto." *International Journal of Housing Policy* 13(4): 408–32. DOI: 10.1080/14616718.2013.840112

CASE STUDY 5.1

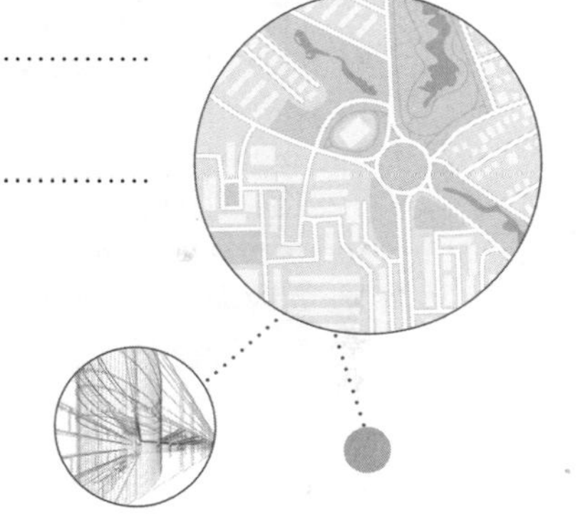

Saskatchewan's Affordable Housing Challenge: Allocation of Public Funding in a Thriving Province

Jenna Mouck

SUMMARY

Saskatchewan has seen unprecedented economic and population growth since 2007, which has resulted in significant housing challenges across the province, including increasing house prices, low vacancy rates, deteriorating public housing stock, and wages that have not kept pace with the cost of living increase. The Government of Saskatchewan is predicting high population increases into the future, which will continue to put pressure on the housing system.

The provincial government has responded in a number of ways to address these housing challenges, since they are, and will continue to be, an impediment to growth. The economic and population growth, coupled with a significant decline in federal funding for low-income housing, has contributed to the need for action in the province.

The Province has taken a holistic view of the **housing continuum** and has begun planning for and making significant changes to the way it does business and in how it is addressing the housing situation across the province. By taking action all across the housing continuum, and collaborating with others involved in the housing sector, the Province has made positive changes at a time when change is prevalent in the province. As programs are revised and public funds are better allocated to address the various needs identified, residents, employers, and investors will begin to see the results of these changes.

OUTLINE

- Rapid population growth in a recent economic boom was instrumental in the development of housing goals to address challenges in the province of Saskatchewan.
- Changes to existing programs and the Province's social housing portfolio include innovative approaches to housing design, targeting housing for those in need, and changes to the criteria for social housing tenants.
- While the province is currently growing rapidly and experiencing unprecedented economic growth, the lessons learned through Saskatchewan's "housing continuum" approach could be applied in other contexts in Canada

The Saskatchewan Context

Since Saskatchewan's economy began booming in 2007, significant housing challenges arose and have persisted across the province. Increasing housing prices, both for rentals and homeownership, low vacancy rates, deteriorating housing conditions, and wages that have not kept pace with the increase in the cost of living are all contributing to a lack of affordable housing in many areas of the province; this situation is most evident in the major centres and **high-growth areas**.

The Government of Saskatchewan is focused on industry and is forecasting a population of 1.2 million by the year 2020, an increase of more than 10 per cent in just eight years, according to a population estimate by the Saskatchewan Bureau of Statistics.

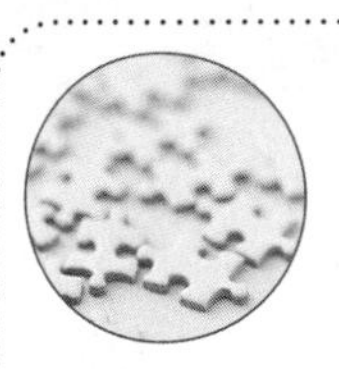

KEY ISSUE

Growth Management

High-growth areas are prevalent across the country. These areas result in a multitude of both opportunities and challenges for communities and regions. Infrastructure challenges, including municipal services, education, health care, and housing, are some of the most prevalent problems in high-growth areas. With a strong influx to the population, housing prices increase rapidly, and supply can no longer keep up with demand. This results in those in the lower end of the housing continuum struggling to afford the private market, and it leaves the public sector struggling to serve the population most in need.

Economic and Population Growth

Saskatchewan's population growth in 2007–8 was the strongest since the early 1970s, with the province leading the country in **interprovincial migration** (Canadian Press 2012). With strong employment and a growing economy, Saskatchewan transformed from a "have not" to a "have" province. Even the recession in 2008 failed to affect Saskatchewan the way it did elsewhere in Canada; the resource sector continued to grow, and the economy continued to boom.

Saskatchewan has a diverse economy based on oil, potash, uranium, farm equipment, wheat, and other crops. Major contributors are the potash and oil sectors. Saskatchewan is the world's largest potash producer and typically accounts for about 25 to 30 per cent of world potash production (Government of Saskatchewan, n.d.). Potash is generally mined in the southeast and south-central areas of the province. Saskatchewan is also Canada's second-largest producer of oil (Canadian Association of Petroleum Producers, n.d.). Oil and gas is generally found in the southeast corner of the province, as well as along its western edge from the southern edge of the province to north of Lloydminster. Farming is a major industry across the province and contributes significantly to the provincial economy. A map of select industries in Saskatchewan is shown in Figure 5.1.1.

Saskatchewan's population dropped throughout most of the 1980s and 1990s. However, a combination of demographic factors has resulted in unprecedented population growth in Saskatchewan since 2007. **Natural growth**, interprovincial migration, and immigration have all contributed to the approximately 1.5 per cent annual growth rate over the past five years (Elliot 2012).

Saskatchewan's population reached the 1-million mark in 2007. It had exceeded 1 million between 1983 and 2001 but had fallen below that mark as of 1 July 2001.

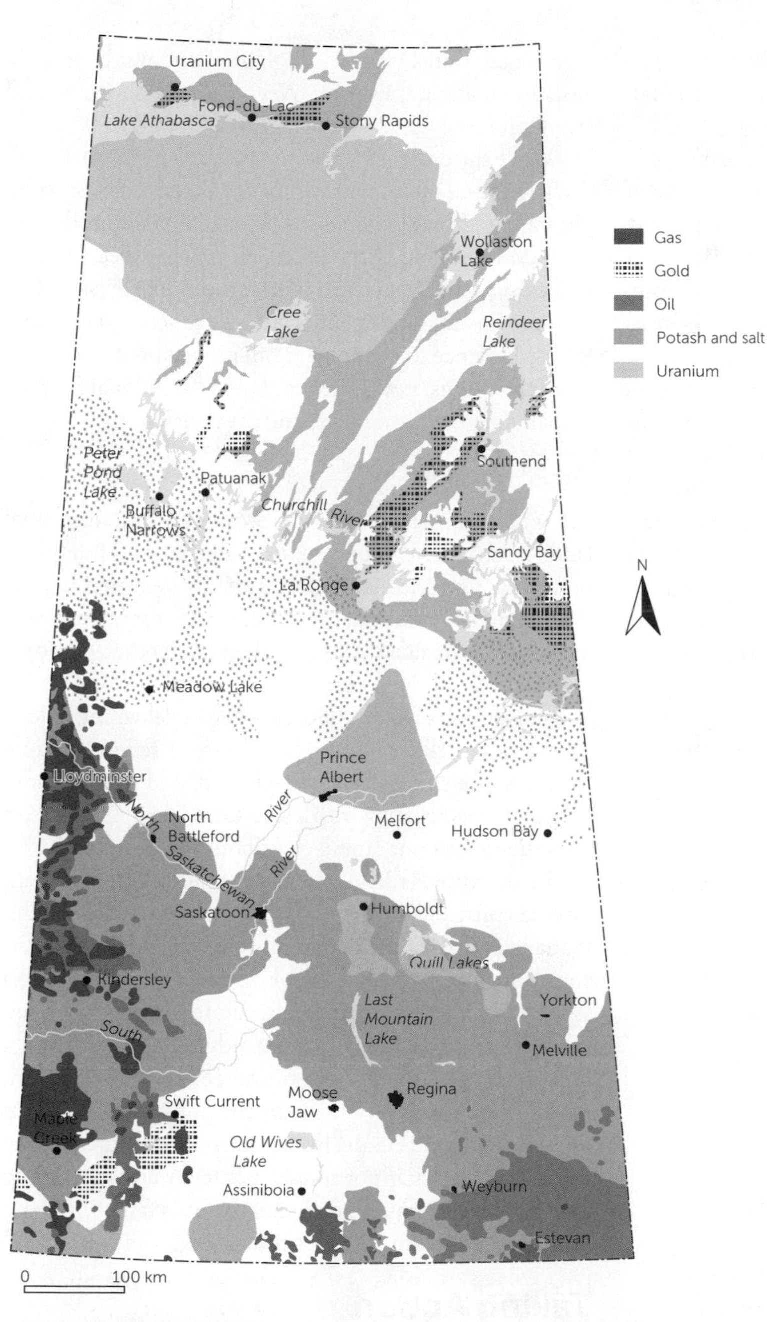

Figure 5.1.1 • Saskatchewan industry.

Source: Adapted from The Encyclopedia of Saskatchewan.

In 2012, the Government of Saskatchewan released its vision for a province of 1.2 million people by 2020. The Saskatchewan Plan for Growth includes a road map for getting there while emphasizing that improved quality of life in Saskatchewan is the purpose of economic growth. The plan states:

After decades of stagnation and decline, Saskatchewan communities are growing and will continue to grow because more young people are staying in Saskatchewan, more families are moving home and more new facilities from across Canada and around the world are choosing Saskatchewan as the place to be. Agriculture and natural resources will continue to drive growth in Saskatchewan's exports. As an export and trade intensive economy, Saskatchewan is a safe and stable supplier of food, potash and energy to emerging economies around the globe.

According to the Canada Mortgage and Housing Corporation (CMHC), Saskatchewan's expanding economy was projected to lead to an increase in employment of 2.6 per cent in 2014, and the unemployment rate was expected to remain low (CMHC 2013a).

Interprovincial migration was largely responsible for the surge in late 2006 and 2007. The population increase was the largest the province had experienced since 1921; in 2010, Saskatchewan hit a historical population high, and it has continued to increase. Saskatchewan now ranks among the national leaders in provincial population and economic growth (Statistics Canada 2007).

Housing Situation

A growing population has increased the demand for housing supply and affected affordability in the province. Saskatchewan has seen a drastic increase in the price of homes and rental prices since 2006; vacancy rates have also fallen, and housing starts have not kept pace with the demand. In 2012 alone, average resale prices increased 5.7 per cent across the province (CMHC 2013a). In April 2013, approximately 51 per cent of homes were valued at more than $300,000 in Regina (CMHC 2013c), and in Saskatoon approximately

22 per cent of homes were valued at more than $400,000 (CMHC 2013d), making them affordable to only a small segment of the population.

Rental prices have also increased substantially; across the province, the monthly rent for a two-bedroom apartment increased by some 4.1 per cent to $977 from 2012 to 2013 (CMHC 2013b). Average two-bedroom apartment rental prices in Regina, Saskatoon, Estevan, and Lloydminster surpassed the $1000 mark, up to as much as $1143 (CMHC 2013b). Rental affordability is a major issue, but vacancy rates are also a source of concern. Average apartment vacancy in the province was 3 per cent in April 2013 (CMHC 2013b), with rates as low as 0 per cent in Estevan and 1.1 per cent in both Weyburn and Lloydminster because of the increase in heavy oil production in the region. However, vacancy rates had increased marginally in both Regina and Saskatoon: Regina's was up from 0.6 to 1.9 per cent, and Saskatoon's increased from 3.1 to 3.3 per cent over the previous year (CMHC 2013b).

Significant housing challenges exist in certain areas of the province because of industrial activity. High-growth areas have seen drastically increased prices and extremely low vacancy rates, which has left those not employed in high-paying industry jobs in need of affordable accommodations. These high-growth areas include Saskatoon, Regina, Estevan, Weyburn, Lloydminster, and Humboldt.

While the provincial government attempts to tackle the challenges associated with rapid growth, other factors complicate the affordable housing realm. Federal funding has been declining on an annual basis and is expected to be non-existent by 2030 when mortgages on the existing federal–provincial housing units are fully amortized. In order for the Province to continue to offer the current affordable housing options, it will have to provide additional funding to maintain, replace, and expand the housing stock.

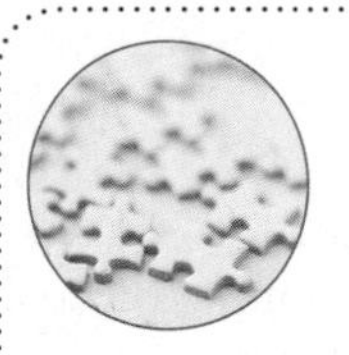

KEY ISSUE

Rental Housing

Renting has traditionally been a more affordable housing option than homeownership, which is generally why young and low-income households rent rather than own. The rental vacancy rate measures the per cent of all purpose-built rentals that are vacant and available for rent. Three per cent is widely accepted as a balanced rental vacancy rate (FCM, n.d.). Lower vacancy rates mean that households have greater difficulty finding a place to rent. This high demand may drive up rents. In 2009, the City of Regina adopted a new housing policy to stimulate an increase in rental housing; the policy included a moratorium on apartment-to-condominium conversions when the vacancy rate falls below 3 per cent.

Social housing

For decades, CMHC and its provincial and territorial counterparts have provided a wide variety of housing programs, some cost-shared and others delivered by one government or the other. From 1973 to 1985, there was increased expansion of the social housing stock. Until 1986, the provinces and territories administered the joint public housing projects, but in 1986, they began delivering federal programs as well. As the 1990s progressed, federal funding for social housing gradually decreased, and new program delivery ended in 1993 with a shift from ongoing long-term subsidies to upfront capital contributions (CMHC, n.d.).

In the 1996 federal budget, the federal government offered provinces and territories the opportunity to manage and deliver existing off-reserve federally funded social housing (CMHC, n.d.). In 1997, Saskatchewan signed the first agreement with the federal government, involving approximately 33,000 federally assisted social housing units. According to the agreement, federal funding will continue to decline annually until it ends in about 2030. Based on this declining subsidy and the current size of the housing portfolio, Saskatchewan will have to reassess the program and ensure that it is targeting those most in need as efficiently as possible.

Taking Action

Based on such strong economic growth and its impact on the housing sector, both in terms of supply and affordability, the Government of Saskatchewan prepared and released "A strong foundation—The Housing Strategy for Saskatchewan" (Government of Saskatchewan 2011). The strategy encompasses the entire housing sector and promotes a housing environment in which all Saskatchewan residents will have access to homes that enhance their well-being, build local communities, and contribute to a growing province.

As part of the Province's action in the first year of the eight-year strategy, it committed additional funding for targeted housing programs. A few months earlier, at the annual Housing and Development Summit in spring 2011, the Province had announced that it would provide additional funding to help increase housing supply across the province.

Thus, as families and individuals in need struggle to afford private-market housing, the Province is attempting to better allocate the limited funding available so that it can best serve those who need it the most.

Housing Continuum

As part of the Housing Strategy, the Province began looking at housing as a continuum—the spectrum of housing in Saskatchewan communities (see Figure 5.1.2). Individuals and families can progress through the housing continuum or move around, depending on their stage in life and the changes that take place. The Province is assessing how targeting funding at certain points along the continuum can alleviate pressure and work to move individuals and families through the continuum.

The different types of housing, from no house at all to private homeownership, have their own unique characteristics but are connected through the broader continuum. Activity or pressure in one area of the continuum affects other areas. For example, a rise in the cost of homeownership can create a backlog in the private rental market if the supply of rental housing is not there, because individuals and families cannot afford to own. Affordable homeownership is normally the mechanism whereby people can vacate private rental housing, which benefits others with lower incomes and less economic and social independence.

The Province views the housing continuum as a tool to understand Saskatchewan's housing infrastructure and the people within it. As a result, program policy professionals are creatively addressing program design issues and taking a proactive approach to lessen the effects of the housing crisis on residents most in need. The Province has begun implementing changes to various housing programs, as well as reallocating funds to new initiatives that have been designed to work in conjunction with existing programs. The bulk of the public funds, including federal and provincial funding, is targeted to low-income residents most in need of affordable accommodation that cannot be met by the private market. However, the Province has also developed initiatives that do more than target low-income residents; it is addressing the entire housing continuum through a range of programs to meet the diverse needs in various communities, thus ensuring a healthy housing continuum for all residents. A number of these programs are discussed in greater detail below.

Planning for Housing

The government is also stepping up planning for housing and has developed a program to encourage municipalities to plan for housing at all points along the continuum. The program will better prepare municipalities for housing development and may provide incentives to developers looking to tap into housing needs in various municipalities across the province. This upfront housing readiness may involve financial incentives, tax breaks, fee waivers, subsidized land, special zoning bylaw provisions, or other innovative methods of increasing housing development within each municipality.

In early 2012, the Province announced two additional sources of funding as part of an existing housing program to allow municipalities, or groups of municipalities, to complete a **Housing Plan**, as well as additional funding to implement strategies contained within a Housing Plan or **needs assessment**. The program is aimed at

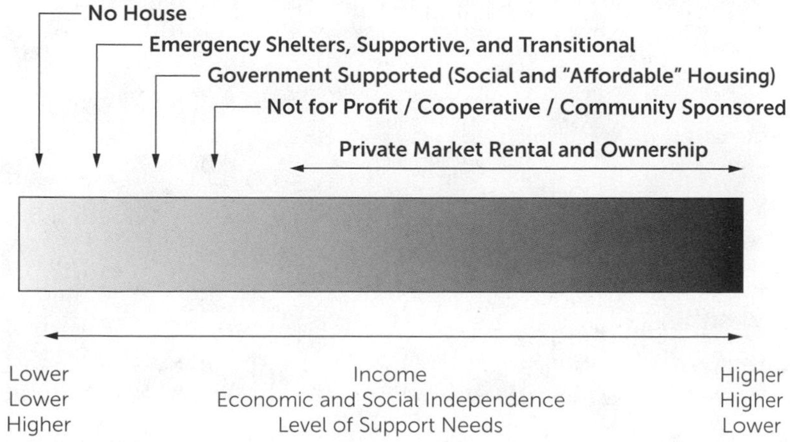

Figure 5.1.2 • Saskatchewan's housing continuum.

Source: Government of Saskatchewan 2011.

helping a greater number of municipalities to meet their housing needs and sustain their population and economic growth.

The Province recognizes the various roles other parties play in the housing sector and is working with municipalities and regions to help address housing needs on a local/regional basis. The funds are intended to ensure that municipalities are in a position to manage current and future housing pressures.

The Encouraging Community Housing Options—Housing Plan and Action Component allocates funds to eligible applicants on a first-come, first-served basis. A minimum population base, as well as a completed or underway **Official Community Plan (OCP)**, is required for an applicant to qualify for funding. The program then provides a cost-matched grant for each of the components: up to $20,000 for a Housing Plan and up to $10,000 for action strategies identified. A total of $400,000 was allocated for the two components.

Since its release, more than half of the available funding has been allocated to 14 municipalities/regions across the province. More than a quarter of the funding has been paid out: seven housing plans and three actions have been completed, with an additional four housing plans currently underway. The municipalities approved include the high-growth areas within the province, which means that there will be further housing work in these areas, ideally involving the private sector.

Spurring Innovation in Housing

The government has also invested in an innovation fund (the Summit Action Fund) to spur new ideas in the development of housing. The hope is that private developers, researchers, housing providers, and others will develop and implement new technologies and concepts that have the potential to affect the housing industry on a broad scale, thereby addressing the housing challenges experienced across the province.

A total of $6 million was allocated to the Summit Action Fund. The goal of the fund is to increase housing supply across the province through creative, flexible approaches and innovation in the housing sector. Delivered through calls for proposals, the fund covers from 30 to 70 per cent of eligible project costs, to a maximum of $500,000. Projects are evaluated on the basis of their impact on housing supply, level of innovation, level of private investment, demonstration value, value to the housing industry, probability of success, urgency, long-term benefit, and alignment with the Summit Action Fund principles. A Selection Committee, made up of representatives of the Saskatchewan Housing Corporation Board and the Canadian Home Builders' Association (Saskatchewan), is responsible for allocating the funds.

So far there have been three calls for proposals. Close to $3.9 million was allocated to 12 projects in the first two intakes; the result of the third intake has not yet been announced. In 2011 and 2012, $1.3 million had already been disbursed to proponents selected in the first and second intakes. The selected projects include providing financial assistance to first-time homebuyers, building new spaces for adults with intellectual disabilities, developing new affordable houses in the inner city, developing a revolving local homeownership program, helping employers offer housing benefits to employees, creating new rental units in a high-growth area through significant municipal financial incentives, and partnering with a community fundraising effort to develop a retirement villa.

Figure 5.1.3 • Tax Sponsorship Program, Hartford Greens, Saskatoon, showing progress of new, affordable homes in communities in need.

Source: Government of Saskatchewan 2012b.

As progress on these initiatives continues to unfold, the hope is that these subsidized models will provide enough information, by way of lessons learned, to pave the way for future housing developments across the province. The major results of the initiative will be evident in the coming years.

Housing Targeted to Those Most in Need

The Province's Social and Affordable Housing Programs, which make up the public housing portfolio, have also been reviewed, and changes are being made to provide access to those most in need. Social housing is subsidized, provincially owned housing targeted to low-income households that would otherwise not be able to afford safe, secure shelter. Social housing tenants pay rent calculated on a sliding scale to a maximum of 25 per cent of their income. Affordable housing is provincially owned with rent set at the lower end of the private market.

Social housing program changes

In a news release in July 2012, the Province announced changes to the Social Housing Program. These changes were intended to help make room "for those truly in need by encouraging people with adequate resources to move into the private market." The news release added that "these changes are necessary when you consider how much Saskatchewan's housing market has evolved over the past several years and factor in our record-level population growth." The decision to implement these changes was based not only on the lack of social housing but on the need to move individuals and families along the housing continuum in order to reduce backlogs in the continuum of available housing.

Under the program changes, new social housing tenants are being selected on the basis of a more balanced approach, taking into consideration their financial circumstances and the safety of, condition of, and crowding in their existing living situation. Consideration is also given to the homeless and victims of domestic violence. Revisions include the introduction of **income and asset limits**; rent set at 30 per cent of adjusted household income, thus creating more transparent rental prices; removal of an $800 rental cap, which will encourage households that would experience a drastic rent increase based on the 30 per cent income calculation to move to the private market and create spaces for those in the greatest need; and an increase of the minimum rent from $100 to $314 in line with the minimum shelter benefit provided by the Ministry of Social Services (Government of Saskatchewan 2012a).

Portfolio changes

The Province's assets are also being evaluated to ensure that programs are meeting their intent and utilizing the depleting funds most efficiently. As a result, the government is selling off single-family detached units (Davis 2012) that require significant repairs. The proceeds of these sales will allow the Province to construct new, more efficient units for a greater number of families and individuals than the current stock can accommodate. The units will be in the form of multi-family dwellings, which are easier to operate and maintain over the long term. The program will also serve as an incentive for families who are capable of moving out of the subsidized multi-family units into private-sector single-family homes for rent or sale.

As part of the Government of Saskatchewan's key actions to target housing programs and services to those in need, the Province plans to sell 300 government-owned single-family housing units in Regina, Moose Jaw, and Prince Albert and undeveloped land in Regina; the Province would then reinvest the proceeds in new medium-density, multi-unit housing. This action will generate additional funding for the development of new low-income rental units. Many of these homes are fairly old, have become quite costly to maintain, and can be very challenging for some household types (e.g., a single parent would be required to ensure that snow is cleared and the lawn mown). The Province believes that it can provide better service to their client group if they live in multi-unit homes and can also maintain the properties more cost-effectively if they are in closer proximity to each other, creating a more energy-efficient and effective public housing system.

The current single-family stock will also offer excellent entry-level homeownership opportunities within the three communities. However the Province is aware that timing is a critical issue in the sale of so many entry-level homes and is committed to not flooding the market, which would affect people trying to sell their homes. Because of sensitivities around divesting certain public properties, the Province is also committed to constructing new units before selling the existing units to ensure that the stock is maintained.

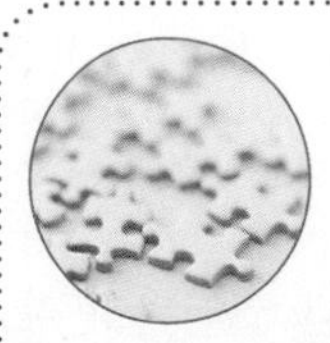

KEY ISSUE

Maintaining Infrastructure

Across Canada, in multiple sectors, the burden of maintaining existing infrastructure is a challenge. Housing, education, health care, and municipal infrastructure are all at various stages of requiring additional investment and/or replacement. As the population continues to grow and the closure of facilities becomes less acceptable, governments are struggling to keep facilities safe and deliver the programming that citizens expect and require.

Offering some of the single-family homes in the public portfolio for sale should encourage more individuals and families to transition to the private rental or homeownership market. Moreover, why should taxpayers, many of whom live in private rental housing, have to contribute to the subsidization of low-income tenants in single-family homes?

The portfolio changes, along with changes to the Social Housing Program, will contribute to Saskatchewan achieving the goal of having those most in need residing in social housing.

Progress in Saskatchewan

While progress on a concept as large as housing supply and affordability in Saskatchewan is difficult to track and measure, positive steps have clearly been taken to address the housing shortage and high prices. As the Housing Strategy clearly states, collaboration across the housing sector must take place to support the province's growing population and ensure that housing is accessible to all residents. The Housing Strategy also states that it is not about quick fixes: it is about making change for the long term.

There has been much action on the part of the private sector, the non-profit sector, and the municipal sector in recent years. All sectors are playing their role in addressing the housing crisis that has affected this growing province.

A Model for Addressing Housing Challenges

The housing challenges in Saskatchewan are not unique. The timing, history, and circumstances around them may be, but housing supply and affordability are challenges across Canada and around the world. The Province of Saskatchewan understood the situation and the future outlook of a growing province, accepted the reality that programs quickly become outdated in a changing economy, and collaborated with others to plan for housing across the province, spurring innovation on the part of those who do innovation best and adapting programming with an eye to the future.

It was all about coming together and taking action to achieve small gains in the right direction and taking a realistic approach that would help to overcome the housing challenges over the long term.

Conclusion

The Government of Saskatchewan is beginning to make changes at a time when change is a constant in the province. As programs are revised and public funds better allocated to address the various needs identified, residents, employers, and investors will begin to see the results of these changes.

KEY TERMS

High-growth areas
Housing continuum
Housing Plan
Income and asset limits
Interprovincial migration
Natural growth
Needs assessment
Official Community Plan (OCP)

REFERENCES

CMHC (Canada Mortgage and Housing Corporation). n.d. "The evolution of social housing in Canada." http://www.cmhc-schl.gc.ca/en/corp/about/cahoob/upload/Chapter_9_EN_W.pdf.

——. 2013a. "Housing market outlook Canada edition—Canadian housing market: Housing starts lower in 2013, increasing modestly in 2014." http://www.cmhc-schl.gc.ca/odpub/esub/61500/61500_2013_Q03.pdf.

——. 2013b. "Rental market report Saskatchewan highlights." https://www03.cmhc-schl.gc.ca/catalog/productDetail.cfm?lang=en&cat=59&itm=15&fr=1382299804641.

——. 2013c. "Housing now Regina CMA." https://www03.cmhc-schl.gc.ca/catalog/productDetail.cfm?lang=en&cat=70&itm=53&fr=1382299750391.

——. 2013d. "Housing now Saskatoon CMA." https://www03.cmhc-schl.gc.ca/catalog/productDetail.cfm?lang=en&cat=70&itm=55&fr=1382299750391.

Canadian Association of Petroleum Producers. n.d. "Saskatchewan's industry." http://www.capp.ca/canadaIndustry/industryAcrossCanada/Pages/Saskatchewan.aspx.

Canadian Press. 2012. "Saskatchewan: The 'it' place to be." *Maclean's Magazine*. http://www2.macleans.ca/2012/12/22/saskatchewan-the-it-place-to-be.

Davis, L. 2012. "120 single family homes may be sold." Discover Moosejaw website. http://www.discovermoosejaw.com/index.php?option=com_content&task=-view&id=25966&Itemid=902).

Elliot, D. 2012. "Recent demographic trends in Regina and Saskatchewan." http://www.sasktrends.ca/Demographics%20for%20Business%20Trends%20Regina%20Chamber.pdf.

FCM (Federation of Canadian Municipalities). n.d. "No vacancy: Trends in rental housing in Canada." http://www.fcm.ca/Documents/reports/FCM/No_Vacancy_Trends_in_Rental_Housing_in_Canada_EN.pdf.

Government of Saskatchewan. n.d. "Potash." http://www.ir.gov.sk.ca/Potash.

——. 2011. "A strong foundation—The Housing Strategy for Saskatchewan." http://www.gov.sk.ca/adx/aspx/adxGetMedia.aspx?mediaId=1513&PN=Shared.

——. 2012a. News release: "Social Housing changes to help more people in need." http://www.gov.sk.ca/news?newsId=fc168456-6fb4-4b3d-98db-834b29081116.

——. 2012b. "Update on the Housing Strategy for Saskatchewan—April 2012." http://www.socialservices.gov.sk.ca/2012Update-HousingStrategy.pdf.

Statistics Canada. 2007. "Canada's population estimates." http://www.statcan.gc.ca/daily-quotidien/071219/dq071219b-eng.htm.

"The Encyclopedia of Saskatchewan." n.d. "Mineral resources." http://esask.uregina.ca/entry/mineral_resources.html.

CASE STUDY 5.2

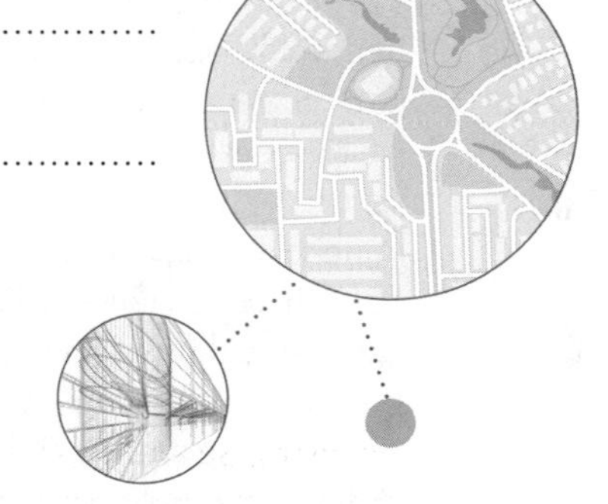

We Call Regent Park Home: Tenant Perspectives on Redevelopment of Their Toronto Public Housing Community

Laura C. Johnson

SUMMARY

Regent Park, Canada's oldest and largest public housing project, located in downtown Toronto, is undergoing redevelopment. In a 15- to 20-year phased public–private partnership initiative, the deteriorating low-density 1950s **modernist** housing development is being razed and rebuilt as a high-density, **new urbanist**, mixed-use, and socially mixed community. Instead of an enclosed island off the grid of city streets, the site is being reintegrated into the surrounding streetscape. A social development plan guaranteed the original low-income Regent Park tenants a right of return to their rebuilt homes. Qualitative longitudinal research has tracked a sample of 52 households from the first phase of those original, low-income public housing residents through their displacement, temporary relocation, and eventual resettlement—or not—into newly built housing in their old neighbourhood. This chapter documents their experiences in this process and their evaluation of the more and less successful aspects of this innovative redevelopment. At the time of writing, halfway into the redevelopment, fewer than half of the sampled households had returned to their original community. Some left by choice, some are still waiting, some feel they have been excluded. Lessons learned can inform subsequent phases of this process as well as public housing renewal in other jurisdictions.

OUTLINE

- This chapter examines Regent Park, Canada's oldest and largest public housing project, during its redevelopment phase.
- The project involved conflicts between tenants' views and the prerogatives of the public–private partnership directing the redevelopment.
- Because similar public housing projects across the country are facing redesign and redevelopment, the lessons from Regent Park could help improve the design and planning processes, easing the return of tenants in other cases.

Introduction

Toronto, like many North American cities, faces the challenge of a deteriorating public housing stock. Early in the new millennium, the Toronto Community Housing Corporation (TCHC) launched a $1-billion, phased 15- to 20-year plan to redevelop Regent Park, Canada's first and largest public housing project. The project was originally built as slum clearance in the 1940s and 1950s, but its reputation quickly became tarnished as buildings fell into disrepair. Regent Park became badly stigmatized (Purdy 2005; Silver 2011), with a reputation as a high-crime area characterized by drug dealing and vandalism. Critics blamed its 1950s low-density modernist design, located off the grid of city streets.

In the mid-1990s, John Sewell, former mayor of Toronto and ex-chair of the Metro Toronto Housing Authority, connected Regent Park's problems to flawed planning of the original streetscape:

> [I]f you design space badly, you have very bad social results. It's the planners who are to blame for this being such a difficult community to live in. . . . they placed these apartment structures

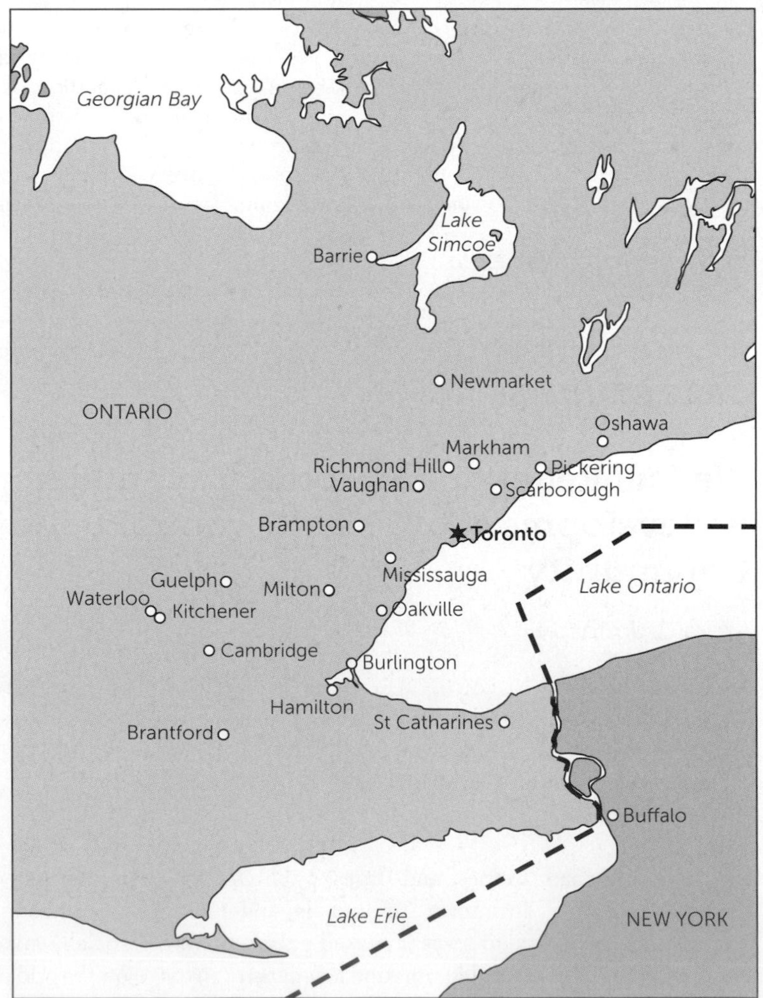

Figure 5.2.1 • Location map.

> among a sea of grass on no streets at all. No one feels that they're connected to anything. They've cut this community off from anything that surrounds it. . . . No one knows who is in control of it and usually what happens in that situation, the strongest take over (Weyman and NFB 1994, cited in Purdy 2005).

Developed as a public–private partnership between TCHC and private real estate developer Daniels Corporation, the current redevelopment plan calls for the demolition of virtually all the buildings, changing the streetscape from a modernist to a new urbanist design. The tools in today's planners' toolbox include introducing **social mix**, increasing density, diversifying land uses on site, and implementing participatory planning and community development initiatives. One of the planners' stated objectives is to improve the quality of life for the public housing community. Tenants were offered two notable opportunities: (1) involvement in planning the redevelopment and (2) right of return to their rebuilt community. This chapter assesses how successfully these commitments were implemented.

The effort is currently at about the halfway mark. This chapter is based on my longitudinal research tracking a sample of 52 resident households through displacement, relocation, and eventual resettlement—or not—into their rebuilt community and exploring how well the housing authority's promises have been met. The chapter will examine tenants' role in the redevelopment process, along with their relationship with the housing authority, their experiences exercising their right of return, and their views on their neighbourhood's redeveloped streetscape. The chapter concludes by considering implications of the first phase of the Regent Park model for subsequent redevelopment phases and for redevelopment of public housing elsewhere.

Figure 5.2.2 • A two-storey mural with a portrait of a Regent Park resident of a building slated for redevelopment. The mural is one of series produced by Regent Park youth supervised by artist Dan Bergeron.

Source: R.E. Johnson.

Context

This is a downtown neighbourhood, an immigrant settlement area that is among the most ethno-culturally diverse in the world. It has a higher proportion of young people and large households than the city population as a whole, with household incomes far below those of most other neighbourhoods. Regent Park has a long tradition of active community involvement as well as a rich collection of innovative community services, including a neighbourhood health centre and a student tutoring program, Pathways to Education, that has been emulated in other communities across Canada. Regent Park is managed by TCHC, Canada's largest **social housing** provider, which oversees 58,500 housing units (TCHC 2013), with 67,704 households on the waiting list for social housing in 2011 (City of Toronto 2011).

Research Methods

The study recruited a volunteer sample of Phase 1 households by various means, including direct screening, posters, and referrals from community agencies. Information was collected by personal interviews. In all, 95 household interviews were conducted, mainly in English. With residents' permission, interviews were audio- and/or video-recorded. NVivo software was used to identify key themes. In addition, 12 key informant interviews were conducted with representatives of the housing authority, the developer, and community organizations. Throughout the research process, my research assistants and I attended public meetings, including community updates organized by TCHC.

"Our New Housing": Resident Participation in Planning and Design

The 15- to 20-year phased redevelopment plan was conceived and implemented via a partnership between TCHC, the municipality of Toronto, the developer (Daniels Corporation), the Regent Park residents' association and its predecessor resident organizations, and other local community organizations. Participation of residents was key to planning the redevelopment, and residents' input has influenced some principles and key design decisions of the redevelopment—for example, the strategy of razing and rebuilding without retaining any of the original buildings. Yet on issues closer to their individual homes, Phase 1 residents sometimes felt their design views were ignored. Overall small sizes of many of the units, insufficient storage space, lack of bathtubs in bathrooms of seniors' units, absence of apartment balconies on some units, and open-plan kitchens throughout are some of the design features to which residents report they objected in the early planning stage.

The issue of kitchen design has cultural significance for many residents, especially Muslims, who believe a kitchen should offer women privacy for cooking. In the new open-plan designs, the person cooking is visible. Some units even give a guest a direct sightline into the kitchen from the front door. One resident who had participated in planning meetings explained his objection to the unit design:

> The living room and the kitchen are not separated.... And 90 to 100 per cent of the Asians, the Africans ... do not like that unless they grow up here. They do not like kitchen and living room that does not have any separation.

The subject is not simple; it raises important issues about how much public housing should cater to the ethno-cultural preferences of specific tenants and how much tenants of public housing should get to choose the design of their housing units. *The Economist* (2013) reported on public opposition in The Netherlands to building "halal flats" to suit immigrants' preferences for dwellings with partitions separating women and men. Nonetheless, many Regent Park tenants who—when consulted—opposed the open-plan kitchens were surprised that no other options were offered. Muslims were not alone in wanting kitchen privacy. Another study participant from a different background described her misgivings:

> I don't like the idea of an open kitchen.... God—someone walks into your door [and] you haven't done your dishes—the whole place looks like, you know!... I'm not the cleanest person, but I do try to be!

Some critics fault the public–private partnership model that delegates major design decisions to the developer (Rahder and Milgrom 2004). But the redevelopment is ongoing; there is an active residents' association, which could organize better communication between tenants and TCHC to address these features of the landlord–tenant relationship in future phases of the project.

At an update meeting organized by TCHC, one tenant asked whether the buildings needed to be so tall and so close together. The TCHC official chairing the meeting said yes: increased density was how the redevelopment was financed. Here too, tenant input was not decisive. Residents subsequently learned that under a revised zoning application, even greater height and density are now being proposed for the subsequent phases.

Choosing Temporary Housing

At the displacement starting in 2005, tenants from the approximately 400 Phase 1 households were asked to line up to choose temporary housing, shifting to vacant units within Regent Park or moving to TCHC housing elsewhere in metropolitan Toronto. They understood that a system of first-come, first-served would also establish tenants' priority for choosing their new housing once it was built. Some tenants objected that they hadn't been given sufficient notice about the timing of the line-up. The line-up was a particular hardship for residents with young children, for those going to work, and for seniors and persons with health challenges. One study participant recalled her experience:

> I just saw people are running, I told my children I just go and look.... So when I go there there was a big line, so I just stand in the line and I told [my daughter] if I get late to come you just phone my sister is there... and you go to school and she will come... and she stand for me in the line.... That's how.

After hours of standing, tenants were given sequential numbers, after which they could leave.

The system was difficult, but many felt it was worth the effort. Some, however, were vociferous in their criticism. TCHC subsequently organized supplementary line-ups according to household size and the number of bedrooms for which households were eligible. With the introduction of a lottery to establish order of choice for newly built housing, some tenants became anxious that the rules were changing.

Moving Out—Redux

Some Phase 1 residents who moved to alternative housing within the old Regent Park faced another unexpected challenge. In a last-minute change, the City of Toronto decided to build an indoor swimming pool in Regent Park. Although most community members applauded this addition, its site was a building and adjacent townhouses to which about a dozen Phase 1 households had been relocated. The demolition of these structures ahead of schedule meant that those households had to move a second time. Although they were assured by TCHC that in exchange for this inconvenience they would be placed atop the list for choosing new housing, some found the experience particularly stressful.

"We're Regent Parkers for Life": Tenants' Exercise of Their Right of Return

In many North American examples of public housing redevelopment, the original population has been scattered (Clampet-Lundquist 2004; Manzo, Kleit, and Couch 2008). Regent Park's planners consulted widely about recent redevelopment initiatives in other jurisdictions. Their stated aim in Toronto—a hallmark of this project—was to guarantee tenants' right of return to a newly built unit inside Regent Park. This point was made clear to tenants in the early consultative meetings and at the time of their displacement from the Regent Park homes. Yet as presented at a Phase 1 community update meeting late in 2010, three buildings in an area designated as "East Downtown"—half a kilometre or more from Regent Park—were to provide some of the Regent Park replacement units, with an overall split of 1583 on-site and 500—subsequently revised to 266—in East Downtown (TCHC 2010; 2015).

These units were offered as a voluntary choice, but some felt pressure to accept or at least consider the off-site option. One woman described her disappointment at being offered housing outside the footprint:

> When we moved they told us we can move back... but in the end they told us we have to move into Carlton or whatever other places—like that's not

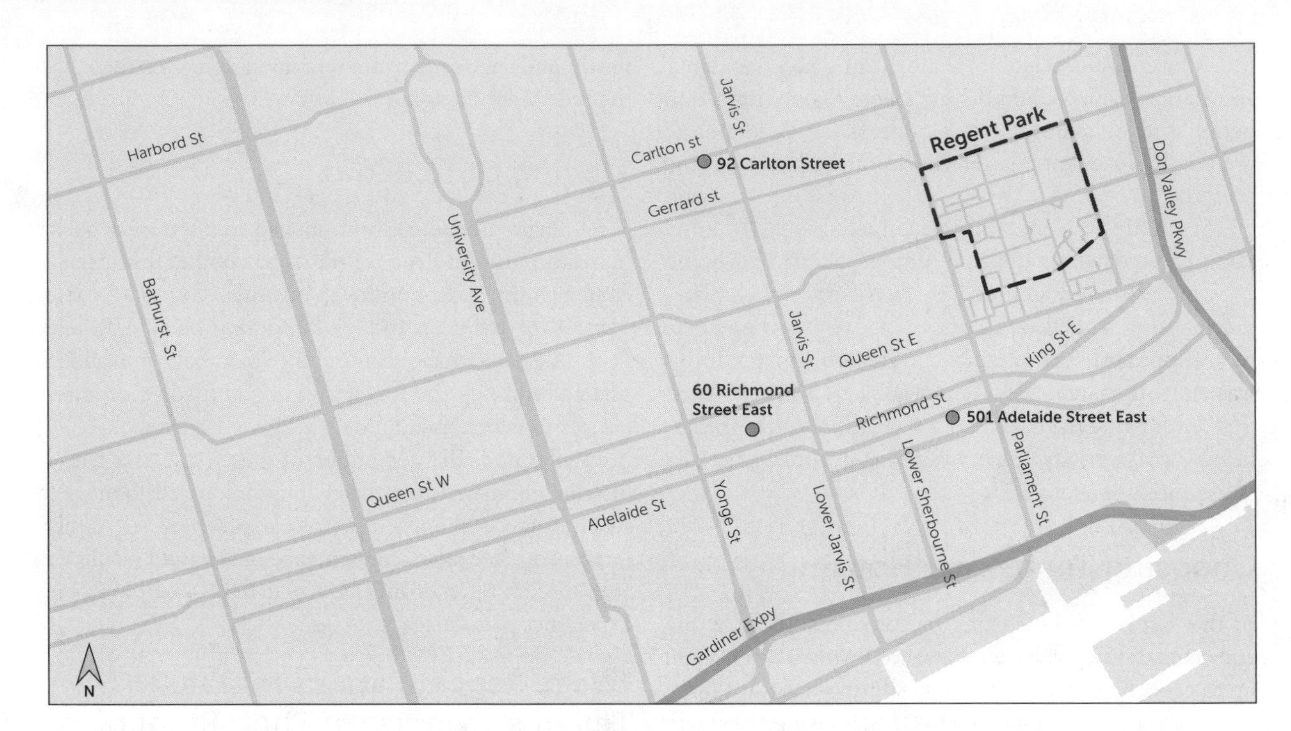

Figure 5.2.3 • Regent Park and surrounding areas, including three alternative housing locations in East Downtown offered to Phase 1 Regent Park tenants.

> basically within Regent . . . that's further, and it's really hard for me.

Another woman explained her preference for returning to her previous neighbourhood:

> I said I wanted to move back to the place I lived before. . . . I wanted to come back in my old area. I said I don't want to move [to] Richmond or something like that—it's not safe. It's far from the grocery and everything.

From our most recent interviews (see Table 5.2.1), only 21 of 52 Phase 1 households have now been resettled inside the Regent Park footprint. Another 10 were moved to new housing in East Downtown outside the footprint; four were still waiting for what they considered suitable new housing inside Regent Park; three had left the TCHC system; six preferred to stay in other alternative housing; and eight were lost from the study. If these participants' experience was typical, the actual rate of return for Phase 1 was somewhere between 40 per cent (counting just those resettled within the Regent Park footprint) and 60 per cent. Data from the almost-complete Phase 2 show a higher rate of return, with an estimated 65 per cent of displaced households already resettled inside the boundaries of Regent Park and an additional 23 per cent still expecting to resettle.

Table 5.2.1 • Housing Status of Sampled Phase 1 Regent Park Resident Households, 2013 (n=52)

Housing Status	Number of Households
New housing inside Regent Park footprint	21
New housing outside Regent Park footprint	10
Still waiting for new housing inside Regent Park	4
Living elsewhere in TCHC system	6
Outside TCHC system	3
Lost from study	8
Total	52

Those who have resettled inside the Regent Park footprint are generally pleased with their new housing; those resettled outside express mixed feelings. Some are pleased to have left Regent Park, but others are disappointed and resentful, feeling that they "didn't make the cut." They report feeling pressured by TCHC relocation counsellors to accept equivalent new housing outside the boundaries of their original neighbourhood. Several, having grown tired of living in temporary quarters, settled very reluctantly in the new accommodations.

Initial plans called for an approximately 40 to 60 per cent split between **Rent Geared to Income (RGI)** units and market condominiums. Since the TCHC Regent Park tenant households tended to be larger than those who would occupy the private-market units, it was anticipated that the population would be evenly divided between RGI and condominium dwellers. But in its newest iteration (May 2013), Phase 3 of the plan proposes to reduce the proportion of social housing to 25 per cent of housing units by adding more condominium units, thereby increasing the population density of the site; even with this change, RGI occupants (because of their generally larger families) will account for an estimated 45 per cent of residents of the completed development. This change, according to TCHC, will make the redevelopment financially feasible. To the low-income tenants, it is another sign that they are losing the Regent Park that they knew and called home. It marks another change in what they understood to be the revitalization plan for their community.

Tenants' Communication with the Housing Authority

A frequent complaint concerned the lack of information given to tenants about unit selection. When choosing new housing units, tenants were shown only floor plans. While some could interpret the plans, others found it difficult to visualize three-dimensional space from them. They relied on children, other relatives, or friends for help. One resident who was proficient at plan interpretation found a creative way to help some of her neighbours visualize the size and layout of new apartments:

> For one lady . . . I took green painter's tape and marked out the floor in the living room because I had a bigger living room than her. I said "This is your living room and kitchen," and I marked out where the sink would be, the fridge . . . the sofa and everything . . . she visualized it that way. . . . They [TCHC] handed out these papers . . . yet they couldn't do a three-dimensional video thing of these apartments? . . . Some people just can't understand it when you look at a piece of paper.

Another tenant complaint concerned the kitchen stove in new units: with a control limiting how hot the electric stove can get, it would take a long time just to boil water. To TCHC management this is a safety feature, but some tenants felt it would prevent them from cooking in their traditional ways.

These experiences left some tenants feeling powerless and frustrated.

Footprint, Screening, and Tenants' Information Deficit

From the very beginning, some Phase 1 tenants interviewed had requested temporary relocation and eventual resettlement alongside their previous neighbours, but they reported that TCHC was unsympathetic to this request. Some who agreed to resettle outside their original neighbourhood felt bitter—as though they had been screened out of their rightful place in a neighbourhood to which they belonged. Sadness was another response—people described missing their former neighbours, their old neighbourhood, and the shops and services available nearby. Some of those outside the footprint held out hope that they would eventually be able to return.

The idea of screening pervaded many resettled residents' narratives. Screening would contravene TCHC policy and violate explicit promises to tenants. Nonetheless, a certain number of Phase 1 residents, including some who did move back into the footprint, believed that a covert screening mechanism was used to determine who would be allowed back inside the footprint.

The pressure to accept alternative housing outside the Regent Park footprint was part of a larger picture whereby tenants felt that rules and assurances were not being followed. The aforementioned lottery system was another example. Residents had previously believed that their priority for selecting new housing would be determined by the timing of their move into temporary housing. Such changes created a climate of insecurity among the relocated resident households, conducive to a belief in the idea of screening. Despite denial by TCHC staff, people began to believe that only certain residents would be resettled inside the Regent Park footprint. Encouragement by relocation counsellors to consider settling outside the footprint made some residents suspect that they were being screened out.

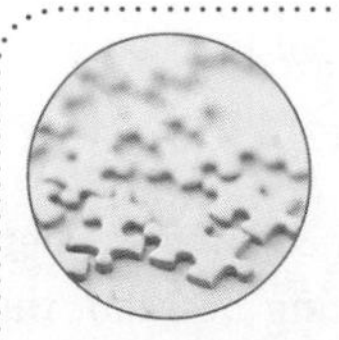

KEY ISSUE

Social Mix

Lacking government support for building or redeveloping public housing, the Regent Park redevelopment was financed by leveraging land on the site for private development. In effect, the spaces between buildings in the original low-density site were sold to finance the rebuilding of the subsidized housing. While there is a financial basis for this approach, it also carries a social rationale—that there is inherent value in mixing populations from different income levels. According to this view, the lower-income tenants are expected to benefit from living in proximity to middle-class neighbours. This view is largely unsubstantiated by empirical research, though numerous housing scholars are investigating the issue (Chaskin and Joseph 2013; Graves and Vale 2012; Tach 2009).

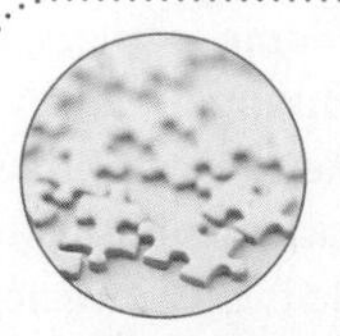

KEY ISSUE

Re-establishing the Street Network

The current redevelopment plan is reintegrating the Regent Park streets with those in the surrounding neighbourhoods, giving residents greater access to amenities and services beyond the project's boundaries. New streetscapes are already in evidence, with buildings oriented toward the street and a range of shops and services that were not available inside the old Regent Park. Entrances to new buildings are well landscaped. Addresses are marked clearly. The remaining old buildings are in sharp contrast, located in fields of trampled grass with garbage dumpsters beside the entrances.

Planners and developers have high expectations that the new streetscape will normalize Regent Park, integrating the neighbourhood into its surroundings and making it more permeable. For some original residents, the issue of permeability isn't as clear-cut as the planners would have it. One Phase 1 resident expressed her surprise that the streets were so open: "anyone could enter" the newly constructed streets. She was uncertain how much safety the new design will offer.

Conclusion

At mid-point in redevelopment, the new Regent Park looks impressive. High-rise, market-rate condominiums and social housing towers are virtually indistinguishable from each other (see Figure 5.2.4). Transformed from an exclusively residential area, the mixed-use neighbourhood now includes a supermarket, a pharmacy, a coffee shop, a restaurant, and the first new bank branch to open in more than 40 years. Redevelopment created more than 1100 new jobs for residents. The municipality, senior levels of government, and Daniels Corporation, the developer, have contributed to funding new amenities on-site, including an indoor aquatic centre, an arts and culture centre, and new premises for local community social and educational organizations.

Some Phase 1 tenants are very pleased with their new homes, others less so. One mother of a large family expressed a wish to return to her previous Regent Park home, a unit that has since been demolished. That place, she said, had plenty of room and a kitchen in which she could cook the way she wanted. She appreciated the privacy it had offered, without picture windows, glass panels by the front door, and open-concept living space.

For these Phase 1 tenants, displacement, relocation, and resettlement were challenging and sometimes alienating experiences. While initial plans promised active tenant engagement, some felt that their views and their attempts at input were ignored. While a right of return was promised, many haven't made it back. Those fortunate (or tough) enough to have resettled into new housing inside their old community are mostly very pleased with their housing situation.

Interviews with original residents revealed the importance of their place attachment to their public housing community. A 14-year-old who had lived in the same

Figure 5.2.4 • New Regent Park housing, with ground-oriented townhouses in the foreground and new high-rise housing in the background.

Source: R.E. Johnson.

Regent Park building for 10 years and had been relocated with her family to a housing unit elsewhere in Regent described her feelings about neighbourly ties and a sense of safety:

> It feels like I spent my whole life there . . . it was like I knew everybody and liked everybody. It was a comfortable place to live. You always felt safe because everybody was around you . . . even though the building was very old, it felt like a good place to be in. . . . I just loved the building [and] the people living in it. . . . We weren't afraid.

The Phase 1 tenants' ties to their neighbours and their neighbourhood are strong, and they want assurance of their right of return. For many, replacement housing in the vicinity of their neighbourhood is unsatisfactory. Residents find particularly stressful the sense that policies and procedures about returning may change during the relocation.

When the proposal was first presented to the community, the rebuilt Regent Park was to have been comprised of approximately 40 per cent RGI and 60 per cent market housing. The latest numbers proposed are 25 per cent RGI and 75 per cent market, with further rezoning changes to increase the number and height of high-rise condominium towers. The new Regent Park will bear little resemblance to the original community the residents had known. Public–private partnership forms the very foundation of Regent Park's redevelopment. The experience of Phase 1 suggests that this corporate model may limit the scope for resident input.

KEY TERMS

Modernism
New urbanism
Rent Geared to Income (RGI)
Social housing
Social mix

REFERENCES

Chaskin, R.J., and M.L. Joseph. 2013. "'Positive' gentrification, social control, and the 'right to the city' in mixed-income communities." *International Journal of Urban and Regional Research* 37(2): 480–502.

City of Toronto. 2011. Quick Fact #16: "Number of households on the active centralized waiting list for social housing in April 2011." Toronto: City of Toronto.

Clampet-Lundquist, S. 2004. "HOPE VI relocation: Moving to new neighborhoods and building new ties." *Housing Policy Debate* 15(2): 415–47.

Graves, E.M., and L.J. Vale. 2012. "The Chicago Housing Authority's plan for transformation: Assessing the first ten years." *Journal of the American Planning Association* 78(4): 464–5.

Manzo, L.C., R.G. Kleit, and D. Couch. 2008. "Moving three times is like having your house on fire once." *Urban Studies* 45(9): 1855–78.

Purdy, S. 2005. "Framing Regent Park: The National Film Board of Canada and the construction of 'outcast spaces' in the inner city, 1953 and 1994." *Media, Culture and Society* 27(4): 523–49.

Rahder, B., and R. Milgrom. 2004. "The uncertain city: Making space for difference." *Canadian Journal of Urban Research* 13(1): 27–45.

Silver, J. 2011. *Good Places to Live: Poverty and Public Housing in Canada*. Toronto: Brunswick Books.

Tach, L. 2009. "More than bricks and mortar: Neighborhood frames, social processes, and the mixed-income redevelopment of a public housing project." *City and Community* 8(3): 269–99.

The Economist. 2013. "Race in the Netherlands: The aftermath of a football tragedy." 12 January.

TCHC. 2010. "Regent Park: Building great neighbourhoods." (Presentation at Community Update Meeting, 28 October.) Toronto: TCHC.

—— 2013. http://www.torontohousing.ca/about.

—— 2015. "Regent Park by the numbers." http://www.torontohousing.ca/regentpark.

Weyman, B., and NFB (National Film Board of Canada). 1994. *Return to Regent Park*. Montreal: NFB.

CASE STUDY 5.3

Energy-Efficiency Retrofits and Planning Solutions for Sustainable Social Housing in Canada

Sasha Tsenkova

SUMMARY

Canada's Economic Action Plan (CEAP) recently provided support for energy-efficiency retrofits in the social housing sector in an effort to create green jobs and provide an effective response to climate change. The social housing sector was targeted as a field of policy intervention in which socially responsible and very professional housing providers had the potential to capitalize on government funding to leverage further investment as well as to showcase the results of transformative change. Within the context of this political commitment, the chapter draws on the first comprehensive assessment of program results with the following objectives:

- To review the results of national and provincial programs targeting energy-efficiency retrofits in the social housing sectors of Ontario, British Columbia, and Alberta, exploring program efficiency and effectiveness in Toronto, Vancouver, and Calgary;
- To identify preferred investment strategies and planning responses by different social housing providers—public, private non-profit, and cooperative—through an in-depth analysis of select case studies.

OUTLINE

- Energy-efficiency retrofits have been completed in a range of social housing projects across Canada.
- The key policies and federal and provincial programs that enabled the retrofits are presented, and their success is evaluated in terms of energy and cost savings and program effectiveness.
- Energy-efficiency retrofits have encouraged experimentation with sustainable design and green technologies such as solar walls/roofs and green roofs, but there are significant market barriers to their effective implementation.

Analytical Framework and Methodology

The general hypothesis advanced in this research is that a more supportive policy framework for energy-efficient transformation of social housing will yield better results. Furthermore, the institutional culture, market share, and commitment to sustainability of social housing providers will put them in a better position to implement innovative strategies for energy-efficiency retrofits in their portfolio (Niebor et al. 2012). The policy environment is deconstructed through analysis of key policy instruments—regulatory and fiscal—to determine the main factors affecting the types of retrofits implemented and investment priorities (Gruis, Tsenkova, and Niebor 2009).

The analytical framework draws on network governance and presents the investment strategies of social housing providers as contextually dependent on the policy environment in which they operate (Van Bortel and Mullins 2009). Network approaches incorporate the explicit recognition of interdependencies and interactions among various actors within policy arenas and the institutional context within which policy and actors operate. Sørensen and Torfing's (2007) concept of network governance is based on interdependent actors where interactions are of a horizontal nature without a dominant party that can control the decision-making process. Approaches of "steering in the shadow of hierarchy" (Scharpf 1997) and coordinative planning can be used to explore the capacity of governments to plan and implement programs in a more complex decision-making environment.

Illustrating the dynamic and messy process of real-life planning and implementation, this chapter describes how interdependent networks of provincial, city, and neighbourhood-based social housing organizations have implemented a program of energy-efficiency retrofits in Canadian cities. Inequalities of power within networks can lead to hierarchical relationships focused on one or more dominant actor, such as a provincial or federal organization, a process that is determined by the institutional and political context for decision-making. The chapter highlights the importance of coordinative planning in which the focus is on ways to deploy organizations so that they can implement mutually agreed-upon goals, objectives, and outcomes. In terms of institutional design, the case studies demonstrate the ability of provincial governments to use their hierarchical powers to steer program implementation in a strategic or comprehensive manner.

The primary research draws on analysis of 18 case studies and more than 45 key informant interviews carried out in the three cities in 2012 (Tsenkova 2013). The chapter profiles three case studies of innovative projects,

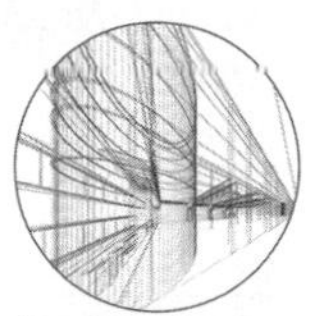

PLANNING THEORY

Coordinative Planning Paradigm

Defined as one of the four main paradigms—rational planning, communicative practice, coordinative planning, and frame-setting—coordinative planning is anticipatory co-ordination about where to go and how to get there (Alexander 2000, 247). The focus is on ways to deploy organizations so that they act at the appropriate time to implement mutually agreed-upon goals and objectives leading to accepted outcomes. In terms of institutional design, the emphasis is on rules and organizational structures to enable co-ordinated action, often through strategic planning. In this context, strategic rationality provides a dynamic framework for the choice of means and the prioritization of objectives and goals. The coordinative planning process is reflective, incorporating inter-organizational networks, large and complex organizational subunits, and institutional constraints in its institutional design (Alexander 1995). The interdependence of participants and their respective orientation may lead to consensus-building and the mutual understanding that defines the normative and methodological aspects of coordinative planning.

identifying major outcomes related to types of energy retrofits, building-envelope improvements, costs, energy savings, and cost recovery. The case study analysis is integrated in a broad comparative evaluation of long-term results focused on program efficiency and effectiveness (Beerepoot 2007; Fuller 2009; Mlecnik, Visscher, and van Hal 2010; Tsenkova 2011). The case study projects were selected on the basis of recommendations from program managers with the purpose of demonstrating innovative practices in energy-efficiency upgrades and some of the more comprehensive measures implemented, including building envelope, technical system upgrades, and installation of renewable energy sources. Data on investment strategies and policy responses by social housing providers were collected through primary research and site visits.

Energy Efficiency Programs in the Social Housing Sector

Canada introduced the Renewable Energy Initiative (REI) in 2009 with $70 million in funding for energy-efficiency upgrades in existing social housing projects and new affordable housing projects. The federal and provincial governments contributed equally to this investment as part of Canada's Economic Action Plan (CEAP). It provided $850 million over two years for the renovation and retrofit of existing social housing, plus a further $475 million to build new rental housing for low-income seniors and persons with disabilities. Federal funding allocation to various provinces and the corresponding number of projects is presented in Table 5.3.1. Overall, CEAP includes $2 billion for new and existing social housing, plus up to $2 billion in loans to municipalities for housing-related infrastructure. It builds on the Government of Canada's commitment in 2008 to provide more than $1.9 billion over five years to help the homeless and improve and build new affordable housing.

The four main objectives of the CEAP Renovation and Retrofit Initiative were to:

- address the demand for renovations and general improvements;
- address the need for energy-efficiency upgrades;
- address the accessibility needs of persons with disabilities; and
- create jobs (CMHC 2010).

In addition to these funds, **Canada Mortgage and Housing Corporation (CMHC)** administers a system of grants to social housing providers who manage social housing under CMHC agreements. The provincial/territorial distribution of these funds—a total of $150 million—is presented in Table 5.3.1. CMHC had specific Canada-wide guidelines and standardized procedures. Priority was given to housing providers with well-managed housing stock and up to $4500 per unit in their replacement reserves. Priority was also given to renovations and retrofits that included critical building systems and services, such as roofs, windows, doors, exterior building envelopes,

Table 5.3.1 • Canada's Economic Action Plan Renovation and Retrofit of Existing Social Housing—Provincially/Territorially Administered

Province/Territory	Funds Allocated (Federal $M)	Projects
Ontario	352.16	5817
British Columbia	88.82	105
Alberta	45.38	1105
All Provinces/Territories	**850.00**	**11,355**
CMHC-Administered		
Ontario	40.50	299
British Columbia	43.20	138
Alberta	11.20	122
All Provinces/Territories	**150.00**	**1312**

Note: Program details are available at http://cmhc.ca/en/inpr/afhoce/fias/fias_017.cfm.

Source: CMHC 2012.

heating systems, plumbing, electrical, and ventilation, as well as renovations required for the health and safety of residents and/or modifications for the disabled. Retrofits were required to meet minimum energy standards, and projects were to be **"shovel-ready,"** with the expectation that they would be completed by March 2011. While there was a formal extension of the CEAP infrastructure stimulus fund until December 2011, the social housing projects were considered "on-schedule" as far as the funds administered by CMHC were concerned, but CEAP funds were administered by the provinces until the end of 2012.

Results

In Ontario, the Ministry of Municipal Affairs and Housing used notional allocation to distribute the funds to the 47 consolidated municipal service providers in the province. If the service manager administered 30 per cent of the total social housing units in the province, they received 30 per cent of the funding available. Service managers had full authority on how they distributed funds to social housing providers (Ontario provincial program manager, personal communication, August 2012). In Toronto, the Shelter, Support and Housing Administration (SSHA) was authorized to submit projects on behalf of the City to the ministry for funding from the Social Housing Renovations and Retrofit Program (SHRRP) and REI.

The total SHRRP allocation of $259 million had a significant impact on the social housing portfolio in Toronto. The Toronto Community Housing Corporation (TCHC) received the largest share (58 per cent), followed by the non-profit housing providers (31 per cent). In terms of the impact measured by the number of units affected, TCHC improved more than half of its portfolio with SHRRP funds, while the units involved in the non-profit and co-op sector accounted for 34 per cent and 11 per cent of the total, respectively. However, a comparison of these statistics against the number of units managed by non-profits and co-ops in the city (see Table 5.3.2) reveals that every non-profit and co-op housing provider received funding and support to upgrade more than 90 per cent of the units in their portfolios. The allocation model used to distribute the REI funding was based on submissions from each social housing provider. The data indicate that REI funding supported more TCHC projects, perhaps because of its institutional capacity and previous experience with renewable energy retrofits since 2006.

In British Columbia, CEAP funding was implemented through the BC Housing Renovation Partnership. The program covered improvements in 20 per cent of the social housing in the province. Because the most significant share was allocated to BC Housing, it is not surprising that the program addressed in a comprehensive manner both energy efficiency and capital improvement needs in the provincial housing portfolio. Most of the funding, $164 million, was directed toward repairs at 84 social housing developments managed by BC Housing and 38 projects managed by the non-profit housing sector (with more than 2000 units). Half of the social housing developments were located in Vancouver.

Table 5.3.2 • SHRRP and REI Funding Allocations and Impact

Proportion of SHRRP and REI Funding Allocated				
	SHRRP		REI	
TCHC	$150,688,073	58%	$21,396,674	70%
Non-profits (other than TCHC)	$79,841,232	31%	$5,797,272	19%
Cooperatives	$28,505,246	11%	$3,478,297	11%
Total	**$259,034,551**	**100%**	**$30,672,243**	**100%**
Proportion of Units Impacted by SHRRP and REI				
	SHRRP		REI	
TCHC	32,419	55%	7,645	70%
Non-profits (other than TCHC)	19,924	34%	2,200	20%
Cooperatives	6,610	11%	1,152	10%
Total	**58,953**	**100%**	**10,997**	100%

Interview data indicate that energy and costs savings were realized, and feedback from tenants was positive. Overall, housing providers believe that the goals of the programs were achieved, but it remains to be seen how much actual energy has been saved across the entire portfolio (Tsenkova 2013). BC Housing developed a business model whereby they partner capital projects with sustainability initiatives from other provincial programs, such as funds for a carbon-neutral public sector through the Public Sector Energy Conservation Agreement (see the Greenbrook sustainability project, below). BC Housing leveraged several funding sources to maximize the reduction in the energy use and greenhouse gas (GHG) emissions of its housing stock, including BC LiveSmart. LiveSmart has more modest targets, but the $2000 per unit subsidy provides an incentive to introduce measures with quick returns.

Figure 5.3.1 • Greenbrook.

Source: Sasha Tsenkova.

Greenbrook Sustainability Project, Surrey, British Columbia

Built in 1974, Greenbrook is a public housing development owned and operated by BC Housing, consisting of 127 units in 28 townhouses that are home to 380 people. The Greenbrook sustainability project combined both building-envelope replacement and energy upgrades to achieve significant energy savings and physical improvements. The use of high-efficiency heating and electrical systems reduced GHG emissions by 86 per cent in 2010 compared to the 2005 baseline. The project boasts the largest residential solar panel installation in western Canada, which offsets about 10 per cent of the site-used electricity and a large portion of the remaining energy consumption, resulting in a housing complex that is very close to being carbon-neutral (Tsenkova and Youssef 2012).

The institutional framework for CEAP program management in Alberta was centralized within the Ministry of Housing and Municipal Affairs, which was responsible for assessment, approval, and disbursement of project funds. The ministry did not have specific guidelines or priorities to complement the federal guidelines; it considered requests for building-envelope improvements, measures addressing safety and accessibility, replacement of heating and ventilation systems, and other specific energy-efficiency retrofits (director, personal communication, November 2012). All of the projects were self-managed by housing providers, with general monitoring and control exercised by the ministry and the CMHC-Prairies Office. Contractors were chosen based on a **tendering process**, following standard procurement guidelines for public works. The disbursement of funds was done on the basis of invoices for completed retrofits in the project-approval document (senior program administrators, personal communication, November 2012).

Interview data provide essential metrics of performance and program results for the province of Alberta. Provincially administered program funds were allocated to 747 projects and involved more than 20,827 units out of a total 26,000 units in the social housing portfolio. Two-thirds of these units are in Edmonton and Calgary, with the largest share of funding allocated to the two largest social housing providers in the province. CMHC program funds of $11.2 million targeted co-ops and non-profit providers in the two most populous cities in the province.

The three case studies present more in-depth analysis of the types of retrofits in Vancouver, Toronto, and Calgary and basic data related to costs, energy savings, and return on investment. The projects include different types of housing—from low- to high-rise developments—built in the 1970s and 1980s. Apartments in Vancouver and Calgary have **Rent Geared to Income (RGI)** whereby tenants pay 30 per cent of their income in rent, while Villa Otthon in Toronto, managed by a non-profit organization, has a mix of RGI and market rental units. In all projects, residents do not pay for utilities, and the cost of heating and hot water is included in the rent.

Types of Retrofits

Grant McNeil Place in North Vancouver is part of BC Housing's portfolio, and its energy retrofits were managed by one of the two energy service companies (ESCOs) selected to implement the program in the province. An energy audit identified a comprehensive package of retrofits, including window and door replacement, sealing to reduce air leakage, replacement of lighting, and provision of low-flow equipment. Larger retrofits were also completed, such as upgrades to space heating, exhaust systems, and domestic hot water systems (see Table 5.3.3). The project dealt with asbestos and mould abatement due to the use of dated building construction methods, poor thermal bridging, and general building lifecycle issues. A public engagement strategy contributed to tenants having a better understanding of the importance and positive effects of retrofits. This helped to improve tenant buy-in and supported effective communication (BC Housing asset manager, personal communication, February 2012).

In Toronto's Villa Otthon, an audit preceded the allocation of SHRRP support. Mechanical upgrades, such as the replacement of heating systems, makeup air units, and cold-water booster pumps, accounted for most of the investment, whereas non-mechanical upgrades, such as general repairs to building facilities and replacement of appliances, represented the smallest percentage of project investment (Tsenkova 2013). Retrofits related to renewable energy technology involved solar thermal installations. Since the retrofits substantially affected all apartments and common areas, management constructed two mock-up apartments to demonstrate the impact they would have. This was particularly helpful in addressing tenant concerns. Building managers worked proactively to minimize the disruption caused by construction work in the building over a 12-month period and had ongoing support from City staff. Tenants reported high levels of satisfaction with the improvement measures (program manager, City of Toronto, personal communication, September 2012).

In Calgary's Baker House, the first priority was window and door replacement, followed by interior modernization (elevator, kitchen, and laundry refurbishment) and access for the disabled (project manager, Baker House, personal communication, April 2012). The window replacement recycled existing bronze metal window frames, while the glass panels were replaced as a cost-reduction measure. Funds used for interior modernization accounted for 20 per cent of the total spending. The high-rise building houses low-income tenants, and current revenue does not allow for the accumulation of sufficient reserve funds to carry out much-needed lifecycle replacement of building-envelope elements and service systems. Other renovations are still needed despite the major improvements resulting from CEAP grant funds. Tenants do not pay for heating costs, which are included in the rent, and even if individual controls are installed in the units to regulate room temperature, there is no real incentive to use them (project manager, Baker House, personal communication, April 2012).

With regard to energy savings, Grant McNeil Place had an annual projected energy saving of 38 per cent, whereas Villa Otthon reported 37 per cent (Table 5.3.3). Cost savings were dependent on the amount of money originally invested and on the type of retrofits implemented. For example, Grant McNeil Place had an original investment of $3.1 million, with annual energy costs savings of $71,330, resulting in a simple cost-recovery time of approximately 43 years. Calculations for return on investment figures equal the annual projected cost savings divided by the original investment. Replacement of the electric heating plant with a gas-fired heating plant in Villa Otthon did cost $2.9 million but was prompted by a disproportionately high bill for heating and utilities that exceeded $400,000 per year. The new system uses natural gas, which resulted in a major reduction in utility bills, but the cost recovery exceeds 77 years (program manager, City of Toronto, personal communication, September 2012).

In addition to the economic benefits of reducing energy use, the feasibility studies claim significant environmental gains resulting from reduced consumption of water and GHG emissions. For example, a reduction of 295 tonnes of GHG is equivalent to growing 7565 tree seedlings for 10 years or taking 54 passenger cars off the road for a year. Such gains are impressive, given the fact that in two of the projects, the GHG reduction

Table 5.3.3 • Case Study Profiles

Case Study	Grant McNeil Place North Vancouver* 	Villa Otthon Lambton, Toronto** 	Baker House Calgary*** 
Project Type and Characteristics			
Year of construction	1976	1989	1971
Building type	Low-rise apartment	Residential tower/ townhouses	High-rise apartment
Number of units	112	194	213
Total area	113,832 sq. ft	n/a	120,319 sq. ft
Project Economics			
Total funding received	$3.1 million	$3.9 million	$1.9 million
Type of rent	RGI	65% RGI	RGI
Rent	$500/month	$1000/month	$500/month
Tenants			
Tenant turnover	Very low	Medium	Very low
Tenants pay utilities	Yes	No	No
Retrofits Completed			
Energy audit performed	Yes	Yes	No
Boiler replacement	Yes	Yes	Yes
Domestic hot water	Yes	Yes	No
Solar hot water system		Yes	
Heat pump (furnace)		Yes	
Unit kitchen and bathroom upgrades	Yes	Yes	Yes
Water pumps	Yes		No
Window replacement	Yes		Yes
Door replacement	Yes		Yes
Lighting	Yes	Yes	No
CPTED (crime prevention through environmental design)	Yes		No
Common area upgrades	Yes	Yes	Yes
Other non-mechanical	Yes	Yes	Yes
Mould/asbestos removal	Yes		No
Tenant Feedback Positive	Positive	n/a	Positive
Energy Costs and Projected Savings*			
Energy costs	$205,412	$455,373	n/a
GJ	3981	n/a	n/a
Energy savings	(38%)	(37%)	n/a
kWh	45,118	3,062,123	n/a
Tonnes GHG/reduction	211	296	n/a
Cost savings	$71,330	$168,244	n/a

*Sources: * BC Housing asset manager, personal communication, February 2012; ** Program manager, City of Toronto, personal communication, September 2012; *** Calgary Housing Company portfolio manager, April 2012.*

Figure 5.3.2 • Major retrofits at Villa Otthon.

Source: Sasha Tsenkova.

is twice and three times the projected amount (Finn Projects 2009).

Program Successes

CEAP offered a major opportunity for the implementation of a comprehensive package of retrofits and improvements in the social housing sector. The two programs—managed by the provinces and by CMHC—provided $972 and $95 million in federal and provincial matching funds, respectively, in BC, Ontario, and Alberta from 2009 to 2012. In terms of efficiency, the investment was critical in addressing the lack of resources to fund capital repairs and system upgrades in the aging social housing stock. The programs were highly relevant, timely, and successful in meeting their broad objectives and accounted for improvements in about 20 to 50 per cent of the social housing in the three provinces. CEAP granted funds for a variety of mechanical, structural, and building-envelope retrofits affecting two-thirds of the social housing in Toronto, Calgary, and Vancouver. The impact, in terms of units upgraded, was particularly significant for the non-profit and cooperative housing providers, which saw on average more than 60 per cent of their portfolio affected by program measures.

Part of the success can be attributed to the efficient management of the programs by existing federal, provincial, and municipal housing institutions. The institutional framework for rapid deployment of program funds (centralized in BC versus decentralized in Ontario) gave the service providers sufficient autonomy to address priority needs. Because of their institutional capacity, large social housing providers in Vancouver, Toronto, and Calgary were able to address both energy efficiency and capital improvement needs in their social housing portfolio in a more comprehensive manner (Tsenkova 2013). A more robust policy framework for energy-efficiency retrofits in BC and Ontario—incorporating a range of regulatory, fiscal, and institutional policy instruments—positively influenced portfolio investment strategies. In Alberta, less emphasis was placed on energy-efficiency retrofits and more on general renovations of the social housing stock, safety improvements, and enhanced accessibility for persons with disabilities. By contrast, BC Housing developed a business model where energy efficiency was systematically pursued through the partnering of CEAP capital projects with provincial sustainability initiatives. In Ontario, housing providers leveraged funding from REI and other utility-managed programs to maximize the reduction in energy use and GHG emissions.

Hubbard Boulevard: Energy Efficiency by Design

SHRRP funding, REI, and other energy incentive programs enabled the rebuilding of a historic asset in Toronto's popular Hubbard Boulevard neighbourhood, providing 27 social housing units. The building's interior was completely rebuilt, with the original stained glass windows and other historic elements incorporated in the new sustainable design. The total cost of the regeneration was $5,894,340, and it included the following green elements:

- energy-efficient heating, air conditioning, and lighting;
- rooftop solar panels to generate electricity;

- a green roof to improve aesthetics, building cooling, and rainwater management;
- a building automation system to manage energy use.

The effectiveness of CEAP programs is difficult to evaluate in the absence of a systematic monitoring and post-retrofit evaluation system (Kaufman and Palmer 2011; Kikuchi, Bristow, and Kennedy 2009). Nevertheless, evidence from 18 case study projects demonstrates substantial improvements in the quality of social housing, targeted approaches to retrofits, integration of both mechanical and building-envelope measures, and high potential for energy savings (20 to 45 per cent). Notwithstanding the emphasis on "best practices" in the analysis, it is evident that the programs have prompted a more strategic approach to asset management and energy retrofits on the part of major public, non-profit, and cooperative housing providers. Another critical success factor was institutional innovation in BC's use of energy service companies (Tsenkova 2013). With increased interest in energy services achieving energy services achieving energy and environmental goals, some companies providing energy to final energy users, including the supply and installation of energy-efficient equipment and building refurbishment, have begun to proliferate. Such companies differ from traditional energy or equipment suppliers in that they can arrange financing for their operations and their remuneration is directly tied to the energy savings achieved. While the ESCO model was more expensive than the project manager/contract services model, it had value-added components, including economies of scale, a "one-stop shopping" approach, and enhanced accountability for project planning, financing, and monitoring.

Figure 5.3.3 • Hubbard Boulevard.

Source: Sasha Tsenkova.

The CMHC-managed program targeted federal housing co-ops. Staff at the federal level worked hard to overcome the constraints of a decentralized model of social housing providers to ensure that program benefits were available to all. Efforts included assistance with project submissions, monitoring of spending, and site inspections to ensure consistency between planned and actual program measures. Some of the most popular retrofits in addition to lighting—"the low-hanging fruit"—were roofing, window replacement, cladding/insulation, and mechanical system upgrades (boilers). Energy-efficiency retrofits were supported, but this was not necessarily a priority. The cooperatives were able to address the tension between short-term affordability goals and the long-term viability of their housing stock using program funds, thus developing much-needed experience with strategic planning.

Figure 5.3.4 profiles the allocation of funds from the CMHC-administered program to federal housing cooperatives in the four cities under review. The data illustrate the investment priorities and choices made by a variety of housing organizations, measured by the cost of major retrofits. Window and door replacement consumed a considerable portion of the budget in the four cities ($11 million). In Edmonton, it accounted for 43 per cent of the total spending and in Calgary for 29 per cent. Roof replacement was the second-most important type of retrofit ($9 million), particularly significant in Vancouver and Toronto. Other major categories of retrofits—interior upgrades and foundation work/exterior cladding/insulation—accounted for $7 and $6 million. The first of these was important for cooperatives and non-profits in Calgary and Vancouver, while the second was critical for social providers in Toronto.

Program Challenges

Program challenges were associated with tight timelines and difficulties in co-ordinating and planning strategic retrofits. Although the projects supported through CEAP were deemed "shovel-ready," housing providers and building managers had to operate within a two-year time frame. Unexpected building envelope problems, such as mould, poor insulation, and asbestos removal, were frequently reported, resulting in cost overruns, project delays, and potential loss of funding if projects were not implemented on time.

The tight deadlines for program management and administration and the sometimes unpredictable nature of construction work led to program extensions and reallocation of funds to other types of measures. The unofficial extension of the program's two-year time frame was mainly due to the fact that extra time was needed for the tendering and contracting of retrofit work, since delays resulting from permit applications and building inspection processes can be lengthy. Some of the smaller social housing organizations reportedly faced capacity constraints and difficulties in managing construction work and contracts and even in qualifying for program funds because of the complex guidelines and procedures.

One of the greatest challenges was the high cost of the program and the lack of sustainability in funding. In more comprehensive improvement and energy-efficiency projects, such as those in the case studies, simple payback periods can be anywhere between 39 and 67 years. Even though the financial viability and cost-benefit of the programs were not the main objectives, they highlight future economic challenges if programs need to be operated on a cost-recovery principle. Limited market penetration of energy efficiency in social housing is constrained by lack of access to capital, high risk, and split incentives (Schüle 2009; Stephenson et al. 2010). Small housing providers face significant challenges in getting access to standard loans and mortgages because of their cash flow and general rent revenue, which makes investments in energy-efficient components much more challenging, given their high upfront costs, lengthy payback period, and uncertainty in energy pricing. Moreover, rents in the social housing sector are often set as a percentage of household income and include utilities, so tenants do not have a direct incentive to reduce their energy consumption. While social housing providers are interested in investing in energy-efficient mechanical systems, tenants

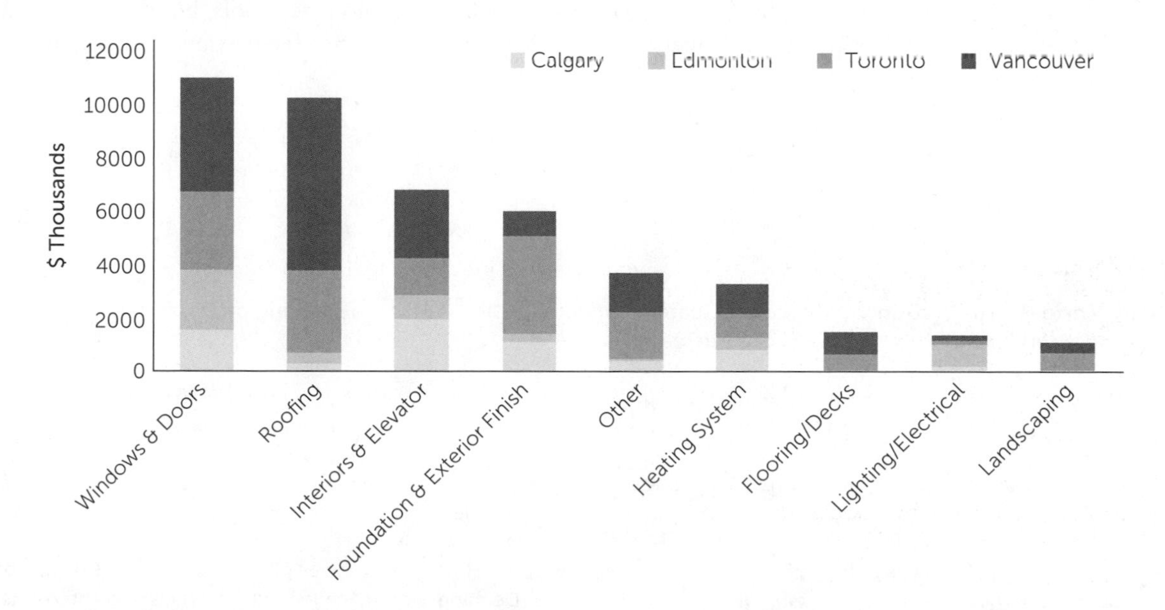

Figure 5.3.4 • A comparison of retrofit and investment priorities, CMHC funding.

often object to such measures because they create temporary inconvenience during the implementation period. Such split incentives, in addition to general behavioural failures and reluctance on the part of consumers to adopt energy-responsible behaviour, hinder the adoption of energy-efficiency measures (Moezzi 2009).

Furthermore, for a number of social housing providers, particularly in Alberta, energy efficiency was a low priority relative to other portfolio considerations (for example, elevator and window replacement or roofing). This is also true for a number of small community-based social housing organizations (non-profits and cooperatives) in BC and Ontario, which lack the institutional capacity to comprehensively plan for energy-efficient retrofits because of limited, asymmetric information and other structural barriers (Leone and Carroll 2010). Market failures and barriers explain the reluctance to adopt cost-effective energy-efficiency measures derived from mainstream economic, organizational, and behavioural theories. Market failures have to do with flaws in the operation of the market, the best known being a lack of complete information and the split incentives problem. The institutional or structural barriers to energy-efficiency implementation result from three conditions: when energy costs are a low priority relative to other factors; when barriers in capital markets inhibit the purchase of energy-efficient technologies; and when energy-efficient markets are incomplete (i.e., energy efficiency is a secondary attribute bundled with other product features) (International Energy Agency 2007).

The physical condition of the social housing portfolio and the lack of adequate reserves to address capital needs significantly affected the implementation of the programs. The trade-off between energy-efficient retrofits and the replacement of deteriorated mechanical and building-envelope components constituted a major challenge to the programs, particularly the segment administered by CMHC. Sometimes there was not enough funding to do both.

The specific retrofit measures in the case studies are diverse and illustrate the significant challenges of such programs in economic terms. If the simple payback of energy-efficiency measures is used as an overall consideration for return on investment, it will be difficult to make the case for green retrofits in the social housing sector. Feasibility studies, however, point to significant environmental benefits resulting from reduced energy and water consumption and reduced GHG emissions (Brophy et al. 2010). Some of these metrics of performance, as well as the social impact measured in tenant satisfaction and improved health and well-being, are difficult to quantify, and any attempt to introduce a system of monitoring and evaluation of results achieved has been very limited, even on a pilot basis. However, key informant interviews with BC Housing policy officials indicate that the monitoring of selected projects is included in ESCO contracts.

While the greening of social housing has many benefits, the installation of green technologies is a strain on capital reserves. CEAP and REI have introduced an important financial boost to experimentation with sustainable design and green technologies such as solar walls/roofs and green roofs, but there are significant market barriers to effective implementation.

KEY TERMS

Canada Mortgage and Housing Corporation (CMHC)
Rent Geared to Income (RGI)
Shovel-ready
Tendering process

REFERENCES

Alexander, E. 1995. *How Organisations Act Together: Interorganizational Coordination in Theory and Practice*. Amsterdam: Gordon and Breach.

——. 2000. "Rationality revisited: Planning paradigms in a post-modernist perspective. *Journal of Planning Education and Research* 19(3): 242–57.

Beerepoot, M. 2007. *Energy Policy Instruments and Technical Change in the Residential Building Sector*. Amsterdam: IOS Press.

Brophy, H., H. Jamasb, L. Platchkov, and M. Pollitt. 2010. "Demand-side management strategies and the residential sector: Lessons from international experience." In

EPRG Working Paper 1034; Cambridge Working Paper in Economics 1060. Cambridge: Electricity Policy Research Group, University of Cambridge.

CMHC (Canada Mortgage and Housing Corporation). 2010. CHS—*Public Funds and National Housing Act (Social Housing) 2009*. Ottawa: CMHC.

——. 2012. "CEAP—Renovation and retrofit projects." Ottawa: CMHC. http://www.cmhc.ca/housingactionplan/reresoho/lipr.cfm.

Finn Projects (Synchronicity Projects Inc.). 2009. *Building Energy Audit for Broadview Housing Co-operative: 1050 Broadview Ave*. Toronto: Finn Projects.

Fuller, M. 2009. *Enabling Investments in Energy Efficiency: A Study of Energy Efficiency Programs That Reduce First-Cost Barriers in the Residential Sector*. Berkeley: University of California, California Institute for Energy and Environment.

Gruis, V., S. Tsenkova, and N. Niebor, eds. 2009. *Management of Privatised Housing: International Perspectives*. Chichester: Wiley-Blackwell.

International Energy Agency. 2007. *Mind the Gap: Quantifying Principal–Agent Problems in Energy Efficiency*. Paris: OECD Publishing.

Kaufman, N., and K. Palmer. 2011. "Energy efficiency program evaluations: opportunities for learning and inputs to incentive mechanisms." *Energy Efficiency* 5(2): 243–68.

Kikuchi, E., D. Bristow, and C. Kennedy. 2009. "Evaluation of region-specific residential energy systems for GHG reductions: Case studies in Canadian cities." *Energy Policy* 37(4): 1257–66.

Leone, R., and B.W. Carroll. 2010. "Decentralisation and devolution in Canadian social housing policy. *Environment and Planning C: Government and Policy* 28: 389–404.

Mlecnik, E., H. Visscher, and A. van Hal. 2010. "Barriers and opportunities for labels for highly energy-efficient houses." *Energy Policy* 38(8): 4592–603.

Moezzi, M. 2009. *Behavioral Assumptions in Energy Efficiency Potential Studies*. Berkeley: California Institute for Energy and Environment.

Niebor, N., S. Tsenkova, and V. Gruis, eds. 2012. *Energy Efficiency in Housing Management*. London: Routledge.

Scharpf, F. 1997. *Games Real Actors Play: Actor-Centered Institutionalism in Policy Research*. Boulder, CO: Westview Press.

Schüle, R. 2009. *Energy Efficiency Watch: Final Report on the Evaluation of National Energy Efficiency Action Plans*. Berlin: Wuppertal Institute.

Sørensen, E., and J. Torfing, eds. 2007. *Theories of Democratic Network Governance*. London: Palgrave MacMillan.

Stephenson, J., B. Barton, G. Carrington, D. Gnoth, R. Lawson, and P. Thorsnes. 2010. "Energy cultures: A framework for understanding energy behaviours." *Energy Policy* 38(10): 6120–9.

Tsenkova, S. 2011. *Investing in Social Housing: Lessons from Ex Post Evaluations of CEB Programmes*. Paris: Council of Europe Development Bank.

——. 2013. *Retrofits for the Future. Renovation and Energy Efficiency Programs in Canada*. Cities, Policy and Planning Research series, January. Calgary: University of Calgary.

——. and K. Youssef. 2012. "Canada: Energy efficiency retrofits—Policy solutions for sustainable social housing." In N. Niebor et al., eds, *Energy Efficiency in Housing Management*, 209–31. London: Routledge.

Van Bortel, G., and D. Mullins. 2009. "Critical perspectives on network governance in urban regeneration, community involvement and integration. *Journal of Housing and the Built Environment* 24(2): 203–19.

INTERVIEWS

BC Housing asset manager, February 2012. In-person interview.

BC Housing project manager, February 2012. In-person interview.

Portfolio manager, Calgary Housing Company, April 2012. In-person interview.

Project manager, Baker House, April 2012. In-person interview.

Program manager, City of Toronto, September 2012. In-person interview.

CMHC program director, August 2012. In-person interview.

Director, Alberta Ministry of Housing and Municipal Affairs, November 2012. In-person interview.

Personal interview, senior program administrators, November 2012. In-person interview.

Ontario provincial program manager, August 2012. In-person interview.

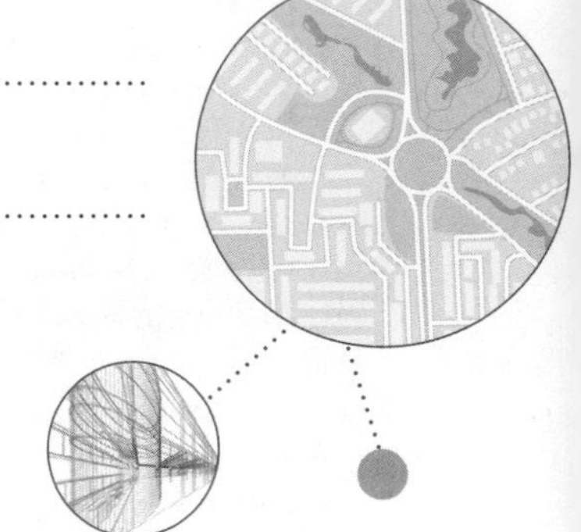

CASE STUDY 5.4

Meeting the Workforce Housing Needs of a Resort Municipality: The Whistler Example

Marla Zucht and Margaret Eberle

SUMMARY

This case study describes the problem of employee housing in Whistler, BC, the evolution of the Whistler response to the shortfall in employee housing, and the current innovative approaches used in Whistler for addressing the critical issue of affordable workforce housing. Through the use of policy, legislative, and regulatory analysis, along with a literature review, the case study highlights the experiences and lessons learned in developing local solutions for the creation of employee housing in Whistler. It emphasizes the challenges, the elements of success, and the lessons that are transferable, along with how replicable this alternative housing approach may be for other municipalities facing similar issues.

OUTLINE

- The development of affordable housing for the workforce in Whistler, BC, a resort municipality, offers a unique approach to persistent affordability issues.
- Among the possible alternatives for affordable housing, Whistler's solution to the community's persistent housing problem shows a unique governance and process involving planners, developers, and the public.
- The affordable housing options in Whistler involve several innovations in workforce housing and governance and have been successful at meeting the goal of 75 per cent local workforce residency.
- Although Whistler has an unusual governance structure as a resort municipality, there are several possible ways to transfer this model to other Canadian cities facing housing affordability challenges.

Introduction to the Planning Problem

Whistler is a unique place, located in the beautiful Coast Mountains two hours north of Vancouver, offering spectacular skiing and mountain biking on two world-class mountains and hosting visitors from around the world. Throughout the 1970s and 1980s, Whistler's population grew and development occurred at an exponential rate. Second-home owners, seasonal workers, and entrepreneurs flocked to Whistler to enjoy the mountain lifestyle. The price of real estate rose dramatically, making it difficult for employees to find affordable accommodation. Along with the increasing popularity and population of the resort community were some other critical limiting factors, including a constrained mountain–valley geography and a community growth management strategy that capped the number of allowable beds in the valley. Because housing is a market commodity, when demand

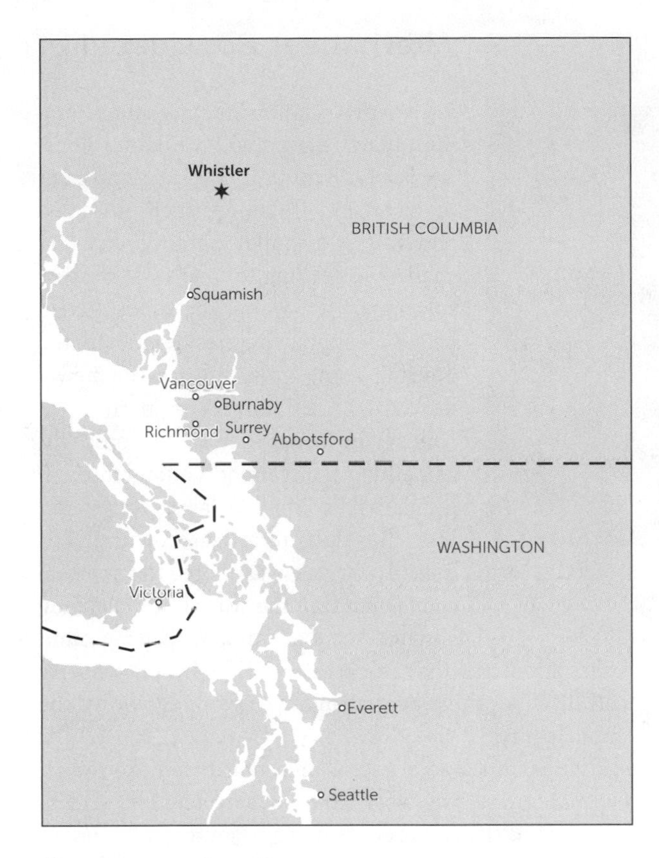

Figure 5.4.1 • Location map.

exceeds supply, prices rise. This created a situation in which local housing costs were increasing faster than local incomes, making housing increasingly unaffordable for local employees.

Because of a shortage of available housing at affordable rates, many local employees were forced to live outside Whistler in neighbouring towns where housing was more affordable and travelled the mountain highway daily to their place of work. It became difficult for local businesses to attract and retain employees. Clearly, the market alone would not create housing for the workforce at workforce wages. **Social housing** was not the right fit for this population. Another solution was needed.

Recognizing this, the Resort Municipality of Whistler (RMOW) commissioned a consultant to investigate the housing needs of the community and to provide recommendations on what might be done to address them (City Spaces Consulting Ltd 1997). The consultant estimated the current and projected housing needs of the community at **build-out** in 2000 and concluded:

> there is a serious and worsening housing affordability problem in Whistler. The problem is structural, not cyclical—it's not a matter of riding out the cycle. Unless the situation changes dramatically, the private market will not meet the housing needs of most local working people in an affordable way (City Spaces Consulting Ltd, i).

Stakeholders and Partnerships

Whistler has a long history of trying to address its workforce housing needs creatively, exploring a number of different avenues and approaches, and utilizing the skills and resources of a number of different stakeholders.

Figure 5.4.2 • Whistler.

Source: David McColm.

Figure 5.4.3 • Whistler Creek Court.

In 1975 the Resort Municipality of Whistler was formed with special resort status and its own legislation—the Resort Municipality of Whistler Act—making Whistler Canada's first resort municipality. Under the Act, the Province retained broad, discretionary land-use powers to regulate forms of development within the municipality's boundaries and the power to initiate and finance a new sewer system.

The RMOW adopted its first **Official Community Plan (OCP)** in 1976. The OCP is a provincially mandated regulatory document and set of high-level plans and policies, such as land-use designations that guide land-use planning, including employee housing, social, economic, and environmental policies, and civic infrastructure investments. Just over a year later, Whistler was granted 53 acres of Crown land from the Province and began developing the Whistler village.

The first formal response to high home prices was initiated by local employers, who banded together in 1983 to form the Whistler Valley Housing Society (WVHS), a not-for-profit housing organization governed by a volunteer board of directors consisting of employers and RMOW. This structure enabled the organization to qualify for funding from the **Canada Mortgage and Housing Corporation's (CMHC)** social housing programs. In 1984, WVHS produced Whistler's first, and to this day only, social housing at Whistler Creek Court with funding from CMHC. These units consist of 20 self-contained townhomes with rents geared to residents' income.

Alternatives Considered

Whistler was not the first mountain community to be faced with workforce housing issues. Banff, Alberta, and a number of resorts in Colorado, such as Aspen and Vail, had similar employee housing challenges leading to their own housing solutions. Throughout the 1980s, WVHS tested an approach to resident-restricted housing using ground leases on newly developed residential properties to control the occupancy of the property, something that cannot be done through municipal zoning bylaws. A ground lease is a long-term lease on the land (usually 60 years or more with renewal rights) limiting the use of any dwelling situated on that land. Whistler ground leases stipulated that an occupant of the leased property and dwelling must be an employee working within the municipality.

While this was a useful mechanism for controlling occupancy, creating affordable housing requires significant financial resources. Two main sources of funding were considered in Whistler.

Senior government funding

Although the WVHS could have applied for more funding through CMHC to build more subsidized housing, this was not the ideal fit. Whistler wanted to be self-sustaining and support its own housing developments rather than dependent upon external funding sources and assistance.

Development linkage fees

Many American communities use inclusionary zoning to create affordable housing. Inclusionary zoning requires developers to build affordable housing as part of their rezoning efforts. With inclusionary zoning, developer contributions toward affordable housing generally take two forms: direct construction of affordable housing units or a monetary contribution (cash in-lieu) to a housing fund. Fortunately, Section 933(2.1) of British Columbia's Local Government Act gives Whistler and other designated resort municipalities the ability to impose development charges to finance the building of resident housing for the employees of resort operations, something that is not available to BC municipalities that are not resort communities.

The Solution and How It Works

The latter approach seemed to be the way forward for Whistler, and in 1990 the RMOW enacted the Employee Housing Service Charge Bylaw, requiring developers of commercial, tourist, and industrial land to either build resident housing or contribute cash in-lieu to a designated housing fund, which the RMOW would then use as capital to build employee housing.

This approach succeeded in getting employers and commercial developers to become part of Whistler's employee housing solution. Following the enactment of the bylaw, contributions by commercial developers were made to the municipal housing fund as a condition of development approval. By 1996, the fund had accumulated $6 million in development levies, which was leveraged to create 326 resident-restricted beds for the local workforce.

With the enactment of the Employee Housing Service Charge Bylaw, RMOW Municipal Council also needed to address the issue of who or what organization would steer the process. Council commissioned City Spaces Consulting Ltd to conduct a housing study and examine alternative organizational approaches for handling the housing functions currently shared by the RMOW and WVHS. The consultants drew upon the experiences of other communities in Canada and the US. They concluded that a standalone housing authority would be the appropriate organizational entity, because it offered an arm's-length operating structure, particularly important since the entity would be involved in somewhat "risky" or unconventional housing development in order to tackle the arduous challenge of creating affordable resident-restricted housing for the resort community.

Figure 5.4.4 • 1060 Legacy Way under construction.

Governance

The Whistler Housing Authority Ltd (WHA) was created in 1997 by council to oversee the creation, administration, and management of resident-restricted housing in Whistler. It is a wholly owned subsidiary of the RMOW, and the municipality (the shareholder) owns all of its shares. RMOW Municipal Council, as the governing body of the shareholder, maintains control of Whistler Housing Authority Ltd through its ownership of all the shares. Customarily, the municipality has appointed members of council and senior RMOW staff to the WHA's board of directors sufficient in number (i.e., four of the seven directors) to form the majority and thus maintain control of the company. Two resident positions on the board guarantee a voice for tenants and owners. The WHA also co-ordinates the efforts of the Whistler Valley Housing Society.

The WHA mission:

> Whistler's long term success as a vibrant resort community is contingent upon retaining a stable resident workforce. We will partner with the resort community to sustain a range and supply of housing options for Whistler's active and retired workforce (WHA, n.d. a, 3).

Among WHA's strategic goals, one stands out:

> Secure a sufficient and diverse supply of resident housing in order to keep at least 75% of Whistler's workforce living in Whistler (WHA, n.d. a, 4).

Although operating at arm's-length from the RMOW, the WHA still collaborates closely with the municipality. Responsibility for various housing-related functions is distributed among RMOW staff and council and WHA staff and the board (WHA, n.d. a, 5). The RMOW is largely responsible for regulatory and municipal policies related to development of private-sector and employer-initiated restricted housing, and WHA is the educator, researcher, liaison, facilitator, advocate, and developer/owner, as well as property manager for public-sector and WHA-initiated restricted housing.

The WHA's financial model ensures that the monthly rents from WHA-owned resident-restricted rental units cover debt servicing of mortgages, property management, contributions to capital reserves, and the WHA's operating costs, thus enabling the WHA to operate self-sufficiently with no assistance from municipal property taxes. Funding for new resident-restricted housing continues to be collected through the Employee Housing Service Charge Bylaw and retained in the municipal housing reserve fund. Contributions to the housing reserve fund vary each year depending on the level of development activity. The balance in the fund by the end of 2012 was 1.7 million.

New resident-restricted units continue to be acquired using a combination of funds from the Employee Housing Service Charge Bylaw reserve fund and debt financing. In addition, some developers provide affordable units in-lieu of cash contributions. Over the years, WHA has also sold older real estate assets and used the sale proceeds to develop additional resident-restricted housing.

Process and Role of Planners and the Public

Many of the functions of WHA were originally carried out by the RMOW planning department, including managing proposal calls and developing employee-restricted housing. Whistler's current resident housing policy evolved over many years, involving many community stakeholders and in concert with major land-use and community planning decisions such as the sustainability plan Whistler 2020 and Whistler's Official Community Plan. Early in Whistler's resort planning, it was determined that in order to preserve and sustain the integrity and diversity of the resort community, most of Whistler's workforce should reside within the community. As a result, a community goal of housing a minimum of 75 per cent of Whistler's workforce within municipal boundaries was entrenched.

The WHA connects with the public through various forms of community engagement in an effort to promote and educate the public about housing options and opportunities. The WHA also meets regularly with the RMOW and council on local housing issues. This is especially important to ensure that the unique housing model with its various restrictions and innovations is well understood. In addition, the WHA website describes the legal agreements that form part of the resident housing portfolio.

Figure 5.4.5 • Public meeting.

Challenges Experienced

Whistler has faced numerous challenges in achieving its employee housing goals, and these challenges have evolved over time. In the early days, the primary challenge was generating sufficient revenue through the Employee Housing Service Charge Bylaw to build employee housing at the scale required. The cash in-lieu fee was originally set at approximately $5500 per employee generated. This amount was quickly found to be too low to provide the equity necessary to build a bed unit. It also became clear that residential development, along with commercial, tourist, and industrial development, contributed to job creation and hence increased the need for additional workforce housing, so in hindsight residential development should have been incorporated in the bylaw.

The challenge of the growing demand for employee housing facing the small but increasingly burdened municipal planning department prompted the RMOW to create a separate entity to specifically focus and be tasked with the creation of employee housing. This arm's-length separate housing entity has enabled the WHA to focus efforts on both public and privately developed rental and ownership employee housing. As of 2012, Whistler's almost 1900 units of resident-restricted housing consist of 57 per cent resident-restricted affordable ownership units and 43 per cent resident-restricted rental units.

Over time, a greater share of Whistler's employee housing has been created as price- and resident-restricted ownership housing as opposed to rental housing. Ownership units are either occupancy-restricted or occupancy- and resale-restricted. Occupancy restrictions are enforced through **covenants** on title and managed by WHA. Resale restrictions are also guaranteed through covenants registered on title and are in place to avoid **windfall profits** for first purchasers as well as to ensure the long-term affordability of the resident-restricted housing. Initially, maximum resale prices were tied to the Royal Bank of Canada prime lending rate and later to the Vancouver Housing Price Index. However, these formulas were abandoned in favour of the current scheme, which links the appreciation escalator to the Canadian Core Consumer Price Index. This formula is now applied to all new properties and any resale units that were originally sold under the old formulas. WHA controls the resale process by calculating the maximum resale price and working with the seller to list and market the property to qualified buyers from the waitlist.

In its early phases of development, the WHA, like many developers of affordable housing, faced **nimbyism** opposition to a number of employee-restricted housing projects. Neighbours voiced concerns about potential impacts on real estate values and speculated about nefarious occupants residing in the employee units. But over time and with the continual development of additional resident-restricted housing in the community, the neighbourhood opposition to employee housing developments eventually disappeared.

During the evolution of the resident-restricted housing program, the WHA has experimented with several different approaches to allocating the restricted units. While both a lottery approach and a points system were tried, the most successful approach, and the one used today, has been a first-come, first-served waitlist approach. Owners of resident-restricted housing units can move to different restricted units but must still go through the WHA's waitlist system. An owner is not permitted to own more than one resident-restricted unit.

Looking forward, the WHA anticipates the new challenge of ensuring adequate funding for ongoing operations. With the focus in recent years on the development of resident-restricted ownership units, which do not generate the same ongoing revenue as the rental inventory, the WHA is looking at additional funding options, such as instituting application fees for its waitlist and other measures to improve its ability to be self-funding.

Because many of the resident-restricted housing units were built within the past two decades, they will soon need repairs and more significant upgrades. The WHA is aware of this aging housing stock and will need to place greater emphasis on sustainable building practices to ensure that the affordable housing inventory continues to respond to the community's housing needs well into the future.

Other current challenges and opportunities include expanding the diversity of housing available in the community to meet the housing needs of seniors and the demand for live–work accommodation.

Figure 5.4.6 • The Terrace.

Innovations

Whistler was one of the first municipalities in Canada to actively address the problem of resident employee housing needs. It created one of the first employee housing organizations in Canada, beginning with the WVHS, formed in 1983, and the W.V. Housing Corporation, founded in 1990. The WHA was incorporated out of these organizations in 1997. Banff Housing Corporation was incorporated in 1993 after many years of grappling with resident housing issues. Incorporation took place at the federal government level because of Banff's national park status and via a non-profit housing body, since Banff did not obtain municipal status until 1990.

Senior governments, once instrumental in Canada in providing the resources to non-profits, including local government non-profits, to create affordable housing for low-income households are no longer playing that role. One of the most important contributions of the WHA is that it pioneered an approach to creating affordable housing for local residents with no senior government subsidy. This has been made possible through the use of linkage fees generated from the Employee Housing Service Charge Bylaw. The fees have been instrumental in creating the equity and the leverage to develop a significant stock of restricted housing. Other municipalities have developed housing reserve funds and implemented inclusionary housing policies requiring a certain proportion of affordable units to be included in projects that involve rezoning, but these measures typically result in few units.

Linked to the WHA's success as outlined above is another innovation. WHA housing prices are occupant- and price-restricted but not allocated on the basis of **income testing**. The premise behind this policy relates back to the WHA's mission to provide "a range and supply of housing options for Whistler's active and retired workforce." Recognizing that a diverse housing inventory integrated throughout Whistler is necessary to meet the assorted housing needs of the Whistler workforce, the WHA does not target any specific income bracket as being exclusively eligible for resident-restricted housing. As a result, the resident-restricted neighbourhoods embody a healthy diversity of residents representing an array of ages and incomes living in a variety of housing forms. Homeowners can build equity as their properties' values increase at the rate of inflation. In Whistler, the analogy used for this housing approach is "a nest, not a nest egg."

Another innovation is the duration of housing affordability. In contrast to some inclusionary housing schemes that fix the price-control period for a set number of years, Whistler's price restrictions remain in place in perpetuity, thus providing a permanent housing resource and community legacy for future generations of Whistlerites. Whistler learned from its own experience that occupancy restrictions on the property through ground leases were not sufficient to keep the properties affordable for the local workforce. The additional price restrictions registered on title proved necessary to prevent speculation in the real estate market and keep the units affordable over the long term.

Most municipal efforts to improve housing affordability have focused on increasing the supply of low-cost rental housing for low-income and vulnerable households, usually in concert with senior levels of government. However, in high-amenity and high-priced real estate communities like Whistler, even moderate-income households face affordability challenges. Political willingness to address employee housing needs has been instrumental in Whistler's ability to address the housing challenge. A lack of political willingness at the local government level often poses a significant obstacle in other communities grappling with similar housing affordability challenges.

What Whistler Has Achieved

Whistler maintains a sophisticated monitoring program associated with Whistler 2020, its long-range sustainability plan. It shows that Whistler has exceeded its target of 75 per cent local workforce residency, with 80 per cent of the Whistler workforce living locally in 2013.

The WHA resident-restricted housing inventory increased from 848 units in 1999 to 1899 by the end of 2013.

With the addition of nearly 500 new resident-restricted units between 2009 and 2012, access and wait

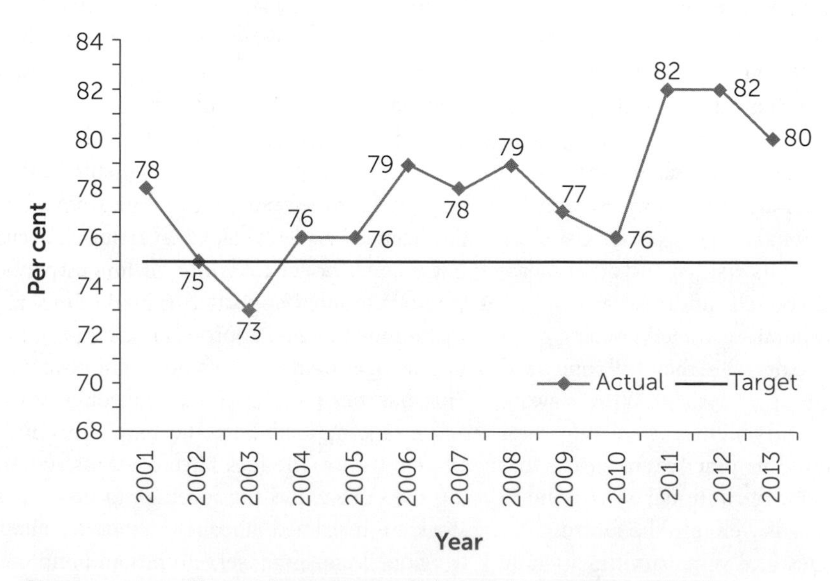

Figure 5.4.7 • Share of employees living and working in Whistler.

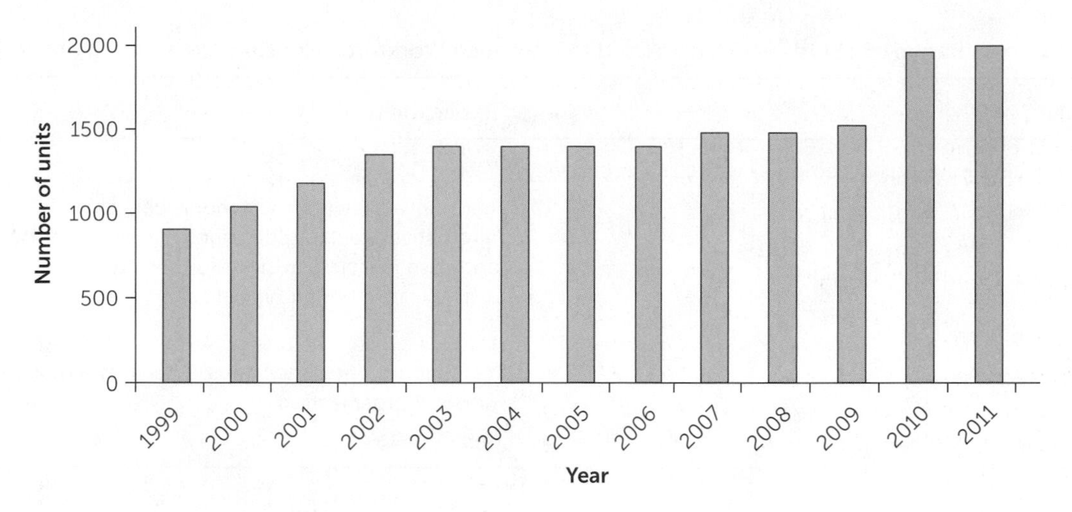

Figure 5.4.8 • Restricted dwelling units.

times have been improving. The WHA still has waitlists for rental and affordable ownership housing, highlighting the fact that there still are families and households in Whistler that are not adequately housed. Although wait times have decreased considerably because of the regular turnover of units, the WHA still had 360 applicants on the purchase waitlist and 204 applicants on the rental list as of December 2013. Average wait times to purchase resident-restricted housing for most unit types is less than a year, but wait times are significantly longer for duplexes and single-family homes.

To better understand outcomes, the WHA surveys both homeowners and tenants when they move out of their resident-restricted unit. The survey's questions relate to the livability of, satisfaction with, and experience of living in Whistler's resident-restricted housing. The following highlights some of the 2012 survey results:

- Ninety-five per cent of the respondents stated that the resident-restricted housing unit they had purchased through the WHA was affordable.
- Respondents highlighted the ease of purchasing and selling through the WHA program, the opportunity to be able to own a home at a price that was affordable, and the ability to live in the community where they work.
- Thirty per cent of respondents, although they were selling their units, indicated that they would remain on the WHA waitlist for an opportunity to purchase another resident-restricted unit in the future.

Lessons and Transferability to Other Municipalities

With no senior government funding available for new affordable housing and with municipalities lacking the resources to subsidize housing, some communities experiencing a mismatch between employee incomes and housing costs are choosing to play an active role to modify outcomes in the housing market. Like Whistler, they have turned to market-based tools and price-control mechanisms to facilitate and preserve new affordable units. The focus is typically workforce housing as opposed to low-income housing (which requires significant senior government subsidies to achieve affordability). A growing number of municipalities and communities have drawn upon the Whistler experience to develop their own solutions to plan for and meet their workforce housing needs, including Vancouver, Tofino, Banff, Salt Spring Island, Bowen Island, Langford, and the University of British Columbia.

Whistler's housing model is ideally suited for high-growth markets with sustained rather than cyclical growth patterns. Otherwise, in low- or no-growth periods, there would be no reliable source of funds. Since financial injections for Whistler's resident housing program come from the development of local businesses, this housing model is also ideal for an area with expanding or redeveloping commercial, industrial, or tourism industries.

An approach to affordable housing like Whistler's also requires strong support from local politicians and local residents. Not all municipal politicians share a belief in an active role for local government in assuring an affordable housing supply, particularly a supply serving

Table 5.4.1 • A Range of Tools That Can Be Used to Introduce Workforce Housing

Tool	Replicability
Covenants and housing agreements controlling occupancy	Yes
Linkage fees	Applicable only in "resort municipalities" in BC. Alternatives such as community amenity contributions and cash in-lieu provisions for density bonuses are available in other municipalities.
Resale price restrictions	Yes
Housing authority	Yes, but not necessary; can be done by a municipality or other organization
Annual housing needs assessments	Yes

moderate-income households. Whistler demonstrates that with sufficient political will, the tools exist and can be implemented so that municipalities can effectively meet their community's employee housing needs. Most of the specific tools, mechanisms, and policies used in Whistler are transferable within BC and probably elsewhere in Canada.

Conclusions

The Whistler experience demonstrates how and what municipalities can achieve with respect to meeting their workforce housing needs independently of senior government funding. Whistler has exceeded its target of housing 75 per cent of the resident workforce locally through the use of a combination of planning, policy, and regulatory mechanisms and with strong political will and community support for its employee housing objectives. Canadian municipalities can and have learned from Whistler's innovative housing model and its experience to develop an alternative housing approach that best meets their own local circumstances.

KEY TERMS

Build-out
Canada Mortgage and Housing Corporation (CMHC)
Covenant
Income testing
nimbyism
Official Community Plan (OCP)
Social housing
Windfall profits

REFERENCES

City Spaces Consulting Ltd. 1997. *Whistler Housing—Discussion Paper #1 Housing Functions and Organizational Approaches*. Vancouver: City Spaces Consulting Ltd.

McKeever, G., and M. Zucht. 2009. *The Whistler Housing Authority Story: A History of Affordable Housing in Whistler*. Whistler, BC: Whistler Housing Authority.

WHA (Whistler Housing Authority). n.d. *Governance and Operational Policy*. Whistler, BC: Whistler Housing Authority.

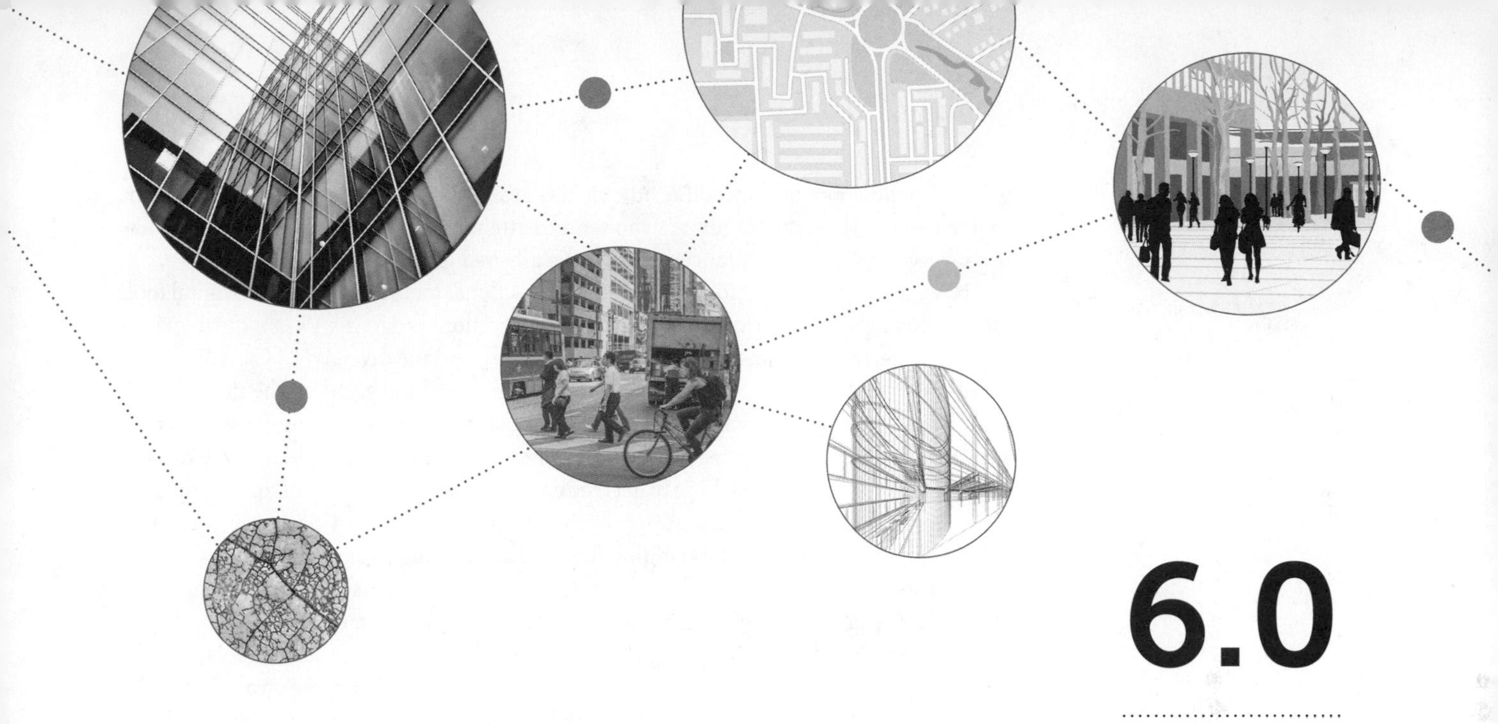

6.0

INTRODUCTION

Participatory Processes

The other sections in this volume have illustrated that planning is a complex process involving many public, private, and non-profit actors. Many municipalities and regions have seen that the process can be just as important as the product (Friedmann 1973). When local people are involved in setting long-term goals for their neighbourhood, suggesting ways to accommodate density or integrate nature into their communities, they feel a sense of ownership over the final plan. However, this input may conflict with planners' broader goals or the views of other interest groups.

Participatory, collaborative, and communicative planning techniques are a major area of inquiry for planning theorists such as Patsy Healey (1997), John Forester (1989), Bent Flyvbjerg (1998), and Susan Fainstein (2000). Many of these authors examine the power dynamics and communicative techniques that can either enable or hinder the engagement of under-represented groups in planning processes. This theoretical focus, beginning in the 1960s as a response to massive citizen protests, completely redirected the discipline of planning from form oriented to process oriented. This is one reason why this volume includes a section on participatory processes, even though many of the case studies in other sections already highlight participation and citizen engagement.

A second reason is that citizen involvement in planning processes has evolved from its origins in the 1960s when citizens demanded a voice in major planning decisions that could potentially destroy the urban fabric. Forms of participation that became commonplace through the following decades included town hall meetings, open houses, and focus groups with key stakeholders such as local business owners. In Canada and other countries, early and ongoing public participation became required under provincial Planning Acts for many planning decisions and processes, including the development of Official Plans. As commonplace as town hall meetings, open houses, and focus groups are now, neighbourhood residents may still be unaware of them or lack the time to participate. Thus, the "regulars" who tend to attend these events time and again may not be giving planners a balanced view of the affected groups.

New forms of participation build on these traditional methods, using new digital tools and increasingly involving marketing and communications programs. With more demands on their time, the prevalence of dual-income families, and the diversity of Canadian communities, new tools allow people to participate from their homes and on their own schedules. Online surveys, participatory budgeting tools, and crowdsourcing have increased planners' abilities to obtain feedback and suggestions from a much wider demographic. They can even be used in teaching to get feedback from high school or university students, since youth and young adults are typically under-represented in planning processes. While nothing will replace the traditional town hall meeting or open house, new digital tools have the potential to broaden participation to youth, recent immigrants, and those facing barriers to participating in an event occurring at a fixed time and place. Marketing and advertising are increasingly being used to attract participants to online tools.

However, issues such as Internet availability among low-income groups, privacy issues, and resistance to participatory decision-making still present barriers to digital processes. Digital participation in planning is still largely limited to comments on a range of options provided, with few examples of participatory budgeting or other processes in which participants' views are incorporated into decision-making. As Pamela Robinson and Michael DeRuyter discuss in their case study on digital tools in Canadian planning, the new tools can merely mimic the old methods of incorporating public feedback on plans, because planners are unwilling to surrender control of the planning process.

Ann McAfee, co-director of planning for the City of Vancouver for 12 years, describes the neighbourhood-based approach to CityPlan beginning in 1992. Public engagement began in the initial steps of plan-making and continued through the decision-making process: City councillors wanted the public to understand first-hand the difficult choices that Vancouver faced with limited land, changing service demands, and inadequate funds. Participants of all ages shared ideas for the plan with the stipulation that each idea should be paired with a solution. On issues where there was no preferred direction, participants were encouraged to choose among different futures for the city. The process resulted in new policies to accommodate future growth and new city-wide plans for industrial lands, greenways, transportation, sustainability, and culture; nine Community Visions grounded CityPlan in neighbourhoods. Although the process took longer than expected, public involvement in CityPlan led to a broader acceptance of controversial decisions, since CityPlan was considered to be "their" plan.

Jeff Biggar tells a very different story in his case study of the development of a cultural plan in Hamilton, Ontario. Using roundtables for cultural leaders, multi-stakeholder and public workshops, and interviews, the City of Hamilton began its cultural planning process. Subsequent phases included a cultural mapping process, the establishment of the city's first-ever Cultural Policy in 2012, and a Culture Plan adopted in 2013. However, communication problems, a non-transparent process, and reluctance to invest in what some consider "low-priority" cultural planning have plagued the development of the Culture Plan. Low levels of social capital, trust, and cooperation among stakeholders were significant issues, and at this point no funding has been allocated for the plan's implementation.

Lisa Brideau and Amanda Mitchell present Vancouver's Greenest City Plan, one of the first to use a crowdsourced approach to idea and solution generation. The process also integrated unique marketing approaches, such as the creation of YouTube videos, visionary advertisements, and a PechaKucha Night, to generate public interest in the plan. A well-planned process framed a common challenge for the future, established ground rules for suggesting ideas, and then allowed voting on the suggestions. The goal was to engage

20,000 people in the process, including website visits, submitted ideas, votes, views of videos, and attendance at events; in the end some 21,000 visited the website, more than 3000 submitted ideas, about 28,000 votes were counted, and some 2100 comments were made. While there are still lessons to be learned, which might help other municipalities in implementing a similar process, the process exceeded planners' expectations.

Juan Torres and Natasha Blanchet-Cohen address a group that is often left out of traditional planning processes: youth. Although past attempts at engaging youth had failed, the municipality of Sainte-Julie, Quebec, teamed with researchers and students from the Université de Montréal and Concordia University to design two participatory processes: evaluating a public transit service for teenagers and upgrading a skate park. As a result of interviews and design workshops with local youth, the municipality of 30,000 residents changed its communication tools and revisited the membership procedures for the public transit service, and design changes were made to the skate park. This is an excellent example of how a small planning department introduced an innovative approach to engagement and produced guidelines for other municipalities seeking to involve youth as partners in the planning process.

Pamela Robinson and Michael DeRuyter delve into the sea of social media in their case, which explores the use of Web 2.0 and its social media tools in municipal planning departments. Comparing 17 Canadian municipalities, their research found that some tools were widely used (such as Facebook, Twitter, and YouTube), while open data is the next technological frontier for Canadian municipalities. However, these tools are used mainly to disseminate and gather information in the same way that traditional methods such as participation at public meetings or writing letters/emails to city councillors are used. At this point, social media does not replace traditional means of participation, because it is not yet being used to generate content and ideas in real time or to devolve power and make decisions in real time. Whether this is a possibility in the future depends on the extent to which municipalities surrender control over planning processes and accept the fluid, "flat" hierarchy of the Web.

As these cases illustrate, public participation is simultaneously the most rewarding and the most challenging task that planners face. Conflicts are inevitable, and the process can be stalled by many factors, including the instability of municipal governments or powerful actors and institutions. However, these cases illustrate that there is something to be learned in the process of involving as many demographics as possible in the strategic planning of our cities and regions: as planners, we need to educate the public on planning goals, concepts, and tools so that they can help make difficult decisions. Participation without education can result in nimbyism, while it is clear that an understanding of planning constraints, the physical characteristics of cities and regions, and land-use policies can help community members share the decision-making process. Community-based organizations may even become partners in plan implementation. Engagement processes can be designed so that individuals feel a sense of control over where their communities will go in the coming decades but still reflect planning principles and values.

REFERENCES

Fainstein, S.S. 2000. "New directions in planning theory." *Urban Affairs Review* 35(4): 451–78.

Flyvbjerg, B. 1998. *Rationality and Power*. Chicago: University of Chicago Press.

Forester, J. 1989. *Planning in the Face of Power*. Berkeley: University of California Press.

Friedmann, J. 1973. *Retracking America: A Theory of Transactive Planning*. Garden City, NY: Anchor Press/Doubleday.

Healey, P. 1997. *Collaborative Planning*. Hampshire, UK: Macmillan.

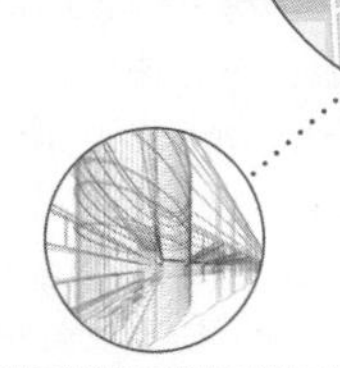

CASE STUDY 6.1

People and Plans: Vancouver's CityPlan Process

Ann McAfee

SUMMARY

Vancouver is internationally recognized as one of the world's most liveable cities. This case study describes the role public engagement played in addressing choices Vancouver faced in moving toward becoming a more sustainable city. The study draws from the author's experience leading Vancouver's CityPlan process. Subsequently, the author has participated in public engagement processes in various cities and under the new realities of the global economy and **social media**. These experiences tested the transferability of Vancouver's planning processes.

This case study is in two parts. The first part describes the three-phase CityPlan program to establish directions for Vancouver. The first phase (1992–5) involved more than 100,000 people in considering the choices and consequences of a wide range of strategic directions. The result was CityPlan, a council-adopted strategic plan (City of Vancouver 1995). The second phase (1995–2008) continued stakeholder engagement in preparing and implementing a variety of policies based on CityPlan directions. The third phase involved three planning initiatives in "single-family" neighbourhoods—Infill Zoning (2003–9), Community Visions (1995–2010), and a Neighbourhood Centre Plan (2003–4)—designed to increase housing choice and make more efficient use of existing services.

The second part of the case study reflects on the outcomes of CityPlan. The study concludes that four features of planning processes—including all city responsibilities, broad public engagement commencing with the initial steps of plan-making, public involvement in choice-making when limited land or funds require trade-offs between city values, and allocating funds for early implementation—contribute to public support for plans. The study also concludes that plans that are expeditiously implemented through regulation or funding benefit from public support. Phased planning processes that require further plan-making prior to implementation can experience approval decay among politicians and citizens.

OUTLINE

- Public involvement in municipal plan-making can lead to public support for city-wide initiatives, such as Vancouver's CityPlan.
- The CityPlan process asked the public to generate ideas and solutions, make choices between the alternatives, and then discuss the options with City staff. This process contributed to the implementation of neighbourhood plans in a way that met local needs.
- The CityPlan process encouraged experimentation with planning processes and funding options. Many of the tools can be adapted for use in other municipalities using new technologies.

Vancouver's Planning Context

In the two decades following the Second World War, Vancouver (population 410,375 in 1966) was described as an unspectacular city in a spectacular setting. Forty years later, Vancouver (population 578,041 in 2006) was the world's most liveable city. What changed?

From the 1930s, Vancouver's growth followed the typical North American model of workers commuting between downtown jobs and suburban homes. In 1972, a newly elected council led by Mayor Art Phillips (1972–6) adopted a new vision for Vancouver. A proposed freeway into downtown was defeated. The 1976 Downtown Plan, followed by the 1991 Central Area Plan, provided a context for a more intensive and environmentally sustainable downtown through the construction of new communities on previously industrial lands. City policies required new communities to provide a mix of housing opportunities and full community services. Vancouver's inner city became an internationally recognized showcase for downtown live–work (Punter 2003). Development on vacant industrial sites was largely unopposed. Planners prepared plans for council approval with minimal public comment.

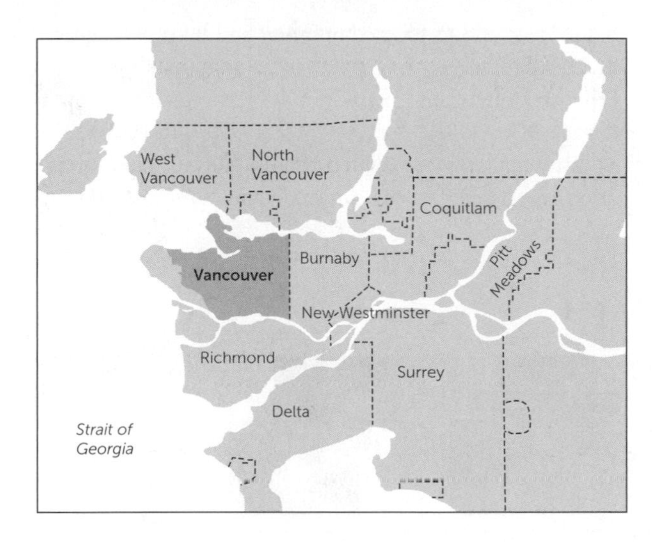

Figure 6.1.1 • Location map.

The relationship between people, politicians, planners, plans, and protests changed in the early 1990s. Questions were raised about the economic consequences of continuing to redevelop inner-city industrial lands for housing. With continued growth and limited revenue sources, Vancouver sought to make more efficient use of the 70 per cent of the city zoned single-family. However, proposals to add density in established neighbourhoods resulted in community outcry.

In 1992, Vancouver Council invited the public to "Walk in Council's Shoes." Council wanted the public to wrestle with difficult choices resulting from limited land, changing service demands, and inadequate funds. Council directed planners to develop a new process to "Hear about All Issues, Hear from New People, and Hear in New Ways." These became the "Prime Directives" of Vancouver's CityPlan process.

CityPlan: Engaging the Public in Plan-Making

The terms "citizen" and "public" and "involvement," "participation," and "engagement" are often used interchangeably. All describe processes by which public concerns, needs, and values are incorporated into decision-making.

An early step in any planning process is determining what role citizens will play. Options as described on the IAP2 Spectrum of Public Participation (International Association for Public Engagement 2007) range from "Inform," "Consult," "Involve,"

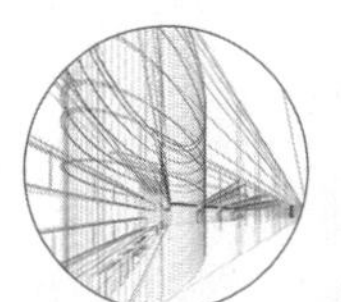

PLANNING THEORY

Steps in the Planning Process

Planning processes that lead to zoning changes or budget allocations typically involve:

1. information-gathering and analysis;
2. making choices between land uses and funding priorities;
3. preparation of a draft plan, leading to
4. public input; and
5. council decisions.

Delaying public input until Step 4 can result in resistance to plans prepared by officials. The CityPlan process involved the public from Step 1. This built support for the plan by engaging the public in identifying issues and advising on choices.

"Collaborate" to "Empower." The CityPlan approach was best described as "Collaborate." Council wanted a process to partner with the public at each stage and incorporate advice from the public into decisions to the maximum extent possible. However, council was clear: the public advises, and the elected council decides. The four-step CityPlan process invited the public to:

1. generate ideas for the plan;
2. review the ideas and recommend those for further consideration;
3. consider issues, choices, and consequences of possible Directions;
4. review the draft plan.

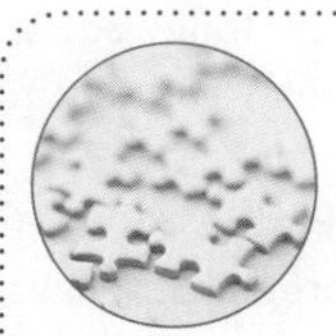

KEY ISSUE

Public Involvement

- Involve public from Step 1 in plan-making process.
- Address sustainability by considering all City responsibilities.
- When land or funds are limited, focus engagement on key choices.
- Following plan adoption, allocate funds for implementation.

Step 1: People Generate Ideas (November 1992–March 1993)

CityPlan started with Vancouver's Mayor Gordon Campbell (1986–93) sending letters to 1000 randomly selected households and to 1000 community and businesses groups inviting people to participate in a City plan-making process. Print and electronic media ran stories explaining how to engage.

Participants were invited to join "City Circles," groups of 10 to 15 people who came together to share ideas for

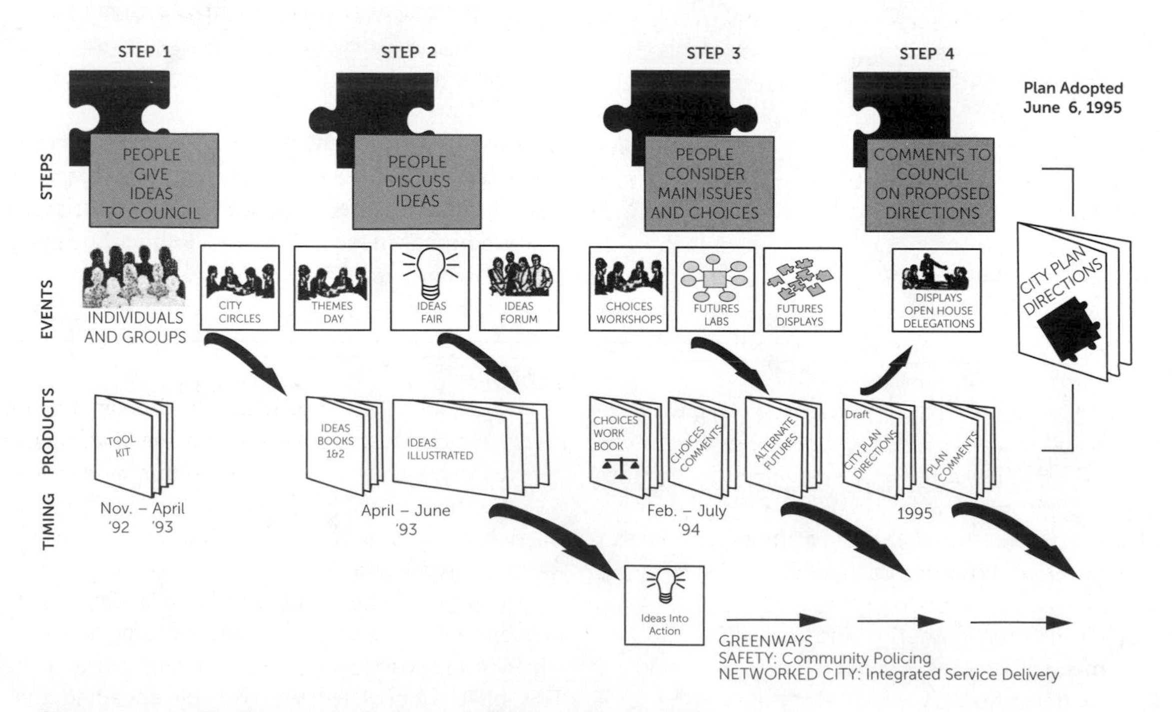

Figure 6.1.2 • CityPlan process: Steps, events, and products.

Source: Adapted from graphics by Brian Riera, courtesy of the City of Vancouver.

CityPlan. Some were members of existing organizations who put CityPlan on their agenda (e.g., the Board of Trade, homeowner groups). Others were individuals with similar interests. The intention was for people to talk to people about ideas for Vancouver's future. Circles were facilitated by citizen volunteers, and tool kits (City of Vancouver 1993b) ensured that all participants had equal access to information, encouraging participants to discuss issues rather than debate facts. A Resource Centre, staffed by representatives from five City departments, provided program support.

Circles were invited to identify ideas for the City's new plan. Most circles focused on one idea or direction. To manage expectations, the only stipulation was that "Ideas and Solutions travel together." For example if a circle was proposing community policing, it had to suggest how this could be achieved by raising or reallocating funds. More than 300 circles prepared ideas.

To engage Vancouver's large multicultural community, tool kit information was provided in six written languages and presentations were delivered in eight spoken languages. Seventy circles worked in languages other than English. Large-format text and recorded information supported those with vision challenges. In addition to "adult" circles, there were 150 Classroom Circles for young children, who drew pictures of their desired neighbourhood. Older students participated in the "adult" program.

To ensure that people's ideas could be shared, all submissions were published in a 477-page *Ideas Book* (City of Vancouver 1993a). Most submissions were written, and some included maps, videos, photographs, and models. More than 3000 submissions were received, some from individuals and others from circles. Staff did not edit any of them. Submissions in languages other than English were printed in the language of submission and translated into English. All participants received copies of the *Ideas Book*, which the circles used to identify key themes and as a permanent record of submissions.

During Step 1, active council involvement, media coverage, and broad public engagement generated a city planning "buzz." Participants said that being invited to engage in the first step of the plan-making process built trust and made them feel their contributions were valued. Councillors who attended discussions observed that they were meeting "new people" and not just "the usual suspects."

From a staff perspective, Step 1 submissions were well thought out, with most tackling the challenge of "Ideas and Solutions travelling together." The CityPlan process could be described as an early example of **crowdsourcing**. Problem-solving was no longer the exclusive activity of City experts: tasks were outsourced to the community.

Figure 6.1.3 • City Circles discuss ideas.

Source: Ann McAfee.

Step 2: People Discuss and Review Ideas (April–June 1993)

A three-day "Ideas Fair" attracted 10,000 people to share ideas. Volunteer artists helped circles display proposals for Vancouver's new plan on coloured six-foot by four-foot panels. The use of artists ensured that the quality of the displays was not influenced by resources: both the Urban Development Institute and the Vietnamese Single Mothers Circle were proud of their submissions.

Participants used an "Ideas Checkbook" to recommend directions the City should pursue. Staff tabulated the results, which were shared with circle representatives at an Ideas Forum. The forum was to publicly confirm ideas that staff would recommend to council for further consideration.

Council reviewed the Step 2 output:

- Some ideas received minimal public support. Council removed them from further consideration.

Figure 6.1.4 • Circles illustrate ideas.

Source: Ann McAfee.

Figure 6.1.5 • Ideas Fair: Mayor Gordon Campbell listens to a Classroom Circle.

Source: Ann McAfee.

- Other ideas received broad public support (e.g., community service delivery through Neighbourhood Integrated Service Teams [NISTs] and reconfiguring selected streets to create a network of walking and cycling "greenways"). In these cases, council did not wait for the plan to be completed and allocated funds to NISTs and greenways. This built creditability for the process: council was seen to be listening and responding to people's ideas.
- On some topics, there was no public consensus. For example, some people supported intensification to increase housing choice, while others wanted to preserve single-family neighbourhoods by continuing to direct growth to industrial "brownfield" and urban fringe "greenfield" sites. Some people supported increasing taxes to pay for new services, but others wanted to redeploy existing budgets to meet new needs. The areas in which there was no clearly preferred City Direction became the focus of Step 3.

Step 3: People Make Choices (February–August 1994)

The ideas from Steps 1 and 2 raised issues and choices. The objective of Step 3 was to engage the public in addressing the difficult choices a council faces when deciding between often equally valid but conflicting directions. Staff assembled a 12-theme, 40-page *Making Choices Workbook* (City of Vancouver 1994). Each theme (housing, jobs, neighbourhoods, movement, services, safety, infrastructure, arts, public places, environment, finance, and decision-making) described choices, the consequence of each choice, and how the choice could be implemented. There were no "right" or "wrong" choices, just different consequences for the city. Translated into six languages, the workbook was distributed to City Circles, to 6000 people on the CityPlan mailing list, and through libraries and community centres. Workshops, media coverage, and speakers provided information about the choices. The circles then discussed the workbook choices and completed questionnaires indicating their recommended City Directions.

The *Making Choices Workbook* responses resulted in the clarification of some directions. However, there were areas of continuing uncertainty. Staff identified four futures (scenarios), which differed on the topics for which workbook responses had failed to provide clear directions: housing, neighbourhood character, jobs, community services, and decision-making. City Circle representatives assisted with illustrating the futures, and a tent displaying

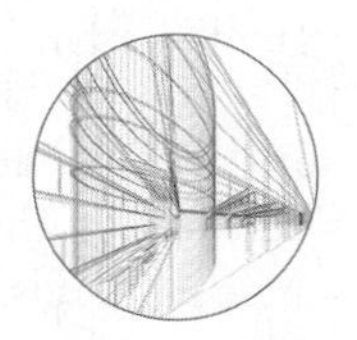

PLANNING THEORY

Governance and Canadian Municipalities

The Canadian Constitution Act (1982) allocates the powers of government between the federal and provincial levels. Section 92 assigns municipal governance to provinces. In British Columbia, the Municipal Act (and the Vancouver Charter) gives local governments the authority to adopt **Official Community Plans (OCP)** and implement **zoning bylaws** with no appeal to a provincial authority. By comparison, in Ontario the Ontario Municipal Board (OMB) operates as a tribunal to hear appeals on planning disputes related to zoning, subdivision, official plans, consents, and variances. Differences in provincial planning legislation mean that the consequences of planning recommendations and council decisions can vary across the country.

the futures toured the city, with more than 15,000 people viewing and discussing the choices. The four futures were also described in an eight-page futures questionnaire mailed to all city households and businesses and inserted (translated) into ethnic newspapers. Results of futures questionnaires were tabulated. Meanwhile, a stratified survey went to 1500 people, asking them for their preferred directions. The survey was designed to "fill in the gaps" in that the sample was focused on areas and for demographics otherwise under-represented in previous responses.

Step 4: People Discuss the Draft Plan with Council (February–June 1995)

Results of the *Making Choices Workbook*, the futures questionnaire, and the random sample were assembled by staff into a draft city plan for public review. Displays at fire halls, a City Hall open house, and posting the draft plan online gave people an opportunity to review and discuss the proposed City Directions.

As a final step, council held public hearings. Most speakers described their role in the process and encouraged council to adopt "their plan." Eighty per cent supported CityPlan, and 20 per cent recommended that council reject it, with most of those who spoke against approval disagreeing with intensification in single-family neighbourhoods. In 1995, council approved CityPlan; under the Vancouver Charter (1953), there could be no appeal of council's decision.

CityPlan in Retrospect

The CityPlan process took 20 months over two-and-a-half years. There were gaps in the public process because of a civic election and staff's need for time to assemble the workbook and draft plan. Throughout the process, people were offered many ways to engage, including purpose-designed programs for youth and ethnic communities. A subsequent survey concluded that more than 100,000 people, representing 40 per cent of city households, participated in the CityPlan process. The four-step Directions process from 1992 to 1995 cost $3.4 million, half of which was covered by reassigning staff from other programs. New costs were incurred for printed materials (including translation) and events. The relationship between council and staff changed through the process. For example, council met regularly with the CityPlan team to discuss progress, and a "Sponsor Committee" of department heads built cross-organization commitment to implementation. An interdepartmental team, managed by the director of planning, included staff from Planning, Engineering, Social Planning, Housing, Parks, Fire, Police, Finance, and Environmental Health. Staffing levels ranged from eight during plan preparation, to 15 to 30 during active outreach, to more than 100 for the Ideas Fair. Consultants provided communication support, but no consultants were involved in managing the process or developing content.

When CityPlan started, people wondered what kind of plan would emerge. Would citizens continue to say "not in my backyard"? This was not the case. CityPlan met the Federation of Canadian Municipalities principles for sustainable planning (FCM 2011) and retained many features people liked about Vancouver (e.g., a vibrant downtown, affordable housing programs). Elsewhere, difficult choices had been made between often equally valued directions. For example, CityPlan changed City policies by preserving remaining industrial lands, increasing housing choice in single-family areas, providing neighbourhood-based service delivery, and requiring that new development pay its way.

To this point, CityPlan was best described by a newspaper headline: "Master plan provides destination, but no route" (Howard 1995). If developing an Official Community Plan could be compared to writing a book, CityPlan was Chapter 1. To tell the rest of the story, the City needed to update or develop a variety of city-wide and area plans.

Implementing CityPlan

City-wide Policy Plans (1995–2008)

During the decade following CityPlan, Vancouver Council further articulated Directions through new city-wide policy plans—e.g., Industrial Lands Strategy (1995), Greenways Plan (1995), Transportation Plan (City of Vancouver 1997b), Financing Growth Policy (City of Vancouver 1997a; 2003), Sustainability Plan (2002), Community Climate Change Action Plan (2003), and Culture Plan (2005–8). All policy plans started with council-approved terms of reference specifying the task, roles, and engagement process.

A retrospective consideration of CityPlan policy plans yields two conclusions. First, processes need to be custom-designed to address the specific circumstances, and not all processes require extensive public consultation. For example, the Transportation Plan continued broad public engagement. Given CityPlan support for retaining industrial lands, staff prepared new industrial zoning schedules, with public review directed to owners and tenants of industrial properties. The Financing Growth Strategy and the Culture Plan were prepared by staff with the assistance of multi-stakeholder task forces. Second, moving expeditiously between policy agreements (e.g., CityPlan) and implementation through zoning and capital improvements increases the likelihood of public support for implementation. As will be seen in the next section, longer planning processes can lose momentum.

Implementing CityPlan in Neighbourhoods (1995–2010)

CityPlan built broad public support for new Directions for the 60 per cent of the city zoned single-family. These Directions were implemented through three programs.

New city-wide residential zoning

New **infill development** provided more housing choices in neighbourhoods and made more efficient use of existing services. Prior to CityPlan, secondary suites (a second dwelling unit in an otherwise "single-family" dwelling) and laneway houses (a second dwelling located off a rear lane) were considered and rejected because of community opposition. Following CityPlan, rezonings to improve the design of apartments above shops (2003), increase housing opportunities through secondary suites (2004), and laneway houses (2009) received community support.

Community Visions (area plans)

Other housing and neighbourhood improvements were to be implemented through nine Community Visions (1998–2010; see City of Vancouver 1998). Visions were designed to ground CityPlan in predominantly single-family neighbourhoods. They provided a context for land use, funding, and local service delivery to further CityPlan in a way that met local needs. Public engagement included resident committees that ensured that the plan included broad community input and reflected community priorities. A "City Perspectives Panel" of residents from other communities monitored the impact of plans on adjacent neighbourhoods. The Vision process was based on the four-step CityPlan process. However, there was a fundamental difference—Community Visions had CityPlan as a context. Visions illustrate a combination of "top-down" and "bottom-up" planning. Broad city-wide directions to increase housing choice were grounded by communities, who identified location-specific land uses and local funding priorities. Each Vision cost $275,000 for public outreach, meetings, surveys, and translation. On average, 55 per cent of area households participated, and all Visions included multicultural outreach and a school program. One unexpected consequence of the Community Vision process was the emergence of 70 community groups involving more than 1000 people in neighbourhood-initiated improvement projects.

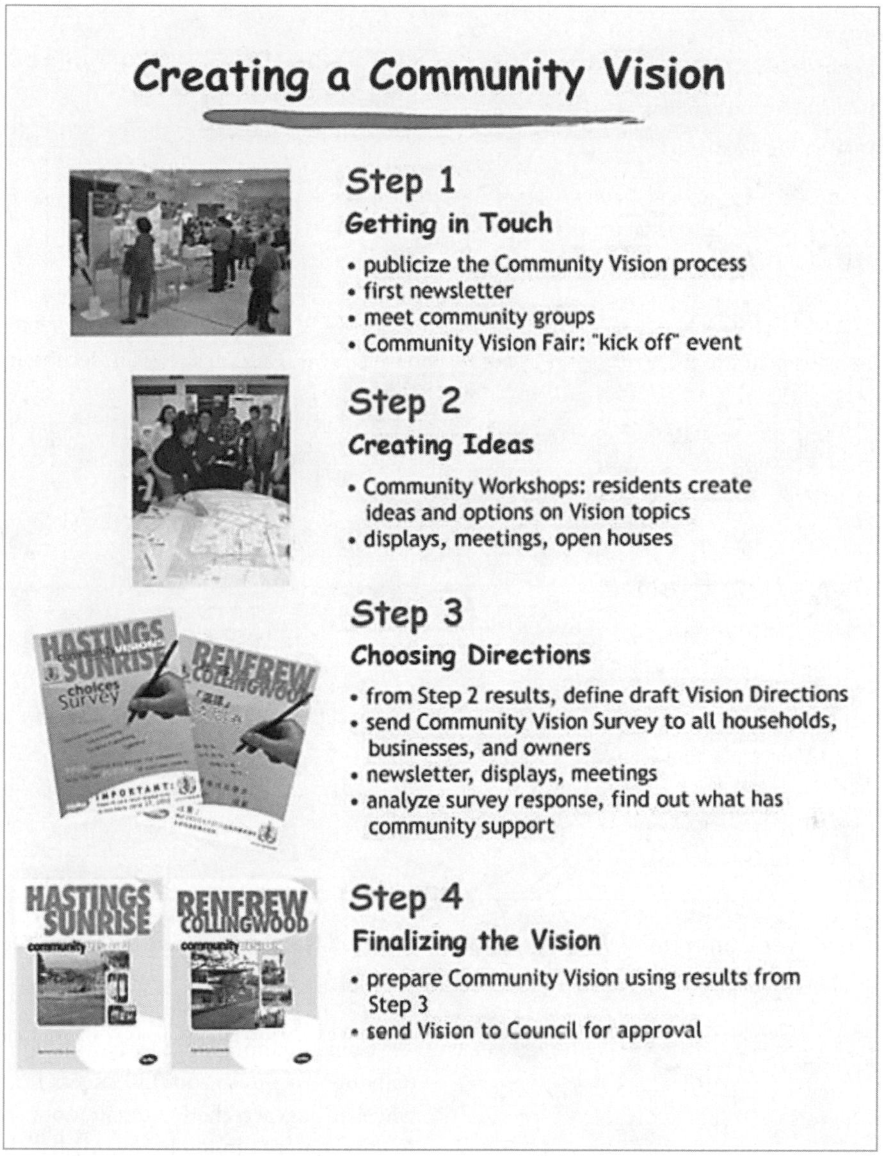

Creating a Community Vision

Step 1
Getting in Touch

- publicize the Community Vision process
- first newsletter
- meet community groups
- Community Vision Fair: "kick off" event

Step 2
Creating Ideas

- Community Workshops: residents create ideas and options on Vision topics
- displays, meetings, open houses

Step 3
Choosing Directions

- from Step 2 results, define draft Vision Directions
- send Community Vision Survey to all households, businesses, and owners
- newsletter, displays, meetings
- analyze survey response, find out what has community support

Step 4
Finalizing the Vision

- prepare Community Vision using results from Step 3
- send Vision to Council for approval

Figure 6.1.6 • Community Vision Process.

Source: Graphics by Nancy Wormald, Courtesy of the City of Vancouver.

Neighbourhood Centres

Community Visions identified 19 Neighbourhood Centres on which to focus future development. Neighbourhood Centre Housing Plans involved planners (with architectural and urban design expertise) working with area residents and developers to prepare new zoning schedules and shopping area improvements.

Because property taxes do not cover all the costs of growth, **Community Amenity Contributions** (applied to rezonings) and **Development Cost Levies** (applied to all new construction) were collected to fund improved community services. The charges were seen by the community as a "good faith" gesture in return for intensification.

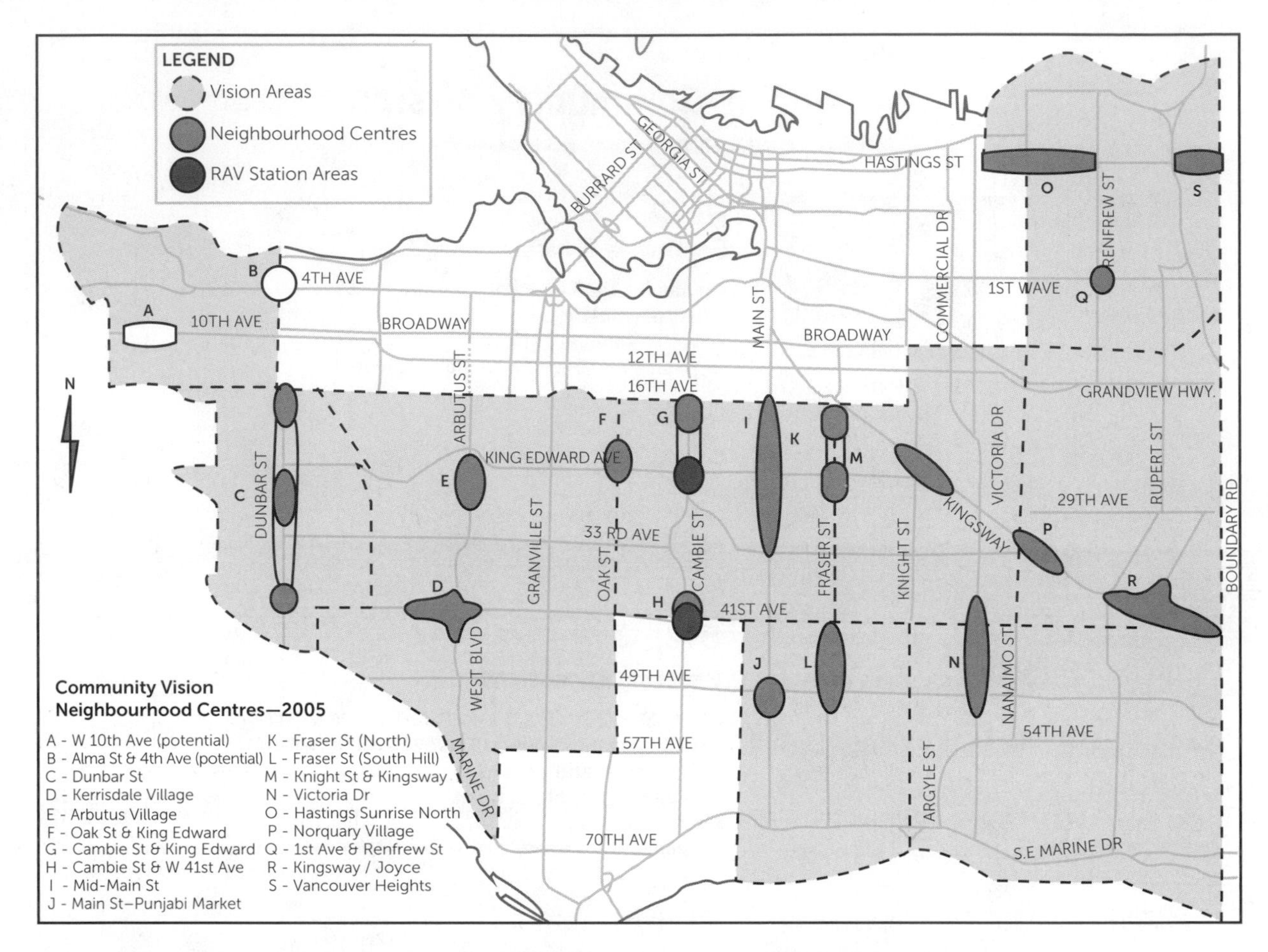

Figure 6.1.7 • Vancouver Community Vision areas and proposed Neighbourhood Centres, 2005.

Source: Adapted from graphics by Nancy Wormald, courtesy of the City of Vancouver.

Figure 6.1.8 • Community-led planning.

Source: Picture by Janice Lowe, courtesy of the City of Vancouver.

Council approved the first Neighbourhood Centre (City of Vancouver 2004) to cheers from the community, who had just watched council adopt "their" centre plan. **Nimbyism** (not-in-my-backyard) had become yimbyism (yes-in-my-backyard). The scale of the first community-supported Neighbourhood Centre rezoning (including a 400-unit high-rise and infill zoning for 1500 area properties) belied the criticism that community engagement reduces the likelihood of bold steps.

Neighbourhoods after the plans

While staged planning processes that obtain broad support for key policies before detailing zoning and budget allocations offer the benefit of focusing public attention on "bite-sized" tasks, they take time to complete. As the years passed, commitment to implementing CityPlan in

Vancouver's single-family neighbourhoods waned. By 2007, proposals to increase density were being met by community resistance. This raises the question of why communities shifted from enraged (1992) to engaged (2004) and returned to enraged (2007). There were several reasons. Resources for CityPlan implementation were considerably less than for plan preparation. The process of preparing nine Community Visions before initiating rezonings delayed implementation. In 2006, the incumbent mayor, concerned that CityPlan was taking too long to implement, proposed a new program to increase densities in single-family neighbourhoods. The *EcoDensity Charter* (City of Vancouver 2008) was seen by communities as ignoring approved Visions. A new mayor and council won the 2008 election with a greener city and affordable housing as their priorities. Both supported CityPlan Directions, although they were not presented as such. Between 2006 and 2010, most senior staff with CityPlan experience retired or relocated. While public engagement continues to be a component of Vancouver's area plans, newspaper headlines "Neighbourhoods in revolt" (Lee 2013) suggest that recent area planning processes are seen as more "top down" than "bottom up."

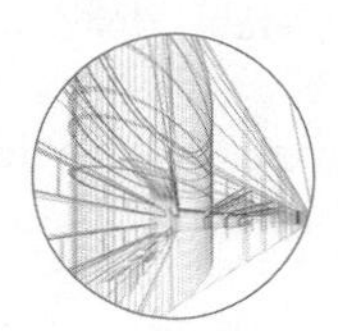

PLANNING THEORY

Addressing Public Cynicism

Tools to address public cynicism include:

- equal access to information;
- engaging the public at each step of the process;
- many ways to participate;
- people speak to people, sharing values and addressing choices;
- people speak to council;
- council allocates funds for implementation.

CityPlan: Lessons for Plan-Making

The rationale for describing case studies such as CityPlan is to draw lessons with potential for transferability to other cities. CityPlan offers several lessons for engaging citizens in plan-making.

The Role of Visionary Leadership: Lead, Listen, Lead

CityPlan demonstrates the role elected officials can play in supporting a plan-making process. Council led by articulating a new planning process, listened to public input, and led by adopting new City Directions. All too often, public processes result in wish lists for which planners and politicians are expected to find answers. Vancouver's CityPlan process engaged participants in both suggesting their preferred directions and advising on how to achieve them. Having "Ideas and Solutions travel together" managed expectations and provided a reality check.

What City Council did not do was claim personal credit for the plan. The completed plan was not the "mayor's plan" or "council's plan." It was "Vancouver's plan." CityPlan was supported by five councils over 14 years. Public support waned when council's commitment to CityPlan wavered.

Including Citizens in Plan-Making: Tools for Change

CityPlan started with council's desire to develop a new process to "Hear about All Issues, Hear from New People, and Hear in New Ways." Hearing about all city responsibilities better ensures that the range of topics that together deliver a sustainable city are integrated into the city's plan. Hearing from new people ensures that the city's many voices are included in the plan-making process. The value of CityPlan was the focus on public choice-making. When a draft plan is based on staff's choices, the public is presented with what appears to be a fait accompli. There is no public ownership of the choices. When choices and their consequences are incorporated into the public engagement process, the public better understands trade-offs resulting from limited land and funds.

Prior to CityPlan, there was considerable distrust of council. People were sceptical about whether council would listen to ideas and implement them. CityPlan demonstrated how "hearing in new ways" offers a variety of tools for addressing public cynicism.

The direction-setting phase of CityPlan occurred before widespread use of the Internet. Today, information and feedback are regularly delivered through an interactive website (see Brideau and Mitchell's chapter on Vancouver's Greenest City Action Plan). However, few studies compare the contributions of "online" and "on-site" engagement. This paper concludes that if CityPlan were being delivered today, the Web would be an excellent platform for disseminating information (e.g., CityPlan Steps 1 and 4) and a less satisfactory way of assembling plan content (e.g., CityPlan Steps 2 and 3). There is no evidence to suggest that online engagement would have significantly improved CityPlan's mix of participants in terms of income, household, age, and ethnicity. Values placed on facts leading to difficult trade-offs benefit from face-to-face discussions, which build understanding, consensus, and cooperative working relationships. This paper echoes Brideau and Mitchell's conclusion that there is a symbiotic relationship between online and on-site engagement tools.

Criticisms of Engagement and the CityPlan Process

CityPlan engendered debate in the planning profession. Vancouver's planners were criticized as being public process pollsters (Seelig and Seelig 1997; McAfee 1997; Seelig, Seelig, and McAfee 2008). While Vancouver's planning processes may appear to be more about communities talking and less about analysis, that was not the case. Key policy directions had a solid analytical basis (e.g., fiscal impact, transportation mode share, and post-occupancy studies).

Would CityPlan have been much different if planners had followed the traditional process of preparing a draft plan for public response? The answers are "no" and "yes." "No," in that most Directions that emerged from CityPlan—increasing housing choice in neighbourhoods, more efficient use of existing services through intensification, and maintaining a diverse economy—would have been proposed in a staff plan. "Yes," in that if past experience is any indicator, the public would have opposed rezoning proposals in residential areas, preserving industrial land, development cost charges, and decisions to fund some services and not others. CityPlan brought a wide range of stakeholders together to agree on shared Directions.

While CityPlan did deliver these benefits, a retrospective look at the process points to several cautions. Public support is fragile, so the time between plan approval and implementation is critical. In the five years following CityPlan, a variety of city-wide plans based on CityPlan Directions were adopted and implemented. However, nine Community Visions took 10 years to complete. Reduced council commitment to CityPlan contributed to loss of community support. By 2010, Vancouver was again embroiled in controversial neighbourhood plans.

Conclusions

CityPlan demonstrates that citizen involvement is not easy. There is no one "public interest." People have various needs, perspectives, and life experiences, all of which enrich public policy. To be successful, public involvement must respect and respond to these differences. Underlying all public involvement should be clear principles for how citizens will be involved and what their role will be in decision-making. Beyond these principles, there is no "right way" to engage the public. Each process needs to fit the task and resources.

Caution is necessary in transferring processes and tools to new circumstances. To some extent, British Columbia's flexible municipal legislation encouraged experimentation with new processes, planning procedures, and funding options. That said, engaging the public in the early phases of preparing a plan and in advising on funding and land-use choices can be implemented under any governance system. Many of the tools for change used in the Vancouver CityPlan process, augmented by opportunities presented by new technologies, are transferrable provided decision-makers are willing to listen and incorporate public input.

Planners learn by sharing experiences—both successful and otherwise. Case studies provide a tool kit of planning practices. The task is to acknowledge the case study context and adapt the process to address the challenges facing tomorrow's planners.

KEY TERMS

Community Amenity Contributions
Crowdsourcing
Development charges (or Development Cost Levies)
Infill development
Nimbyism
Official Community Plan (OCP)
Social media
Zoning bylaws/schedules

REFERENCES

City of Vancouver. 1993a. *CityPlan Ideas Book*. Vancouver: City of Vancouver.

——. 1993b. *CityPlan Tool Kit*. Vancouver: City of Vancouver.

——. 1994. *CityPlan Making Choices Workbook*. Vancouver: City of Vancouver.

——. 1995. *CityPlan: Directions for Vancouver*. Vancouver: City of Vancouver.

——. 1997a. *Financing Growth: Paying for City Facilities to Serve a Growing Population*. Vancouver: City of Vancouver.

——. 1997b. *Transportation Plan*. Vancouver: City of Vancouver.

——. 1998. *Kensington–Cedar Cottage Community Vision*. Vancouver: City of Vancouver.

——. 2003. *Financing Growth: Paying for City Facilities to Serve a Growing Population*. Vancouver: City of Vancouver.

——. 2004. *Kingsway and Knight Neighbourhood Centre: Housing Area Plan*. Vancouver: City of Vancouver.

——. 2008. *EcoDensity Charter*. Vancouver: City of Vancouver.

FCM (Federation of Canadian Municipalities). 2011. "Policy statement on environmental issues and sustainable development." Ottawa: FCM. http://www.fcm.ca/home.htm.

Howard, R. 1995. "Master plan provides destination, but no route." *Globe and Mail* 22 April.

International Association for Public Engagement. 2007. http://www.iap2.org.

Lee, J. 2013. "Relations with Vancouver city hall sour in four communities angered by planning process." *Vancouver Sun* 24 September.

McAfee, A. 1997. "When theory meets practice—Citizen participation in planning." *Plan Canada* 37(3): 18–22.

Punter, J. 2003. *The Vancouver Achievement: Urban Planning and Design*. Vancouver: University of British Columbia Press.

Seelig, M., and J. Seelig. 1997. "Participation or abdication?" *Plan Canada* 37(3): 18–22.

——. and A. McAfee. 2008. "An update on *CityPlan*." In J. Grant, ed., *A Reader in Canadian Planning: Linking Theory and Practice*, 116–18. Scarborough, ON: Thomson Nelson.

Vancouver Charter [1953-55-1].

CASE STUDY 6.2

Cultural Planning and Governance Innovation: The Case of Hamilton

Jeff Biggar

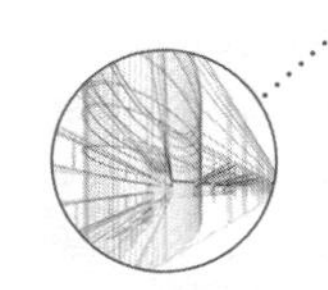

SUMMARY

This paper examines the challenges associated with developing a **Cultural Plan** in Hamilton, Ontario. It considers what actors were consulted in the process as well as how the consultation process was organized and the techniques used to foster dialogue between various **stakeholders**. It also examines the opportunities and challenges associated with undertaking a municipal Cultural Plan for stakeholders who were involved in the consultation process as well as the broader community.

OUTLINE

- Governments are becoming more entrepreneurial in planning for culture, which has become more closely tied to community economic development in Canadian municipalities.
- The development of cultural planning in Hamilton has occurred as the city has transitioned from a manufacturing base to a region with a post-industrial focus. The process of public participation in its Cultural Plan encountered a number of challenges.
- With several barriers to implementation of the plan, this case offers several lessons for other municipalities in re-imagining the relationships between culture, place, and work.

Introduction

The ascendance of urban strategies related to **culture** and creativity has captured the attention of policy-makers and academics both globally and locally. Canadian cities, in particular, have capitalized on the contribution of art and culture to local economic development. Planning is no longer restricted to decisions involving land use and regulation but is increasingly focused on the cultural characteristics of an urban region. This chapter evaluates urban cultural planning in the City of Hamilton. The chapter begins with a background on the relationship between culture and local governance. The details of the research project are then presented, followed by a discussion of the process, techniques, and actors involved in the formation of Hamilton's Cultural Plan.

Culture and Planning

Public-sector reforms in the UK and North America have transformed urban governance over the past three decades. Broad structural changes have led to a shift from a top-down, hierarchical mode of governance to a partnership mode (Harvey 1989; Hastings 1996). City governments are becoming more entrepreneurial in how they respond to urban problems, increasingly relying on external partners like businesses and communities to deliver public programs.

Many scholars have called for a critical analysis of modern governance arrangements, developing theories that assess the interactive nature of planning, which includes increased collaboration between departments as well as between state and non-state actors (Healey 1995, 1997; Forester 1999; Sager 1994; Booher and Innes 2002; Hajer and Wagenaar 2003; Goldsmith and Eggers 2004; Agranoff 2007). Bringing different actors together has the potential to enable new learning opportunities and to enhance the ability of stakeholders to make active decisions about their city. However, in order for positive relationships between state and non-state agencies, citizens, and other stakeholders to take shape, a degree of trust must underpin these relations. The literature has shown that the effectiveness of local governance is linked to the strength of networks and high levels of social capital (Amin and Thrift 1995; Putnam and Leonardi 1993). Commonly thought of as the overall resources that individuals and groups possess, social capital emphasizes the inherent value of trust, norms, and social relationships to both individuals and society at large (Bourdieu 1986; Coleman 1988).

Strong networks and high levels of social capital combine to provide the conditions for a strong governance dynamic to develop. However, power relations invariably contribute to an uneven distribution of power within governance (Flyvbjerg 1998). Participation tends to favour the resource-rich: those of similar demographic characteristics who possess political know-how, time, and professional connections (Agger 2012; Quick and Feldman 2011). The resource-poor (e.g., immigrants and low-income residents) are at a disadvantage. Thus, the social worlds that come together through planning exercises are not always representative of the broader public.

One recent trend within planning is to adopt a cultural planning approach, which tries to include a wide range of community actors in the planning process. Hawkes (2001, 3) argues that culture is the "bedrock of society," for it "covers both the values upon which a society is based and the embodiments and expressions of these values in the day-to-day world of that society." Culture is not simply about galleries, theatres, and other traditional elements of high art, nor is it only about popular culture, entertainment, or heritage. It encompasses a

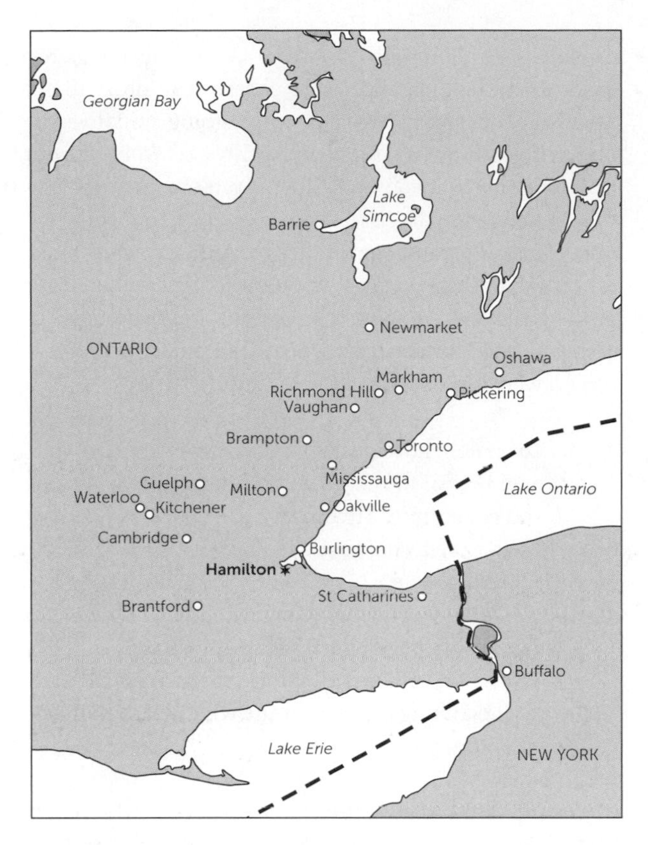

Figure 6.2.1 • Location map.

range of values and processes, including participation, inclusion, innovation, education, tolerance, and well-being, as well as respect for the environment (Duxbury and Jeanotte 2012).

Since the late 1980s, cultural planning has been promoted as an important tool in helping cities to mobilize opportunities for economic growth, revitalizing downtowns, and improving quality of life (Bianchini 1996; Evans 2001; Stevenson 1992). Reflecting the popularity of creativity discourse in policy circles, cultural planning has been promoted as a tool for cities to build a culture-rich environment for residents and tourists (Grodach 2013) but also for potential residents, knowledge workers, and members of the "creative class" (Florida 2002, 69). Florida defines this class as including occupations in science, engineering, higher education, the arts, design, and entertainment, as well as knowledge workers in health care, high-tech, financial services, and the legal field.

A cultural approach to planning is positioned as a means of breaking down the silos between planning departments, improving inter-organizational relations. It is also seen as a means of bringing together municipal government actors with those in the private and voluntary sectors (Montgomery 1990). The rationale is that applying a cultural lens to social, environmental, economic, and community development objectives will strengthen existing services and programs while at the same time alleviating barriers to collaboration (Hume 2009). In this way, cultural planning is viewed as a bridge-building strategy for municipal governments. Unlike cultural policy, which is concerned with supporting more elite forms of art (such as museums, film, and publishing), cultural planning is focused on cultivating local cultural activity, including anything from cultural programming to local events and street festivals (Evans 2005; Stevenson 2004). The next section illustrates how cultural planning is understood by stakeholders involved in the planning process.

Cultural Planning in Hamilton

Hamilton is a mid-sized city with a population of approximately 500,000. Historically, Hamilton was dominated by employment in the steel industry but has recently begun to diversify its economy, focusing on biotechnology, arts, and culture (City of Hamilton 2010). Hamilton has a rich cultural fabric and is home to one of the largest communities of newcomers in Canada. The city has been identified as a key "urban growth centre," according to Ontario's Places to Grow strategic plan (Province of Ontario 2006). Urban growth centres are considered a priority for business attraction. To accompany this growth directive, there has been a surge in government-funded workshops across Ontario to promote the benefits of "municipal cultural planning." Since 2009, the Province of Ontario has invested $3 million in municipal cultural planning to help municipalities identify how their cultural resources could better generate opportunities for economic growth (Kovacs 2011). Hamilton received $50,000 from the Province's "creative community prosperity fund" to support their cultural planning process, particularly to initiate community and stakeholder engagement sessions and establish a cultural policy framework.

Led by the City's Culture Division, which is housed in the Planning and Economic Development Department, the City launched its first Culture Plan in 2008 (City of Hamilton 2013). Dubbed "Love Your City," the initiative entailed a number of phases, ranging from spatial mapping to public engagement. The overall goal was to make culture an instrumental tool in city-building. This included developing cultural businesses, increasing tourism, and rejuvenating the downtown core (City of Hamilton 2013). Table 6.2.1 lays out progress the City has made to date.

Between Phase 1 and Phase 2, the City undertook a number of steps to include the public in the planning process. First, the City established cultural leaders' roundtables (artists, cultural and creative workers) and multi-stakeholder workshops (staff from the City's culture office, staff from different departments, and non-profit and community organizations). Second, the City held public workshops, including the "citizens' reference panel," a series of workshops and participation exercises that convened members of the public over a period of three weekends. This produced the City's definition of culture, which was mutually determined by City staff and the public. Taken together, these participatory activities generated recommendations currently found in the Cultural Plan (City of Hamilton 2013).

To gain information about the planning process, 20 semi-structured interviews were conducted during July and August 2012 with key stakeholders (city staff, artists, planners, cultural organizations, and other non-profit groups) involved in Phase 1 and 2 of the project, specifically those who attended the cultural leaders' roundtable and the multi-stakeholder workshops. However, participants who took part in public consultations, such as the citizens' reference panel, were not interviewed. Public consultations, workshops, and City Council meetings relevant to cultural planning were observed in order to get a sense of the overall reaction to ongoing initiatives. A research goal was to learn how the City was working with people, who was involved, and what their response was to this initiative. As discussed earlier, it is evident from the literature that planning includes some actors and excludes others. Accordingly, it was important to get a better sense of the goals and vision guiding the cultural planning process and how participation works. The objectives for the project were fourfold:

1. To explore how local government works with members of the cultural community, City staff, and community organizations
2. To analyze techniques used by government to foster dialogue
3. To assess how the role of culture is understood
4. To examine the opportunities and challenges associated with undertaking a municipal cultural planning process for the wider Hamilton community

After conducting interviews with participants, responses were analyzed, coded, and categorized by theme. The table below represents the results of the study.

Many non-governmental respondents felt that the City did a poor job in communicating with stakeholders and keeping lines of communication open throughout the process. Respondents representing the City felt

Table 6.2.1 • Cultural Planning Phases

Milestone	Description	Timeline
Phase 1	*Cultural Mapping Exercise:* The City hired consultants to conduct an inventory of the city's cultural assets and established benchmarks, such as the number of cultural industries and employment in the arts and culture, to illustrate the spatial configuration of cultural resources.	Spring 2010
Phase 2	*Cultural Policy:* The City established their first-ever Cultural Policy, which consisted of a vision statement, definition of culture, and guiding principles. This document was directly informed by input gathered from public workshops.	Summer 2012
Phase 3	*Cultural Plan:* A formalized plan was developed combining the above phases into a set of actions and deliverables. It consists of strategic goals and actions distilled from 250 recommendations. The Cultural Plan was approved by City Council but without explicit financial commitments.	October 2013

Table 6.2.2 • Analysis of Interviews

Area of Investigation	Characteristics of Local Governance	Planning Challenges
Local Planning Culture	• Misunderstanding of roles • Poor communication • Lack of trust	• Unchallenged assumptions regarding roles • City not seen as reliable partner • Fear of "over-planning"
Planning Techniques and Activities	• Long gaps between planning events • Outmoded planning procedures • Engaging with "usual suspects"	• Participation fatigue • Disengaged groups • Lack of involvement in decision-making, missing voices
The Role of Culture	• Having a framework in place • Competing visions of "culture"	• Leveraging Cultural Plan to build on existing cultural activity • Working with a divided cultural community
Implications for Hamilton	• Proving that culture matters • Political uncertainty • Adversity • Resistance to change	• A change in political will • Re-imagining the set of relationships between place, culture, and work

that they did a sufficient job at communicating but suggested the need to establish relationships with groups not previously consulted. Obstacles to overcoming communication gaps include a lack of staff, resources, and training. A lack of transparency contributed to a gap between how the City positions its role and how non-governmental stakeholders perceive this role. While the City has been forthcoming about its role as catalyst and community partner, the presentation of their role as a "reliable partner" was met with apprehension by some respondents. Some respondents felt that government intervention in the cultural sphere could lead to a narrow focus on aesthetic changes and a "contrived" notion of culture.

Many respondents, both government and non-government, felt that City staff from the culture office were ambiguous about what the future involvement of stakeholders would entail. Culture staff, on the other hand, were held back from making commitments, such as assigning tasks to groups and establishing timelines, until they receive approval from City Council. Government respondents found these procedural delays frustrating, because it resulted in missed opportunities to build trust with stakeholders inside and outside of government. In terms of participation, some non-governmental respondents felt that the multi-stakeholder workshop was tokenistic—a symbolic gesture toward the city's diverse cultural communities that were not equal to mainstream groups in planning meetings. In addition, a few respondents felt that certain narratives regarding diversity—for example, an account of the city's Aboriginal history—were left out of the planning stages.

Most respondents indicated that cultural planning is a low priority in relation to other municipal policy areas, such as social services and health care. Because culture is a non-essential service, its strategic importance is not prioritized. Many respondents had divergent opinions on what culture means and the role that culture plays in urban development. Some referred to culture as a way of life, while others defined culture in terms of arts for art's sake (recognizing the intrinsic value of art). Others emphasized the role of culture in innovation and the knowledge economy (market-based understandings). The majority of respondents stated that Hamilton's ability to move forward on major city-building initiatives was plagued by the stigma of its reputation as a "have-not" city. In this sense, culture may have a role to play in boosting civic confidence, but it must be understood as one part of a larger revitalization effort.

Most respondents were in agreement about the need for better leadership. There was a feeling among non-governmental respondents that City staff ultimately yields to council, which will likely favour cultural development

Figure 6.2.2 • Site map, James Street North, Hamilton.

with a strong economic rationale and act less favourably toward non-revenue-generating cultural activities such as community arts.

Amid planning challenges, some respondents felt that establishing a Cultural Plan would bring positive benefits. On a policy level, having a framework in place would legitimize culture and hold decision-makers accountable. On a civic level, a Cultural Plan would help to improve the outside perception of the city's image and boost confidence among residents through programs that actively celebrate community life. Lastly, many respondents cited a large disconnect between working-class communities and creative communities.

Interviews with key stakeholders revealed many insights into the characteristics of local governance. City officials collaborated with external stakeholders and were effective in cultivating interest around the plan; however, much of the momentum languished because of delays in process. This outcome can be attributed to outmoded methods of stakeholder consultation. Many non-governmental respondents perceived the City's participation efforts as a series of one-off consultations and forums with no common thread holding these procedures together. Consultations did not build on the progress of preceding events. One respondent noted:

> The City could do a much better job of soliciting input and really try better ways to get the stakeholders involved in policy development in a more meaningful way. It is still generally the same format used 20 years ago with public meetings. . . . Particularly, in the creative industries and the cultural sector, you have a diversity of personalities. It is difficult . . . to get consensus on issues because some people are fiercely independent, which makes it a little bit like herding cats (urban planner, non-profit sector, 26 July 2012).

The cultural planning process had many challenges in forming an integrative network. While stakeholders had opportunities to share their knowledge of local practices, many felt that this knowledge exchange would have little significance in outcomes. This dynamic contributed to low levels of trust, leaving many to adopt a "wait and see" approach. The following

© Mike Lalich

© Sheryl Nadler

Figure 6.2.3 • Grassroots cultural scene on James Street North.

quote illustrates the perception of the City not being "a reliable partner":

> I felt as though the City's interest in us [organization] was tokenistic, not genuine. They come to us only when they need something, but our calls go unreturned. There have been no signs that our feedback has been reflected thus far, and as a consequence of that, we do not have a lot of interest in the implementation phase (executive director, cultural organization, 12 July 2012).

Healey (1998) stresses the importance of establishing a robust and transparent agenda from the start of planning processes in order to create trust among stakeholders, account for social differences, and create pathways for social learning and collaboration. Without such an agenda, the implementation potential of projects deteriorates. Based on the interviews, I can suggest that the city faced challenges in terms of generating the institutional capacity to build sustainable partnership opportunities.

Reaching consensus proved difficult. While the "definition of culture" was jointly determined by members of the public who participated in the citizens' reference panel, the City received meagre support from the cultural community about their vision for culture. Much of the mistrust of City staff on the part of members of the community derived from its focus on design initiatives. One respondent called attention to the negative outcomes that could be brought on by planning interventions:

> James Street has been touted as an organic movement because people got together and did something. It wasn't a plan where the city came in and said we have made this creative zone because this is where all the artists have been. We don't want the banners going up calling this the arts district. We don't want the big gates. We want the neighbourhood to be allowed to determine itself (small-business owner, 2 August 2012).

Interestingly, many respondents from the City were consciously aware of the sensitivities surrounding planning for culture:

> You have to be careful about destroying what you're trying to create. You can really over-plan something.... It's a delicate flower (cultural heritage manager, City of Hamilton, 23 July 2012).

The above tensions speak to whether cultural development can be improved through planning. For Hamilton, it is still premature to assess whether City-led planning initiatives will interfere with the organic, bottom-up nature of development. In order for it to work, the city should avoid actions that lay claim to ownership over cultural amenities and spaces established through previously unplanned projects.

While the hopeful rhetoric found in the City's Cultural Plan positions culture as a catalyst for revitalization, the evidence from this study suggests that other investments must also be prioritized or the City will fall short in addressing social stresses such as poverty and unemployment.

Many respondents cautioned that a City-led initiative about culture was merely educating receptive minds, such

Figure 6.2.4 • "Art is the New Steel." Why might some consider this controversial?

as those who work in cultural industries. The concern is that ordinary citizens do not identify with the broadly conceived vision of culture outlined by the City. As one respondent noted:

> Citizens have much more basic demands of their councillors, such as a pothole down the street. Most people don't talk to councillors about a larger vision for the city. They don't make statements like "I want Hamilton to be a cultural centre" (project director, City of Hamilton, 8 August 2012).

Many respondents also noted a lack of understanding of the arts by local residents, many of whom have been displaced from manufacturing-based jobs and have little knowledge of cultural sectors. As one participant noted, it has been difficult to engage the local community:

> I'm not convinced there was a lot of work done in the planning process to connect cultural industries to unions and their members. We [the public] need to understand more fully what it means to practise as an artist. . . . We are not there yet (social planner, non-profit sector, 26 July 2012).

While the Cultural Plan has been officially adopted, no budgetary funds have been guaranteed. A lack of broader support for the plan may encourage City Council to question the Cultural Plan's value in relation to other municipal areas. Considering culture is not prioritized by some residents and councillors, City staff may receive pushback from City Council regarding the justification for allocating resources to culture. With regard to the stakeholders involved, lack of ongoing exchange of knowledge throughout the planning process may constrain relations between City staff and the cultural community in the implementation phase. Consultation with stakeholders was largely confined to the beginning of the planning process, with sporadic moments of input throughout. This method may hinder those leading the planning process in becoming aware of the evolving needs of the arts and cultural communities, and, most important, they may lose support in implementation.

Conclusion

This chapter evaluated the planning process related to the development of Hamilton's Cultural Plan. It assessed the conditions (institutional structure and political culture), actors (City staff, planners, cultural workers, community and non-profit groups), and channels (planning meetings) that shaped processes related to the Cultural Plan. It found that cultural planning creates opportunities for future involvement and capacity-building around the plan. However, the plan's vision of providing mutual benefit is yet to be achieved. Many of the interviewees expressed concerns regarding the fait accompli nature of public participation whereby cities offer consultation exercises to garner public support but then go ahead with their pre-existing vision. These results are consistent with those of studies examining culture-led urban development projects in which consultants, city staff, and architects retain much of the decision-making power (Evans 2005). However, considering that the long-term outcomes of the plan have yet to be seen, it is premature to comment on who will drive the cultural agenda and whether or not the cultural planning process will translate into meaningful action for those involved.

Thus far, however, I found that the local governance system is characterized by low levels of social capital, contributing to weak relations between government and non-governmental stakeholders. Cities that demonstrate low levels of social capital may be hindered in developing strong relationships with stakeholders. Prior to beginning a planning process, cities should consider whether previous planning processes have weakened levels of trust and cooperation between citizens and stakeholders. If so, this tension should be addressed as a way of learning from past mistakes.

In Hamilton, non-governmental stakeholders were sceptical about the City's intervention into the cultural sphere. There was a fear of "over-planning" and a loss of interest in the planning process as a result of people feeling disengaged. In cautioning against using culture as a solution to a range of other policy problems, Stevenson (2004, 129) suggests that "culture must actually mean *something*, but it cannot mean everything." This case study indicates that the term "culture" is a highly contested concept meaning different things to different people. This makes it difficult to use culture as a means of bridging the gap between diverse social groups. While respondents did not agree on the meaning of culture, they acknowledged that in a time of economic transition, any forward-thinking initiatives, such as a Cultural Plan, were welcome. What was questioned was culture's potential to lead the transition.

While the procedural norms embedded within the planning process left many frustrated, they provided learning opportunities for thinking about the wider implications of cultural planning. The cultural planning process prompted many to clarify the challenges Hamilton faces transitioning from an industrial to a post-industrial city, including the knowledge gap between more traditional working-class communities and the arts and culture communities. Ultimately, the degree to which cultural investment works for diverse social groups—such as Hamilton's working-class communities—remains unclear. While these narratives were missing from stakeholder conversations, they should not be overlooked by decision-makers when determining the future cultural agenda. As the City takes broader efforts to re-tool its economy, there is a serious need for stakeholders and the public to re-imagine relationships between culture, place, and work. Unless they do so, a Cultural Plan may have little relevance for those residents who give meaning to the place. Exploring new methods of public consultation may create the pathways for less resource-rich participants to be involved, creating opportunities to seek new members, new understandings of public issues, and new opportunities to work together.

KEY TERMS

Cultural Plan
Culture
Stakeholder

LINKS

Learn how the City of Hamilton defines "culture": http://www.youtube.com/watch?v=NSCFxDKJWwo.

Learn how the City of Hamilton is positioning "culture" and "creativity" as a local economic development strategy: http://www.youtube.com/watch?v=hpfbhv373n4.

REFERENCES

Agger, A. 2012. "Towards tailor-made participation: How to involve different types of citizens in participatory governance." *Town Planning Review* 83(1): 29–45.

Agranoff, R. 2007. *Managing within Networks: Adding Value to Public Organizations*. Georgetown: Georgetown University Press.

Amin, A., and N. Thrift. 1995. "Globalisation, institutional 'thickness' and the local economy." *Managing Cities: The New Urban Context* 12: 91–108.

Bianchini, F. 1996. "Cultural planning: An innovative approach to urban development." In J. Verwijnenand and P. Lehtovuori, eds, *Managing Urban Change*, 18–25. Helsinki: University of Art and Design Helsinki.

Booher, D.E., and J.E. Innes. 2002. "Network power in collaborative planning." *Journal of Planning Education and Research* 21(3): 221–36.

Bourdieu, P. 1986. "Forms of capital." In J.G. Richardson and C.T. Westport, eds, *Handbook of Theory for the Sociology of Education*, 241–58. Westport, CT: Greenwood Press.

City of Hamilton. 2010. "City of Hamilton Economic Development Strategy, 2010–2015." http://www.investin-hamilton.ca/wp-content/uploads/2011/06/Hamilton-EcDev-Strategy2010.pdf.

——. 2012. "Phase 2—Cultural Policy: Report to Council." http://www.hamilton.ca/NR/rdonlyres/CA29FB48-0C1E-4C3B-A908.

——. 2013. "Draft Cultural Plan." Appendix A to Report PED13045. http://www.hamilton.ca/NR/rdonlyres/B50B8AEB-7F26-4937-9C64-92241E167099/0/CulturalPlan_DRAFT_CityOfHamilton.pdf.

Coleman, J.S. 1988. "Social capital in the creation of human capital." *American Journal of Sociology* 94: S95–S120.

Duxbury, N., and M.S. Jeannotte. 2012. "Including culture in sustainability: An assessment of Canada's Integrated Community Sustainability Plans." *International Journal of Urban Sustainable Development:* 4(1): 1–19.

Evans, G. 2001. *Cultural Planning: An Urban Renaissance?* London: Routledge.

——. 2005. "Measure for measure: Evaluating the evidence of culture's contribution to regeneration." *Urban Studies* 42(5–6): 959–83.

Florida, R. 2002. *The Rise of the Creative Class and How It's Transforming Work, Leisure, Community and Everyday Life*. London: Basic Books.

Flyvbjerg, B. 1998. *Rationality and Power: Democracy in Practice*. Chicago: University of Chicago Press.

Forester, J. 1989. *Planning in the Face of Power*. Berkeley: University of California Press.

Goldsmith, S., and W.D. Eggers. 2004. *Governing by Network: The New Shape of the Public Sector*. Washington: Brookings Institution Press.

Grodach, C. 2013. "Cultural economy planning in creative cities: Discourse and practice. *International Journal of Urban and Regional Research* 37(5): 1747–65.

Hajer, M.A., and H. Wagenaar, eds. 2003. *Deliberative Policy Analysis: Understanding Governance in the Network Society*. Cambridge, MA: Cambridge University Press.

Harvey, D. 1989. "From managerialism to entrepreneurialism: The transformation in urban governance in late capitalism." *Geografiska Annaler, Series B, Human Geography* 71(1): 3–17.

Hastings, A. 1996. "Unravelling the process of 'partnership' in urban regeneration policy." *Urban Studies* 33(2): 253–68.

Hawkes, J. 2001. *The Fourth Pillar of Sustainability: Culture's Essential Role in Public Planning*. Melbourne, Australia: Common Ground.

Healey, P. 1997. *Collaborative Planning: Shaping Places in Fragmented Societies*. London: Macmillan.

——. 1998. "Building institutional capacity through collaborative approaches to urban planning." *Environment and Planning A* 30(9): 1531–46.

Hume, G. 2009. *Cultural Planning for Creative Communities*. St Thomas, ON: Municipal World.

Kovacs, J.F. 2011. "Cultural plan implementation and outcomes in Ontario, Canada." *Cultural Trends* 19(3): 209–24.

Montgomery, J. 1990. "Cities and the art of cultural planning." *Planning Practice and Research* 5(3): 17–24.

Ontario Ministry of Infrastructure. 2006. *Place to Grow: Growth Plan for the Greater Golden Horseshoe*. https://www.placestogrow.ca/index.php?option=com_content&task=view&id=9&Itemid=12.

Putnam, R.D., and R. Leonardi. 1993. *Making Democracy Work: Civic Traditions in Modern Italy*. Princeton, NJ: Princeton University Press.

Quick, K.S., and M.S. Feldman. 2011. "Distinguishing participation and inclusion." *Journal of Planning Education and Research* 31(3): 272–90.

Sager, T. 1994. *Communicative Planning Theory*. Aldershot, UK: Avebury.

Stevenson, D. 2004. "'Civic gold' rush: Cultural planning and the politics of the third way." *International Journal of Cultural Policy* 10(1): 119–31.

——. 1992. "Urban re-enchantment and the magic of cultural planning." *Culture and Policy* 4: 3–18.

INTERVIEWS

Urban planner, non-profit sector, 26 July 2012. Telephone interview.

Executive director, cultural organization, 12 July 2012. In-person interview.

Small-business owner, 2 August 2012. In-person interview.

Cultural heritage manager, City of Hamilton, 23 July 23 2012. In-person interview.

Project director, City of Hamilton, 8 August 2012. Telephone interview.

Social planner, non-profit sector, 26 July 2012. In-person interview.

CASE STUDY 6.3

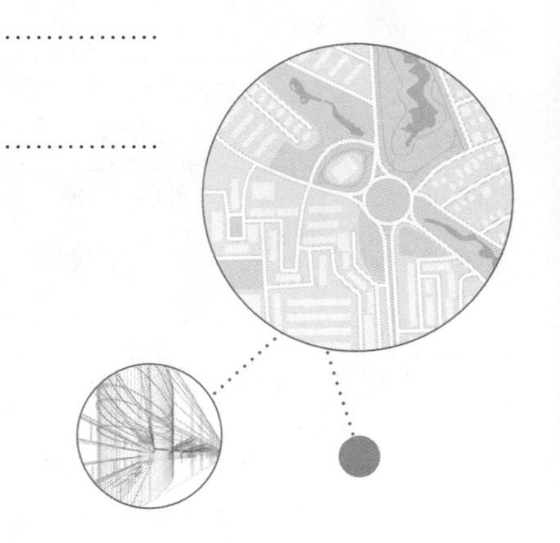

Taking It Online: How the City of Vancouver Became Comfortable with Engaging Residents in Their PJs: Vancouver's Use of Online Crowdsourcing to Engage Residents during the Development of the Greenest City Action Plan

Lisa Brideau and Amanda Mitchell

SUMMARY

Public engagement is of critical importance for municipalities, especially when they are developing plans that will lead to significant changes. It can be hard for municipalities, with limited staff and budgets, to justify the risk of trying new engagement techniques. Sometimes, however, the risk is worth it. To engage Vancouverites in an important dialogue about sustainability, the City asked residents to "Talk Green to Us" and share their ideas on how to make Vancouver the greenest city in the world. This case study reviews the use of an online **ideation tool**: a forum to **crowdsource** and discuss green ideas—a first for the City and a brave departure from more traditional engagement methods. The City stepped out of its comfort zone and used new digital tools to create an engagement process in which residents could participate while at home in their pyjamas. The Talk Green to Us forum helped to change how the City of Vancouver does engagement by proving the value and utility of digital tools. It also taught City planners key lessons about how to make online engagement successful; these lessons are shared in this chapter.

OUTLINE

- The City of Vancouver recently took an ambitious approach to developing a plan that would make it the greenest city in the world by 2020.
- The development of the engagement plan began with 10 goals and asked the public how to meet these goals. Through an online forum as well as in-person events, people were asked to suggest ideas, review all the options, and vote for a limited number. This created a broader dialogue between members of the public and the planning team.
- The results of engaging a broader public included a higher-than-anticipated number of responses and ideas generated, a better understanding of how to use social media in planning processes, and the ability to explain the viability of an idea in terms of implementation.

Introduction

The City of Vancouver's Greenest City planning effort had an ambitious goal—to create a plan that would make Vancouver the greenest city in the world by 2020. To achieve a goal this bold requires an equally bold engagement process. This case study details the City of Vancouver's brave foray into online engagement, how an online crowdsourcing forum and **social media** networks allowed the City to tap into the ideas and creativity of a passionate public, and how the success with this process changed the face of public engagement at the City of Vancouver.

The Context

The City of Vancouver is a municipality on the west coast of Canada. With a population of 603,000, it is

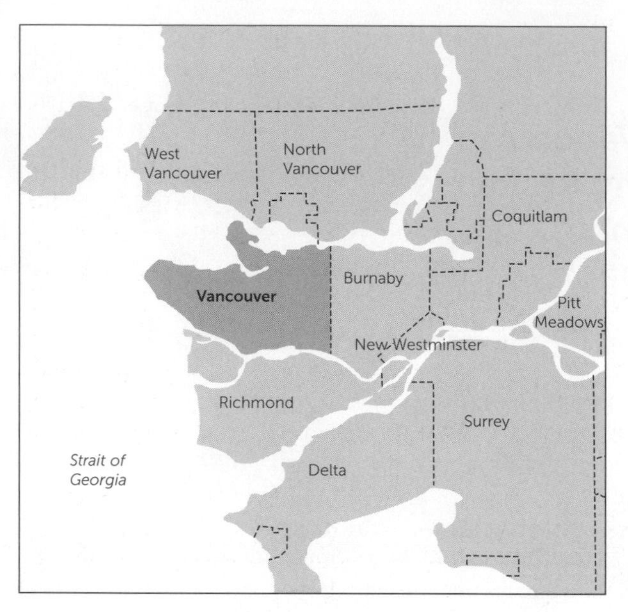

Figure 6.3.1 • Location map.

the largest city in British Columbia, the eighth-largest municipality in Canada, and the core of the third-largest metropolitan region in Canada. Vancouver is governed by 10 councillors and a mayor elected at large. Vancouver has a diverse population; according to 2011 census data, 49 per cent of Vancouver residents speak English, and 25 per cent speak Chinese as their first language.

The Start of the Greenest City Vision

In 2009, building on Vancouver's long history of action on sustainability, Mayor Gregor Robertson set the ambitious vision of making Vancouver the greenest city in the world by 2020. To determine what needed to be done to get there, the mayor convened a task force of local thought leaders in sustainability, including Dr David Suzuki. This Greenest City Action Team (GCAT) established appropriate goals and targets necessary to succeed.

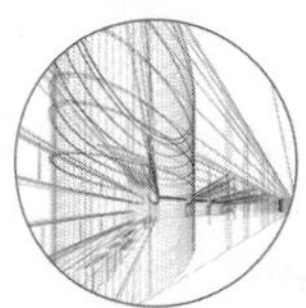

PLANNING THEORY

Diversity

Diversity of languages and cultures in cities can stretch public engagement budgets very thin just in terms of the costs of basic translation services. While online engagement in multiple languages is currently a challenge, online engagement on an English platform does potentially expand participation to those comfortable *writing* in English who might not be comfortable speaking it or attending a workshop. Developing creative ways to meaningfully engage across many cultures and languages (online and off) will be an ongoing challenge for planners as Canadian communities continue to diversify. Arnstein (1969), Davidoff (1965), and Forester (1999) provide broader discussions on meaningful engagement and the importance of including a diversity of voices.

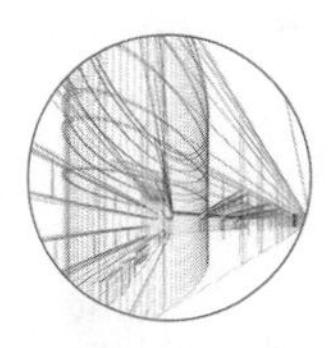

PLANNING THEORY

The Balance of Power

While other cities in the Province of BC get their governing authority from the Province via the Local Government Act and the Community Charter, Vancouver gets its authority from the Vancouver Charter (1953). The Vancouver Charter provides powers that other municipalities do not have, such as a separately elected Park Board, a separate building code, the ability to impose special development levies, and the ability to delegate to the director of planning the authority to approve development permits. Vancouver is a large city of national significance with unique governance powers granted by the Province. It is worthwhile to consider how this may be relevant to Vancouver's policy directions and status as an innovator in urban sustainability (Forester 1989).

The Greenest City Planning Challenge

In 2010, City Council accepted the GCAT's recommended 10 goals (see Table 6.3.1) and directed staff to report back with a detailed Greenest City Action Plan (GCAP) that would act as a road map for achieving the targets.

The Greenest City Action Plan would need to identify bold actions and significant policy changes for the City in order to meet the adopted targets, so a robust engagement process would be needed. The planning challenge was to engage a broad spectrum of people, gather ideas to inform the plan, and build support for implementation.

The Greenest City Planning Team

To lead development of the **action plan**, a Greenest City Planning Team (GCPT) was created in the Sustainability Department. This team of one part-time and two full-time planners designed, organized, and facilitated the public engagement work and connected public feedback with the other staff who were developing the plans, among other duties.

The GCPT was not responsible for writing the technical content of the plans. For each of the 10 goals, a staff working group was created to come up with a plan that met the specific targets. These cross-departmental and

Table 6.3.1 • Greenest City Goals

Goals and Targets
Green Economy
• Target 1: Double the number of green jobs over 2010 levels by 2020.
• Target 2: Double the number of companies that are actively engaged in greening their operations over 2011 levels by 2020.
Climate Leadership
• Target: Reduce community-based greenhouse gas emissions by 33% from 2007 levels.
Green Buildings
• Target 1: Require all buildings constructed from 2020 onward to be carbon-neutral in operations.
• Target 2: Reduce energy use and GHG emissions in existing buildings by 20% over 2007 levels.
Green Transportation
• Target 1: Make the majority of trips (more than 50%) by foot, bicycle, and public transit.
• Target 2: Reduce average distance driven per resident by 20% from 2007 levels.
Zero Waste
• Target: Reduce total solid waste going to the landfill or incinerator by 50% from 2008 levels.
Access to Nature
• Target 1: Ensure that every person lives within a five-minute walk of a park, greenway, or other greenspace by 2020.
• Target 2: Plant 150,000 additional trees in the city between 2010 and 2020.
Lighter Footprint
• Target: Reduce Vancouver's ecological footprint by 33% over 2006 levels.
Clean Water
• Target 1: Meet or beat the most stringent of British Columbian, Canadian, and appropriate international drinking water quality standards and guidelines.
• Target 2: Reduce per capita water consumption by 33% from 2006 levels.
Clean Air
• Target: Meet or beat the most stringent air quality guidelines from Metro Vancouver, British Columbia, Canada, and the World Health Organization.
Local Food
• Target: Increase city-wide and neighbourhood food assets by a minimum of 50% over 2010 levels.

Source: Lisa Brideau; Amanda Mitchell; City of Vancouver, Corporate Communications Department.

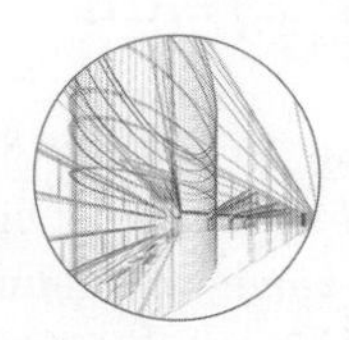

PLANNING THEORY

Meaningful Engagement

Creating a sense of ownership over plans and targets is critical to ensuring that support continues into implementation; this is true of staff and the public. Involving key staff people who will be responsible for implementation of the plan will ensure that their concerns are addressed and that they will be more likely to adhere to and champion the plan during implementation. Meaningfully incorporating and reflecting public input in the plan will help to ensure that public support exists when difficult steps are taken in the implementation phase. Depending on the tools used, online engagement can have the added benefit of the ability to follow up with people at a later phase in the project, which can be useful in maintaining a sense of ownership throughout a project's life (Arnstein 1969).

interdisciplinary working groups were each chaired by a senior staff member. Led by the deputy city manager, a steering committee comprised of these senior staff members met monthly to discuss progress. In total, more than 80 staff members from across the organization participated in these working groups.

The Engagement Plan

The City recognized that it could not achieve the Greenest City targets without support from community organizations, residents, and businesses, so the **engagement plan** was developed with the dual intent of getting public input while also building a base of interested and active citizens who would support and participate in future work. Additionally, the GCPT wanted to reach people who do not normally participate in the City's engagement efforts and to deepen the public conversation about sustainability.

The engagement plan had two phases: collecting ideas and getting feedback on the draft plan. This case study focuses on the first phase.

Taking It Online

Key staff who manage public engagement in various departments at the City felt strongly that the City should incorporate online consultation into its suite of engagement tools. With that support, the GCPT advocated for the use of an online ideation tool to brainstorm (crowdsource) ideas. The desire was to prove that online engagement could be as strong as, or even stronger than, traditional methods.

An online approach would:

- allow people to contribute at a time and place convenient to them—even at home in their pyjamas;
- allow people to see what others were saying or supporting;
- allow staff to provide responses to questions that everyone could see;
- potentially reach a younger age group than traditional engagement approaches.

Staff were searching for an ideation tool that would have the ability to:

- submit ideas;
- support ideas (through voting or endorsements);
- comment on ideas (ideally with threaded conversations);
- categorize ideas;
- update the status of an idea;
- customize the design to blend with the look and feel of the rest of the engagement materials.

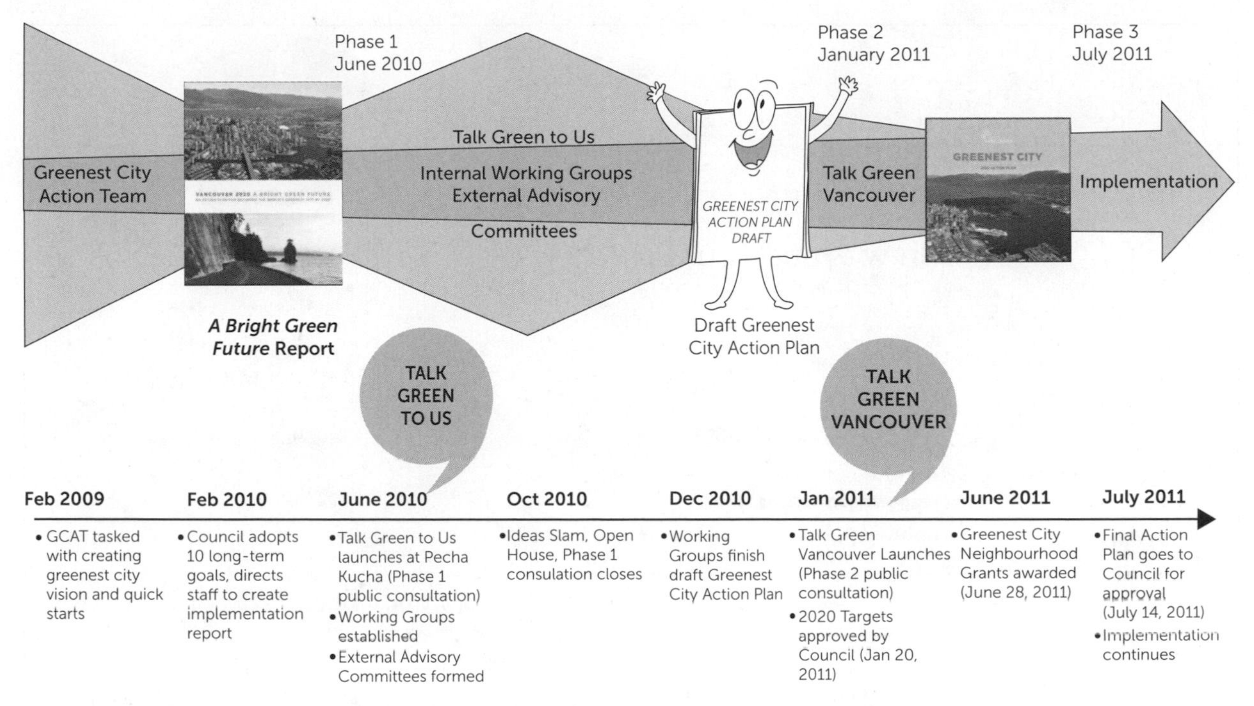

Figure 6.3.2 • Greenest City project timeline.

Source: Courtesy of the City of Vancouver.

Once it was confirmed that the desired functionality could be supplied at a modest cost (using UserVoice), the director of corporate communications championed the online forum proposal with senior management and gained approval.

Crowdsourcing Ideas

Doing something for the first time, especially in a bureaucratic context that is sometimes resistant to change, means it is important to select the right project to use as a pilot. The Greenest City plan was a good testing ground for the online forum for several reasons. It was a high-profile, city-wide initiative with potentially wide-ranging impacts requiring broad public input; it had ample resources to support the engagement effort (budget and staff); and it was a positive subject in need of ideas on how to achieve aspirational goals.

To orient the program for success, the forum was set up with the following directions in mind.

Use Aspirational Branding

To have some fun and get people interested in participating, less "corporate" branding was used. This was a direction supported by the director of corporate communications. A creative firm developed the "Talk Green to Us" campaign, which included a standalone, non–City-branded website and a provocative ad campaign.

Frame around a Shared Challenge

The focus of the consultation was not to get input on the goals or targets but rather to focus on how these targets could be met. It was framed as a shared challenge, asking people: "How can we reach our 2020 Greenest City targets?" To provide context, a video was created that outlined the 10 goals and invited people to add their ideas on how to reach them. Keeping the focus of the dialogue tight on solutions was key to getting the specific ideas that were needed.

Courtesy of the City of Vancouver

Figure 6.3.3 • Talk Green forum screen capture.

Establish Ground Rules and Follow Them

Discussion guidelines were established at the outset to set expectations for participant behaviour. Content that failed to meet these expectations was not published. In addition, the forum was pre-moderated to avoid the posting of any inappropriate material on a City website.

Set Expectations

The forum included a backgrounder on how ideas would be used. The backgrounder explicitly stated that the City was not obliged to implement the ideas—even if an idea received many votes. What was promised was that staff working groups would

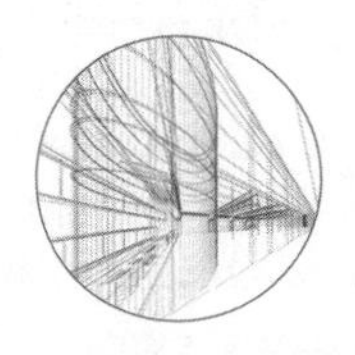

PLANNING THEORY

Unique Issues in Digital Participation

Planners using digital media tools are responsible for understanding the unique issues related to online consultation, such as data privacy (e.g., that different laws govern privacy of data stored on servers in Canada versus those in the US) and the importance of creating a space free from abusive language and harassment (e.g., by establishing and enforcing clear terms and guidelines for participation). Daren C. Brabham (2009) discusses additional issues to consider with respect to online engagement, while Seltzer and Mahmoudi (2013) discuss issues specific to crowdsourcing.

Figure 6.3.4 • Talk Green to Us advertisement.

Source: Courtesy of the City of Vancouver.

consider every idea submitted as they created the draft action plan.

Seed the Conversation

To start the conversation, staff pre-populated the forum with ideas that the working groups were initially considering. By doing so, the City was transparent about the ideas that were being considered, and staff got direct feedback on these strategies.

Participate

While the forum was open, staff logged in to answer questions, guide conversation, explain jurisdictional authority, and share relevant resources (e.g., links to existing programs that were similar to the proposed ideas).

Launch

The GCPT struggled with how to tell the public about the Talk Green to Us forum and ask for participation. The City did not have a particularly strong social Web presence at the time, and the risk with the online forum was lack of participation.

As part of trying to make public engagement fun and easy, and embracing the concept of going where the people are (rather than making them come to you), the City partnered with the University of British Columbia and the organizers of the Vancouver Pecha Kucha Night to launch the Talk Green to Us campaign. Pecha Kucha events take place globally, offering an evening of inspirational local speakers. Each speaker has 20 slides that self-advance after 20 seconds, for a total of 6 minutes 40 seconds per speaker. It is a fast-paced and fun format and a popular event in Vancouver. The very successful "Walk the Talk: Green Your City" Pecha Kucha Night was held at the Queen Elizabeth Theatre. Tickets were kept at an affordable $10, with free tickets available to those unable to pay. More than 2000 people attended this sold-out event, and people were even reselling tickets at the door—a first for a City of Vancouver consultation event.

Getting Social

The Talk Green to Us online forum (http://vancouver.uservoice.com/forums/56390-gc-2020) was open from June through September of 2010, a long time in terms of keeping people interested and engaged. Throughout the campaign, social media (dedicated Facebook [https://www.facebook.com/greenestcity], Twitter [https://twitter.com/greenestcity], Flickr [http://www.flickr.com/photos/greenestcity2020], and YouTube [http://www.youtube.com/playlist?list=PLDDB290AB0A400EBD&feature=plpp] accounts) were used to drive people to the online forum, share ideas, promote events, and encourage participation. To keep people interested and collect more followers:

- compelling ideas from the forum were cross-posted on Facebook and Twitter;
- presentations from in-person events were filmed and posted on YouTube so that those unable to attend could watch online;
- the GCPT collaborated with influential supporters with large social networks; and
- Facebook advertising was purchased to increase awareness among local Facebook users.

Closing the Forum

The online forum ended with a contest. Four of the top ideas would be presented to the mayor, senior staff, and sustainable businesses leaders in a "Dragon's Den"–style Ideas Slam. For each of the ideas selected, a local advocate had five minutes to sell his or her idea to the panel. The panel then discussed what they liked about the idea. This sold-out event was hosted by CBC Radio's Bill Richardson and included well-known local musical guests. To extend the reach, the event was live-tweeted and broadcast on cable TV. This was an extremely useful event for driving traffic to the online ideas forum, and participation on the forum spiked when the event was announced.

Considering Ideas

Staff working groups reviewed every idea submitted as they developed the draft action plan. After ideas were reviewed, their status on the forum was updated to indicate whether they were under consideration, started, planned, completed, or declined. When the status of an idea changed, the system automatically sent an email to every person who voted, commented on, or submitted the idea. This simple feedback loop let people know that the City appreciated their participation. So often in consultations, this follow-up step is not taken. It made participants feel that their time was valued, that the City was listening, and that the City thought their input was worthwhile. It also helped to build support for and ownership of the plan.

These sentiments are reflected in the following comments from a post-consultation survey:

> Too often I find that organizations solicit my feedback, and then feed it into a black hole. Was I heard? Did it make a difference? Is anyone really listening? Who knows! The Talk Green to Us team has been sharing updates on the progress of the public's ideas, both on ideas I submitted and those I voted on. I'm seeing tangible outcomes from my participation. . . . Which is remarkable and re-energizing. Hooray for not feeling like you've wasted your time!

> Timely feedback from the City on proposed measures made my involvement feel worthwhile. Looking at the status of submitted Talk Green to Us ideas, and seeing the City actually take on new initiatives as a result. One of my ideas was accepted and acted upon!

Results

The original target for the first engagement phase was to have 20,000 people engaged (a combination of website visits, ideas submitted, votes on ideas, video views, and attendance at events). When the online forum closed, more than 21,000 people had visited the Talk Green to Us website, and more than 3000 users had submitted 726 unique ideas, 28,000 votes, and 2100 comments. This exceeded all expectations. Many ideas from the forum are in the final plan, and some took on a life of their own—such as the CityStudio, a collaboration between the City and six post-secondary institutions, one of the most popular ideas on the forum.

Table 6.3.2 • Statistics on the Number of Participants Involved

Ideas	726*
Registered users	3,154
Comments	2,140
Votes	28,026
Visits	21,486
Absolute unique visitors	13,747
Page views	38,421
Percent new visits	65%
Countries visiting	123
Visitors from Canada	17,914
Percent visitors from Canada	84%
Visitors from Vancouver	12,774 (59%)
Views on Talk Green to Us video	7,834
Attendance at events	3,728
New Twitter followers	1,177 (2,427 total)
New Facebook fans	1,164 (1,490 total)

* The total number of unique ideas does not reflect the total number of ideas submitted. More than 70 ideas were merged because of their similarity.

Wins and Fails

Typically in open-house–style engagement events, members of the public speak to members of staff or simply fill out comment sheets that only staff see. One of the biggest benefits of the online forum was that it allowed people to see and comment on what others were saying; it became a dialogue among many people instead of a closed conversation between two. The forum allowed staff to openly respond to comments and suggestions, which can make subsequent dialogue more informed and relevant. The ability to loop back and identify which ideas would be included in the plan—and why—is a level of transparency that simply cannot be achieved with traditional open-house engagement styles in which the public anonymously submits comment forms and the input is aggregated and incorporated into the plan in a way that is a mystery to the public.

The Greenest City engagement efforts benefitted from the fact that sustainability is a passionate topic for Vancouverites, and many were keen to engage. It is possible that other planning initiatives may struggle to develop the interest and critical level of discussion activity necessary to make a crowdsourcing platform like this work; it is not an appropriate tool for every public engagement program.

There are some issues with an online forum format; below are several that the Greenest City Planning Team encountered.

Exclusion

Digital engagement tools potentially exclude some people from participation because of their limited access to the Internet or a lack of proficiency in navigating websites. At the same time, in-person events potentially exclude those who cannot attend at the chosen time. The two approaches can work together to maximize access to the engagement process.

Determining Who Was Engaged

Google Analytics showed that visitors to the Talk Green to Us website came from all over the world and 59 per cent of visits originated in Vancouver. Since demographic information on users was not collected, it was difficult to know who was—and who wasn't—participating.

Language Barrier

Since the website was in English only, participation was limited to those comfortable with communicating in written English. This was a definite shortcoming. To compensate, the subsequent phase of engagement had a particular focus on engaging multicultural communities.

Key Lessons Learned

Online and Off-Line Engagement Are Symbiotic

During the engagement process, it became clear that the relationship between social media, online engagement, and in-person events is synergistic. The cultivation of an online community helped to create an interested group to spread the word about workshops and events. In-person events supported the off-line engagement while strengthening the online community. This symbiotic relationship between online and off-line engagement activities was demonstrated when website hits peaked around the campaigns major in-person events.

Online Engagement Works

From the survey responses, blog comments, open-house comments, and the level of participation on the Talk Green to Us website, it was clear that Vancouverites were interested in engaging online. In fact, after the forum closed, participants lamented that it was no longer open.

Social Media Are Legitimate Engagement Tools

The City learned that using social media is not a mere trivial pursuit but a serious and efficient way to develop community and keep people engaged. The social media networks helped to develop a sense of community around the Greenest City planning process. They provided a space where people could get updates on the plan, contribute to the conversation, and feel connected to the process and each other. Additionally, the people engaged with the City on social media often became involved in other consultation efforts, online and off.

Internal Champions Can Assist Implementation

Involving staff from across the organization ensured that the departments ultimately responsible for delivering on the targets were actively involved in creating the action plan. Having senior staff and management involved meant that actions identified in the plan could be integrated into department work plans and budgets, which helped to jump-start implementation.

Conclusion

The use of online tools is not limited to larger cities like Vancouver with substantial planning departments and resources. Although the Greenest City example was well resourced and large scale, smaller municipalities have also had success with online engagement; the key is to be realistic about the resources required to do it well. Provided the community is comfortable online, it would be worthwhile for a municipality to consider an online component of some kind for engagement, especially for the idea-generating phase of an important city-wide initiative. To ensure success, the municipality should set expectations, have a clear plan for how input will be used, and allocate adequate staff. Above all else, online engagement should not replace in-person events but provide another avenue for people to get—and stay—involved. Online engagement is not a cheap shortcut or a substitute for in-person consultation: it is a way to expand and deepen engagement (Seltzer and Mahmoudi 2013).

Planners in the GCPT played a critical role in this project by suggesting and championing digital engagement techniques. They researched proposed tools, sought the support of management to proceed, implemented the engagement plan, and shared lessons learned with other staff.

The campaign profoundly affected the way public engagement is done at the City of Vancouver. Based on its success,[1] the City has increased its comfort level with online engagement to the point where the aim is for every engagement process to have an online component. Sometimes it could simply be an online questionnaire, but it is often more complex. The Greenest City experience has opened the door to experimenting with other online tools, including discussion forums, crowdmapping, budget calculators, and the use of Twitter for collecting and responding to service issues. In 2013, the City launched Talk Vancouver, an online community panel where Vancouverites can provide feedback to the City in an ongoing way—a significant investment in ongoing online engagement.

The lessons learned from the social media outreach have benefitted other City accounts and projects and have helped shape the City's social media approach. Greenest City social networks continue to be active with an ever-growing audience, providing updates on the Greenest City implementation and the City's sustainability programs.

The new approach for engagement efforts in Vancouver is to find ways to make engagement fun and easy whenever possible. This means meeting people where they are—at local events or online at their computer—and making it as convenient as possible for them to participate. Its willingness to try social media and

1 The Greenest City Action Plan won a Sustainable Communities Award for Planning from the Federation of Canadian Municipalities in 2012. http://www.fcm.ca/home/awards/fcm-sustainable-communities-awards/2012-winners/2012-planning-co-winner-1.htm.

online tools has led the City to embrace new engagement methods and communication techniques. New people are engaging with the City, and the City has new ways to gather public input to guide program and policy development.

The Greenest City Action Plan is still guiding City efforts to be more sustainable. Every year the City releases an update report on the progress made, highlighting new policies and programs, a report Vancouverites can read at home, in their pyjamas.

KEY TERMS

Action plan
Crowdsourcing
Engagement plan
Ideation tool
Social media

REFERENCES

Arnstein, S.R. 1969. "A Ladder of Citizen Participation." *Journal of the American Planning Association* 35(4): 216–24.

Brabham, D.C. 2009. "Crowdsourcing the public participation process for planning projects." *Planning Theory* 8(3): 242–62.

Davidoff, P. 1965. "Advocacy and pluralism in planning." *Journal of the American Institute of Planners* 31(4): 331–8.

Forester, J. 1989. *Planning in the Face of Power*. Berkeley: University of California Press.

—— 1999. *The Deliberative Practitioner: Encouraging Participatory Planning Processes*. Cambridge, MA: MIT Press.

Seltzer, E., and D. Mahmoudi. 2013. "Citizen participation, open innovation, and crowdsourcing: Challenges and opportunities for planning." *Journal of Planning Literature* 28(1): 3–18.

Vancouver Charter [1953-55-1].

CASE STUDY 6.4

Reaching Youth: Tools for Participating in the Upgrading and Evaluation of Municipal Equipment and Services

Juan Torres and Natasha Blanchet-Cohen

SUMMARY

This chapter presents the experience of youth participation in enhancing local services and equipment in Sainte-Julie, a suburb of Montreal. As an accredited Child Friendly Municipality, Sainte-Julie was committed to youth participation, but past attempts at involving youth had failed. Municipal staff felt ill-equipped and uncertain about the process and usefulness of involving youth. In 2011, they welcomed collaborating with planning and community development researchers and students in designing and carrying out two participatory processes: (1) evaluating an innovative public transit service for teenagers (Taxi 12-17) and (2) upgrading a skate park. For the Taxi 12-17 service, we used interviews, focus groups, and surveys with users, other youth, and parents. As a result, the municipality changed its communication tools and revisited the membership procedures. To upgrade the skate park, we began by connecting with youth at the site and holding interviews and then organized design workshops so that youth could make suggestions for changes. Youth input informed the municipal budget allocation. The two case studies point to the importance of municipal staff's openness to different strategies to reach out to youth, to being transparent about the level and purpose of youth's involvement, and to revisiting conventional communication tools. The case studies also shed light on the importance of collaboration among municipal services and local institutions.

OUTLINE

- While youth participation in decisions at the municipal level is important, municipal planning departments may consider youth engagement difficult and time-consuming.
- Youth participation in evaluating a public transit service called Taxi 12-17 and in upgrading a skate park with the Sainte-Julie Parks and Recreation Department allowed staff to make key changes to better meet the needs of their target demographic group.
- The Sainte-Julie cases highlight the benefits of youth participation for informing municipal services and programs and the need for enhancing collaboration across all municipal departments involved in planning that affects youth.

Introduction: Involving Youth in Municipal Planning

In Canada, municipalities are local governments in direct contact with residents. They provide infrastructure and services that have a large impact on youth's everyday lives. But usually **youth** are not involved in the design of these infrastructures and services, nor consulted on how municipalities are meeting their needs or could better meet them (Freeman and Tranter 2011; Hörschelmann and van Blerk 2012).

Lack of youth participation at the municipal level stems from several factors, including paternalistic attitudes and lack of training among professionals (Driskell 2002; Freeman, Nairn, and Sligo 2003). In some cases, municipal staff are concerned whether the youth they reach out to are representative. Youth themselves are rarely asked to have a voice and unsure of how they could become involved in municipal decisions that affect them. There is thus a shared perception that participatory processes are difficult and time-consuming and the outcomes modest and not necessarily worthwhile.

However, the imperative to include young people in municipal planning arises from the recognition of the right to participate and from evidence of the benefits of youth participation (Blanchet-Cohen 2006; Torres and Lessard 2007). The UN Convention on the Rights of the Child recognizes in Article 12 that all young people are entitled to express their views on matters that concern them (UN 1989). Agenda 21 of the Earth Summit identifies children and youth as a group that must be involved in participatory processes to create more people-friendly and sustainable environments (UN 1992). Supporting child rights to participate is also an aim of UNICEF's Child Friendly Cities initiative and Quebec's unique accreditation program for child-friendly municipalities, managed by Carrefour action municipale et famille (CAMF). Criteria for accreditation include environmental quality and diversity of services, as well as the capacity of children and youth to "influence decisions about their city" and "express their opinion on the city they want" (UNICEF 2013).

Evidence of the benefits of youth participation in planning has increased since the pioneering work of Kevin Lynch (1977), who recognized the need and tools for working with young people in cities. With youth participation, places and programs can better respond to youth expectations, and youth in turn are more likely to make a commitment to places and programs when they have been involved in their design and implementation (Chawla 2002a; 2002b). In Canada, many municipalities support youth participation in community planning, although implementation remains difficult.

The Sainte-Julie Cases

Context

Sainte-Julie is a municipality with 30,000 residents, an off-island suburb of Montreal situated six kilometres east of the St Lawrence River. Primarily French-speaking, the community has a median income and an educational level above the provincial and metropolitan averages (Statistics Canada 2011). It is recognized as family-oriented and ranked several times as Quebec's "best municipality to live in" by the Relative Happiness Index (a private ranking program). In 2009, following the mayor's support, Sainte-Julie received the Child Friendly Municipality accreditation, one of the first in the province to do so.

Like other municipalities of its size, Sainte-Julie has several departments that deal with youth programs and services. The Parks and Recreation Department (PRD) is responsible for a range of places and activities that occupy youth after school, including sport fields, parks, libraries, and youth centres. The department is also a hub for community groups that provide services and activities to citizens of all ages throughout the year. Given this direct contact with children and youth, the PRD had naturally taken the lead in obtaining and maintaining Sainte-Julie's Child Friendly Municipality accreditation.

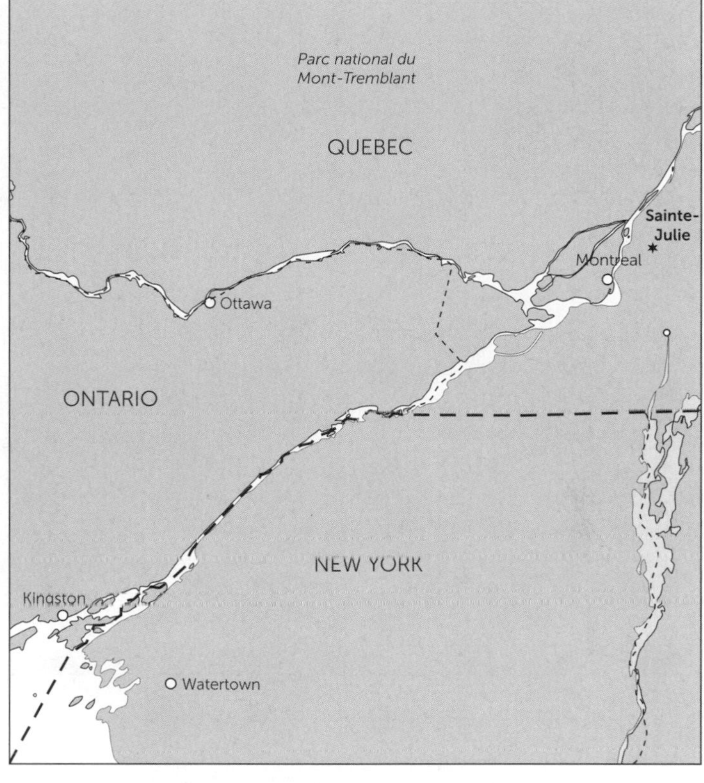

Figure 6.4.1 • Location map.

Local Openness . . . and Challenges

As a Child Friendly Municipality, Sainte-Julie is open and committed to youth participation, but in 2011, PRD's staff felt dissatisfied and disappointed with their earlier attempts to engage youth. They identified three main obstacles to youth participation:

1. The constant renewal of the youth population
2. The legal status of adolescents as minors, which complicated procedures, including the requirement for parental consent
3. The small size of the PRD staff (four, including the director) given the municipality's needs

Previously, the department had wanted to upgrade the local skate park. It had made a call for bids and included youth on the selection board. Staff tried to contact youth by telephone and mail based on user-lists from various municipal services, but response rates were low and feedback limited. Despite this, some youth joined the selection board, but their vote counted for only 10 per cent of the global evaluation and their selected bid was rejected by the rest of the board. While it was a transparent process, youth felt frustrated that their opinion had little weight in decision-making. Staff felt that youth were disengaged and lacked capacity for reflection and insight. They were unsure how to proceed, confirming their feeling that youth participation was difficult and time- and resource-consuming. The process was put on hold, and the skate park remained unchanged.

This is the context in which our multidisciplinary urban planning and community development team from local universities, supported by CAMF, worked with Sainte-Julie's PRD over a 10-month period. The municipality welcomed the partnership. Our goal was twofold:

- to support Sainte-Julie in developing adapted participatory processes;
- to document the experience to inform a guide for municipalities in Quebec.

For our team of two professors and three research assistants from the Université de Montréal and Concordia University, the collaboration was also an opportunity to develop practice and knowledge as well as provide for a stimulating environment for future professionals. For the PRD, the partnership was an opportunity to discover new ways to facilitate participatory processes and to fulfill their municipal commitment toward children's and youth's rights to participate in local decision-making.

At the outset, the PRD identified two areas that would benefit from youth participation: evaluating a

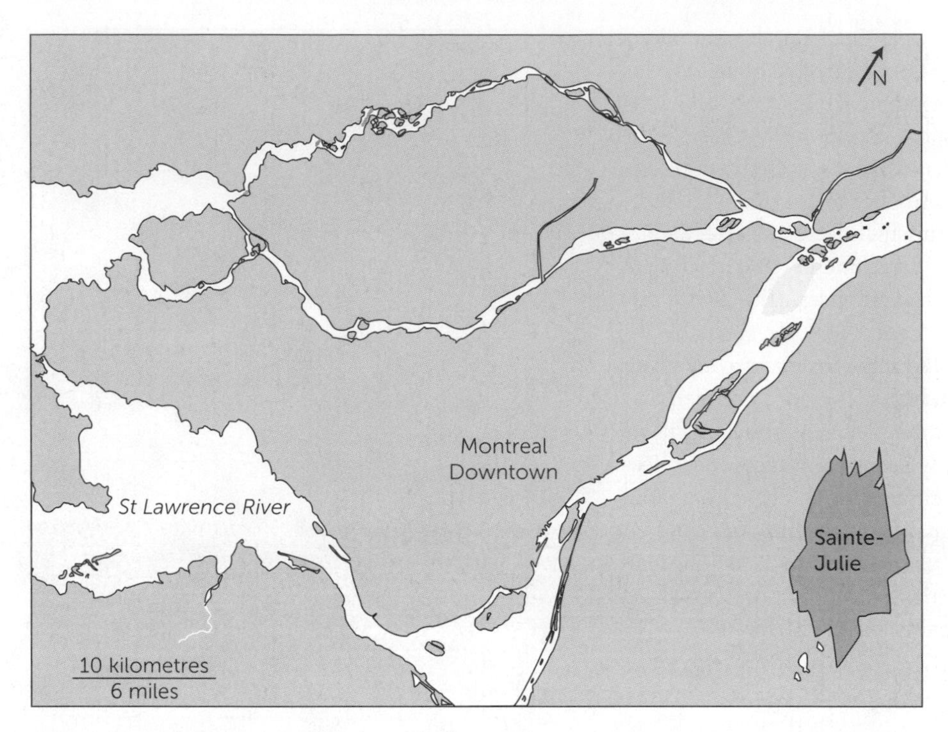

Figure 6.4.2 • Location of Sainte-Julie.

recently implemented public transit service for teenagers (Taxi 12-17) and upgrading the skate park.

Case 1: Engaging Youth in Evaluating a Public Transit Service

The transition from primary to high school often brings deep changes to the adolescent's daily life, especially in terms of mobility. Adolescents may gain in autonomy (Clifton 2003), but since high school zones reach out more broadly, they are usually located farther away from students' homes, discouraging **active transportation** (Torres, Lewis, and Bussière 2010). According to an origin-destination survey (AMT 2008), commuting between home and school represents around 80 per cent of the adolescents' daily trips in Sainte-Julie. Fourteen-year-old adolescents make only 12 per cent of these trips by foot, the rest by school bus. Those aged 17 walk even less: most use a car as passengers (27 per cent of the trips) or as drivers (29 per cent).

Following a paper-based survey conducted by the municipality in local schools among 400 teenagers, which identified a need for public transportation, the PRD implemented the Taxi 12-17 service in 2009. The concept was simple: any adolescent aged between 12 and 17 could become a member and call a taxi to travel with other teenagers within the municipality (for a cost of $5) or to a neighbouring municipality (for a cost of $10). The price could be shared among a maximum of six teenagers. The affordable rate and the availability of the service (4 pm to midnight on weekdays and 10 am to midnight on weekends) allowed adolescents to travel throughout the year without depending on their parents. For the municipality, partnering with a local taxi company and paying the difference between the member fee and the actual cost of the trips was less expensive than establishing a conventional public transportation service.

A recipient of awards for its innovative approach, Taxi 12-17 inspired other municipalities in Quebec and abroad. However, the service remained underused, even after publicity campaigns in strategic places like the high school and the nearest ski centre: there were only 150 registered members in 2011 and an average of less than one trip per day since the creation of the service (Ville de Sainte-Julie 2011). The municipality recognized that they needed to involve young people, the main stakeholders, in order to understand why the service was underused and find ways of improving it.

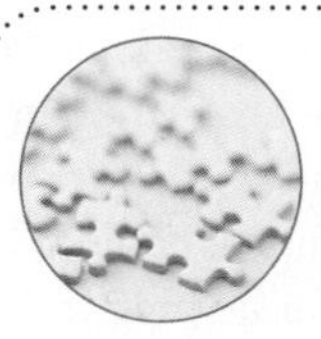

KEY ISSUE

Transportation Issues for Youth in Suburban Municipalities

In suburban municipalities like Sainte-Julie, car dependency may result not only from long distances to daily destinations—such as schools, shopping, or the cinema—but also from the lack of public transit. In Quebec, low-density and low-functional-diversity municipalities often partner with neighbouring municipalities to develop shared public transportation services. Sainte-Julie has its own public transportation service, but like its neighbouring municipalities, service operates only during the weekday rush hours and mainly to downtown Montreal and the metro station on the south shore. For children and youth under the age of driving (16), this limited service raises the issue of accessibility as a condition for social justice. Furthermore, it reduces the opportunities for active transportation, a useful form of physical activity that protects young people from contemporary health problems like overweight and obesity.

Municipality of Sainte-Julie

Figure 6.4.3 • Taxi 12-17 flyer.

Tools with youth

Evaluation activities included focus groups with adolescents, interviews with parents, and a survey of users. First, we carried out two focus groups identified from the sports activities they attended on weekdays evenings at the local high school (the school and the municipality share their sport facilities). The focus groups, with a total of 11 teenagers, took place in a classroom. Young people were asked to talk about the Taxi 12-17: how well they knew it, how often they used it, what made it adapted or not adapted to their needs, and what they would recommend to increase usage. Then we interviewed 11 parents while they were waiting for their children at the same sports facilities. We asked them about their familiarity with the service, about how well it responded (or not) to their needs, and for their ideas about increasing usage. Focus groups and interviews were audio-recorded and transcribed for analysis. Finally, a telephone survey allowed us to gather the point of view of 17 users of the service on their degree of satisfaction, their parents' support, and their recommendations for improvement.

Outcomes

To share results with the municipality, a concise four-page summary was produced, and a presentation was made to key staff. Results showed that teenagers and parents both considered Taxi 12-17 a great idea, especially useful for single-parent families. Some parents, however, were unaware of the program; others expressed concern with having 12-year-old adolescents on their own; they preferred the idea of young people travelling together. Some parents even questioned the role of municipalities, considering "chauffeuring" their own responsibility. While enthusiastic about the service, they saw it as a last-resort option.

Participants identified some weaknesses in the service. Registration was perceived as cumbersome, since adolescents needed parental authorization and the procedure had to be done in person with parents present during office hours at the City Hall. Also, operation rules appeared complicated, especially with regard to fares, the number of persons by type of trip, and the service hours.

For the PRD, this feedback was useful: it confirmed that despite its modest use, the service was relevant. The municipality was surprised about single-parent families' interest in the Taxi 12-17 and the perspective of it being a "last resort." In response to the need for more awareness of the service, municipal staff redesigned their communication and promotion tools—e.g., flyers (Figure 6.4.3) and a web page. To ease membership procedures, the places and schedule for registration were revisited. Most significant was realizing the need to target parents as well in order to foster their trust, a position that was unusual for a municipality but proved to be necessary to address concerns about safety and parents' perhaps overprotective mentality.

Case 2: Upgrading a Skate Park

Skate parks are popular facilities among adolescents, particularly males. Public ones usually include a combination of prefabricated obstacles. Their design can be complex, given their multiple users (skateboard, BMX, or inline skating) and the diversity of user skills, especially depending on age and experience.

Sainte-Julie has a public skate park in the heart of its residential area (at the Jules-Choquet Complex). Since its construction more than 10 years ago, it has been popular among young people. The municipality had a $25,000 budget for renovation, untouched since it aborted the bid procedure mentioned earlier. PRD staff were interested in exploring new strategies for reaching youth; they saw youth involvement as necessary for an optimal investment of the budget in buying the most suitable prefabricated obstacles. Staff recognized that they did not have the competence to make such decisions without youth's knowledge and experience, yet they were unsure as to how to go about reaching and engaging youth in helping them make a decision.

Tools with youth

The participatory process included several activities carried out over one month in the fall. First, to build a relationship and an understanding of how the space was used and by whom, we scheduled informal "hanging around" periods over two afternoons at the skate park. While teenagers shared their tricks, we explained the reasons for our presence: the municipality had a budget to upgrade the park but wanted youth input. We invited the teenagers (in person and through posters placed at the park) to a special activity and to visit the project's page on Facebook.

The special activity held the following week was a social event at the skate park. The PRD brought a tent, hot chocolate, and a sound system to plug into the adolescents' iPods. During the event, we interviewed six adolescents, seeking age and interest diversity (skateboarders, inline skaters, cyclists). Semi-structured interviews explored their experience as skate park users. The interviews were recorded and transcribed for analysis to provide context on skate park use.

Once relationships established an understanding of the current usage, three design workshops were organized at the park services building in the vicinity of the skate park (Figure 6.4.4). Each workshop was held with a group of four or five adolescents. The workshop began with a discussion about the issues of the actual skate park. Then around a table with printed plans of the skate park and its surroundings,

Figure 6.4.4 • Design workshop with youth.

youth marked the main problems and sketched proposed solutions. A staff member from PRD supplied information about the shapes, sizes, and costs of prefabricated obstacles, giving feedback to youth as they shared their ideas.

Outcomes

As was done for Taxi 12-17, the results were summarized in a four-page report and presented to the PRD staff. They were also published on the Facebook page of the project and made accessible to the youth participants. We found that the skate park was a meaningful place for adolescents, regardless of age or skills. For many of them, it was actually the place they enjoyed most in Sainte-Julie. Its accessibility and the opportunity to gather with other teens were valued in ways that surprised the PRD staff.

Youth proposals for upgrading or remodelling the skate park were both creative and realistic (Figure 6.4.5): besides expressing their preferences among affordable prefabricated obstacles, they made suggestions for improving the conviviality of the place by increasing the vegetation and including a rest area for users and their friends. They highlighted the importance of the skate park as a sports facility and as a place for socializing; this was a revelation to the staff.

Social media played an important role during the process. As in other municipalities with institutional restrictions on the use of Web-based communication tools, the PRD initially did not see the relevance of social media for them, but after our success in communicating with youth they requested a change in policy. The PRD staff began using the project's Facebook page to further consult with the park users on issues such as material or design details. They also used this medium to announce the completion of the redesigning and to extend an invitation to a reopening celebration eight months later.

Learning from the Cases

For the PRD, the two cases were successful in that the planning decisions better reflected youth needs: in the case of the skate park, the choice of prefabricated obstacles; in the case of Taxi 12-17, a revision of registration procedures as well as revamping communication and promotion materials. Engaging with youth provided the PRD with an understanding of their multiple perspectives on mobility and public spaces, especially as means of socializing and feeling part of the municipality. Most important, the cases

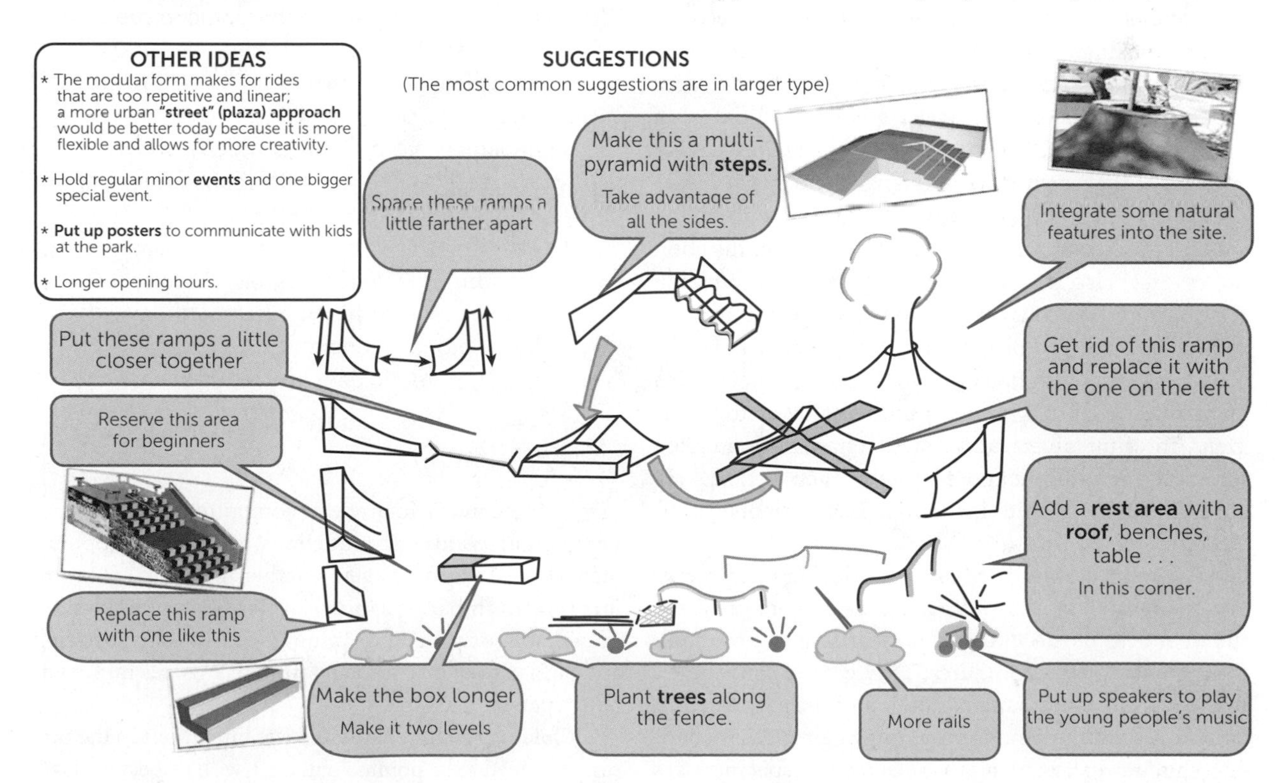

Figure 6.4.5 • Sketched proposals.

illustrated the pertinence and value of adolescents' contribution in both participatory situations. "Not to underestimate youth knowledge and potential contribution" captured the PRD staff's advice to other municipalities.

More broadly, our work enabled the municipality to develop new strategies for reaching youth and maintaining momentum in participatory activities. There was **capacity-building**, illustrated by the replication of the approach and tools in other municipal activities, such as organizing the adolescent annual party during the summer and continuing the renovation process for local parks.

According to the PRD staff, the collaboration with university researchers and students gave them ideas and tools on how to approach youth. Staff saw the partnership as a form of professional development, a reminder of the importance of a new openness for professionals in municipalities. Creating partnerships with local universities or other post-secondary educational institutions was seen as a win/win: they provide field experience to students and opportunities for relevant research, while at the same time they offer significant assistance to small municipalities with limited resources. That said, while partnerships can help, they require transparency and reciprocal trust, since they often challenge well-established ways of working.

Another outcome for the municipality was recognition of the importance of diversifying communication channels. Traditional approaches (e.g., posters, messages on the municipality web page) were appropriate but insufficient. The PRD staff realized they had to "reach adolescents at their own places"; this is where social media proved a useful tool for keeping in contact with youth, resulting in the municipality revising its communication policy. Staff also recognized the importance of building relationships and connecting informally, gaining trust by spending time at the places youth use. Organizing festive activities (with music!) in order to "break the ice" was also an appreciated strategy—fun, fruitful, and easy to implement. These tools were quite different from those adopted for adults, who can be more aware of their rights as citizens and responsive to invitations in newspapers, posters, or radio.

A very important success factor during this process was the municipality's engagement and, more precisely, the openness of the PRD staff to learning and to modifying policies and procedures. They greatly contributed to the planning and implementation of the participatory activities with the youth, providing their ideas, experience, and resources. They also selected two specific cases in which the municipality was committed to seeing that youth input led to concrete actions, avoiding disappointment. It has yet to be seen whether municipal commitment to youth participation can translate into planning on regional or provincial issues (such as transportation and land uses), since no regulation recognizes minors as participants.

While successful, there were two limitations to the process that need to be pointed out. First, with respect to Hart's

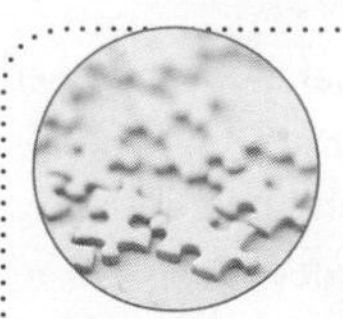

KEY ISSUE

Guidelines for Engaging Youth for Municipalities

Although planning increasingly recognizes public participation as an important condition for better processes and places, manuals and curricula dealing with youth participation remain scarce (Blanchet-Cohen, Mack, and Cook 2011). On the basis of experiences like those presented here, guidelines for municipal staff were produced (CAMF 2013), including:

- The need for, and benefits of, involving youth in the planning design and implementation of municipal services and programs. As users, they have valuable views and insights, and their involvement provides for greater effectiveness.
- The importance of being transparent and clear about the purpose and degree of youth participation. From the beginning, sharing the scope of possibilities will make the outcome more successful for both the youth and the municipal professionals.
- The use of appropriate tools for engaging youth. Tools need to be adapted to the specific realities and diversity of youth.
- The value of strengthening collaboration at different levels (between municipal services, government levels, sectorial agencies, etc.) given their effect on young people's everyday lives.

(1997) classification, adolescents' participation remained at the level of **consultation**. Unlike **tokenism**, consulting can be an authentic participatory approach and an important step forward for young people rarely involved in the local decision-making process. However, diversity in the forms of participation allows community members to find more and better ways to collaborate (Hart 2008). Therefore, it would be important in the future for youth to participate not only in consultations concerning the places and services they use but also in defining the topics to be discussed and the forms of collaboration to be implemented. In other words, an important challenge, even for proactive municipalities like Sainte-Julie, is to collaborate with their young citizens as *partners* and not only as *users* of services or places. One can expect that ongoing participation increases awareness and engagement among authorities and youth, but proactive initiatives by local leaders are essential to enable and maintain partnership.

This challenge raises another issue: youth participation activities often remain attached to a single department rather than extending across all municipal dimensions. Youth concerns go well beyond recreation facilities; they also touch on education, public safety, environment, and so on. Furthermore, to authentically involve children and youth in the decision-making process affecting their community, other government levels (provincial, federal) and actors (e.g., school boards, local associations) are essential partners. Enhancing living environments *with* youth calls for broad and meaningful collaborations.

Conclusion

The two consultation processes presented in this chapter shed light on several obstacles to and opportunities for youth participation in municipal contexts. Many municipalities do not have the resources (particularly in terms of staff), nor the experience, to conduct participatory processes. Lack of tools and apprehension about the usefulness and difficulty of participation appear as obstacles. These case studies show that given some support and openness, municipalities can effectively engage youth in planning for more effective programs and services.

In Sainte-Julie, youth participation in a process for evaluating and upgrading youth-oriented services and equipment proved to be feasible and useful, leading to concrete changes, making municipal work more effective, and showing youth the impact of their engagement. Evidence about the transferability of the process is as promising as the publication of guidelines for youth involvement for municipalities in Canada. However, developing a participatory culture involves the different municipal departments collaborating with youth not only as mere users of local services and facilities but as real partners, together deciding the issues to address and the ways of resolving them. Like other citizens, youth are concerned with issues that go well beyond recreation activities and beyond municipal jurisdiction.

KEY TERMS

Active transportation
Capacity-building
Consultation
Tokenism
Youth

REFERENCES

AMT (Agence métropolitaine de transport). 2008. *Mobilité des personnes dans la région de Montréal, Enquête Origine-Destination 2008*. Version 08.2a. Montreal: AMT.

Blanchet-Cohen, N. 2006. "Young people's participation in Canadian municipalities: Claiming meaningful space." *Canadian Review of Social Policy* 57: 71–84.

——. E. Mack, and M. Cook. 2010. "Changing the landscape: Engaging youth in social change." http://www.mcconnellfoundation.ca/en/resources/report/youthscape-guidebook-changing-the-landscape-involving-youth-in-social-change.

CAMF (Carrefour action municipale et famille). 2013. «Municipalité amie des enfants : Établir une démarche participative avec les jeunes.» http://www.amiedesenfants.ca/attachments/article/88/Guide_FINAL_PDF.pdf.

Chawla, L. 2002a. *Growing up in an Urbanising World*. Paris and London: UNESCO, Earthscan.

——. 2002b. "Insight, creativity and thoughts on the environment: Integrating children and youth into human settlement development." *Environment and Urbanization* 14(2): 11–21.

Clifton, K.J. 2003. "Independent mobility among teenagers: An exploration of travel to after-school activities." *Journal of the Transportation Research Board* 1854: 74–80.

Driskell, D. 2002. *Creating Better Cities with Children and Youth: A Manual for Participation*. Paris and London: UNESCO, Earthscan.

Freeman, C., K. Nairn, and J. Sligo. 2003. "'Professionalising' participation: From rhetoric to practice." *Children's Geographies* 1(1): 53–70.

Freeman, C., and P. Tranter. 2011. *Children and Their Urban Environment: Changing Worlds*. London and Washington: Earthscan.

Hart R. 1997. Children's Participation: *The Theory and Practice of Involving Young Citizens in Community Development and Environmental Care*. London: Earthscan, UNICEF.

——. 2008. "Stepping back from 'the ladder': Reflections on a model of participatory work with children." In A. Reid, B. Jensen, J. Nikel, and V. Simovska, eds, *Participation and Learning*, 19–31. New York: Springer.

Hörschelmann, K., and L. van Blerk. 2012. *Children, Youth and the City*. Abingdon, UK, and New York: Routledge.

Lynch, K. 1977. *Growing up in Cities: Studies of the Spatial Environment of Adolescence in Cracow, Melbourne, Mexico City, Salta, Toluca, and Warszawa*. Cambridge, MA, and Paris: MIT Press, UNESCO.

Statistics Canada. 2011. *National Household Survey*. Ottawa: Statistics Canada.

Torres, J., and M. Lessard. 2007. "Children and design students as partners in community design: The Growing Up in Cities projects in Montreal and Guadalajara." *Children, Youth and Environments* 17(2): 527–40.

Torres, J., P. Lewis, and Y. Bussière. 2010. "Schools' territorial policy and active commuting: Institutional influences in Montreal and Trois-Rivières." *Journal of Urban Planning and Development* 136(4): 287–93.

UN (United Nations). 1989. "Convention on the Rights of the Child." UN, Office of the High Commissioner for the Human Rights. http://www.ohchr.org/en/professionalinterest/pages/crc.aspx.

——. 1992. *Agenda 21*. UN, Department of Economic and Social Affairs, Division for Sustainable Development.

UNICEF. 2013. "What is a Child Friendly City?" http://childfriendlycities.org/overview/what-is-a-child-friendly-city.

Ville de Sainte-Julie. 2011. "Taxi 12-17." Paper presented at the Annual Conference on Municipal Recreation (Conférence annuelle du loisir municipal), Rimouski, Québec. 5–7 October.

CASE STUDY 6.5

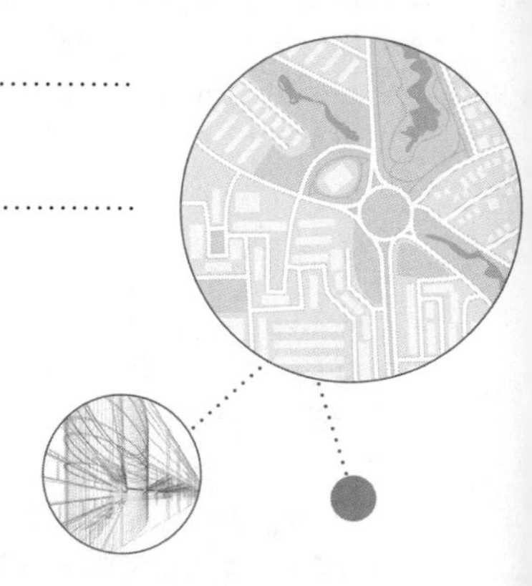

I "Like" You, You Make My Heart Twitter, But . . . : Reflections for Urban Planners from an Early Assessment of Social Media Deployment by Canadian Local Governments

Pamela Robinson and Michael DeRuyter

SUMMARY

Research findings show that social media are being used to increase the amount of participation, to offer new venues for participation, and to reach people who would not normally participate through more traditional or old-fashioned methods (e.g., sending email, attending a public meeting). However, the local governments scanned were not, thus far, capitalizing on the collaborative, crowd-sourced, or volunteered geographic information capabilities of these tools, even though they have potential applicability for municipal land-use planning activities.

In Canada in 2013, land-use planning and the consultation processes that support it were still practices deeply rooted in real life, with planners consulting and engaging with real people in real places. Yet the geographically dispersed and sometimes anonymous contributions of participants enabled through Web 2.0 and its suite of **social media** tools present practical challenges for our current planning processes. Finally, municipal planners work in institutions that are risk-averse in political contexts where the opportunities to experiment are limited. So despite the exciting promise of Web 2.0 tools, moving beyond using their functionality for information-gathering and dissemination to real-time interactive, collaborative, decision-making is a tremendous challenge.

OUTLINE

- A growing number of Canadian municipalities use Web 2.0 tools, particularly Facebook and Twitter.
- However, the potential to involve the public in planning processes using Web 2.0 tools is still unrealized as planners struggle to move beyond the familiar, place-based methods of participation.
- Planners and the public are still not equal decision-makers in planning processes, and Web 2.0 tools highlight the flat hierarchies innate in technological approaches to participation.

Introduction

Now more than ever, the work of municipal government, and of planners in particular, is unfolding in a digitally enabled, connected world. The initial rise of the Internet created the promise of electronic government, then next, with the evolution to interactive social media also known as Web 2.0, came a whole new suite of possibilities. The promise and expectation of real-time, online, interactive access to information, data, government staff, politicians, and other interested people is alluring in the practice of urban planning in which planners consult and engage the public as part of their long-range and development control activities.

The *potential* for social media to make meaningful contributions to local democracy and, more specifically, planning is widely accepted. This belief in the transformative potential of social media places real-time pressure on local governments and municipal planners to adopt these tools as part of efforts to consult and engage the public. Yet research that *actually* assesses the context and intended purpose of local government's and municipal planners' use of social media is only just emerging. Drawing from a scan of how Canadian municipal governments are using social media platforms like Facebook, Twitter, wikis, and YouTube, this chapter begins with the question: *What role do these tools have in formal land-use planning processes?*

Urban planning is a field in which professionals are compelled to use evidence to inform practice. Therefore, it is important for planners and planning students to differentiate between the potential, the promise, and the hype when new technologies present new opportunities for planning practice. This chapter reveals that thus far Canadian municipal governments, and planners working for them, are using social media to gather more public input and to more efficiently disseminate information from municipal governments to their residents. These findings mean that the transformative potential of social media has not yet been realized. In the late summer of 2012, a scan of 17 Canadian municipalities was conducted to inventory and assess the 10 most common social media tools being used by local governments for land-use planning activities in development control, urban design, and long-range planning contexts. The findings are presented, and then reflections about the implications of the findings for land-use planning are shared.

Methodology

This case study emerges from a scan of Web 2.0 tool use by 17 municipal governments in Canada in July and August 2012 (DeRuyter 2012). The 17 cities were chosen on the basis of cross-Canada representation: 13 are capitals of their respective province or territory, and the other four (Mississauga, Saskatoon, Calgary, and Ottawa) were chosen on the basis of their size and importance to their respective provinces. The 17 municipalities combined are home to 10,320,962 citizens, or approximately 30.8 per cent of Canada's population (Statistics Canada 2011b). The average size of the municipalities was 607,115, with their populations ranging from 6,800 (Iqaluit) to 2,615,060 (City of Toronto).

The use of Web 2.0 tools was determined by visiting the municipality's website. The home page of each site was scanned to determine whether Web 2.0 tools would be noticeable to people visiting the websites. If no tools were found on the website's homepage, a Google search was completed using the terms "social media" or "Web 2.0." If no Web 2.0 tool use was detected through these methods, the individual tools were then searched by name on a systematic basis. If no Web 2.0 tools were detected after these searches, it was determined that they were not used by the municipality. The scan approach was designed to mimic the experience of a member of the public visiting the website.

This snapshot-style search may have had an impact on the number of accounts listed in this chapter. For instance, the City of Toronto is listed as having 31 Facebook accounts, because this was the number listed on its official city website. There may be additional accounts either directly tied to the City or affiliated with the City government; however, presence and notification on the City government's official website was used as a proxy to identify City accounts.

The 10 types of Web 2.0 tools were chosen following an extensive literature review of Web 2.0 use at the local government level to determine which tools were most often mentioned. The most commonly used are the "Big Three" of Web 2.0 tools: Facebook, Twitter, and YouTube. These three tools can be considered the cornerstones of Web 2.0, because they each own the largest market share of their respective uses (social networking, **microblogging**, and video hosting). After these three, seven additional tools used were selected. Flickr was chosen because it is the largest and most common photo-sharing website on the Internet. RSS was chosen because it is the largest syndication and aggregate tool. The other five tools (blogs, **mobile apps**, **podcasts**, **open data**, volunteered geographic information) were chosen without having any specific website or brand name attached because they are as yet not dominated by one specific company (although volunteered geographic information is largely dependent on Google Maps).

Findings

The scan sought to answer the question: *Are Canadian municipalities adopting Web 2.0 technologies, and, if so, in what ways are these tools used?* The scan revealed that thus

Table 6.5.1 • Canadian Municipal Adoption of Web 2.0 Tools (Summer 2012 Baseline Data)

City	Facebook (#)	Twitter (#)	YouTube (#)	Flickr (#)	Blogs (#)	Podcasts	RSS (#)	Open Data	VGI	Apps (#)
Calgary	Y(1)	Y(1)	Y(l)	N	Y(1)	N	Y	N	N	Y(5)
Charlottetown	Y(1)	Y(1)	N	N	N	N	Y	N	N	N
Edmonton	Y(10)	Y(9)	Y(1)	Y(1)	Y(3)	N	Y	Y	Y	Y(1)
Halifax (RM)	Y(8)	Y(9)	Y(2)	N	Y(2)	N	Y(3)	N	N	N
Iqaluit	Y(2)	Y(1)	N	N	N	N	N	N	N	N
Mississauga	Y(8)	Y(7)	Y(1)	Y	Y(1)	N	Y(7)	Y	N	Y(1)
Montreal	Y(28)	Y(22)	Y(10)	Y(4)	Y	N	Y	Y	N	N
Ottawa	Y(1)	Y(3)	Y(1)	Y(1)	Y	N	Y(7)	Y	Y	N*
Regina	Y(1)	Y(1)	Y(1)	N	N	N	N	Y	N	Y
Saint John	Y(1)	Y(1)	Y(1)	N	N	N	Y(5)	N	N	N
Saskatoon	Y(1)	Y(1)	Y(1)	Y	Y(1)	N	N	N	N	N
St John's	Y(1)	Y(1)	Y(1)	N	Y(1)	N	N	N	Y	N
Toronto	Y(28)	Y(31)	Y	Y(2)	Y(3)	Y	Y	Y	Y	Y(2)
Vancouver	Y(13)	Y(15)	Y(5)	Y(5)	Y	Y	Y	Y	Y	Y
Whitehorse	N	Y(1)	Y(1)	N	N	N	Y	N	Y	N
Winnipeg	Y(10)	Y(8)	Y(8/2)	N	Y(5)	N	Y(3)	Y	N	Y
Yellowknife	N	N	N	N	N	N	Y	N	Y	N
Total	15 of 17	16 of 17	14 of 17	7 of 17	11 of 17	2 of 17	13 of 17	8 of 17	7 of 17	7 of 17

far, municipalities are primarily using Web 2.0 to collect and disseminate information from and to the public.

Table 6.5.1 shows that all of the municipalities scanned have at least one Web 2.0 tool, with the majority of municipalities adopting more than one. With few exceptions, the "Big Three" are universally adopted, which is not surprising given that they are the most established, most popular, and easy to use. The purpose and style in the way that the municipalities use these tools varies greatly, as does the number of sites/accounts maintained by each city. In terms of the scale of uptake, it is clear that the size of the city plays a major role in that Montreal, Toronto, and Vancouver (traditionally considered Canada's most important and international cities) have the most accounts listed on their sites.

Much as in private society, Facebook, Twitter, and YouTube are now commonplace in municipal service tool belts. The overall scan showed that 15 of 17 municipalities (88 per cent) had adopted at least one official Facebook account, 14 of 17 (82 per cent) had adopted a YouTube account, and 16 of 17 had at least one Twitter account (94 per cent). Of the Big Three, Twitter seems to be the most effective at gaining support, with by far the largest number of followers (a common measurement of a site's popularity) than the other two tools. There appears to be a direct, and not surprising, correlation between content and followers: the more content released via the tool, the higher the number of followers. While some Facebook accounts are updated fairly regularly, the speed at which a microblogging site like Twitter can disseminate information makes it the easiest tool to quickly update, and this rapidity might account for the fact that 16 of the 17 municipalities deploy it. These tools, this scan discovered, were also being used to deliver service and program information and issue-related material (e.g., background reports) to the public as well as to gather feedback and information from the public.

The research also revealed that open data is quickly becoming the next technological frontier for Canadian municipalities. Across Canada, cities like Toronto, Montreal, Calgary, and Regina have vast open data catalogues (Currie 2013). These open data catalogues are intended to spark a flurry of app development with the goal of encouraging the public and private developers to make use of the data. While other Web 2.0 tools can also result in some significant returns, the municipal service–focused capabilities of open data application mashups makes them particularly noteworthy. This is evident in the proliferation of applications allowing the public to tell their municipal government where potholes and graffiti are located. Across Canada, numerous mobile apps enable transit users to find out when the next bus, streetcar, or subway train will arrive. Given the public demand for better municipal service provision, making public service–related open data available enables municipal governments to appear responsive and transparent while also technologically savvy.

From this scan we can conclude that Web 2.0 tools and technologies have become popular among Canada's municipalities. The question is no longer whether Canadian municipalities will adopt Web 2.0 tools and technologies, since every municipality studied deploys at least one Web 2.0 tool and the majority have many. Now that Canadian municipalities are adopting Web 2.0 tools, it is important to consider the implications of these tools in land-use planning practice.

Implications for Planning Practice

Thus far, our research shows that in practice, Web 2.0 tools are still mainly used to disseminate and gather information. This finding is interesting in that it draws attention to the gap between the potential functionality of Web 2.0 tools and the reality that they are not being used for their collaborative idea-generating and decision-making capacity.

Yet despite all this uptake, we still have not yet seen the infiltration of social media values into the land-use planning process. Facebook is being used to initiate discussions, Twitter is being used to disseminate information, wikis and other crowdsourcing tools are being used to begin discussions, apps are being used as open data to influence municipal service, and crowdsourced input is having a direct impact on municipal service provision. SeeClickFix.com is a good example of the public flagging concerns about potholes and other municipal problems, with local governments receiving feedback faster and thus able to more readily respond. But thus far, crowdsourced ideas are being fed into the planning process in the same way that ideas coming from participation at public meetings and letters or emails to councillors and staff are. They are part of the total input and ideas, all of which are still mediated by professional planners and ultimately decided upon within the standard political process. To date, we have not discovered a planning consultation process in which crowdsourced input has significantly influenced a planning decision-making process.

Time may be a factor here. Web 2.0 tools only emerged in recent years, and their popularity is quite new, which means that municipalities are still in the relatively early stages of Web 2.0 adoption. This emphasis on information

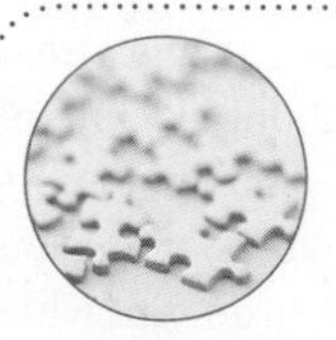

KEY ISSUE

Information Disseminating and Gathering

Many Canadian cities use Facebook as part of their public participation processes. But how do we use the "data" gathered from these sites in our planning processes? The City of Edmonton has a Facebook page (https://www.facebook.com/cityofedmonton). The 2014 population of Edmonton was 877,926, and its Facebook page had 41,123 likes by 30 May 2015. But what do these likes mean? Because of the diffuse geography of Facebook users, we don't know where the people who "liked" Edmonton live. And what does it mean to "like" something on Facebook? Is clicking an act of civic participation?

Facebook pages, like other forms of social media, can help local governments by functioning as an ordinary call for public input: people share their input into a process that then reviews the information gathered and makes a decision. This is a good example of how social media is being deployed in planning practice today: to deliver information quickly to people in places where they can engage online and to allow people in these places to input their information easily using multi-media submissions.

delivery parallels the way that ordinary people use the Internet: primarily for information purposes (Statistics Canada 2011a). Thus, local governments are neither ahead of nor behind the wave of implementation but rather sitting firmly on top of it.

Public administrations and planners are often criticized for being too insular (Fyfe and Crookall 2010; Sossin 2010; Tapscott, Williams, and Herman 2008) and for not pushing far enough to engage or consult the public in their decision-making processes. Here the populist appeal of social media presents an obvious opportunity to go *to* the public instead of asking the public to come to local government by means of a public meeting. But the dispersed and decentralized nature of social media use makes it interesting and important to consider a range of new potential deployment options through the lens of governance issues.

Planners, when working for local governments, have limited opportunities to "fail" when consulting or engaging the public. For example, in the case of a mandatory public meeting, there are legally specific requirements to confirm the successful circulation of public notices. A breach of this process can lead to an automatic appeal of a council's decision by the developer based on improper circulation. More specifically, in Ontario there is a 120-metre meeting notification circulation radius for the statutory required public meeting for Official Plan amendments, zoning bylaw amendments, and minor variance applications. Meeting notifications are sent to the property-owner by first-class mail. But when meeting notices are circulated on Twitter and Facebook, planners cannot confirm that the necessary members of the public received the notice. So despite good intentions, the widespread reach and dispersed nature of Web 2.0 tools can lead to failure for planners.

The hierarchical structure of government has also been cited as an impediment to Web 2.0 adoption. This obstacle is repeatedly mentioned in Fyfe and Crookall's study of Web 2.0 use in Canada's governments:

> The most significant impediment to government use of social media is the clay layer in management and the hierarchical public service culture. ... Social media demands a new paradigm regarding the ownership, use and management of information (Fyfe and Crookall 2010, 82).

The agile, fluid, and "flat" hierarchy of social media communities is markedly at odds with the hierarchical cultures and practices within government, including at the municipal level. Adding to this situation is the fact that urban planners in municipal government are not typically the final decision-makers in planning processes: they provide sound professional advice to municipal councillors, who ultimately make these decisions on the council floor. Yet as social media are used more and more by the private sector, the not-for-profit sector, and citizens themselves, local governments, as the closest level of government to the people, run the risk of being perceived as lagging behind because of these clashes of culture.

Local governments also have an implicit spatial hierarchy when it comes to the value placed on contributions from the public: typically, priority is given to the perspectives of the people who are most affected by a particular

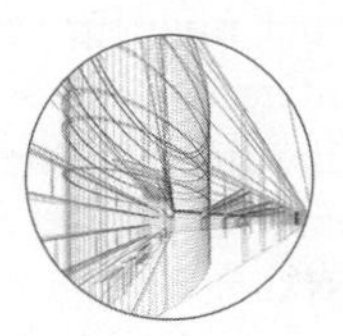

PLANNING THEORY

Digital Participation Tools

Do we need a new theory of planning to help explain digitally enabled public consultation? Or does Web 2.0 just provide new mechanisms and channels for input from the public into the public consultation processes that planning has used for the past 40 years? There is a lively debate ongoing right now among planning scholars questioning whether public participation enabled via Web 2.0 tools is really any different from the kinds of public consultation enabled through a room full of chairs in a town hall or letters to municipal councillors and planners (Seiber et al. forthcoming). Does Arnstein's (1969) Ladder of Citizen Participation account for the dynamics of digitally enabled participation? Is crowdsourcing activity consistent with the type of collaborative planning Healey (2005) proposes? Is volunteering geographic information or joining a hackathon consistent with rational choice theory? As our understanding of how Web 2.0 tools are used in planning expands, it is important to keep raising these questions and others that challenge planning theory.

development. But given that anyone in the world can contribute to a planning process when consultation is enabled through social media, how can planners differentiate between who is local and who is not? Furthermore, many social media users participate anonymously or use a pseudonym, a practice that is less likely in traditional planning consultation processes where participants sign in and often preface their comments by stating their name and address in front of a room full of people, including their neighbours. Anonymous or masked participation makes it difficult for planners to ascertain who is contributing to the process, who is not contributing, and whether the ideas generated represent a variety of perspectives or those of a single user. For planners, who have long sought to engage as many people as people as possible, the broad reach of Web 2.0 tools presents some functional challenges.

What is clear is that Web 2.0 tools present an opportunity and a challenge to local governments and their planners, because these tools fundamentally challenge the hierarchies and norms of practice of these institutions and the professionals who work there. Social media were designed to capture "the wisdom of the crowd," yet urban planning is a profession with its own claim to expertise. Listening to, considering, and mediating crowd-based expertise is a new challenge for planners and others on the receiving end of crowdsourced community input (Seiber et al. forthcoming). Although the planning process has long consulted the public, crowdsourced ideas and solutions are sometimes the result of public wisdom assembled outside of formal planning consultation processes, and thus planners are not yet sure what to do with this type of input.

Conclusion

Thus far in planning practice in Canada, we observe that social media are helping municipal staff and planners disseminate information faster on online platforms where residents might already be engaged, using tools that traditionally marginalized groups of people (e.g., youth) are already using. To date, social media are a complement to and not a replacement of traditional means of public consultation and engagement—planners still hold public meetings and still post municipal notices in newspapers of record, and sometimes they tweet, post on Facebook, and use YouTube to introduce citizens to planning issues or problems. Planners in Canada are not, so far, using social media's potential to generate planning content and ideas in real time or to devolve power or make decisions in real time. The potential for decentralized decision-making facilitated through Web 2.0 tools has not yet taken hold.

Ultimately, this research shows us that in order for more of the potential of Web 2.0 tools to be realized, all levels of government have to come to terms with the notion that some surrendering of their complete control over the situation is needed. The challenge lies in government inertia: "elected leaders and the public service infrastructure that supports them must overcome the cultural barriers that stand in the way of a more open, free-form style of collaboration" (Deloitte Touche Tohmatsu 2008, 2). Some researchers argue that this, rather than a lack of understanding or resources, is why many governments have failed to move beyond the communication-focused style of e-government information with Web 2.0 tools. Through their research, Deakins et al. remind us that

> [p]ublic sector organizations also continue to favour e-government services that offer even more information to citizens, rather than providing collaborative service channels that even-handedly engage every stakeholder. . . . [M]odern technology is being used by local government largely because it is available, and for interactions with stakeholders that are both directive and maintain control over citizens (Deakins et al. 2010, 530).

Yet Lisa Brideau and Amanda Mitchell's chapter, "Taking It Online," presents the example of the City of Vancouver's experiment with crowdsourcing for the Greenest City project. At first read, this case could be interpreted as contradicting the findings of our research results. Instead, Vancouver's experience actually coincides with social media deployment across Canada in that a municipal government used social media tools in a civic engagement process. They disseminated information faster to more people, and their team received input from more people more efficiently. In these ways Vancouver's experience is similar to that of the other municipalities discussed here. The staff team in Vancouver also began to experiment with the interactive nature of social media, allowing people who were geographically distributed (not in a public meeting room, all together at the same time) to review each other's feedback and to respond in real time. When municipal governments begin to use these kinds of interactive processes in formal decision-making processes, the shift will signal a fundamental change in how the public is consulted and engaged in municipal governance across Canada.

In flagging when real change may take place, this chapter contributes to a tendency in academic literature to want governments to "go big" with their Web 2.0 initiatives in the hope of achieving deep and integrated Web services in a short time through a greater focus on interaction and service-based tools. This is understandable, given the speed at which Web 2.0 has risen and the exponential rate of technology development in the field, which has augmented the expectations of researchers and the public about how quickly municipalities and other governments should adopt a robust initiative. Here the public appetite for more rapid adoption is at odds with public pressure on governments at all levels to "get it right the first time." In the case of Web 2.0 adoption in local planning processes, the public cannot have its cake and it eat it too.

This case study differs from others in this book. Instead of highlighting lessons learned from reflections about completed planning projects, it signals the challenges of trying to capitalize on new opportunities. The findings speak directly to the challenges municipal planners in Canada face when seeking to use new tools as part of the "same old" land-use planning processes. The deployment of Web 2.0 tools is an evolving issue in contemporary planning practice. Ultimately, this case study raises questions about the capacity of municipal planning as we currently practise it to meet the needs of increasingly complex communities in which our publics are seeking and sometimes demanding planning processes that are more open, connected, transparent, fluid, accessible, and effective.

KEY TERMS

Microblogging
Mobile apps
Open data
Podcasts
Social media

REFERENCES

Arnstein, S.R. 1969. "A Ladder of Citizen Participation." *Journal of the American Institute of Planners* 35(4): 216–24.

Currie, L. 2013. *The Role of Canadian Municipal Open Data Initiatives: A Multi-City Evaluation.* http://www.slideshare.net/LiamCurrie/canadian-municipal-open-data-initiatives.

Deakins, E., S. Dillon, H. Al Namani, and C. Zhang. 2010. "Local e-government impact in China, New Zealand, Oman and the United Kingdom." *International Journal of Public Sector Management* 23(6): 520–34.

Deloitte Touche Tohmatsu. 2008. *Change Your World or the World Will Change You: The Future of Collaborative*

Government and Web 2.0. Toronto: Deloitte Touche Tohmatsu.

DeRuyter, M. 2012. "Entering the fray: Canadian municipalities' use of Web 2.0." Master's thesis, School of Politics and Public Administration, Ryerson University, Toronto.

Fyfe, T., and P. Crookall. 2010. *Social Media and Public Sector Policy Dilemma*. Toronto: Institute of Public Administration of Canada.

Healey, P. 2005. *Collaborative Planning: Shaping Spaces in Fragmented Societies*. 2nd edn. New York: Palgrave Macmillan.

Seiber, R., P. Robinson, P. Johnson, and J. Corbett. Forthcoming. *Defining Participation on the Geospatial Web 2.0: Empirical Evidence from the Field*.

Sossin, L. 2010. "Democratic administration." In Christopher Dunn, ed., *The Handbook of Canadian Public Administration*, 2nd ed., 364–80. Toronto: Oxford University Press.

Statistics Canada. 2011a. "Canadian Internet use survey." http://www.statcan.gc.ca/daily quotidien/110525/dq110525b-eng.htm.

——. 2011b. "Population and dwelling counts, for Canada and census subdivisions (municipalities) with 5,000-plus population, 2011 and 2006 censuses." http://www12.statcan.gc.ca/census-recensement/2011/dp-pd/hlt-fst/pd-pl/Table-Tableau.cfm?LANG=Eng&T=307&S=11&O=A&RPP=100.

Tapscott, D., A.D. Williams, and D. Herman. 2008. *Government 2.0: Transforming Government and Governance for the Twenty-First Century*. Toronto: New Paradigm Learning Corporation.

7.0

INTRODUCTION
Urban Design

Public and private spaces are integral to communities: urban parks, public squares, and streetscapes enable us to connect with each other and with nature. In many countries, these spaces have allowed people to have conversations with neighbours, randomly encounter strangers, and assemble for community events or political protests. Streets, avenues, and boulevards are integral in the social life of communities, integrating shops, community services, play spaces, and markets. The interface between buildings and the street is critical in human perception and affects the way people use sidewalks and urban infrastructure. In Section 3, we saw how important the physical design of towns and cities can be to healthy communities. Here we focus on the ways in which urban design elements can be integrated into projects, plans, and regulations to physically reshape our cities and manage population growth.

The use of public spaces is often highly contested, and powerful actors try to control the ways in which people can assemble and occupy spaces, as seen in the Occupy Wall Street phenomenon, a public response to corporate practices during the devastating recession in the United States, which spread quickly to other countries. In Baron Haussmann's Paris, one response to revolutionary actions among the citizenry was the destruction of parts of the city to build broad avenues that would allow military manoeuvring (Harvey 2003). Some authors, such as Yiftachel (1998) and Sandercock (1998), argue that controlling the social aspects of urban spaces, particularly those that are considered threatening to societal norms and expectations, has always been a fundamental goal in planning practice.

Another theme in urban design since the 1980s has been the interplay between public use and private ownership. Public spaces, such as streets, parks, and plazas, allow urban life to penetrate them and generally enable human activities to occur, even those that might be considered less desirable. Private spaces are of necessity controlled and managed by their owners and may have certain hours of operation or allow the removal of disruptive people or groups (e.g., teenagers). More troubling are those public–private spaces perceived as public but which are actually highly regulated. One example is indoor "streets" within shopping malls that mimic those in historic city centres but are actually owned and controlled by private corporations. Many argue that the privatization of public space is of increasing concern with the development of public–private partnerships and funding agreements—e.g., the practice of allowing fans inside a hockey arena to consume only food and drink purchased at the venue.

Sustainability concerns have recently had a major impact on urban design: many municipal and regional governments aim to create more complete streets (Filion 1999), make better connections for pedestrians and cyclists, include higher densities to support public transit (e.g., Calthorpe 1993), and incorporate sustainable materials. Contrasted with the postwar urban design aesthetics (e.g., Jacobs 1961), streets are understood to be much more than conduits to quickly move traffic, and buildings are more human-scaled; new urbanism, Smart Growth, and sustainable development principles have resulted in some improvements on the postwar built form. Many cities and rural communities have begun to plan for intensification along their existing road networks, public spaces, and public transit infrastructure (Grant 2005). However, planners and developers continue to struggle with implementation within institutions and governance models that have more traditional views on the built form (Grant 2009).

Urban design, perhaps more than any other sub-discipline of planning, has also struggled with its normative values; urban designers and planners strongly believe that they know the principles and elements of the built environment that work best. Some, such as Talen and Ellis (2002), have argued for more emphasis on physical planning and specifically, a "good city form." Further, urban designers and planners often believe that there are social problems that can be alleviated with specific design approaches. But history has shown us that design values can be dangerous: the postwar planning agenda, bent on progress and breaking from tradition, emphasized separation of land uses, monolithic buildings that presented blank facades to the street, high-rise apartments surrounded by empty lawns, and car-based transportation infrastructure that had major negative effects on pedestrians and cyclists. The consensus on the built form that many allude to (Talen and Ellis 2002; Filion 1999) may prove to have some unforeseen negative impacts on future populations.

The following cases integrate these ideas and many more in their exploration of urban design projects, guidelines, and processes from a planning perspective.

Leslie Shieh's case study on the redesign of the Westminster Quay Public Market into River Market shows how design can put local residents' needs ahead of tourists' interests. Like other public markets in Canada, this gem in New Westminster, BC, suffered from increased retail competition by the early 2000s and needed a new direction. Using a public consultation process to find out what locals wanted, River Market aimed to create a food precinct that would allow neighbours to do things together, such as take cooking classes or plant vegetables in an edible garden. The market features local eateries and entrepreneurs, enabling visitors to invest in local businesses through a founders program and a limited number of subsidized rents to encourage start-ups. This transition from a festival marketplace oriented toward private profit and tourism to a locally oriented enterprise supporting the local neighbourhood illustrates the public–private dualism in urban spaces.

James White explores the case of the East Bayfront Precinct in Toronto, part of the City of Toronto's secondary plan for the waterfront. As a response to prescriptive bylaws, the City decided to emphasize placemaking and high-quality design in the development of this detailed plan. Masterplanning for the site began in 2003, integrating public forums and developing shared goals for the precinct: improving the north–south connections between the city and waterfront, better pedestrian spaces, flexible spaces, and addressing the need for affordable housing. The East Bayfront Precinct Plan was adopted in 2005, and a special district zoning amendment ensured the size and location of public space, distribution of roads, and the configuration and heights of buildings. It also stipulated ground-floor commercial units and affordable housing targets. Implementation of the plan began in 2009 and will continue until the 2020s.

For a rural example of an urban design plan, we turn to Beverly Sandalack and Francisco Alaniz Uribe, who present the case of Benalto, Alberta, a hamlet of just 300 residents. Benalto sought to preserve its small-town character from future development through a new Area Redevelopment Plan and urban design guidelines. Using townscape analysis to identify trends and patterns (morphology, typology, spatial structure, urban qualities) and a workshop with residents, a plan was developed using an open-space system as a framework. Urban design guidelines were produced in a graphic poster format to enable residents, planning administrators, and developers to understand the plan concepts. The Area Redevelopment Plan was approved by council in 2006 and continues to guide land use and development in the hamlet.

In the fascinating story of the development of the American Embassy in Ottawa, Jason Burke shows how security concerns overrode urban design in the public interest. Security considerations, many of which contradict each other, were instrumental in every aspect of the project, from choosing the site to the building materials and the design of temporary and permanent barriers on Sussex Drive. Burke argues that these measures represent a form of exclusionary zoning and the increasing privatization of public space; residents' and businesses' concerns were essentially trumped by this "securitization" throughout the design process. The traditional ceremonial parade route of the capital has been seriously undermined through the urban design elements and lane closures following 9/11. This case also emphasizes the lack of co-ordination and agreement between the federal government, which owned the land, and the municipal government, which would ultimately be affected by the development.

Cal Brook and Matt Reid offer a case study on a critical topic: urban intensification. As cities like Toronto continue to grow, main streets such as the city's Avenues can be targeted for increased density. These important corridors already offer pedestrian and transit infrastructure and were identified in the City's 2006 Official Plan. Mid-rise buildings have a great potential to accommodate new business and residential growth, but after adoption of the Official Plan, developers still struggled to understand the policies, process, and design elements involved. The Toronto Avenues and Mid-Rise Buildings Study analyzed the existing conditions on the Avenues, and consultants held focus group sessions with the public and development community. The final result, a set of 36 Performance Standards for mid-rise buildings, was adopted by City Council and incorporated into the City-wide zoning bylaw in August 2010. The clear, simple illustrated guidelines were designed to be easily understood by developers, planners, residents, and business owners on the Avenues.

These cases illustrate how municipalities, developers, businesses, and residents alike can participate in the urban design process and contribute to innovative guidelines and regulations. In most cases, the integration of local needs into the process had a beneficial impact on adoption and implementation; Jason Burke's case of the American Embassy provides a counterpoint. Urban design trends come and go, but in recent years the combination of Smart Growth, new urbanism, and sustainable development principles seems to have gained traction, perhaps because so many of the design principles are drawn from historical examples. The human-scaled built form popular in many early Canadian cities and prevalent until the 1950s may have returned for good. However, urban design remains contentious because it largely concerns aesthetics, and there are many differences in perception in this area. For urban designers, the end result may be unpredictable among populations who do not share their predominant values. Visual explanations of urban design principles, seen in the Benalto and Toronto mid-rise cases, could be an important step toward the development of shared design values.

REFERENCES

Calthorpe, P. 1993. *The Next American Metropolis*. New York: Princeton Architectural Press.

Filion, P. 1999. "Rupture or continuity? Modern and postmodern planning in Toronto." *International Journal of Urban and Regional Research* 23(3): 421–44.

Grant, J. 2005. "Rethinking the public interest as a planning concept." *Plan Canada* 45(2): 48–50.

——. 2009. "Theory and practice in planning the suburbs: Challenges to implementing new urbanism, Smart Growth, and sustainability principles." *Planning Theory and Practice* 10(1): 11–33.

Harvey, D. 2003. "The right to the city." *International Journal of Urban and Regional Research* 27(4): 939–41.

Jacobs, J. 1961. *The Death and Life of Great American Cities*. New York: Random House.

Sandercock, L., ed. 1998. *Making the Invisible Visible: A Multicultural Planning History*. Berkeley and Los Angeles: University of California Press.

Talen, E., and C. Ellis. 2002. "Beyond relativism: Reclaiming the search for good city form." *Journal of Planning Education and Research* 22: 36–49.

Yiftachel, O. 1998. "Planning and social control: Exploring the dark side." *Journal of Planning Literature* 12(4): 395–406.

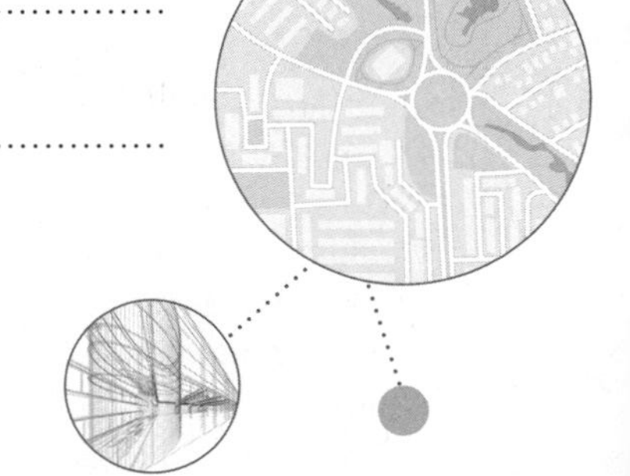

CASE STUDY 7.1

What's Public about Public Markets? Beyond Public and Private Space in Community-Building

Leslie Shieh

SUMMARY

The distinction between public and private is fundamental to the conceptualization and organization of urban spaces. In everyday speech, **public spaces** refer to places accessible to everyone, with little or no restrictions placed on the users. Public spaces are where people go for individual or group activities and are defined by amenities that support gathering and encourage staying, such as seating, vegetation, and children's play structures (Carr et al. 1992, 50). **Private spaces** are controlled to give users privacy, exclusive use, and protection from the intrusion of others (Madanipour 2003, 68). In the contemporary city, however, the distinction between public and private space is often blurred and complex.

Many **public markets** are privately owned and managed. What, then, is public about the public market? How does a public market remain relevant in community life? These were key questions in the redesign of the 22-year-old Westminster Quay Public Market (WQPM) in New Westminster, BC. The redevelopment of WQPM in 2008 as River Market explored the commonly disregarded topic of **public life** in private spaces (Kirby 2008). I worked as the project planner in the public market's redevelopment. My responsibilities included overseeing the project finances and participating in the development process, which included public consultation, design, and programming decisions.

This case study begins with a brief overview of the public market tradition and the rise of the **festival marketplace** development model, of which the 1986 WQPM is an example. It then examines the festival marketplace through its function as an urban public space. Expanding on this, the last section discusses how design considerations in the redevelopment of the market have sought to strengthen River Market's public space function.

OUTLINE

- The public market has a long tradition in North American cities. In recent years, the festival marketplace development model attempted to integrate mixed-use and leisure-based development with the reurbanization of urban waterfronts.
- Westminster Quay Public Market, an example of a festival marketplace, was recently redesigned as River Market, which functions as an urban public space that meets the needs of local residents rather than tourists.
- The approaches undertaken in the redevelopment of the market, including a focus on local businesses, local food, and community spaces, have strengthened its public space functions.

Public Market Tradition and Festival Marketplace

For nearly three centuries, public markets were the primary retail food source for American and Canadian colonial towns and cities (Mayo 1993; Basil 2012). Local authorities established public markets in order to provide fresh food for growing urban populations. By the late nineteenth century, privately owned market houses were competing with and replacing the publicly owned markets. Some sought to attract families and middle-class shoppers and boasted the latest in construction, lighting, ventilation, and refrigeration (Tangires 2003; Basil 2012).

By the mid-twentieth century, many public markets had failed. One insurmountable challenge they faced was competition from grocery stores, and later supermarkets, that were conveniently located in growing suburban areas. Grocery stores were also quicker at responding to customer preferences and changing their product mix, whereas public market operators were constrained by leases and vendors' independent choices. Moreover, some farmers preferred selling to a grocery store, which guaranteed the purchase of their entire stock, as opposed to setting up a stall at a public market. Later, with the rise and dominance of chain stores and large national corporations in food distribution, processing, and retailing, public markets struggled to remain relevant (Mayo 1993).

In the 1960s, there was some renewed interest in public markets, specifically farmers' markets, where customers bought directly from growers. This return to a more traditional food system grew out of consumer concerns about the negative impact of the industrial food system on health and the environment (e.g., Carson 1962). This was the early rise of the local and organic agricultural movement that continues today to promote safer farming practices, higher product quality, fair exchange, and interaction between producers and consumers.

What received the most attention from city mayors and planners were marketplaces in new or renovated buildings undertaken as part of **urban revitalization** efforts. The concept of festival marketplaces was pioneered at San

Francisco's Ghirardelli Square. The project, developed in 1964, converted an old chocolate factory into a series of small retail and craft shops, cafés, and restaurants surrounding a square where entertainers performed. The concept was formalized by Baltimore developer James Rouse. His two projects, Faneuil Hall Marketplace in Boston and Harborplace in Baltimore, became showpieces of American inner-city regeneration through public–private partnership (Olsen 2003; Bloom 2004). In both cities, large derelict industrial port areas were redeveloped with the then-novel combination of restored buildings, boutique shopping, bars, restaurants, and hotels and a festival atmosphere created by free events and live entertainment. Revitalization-linked marketplaces in Canada, such as Ottawa's Byward Market and Vancouver's Granville Island, have also been extraordinary successes (Tunbridge 1992, 2001; Ley 1996); Tunbridge (1992) uses the case study of Byward Market to discuss how Canadian festival marketplaces differ from American ones. Given the immense appeal of these early projects, marketplaces became the centrepiece of downtown and waterfront revitalization strategies in cities across North America during the 1970s and 1980s. City mayors saw them as a means to involve the private sector in financing the restoration of historic buildings and as catalysts to drive real estate development in decaying downtown districts (Martin 1984; Frieden and Sagalyn 1989; Sawicki 1989; Goss 1996).

The WQPM was developed during this time of interest in a mixed-use and leisure-oriented approach to urban waterfront revitalization. Downtown New Westminster was a thriving commercial centre in the early 1900s. Located on the Fraser River, it was a designated world port and linked to the Canadian Pacific Railway network. By the 1960s, however, the once-thriving downtown was facing a steady decline. Retail and commercial activity would relocate to new shopping malls. The port's infrastructure was insufficient to support the shift from cargo to container shipments, and it would eventually close. Throughout the 1980s, the municipal government undertook a series of major projects in a concerted effort to revitalize and to give the waterfront and the downtown core a new function. This action was supported by regional planners of the Greater Vancouver Regional District (now Metro Vancouver), who had concluded that New Westminster should become a regional town centre where residents of New Westminster and neighbouring suburbs could work and shop closer to their homes rather than commuting into downtown Vancouver (FCCDC 1982).

A joint municipal–provincial agency, First Capital City Development Corporation (FCCDC), was formed in

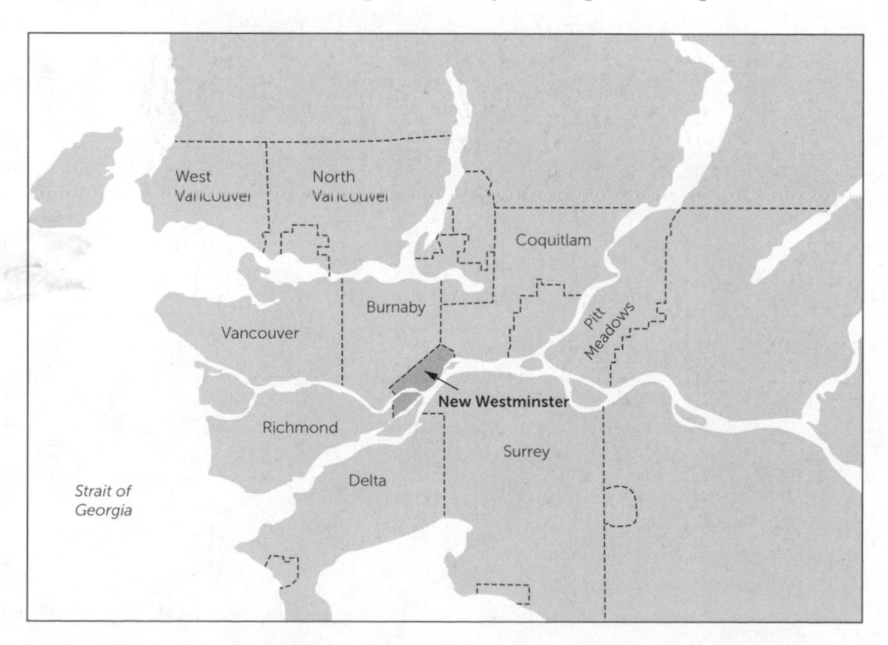

Figure 7.1.1 • New Westminster, a city of 60,000 residents, is located in the geographical centre of Metro Vancouver, a region with 2.3 million people. It was the first capital of British Columbia (1859 to 1866), named after the London borough by Queen Victoria.

Source: L. Shieh based on Metro Vancouver 2010.

1977 to realize the objectives of the downtown community plan for New Westminster. One of its principal functions was to facilitate the conversion of former industrial sites along the Fraser River into a new mixed-use residential and commercial district. Referred to as Westminster Quay, the 47-acre waterfront development would include 1200 condominium units, a hotel, a public market, an office tower, and a riverfront esplanade. FCCDC undertook the tasks of assembling the land, preparing the development plan, constructing the public amenities, and marketing the individual properties to private developers (FCCDC 1982; 1984). This method of funding was common at the time whereby a redevelopment agency converted underutilized land to higher uses, resold land parcels to private-sector developers, and used the sales revenue to defray the costs of the project (Gordon 1997; Shaw 2001). The completion of the rapid transit connection (SkyTrain) from downtown Vancouver to downtown New Westminster in 1986 further improved the convenience and attraction of the waterfront project.

The WQPM was initially built as a partnership between FCCDC and Laing Property Corporation, a local developer of shopping centres that later acquired full ownership. The two-storey public market opened on 19 July 1986 to coincide with the Vancouver World Exposition. The design of the market building was food-oriented, with booths and stalls on the ground floor and a food court, restaurants, and specialty retail shops on the second floor.

Festival Marketplaces as Public Spaces

The marketplace is a fitting entry point into the examination of public spaces. The notion of urban public space can be traced back at least to the ancient Greek agora. The agora had multiple functions but evolved principally into a secular marketplace open to all (Hall 1998, 39). This function as a non-exclusive place for commerce, but also

Figure 7.1.2 • Major development projects along the New Westminster waterfront from 1980 to 2013.

Source: Aerial photograph courtesy of River Market.

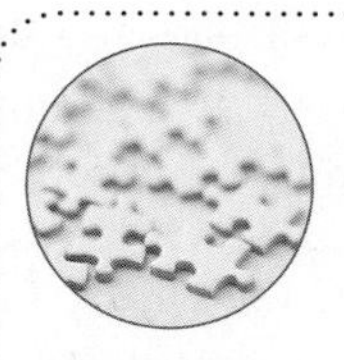

KEY ISSUE

Public Spaces

The planning literature on public urban spaces consists of three related discourses. The first examines the significance of public spaces in our cities and is concerned with the social, political, and economic factors leading to their decline (Mitchell 2003; Kohn 2004; Low and Smith 2006). The second, coming in particular from urban design literature, focuses on the creation of new public spaces and placemaking strategies. The third area of inquiry seeks to provide tools for analyzing public spaces and their publicness (Varna and Tiesdell 2010). The planning and designing of the revitalized market drew on the thoughts and ideas of these fields.

for informal and formal public assembly, would become qualities associated with the idealized public space. These qualities were not lost on the festival marketplace developers as well as their critics. Their opposing perspectives underscore the planning debate on the privatization of urban public spaces.

James Rouse, the developer credited with the festival marketplace concept, saw the marketplace as a public space. In a memo to his advisors, Rouse wrote in 1978:

> Banks and office buildings don't belong to the people who use them. They stand apart from the people and so do museums and universities and hospitals. But the marketplace is a common place. It is meant to serve people, to make them feel at home and comfortable (cited in Olsen 2003, 271).

Rouse was writing at a time in the United States when most people who had the means were moving to single-family homes in the suburbs. Downtown department stores, retail districts, and entertainment venues either kept shorter hours, because there were fewer people downtown after working hours, or closed their operations entirely (Frieden and Sagalyn 1989). Festival marketplaces drew staggering crowds back to downtowns, and commentaries written at the time celebrated the return of public life to the city (Goss 1996). One architecture critic wrote of the marketplace as "a halfway house for people from the car culture who are trying to learn to love cities again" (Campbell 1981, 25).

Through architecture and design, festival marketplaces suited the middle-class taste for fresh, ethnic, and gourmet food and an interest in history and historical preservation (Frieden and Sagalyn 1989). They provided an escape from places like supermarkets where people go for the sole purpose of buying food. They were designed to be "hang-out" places where the main activity was *flanerie*—the activity of strolling and looking (Banerjee 2001). Rouse observed that "people don't come to a festival marketplace for the purpose of shopping or eating... they come for the delight. They come and they walk slowly and they smile and they sit and look at other people. There are very few places in a city that people can go with no purpose" (cited in Olsen 2003, 270).

Festival marketplaces are not without critics who question the richness and authenticity of this public life. They argue that as public spaces, festival marketplaces support a public life increasingly shaped by a consumer culture (Banerjee 2001). They are concerned that urban public spaces are not only increasingly becoming places for leisure and consumption but that they are designed like theme parks to evoke nostalgia (Sorkin 1992), such as suburban shopping malls modelled after traditional downtown main streets and festival marketplaces designed to recreate the atmosphere of town market squares:

> Such nostalgia is rarely innocent, however. It is, rather, a highly constructed, corporatized image of a market quite unlike the idealization of the agora as a place of commerce and politics. In the name of comfort, safety, and profit, political activity is replaced . . . by a highly commodified spectacle designed to sell—to sell either goods or the city as a whole (Mitchell 2003, 138).

For Mitchell and other public space scholars, one of the most important impacts the decline of public space has on cities is diminished public life and civic engagement. A true public space, accessible by different social groups, is critical for informal and formal social interactions, political conversations, and in essence the working of democracy. Scholars bemoan a society that is less interested in public matters and more driven by private interests and personal desires (Sennett 1992/1977; Mitchell 2003; Kohn 2004; Low and Smith 2006). Festival marketplaces are part of a wider trend toward **privately owned public spaces** that include shopping malls, plazas, and gallerias where design creates "an illusion of public space, but from which the risks and uncertainties of public life are carefully edited out" (Banerjee 2001, 13).

The opposing perspectives on the public space function of festival marketplaces point to different aspects of what constitutes a public space. Synthesizing the public space literature, Varna and Tiesdell (2010) identify five core dimensions of "publicness": ownership, control, civility, physical configuration, and animation. Defining public spaces through these dimensions is useful, because it provides a range of possible criteria to qualify a place as a public space. A place derives its degree of publicness from the interaction between the five dimensions. Failing to meet one criterion would not necessarily categorize a place as a private space.

Table 7.1.1 examines the traditional public market and festival marketplace through each of these dimensions. Examining these two types of markets in this manner shows how their publicness differs in two main respects. First, traditional public markets are predominantly publicly (municipally) owned and operated to provide a public service. Festival marketplaces are usually owned privately, though many were developed under public–private partnerships with the public goal of revitalizing industrial waterfronts. As such, management priorities are often financially driven. Second, the main function of traditional public markets is to provide a place for sellers and buyers to congregate. Accessible to anyone, traditional public markets bring together and facilitate the interaction between strangers and different social groups. Festival marketplaces, as places of leisure, encourage stay and consumption through their architectural design and entertainment. Their regulations and programming cues seek to create a place attractive to more narrowly targeted social groups—namely, tourists and urban middle-class consumers.

Rethinking the Public Market Concept

By the late 1990s, many festival marketplaces were closing or re-tenanting to national franchises. They have been most successful in cities with large numbers of tourists and where they are close to large populations of office workers. In locations lacking one or both of these supports, they have been economic failures (Metzger 2001). Moreover, their once-novel setting increasingly lost appeal. Shopping malls and grocery stores soon co-opted the public market's look and feel. Internal documents show that by the early 2000s, WQPM was struggling financially. By the time it was closed for redevelopment in 2008, two-thirds of the stalls were vacant.

Learning from WQPM's challenges as a festival marketplace, its redevelopment as River Market considered the public market concept in a new way, examining the relevance of public markets in urban communities when there are now numerous options for fresh and gourmet food. To determine the needs of the local community, the project began with a month-long community consultation. More than 1000 residents, particularly those who lived in the immediate downtown neighbourhood, participated in focus groups and completed surveys and mail-in response forms. The overwhelming response and willingness to participate demonstrated that the community indeed wanted a public market. When asked what they would like the market to be, the comments received often spoke about a place that mattered, such as "a place more for local residents than tourists," "a place [where] I would be proud to bring out-of-town visitors," and "a place I would have reasons to visit several times a week." The new market, renamed River Market to give greater reference to its location by the Fraser River, opened in 2011. The first floor is designed for food and retail, and second-floor uses include a circus school, a music school, and a co-working facility. Three community functions that have been central to the market's redevelopment are discussed below.

An Activity Precinct around Food

Markets have always been food-oriented, be it the selling of food in traditional public market halls or the consumption of food in the food courts of festival marketplaces. River Market's planning considered food more comprehensively, exploring how to design for food-oriented activities. The appeal of the festival marketplace falls under the "experience economy" where memorable events are orchestrated as the product being sold (Pine and Gilmore 1999). In contrast, River Market seeks to strengthen an "activity economy" where value comes from people doing things together in one place. The public market is a place where we buy and sell things, of course, but there are other venues,

Table 7.1.1 • Public Markets and Publicness

Dimensions of Publicness	Traditional Public Market	Festival Marketplace	River Market
Ownership: • More public: owned by public body for public use • Less public: owned by private entity for private use	• Publicly owned for public use and to provide a public service (i.e., better access to fresh food)	• Privately owned space for public and private uses	• Privately owned space for public and private uses
Control: • More public: no visible control presence • Less public: explicit control presence (e.g., security guards)	• Accessible to anyone. Regulations protect public/ community interest	• Regulations protect people and property from harm • Subtle expression of control by market management and businesses	• Regulations protect people and property from harm • Subtle expression of control by market management and businesses
Civility: • More public: managed for the needs of many and different social groups • Less public: over/ under-managed deters use by certain social groups	• Amenities and facilities for sellers and buyers (mainly men and later women and family), maintained by a caretaking staff	• Maintained by a caretaking staff for tourists and nearby office workers • Provision of basic facilities (e.g., seating, washrooms, lighting)	• Maintained by a caretaking staff for local residents, students, nearby office workers, and tourists • Provision of basic facilities (e.g., seating, washrooms, lighting)
Physical Configuration: • More public: accessible, no obvious barriers to entry • Less public: off beaten path	• Located and designed to be easily accessible	• Located and designed to be easily accessible by foot, car, and transit	• Located and designed to be easily accessible by foot, car, transit, and bike
Animation: • More public: support for a wide range of potential uses and activities; opportunities for engagement and discovery • Less public: dead space, support for limited range of potential use and activity	• Space design and programming primarily support the buying and selling of goods • Opportunities for social interaction	• Space design and programming support the buying and selling of goods and leisure visits • Opportunities for social interaction and passive engagement (e.g., people-watching)	• Space design and programming support more than the buying and selling of goods • Programming for different age groups and interests • Adaptable spaces for ad hoc and planned programming • Activities animate the space at different times of day (e.g., daytime shopping, and evening music lessons) • Affordable space for community groups and residents to host events and gatherings (e.g., birthday parties, book sales, fundraisers)

Source: Based on the five dimensions of publicness outlined in Varna and Tiesdell 2010.

both physical and virtual, where such commerce can take place. What is key for the public market is to engage the public. Specifically, it can provide a place where neighbours do things together. The activities are shopping for food and consuming food, yet the shift is toward deeper engagement and participation, like learning and playing. Comparing River Market with the traditional public market and festival marketplace (Table 7.1.1) highlights its greater publicness in the animation dimension.

Identifying the kinds of activities that bring people together has been at the core of community development (Putnam 2000). People mix in activities, often bridging cultural and demographic differences. The central objective for the new market is to create an activity **precinct** around food, supporting shared activities—from growing food in edible gardens to learning about food in cooking classes to buying food from local growers and businesses. As a planning concept, a precinct is a place developed around activities for a targeted group. In the case of the new market, the targeted group is people who live in the local community. If a precinct is successful, businesses that service the targeted group will want to locate there. As well, people outside of the targeted group will want to spend time there to people-watch and experience the place, just as tourists want to visit places that locals frequent (HB Lanarc Consultants 2008).

Figure 7.1.3 • These before and after redevelopment photos are examples of two design considerations. One, the landscape around the market was converted to edible gardens (top row). Two, the interior of the first floor was designed to give shops a storefront-like space with a commercial kitchen and a central food hall that functions as a seating area but can also be transformed into an open space for gatherings (bottom row).

Sources: Pre-redevelopment photographs courtesy of River Market Inc.; post-redevelopment photographs L. Shieh.

An Assembly of Homegrown Businesses

Another unique aspect of River Market is its tenant mix of locally owned businesses. At the municipal level, land-use policies can control the entrance of big-box retailers but do not necessarily facilitate the entrance of small, locally owned businesses. At large-scale retail centres, the developer's leasing strategy typically focuses on chain stores for their brand recognition and covenant strength. Though the approach carries greater financial risk, River Market engages with independent owner-operators. The intent is not only to lease to local businesses but also to support their growth. River Market's goals have not only attracted the region's best independent eateries but also brought together like-minded entrepreneurs in realizing and shaping the activity precinct around food. For instance, the owner of the bakery teaches a weekly baking class for children, and the chefs at two of the restaurants hold regular cooking demonstrations as part of the dining experience. Another shop-owner collects kitchen scraps from residents and other market shops to feed the pigs at a local farm.

Taking a different approach from festival marketplaces where vendors are allocated small stalls, River Market's redesign gives shops a storefront-like space outfitted with a commercial kitchen. Consequently, the market can accommodate fewer shops and businesses than its 1986 design. This decision to reconfigure may disappoint seasonal visitors who come expecting a market of small stalls like bazaars or farmers' markets. But importantly, this design has provided the owner-operators with the opportunity to expand their businesses beyond food retailing at the market. For example, two of the shops prepare food for catering and another for a food truck business in their market kitchens.

In this cluster of independent businesses, some are more established, with two or three store locations. Others are new businesses that require financial assistance. Two support programs are currently being tested at River Market. The founders program provides a structure through which people can invest in local business for $500 or $1000. These small loans provide the needed business start-up costs that are increasingly more difficult and costly for small businesses to acquire from banks. The loans are to be paid back in five years' time, during which the founding members receive a certain amount of food credits annually. The second program provides small, affordable spaces to incubate new start-ups. There is a 200-square-foot storefront and a shared 800-square-foot retail studio space for artisans with shorter lease terms and subsidized rents. Here, new entrepreneurs can spend a few months or a year testing ideas, setting up, and building a customer base before deciding to move into a commercial retail space, which typically has a three- or five-year lease term.

A Community Exchange

The redesign of the market building includes several flexible, multi-use spaces for formal and informal gatherings. One such space is the landscaping around the market, which has been transformed into an edible community garden. The plantings in the garden change every year with the volunteers. Another space is the central food hall, which was designed to function as a seating area but also as an open space for gatherings. It has been transformed to provide space for a community lecture series, fundraisers, and wedding receptions. During the winter months, it provides the city's farmers' market with an indoor location. A third multi-use space is a square on the second-floor landing, which has been used for art exhibitions and workshops. Lastly, there is a conference room that market shops and community groups can reserve for meetings and workshops. These activities and events have greatly contributed to animating the market.

As members of the community, the owners of market shops are also contributors to community life. For instance, the grocer at River Market contributes 1 per cent of the dollars spent by customers to a fund to kick-start local projects. One of the shop-owners regularly convenes community dinners in the food hall where friends and strangers sit and enjoy a meal together at a long table.

Knowledge-sharing has been equally important in strengthening the relevance of the new public market in the community. Members of the management team serve on boards and advisory committees of public and non-profit organizations whose purposes range from tourism and economic development to sustainability. In 2013, River Market participated in a pilot study that examined the challenges of organic food waste collection in public spaces, done in anticipation of a region-wide ban of organic food waste in landfills. And to support the building of public markets in other communities, lessons learned from the redevelopment of WQPM into River Market have been shared informally and formally with designers, developers, and other organizations.

Figure 7.1.4 • This open square on the second floor of the market building was designed to be an adaptable space to support a variety of activities, such as (clockwise) a puppet workshop for children, a community dinner, a craft market, and a nutrition workshop for seniors.

Source: L. Shieh.

Conclusion

Project for Public Space, a non-profit corporation founded in 1975 to help create and sustain public spaces that build stronger communities and a strong advocate for public markets, has found that successful public markets are those that benefit their community (Spitzer and Baum 1995). Many festival marketplaces were built as part of efforts to revitalize decaying downtown districts, and the popular response to them at the time brings attention to the growing appeal of leisure spaces different from standardized suburban shopping malls. As public spaces, however, they are retail centres managed akin to such shopping malls—publicly accessible with restrictions and aimed at the interests of middle-class shoppers (Bloom 2004, 158). In rethinking the marketplace model, the redevelopment of WQPM as River Market focused on three functions a public market can have in a community: an activities precinct, a supported network of local businesses, and a community exchange.

This case study of a public market's revitalization demonstrates that a privately owned public space is not necessarily a space in which social interaction and civic engagement are diminished. Working at a project level and as part of a multidisciplinary team in the private sector, the urban planner can draw attention to the nuanced understanding of urban spaces, such as what makes a public space. In this work, the planner must understand the economics of development and also relate development and the built environment that is being created to the broader context of the city, including its history, culture, and social change (Wadley 2004). The often-held misconception is that developers and municipal planners work in opposition to one another. However, planners can have a positive impact on public as well as private real estate development projects (Frieden 1990; Peiser 1990; Coiacett 2000). A good development, whether undertaken by a private development firm or a public agency, is characterized by innovation and efficient mobilization of land, labour, and capital, as well as local commitment and its ability to address community needs (Wadley 2004, 178).

KEY TERMS

Festival marketplace
Precinct
Privately owned public space
Private spaces
Public life
Public markets
Public spaces
Urban revitalization

REFERENCES

Banerjee, T. 2001. "The future of public space: Beyond invented streets and reinvented places." *Journal of the American Planning Association* 67(1): 9–24.

Basil, M. 2012. "A history of farmers' markets in Canada." *Journal of Historical Research in Marketing* 4(3): 387–407.

Bloom, N.D. 2004. *Merchant of Illusion: James Rouse, America's Salesman of the Businessman's Utopia*. Columbus: Ohio State University Press.

Campbell, R. 1981. "Evaluation: Boston's 'upper of urbanity.'" *American Institute of Architects Journal* 70: 24–31.

Carr, S., M. Francis, L.G. Rivlin, and M.S. Andrew. 1992. *Public Space*. London: Cambridge University Press.

Carson, R. 1962. *Silent Spring*. Boston: Houghton Mifflin.

Coiacett, E. 2000. "Places shape place shapers? Real estate developers' outlooks concerning community, planning and development differ between places." *Planning Practice and Research* 15(4): 353–74.

FCCDC (First Capital City Development Corporation). 1982. Answers about New Westminster's downtown and waterfront redevelopment program." Pamphlet. New Westminster, BC: FCCDC.

——. 1984. The New Westminster Downtown Redevelopment Program. Press release, 19 September. New Westminster, BC: FCCDC.

Frieden, B.J. 1990. "City centre transformed: Planners as developers." *Journal of the American Planning Association* 56(4): 423–8.

——. and L.B. Sagalyn. 1989. *Downtown Inc*. Cambridge, MA: MIT Press.

Gordon, D.L. 1997. "Financing urban waterfront redevelopment." *Journal of the American Planning Association* 63(2): 244–65.

Goss, J. 1996. "Disquiet on the waterfront: Reflections on nostalgia and utopia in the urban archetypes of festival marketplaces." *Urban Geography* 17(3): 221–47.

Hall, P. 1998. *Cities in Civilization*. New York: Pantheon Books.

HB Lanarc Consultants. 2008. "The power of precincts: Overview of core concepts." Document prepared for River Market New Westminster Quay.

Kirby, A. 2008. "The production of private space and its implications for urban social relations." *Political Geography* 27: 74–95.

Kohn, M. 2004. *Brave New Neighborhoods: The Privatization of Public Space*. New York: Routledge.

Ley, D. 1996. *The New Middle Class and the Remaking of the Central City*. Oxford and New York: Oxford University Press.

Low, S., and N. Smith, eds. 2006. *The Politics of Public Place*. New York: Routledge.

Madanipour, A. 2003. *Public and Private Spaces of the City*. New York: Routledge.

Martin, J.A. 1984. "If baseball can't save cities anymore, what can a festival market do?" *Journal of Cultural Geography* 5: 33–46.

Mayo, J.M. 1993. *The American Grocery Store: The Business Evolution of an Architectural Space*. Westport, CT: Greenwood Press.

Metro Vancouver. 2010. Local Governments of Metro Vancouver Map.

Metzger, J.T. 2001. "The failed promise of a festival marketplace: South Street Seaport in Lower Manhattan." *Planning Perspectives* 16(1): 25–46.

Mitchell, D. 2003. *The Right to the City: Social Justice and the Fight for Public Space*. New York: Guilford Press.

Olsen, J. 2003. *Better Places, Better Lives: A Biography of James Rouse*. Washington: Urban Land Institute.

Peiser, R. 1990. "Who plans America? Planners or developers?" *Journal of the American Planning Association* 56(4): 496–503.

Pine, B.J., II, and J.H. Gilmore. 1999. *The Experience Economy: Work Is Theatre and Every Business a Stage*. Boston: Harvard Business School Press.

Putnam, R. 2000. *Bowling Alone: The Collapse and Revival of American Community*. New York: Simon and Schuster Paperbacks.

Sawicki, D. 1989. "The festival marketplace as public policy: Guidelines for future policy decisions." *Journal of the American Planning Association* 55(3): 347–61.

Sennett, R. 1992/1977. *The Fall of Public Man*. New York: W.W. Norton.

Shaw, B. 2001. "History at the water's edge." In R. Marshall, ed., *Waterfronts in Post-industrial Cities*, 160–72. London: Spon Press.

Sorkin, M., ed. 1992. *Variations on a Theme Park: The New American City and the End of Public Space*. New York: Hill and Wang.

Spitzer, T.M., and H. Baum. 1995. *Public Markets and Community Revitalization*. Washington: Urban Land Institute and Project for Public Spaces, Inc.

Tangires, H. 2003. *Public Markets and Civic Culture in Nineteenth-Century America*. Baltimore, MD: Johns Hopkins University Press.

Tunbridge, J.E. 1992. "Farmers'/festival markets: The case of Byward Market, Ottawa." *Canadian Urban Landscapes* 36(3): 280–5.

——. 2001. "Ottawa's Byward Market: A festive bone of contention?" *Canadian Urban Landscapes* 45(3): 356–70.

Varna, G., and S. Tiesdell. 2010. "Assessing the publicness of public space: The star model of publicness." *Journal of Urban Design* 15(4): 575–98.

Wadley, D. 2004. "Good development, better planning: The nexus revisited." *Planning Research and Practice* 19(2): 173–93.

CASE STUDY 7.2

Planning as Placemaking? The Case of the East Bayfront Precinct Planning Process

James T. White

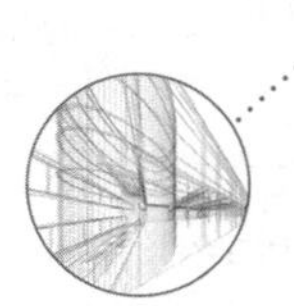

SUMMARY

Influencing the design of the built environment is one of many important tasks that planners undertake in Canada. The land-use policies and regulatory controls they administer determine how land is developed, but in many cases their efforts are restricted by zoning bylaws that identify the permitted use of each and every land parcel in the municipality (Hodge 2003). The exacting requirements of zoning often leave little room for Canadian planners to have a qualitative influence on the design and form of the built environment (Grant 2000), and as a result, real estate developers tend to shape large parts of the Canadian urban landscape. This has resulted in repetitive low-density development, unimaginative architecture, poorly conceived public space, and a subsequent sense of "placelessness" in towns and cities across the country (Relph 1987). In this chapter, a possible remedy for this endemic planning problem is introduced. The chapter begins with a brief overview of the historical relationship between zoning and land-use planning in Canada before demonstrating some of its negative impacts. It then explains the concept of "masterplanning" and argues that it can be used to encourage more sophisticated land-use zoning. To understand the concept better, the paper turns to the case of Toronto's waterfront and reveals how masterplanning and zoning have been used together to generate a more design-sensitive framework for planning decision-making. The chapter concludes that Canadian planning practitioners might consider masterplanning as a positive placemaking tool to reconcile some of the deficiencies of zoning.

OUTLINE

- Land-use policies and regulatory controls, such as zoning bylaws, have had a major impact on the physical design of Canadian cities.
- Masterplanning can help to reconcile some of the negative spatial outcomes of zoning, creating places with better design qualities. The East Bayfront Precinct Plan can be considered an example of masterplanning in practice.
- Successful elements of the precinct planning process included "codifying" the size and location of public

space, the distribution of roads and other public rights-of-way, and the configuration and heights of buildings. Rules were set out for the provision of affordable housing and funding of public amenities established in the master plan. Other Canadian municipalities can use this design-sensitive approach to physical planning in their own jurisdictions.

Zoning and Canadian Land-Use Planning

Zoning is the regulatory mechanism used across Canada to organize and distribute new development. Land is divided into parcels, and each parcel is assigned a different zone. The rules and restrictions for each zone are then detailed in a land-use **bylaw**. This is a heavily legalistic document that prescribes the permissible land use, lot size, building height, and building size allowed on each parcel of land in the municipality (Cullingwood and Caves 2009). On undeveloped or vacant sites, the **zoning bylaw** is typically accompanied by a subdivision plan that provides further regulatory requirements related to the layout of highways and sidewalks and the arrangement of land parcels. By determining the type, scale and organization of buildings, streets, and blocks, the zoning bylaw and subdivision plan act together as a physical design control mechanism (Barnett 2011).

The process of zoning and **subdivision** was popularized in the United States during the early twentieth century as local government officials sought to manage urban growth and migration in increasingly polluted towns and cities. Zoning allowed incompatible land uses to be separated while protecting the rights of new property-owners as cities expanded (Grant 2002). Facing similar challenges, and keen to protect the property ownership rights of wealthier citizens, Canadian provinces and municipalities also began to experiment with zoning (Grunton 1991). The first zoning bylaw was established in Kitchener, Ontario, in 1924, and ever since, zoning has remained the core mechanism for managing the future use of land in Canada (Grant 2000).

Zoning was not always destined to be such a dominant force in Canadian land-use planning. During the early twentieth century, the British garden city movement, with its aesthetic emphasis on comprehensive "plan-based" approaches to urban growth, began to gain traction (Grant 2000). A Town Planning Institute of Canada was formed, and British planners were hired to draw up visionary town plans for cities like Toronto, Calgary, and Vancouver (Grunton 1991). In most instances, however, the grand concepts that were envisioned in these plans never caught on, and by the early 1920s, Canadian municipalities were quick to adopt the simpler and more prescriptive zoning controls favoured in the United States (Grunton 1991). These controls reinforced the protection of property rights and encouraged a culture of single-use development. That being said, Canadian planners did retain slightly more influence over the scope and scale of real estate development than their American counterparts. In the United States, the US Constitution specifically protects an individual's right to private property, whereas in Canada, the Canadian Charter of Rights and Freedoms makes no such allowance. Zoning and subdivision decisions in the United States therefore tend to rest on the extent to which private property rights are infringed. In contrast, Canadian planners have more opportunities to make decisions that might be deemed "in the public interest" (Hodge 2003).

Regardless of this more discretionary power, zoning has remained the stalwart of planning and development decision-making in Canada, and although planning has become more responsive to community desires over the past 50 years, it is still a largely "passive activity accommodating the initiatives undertaken by private entrepreneurs" (Grunton 1991, 101). The result is that Canadian planning and zoning has allowed real estate developers to create an urban landscape of strictly subdivided land uses, boxy downtown towers, concrete plazas, strip malls, and "cookie cutter" residential sprawl (Grant 2000).

Dissatisfied with the places and spaces created by more than 50 years of zoning-led planning, citizens and planning professionals alike have begun to question the long-held assumption that zoning is the best method for regulating development in Canadian municipalities. On the whole, Canadians have a "love/hate relationship with planning and zoning" (Grant 2000, 453); some of the time they appreciate the predictability of zoning, while on other occasions they find that the environments that are created lack a **sense of place**. As Carmona (2009, 2649) more broadly contends, "The evidence overwhelmingly suggests that coding in the form of non–site-specific development standards is unlikely to provide the answers to delivering better urban design."

Masterplanning: Beyond Zoning and toward Placemaking

With this in mind, it is important to consider how Canadian planners might reconcile some of the negative spatial outcomes created by zoning and engineer a shift toward more "place-focused" planning. Two of the most successful methods introduced in recent years to encourage this are mixed-use zones and heritage conservation areas. They have helped some municipalities to create more "sympathetic building forms and patterns" (Grant 2008, 341) and foster greater social and economic diversity, particularly in inner-city neighbourhoods. Nevertheless, more can still be done to generate a broader shift away from the straitjacket of zoning and subdivision. One possible way is to recast planning as an activity in **placemaking** rather than a rigid regulatory control mechanism. Under such a definition, the process of planning neighbourhoods is understood as a conscious process of design intervention that emphasizes the relationship between behaviour and physical space (Carmona et al. 2010; Adams and Tiesdell 2013). Regulation still plays a core role in the decision-making process, but the tools employed by planners are more focused on enabling sustainable urban development practices that promote spatial design principles like density, permeable blocks, and active mixed streets (e.g., Jacobs 1961) rather than separated land-use zones.

The planning instrument that is perhaps best equipped to facilitate this shift is the **master plan**. Masterplanning provides planners with a strategic document that establishes an integrated framework of the streets, blocks, buildings, and spaces in a given geographic area (CABE 2004; Tiesdell and MacFarlane 2007). Most master plans include a spatial site layout, three-dimensional visualization of the anticipated form of development, and accompanying codes, which, rather like zoning bylaws, set out detailed rules on building types, landscape design, and architectural treatment (Tiesdell and MacFarlane 2007). The crucial difference between the codes that accompany a master plan and a traditional zoning bylaw is that master plans always have a distinct spatial focus. This can help to guarantee better planning and design outcomes while, at the same time, encouraging different actors involved in the development process to work together around a shared design-led vision of the future (Bell 2005; Tiesdell and MacFarlane 2007).

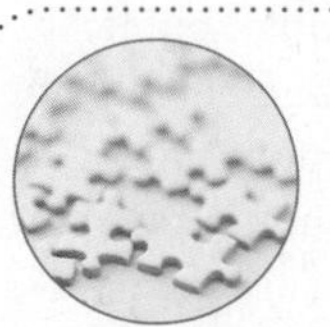

KEY ISSUE

The Planning Challenges Created by Land-Use Zoning

- Zoning is a legalistic method of co-ordinating future lands uses.
- It can lead to sprawling development and separated land uses.
- It has a tendency to generate "placelessness" and poor urban design.

The potential for masterplanning to generate a culture of "placemaking":

- Masterplanning provides a strategic approach to co-ordinating future land uses.
- It allows for planning and design ideas to be visualized and phased over time.
- It establishes a framework of streets, blocks, buildings, and spaces.
- It can be used to inform the language contained in a traditional land-use zoning bylaw.

Placemaking through Precinct Planning on Toronto's Waterfront

To understand how masterplanning might improve the design sensitivity of land-use zoning, the chapter now turns to the case of Toronto's waterfront redevelopment. The findings presented in the following paragraphs are drawn from a two-year qualitative study of urban planning and design that employed a series of semi-structured interviews, archival research, and direct observation of the built environment to examine the design decision-making processes on the waterfront between 1999 and 2010. Since 1999, the redevelopment of the city's Lake Ontario waterfront has been led by a semi-autonomous government agency called the Toronto Waterfront Revitalization Corporation (TWRC) (now referred to as Waterfront Toronto) that was established by the City of Toronto, the Province of Ontario, and the federal government (City of Toronto 2004). One of the primary challenges the

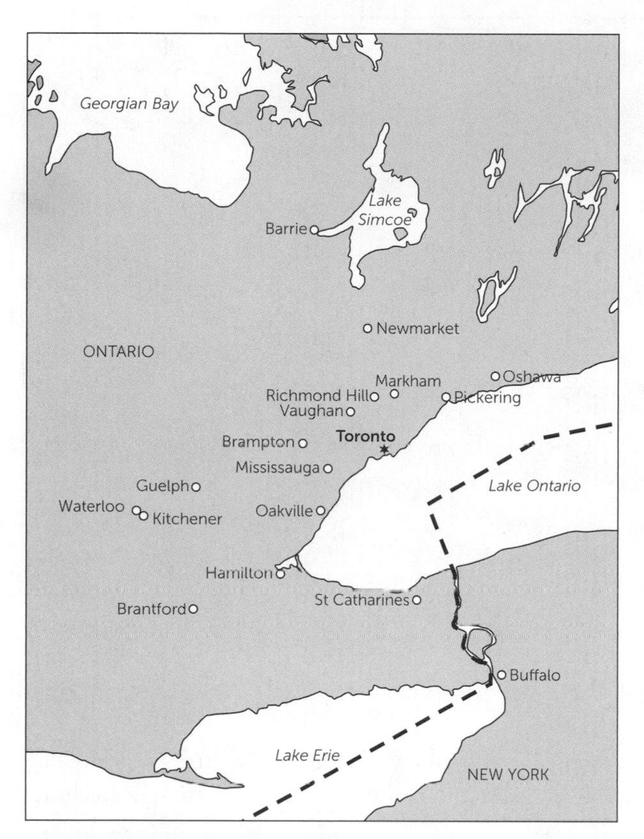

Figure 7.2.1 • Location map.

corporation faced after it was created was how to effectively plan and regulate new development on the 800-hectare post-industrial site (see Figure 7.2.2). Recognizing the corporation's predicament, the City of Toronto Planning Department selected the waterfront as a potential testing ground for a new type of design-led planning framework that aimed to strip away some of the heavy regulatory language found in land-use policies and give the TWRC more planning and design freedom (senior planning manager, City of Toronto, personal communication, 2011). Planning in Toronto is governed by the Province of Ontario's Planning Act (1990), which stipulates that all Ontario municipalities have to prepare an **Official Plan** and accompanying zoning bylaw to regulate the development of land. The Act states that an Official Plan must contain the goals, objectives, and policies for future physical change in the municipality and outline the anticipated social, economic, and environmental impacts of the proposals (Province of Ontario 1990, Sec. 16). Although not explicitly called for in the Act, larger municipalities also tend to prepare Part II Official Plans, or secondary plans, to accompany the primary Official Plan. These plans offer more detailed planning principles for specific geographical areas within the municipality—such as a waterfront—where major development or physical changes are anticipated.

A secondary plan on Toronto's waterfront was created in 2001 (City of Toronto 2001). Produced by the City of Toronto Planning Department in consultation with the new redevelopment corporation, the plan was written in the more flexible and open format imagined by the City of Toronto. As a result, the new plan differed from the City's regular secondary plans, which typically list a series of policies for an area in conjunction with simple two-dimensional maps and supporting zoning regulations. The waterfront secondary plan was, by contrast, a colourful document complete with photography, visionary plans, and three-dimensional visuals. It aimed to enable high-quality development rather than regulate land parcel by parcel and offered no specific development proposals or supporting zoning bylaws.

As part of the secondary plan process, the City of Toronto and the TWRC also recognized that the waterfront could not all be planned in one go: it was just too large. In an effort to find a solution to this problem, the plan's authors proposed to subdivide the redevelopment area into a series of manageable "bite-sized chunks" (senior executive and urban designer, TWRC, personal communication, 2011) that could be individually masterplanned under the banner of the vision established in the secondary plan. Termed the "precinct plans," they were described as follows:

> The precinct development strategies will deal with street and block patterns and building heights, urban design, community services and facilities including schools and local parks, and a strategy for achieving affordable housing targets in the Central Waterfront. The precinct development strategies will also address business relocation requirements and financing options (City of Toronto 2001, 21).

Rather like the "new look" secondary plan that governed them, the precinct plans were an entirely new concept within the City of Toronto hierarchy of plans. However, there was one crucial difference: unlike the secondary plan, the precinct plans were not legally binding components of the Official Plan (City of Toronto 2002) and therefore floated in a "grey area" between the official

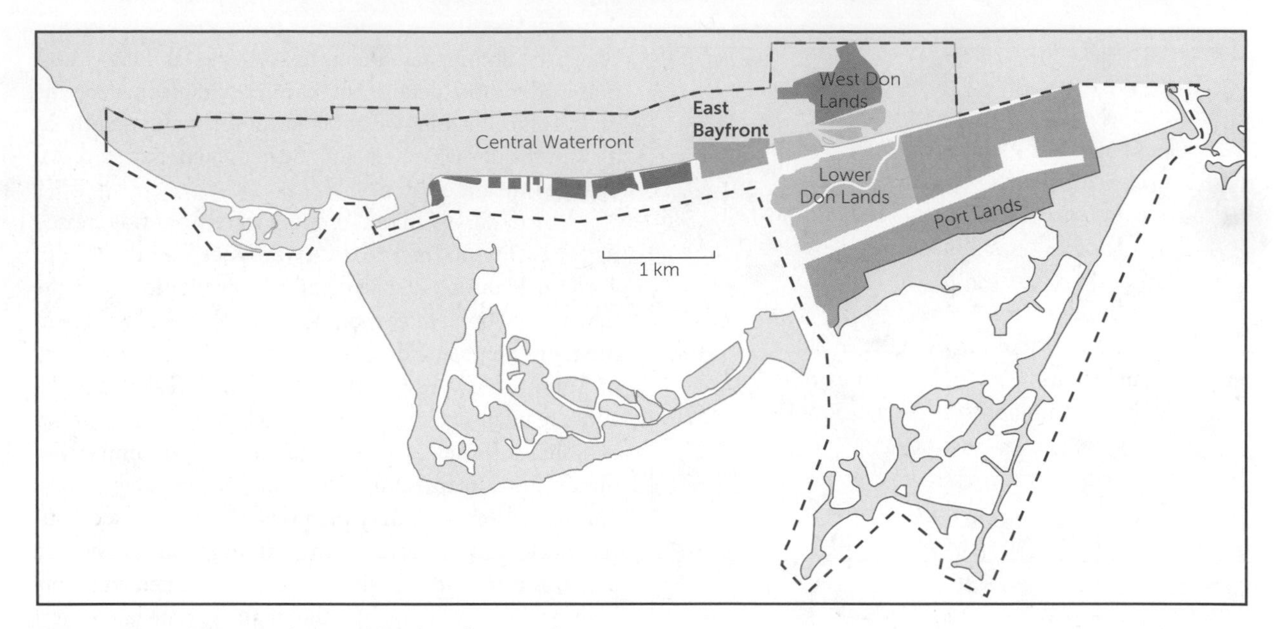

Figure 7.2.2 • The precincts of Toronto's waterfront.

Source: With kind permission of Waterfront Toronto.

secondary plan and the existing City of Toronto zoning bylaw, which was still to provide the legal framework for land use and development on the waterfront. The City of Toronto and the TWRC reconciled this legal impediment by designating the precinct land parcels as special districts within the wider City of Toronto Zoning Bylaw (City of Toronto 2010). With this approach in place, the new precinct plans had the potential to act as "bridges" between the waterfront secondary plan and a series of new amended zoning bylaws and subdivision codes that would reflect the design ambitions of the precinct plan (TWRC 2005a). To explore how this idea was put into practice, let us examine one of the waterfront precincts in more detail.

The Easy Bayfront Precinct Plan

Covering an area of 23 hectares, the East Bayfront precinct is located close to Toronto's downtown core on the eastern side of the waterfront redevelopment area. The precinct extends from Lower Jarvis Street in the west to Parliament Street in the east. On its northern edge, the precinct is bounded by a high-volume elevated expressway and a raised railway corridor, while to the south, the site extends all the way to Lake Ontario. Located in a city that is famed for its vibrant urban neighbourhoods, both the East Bayfront site and the wider waterfront lands present a stark contrast. Although Toronto's downtown core is located only a few hundred metres away, the railway corridor and elevated highway create a deep chasm between the city and the lake. Pedestrians and road-users have to traverse underpasses, intersections, and deafening highway noise before reaching the waterfront, and when they arrive, the area feels like a forgotten segment of the city.

Following the publication of the new secondary plan, the TWRC initiated the masterplanning process for the East Bayfront in 2003 (Waterfront Toronto 2012) on a segment of land that was created during an extensive landfill program led by the federal Toronto Harbour Commission (THC). Throughout the early and mid-twentieth century, the THC was the custodian of all the public land on the waterfront, and in 1952 it selected the East Bayfront as the final portion of the waterfront to be reclaimed and amalgamated into the city's growing port. Its time as a site of industry, however, was short-lived. During the 1960s, containerization became the norm, and shipping almost entirely ceased on Lake Ontario because the new class of "supercontainer" ships were too large to navigate beyond Montreal (Desfor 1993; Laidley 2007). Subsequently, the East Bayfront, like much of the industrial land on Toronto's waterfront, became underused. When the precinct planning process was initiated, few active land uses remained.

An experienced American urban design firm, Koetter Kim and Associates, was commissioned to produce the precinct plan for the East Bayfront (Hume 2003). At an introductory event held by the TWRC in August 2003, the project leader, Fred Koetter, spoke of his desire to develop a flexible master plan that would respond to changing uses over time. Koetter also emphasized that the key to success lay in a transparent planning process. Getting the process right, he stated, would improve the chances of creating a great place (Hume 2003). A series of iterative public engagement events were therefore incorporated into the planning process (TWRC 2002).

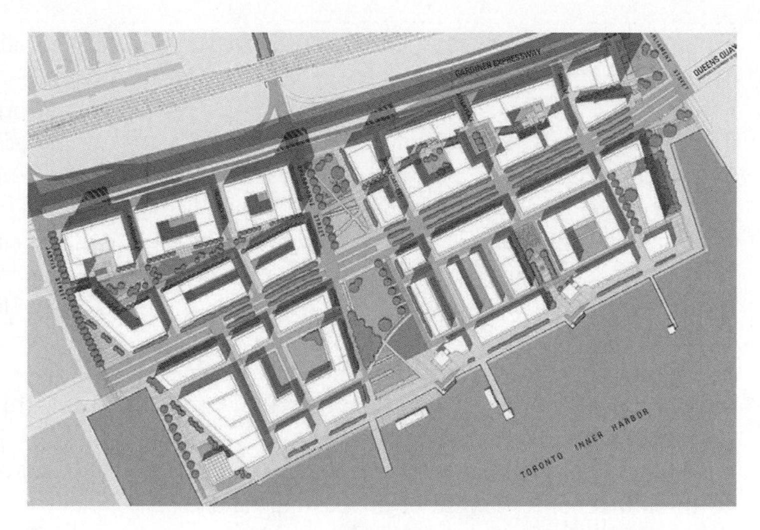

Figure 7.2.3 • East Bayfront precinct plan final master plan.

Source: With kind permission of Waterfront Toronto.

Working with local people, TWRC completed the master plan in four stages. The design team started the process by hosting a public forum to solicit general thoughts and ideas from the community. A few months later, they shared a range of initial ideas at a second public forum. Feedback from this session was then used as the basis for producing a series of focused design concepts, which were presented at a third public forum. Finally, using detailed feedback from forum participants, the design team produced a complete version of the master plan (TWRC 2005a). The entire process took six months, and approximately 200 to 250 people attended each of the events (TWRC 2003). The majority of attendees were from local neighbourhood associations, but numerous other people, including local government officials, members of the local design community, advocacy groups, and graduate students, also took part. In addition, a series of smaller stakeholders meetings were convened. These meetings provided an opportunity for representatives from local community groups and businesses located on the waterfront to have a more focused discussion with the design team about the masterplanning process.

The precinct master plan that emerged from the participatory design process ran to 50 pages and included a mixture of accessible text, plans, three-dimensional graphics, watercolours, and photographs to convey the vision for the area's transformation (TWRC 2005b). As illustrated in Figure 7.2.3, it established the arrangement of the proposed blocks, road layout, and pedestrian network and included a mixture of residential and commercial buildings that explored the potential proportions of the three-dimensional blocks imagined on the site (Love 2009).

The plan also clearly reflected the outcomes from the four public forums. Since the beginning of the masterplanning process, for example, both the community and the design team had developed a shared wish to improve the poor north–south connections between the city and the waterfront. The final master plan thus employed a series of diagonal sightlines and public spaces to reinforce the connection of the north–south streets to Lake Ontario. In the finished plan, the design team also emphasized the crucial role that the water's edge promenade and Queens Quay—the main road running through the site—would play as pedestrian places, a suggestion that participants had made numerous times during the public forums. The interface with the lake would be car-free and continuously animated at street level by ground-floor retail. And to increase its versatility, the promenade would also be tiered, creating a protected zone for walking by the water and a more flexible space for outdoor seating and restaurants behind. Similarly, the design team had undertaken a considerable amount of detailed design work to ensure that Queens Quay would become an inviting boulevard that could be shared by pedestrians, cyclists, streetcars, and vehicles. The design team also dealt with the concerns that participants had raised about building height in the final master plan. Taller buildings would be limited to strategic locations along the northern perimeter of the precinct, and as a general rule, all of the buildings would step down toward the lake. Finally, the precinct plan addressed affordable housing. As a minimum, the master plan stated that 20 per cent (1250) of the 6300 proposed units on site would be affordable, and an additional 5 per cent (315) would be allocated for **social housing** (TWRC 2005b). The precinct plan offered a design-led strategic response to the site, and it was presented in an accessible format that experts and laypeople alike could understand and interpret.

Masterplanning as a Design-Sensitive Blueprint for Zoning

The City of Toronto endorsed the East Bayfront Precinct Plan in December 2005 and soon after instructed City staff to begin the process of preparing an amended zoning bylaw for the precinct (City of Toronto 2005). The special district zoning amendment "codified" the built-form components of the plan, including the size and location of public space, the distribution of roads and other public **rights-of-way**, and the configuration and heights of buildings (see Figure 7.2.4). In addition, it set out rules for the provision of affordable housing and funding of public amenities established in Koetter Kim's master plan (City of Toronto 2006b). Since all of the planning and urban design principles for the East Bayfront had been agreed upon in advance during the precinct planning process, the City of Toronto Planning Department was able to produce the amended zoning bylaw in less than a year. Moreover, the process demanded only minimal public participation, because no substantive changes to the planning vision and design principles were made.

The amended zoning bylaw was enacted by the City of Toronto Council in September 2006 (City of Toronto 2006a). All of the developable land parcels in the precinct were zoned "CR" (commercial–residential), while all of the proposed public spaces were zoned "G" (public open space). To achieve the street-level vibrancy imagined in the precinct plan, the bylaw also banned residential units at ground level and stipulated that all ground-floor uses within CR zones had to be animated with commercial units. Through a series of strict **setbacks** and maximum height rules, the bylaw also determined the scale of the perimeter blocks proposed in the precinct plan and confirmed the community's wish to see any new buildings in the precinct step down to the lake. To meet the masterplanning objective that affordable housing be a core component of the East Bayfront precinct, the bylaw further required that at least 20 per cent of the dwelling units constructed on each land parcel be maintained as affordable rental units for no less than 25 years or gave developers the option to make an in-lieu payment to the City of Toronto for affordable housing. The precinct plan and zoning bylaw amendment was also supported by an official "plan of subdivision" that regulated the specific orientation of roads and building configurations within each of the land parcels in the precinct.

Zoning bylaws can often end up being cumbersome planning and design tools that generate regimented and unimaginative urban form (Barnett 1982). The East Bayfront precinct masterplanning process created the conditions for a design-sensitive planning practice that extended the limits of traditional zoning. The precinct plan provided a sense of security for the TWRC, the City of Toronto, and community stakeholders about how the area might look and feel in the future. The next stage in the East Bayfront story, however, is the process of implementation. It began in 2009 and will likely be ongoing until the mid-2020s. Planning for future land uses is never a watertight process, and already some of the new

TORONTO City Planning Division
Map B: Maximum Heights Plan
East Bayfront - West Precinct

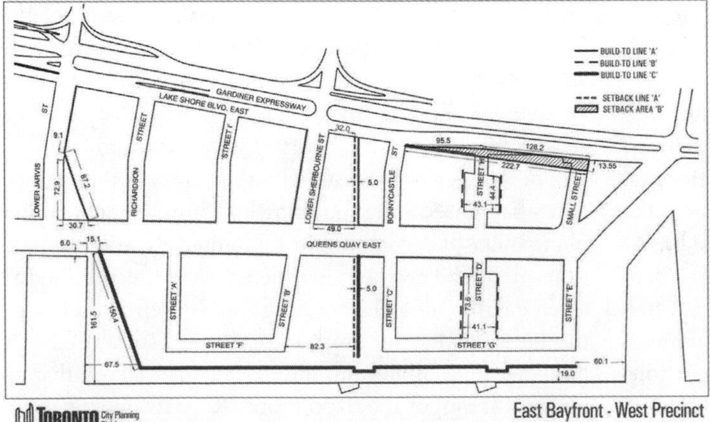

TORONTO City Planning Division
Map 4: Build-To/Setback Plan
East Bayfront - West Precinct

Figure 7.2.4 • Sample requirements of the East Bayfront zoning bylaw.

Source: City of Toronto 2006a, 18 (top), 23 (bottom).

buildings and public spaces constructed in the precinct have forced changes to the vision and principles contained in the master plan. As developers have proposed these alterations, often as a result of perceived market imperatives, the TWRC has been careful to keep local people informed and engage them in the process of amending the zoning provisions. While it is almost always inevitable that real estate developers will seek to alter the existing planning framework to suit their commercial needs, a master plan can still act as a powerful visual tool that outlines the long-term planning and design vision for the site. With such a document in place, it is hoped that any proposed changes can be subject to more rigorous and informed review by the governing authority and local stakeholders.

Conclusion

This chapter has shown how masterplanning can lead to a more design-sensitive approach to land-use zoning. The concept of zoning emerged in the early twentieth century in Canada as a tool to separate incompatible land uses into single-use districts while protecting the property rights of wealthy citizens. However, zoning decisions are often made at the expense of neighbourhood design and long-term planning considerations, leading to urban environments that are dominated by strip malls and low-density housing. The chapter introduced the concept of masterplanning as a strategic planning tool that creates a more design-led framework for new development.

The East Bayfront precinct on Toronto's redeveloping waterfront is an example of masterplanning in practice. While every jurisdiction in Canada faces its own unique land-use and urban design challenges, this case study has aimed to demonstrate that new and innovative mechanisms should always be tried and tested. Developing a master plan to *inform* the shape and scope of zoning and subdivision plans is one way in which this can be achieved. For Canadian practitioners struggling to encourage more design-sensitive planning, masterplanning can be seen as a way to recast the often market-led process of city-building as a positive exercise in "placemaking" and an opportunity to generate a more rigorous focus on space and place in Canadian practice.

KEY TERMS

Bylaw (or by-law)
Master plan
Official Plan
Placemaking
Right-of-way
Sense of place
Setbacks
Social housing
Subdivision
Zoning bylaws/schedules

REFERENCES

Adams, D., and S. Tiesdell. 2013. *Shaping Places: Urban Planning, Design and Development*. Abingdon, UK: Routledge.

Barnett, J. 1982. *An Introduction to Urban Design*. New York: Ledgebrook Associates, Inc.

———. 2011. "How codes shaped development in the United States, and why they should be changed." In S. Marshall, ed., *Urban Planning and Coding*. Abingdon, UK: Routledge.

Bell, D. 2005. "The emergence of contemporary masterplans: Property markets and the value of urban design." *Journal of Urban Design* 10(1): 81–110.

CABE (Commission for Architecture and the Built Environment). 2004. *Creating Successful Masterplans: A Guide to Clients*. London: CABE.

Carmona, M. 2009. "Design coding and the creative, market and regulatory tyrannies of practice." *Urban Studies* 46(12): 2643–67.

———. et al. 2010. *Public Places Urban Spaces*. Abingdon, UK: Routledge.

City of Toronto. 2001. *Making Waves: Central Waterfront Part II Plan*. Toronto: City of Toronto.

———. 2002. *Toronto Official Plan*. Toronto: City of Toronto.

———. 2004. *Governance Structure for Toronto Waterfront Revitalization Corporation*. Toronto: City of Toronto.

———. 2005. *Minutes of the Council of the City of Toronto, December 5, 6 and 7, 2005*. Toronto: City of Toronto.

———. 2006a. *Bylaw No. 1049-2006*. Toronto: City of Toronto.

———. 2006b. *City of Toronto Staff Report: Proposal to Amend Zoning Bylaw 438-86 and Modify the Central Waterfront Secondary Plan for the Lands between Lower Jarvis Street and Small Street to the South of Lake Shore Boulevard East*. Toronto: City of Toronto.

———. 2010. *City of Toronto Zoning Bylaw*. Toronto: City of Toronto.

Cullingwood, B., and R.W. Caves. 2009. *Planning in the USA: Policies, Issues and Processes*. Abingdon, UK: Routledge.

Desfor, G. 1993. "Restructuring the Toronto Harbour Commission: Land politics on the Toronto waterfront." *Journal of Transport Geography* 1(3): 16–181.
Grant, J. 2000. "Planning Canadian cities: Context, continuity and change." In T. Bunting and P. Filion, eds, *Canadian Cities in Transition: The Twenty-First Century*, 2nd edn., 443–61. Don Mills, ON: Oxford University Press.
——. 2002. "Mixed use in theory and practice: Canadian experience with implementing a planning principle." *Journal of the American Planning Association* 68(1): 71–84.
——. 2008. "Tools of the trade." In J. Grant, ed., *A Reader in Canadian Planning: Linking Theory and Practice*, 339–43. Toronto: Thomson Nelson.
Grunton, T. 1991. "Origins of Canadian city planning." In K. Gerecke, ed., *The Canadian City*, 93–114. Montreal: Black Rose Books.
Hodge, G. 2003. *Planning Canadian Communities*. 4th edn. Toronto: Thomson Nelson.
Hume, C. 2003. "Bayfront vision moves forward." *Toronto Star* 23 August.
Jacobs, J. 1961. *The Death and Life of Great American Cities*. New York: Vintage Books.
Laidley, J. 2007. "The ecosystem approach and the global imperative on Toronto's Central Waterfront." *Cities* 24(4): 259–72.
Love, T. 2009. "Urban design after Battery Park City: Opportunities for variety and vitality." In A. Krieger and W.S. Saunders, eds., *Urban Design*, 208–26. Minneapolis: University of Minnesota Press.
Planning Act [R.S.O. 1990, c-P.13].
Relph, T. 1987. *The Modern Urban Landscape: 1880 to the Present*. Baltimore, MD: John Hopkins University Press.
Tiesdell, S., and G. MacFarlane. 2007. "The part and the whole: Implementing masterplans in Glasgow's New Gorbals. *Journal of Urban Design* 12(3): 407–33.
TWRC. 2002. *Waterfront Revitalization Corporation Public Consultation and Participation Strategy*. Toronto: TWRC.
——. 2003. *East Bayfront Precinct Planning: Summary of Public Forum 1*. Toronto: TWRC.
——. 2005a. *East Bayfront Precinct Plan*. Toronto: TWRC.
——. 2005b. *East Bayfront Precinct Planning: Draft Summary of Public Forum #4*. Toronto: TWRC.
Waterfront Toronto. 2010. *Report to the Community 2010: Our New Blue Edge*. Toronto: Waterfront Toronto.
——. 2012. "Planning East Bayfront." http://www.waterfrontoronto.ca/east_bayfront/planning_the_community.

INTERVIEWS

Senior planning manager, City of Toronto, 7 March 2011. In-person interview.
Senior executive and urban designer, TWRC, 18 March 2011. In-person interview.

CASE STUDY 7.3

Developing the Community Plan and Urban Design Plan for Benalto, Alberta

Beverly Sandalack and Francisco Alaniz Uribe

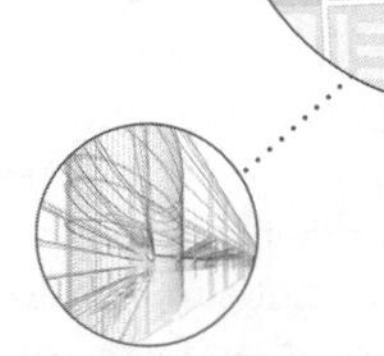

SUMMARY

The Hamlet of Benalto in central Alberta has a traditional town form. However, with obsolescence of the railway and related infrastructure, the structure and function of the town have changed. Benalto's location makes it a desirable place to live in, combining small-town living with big-city convenience, and it is identified by Red Deer County's Municipal Development Plan as an area likely to experience growth. Recent development has been more typical of generic suburbs, and the quality of life, character, and sense of place of Benalto are threatened. The County therefore contracted the Urban Lab at the University of Calgary, an interdisciplinary research group, to prepare a sustainable land-use strategy and design guidelines for Benalto. An Area Redevelopment Plan (ARP) was prepared to provide a framework for Benalto's form and function. The overall aim was the conservation of the hamlet as a cultural landscape that would provide a high quality way of life while responding to contemporary needs.

A theoretical framework, approach, and methods, which had been developed over the course of several projects for considering complex planning and urban design problems, was refined and applied to Benalto. This theoretical framework includes:

- a conceptual approach to understanding the relationships of various components of the (ideal) built environment and their degree of permanence;
- the elaboration of Townscape Analysis as a comprehensive methodology integrating environmental analysis, historical evolution analysis, spatial structure analysis, morphology and typology studies, economic and demographic analysis, and visual analysis as a means of identifying constraints and opportunities and providing direction for land-use decisions in the existing and proposed settlements;
- a compact and inclusive process for community engagement.

The intent of the ARP was to satisfy three interrelated components of Benalto urban process and form:

- small-town character;
- quality of life;
- sense of place.

The Area Redevelopment Plan, which was ultimately adopted by Red Deer County Council (2006), was prepared in a legible and accessible poster format. The project received a National Honour Award in the Canadian Society of Landscape Architects 2007 Professional Awards Program, and the innovative approach and highly graphic nature of the plan were commended.

OUTLINE

- The process of creating an Area Redevelopment Plan for the Hamlet of Benalto, Alberta, included identifying areas for growth, developing design guidelines, and preserving the cultural landscape using a townscape analysis and public workshop.
- The project presented a unique opportunity for the consideration of urban–rural interface issues, for development of a sustainable regional plan, and for the design of a high-quality urban settlement according to sustainable design principles.
- The case also illustrates how this approach to analysis, planning, and design can provide the framework for various scales of urban development, from the regional plan, to the town, to the neighbourhood, and to the street.

Introduction and Context

Red Deer County (RDC) in the province of Alberta in western Canada is one of the fastest-growing regions in North America. People are attracted by the buoyant economy, the high standard of living, and the quality of the environmental setting. While much of Alberta's population growth has been concentrated in the cities of Calgary, Edmonton, and Red Deer, exurban areas have also undergone recent and extreme change.

The Hamlet of Benalto is located just west of Sylvan Lake on rolling parkland. It has a traditional town form. However, with obsolescence of the railway and related infrastructure, the structure and function of the town have changed.

The proximity of Benalto to Sylvan Lake and the city of Red Deer make it desirable as a place to live in, combining small-town living with big-city convenience, and the hamlet is identified as an area likely to experience growth. Its population at the time of this project was just under 300. As of the 2006 census, 175 people resided within the settlement of Benalto, plus 252 in Kountry Meadow Estates, an adjacent mobile home park.

Benalto is distinguished by a compact form, a town centre facing remarkable mountain views, a grid fabric of **walkable** residential streets, a network of open spaces and paths, and memorable tree-lined entries. However, some recent development has been more typical of generic suburbs, and the quality of life, character, and sense of place of Benalto are threatened. New development and redevelopment are expected, and it is desired that the character and quality of life of Benalto be sustained while accommodating appropriate growth. The population of the town and region is now composed of a diverse mix, including many artists and artisans in addition to the original agricultural community, and there is the expectation that as the hamlet evolves, it will retain its small-town character and qualities and offer economic opportunities and a high quality of life. A strategy for this growth was required to provide guidance for sustainable development, revitalization of the town core, and plans for open space and other land uses.

While identity was once largely a product of the unique interaction between local culture and traditions with environmental constraints, the visual qualities and spatial structure of places are now strongly influenced by external political and economic forces, and as a result, many places now tend to look more and more alike. As Michael Hough observed, "the question of regional character has become a question of choice and, therefore, of design rather than of necessity" (1990, 2).

The main instruments for typical development control are regional and municipal plans, which rely heavily on **zoning**, **subdivision** regulation, and **bylaws**. Although they provide a rational basis for land development and subdivision, they emphasize functional efficiency in transportation planning and street layout while often neglecting both the social and aesthetic consequences of these decisions as well as the potential value of local history and culture in providing design and planning direction. **Design guidelines** are a recent instrument of development control in Canadian towns and are statements of intent as to the preferred use, density, and general appearance of new development. Red Deer County wanted a more responsive process and plan, combined with appropriate guidelines, to address the issues facing Benalto.

The authority for municipal planning, subdivision, and development control was established under the Alberta Municipal Government Act (2000), and it provides for a series of statutory and non-statutory plans. The **Municipal Development Plan** (MDP) is the primary planning policy document for use at the municipal level

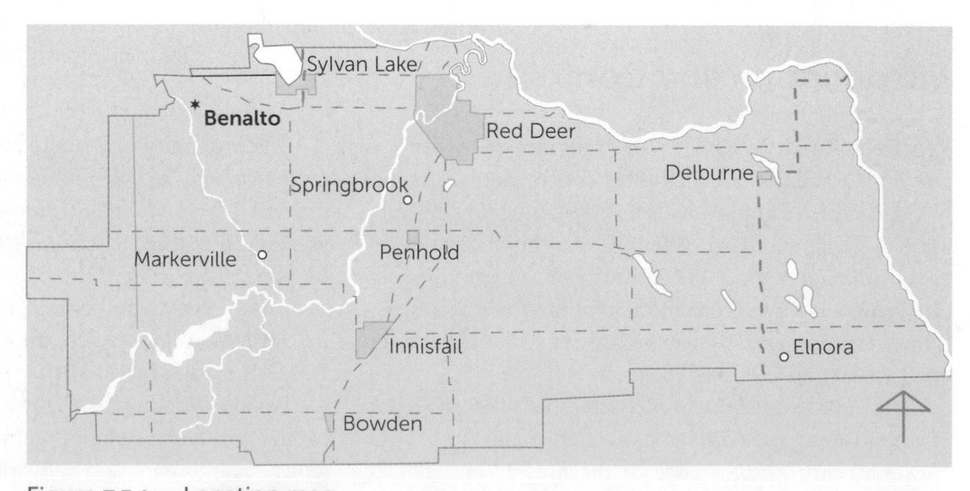

Figure 7.3.1 • Location map.

under which fall various statutory plans (legal documents that must go through three readings and a public hearing before adoption), such as Inter-municipal Development Plans (IDPs), Area Structure Plans (ASPs), and Area Redevelopment Plans (ARPs), and non-statutory plans, which are bylaws passed by resolution to help provide direction, such as Concept Plans and Outline Plans. Red Deer County's MDP (Red Deer County 2006) advocates for Smart Growth and recommends concentration of growth development in nodes in order to support and take advantage of existing infrastructure. Several settlements were identified as Priority Growth Hamlets in the MDP, including Benalto because of its proximity to the city of Red Deer and the town of Sylvan Lake and its location near Highway 11, which connects Red Deer and Sylvan Lake to the west country. Benalto consequently required an Area Redevelopment Plan to provide direction for this anticipated future growth.

The Urban Lab at the University of Calgary was contracted by Red Deer County to prepare an Area Redevelopment Plan that would provide a framework for the hamlet's form and function, to identify potential lands for growth, and to establish guidelines for several specific aspects. The Urban Lab is an innovative research group in the Faculty of Environmental Design and connected to its graduate professional programs. It was established and led by Beverly A. Sandalack, and projects are co-managed with Francisco Alaniz Uribe. Members of the Benalto Project team included graduate research assistants Blair Marsden, Kristina Meehan-Prins, Nathalie Woodhouse, and Natalia Zoldak.

The process for preparing the ARP included several unique aspects:

- The "townscape analysis" approach (described on page 294) was used as a means of identifying constraints and opportunities and of providing direction for land-use decisions in the existing and proposed settlements.
- A two-day workshop with residents was held, consisting of a series of presentations and discussions and including participation by local schoolchildren, who were able to learn more about their town through the physical model and drawings.
- Development of a plan that would guide town form at several nested scales—the town as a whole (within a regional context), the block, the street, and the lot, including development of an open-space system as the framework for the public realm and development of urban design guidelines to guide the three-dimensional realization of the plan concepts.
- Preparation of the plan in a highly graphic poster format that would be accessible and understandable by residents and practical for planning administration, designers, and developers. The County wanted a compact process; therefore, a timeline was developed that would progress efficiently from project start-up in May 2006, to analysis during April and early May, to various concept stages and community input during mid-May and June, to development of the final plan during the summer of 2006. Various consultations with County Council and staff and subsequent plan document revisions were to take place over the following months, and it was anticipated that the final plan would be submitted for approval in the fall of 2006.

Approach and Methodology

A conceptual diagram (Figure 7.3.2) illustrates the relationships of various components of the (ideal) built

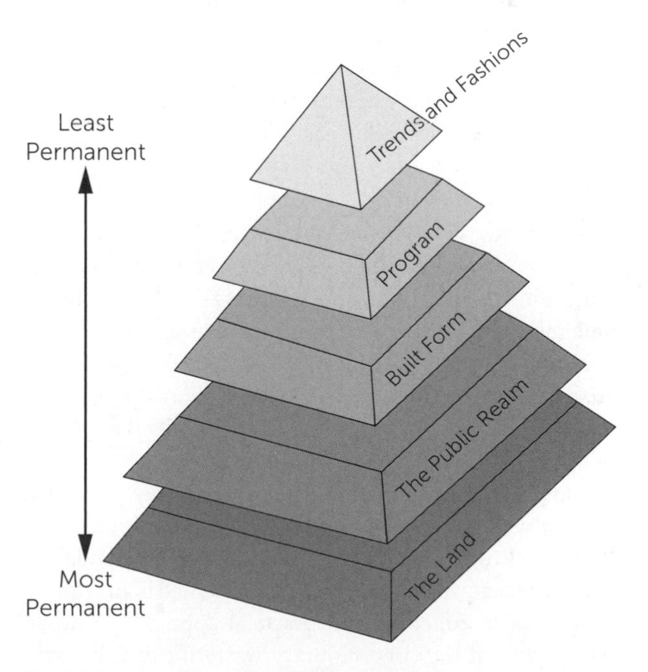

Figure 7.3.2 • The pyramid concept.

environment and their degree of permanence (Sandalack and Nicolai 2006).

- The land, as well as landscape character, is the most permanent aspect of the built environment, with the greatest potential to contribute to a sense of place.
- Much of our everyday urban existence occurs within the shared city space made up of the streets, sidewalks, parks, squares, and plazas. Collectively, this is known as the public realm, and it constitutes the next most permanent component of the built landscape.
- Several generations of built form will come and go within the lifecycle of the urban structure. However, if the infrastructure of the public realm is intact, then the built form has a sense of continuity and meaning over time.
- Each individual building, if the form is resilient, may be used for various activities or programs. This **resilience** of built form contributes further to the establishment of a sense of place through continuity of form.
- The least permanent aspects of the built environment, and of design activity, are the transient and ephemeral trends and fashions. While they frequently add the qualities of delight and contemporariness to the built environment, there is normally a built-in obsolescence to them, and they should be understood as the least permanent, although not necessarily the least important, aspects of environmental design.

Currently, more design intent is applied to individual buildings, to programming, and to trends and fashions than to the most permanent elements—the public realm and the landscape. The less permanent elements come and go, but the public infrastructure—the integrated system of public spaces—persists and can give a sense of continuity and quality to a place. This is typically the domain of physical planning and urban design.

An approach to planning and urban design called "townscape analysis" was developed from theoretical (Sandalack 1998; Sandalack and Nicolai 1998; Sandalack and Nicolai 2006) and professional work and is an attempt to develop a coherent and practical approach to urban design within the contemporary western city that could inform planning education and practice. This approach emphasizes a methodology for understanding, in detail, the makeup and functioning of a place. It assumes that each place is the product of the interrelationship of natural form and process with built form and community over time.

This approach draws from the theories and methods of landscape architecture, urban planning, urban morphology, and urban design. It includes environmental analysis, morphology/typology, spatial structure, and urban qualities.

Environmental Analysis

Environmental analysis is essential to the development of sustainable places and can also provide design determinants. There is an inherent logic in the evolution of any city or town, and often this has something to do with landscape, topography, and hydrography. Environmental analysis allows better understanding of natural form and process and provides information that can contribute to the development of cities and towns with greater environmental responsiveness, authentic identity, and sense of place.

Ian McHarg (1992) changed the way that environmental planners and designers gathered and analyzed environmental data, effectively demonstrating that physical planning and the design of sites should be based on a thorough understanding of the ecology of the area together with human values. Through an understanding of environmental conditions and characteristics, more locale-specific and ecologically sound planning and design would be produced, and there would be a greater likelihood of producing designs that would be more harmonious with their environmental context and more appropriate to human values and needs. The field of landscape ecology (Forman 1995; Forman and Godron 1986) provides a broad theoretical base to urban ecology and contributes principles upon which urban planning and design decisions can be made.

Morphology/Typology

A key issue of authenticity and identity is the maintenance of continuity. The use of local history in developing an understanding of places offers a reference point for decisions.

Urban morphology is an approach to studying urban form that considers both the physical and spatial components of the urban structure: the physical components of plots, blocks, streets, buildings, and open spaces (Moudon 1997) and their relationships to each other. The morphological approach is useful in understanding the cause-and-effect relationships between urban process and form and between urban form and spatial structure. **Precedents** can be found for more appropriate design, and design solutions can be sanctioned according to whether they correspond to and reinforce the character and typological conditions that have evolved in that place and the qualities that are believed to be desirable.

Spatial Structure

The analysis of a town or city's spatial structure considers land utilization and the pattern of activities that parts of a town or city generate. It describes the location and distribution of particular uses and the functional relationships between them. A number of theories of spatial structure have been formulated, with various ways of conceptualizing space.

Lynch (1960) saw the city image as a system composed of five basic elements: paths, edges, districts, nodes, and landmarks (or monuments) by which urban form can be analyzed and used as a basis for design. This organizing structure has significance for the inhabitants, who form a mental map in which the urban elements provide physical and psychological orientation. Trancik (1986) expanded this framework and discussed the importance of identifying the gaps in the fabric (such as spaces that make no positive contribution to their surroundings or to the experience of the users) and of considering the overall pattern of development. Social processes should also be considered, since spatial structure analysis can only increase the understanding of how towns function when it considers the interrelationships of urban elements with human perceptions and social patterns. Studies of the public realm elements—the streets, squares, public buildings, and open spaces—constitute an important part of this analysis.

Urban Qualities

The position that one takes as a planner is based on a conviction regarding the qualities that are important to urban form and urban life. Various authors (Lynch 1960; Hough 1990; Sandalack and Nicolai 1998) proposed lists of qualities that good places should embody. A number of authors have agreed on the importance of legibility, permeability, human scale, continuity, variety, and environmental responsiveness as universally desirable qualities of urban form. Qualities specific to individual places or regions should also be identified.

Analysis

Environmental Context

The Benalto townsite lies within the Medicine River subbasin, which comprises part of the Red Deer River basin. A major wetland complex is located northwest of Benalto and drains into the Medicine River. Benalto has good drainage, with few sloughs or permanent water bodies.

The land slopes to the west toward the Medicine River, with some areas of steep slope greater than 15 per cent (these are potential hazard areas, according to the RDC Municipal Development Plan). The Benalto townsite slopes to the southwest, providing excellent views of the mountains.

Most of the land surrounding Benalto is Class 3 (suitable for agriculture), with Class 4 (marginal) along the Medicine River and Class 6 (unsuitable) to the northwest. Rural land in Canada is assessed by the Canada Land Inventory according to its capability for various uses. Soil classification indicates the suitability for agricultural crops and considers limitations due to climate, soils, and landscape—this can assist in determining land value and management practice and should be considered a factor in suitability for development.

Historical Evolution

Some of the accompanying images summarize the historical evolution of Benalto within the regional context.

Vision and principles

The following interdependent principles, developed in collaboration with the residents and RDC planners, attempted to address the vision and provide a framework for development.

Strengthen the town centre:

- Reinforce the 50th Avenue Main Street area as the commercial and social core.
- Encourage a mix of uses and live/work opportunities.

- Establish architectural and urban design guidelines.
- Provide guidance for development of the former railway lands to support the town centre, preserve important views, and contribute to public open-space needs.

Preserve "natural capital":

- Conserve and/or frame important views.
- Preserve tree stands and develop a replanting strategy.
- Capitalize on tree-lined roads as memorable entries.
- Link Benalto to the regional path system as a way of connecting to the landscape.

Establish an open-space system:

- Provide a range of green spaces for all areas/all ages.
- Formalize the initial efforts to develop a path system, and establish guidelines for acquisition, design, and management.
- Consider all streets as part of the open-space system, and develop appropriate streetscape guidelines that preserve the town character and encourage walkability.
- Develop a street tree-planting program as a beautification priority.
- Require dedicated green space and path linkages in all new development.

Establish guidelines for new development:

- Build on the established Benalto grid block pattern and street standards in all new growth areas (discourage suburban standards, e.g., curb and gutter, wide carriageway, lack of sidewalks).
- Require new development to provide public open space and linkages to the existing path system.
- Provide a range of house and block types to encourage variety and affordability.
- Establish architectural and urban design guidelines for new development and infill, including modular homes.
- Require site planning and house design to contribute to a high-quality and pedestrian-friendly streetscape and discourage front garages, deep setbacks, and unnecessary grading.
- Require outline plans for new development to include street and block layout, streetscape and landscape plan, and three-dimensional illustrations of overall development.

Workshop with stakeholders

The workshop gave residents along with County planners opportunities to contribute to development of a vision for the plan, for input on several specific aspects of the plan, and for discussion and feedback to the project team. The Benalto School students contributed a series of drawings that expressed what they liked most about Benalto and what they felt was missing.

Land use and concepts

The opportunities and constraints identified in the analysis were synthesized with the input from residents, and an overall plan and guidelines were developed that would respond to the vision and principles. The urban structure plan is intended to reinforce the way of life and quality of life of the hamlet and includes concepts that will encourage a more sustainable form. During this process, several conflicts emerged that the plan attempted to help resolve. For example, a land developer had recently constructed a road and houses according to a form that was more

Figure 7.3.3 • A student workshop.

indicative of a city suburb than Benalto, and residents were anxious that additional growth would not take this form. Accordingly, guidelines were prepared that would influence the form and quality of block patterns, streetscapes, house type, and landscape.

Entries

Benalto currently has two stunning entries (from the west and southeast) through tree-lined roads connecting to 50th Avenue, and these entries help to distinguish Benalto and contribute to its sense of place. A concept for enhancing these entries was developed, and a tree-planting program was proposed for the Range Road 25A approach from Highway 11 in order to develop a third high-quality entry.

Hamlet centre

Benalto's Main Street (50th Avenue) is the commercial and social core of the town, but it currently has several gaps and few businesses. Provision for mixed-use development can help to revitalize it and create economic opportunities. Urban design guidelines will also help to establish the physical form. The 50th Avenue Main Street area forms the visual and functional heart of Benalto and should be supported, reinforced, and enhanced.

Residential development

Benalto's small-town character and quality of life is supported by a grid block pattern, walkable streets, street trees, and appropriate house form. Guidelines for block, street, lot, and house placement build on what is already distinctive.

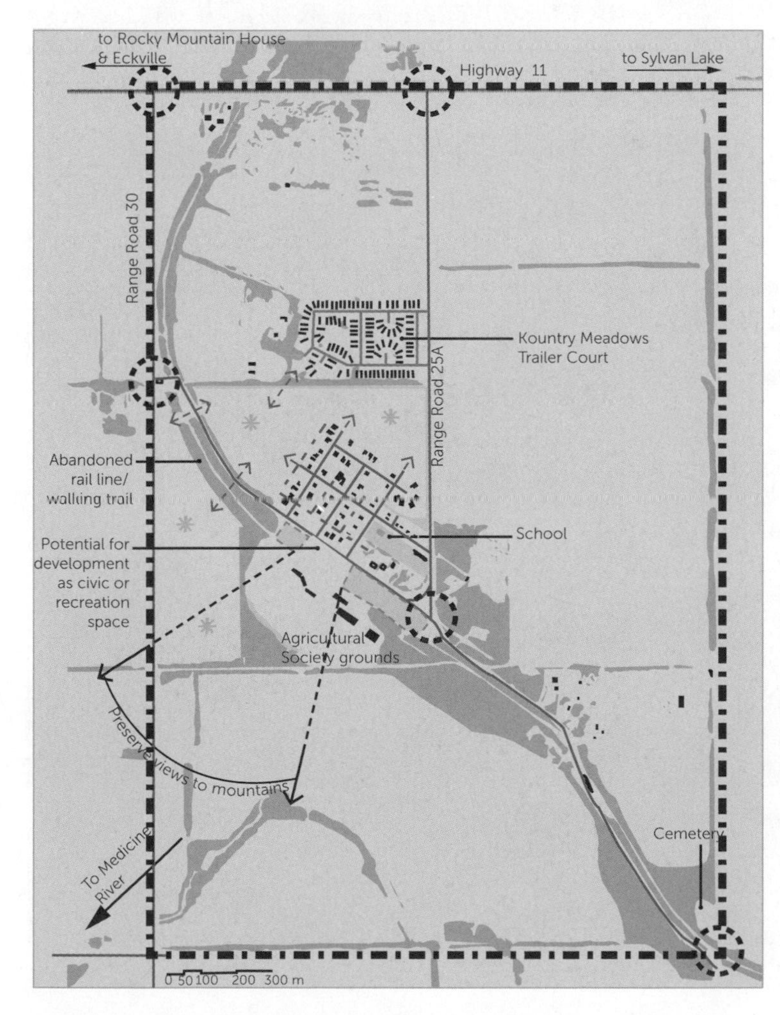

Figure 7.3.4 • Urban structure plan.

Potential areas for growth

Future Benalto growth may take place to the north and west, as shown in the urban structure diagram. New development must conform to the Benalto block and street form (Section 7). Kountry Meadows Trailer Court should be linked to Benalto through streets and trails. Land adjacent to the built-up area will remain in agricultural use during the lifecycle of the plan. The former railway lands may be also be considered as a potential area for residential growth.

Former railway lands

The Benalto former railway lands will provide for public open space and support 50th Avenue Main Street as the town centre. The maximum height of all buildings will be 3.5 metres (measured on the 50th Avenue facade) in order to preserve the views to the west. Buildings proposed within the viewplane indicated on the urban structure diagram will not exceed seven metres in height. Although the primary frontage is toward 50th Avenue, **secondary frontages** are encouraged on other sides facing the **public realm** in order to create multiple active edges.

Block structure

The hamlet of Benalto's small-town character and quality are partly a product of

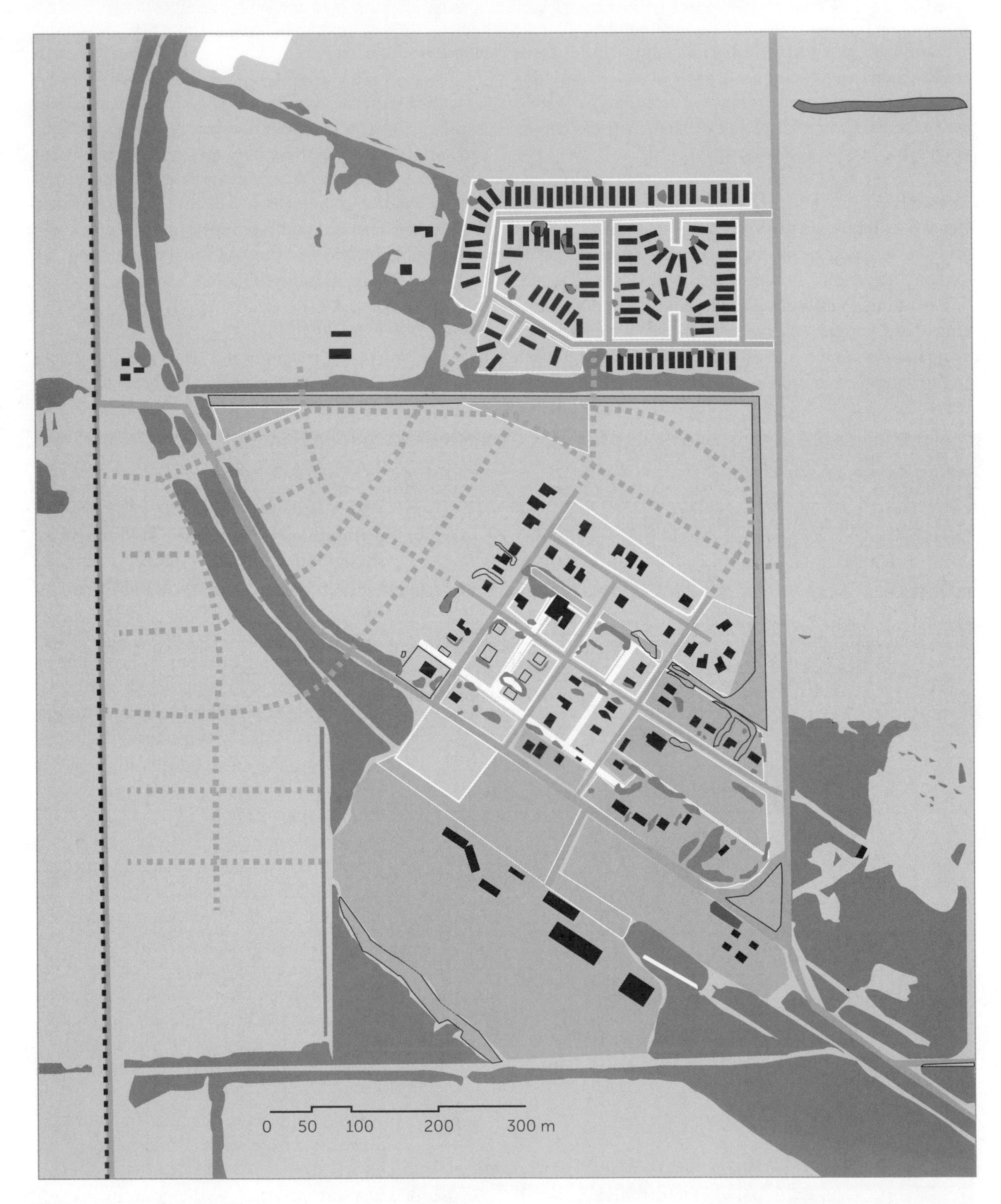

Figure 7.3.5 • Block pattern.

block and street form. Block and street form determines permeability and choice of routes and mode of travel (walking, biking, driving). Benalto is composed of a compact and walkable grid. This contributes to the quality and way of life and is an important component of its sense of place. The grid pattern should be extended and the established block size repeated.

- Maintain street and lane connections into new development areas.
- Maintain existing lane allowances as public thoroughfares. Conversions to public paths are permitted.
- Provide back lanes where possible.
- Curvilinear suburban block patterns will not be permitted.
- Cul-de-sacs will not be permitted.

Open space and trails

The open-space system provides a framework for town structure and is based on concepts of walkability, accessibility, and sustainability. Benalto currently has two significant open spaces—the school grounds and the agricultural society lands. It also has the beginnings of an excellent path system. Walking trails should be developed as a continuous system linking all parts of Benalto and connecting to a regional network.

- Provide a range of green spaces for all areas/all ages.
- New development must include dedicated green space and path linkages.
- Establish a public civic space at the town centre. Acquire land on former railway lands and dedicate it for recreation and open space use.
- Preserve existing tree stands, and develop a management plan and replanting strategy.
- Connect Benalto to the regional path system.

Streets

All Benalto streets should be considered as part of the public realm and included as a component of the open-space system. Back lanes should be provided in all new developments wherever possible, and utilities should be placed in lanes to maximize the plantable area in front.

Street tree strategy

The street tree-planting program should be developed as a beautification priority. Street trees should be planted where possible. In the fall of 2006, more than 150 trees were planted in Benalto as a direct outcome of the plan.

Framework for sustainability

The ARP aims to achieve high standards in land-use patterns, neighbourhood design, site planning, and technology. Leadership in Energy and Environmental Design, or LEED, is a set of standards aimed at raising environmental quality in the built environment and providing guidance for development. LEED-Neighbourhood Design, or LEED-ND, combines principles of good neighbourhood planning, good urbanism, and Smart Growth.

This section of the plan includes concepts for arrangement of land uses as well as site planning practices at the scale of the hamlet, block, street, and lot. They are integrated with the land-use plan and open-space plan and are also elaborated in some specific proposals.

Groundwater Absorption Coefficient (GAC)

This aspect of development was identified as having effects on both sustainability and small-town character/quality. Typical practices of urbanization tend to increase the amount of paved surfaces (driveways, parking lots, walkways) and reduce the amount of infiltration. Permeable surfaces, especially landscaping, allow greater infiltration of rain and storm runoff, recharging of groundwater, and contribution to a more comfortable micro-climate. The Groundwater Absorption Coefficient (GAC) is the percentage of a lot that is required to be free of impervious material (e.g., concrete, asphalt). All new developments are required to have a minimum GAC of 45 per cent. This practice is intended to limit the amount of impervious surface and, as a consequence, to enhance many of the small-town characteristics and qualities.

Policies and urban design guidelines

Hamlet Mixed Use

The existing County Land Use Bylaw does not include a category that allows a mix of uses or that describes the current Main Street. A new land-use designation is

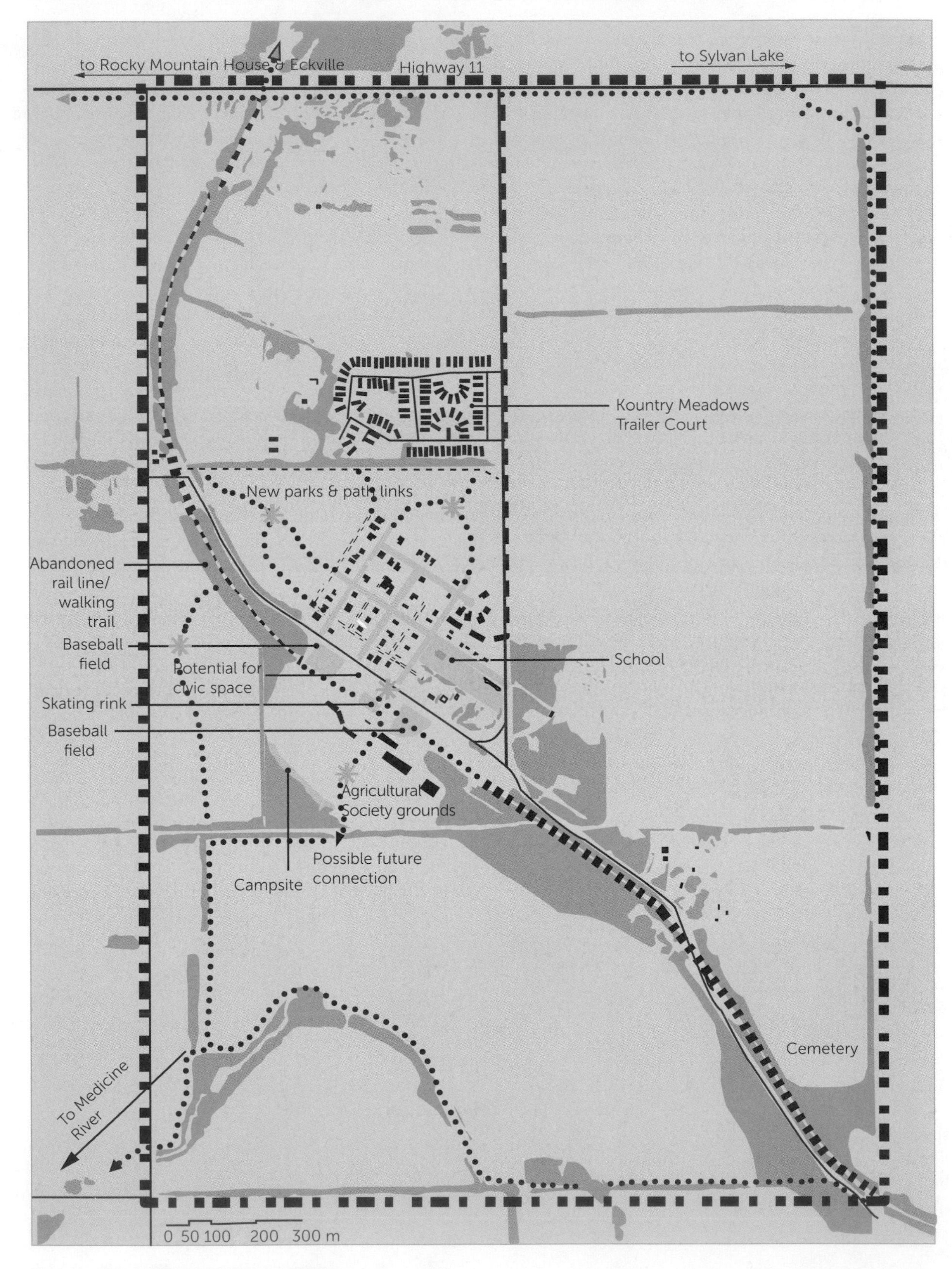

Figure 7.3.6 • Open-space concept.

proposed to facilitate a range of commercial, retail, and office uses together with residential uses, consistent with the small-town character in which these uses are blended. Hamlet Mixed Use land-use areas are contained within the town boundaries.

- There will be no front-yard setback.
- All buildings will be set on the front property line to form a continuous street wall.
- Side-yard setbacks are to be used for active uses such as food service patios, display areas, or public realm enhancements.
- Provision for services and deliveries will be at the rear yards, with appropriate screening to adjacent properties and public space.
- All buildings must have front entries. In the case of multiple-lot buildings, entries must be provided at one per 25 feet of lot frontage.

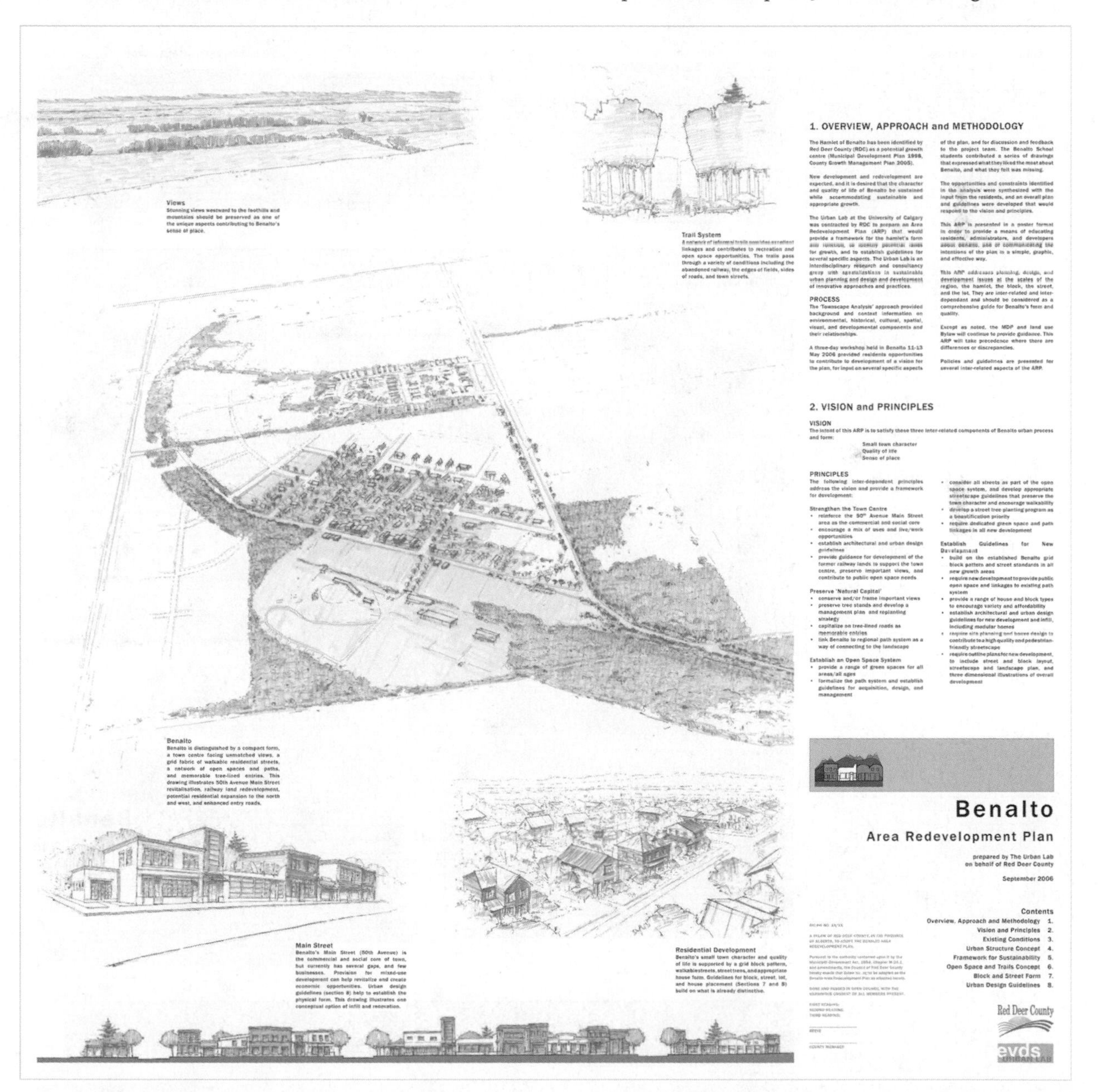

Figure 7.3.7 • ARP poster, side 1.

- Buildings on corners must have entries, windows, and an active street presence on the two public facades.
- The maximum height of all buildings will be 10 metres.
- A mix of uses within two-storey buildings is encouraged, ideally commercial or retail uses on the ground floor and residential or office uses above.

Documentation and Communication of the Plan

The final ARP was presented to County Council for adoption and to the public in an open house. The plan was widely accepted by council and the public. It was prepared as a two-sided poster, printed at various sizes—18" × 18" for wide distribution (residents, developers), and 30" × 30" for County administration. This format was believed to

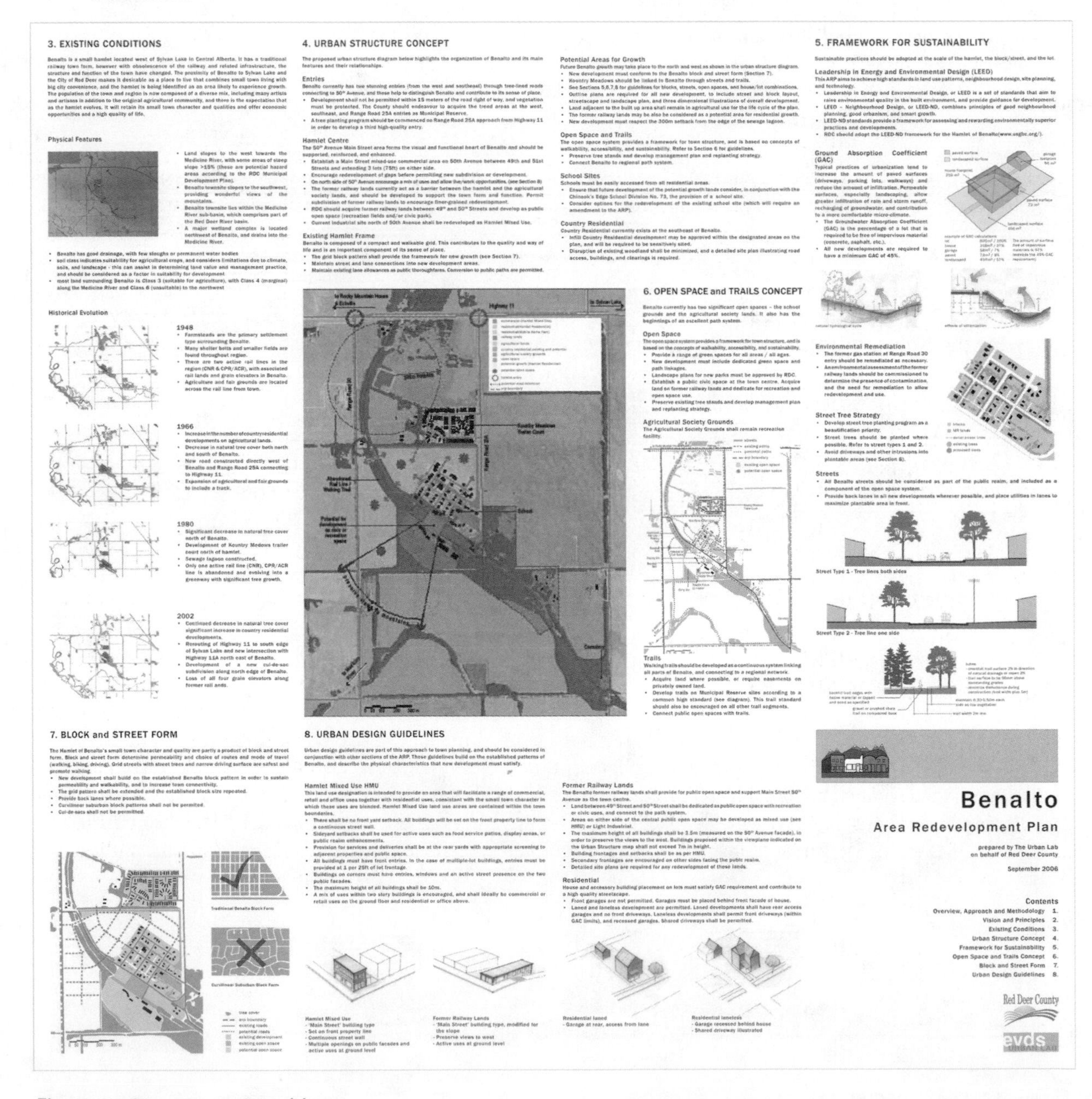

Figure 7.3.8 • ARP poster, side 2.

be more appropriate for a small place, more accessible to residents, and more effective for implementation by planners.

Conclusions

This project has had impacts on municipal planning practice in the RDC and elsewhere, as well as on the approach and methods utilized by the Urban Lab and taught in the Master of Planning Program at the University of Calgary. The project received a National Honour Award in the Canadian Society of Landscape Architects 2007 Professional Awards Program, and the innovative approach and highly graphic nature of the plan were commended by the panel of professional peers.

Benalto's Area Redevelopment Plan was approved by council as a statutory document in the fall of 2006 (Red Deer County 2012a), and it continues to be the guiding land-use and development document for the Hamlet. The County initiated some immediate implementation actions following approval, including planting 140 street trees. The plan has also been identified as a catalytic project in terms of mobilizing and focusing community engagement, and several community members have remained active in local initiatives, including a project to move Benalto's original railway station back from Taber, Alberta, where it had been relocated. The compact process and poster format of the Benalto plan has subsequently been used in several other projects in Red Deer County as well as in other regions in Alberta where former RDC planning staff are now working, since it provides a method and a document that are more accessible and appropriate to the scale of smaller communities.

The townscape analysis methodology has been utilized and refined by the Urban Lab in several other projects of various scales (Urban Lab 2013) and has proven valuable as a pre-design process for understanding a place, for educating those living and working within it, and for contributing to decisions that have a greater likelihood of leading to a higher-quality public realm and to a sense of place. It also constitutes the framework for several courses in the planning program at the University of Calgary.

Morphological analysis, when combined with spatial structure analysis, visual analysis, and an understanding of the environmental conditions and context, provides the comprehensive approach necessary to understanding places as multi-faceted and complex processes. This methodology is useful in helping to identify and anticipate change and to qualitatively analyze those changes. Planning can then be considered as an activity concerned with directing processes, and design can build upon that understanding. It is an action-oriented approach and focuses on continuity and identity and on the public realm as a receptacle for civic life and civic engagement.

The urban design approach proposed in this paper (and introduced in the courses) relies on the town or city as the most important source of information. It requires students to observe—to understand the pattern of activities that take place within it, to understand its visual qualities, and to understand how the urban form developed over time and how particular processes influenced the development of spatial qualities and built form. The approach uses two- and three-dimensional representation as a way of understanding places, as a way of communicating that information to others, as a way of developing design proposals, and as a way of planning and directing future development, which suggests that the ability to work in this way should be considered a necessary skill and competency.

This project presented a unique opportunity for the consideration of urban–rural interface issues, for the development of a sustainable regional plan, and for the design of a showpiece urban settlement according to sustainable planning principles. It also illustrated how landscape analysis, landscape planning, and landscape design can provide the framework for various scales of urban development, from the regional plan, to the town, to the neighbourhood and street. The project offered an innovative prototype for ARPs for hamlets in Red Deer County and serves as an ongoing model for other areas undergoing growth and experiencing rural–urban conflicts.

KEY TERMS

Bylaw (or by-law)
Design guidelines
Environmental analysis
Municipal Development Plan
Precedents
Public realm
Resilience
Secondary frontage
Subdivision
Walkability
Zoning Bylaws/Schedules

REFERENCES

Alberta Municipal Government Act [2000]. Chapter M-26. http://www.qp.alberta.ca/documents/acts/m26.pdf.

Forman, R.T.T. 1995. *Land Mosaics: The Ecology of Landscapes and Regions*. Cambridge: Cambridge University Press.

——. and M. Godron. 1986. *Landscape Ecology*. New York: John Wiley.

Hough, M. 1990. *Out of Place: Restoring Identity to the Regional Landscape*. New Haven, CT: Yale University Press.

Lynch, K. 1960. *The Image of the City*. Cambridge, MA: MIT Press.

McHarg, I. 1992. *Design with Nature*. New York: Natural History Press.

Moudon, A.V. 1997. "Urban morphology as an emerging interdisciplinary field." *Urban Morphology* 1: 3–10.

Red Deer County. 2006. "Council minutes October 24, 2006 Bylaw No. 2006/47." https://reddeercounty.civicweb.net/Documents/DocumentList.aspx?ID=355.

——. 2012a. "Benalto Area Redevelopment Plan." https://reddeercounty.civicweb.net/Documents/DocumentList.aspx?ID=784.

——. 2012b. "Municipal Development Plan." https://reddeercounty.civicweb.net/Documents/DocumentList.aspx?ID=10747.

Sandalack, B.A. 1998. "Continuity of history and form: The Canadian prairie town." Unpublished PhD thesis, Oxford Brookes University, Oxford, UK.

——. and A. Nicolai. 1998. *Urban Structure—Halifax: An Urban Design Approach*. Halifax: Technical University of Nova Scotia.

——. 2006. *The Calgary Project: Urban Form/Urban Life*. Calgary: University of Calgary Press.

Trancik, R. 1986. *Finding Lost Space: Theories of Urban Design*. New York: Van Nostrand Reinhold.

Urban Lab. 2013. "Project descriptions." Calgary: Faculty of Environmental Design, University of Calgary. http://www.ucalgary.ca/urbanlab.

CASE STUDY 7.4

Variations on Empire: Planning the US Embassy in Ottawa

Jason R. Burke

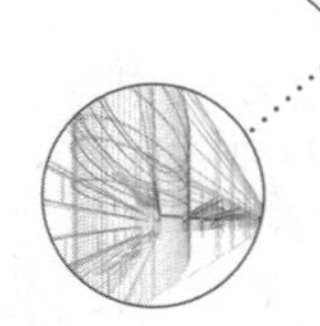

SUMMARY

Terrorism is an ancient phenomenon that in its present form is threatening the way of life for people in cities around the world. In order to address this threat, extensive security measures have been implemented, significantly altering urban environments. This chapter examines these alterations and questions the role of planning in addressing this predicament. The development of the US Embassy in Canada's capital, Ottawa, is used as a case study to demonstrate the impact of anti-terrorism security measures on the urban milieu and to critically examine how planning was involved in building and securitizing the embassy. Through archival research and interview data, it is demonstrated that the processes of planning have been co-opted in the name of security and that the embassy and its security measures have had a detrimental impact on the urban environment.

OUTLINE

- The development of the US Embassy in Canada's capital city is an example of the recent trend of securitization in cities, particularly the transition after 9/11.
- The process of choosing the site, security considerations involved in making the site decision, and consideration of neighbouring residents and businesses raise a number of issues in terms of balancing residents' rights and government agendas.
- In this case, the security concerns seemed to take precedence over all other planning considerations, prompting the question as to whether planning is always in the public interest.

Planning the Secure City: The US Embassy in Ottawa

On 19 February 2009, Barack Obama travelled to Ottawa on his first official international visit. When an American president visits Canada's capital, the downtown core is transformed. Buses are rerouted, streets are closed, and barriers proliferate throughout the public spaces of the city. During Obama's visit, the city was altered into a security fortress for three days, with thousands of makeshift barricades lining the streets, snipers in helicopters and on top of strategic buildings, and bulletproof glass encasing the front entrance to Parliament. Even though the city returned to "normal" following President Obama's departure, the relationship between **security** and urbanism is not limited to such events. Indeed, it has become increasingly integral to urban management and planning in many of the world's prominent cities.

In recent years, the threat of terrorism has significantly altered the way urban environments are planned and managed. Some authors have focused on the negative impact of increased security measures in cities (Coaffee 2004; Graham 2004a; Marcuse 2006; Savitch 2008), while others have demonstrated that security measures are being justified at the great expense of civil liberties (Gregory and Pred 2007; Weizman 2007). From barriers protecting high-risk sites to surveillance cameras recording public activities, many governments and security professionals have come to assume that urban space can be altered to defend against a terrorist attack. Regardless of whether or not these measures are indeed creating safer urban environments, planners have a key role in negotiating between the **public interest** and the demand for increased urban safety and security. Therefore, addressing the need to build more secure cities while maintaining civil liberties is a key dilemma facing urban policy-makers today (Coaffee 2005).

In recent history, government buildings, and especially embassies, have been subject to terrorist attacks. As a result, enhanced security measures surrounding such buildings have proliferated in many countries. The US Embassy on Ottawa's Sussex Drive represents one of the major sites among Canadian cities to have undergone a profound transformation as a result of the threat of terrorism (Figure 7.4.2 and Figure 7.4.3).

Opened in 1999, the US Chancery is a symbolic building, and it stands out as a target among the smaller, diverse building stock of the neighbouring Byward Market. Upon its completion, the embassy was met with tremendous criticism from local residents, who claimed that the building did not fit in with the surrounding milieu (Figure 7.4.4). After the terrorist attacks of 11 September 2001, new obtrusive security measures were implemented surrounding the embassy, furthering the distaste for the embassy and its location in the heart of Ottawa's lively Byward Market neighbourhood. As a prime example of contemporary securitization of urban environments, this case study adds a much-needed "security" dimension to Canadian planning by demonstrating how security considerations have co-opted the **democratic planning** process and how security logic has permeated the field of planning.

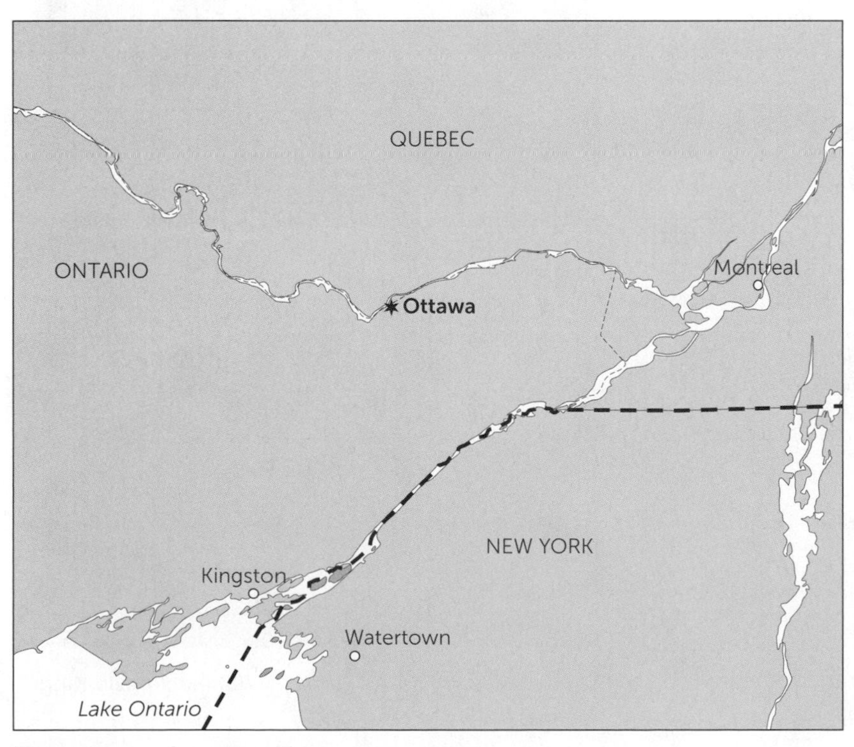

Figure 7.4.1 • Location Map.

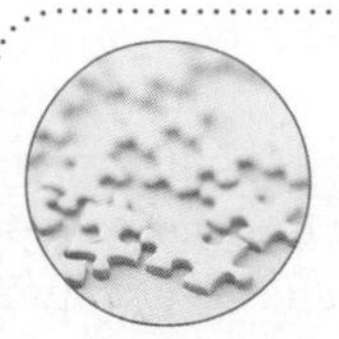

KEY ISSUE

Urban Securitization

Security has become a key concern for Canadian cities. Fear of terrorism and violence from protests has led to increasing forms of surveillance and social control in most Canadian cities.

In recent years, there has been a substantial increase in physical security around ports, government buildings, and sport venues as well as an increase in video surveillance in public spaces and throughout public transit systems. As contemporary forms of urban securitization become increasingly integral to planning, planners will be faced with a crucial dilemma: How can cities become more secure without sacrificing the fundamental qualities of openness and privacy that characterize most cities? As this case shows, the power of security rhetoric tends to trump all other planning considerations, leading to the possibility that planners may have to compromise their communicative role as a negotiator between the state, the public, and the private sector in order for security precautions to ensue.

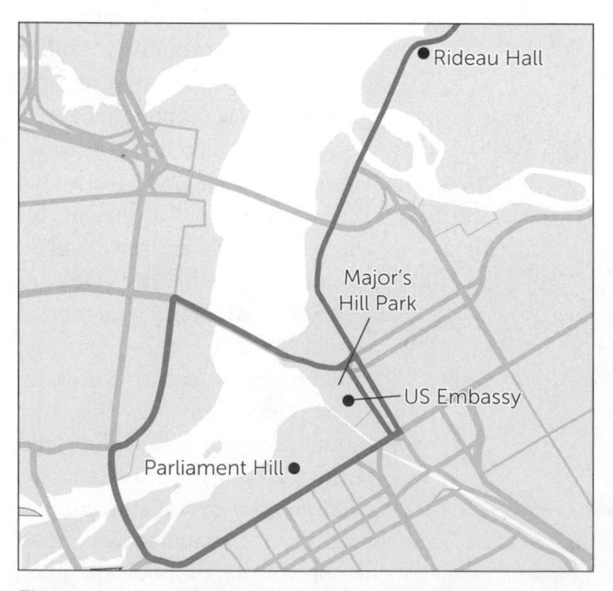

Figure 7.4.2 • The ceremonial parade route in Ottawa; Confederation Boulevard.

Based on archival records and on media sources, this chapter investigates the planning history of the US Embassy, paying particular attention to how fluctuating security considerations and complicated governance structures dictated the planning process. This case study will be used to consider two major questions. First, how did the US Embassy come to be built in this location? In other words, what was the role of planning in deciding to place the embassy on its vulnerable location? Second, how were post-9/11 security measures at the embassy site legitimized and implemented, and what was the role of planners in these developments? By exploring these questions, it is possible to understand how planning has been co-opted by security practices.

Planning and Security

Planning is clearly part of the larger process of securitization. Historically, cities were planned with fortifications

Figure 7.4.3 • The US Embassy: (a) Mackenzie King facade; (b) Sussex facade.

Figure 7.4.4 • The surrounding urban environment.

as a way of protecting the city from outside aggressors and to maintain control and surveillance over populations within the city (Mumford 1961). Planning was also used to reconfigure urban environments in line with militaristic and security principles. A prime example of this is Baron Georges-Eugène Haussmann's nineteenth-century redesign of Paris. To respond to insurrections in the more revolutionary neighbourhoods of the city, parts of the city were destroyed to make way for broad avenues that could facilitate military manoeuvrings in the urban landscape (Harvey 2003; Misselwitz and Weizman 2003; Scott 1998).

By the 1970s, a technique of planning known as "crime prevention through environmental design" (CPTED) became a popular tool for addressing "undesirable" behaviour (Cohen 1985; Jeffrey 1971; Newman 1972). Planners and policy-makers assumed that understanding principles of defence could lead to better planning and safer spaces. This method, developed mostly by criminologists and sociologists, emphasized the idea of a spatial fix to crime and ignored many larger social problems. Nevertheless, CPTED principles have since become an essential element in many planning or redevelopment projects.

Many of these methods of controlling and ordering urban environments have been reworked and reconfigured to suit contemporary cities. For example, the urban fortress method, aptly labelled "citadelization" by Marcuse (2004), has become a common practice for isolating and protecting high-profile spaces and buildings from security threats. Coaffee's (2004) detailed account of how the central city of London has been securitized in response to years of Irish Republican Army bombings presents one of the clearest and most dramatic examples of this form of planning. Obstructions such as **blast walls**, **Jersey barriers**, and bollards have become permanent fixtures around government buildings, embassies, and commercial establishments as part of a range of anti-terrorism security measures. These measures rework urban environments in a way that restricts access to public spaces and hinders freedom of movement (Boddy 2008; Coaffee 2004; Graham 2004b; Marcuse 2004, 2006; Savitch 2008). Sorkin (2004, 258) describes their effects on the urban environment as "the recoding of the landscape in the language of the bomber." Many authors have written on the subject of vulnerabilities and why cities are at risk of terrorism (Beall 2006; Savitch 2008; Molotch and McClain 2003; Dezzani and Lakshmanan 2003; Fraser and Mabee 2004; Gray and Wyly 2007).

Essentially, the use of spatially restrictive security measures can represent a form of **exclusionary zoning** whereby the nature and use of public space is altered and prescribed. Security concerns have perpetuated the increasing privatization of public space (Németh and Hollander 2010), while in other cases, such as gated communities, exclusionary planning is premised on security principles (Blakely and Snyder 1997; Low 2003). For the spaces of the city that remain nominally public, surveillance technology and other forms of public surveillance have become the dominant form of security (Lyon 2004). In other words, there is a long-established link between security, spatial control, and planning, and the case of the US Embassy in Ottawa exemplifies this problematic in the Canadian context.

Description of the US Embassy

The embassy is located on the western edge of the Byward Market, Ottawa's most vibrant neighbourhood. Immediately surrounding the embassy are buildings with major historical and cultural value and a large park with a superb view of Parliament Hill. Adjacent to the building are several layers of security. Surrounding the building are low-lying shrubs and a tall fence. There are security cameras and two visible security booths: one for the pedestrian entrance (located in the centre of the building on Sussex Drive) and one for the automobile entrance (located beside the building on Sussex Drive). There are large retractable barriers that move up from the ground to prevent any vehicle from entering the garage without first being thoroughly inspected. A sidewalk surrounds the embassy and is lined with bollards, street lamps, and trees. Beyond this, one lane of traffic is closed off, with Jersey barriers surrounding the entire building (Figure 7.4.5). The lane closures and Jersey barriers were only implemented after 9/11 and remained in place until 2012 when the ambassador requested that the Jersey barriers along Sussex Drive be replaced by bollards. Finally, the end of Clarence Street, where it intersects with Sussex Drive, was reconfigured to prevent a vehicle from driving directly toward the embassy.

Planning Background of the Embassy

The planning of the US Embassy in Ottawa began in 1972 when the US State Department decided it needed a larger site (National Capital Commission 1986). Its former embassy, opposite Canada's Parliament buildings, was too small and became a security concern, given its full frontage onto a major arterial street in the downtown core. In 1985, the US was offered a prominent site on Sussex Drive, part of the capital's ceremonial parade route. Although this site would eventually become the embassy's location, it was initially rejected outright because of security concerns and size constraints. At the time, this was a bonus for the Sussex/Byward Market redevelopment plan, since it was believed that an embassy built adjacent to the market would have serious repercussions for the marketability of Sussex Drive (National Capital Commission 1985). Furthermore, an embassy on this location would create a wall between the market and Major's Hill Park, a site with impressive vistas of the capital.

Figure 7.4.5 • Security measures for the embassy.

The National Capital Commission (NCC), the planning authority for the federal government's properties in Ottawa-Gatineau, decided that a large empty site in Rockcliffe Park called the Mile Circle would be an ideal location for the US Embassy along with several other embassies. This site's proximity to an exclusive residential area and the residences of the prime minister, the governor general, and the US ambassador made it particularly prestigious. Consultations were conducted with the Royal Canadian Mounted Police (RCMP, the federal police), embassy representatives, the mayor of Ottawa, the Council of Rockcliffe Park, one member of Parliament, the Rockcliffe Flying Club, and the Save the Mile Circle group. During these consultations, the RCMP, the embassy representatives, and the mayor of Ottawa simply stated their requirements for an embassy enclave in the Mile Circle, while the other stakeholders strongly opposed the project on the grounds of accessibility, traffic, environmental preservation, and personal security (National Capital Commission 1986). After an environmental impact assessment, the Mile Circle project was postponed. To this day, the Mile Circle site remains a vacant and underused site.

In 1987, a second round of site selection evaluations for the future location of the US Embassy began (National Capital Commission 1988). More than 30 sites were reviewed, and once again the Sussex Drive location was rejected because of increasing security concerns. The purpose of this site selection process was to find an optimal site that would accommodate the size of the embassy, meet its security needs, conserve the surrounding environment, and minimize social disruption (National Capital Commission 1988). In the first public consultation, a poorly advertised open house, many attendees feared that the embassy would look like a bunker as a result of what they perceived to be exaggerated security concerns (National Capital Commission 1988). The central concern emanating out of the public consultation was that the embassy should not be constructed as though

Table 7.4.1 • Site and Building Requirements for the US Embassy, 1988

Site Requirements

1. Accessible by car and public transport
2. Away from tall buildings
3. Distant from adjacent properties to ensure the security of the surrounding area
4. Little impact on the community, including traffic and noise
5. Located in the core of city and compatible with existing land uses
6. The potential to become an embassy enclave
7. Possibility of expansion in the long term
8. Respect for the safety of the surrounding community
9. Aesthetically pleasing view to and from the site
10. Away from busy crowded areas
11. Less than 300 residents, no hospital or school within 300 metres of the site
12. Not near large shopping malls where hostiles can gather undetected
13. Near high-quality area in a prestigious setting

Building Requirements

1. Prestige, image, and visibility
2. Supportive of capital objectives
3. Security buffer zone surrounding the building in the form of blast walls
4. Narrow windows; glass no more than 15 per cent of building
5. Blast-resistant material used in building
6. Should not devalue surrounding properties
7. Underground parking not directly under building
8. At least a 100-metre setback from the street

Note: These building and site requirements are a combination of requirements specified in both the 1986 and 1988 site selection processes (National Capital Commission 1986, 1988).

Ottawa were a high-risk location. The US had specific site requirements and building requirements for its embassy (see Table 7.4.1).

In the end, a site was finally chosen near the previous Mile Circle site. This site met all of the US State Department's requirements but was opposed by the powerful and wealthy Rockcliffe Park residents. They believed the construction of the embassy in their neighbourhood would be a security risk and would cause increased congestion. The project was eventually postponed. While the rationale behind the final decision to build on the Sussex Drive site is still unclear, Boddy (2008) suggests that the push actually came from a deal between US and Canadian politicians and not the State Department. In fact, there was a reciprocity agreement between the Canadian and US Governments to ensure each country's embassy was guaranteed unobstructed viewing lines to the most important government institutions (National Capital Commission 1993). Even though the US had rejected the Sussex Drive site numerous times, it was finally accepted after almost 10 years, and construction began in 1998.

Compromising Site and Building Requirements for the Sussex Drive Site

Looking at the requirements laid out in Table 7.4.1, it is clear that some of these requirements were sacrificed or compromised in the decision to build on Sussex Drive. One of the most glaring contradictions is the requirement that the embassy be built in the core of the city but also be away from busy places. Of the 13 site requirements, eight were compromised. First, the site fails to ensure the safety of the surrounding area by maintaining a safe distance from surrounding properties. If there were any attack on the embassy, many nearby heritage buildings would likely be damaged. In addition, the closure of one lane of traffic as a security precaution has added to traffic congestion, and the embassy has become a key location for political demonstrations or bomb threats, which results in the closing of the street and a loss of business for merchants along the street. Therefore, building the embassy in this location has affected the economic viability and safety of the surrounding community.

Given the spatial limitations of the site, the embassy will be unable to expand in the long term, and there is no chance of the area becoming an embassy enclave. The embassy was also built in the busiest and most crowded area of the city, something the US State Department initially tried to avoid. Major's Hill Park is the site of numerous festivals, and the Byward Market attracts thousands of people a day to its many restaurants, shops, and clubs. More than 300 people live within a 300-metre radius of the site, residing either above the shops on Sussex Drive or in the proximate condominium buildings. There is also a fashion design school directly across the street from the embassy and a hospital a few blocks away. All these elements contradict the State Department's initial priorities and requirements for the embassy. Finally, the US mission stipulated that the embassy should be built away from large shopping malls where hostiles can gather undetected. Again, this priority was compromised. The embassy was built just two blocks away from Ottawa's largest shopping mall, the Rideau Centre, which has an entrance directly onto Sussex Drive (Figure 7.4.6).

Conversely, the building's features do conform with many of the initial building requirements. It is in a prestigious location, and the building is extremely visible. It supports the objective of the NCC to reinforce the ceremonial parade route of the capital with monumental embassies. The building was constructed with blast-resistant materials, protecting it from a bomb attack. There is no evidence that the surrounding buildings have been devalued in any way. And finally, the parking area is not directly under the building.

Of these eight requirements, only two were compromised and probably for the overall benefit of the urban environment. First, although a security buffer zone was created around the building, it did not come in the form of a blast wall, since this feature was integrated into the structure itself. Second, there is no 100-metre setback from the sidewalk on the Sussex Drive side of the building. According to US Embassy building requirements, all US embassies must have at least a 100-metre setback. Given the narrowness of the site, however, this setback was impossible. Following the events of 11 September 2001, one lane of traffic was closed and barricaded around the embassy, ensuring a larger setback for the building than was initially possible.

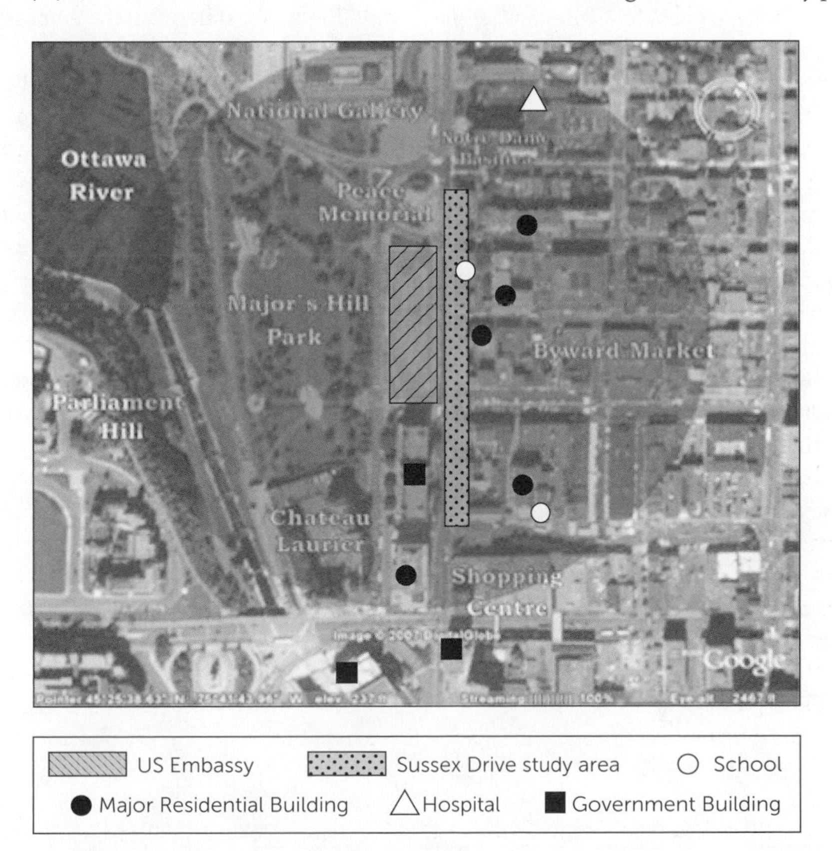

Figure 7.4.6 • Map of security requirements.

Increased Security and Public Consultation

The lane closures were implemented at the request of the RCMP, which is mandated to ensure the safety of diplomatic missions in Canada. In a hasty response, old, decrepit Jersey barriers were temporarily placed around the embassy until a more suitable security measure could be found. In 2002, the City of Ottawa commissioned Dillon Consulting to carry out an environmental assessment of these security measures, and the findings were made public in 2003. Essentially, the justification for the lane closures and the Jersey barriers was to increase the distance between the embassy and the street in the case of a car bomb attack and to render the embassy a less attractive target. The measures were implemented with removable materials so the street could one day be restored.

The Dillon Consulting assessment began with a stakeholders meeting between the US Embassy staff, the Ottawa police, the NCC, the City of Ottawa, and the RCMP. The notes from this meeting are restricted and therefore not accessible to the public. In addition, an open house was organized and attended by 40 people. Citizens were presented with three options for the area, two of which had already been rejected before the consultation. The first option was to do nothing, and the second was to increase security without encroaching onto the street by relocating the embassy or by building a blast wall around it. Relocating the embassy would have been ideal, but the cost of such an undertaking was prohibitive. Finally, the blast wall would be costly, would detract from the aesthetics of the area, and would not be removable.

The third and only realistic option was a set of security measures that would encroach onto the street (see Table 7.4.2). The City of Ottawa suggested closing one lane and replacing it with an elevated sidewalk with hardened street furniture. This idea would have maintained the space for public use and provided an aesthetically secure alternative to the others being proposed, but it was never considered.

The citizens who participated in the public consultations for the environmental assessment had their voices heard but their opinions ignored. Public comments were collected on paper and presented in the resulting publication. From these comments it is clear that many participants were disappointed by the poor advertisement for the event and upset by the limited solutions being proposed before the public was even consulted. Some participants also believed that the extra security would have a negligible impact, since a few extra feet of asphalt would do little to deter a terrorist.

The doubtful public also felt that their rights as taxpayers were being ignored as a result of the lane closure and the loss of public space. One person made a strong statement about the barriers: "If the barriers stay, it will be an encroachment of public property by a foreign power and a precedent for other embassies" (comment in City of Ottawa 2003). Many believed that the embassy had no right to take up public space and that if there was a security concern, it should be relocated to another part of the city. The RCMP stated that the embassy had been built on this site prior to 9/11 and that the "current reality must be addressed" (City of Ottawa 2003). Essentially, the RCMP was suggesting that it was not the embassy's fault that security had become a more serious issue. Back in the 1980s, however, the Sussex Drive location was never an option because of security concerns. Furthermore, construction proceeded after the bombings of two US embassies in Africa in 1998. Surely embassy security was an issue prior to 9/11, and it was not simply that event that made the embassy's site more vulnerable.

As a result of the barriers, the ceremonial parade route of the capital lost some of its grandeur. Furthermore, the closure of one lane on each side of the embassy has resulted in a lengthier commute between Gatineau and Ottawa and more cars idling in front of the embassy, which

Table 7.4.2 • Evaluation of Street Encroaching Security Measures

Security Measure	Addresses Security	Removable	Aesthetic	Cost
Jersey barriers	Yes	Yes	No	Low
Concrete planters	Yes	Yes	Yes	Med.
Bollards	Yes	No	Mod.	Med.
Road closure	Yes	Yes	Mod.	High

must be a security concern. In the end, most of the doubts and disagreements expressed during the consultations were ignored, suggesting that public participation was simply a formality in the planning process; the decision to close one lane of traffic with Jersey barriers had already been taken. However, in March 2010, the US ambassador to Canada announced that the unsightly Jersey barriers would be replaced by an elevated sidewalk and "well-designed" bollards, an idea similar to the proposal made by the City of Ottawa back in 2003 (CBC News 2010).

The Murky World of Security Planning

Given the rejection of the Sussex Drive site in the 1980s, it remains unclear as to why the embassy was eventually built on this seemingly insecure site. The site selection process continued into the early 1990s until, finally, the US government reconsidered its security requirements and decided that the Sussex Drive location would be a suitable location for its new Chancery (National Capital Commission 1993). The federal government rarely enters into analyses of urban planning. However, in this case it played a central role in shaping the development and securitization of this site and the surrounding area. The NCC approved the location of the embassy and then spearheaded the planning, often omitting public consultation and dictating what the building should look like. Since it was on federal land, the municipality had little say in how the space was planned and designed. In other words, the provincial Planning Act and the Official Plan for the City of Ottawa had no jurisdiction over what happened on this parcel of land despite the clear repercussions a highly securitized building would have on the surrounding urban environment.

A major task in planning is to evaluate the impact of a particular project on its surrounding environment. The presence of the embassy on this site posed two problems. First, it did not reinforce the commercial nature of the Byward Market area. Second, it presented a great risk to the merchants and residents of the Byward Market; should the embassy ever be targeted by a terrorist attack, the merchants and the historic buildings nearby will likely be harmed, whereas the embassy building, with its advanced security precautions, would likely withstand the attack. In other words, the planning process had failed to recognize the possible negative repercussions of the embassy and its security measures on the surrounding neighbourhood.

This case also depicts the contradictions that exist within the world of planning. Initially, security was of paramount concern for the embassy, but in the end, the US State Department was willing to circumvent several security concerns for the embassy to be built on this prestigious site. The NCC's Advisory Committee of Design believed that more "sophisticated" security measures could be integrated into the design of the building that would allow it to "sit in an urban environment" (National Capital Commission 1994). After 9/11, however, security suddenly became a major concern once again, requiring the alteration of the urban environment. These alterations received little public scrutiny, because the security imperative behind the lane closures trumped all other considerations. In other words, security rationale became a key force for implementing even the most controversial security measures without any public recourse. From this scenario, it is clear that the planning process was co-opted by security decisions made prior to the public consultation.

Conclusion

Security has always played an integral role in the development of cities. It is as much about protecting people as it is about protecting a system of operation. One of the bleakest examples of urban securitization can be seen around American embassies throughout the world, and the heavily fortified US Embassy in Ottawa is no exception. It stands out as one of the most securitized spaces in Canada and lies in stark contrast to the Canadian Parliament buildings, which still remain relatively accessible.

While there was controversy surrounding the placement and the securitization of the embassy, the US State Department disregarded several site and building requirements in deciding that it would be built on a prestigious site. In the end, the embassy detracted from the character of the neighbourhood and added an element of insecurity that did not previously exist. In this case, security considerations trumped a more democratic and open process of planning. As contemporary forms of urban securitization become increasingly integral to planning, planners will have to either compromise their communicative role as a negotiator between the state, the public, and the private sector or find new avenues to defend the public interest. A proper assessment of the Sussex Drive site and more earnest consultations with surrounding residents and businesses could have resulted in a more considerate use of this space and a more appropriate location for the embassy.

As cities continue to securitize high-profile buildings and public spaces, planners have a crucial role to play in mediating between the desire for more security and the loss of civil liberties. Today, the urban planning process is likely to employ democratic strategies such as participation and consultation. These strategies are currently being displaced by the growing hegemony of security logic, which prioritizes security over all other considerations. While a more democratic security planning process may not guarantee more considerate and accountable security plans, more community involvement could serve to highlight that not all forms of security are useful or required. The potential for better urbanism lies in accepting that the best "security" might be a result of increased democracy.

KEY TERMS

Blast walls
Democratic planning
Exclusionary zoning
Jersey barriers
Public interest
Security

REFERENCES

Beall, J. 2006. "Policy arena: Cities, terrorism and development." *Journal of International Development* 18: 105–20.

Blakely, E.J., and M.G. Snyder. 1997. *Fortress America: Gated Communities in the United States*. Washington: Brookings Institution Press.

Boddy, T. 2008. "Architecture emblematic: Hardened sited and softened symbols." In M. Sorkin, ed., *Indefensible Space: The Architecture of the National Insecurity State*, 277–304. New York: Routledge.

CBC News. 2010. "US Embassy barriers to be replaced." 8 March. http://www.cbc.ca/canada/ottawa/story/2010/03/08/embassy-barriers.html.

City of Ottawa. 2003. *Sussex Drive and Mackenzie Avenue Traffic and Safety Measures Environmental Assessment Study, Final Report*. Ottawa: Dillon Consulting.

Coaffee, J. 2004. "Rings of steel, rings of concrete and rings of confidence: Designing out terrorism in central London pre and post September 11th." *International Journal of Urban and Regional Research* 28(1): 201–11.

——. 2005. "Urban renaissance in the age of terrorism: Revanchism, automated social control or the end of reflection." *International Journal of Urban and Regional Research* 29(2): 447–55.

Cohen, S. 1985. *Visions of Social Control*. Cambridge: Polity Press.

Dezzani, R.J., and T.R. Lakshmanan. 2003. "Recreating secure spaces." In S.L. Cutter, D.B. Richardson, and T.J. Wilbanks, eds, *The Geographical Dimensions of Terrorism*, 169–78. New York: Routledge.

Fraser, E.D.G., and W. Mabee. 2004. "Identifying the secure city: Research to establish a preliminary framework." *Canadian Journal of Urban Research* 13: 89–99.

Graham, S. 2004a. "Introduction: Cities, warfare, and states of emergency." In S. Graham, ed., *Cities, War, and Terrorism: Towards an Urban Geopolitics*, 1–25. Oxford: Blackwell.

——. 2004b. "Postmortem city: Towards an urban geopolitics." *City* 8(2): 165–86.

Gray, M., and E. Wyly. 2007. "The terror city hypothesis." In D. Gregory and A. Pred, eds, *Violent Geographies: Fear, Terror, and Political Violence*, 329–48. New York: Routledge.

Gregory, D., and A. Pred. 2007. "Introduction." In D. Gregory and A. Pred, eds, *Violent Geographies: Fear, Terror and Political Violence*, 1–6. New York: Routledge.

Harvey, D. 2003. *Paris: Capital of Modernity*. New York: Routledge.

Jeffrey, C.R. 1971. *Crime Prevention through Environmental Design*. Beverly Hills, CA: Sage.

Low, S. 2003. *Behind the Gates: Life, Security, and the Pursuit of Happiness in Fortress America*. New York: Routledge.

Lyon, D. 2004. "Technology vs. 'terrorism': Circuits of city surveillance since September 11, 2001." In S. Graham, ed., *Cities, War, and Terrorism: Towards an Urban Geopolitics*, 297–311. Oxford: Blackwell.

Marcuse, P. 2004. "The 'war on terrorism' and life in cities after September 11, 2001." In S. Graham, ed., *Cities, War, and Terrorism: Towards an Urban Geopolitics*. Oxford: Blackwell.

——. 2006. "Security or safety in cities? The threat of terrorism after 9/11." *International Journal of Urban and Regional Research* 30(4): 919–29.

Misselwitz, P., and E. Weizman. 2003. "Military operations as urban planning." In A. Franke, ed., *Territories*, 272–85. Berlin: KW Institute of Contemporary Art.

Molotch, H., and N. McClain. 2003. "Dealing with urban terror: Heritages of control, varieties of intervention, strategies of research." *International Journal of Urban and Regional Research* 27(3): 679–98.

Mumford, L. 1961. *The City in History: Its Origins, Its Transformations and Its Prospects*. New York: Harcourt, Brace and World.

National Capital Commission. 1985. *Sussex/Market Redevelopment Plan—Background Brief*. Ottawa: McKellar Architects.

——. 1986. *The Mile Circle—Site Analysis and Planning Study*. Ottawa: Project Planning Canada Ltd.

——. 1988. *US Embassy Site Selection: Initial Environmental Evaluation, Final Report*. Ottawa: Landplan Collaborative Ltd.

——. 1993. *Sussex-Mackenzie North Development: Initial Environmental Evaluation (IEE) Draft Report*. Ottawa: National Capital Commission

——. 1994. *Comments of the Advisory Committee on Design—Sussex/Mackenzie North U.S. Chancery*. 12 May. Ottawa: National Capital Commission.

Németh, J., and J. Hollander. 2010. "Security zones and New York City's shrinking public space." *International Journal of Urban and Regional Research* 34(1): 20–34.

Newman, O. 1972. *Defensible Space: Crime Prevention through Urban Design*. New York: MacMillan.

Savitch, H.V. 2008. *Cities in a Time of Terror: Space, Territory and Local Resilience*. Armonk, NY: M.E. Sharpe Inc.

Scott, J. 1998. *Seeing Like a State: How Certain Schemes to Improve the Human Condition Have Failed*. New Haven, CT: Yale University Press.

Sorkin, M. 2004. "Urban warfare: A tour of the battlefield." In S. Graham, ed., *Cities, War, and Terrorism: Towards an Urban Geopolitics*, 251–62. Oxford: Blackwell.

Weizman, E. 2007. *Hollow Land: Israel's Architecture of Occupation*. London: Verso.

CASE STUDY 7.5

The Toronto Avenues and Mid-Rise Buildings Study

Cal Brook and Matt Reid

SUMMARY

As the city of Toronto's population continues to increase over the next 20 years, a significant amount of this growth will be accommodated through mid-rise infill buildings along the city's Avenues—major streets where reurbanization is anticipated. Between 5 and 12 storeys, **mid-rise buildings** offer a significant concentration of new employment and residential uses at a scale that supports attractive, pedestrian-friendly streets.

The Toronto Avenues and Mid-Rise Buildings Study (Brook McIlroy Planning + Urban Design, ERA Architects, Quadrangle Architects Limited, and Urban Marketing Collaborative 2010) analyzed the existing conditions along all 162 kilometres of the Avenues, including **right-of-way** width, lot width and depth, existing and adjacent uses, and transit accessibility. Through four public consultation sessions and smaller focus groups with members of the development community, the consultant team prepared 36 Performance Standards—measurable design criteria (i.e., height, massing, streetscape design) to ensure that new buildings on the Avenues support healthy, liveable, and vibrant main streets.

The Performance Standards were adopted by council in July 2010, with five key standards incorporated into the new City-wide zoning bylaw in August 2010. Since the adoption of the study, there has been a surge of development along the Avenues, with local architects interpreting the Performance Standards to create unique, attractive, and feasible mid-rise buildings.

The Toronto Avenues and Mid-Rise Buildings Study was inspired by the mid-rise, **pedestrian-oriented** streetscapes found in great European cities. As other metropolitan areas in Canada face unprecedented population growth, they too should look to mid-rise buildings and a process similar to the Toronto Avenues and Mid-Rise Buildings Study to ensure sensitive, context-specific intensification.

OUTLINE

- The development of Performance Standards for mid-rise buildings in the City of Toronto has been instrumental in encouraging medium-density development along the Avenues, which were identified in the 2006 Official Plan.
- Mid-rise buildings can be conceived as an important building type for infill development on major city streets, and their design elements can harmonize with the existing built form in the community.
- Following the Toronto Avenues and Mid-Rise Buildings Study, City Council adopted the Performance Standards, incorporating five key standards into the new City-wide zoning **bylaw**. This has resulted in new development along the Avenues, major city streets designated for future growth in Toronto's Official Plan.

Introduction

The population of the city of Toronto is expected to reach 3.08 million by 2031, representing a growth in population of approximately 500,000. Comprehensive strategies are required to direct the location and form of this growth in order to mitigate the impacts on existing stable neighbourhoods. The City's Official Plan (2006), which determines how land in the municipality should be used, directs new growth toward intensification areas, including the Downtown and Central Waterfront, Centres, Employment Districts, and the subject of this case study—the Avenues. Avenues are defined in the Official Plan as:

> Important corridors along major streets where reurbanization is anticipated and encouraged to create new housing and job opportunities while improving the pedestrian environment, the look of the street, shopping opportunities and transit service for community residents.

Figure 7.5.1 • Location Map.

This chapter describes the process of developing a vision and implementation plan for the Avenues, including the inventory, analysis, and community consultation used to develop Performance Standards.

As mixed-use corridors, the Avenues function as the social, economic, and cultural spine of communities and are intimately linked to the identity and vitality of the neighbourhoods that surround them. Despite this important role in the city, many Avenues are characterized by one- or two-storey commercial buildings, vacant or underutilized lots, and surface parking areas, making them prime candidates for intensification (a level of growth that is greater than what exists at present). As the Avenues intensify, it is important that new development occurs in a form that will accommodate a significant concentration of new employment and residential uses while ensuring pedestrian-friendly streets and minimizing adverse impacts on adjacent residential neighbourhoods. This can most easily be achieved through a mid-rise building form as expressed in the City's vision for the Avenues:

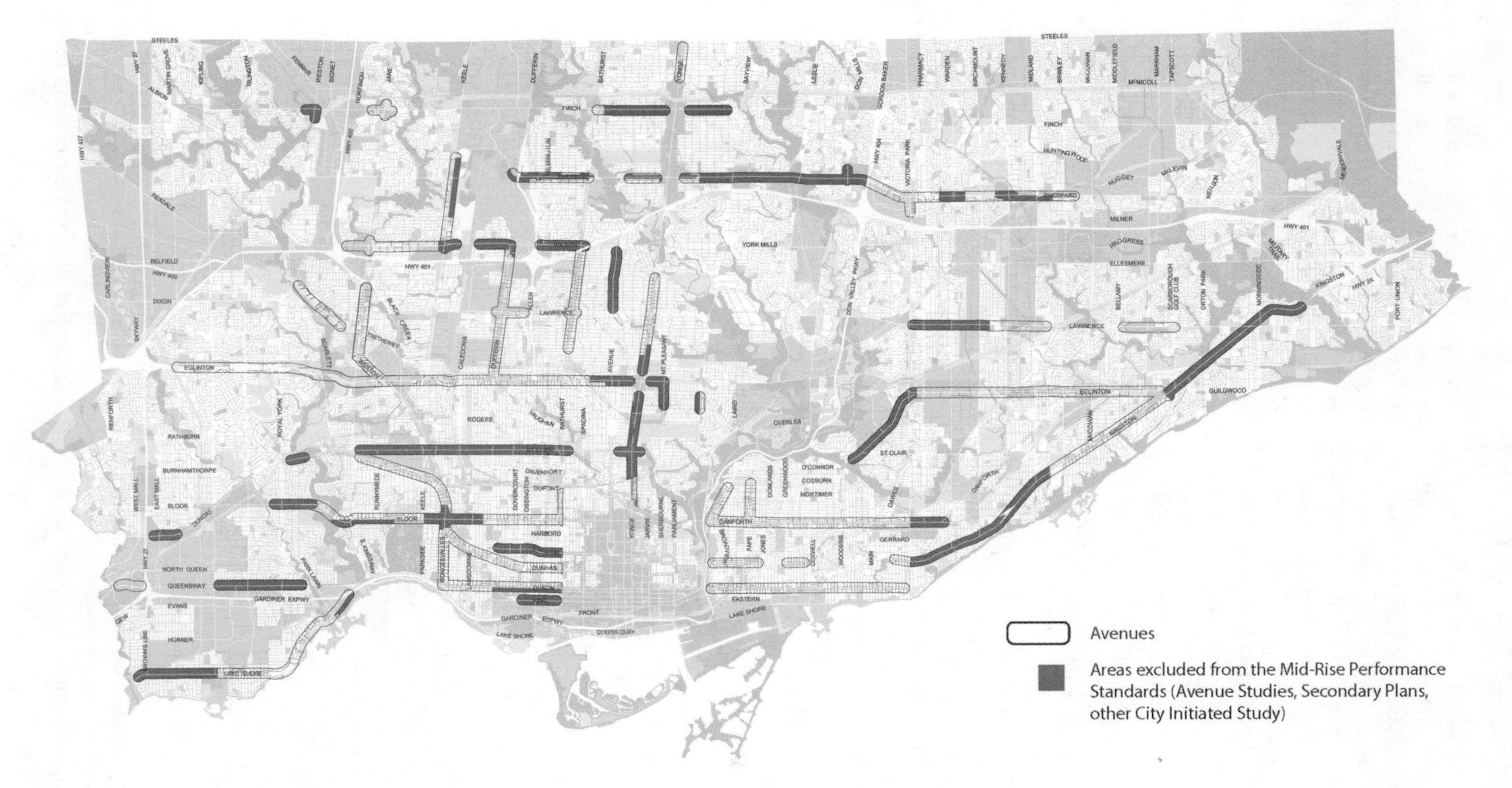

Figure 7.5.2 • Schedule 2: Urban structure map.

Source: Cal Brook and Matt Reid.

> The Avenues vision calls for beautiful tree-lined streets and sun-lit sidewalks, framed by carefully articulated mid-rise buildings providing a multiplicity of retail and community uses at the sidewalk level, with residential and commercial units above. As better transit service is incrementally introduced, and the population increases, the Avenues will be re-energized, supporting improved levels of commercial, retail and community services. Combined with investments in the streetscape and public realm, the setting for a vibrant community life will emerge.

Mid-rise buildings are an "in-between" scale of building. Generally ranging between 5 and 12 storeys, mid-rise buildings accommodate significant useable space in a form that is tall enough for a city feel but low enough to maintain sunlight on the sidewalk and views to the sky. Mid-rise buildings support a comfortable and attractive pedestrian environment through carefully designed ground-floor uses characterized by large windows, attractive **facades**, and active uses (e.g., restaurants, cafés, stores).

The Avenues amount to approximately 324 kilometres of property frontage, about 200 kilometres of which has the characteristics (i.e., lot width and depth) to accommodate mid-rise buildings. If mid-rise development were to occur on half of these properties by 2031, the Avenues would accommodate 250,000 of the city's anticipated new residents. Mid-rise development of the Avenues therefore has the potential to address a significant portion of the city's anticipated growth needs over the next 20 years.

Following the adoption of the City's Official Plan in 2006, the development community was slow to support the mid-rise building form on the Avenues. This was primarily due to an imbalance between the time and capital investment required to rezone properties on the Avenues during the application and approvals process and the financial return on a lower building with fewer units. The Toronto Avenues and Mid-Rise Buildings Study recommends the removal of many of the barriers to mid-rise development by translating the broader policies of the Official Plan into clear, implementable Performance Standards that illustrate the City's expectation for mid-rise buildings on the Avenues. The Performance Standards provide clear, simple, and straightforward design directions for new buildings that can be easily understood by developers, planners, residents, and business owners. In addition, the Performance Standards provided the foundation for new zoning regulations that permit mid-rise development **as-of-right**. These elements remove ambiguity in design and allow the City to expedite approvals when applications are consistent with the design directions, resulting in a more streamlined planning process.

The Study Process

Each Avenue in the City of Toronto has unique characteristics, including lot width and depth, existing and adjacent uses, transit accessibility, and streetscape potential. Recognizing that there is no "one size fits all" solution, the Official Plan recommended individual Avenue Studies for each Avenue, or segments of the Avenue, in order to determine a vision and implementation plan that responds to the context-specific conditions of the Avenue. By 2010, having undertaken 19 Avenue Studies, the City determined that individual studies were slow and expensive and generally yielded similar results. It was decided that all studies completed to date would be analyzed to see where the findings could be applied on a city-wide scale.

The consultant team, led by Brook McIlroy and including ERA Architects, Quadrangle Architects Limited, and Urban Marketing Collaborative, together with City of Toronto planning staff, reviewed and analyzed the Avenue Studies completed to date. This exercise provided a general understanding of the state of the city's Avenues, and the findings were instrumental in the preparation of the Performance Standards. Following this, all 162 kilometres of Toronto's Avenues were inventoried and analyzed to gain a comprehensive understanding of the conditions along the Avenues, ensuring that all development scenarios were accounted for. While this does not preclude the completion of future individual Avenue Studies, the approach allows the City to achieve the Official Plan's vision for the Avenues on a city-wide scale without compromising the unique character of each Avenue.

In addition to extensive background research and analysis, the Toronto Avenues and Mid-Rise Buildings Study aimed to build on the significant public consultation that was undertaken in preparing the initial Avenue Studies. Recognizing the significant impact that redevelopment of the Avenues would have on adjacent neighbourhoods, a unique public outreach strategy was developed to ensure that extensive feedback was received. This included four public open-house sessions to receive feedback on the recommended Performance Standards.

As the Performance Standards were being developed, the consultant team hosted smaller focus groups with members of the development community, the public, and Local Advisory Committees (community representatives) from previous Avenue Studies. Both groups were generally in favour of the Performance Standards and supporting implementation recommendations, though for different reasons. Members of the public felt that mid-rise development would be a catalyst for streetscape improvements and the general health of the neighbourhood, while the development community appreciated the clear vision and certainty achieved through the Performance Standards, which would allow them to make an informed decision about the feasibility of a project at the beginning of the process. This process helped to ensure that the design directions received buy-in from the community while still being feasible and financially implementable from a development standpoint.

In addition, a study website and a Facebook group were created to enable interested parties to obtain information and learn about the study. It provided information about the public meetings, links to various presentations, and the display boards used at the open houses and community meetings.

Many of the concerns raised during the consultation sessions were addressed through minor adjustments to the Performance Standards. When concerns went beyond the scope of the study, they were outlined in detail as the next steps for implementation. This included further amendments to City policies and processes to allow as-of-right zoning for mid-rise buildings on the Avenues, reduced parking and loading standards to reflect the constrained site dimensions that characterize many Avenues, and dedicated City resources to ensure an **expedited approvals process**. As new mid-rise buildings are built, these elements will be further refined as part of an ongoing monitoring period by City staff.

The Performance Standards

Performance Standards are an integrated set of measurable criteria used to establish how existing and planned buildings behave toward each other or "perform" in relation to a set of criteria or principles in an area-specific setting or context. Thirty-six Performance Standards were prepared to transform the Avenues into healthy, liveable, and vibrant main streets while protecting the stability and integrity of adjacent neighbourhoods. The Performance Standards are clearly illustrated and supported by precedent images and provide clear guidance for those seeking to develop mid-rise projects on the Avenues. Key provisions include:

- Buildings are moderate in height—no taller than the right-of-way (property line to property line) is wide.
- Buildings provide an appropriate transition in scale to adjacent neighbourhoods.
- Sidewalks are wide enough to include and support trees, generate a lively pedestrian culture, and ensure accessibility for all.
- Sidewalks receive at least five hours of sunlight per day from the spring through to the fall.
- The ground floors of buildings provide uses that enliven sidewalks and create safe pedestrian conditions.
- The **public realm** is protected and enhanced by limiting vehicle access from the Avenue, encouraging shared access, and creating a public laneway system that is accessible from side streets.
- Streetscape and building design reflects excellence in sustainability, urban design, and architecture, recognizing the important public role of the Avenues in defining the quality of life for the city and its neighbourhoods.

To address the uniqueness of each Avenue, application of the Performance Standards varies according to the physical characteristics of the site (i.e., lot depth and width, topography) and the site location (i.e., width of right-of-way, corner or mid-block site).

Sample Performance Standards

In the following sections, sample Performance Standards are described in greater detail to demonstrate how they guide the design of mid-rise buildings and adjacent streetscapes along the Avenues. Each sample guideline includes the key recommendations, the rationale in relation to the study objectives, and the relevant references to the existing planning framework.

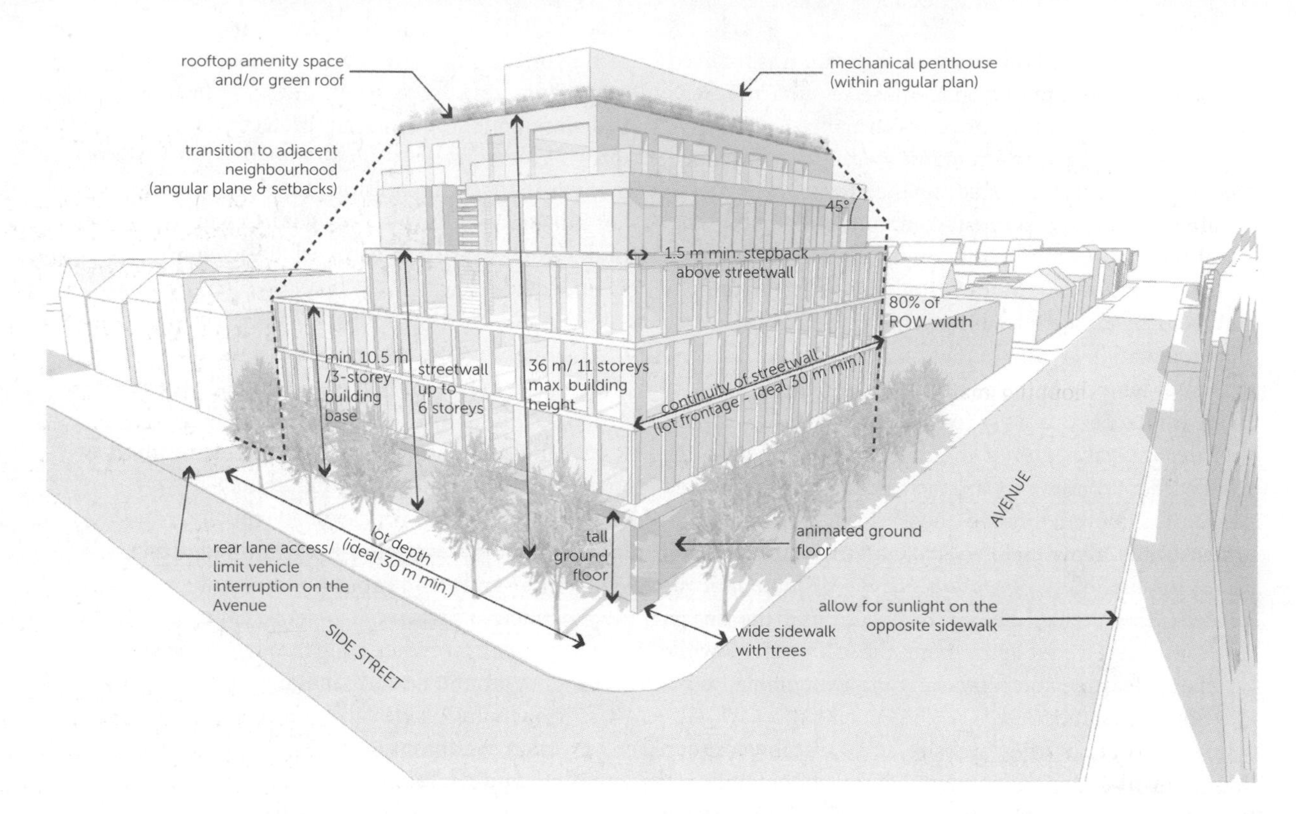

Figure 7.5.3 • Performance Standard diagram.

Sample Performance Standard #1: Maximum Allowable Building Height

The maximum allowable height of buildings on the Avenues will be no taller than the width of the Avenue right-of-way, up to a maximum mid-rise height of 11 storeys (36 metres).

Rationale—The City of Toronto defines a mid-rise building as "taller than a typical house or townhouse, but no taller than the width of the street's public right-of-way." Similarly, the City's Tall Building Guidelines (Urban Strategies Inc. and Hariri Pontarini Architects 2010) define tall buildings as those that are taller than the right-of-way on which they are located. Map 3 in the City's Official Plan identifies the right-of-way widths associated with existing major streets. The majority of the Avenues have a right-of-way width of 20, 27, 30, or 36 metres. Applying Performance Standard #1, a mid-rise building on a 20-metre right-of-way could be 20 metres tall (approximately six storeys). For a site to be developed at this maximum height, it must meet further requirements determined by additional Performance Standards related to the form of the building (i.e., front, side, and rear-yard **stepbacks**).

Sample Performance Standard #2: Minimum Ground Floor Height

The minimum floor-to-floor height of the ground floor should be 4.5 metres to facilitate retail uses at grade.

Rationale—Floor-to-floor heights for commercial uses are generally higher than a typical residential floor. A taller floor-to-floor height at grade will provide for flexibility of grade-level uses and increase the marketability of retail spaces. A taller floor-to-floor height emphasizes this portion of the building and thereby increases the visibility of any developed retail. A floor-to-floor height of 4.5 metres also provides clearance for loading spaces and trucks into the internal spaces of a building (i.e., it would not require double-height garage door openings), which should be at the rear of the site.

Sample Performance Standard #3: Front Facade: Angular Plane

From the front property line, a 45-degree angular plane should be applied at a height equal to 80% of the right-of-way width. The upper storeys of the building should not penetrate this angular plane.

Rationale—The success of the Avenues is contingent on the creation of great main streets with comfortable pedestrian spaces—especially sidewalks. Extensive research about the effects of sunlight (Bosselman et al. 1990) suggests that new buildings should maintain five hours of sunlight per day on Toronto's Avenues between the spring and fall equinox.

Sample Performance Standard #4: Rear Transition to Neighbourhoods: Deep Properties

The location of the building should be determined by a minimum setback of 7.5 metres from the rear property line. Articulation at the rear of the building must conform to a 45-degree angular plane from the rear property line.

Rationale—The classification of a property as shallow or deep is directly related to the width of the right-of-way as well as to the potential of the property to accommodate underground structured parking. A 7.5-metre **setback** allows for a two-way lane (6 metres) and a pedestrian walkway (1.5 metres), which allows access to parking and delivery facilities to be provided from the rear of the building to minimize disruptions to the primary streetscape. A 45-degree angular plane ensures a lower building at the rear and a gradual transition to the rear property line, minimizing shadow impacts and **overlook** on adjacent properties.

Sample Performance Standard #5: Minimum Sidewalk Zones

Mid-rise buildings should be set back from the property line to accommodate a minimum sidewalk width of 4.8 metres. Where right-of-way widths are greater than 30 metres, setbacks should accommodate a 6.0-metre sidewalk.

Rationale—The Avenues and Mid-Rise Buildings study is as much about creating an attractive, welcoming, and safe pedestrian realm as it is about creating mid-rise buildings for people to live and work in. A 4.8-metre sidewalk width is consistent with the recommendations of the City's *Vibrant Streets Manual* (R.E. Millward and Associates 2012) to create enough space for street trees to grow to maturity while maintaining an uninterrupted pedestrian clearway. Properties with right-of-way widths greater than 30 metres can develop taller buildings (up to nine storeys), which are likely to generate additional pedestrian traffic and should therefore ensure a 6.0-metre sidewalk.

The 36 Performance Standards set the framework for development that is appropriately scaled and designed to reflect the unique characteristics of the Avenues. While the directions are highly prescriptive, it is not the intent to limit creative design solutions. Within the Performance Standards, there is significant flexibility and opportunity for architectural interpretation, as demonstrated in the sample projects below.

Implementation

In July 2010, Toronto City Council adopted the Performance Standards, thereby authorizing staff to use them in the evaluation of all new and current mid-rise development proposals on the Avenues and in the implementation of future Avenue studies. In August 2010, five of the 36 Performance Standards were incorporated into the new city-wide zoning bylaw. In so doing, the City has been able to expedite the implementation of key Performance Standards. To ensure that the Performance Standards achieve the City's vision for the Avenues in a consistent and feasible manner, their application is being monitored to identify whether additional standards are needed or the existing standards should be modified.

To further ensure that development on the Avenues is consistent with the City's vision, any development application located along the Avenues must be reviewed by the City's Design Review Panel, whose goal is to improve people's quality of life by promoting design excellence within the public realm, including the pursuit of high-quality architecture, landscape architecture, urban design, and environmental sustainability.

Sample Mid-Rise Buildings

Since the adoption of the Toronto Avenue and Mid-Rise Buildings Study, there has been a surge of mid-rise development along the Avenues. A number of new mid-rise buildings have already been constructed, while many more are in the application or construction phase. The examples below demonstrate how local architects and developers have interpreted the Performance Standards to create unique and attractive mid-rise buildings that have proved to be more feasible because they move quickly through the approvals process.

Sample building #1: Bellefair Kew Beach Residences, Queen Street East (at Bellefair Avenue)

Bellefair Kew Beach Residences is a five-storey **mixed-use building** currently under construction on Queen Street East at Bellefair Avenue. The project includes the adaptive re-use of a former church, which will be retained and updated and will form a significant component of the building's design. At five storeys, the building's height is within a 1:1 ratio with the width of the right-of-way (20 metres). The retail uses on the ground floor are highlighted by a taller (4.5 metres) floor-to-floor height, large windows, and wide sidewalks. Facade **articulation** and a variety of materials (e.g., brick, glass, stone) help to break up the large street frontage while reflecting the smaller-scale commercial uses found on the remainder of Queen Street East. At the rear of the building, six three-storey townhouses provide a transition to the adjacent stable residential neighbourhood.

Sample building #2: Lakehouse Beach Residences, 1960 Queen Street East

Lakehouse Beach Residences is a six-storey mixed-use building replacing a single-storey restaurant at 1960 Queen Street East. At six storeys, the building's height is within a 1:1 ratio with the width of the right-of-way (20 metres). Above the fourth storey, the building steps back within a 45-degree angular plane (from 80 per cent of the right-of-way width) to minimize shadows, allow sun penetration, and maintain a building base height that is complementary to adjacent two-storey buildings. Because the building is on a corner site, this treatment is reflected on both street frontages. The brick applied to the building's initial four storeys reflects the adjacent buildings on Queen Street East, while the extensive glass used on the upper two storeys creates transparency, helping to mitigate the building's height. At the rear of the building, the upper storeys step back to conform to a 45-degree angular plane, providing a significant amount of private amenity space and an appropriate transition to the adjacent residential neighbourhood.

Sample building #3: Cube Lofts, 799 College Street

Cube Lofts is a six-storey residential building located at 799 College Street. At six storeys, the building's height is within a 1:1 ratio with the width of the right-of-way (20 metres). While the building has residential units at grade, it has a taller ground floor height (4.5 metres) to maintain flexibility and accommodate future conversion to retail should the market allow. Individual entrances are provided on the ground floor, helping to break up the building's large frontage while reflecting the adjacent uses. Above the fourth storey, the building steps back to fit within a 45-degree angular plane (from 80 per cent of the right-of-way width), creating significant opportunities for outdoor amenity space. At the rear of the building, rather than conforming to the extent of the 45-degree angular plane envelope (as in the example above), the entire building is set back to accommodate a significant ground-floor amenity space.

Figure 7.5.4 • Precedent rendering of Bellefair Kew Beach Residences.

Source: RAW *Design*

Figure 7.5.5 • Precedent rendering of Lakehouse Beach Residences.

Source: RAW *Design.*

Figure 7.5.6 • Precedent rendering of Cube Lofts.

Source: RAW Design.

Conclusions: Canadian Application of the Study

The Toronto Avenues and Mid-Rise Buildings Study was inspired by the mid-rise, pedestrian-oriented streetscapes found in European cities such as Copenhagen and Paris. Similarly, the recommendations of the study, and the successful mid-rise developments that have resulted since its implementation, can be used to inspire sensitive, context-specific intensification across the country.

Significant growth in urban population is not unique to the city of Toronto. Major metropolitan areas such as Calgary, Edmonton, and Vancouver face the similar challenge of trying to promote intensification while minimizing negative impacts on stable residential neighbourhoods. Smaller cities are not exempt, with cities such as Saskatoon, Oshawa, Moncton, and St John's seeing significant growth. Brampton, one of the fastest-growing cities in Ontario, is in the process of preparing its own mid-rise guidelines.

Each of these cities has a document similar to Toronto's Official Plan, which aims to achieve the intensification goals mandated through broader provincial policy documents, such as Ontario's *Places to Grow: Growth Plan for the Greater Golden Horseshoe* (Ontario Ministry of Infrastructure 2006). This includes Calgary's Municipal Development Plan (2009), Saskatoon's Official Community Plan (2009), and Edmonton's Municipal Development Plan (*The Way We Grow*) (2010). These high-level documents direct growth to the most appropriate and sustainable locations without sacrificing the long-term sustainability of stable residential areas. As these cities explore strategies to accommodate intensification within their cores and along their primary streets, mid-rise development should be encouraged in order to concentrate growth in key locations and reurbanize key areas without relying on taller buildings.

For mid-rise development to be successful in other cities, a guiding document similar to the Avenues and Mid-Rise Building Study is required. However, simply adapting Toronto's Performance Standards to respond to the local context is unlikely to yield such successful results. Instead, municipalities must adopt *the process* that defined the Toronto Avenues and Mid-Rise Buildings Study. This includes background analysis to determine the range of characteristics that define potential development sites and extensive testing and analysis to determine the most appropriate way to develop these areas. Perhaps most important, aspiring municipalities must work closely with the public and the development community to determine a solution for mid-rise intensification that strengthens the existing community character while providing a realistic and feasible response to local development realities.

KEY TERMS

Articulation
As-of-right
Bylaw (or by-law)
Expedited approvals process
Facade
Mid-rise building
Mixed-use building
Overlook
Pedestrian-oriented
Public realm
Right-of-way
Setback
Stepback

REFERENCES

Bosselman, P., E.A. Arens, K. Dunker, and R. Wright. 1990. *CityPlan '91: Sun, Wind, and Pedestrian Comfort: A Study of Toronto's Central Area*. Toronto: City of Toronto.

Brook McIlroy Planning + Urban Design, ERA Architects, Quadrangle Architects Limited, and Urban Marketing Collaborative. 2010. *Toronto Avenues and Mid-Rise Buildings Study*. Toronto: City of Toronto.

City of Calgary. 2009. *Calgary's Municipal Development Plan*. Calgary: City of Calgary.

City of Edmonton. 2010. *Municipal Development Plan (The Way We Grow)*. Edmonton: City of Edmonton.

City of Saskatoon. 2009. *Official Community Plan*. Saskatoon: City of Saskatoon.

City of Toronto. 2010. *Official Plan: Office Consolidation*. Toronto: City of Toronto.

Ontario Ministry of Infrastructure. 2012. *Places to Grow: Growth Plan for the Greater Golden Horseshoe: Office Consolidation*. Ontario: Ministry of Infrastructure.

R.E. Millward Associates Ltd. 2012. *Vibrant Streets: Toronto's Coordinated Street Furniture Program*. Toronto: R.E. Millward Associates Ltd.

Urban Strategies Inc. and Hariri Pontarini Architects. 2010. *Tall Buildings: Inviting Change in Downtown Toronto*. Toronto: Urban Strategies Inc.

8.0

INTRODUCTION

Urban Redevelopment

Urban redevelopment has been one of the most contentious sub-disciplines of planning for more than 60 years. There are always areas within towns and cities that are in decline, just as there are always neighbourhoods growing more attractive to residents and businesses. Municipalities, developers, local residents, and business owners all play a role in reshaping urban areas to redirect growth and replace or rehabilitate elements of the built form. Redevelopment may occur for a number of reasons, including the desire for economic growth, more sustainable development patterns, increasing a city's competitiveness, or creating more opportunities for international tourism.

In the discipline of planning, urban redevelopment has some negative connotations because of the massive destruction of low-income neighbourhoods in the United States, and to some extent in Canada, in favour of new public housing developments, arterials, and expressways during the postwar decades. For example, Montreal lost thousands of homes during rapid redevelopment in the 1960s, seriously affecting working-class neighbourhoods (Artibise 1998). Toronto's Parkdale, once a stable middle-class neighbourhood, experienced disinvestment following construction of the Gardiner Expressway in the 1960s (Slater 2004). In mid-sized cities like Sudbury and Winnipeg, demolition of historic buildings and disinvestment in the city centre continued well into the 1980s. The temptation among planners and developers to erase the problems of the past and replace them with shiny new upscale alternatives (Breitbart 2013) can be difficult to overcome in a neo-liberal economic paradigm.

Redevelopment efforts turned toward rehabilitation of buildings rather than demolition in Canadian cities, largely spurred by changes in the National Housing Act in 1964 and programs funded through Canada Mortgage and Housing Corporation (Wolfe

2004). In the 1970s, redevelopment of former industrial lands became instrumental in preserving vital urban neighbourhoods (e.g., False Creek and Granville Island in Vancouver, Byward Market in Ottawa). Rural redevelopment programs also had positive effects (Robinson and Webster 1985). During the 1980s, redevelopment of inner-city precincts became common in Canadian cities, illustrating the spatial effects of the post-industrial economy (e.g., growth in industries such as finance, business development, and real estate led to development of high-end luxury condominiums reflecting the desires of high-income service-sector employees).

Redevelopment has also been driven by community members: in many of Canada's urban centres, immigrant communities have been instrumental in establishing businesses, making renovations to existing housing, establishing community organizations, and establishing ethnic neighbourhoods. Much higher immigration rates in postwar Canada have contributed to the vitality of urban neighbourhoods compared to American cities, whose inner-city neighbourhoods often continue to show the spatial effects of residential segregation and persistent employment problems associated with the transition to post-industrial economies. Canadian inner cities are often vibrant and desirable places to live in. A critical dilemma in urban redevelopment involves reconciling the needs of long-term local residents and businesses with planners' or politicians' goals for the future (e.g., economic growth, supporting creative industries). This is particularly the case because the large scale of redevelopment efforts often involves several neighbourhoods or urban districts.

Several cases in this section explain that significant efforts from planners, developers, and community members are required in redevelopment processes. The cases also illustrate the contributions of all three levels of government and partnerships with community organizations in order to implement redevelopment plans. In municipalities with small budgets or few resources, these partnerships are often necessary and help to develop a sense of ownership of the plan, which contributes to greater acceptance on the part of the different actors. Conflicting views on the future vision for an area or district, conflicts between political and community views, and lack of engagement with the local community present significant challenges to urban redevelopment efforts.

In their case study, Alison Bain and Ross Burnett examine the development and implementation of Greater Sudbury's Downtown Master Plan. The authors found that for a mid-sized city with budgetary and resource constraints, such a plan needs to be operationalized quickly, have strong political commitment, a community that will hold the municipality accountable, and stakeholders to deliver part of the plan independently. Extensive public engagement included stakeholder interviews, a community forum, a collective visioning process, targeted outreach to youth, Aboriginals, and francophones, design charrettes, and roundtable discussions. The plan includes 52 projects and an Action Strategy separating out short-term, medium-term, long-term, and ongoing delivery; 17 projects will be implemented within the next decade. Since 40 per cent of Canadians live in mid-sized cities like Sudbury, this case could offer important lessons to other municipalities.

The theme of building a sense of ownership through the plan-making process continues with Gerry Couture's case of Winnipeg's CentrePlan. Like Sudbury, Winnipeg suffered serious decline in the downtown area in the 1970s: boarded-up and vacant buildings, endless surface parking, and persistent social problems spurred an intervention from all three levels of government. CentrePlan was launched in 1993, integrating community representation on its steering committee, advisory committee, and strategy teams. The group used an iterative process to determine what local people liked about downtown, what changes they would make, and their views on the proposed solutions. The end result, a list of 15 opportunities and 65 strategies, represented the community's vision for the future, while an Action Plan identified 35 actions, required resources, benefits, and steps required. Many of the actions were fully implemented in 1995–6, and others were carried forward into the new Action Plan. Almost 20 years later, the momentum and commitment generated by the plan is still evident.

Laurie Loison and Raphaël Fischler tell a very different story in their case study of the redevelopment of Montreal's Quartier des spectacles. The case perfectly illustrates the tension between local interests and outward-focused economic redevelopment. After the mid-1990s economic recession, the City of Montreal launched a series of plans and projects to stimulate private development in the Quartier des spectacles, which was home to many arts and entertainment venues but was riddled with empty lots and illicit activity spurred by decades of postwar urban regeneration efforts. Working with local property-owners and the provincial and federal governments, the City envisioned new public buildings and spaces aimed at drawing international users. Shifting values and perceptions of the neighbourhood contributed to conflicts between the large number of actors in the process as artists, informal performance spaces, and long-time renters began to lose out to large homogenous arts venues and a new population of young professionals.

Jason Thorne presents a large-scale redevelopment effort: the Province of Ontario's Places to Grow policy. Ontario's Growth Plan for the Greater Golden Horseshoe (2006) attempted to increase density, intensify development, and manage growth within the municipalities in this region, which includes 10 of the country's largest cities. Plan development began with a series of **Smart Growth** Panels, including local mayors and environmental and agricultural interest groups, and the establishment of a Smart Growth Secretariat. The Places to Grow Act (2005) requires all local planning decisions to conform with the provisions of the provincial plan and all local Official Plans to be brought into conformity with the Growth Plan for the Greater Golden Horseshoe. Twenty-five urban growth centres identified in the plan are targeted for future employment and population growth. Although plan conformity has taken longer than expected and plan interpretation varies across municipalities, significant shifts in development can be seen (e.g., an increase in multiple-unit housing).

These cases are examples of historical and current efforts in Canadian cities. The variety of approaches, from top-down policies initiated at the provincial level to locally driven plans, is illustrative of the flexibility of urban redevelopment and the need to involve a wide range of stakeholders. Several of the cases, notably Sudbury and Winnipeg, show how important it is to include local actors not only in the planning process but as implementation partners as well. As municipalities with limited financial and human resources (e.g., planning staff), these cities offer valuable lessons for other mid-sized cities in Canada.

REFERENCES

Artibise, A.F.J. 1988. "Canada as an urban nation." *Daedalus* 117(4): 237–64.

Breitbart, M.M., ed. 2013. *Creative Economies in Post-industrial Cities: Manufacturing a (Different) Scene*. Farnham, UK: Ashgate.

Robinson, I.M., and D.R. Webster. 1985. "Regional planning in Canada: History, practice, issues, and prospects." *Journal of the American Planning Association* 51(1): 23–33.

Slater, T. 2004. "Municipally managed gentrification in South Parkdale, Toronto." *The Canadian Geographer* 48(3): 303–25.

Wolfe, J. 2004. "Our common past: An interpretation of Canadian planning history." *Plan Canada 75th Anniversary Special Edition*.

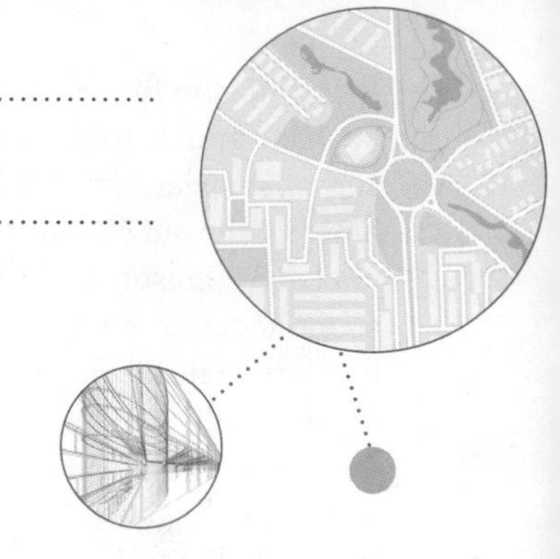

CASE STUDY 8.1

Urban Regeneration in a Mid-sized City: A New Vision for Downtown Sudbury

Alison Bain and Ross Burnett

SUMMARY

In Canada, more than 40 per cent of the population live in mid-sized cities. Within the parameters of a twenty-first-century post-industrial economy, civic leaders of mid-sized cities face the challenge of differentiating their city and defining a competitive advantage. A conventional approach to mid-sized city survival involves erasing signs of physical decline through downtown redevelopment. This chapter offers a case study in the initiation, development, and implementation of a Downtown Master Plan for the City of Greater Sudbury. It critically examines the planning process from vision to implementation in order to reveal the centrality of consultation and education in fostering ownership of the Downtown Master Plan by community and stakeholder partners. This chapter argues that for a plan to be swiftly operationalized to realize significant material changes to the urban fabric, it requires strong political commitment and a community willing to hold the municipality accountable, as well as stakeholders able to independently deliver elements of the plan.

OUTLINE

- The development and implementation of Sudbury's Downtown Master Plan after redevelopment in the 1970s and decades of disinvestment can be considered typical of mid-sized municipalities in Canada.
- The planning process included developing a vision through extensive public engagement and developing a series of land-use recommendations, design guidelines, and 52 targeted revitalization projects.
- The vision for the downtown became a reality with the contribution of project partners, who helped to implement projects, assured buy-in, and ensured additional funding for the plan. The stakeholder relationships have been critical in the successful implementation of a shared vision and strategic plan for improvement.

Introduction

Small and mid-sized cities are often overlooked in discussions of planning policies and processes, which tend to focus on the strategies and plans implemented in large city-regions (Garrett-Petts 2005; Bell and Jayne 2006; van Heur 2010; Nelson 2012; Breitbart 2013). Yet more than 40 per cent of Canadians live in mid-sized cities with populations between 50,000 and 500,000 (Bain and McLean 2011). Thus, it is important to attend to the planning challenges and opportunities that are unique to mid-sized cities. Key challenges facing civic leaders in mid-sized cities are how to differentiate their city and how to define a competitive advantage within the parameters of a new post-industrial economy.

The conventional mantra of mid-sized city survival in the twenty-first century is predicated on its reinvention as a place that is attractive to new businesses, residents, investors, and visitors alike (Bain and McLean 2013). Such reinvention is often focused on the redevelopment of the downtown core with an emphasis on a diversity of culture-imbued economic opportunities (Short 1999). The intent is usually to erase physical reminders of decline in order "to provide the bedrock for a new prideful reconstructed image of place" (Breitbart 2013, 11). Post-industrial visions embraced by local politicians, urban planners, and economic developers in mid-sized cities across North America and Europe have generally "focus[ed] on a renewed physical landscape that includes a revitalized pedestrian-lively downtown, with a refurbished historically-preserved architecture, and noticeably fewer vacant lots and abandoned shops" (Breitbart 2013, 12). Such a mainstream approach to regeneration has been proposed for the historic downtown of the City of Greater Sudbury.

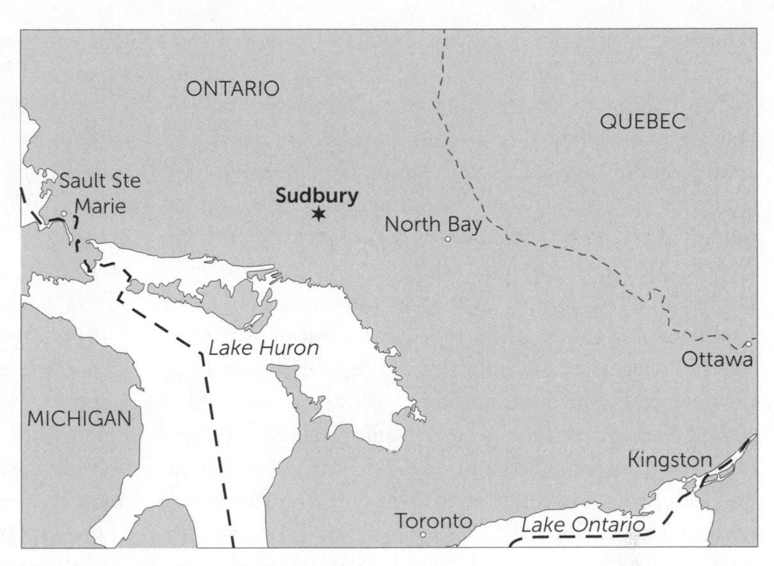

Figure 8.1.1 • Location map.

This chapter offers a case study in the initiation, development, and implementation of a Downtown Master Plan. Drawing on interviews and participant observation, it provides a critical examination of the plan development process in Sudbury from vision to implementation. In a mid-sized city facing municipal budgetary and planning staff resource constraints, this chapter argues that for a Downtown Master Plan to be swiftly operationalized, it requires strong political commitment and a community willing to hold the municipality accountable, as well as stakeholders able to independently deliver elements of the plan. The case study demonstrates that a multivalent plan delivery model that combines political support with urban planning leadership, institutional initiative, and community action can realize significant material changes to the urban fabric in a relatively short period of time.

Mid-sized City Reinvention: The City of Greater Sudbury

Sudbury is located 390 kilometres north of Toronto and 480 kilometres west of Ottawa where Highway 69 and the Trans-Canada Highway intersect. It is the regional capital of northeastern Ontario. Today, Sudbury has a population of more than 160,000 and covers an area of 3267 square kilometres, making it geographically the largest municipality in Ontario and the second-largest in Canada.

A third of Sudbury's population is francophone; the Franco-Ontarian flag was created in Sudbury. It flies atop a pole at city hall, celebrating Sudbury as the birthplace of Franco-Ontarian culture and a hub of French services and education.

Prior to European colonization in the 1600s, the Sudbury area was Aboriginal territory named N'Swakamok, which means "where the three roads meet" (Leadbeater 2008). Two of the nearest Aboriginal territories were Whitefish Lake First Nation and Wanapitei First Nation. To facilitate European resource exploitation, settlement, and expansion, Aboriginal people were displaced from their lands and concentrated in reserves, 20 of which remain within a 300-kilometre radius of Sudbury. The number of Sudbury residents reporting Aboriginal (8160) and Métis (6315) ancestry in the 2006 Canadian census together exceeds that for individual ethic groups of Finns, Germans, Italians, Poles, or Ukrainians. Today, an Aboriginal cultural presence in Sudbury is reinforced through the N'Swakamok Friendship Centre and the Shkagamik-Kwe Health Centre, along with the Native Studies Program at Laurentian University, the Métis and Aboriginal Centre at Collège Boréal, and the First Nations College Experience Program at Cambrian College.

In addition to its rich Indigenous and French heritage, Sudbury is a city defined by its geography. Located on the Precambrian Shield, the city's economy has been closely tied to the mining industry, resulting in more than 125 years of boom and bust cycles of growth. Sudbury began as a company village of the Canadian Pacific Railway (CPR) (Wallace 2004). During construction of the transcontinental railway, copper ore was discovered in the Sudbury Basin, and mining began in 1886. Sudbury gradually developed from a lumber and railway camp into a masterplanned mining town and ultimately the largest hard-rock mining centre in North America. The world's leading nickel-mining corporations, Inco Ltd (formed in 1886 as the Canadian Copper Company, now owned by the Brazilian mining transnational Vale Rio Doce) and Falconbridge Ltd (established in 1928, now owned by the Swiss mining transnational corporation Xstrata Nickel), are located here. Mining caused extensive environmental degradation through acid rain, deforestation, and soil erosion and contributed to the creation of Sudbury's notorious moonscape of the 1940s to 1970s (Edinger 2008). Since the 1980s, however, the city has spearheaded a program of environmental repair and reforestation. Today, the city is recognized as a world leader in environmental renewal for land, air, and water. At the 1992 Earth Summit, the United Nations honoured Sudbury with an award recognizing the city's community-based environmental reclamation strategies. By 2010, re-greening programs had successfully rehabilitated 3350 hectares of land in Sudbury; however, 30,000 hectares of industrial land still needs to be rehabilitated (Bradley 2010).

Sudbury formally became a city in 1930, and its colonial–frontier character gradually changed after World War II (Saarinen 1990; 2013). Population growth, combined with investment support provided by the attainment of remoteness and hinterland status, helped to develop a more mature, service-oriented economy and to enhance the city's role as a regional service centre. In the 1950s, a highway to Toronto was completed, the Sudbury Airport was established, and the city's first Official Plan was approved. In addition to the acceptance of planning as a development tool and the expanded transportation network, medicine and education also helped to extend Sudbury's regional presence. Outside of the downtown core, three hospitals were established in the 1950s, and Laurentian University and Cambrian College of Art and Technology were established in the mid-1960s (the francophone college of applied arts and technology, Collège Boréal, opened in 1995). Lower land costs on the periphery and increasing automobile ownership encouraged urban growth in surrounding townships. Municipal planning in Sudbury, Leadbeater (2008, 34) argues, "has been aggressively oriented toward a retrograde corporate developer–driven commercial and residential expansion that encourages wasteful sprawl, suburban malls at the expense of downtowns, [and] drive-throughs galore."

In an effort to attract reinvestment in the downtown core and to address perceived urban blight, an aggressive urban renewal program was implemented in the 1970s, jointly funded by municipal and provincial governments. In the north end of downtown, 24 hectares was comprehensively redeveloped with a new system of streets, parks, and open spaces. Heritage properties were expropriated and demolished to create large-scale revitalization projects (e.g., the Rainbow Centre shopping mall and the Civic Square office complex). The formation of the Regional Municipality of Sudbury (RMS) in 1973 was the political and planning catalyst for much of this redevelopment. However, none of the eight former towns and cities of the RMS, which were amalgamated in 2001 to form the City of Greater Sudbury, took ownership of the downtown or saw it as their responsibility to maintain.

In the absence of a comprehensive plan for the downtown and with fewer than 600 residents to mobilize for change, the downtown fell into decline and was treated as little more than a political boundary. Buildings were demolished and not replaced, land was underutilized, and people drove through rather than to it. City officials recognized the need to reverse decline and to create an up-to-date framework to guide urban development decision-making. In 2009, motivated by a proposal to establish a new School of Architecture at Laurentian University, the city leveraged provincial and federal regional development funding to retain private urban planning consultants. The urban planning firm Urban Strategies was hired to prepare a 30-year Master Plan and Action Strategy to revitalize the city's downtown core. This Downtown Master Plan is part promotional document for why downtowns matter and part recipe for how to create a "successful" downtown.

Success is framed in the Downtown Master Plan in economic terms as: offering a positive image for the region; providing a strong tax base; incubating business growth; creating jobs; protecting property values; becoming a tourist destination and meeting place; and fostering sustainable growth. The urban boosterist mandate is conveyed at the outset of the document with an agenda set to make "recommendations to reinforce the Downtown's role as the biggest, brightest, and best downtown in Northern Ontario" and to position Sudbury as "the true and distinct 'capital of the North'" (Urban Strategies Inc. 2011, 6). Such ambitions are intended to deliver the broader provincial planning agenda as articulated in the Provincial Policy Statement (2005) and the Growth Plan for Northern

Ontario (2011). These policy documents offer high-level guidance to municipalities on matters of land-use planning and development. All statutory municipal planning documents must support the intention of these provincial policy directives, which emphasize the importance of creating long-range community visions that foster social, economic, environmental, and fiscal sustainability in order to support a good quality of life for residents. This is to be achieved through the efficient management of land and infrastructure, the promotion of a healthy environment, the creation of a diverse, productive, and well-educated workforce, and the provision of opportunities for employment and residential development. The following section explores in greater detail what a Downtown Master Plan is, how it works, and what the development goals and project priorities of Sudbury are.

Developing a Downtown Master Plan for Sudbury

A **master plan** is a highly visual document that presents a large-scale vision of physical change for what the future of a geographically defined area can be. In the UK, the Commission for Architecture and the Built Environment (CABE 2008) asserts that master plans should "set out proposals for buildings, spaces, movement strategy and land use in three dimensions, and match these aspirations with an implementation strategy." A successful master plan should, according to the Urban Task Force (cited in Ardron, Batty, and Cole 2008, 9), be:

- *Visionary:* raise aspirations and provide a vehicle for consensus-building and implementation
- *Deliverable:* define the most appropriate implementation strategy
- *Enabling:* integrate with land-use planning systems and enable new uses and market opportunities to exploit development potential
- *Flexible:* provide the base for negotiation and dispute resolution
- *Participatory:* provide all stakeholders with the means to express their needs and priorities

A master plan is a strategic document used to guide decision-making on where investment should be spent (geography), when investment should be made (timing), and what it should be spent on (projects). It functions as a "call to action" to bring different partners together to achieve their common goal of turning an agreed-upon planning vision into reality. The implementation component of a master plan guides the city and its partners on needed actions, partnership opportunities, cost initiatives, sources of funding, required policy changes, and project-delivery phasing.

The Downtown Master Plan for Sudbury establishes a strategy to transform the core into an active, safe, and diverse destination for people and investment. It strives to create a downtown that is mixed in its character and function, is well integrated with surrounding communities, and supports a unique place identity. The plan provides a series of land-use recommendations, design guidelines, and 52 targeted revitalization projects clusters under three complementary directives to augment the function of the downtown:

- *Activity and Growth:* a downtown that is a node of economic and cultural growth and a destination for residents and visitors
- *Access and Connectivity:* a downtown that is connected to the rest of the city, is easy to navigate, and is accessible to all
- *Beauty and Pride:* a downtown that is beautiful, celebrates the "Spirit of Sudbury," and creates a statement of beauty and northern identity

To deliver these directives, each of the plan's 52 projects is included in the Action Strategy. To help prioritize delivery, the Action Strategy separates projects into short-term delivery (one to five years), medium-term delivery (five to 10 years), long-term delivery (10+ years), and ongoing (anytime). Although one of the primary objectives of the Action Strategy is to provide a time frame for sequencing downtown development projects, inevitably the order of implementation depends on the availability of funding, private-sector partners, technical requirements, and community priorities.

In the absence of a capital commitment for many of the projects, and to help maintain implementation momentum, the consultants identified 25 low-cost, high-impact actions that could be undertaken within the first 12 months of the Downtown Master Plan (e.g., endorsement of the plan, completing feasibility studies, introducing pilot studies, building partnerships, and preparing for design competitions). The intent of this short list is to establish a foundation upon which future change can be built. Such change is also dependent on a supportive policy framework. Thus, recommendations are made

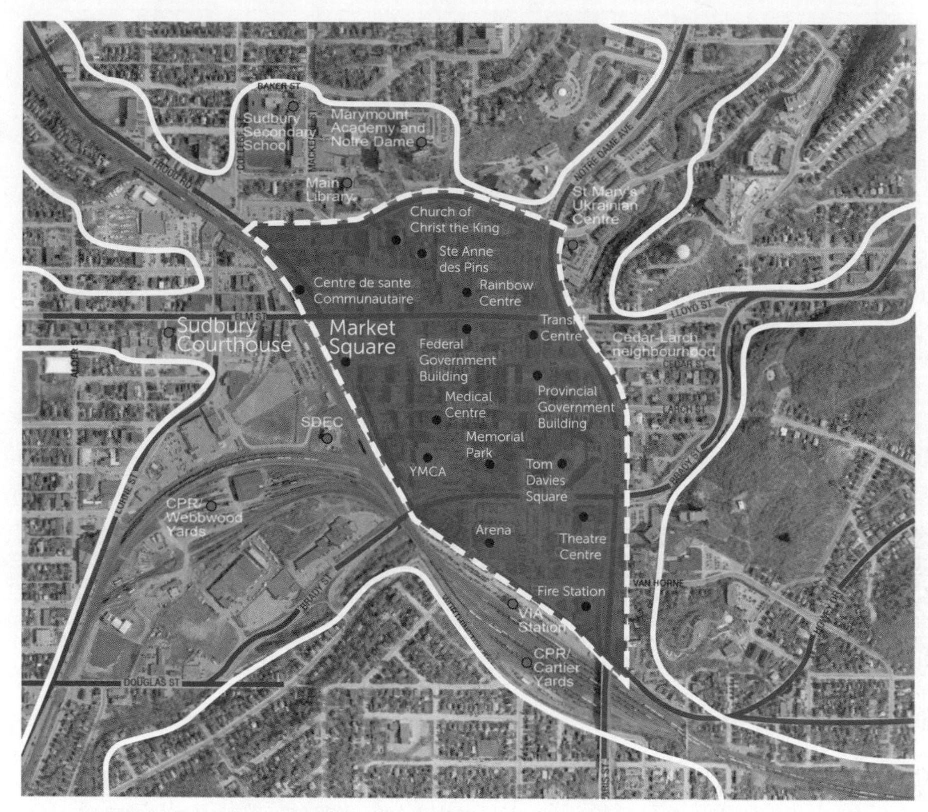

Figure 8.1.2 • The city of Greater Sudbury.

city staff, local businesses, the development industry, landowners, the university and colleges, arts and cultural organizations, police, community interest groups, and residents. These interviews provided background information on the study area and delineated differential stakeholder priorities in order to determine the scope of the project. This phase also included in-depth physical analysis of the study area to establish a baseline condition and identify issues to be addressed and opportunities to be built upon. The consultation cornerstone of this first phase of work was the first Speak Up! Sudbury community forum. This bilingual session presented participants with activity stations allowing residents to engage directly with the project team and have their views documented on the current condition of the downtown, its future opportunities, and their priorities for change. The format also allowed participants to read the views of others and to appreciate that an important part of the plan process is negotiating conflicting priorities and opinions.

The second phase involved a collective visioning process to imagine the many possibilities for the downtown. More than 1000 creative ideas were shared and documented. This stage culminated in the second Speak Up! Sudbury hands-on workshop. At eight workstations, groups of participants had an opportunity to critique and augment components of the emerging vision for downtown Sudbury.

A particular effort was made to collect feedback from harder-to-reach community groups, including youth, Aboriginals, and francophones. Two separate sessions were held with young people at the N'Swakamok Native Friendship Centre and the Sudbury Action Centre for Youth. These informal sessions not only collected the views of at-risk young people on how they use the downtown, what problems they feel exist in the downtown, and

about regulatory policies and zoning controls that need to be amended in the Official Plan, the Metro Centre Community Plan, and the zoning bylaw.

Activating Community through Consultation and Education

The Downtown Master Plan for Sudbury was completed in four phases from September 2010 to February 2012. As Figure 8.1.4 illustrates, community engagement was the core of the plan. Each phase in the planning process served a different purpose:

- Phase 1 (understanding)
- Phase 2 (visioning)
- Phase 3 (planning)
- Phase 4 (action)

The first phase of the Downtown Master Plan sought to understand local issues and opportunities. Stakeholder interviews were undertaken in English and French with

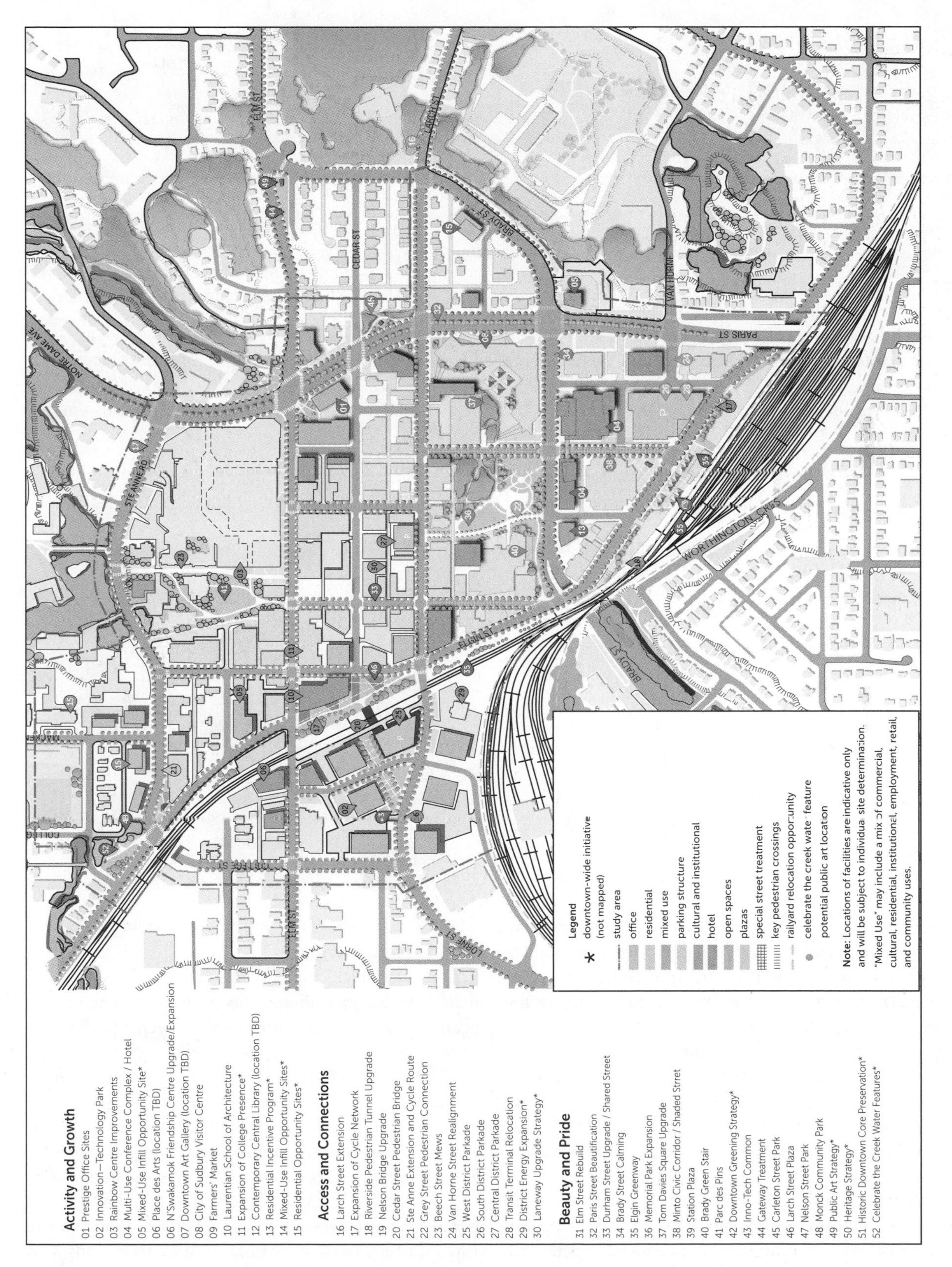

Figure 8.1.3 • Ten-Year Downtown Master Plan map.

Source: Adapted from Urban Strategies Inc. 2011.

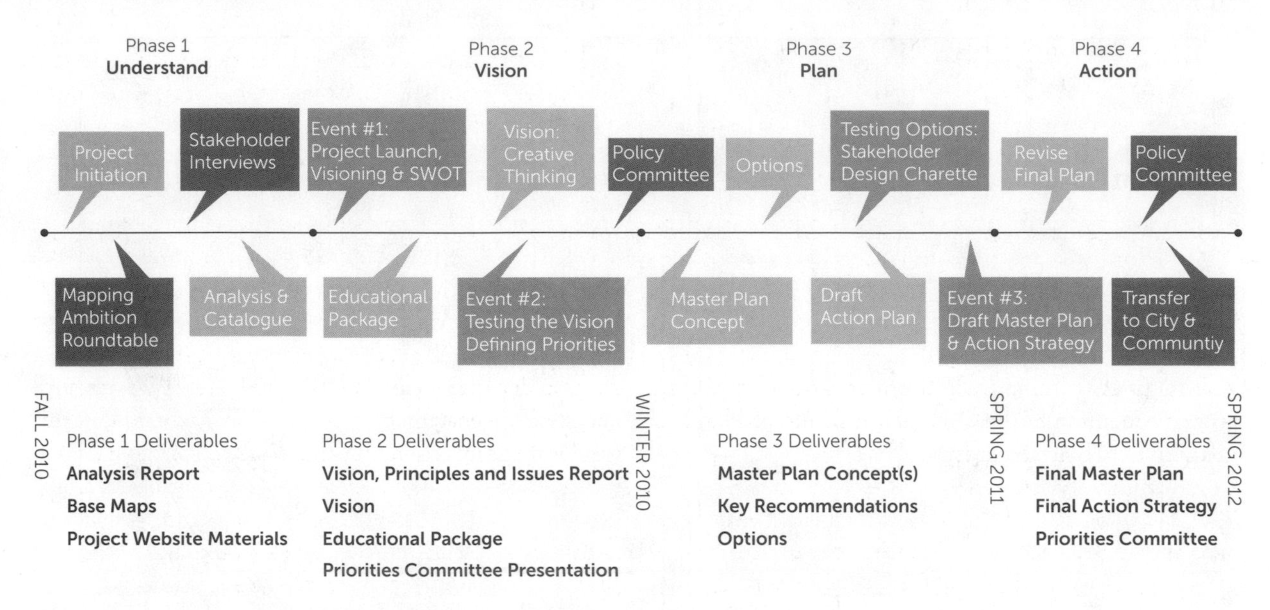

Figure 8.1.4 • Summary of the four phases of the Master Plan process.

Source: Adapted from Urban Strategies Inc. 2011.

what the best things are about the downtown but also educated youth on the importance of a healthy downtown and what urban planning is.

Because Sudbury is a fully bilingual city, securing the input of its francophone community, one of the most visible and active groups in the downtown, was essential. Consultants conducted several separate meetings with francophone groups in French. To ensure full participation by community members, all open-house and community workshop materials were produced in French as well as English. The Downtown Master Plan process was the first time that the municipality had translated non-technical documents into French for public distribution.

Central to stakeholder engagement and education, especially in Phase 1 and Phase 2, was a Community Liaison Group (CLG) made up of 30 members who represented a cross-section of associations, organizations, and institutions. The CLG initially came together as a small group of volunteers to help the city scope the plan. As the planning vision solidified and new projects emerged, the CLG volunteer membership base expanded to capture new project partners and to ensure full stakeholder representation. Inevitably, CLG involvement was driven by project-specific interests and a desire to see particular development agendas realized; however, in an effort to tap into wider community networks, CLG members were also trained as educators to run smaller, local community visioning sessions. This second phase involved a prioritization process in which projects were assessed against the agreed vision and goals of the plan. Projects that offered the maximum benefit were prioritized, along with projects that were the furthest developed and had a lead partner.

Figure 8.1.5 • Outreach event.

Source: Urban Strategies Inc. 2011.

The third phase of the process rendered the contradictions and competing ideas of the visioning exercise into a draft plan with concepts and recommendations that could be tested in hands-on design charrettes and roundtable discussions with stakeholders. With stakeholder feedback on the draft obtained and incorporated into revisions, the fourth phase delivered the final Master Plan and Action Strategy to the city and the community.

Making a Vision a Reality

Together, the vision, plan, and action strategy provide a portrait of what downtown Sudbury could look like in 30 years. While ambitious in scope, the vision and plan are also supported by "actionable projects that can spark Downtown's latent potential and create positive change" (Ferrigan 2012, 1). Unanimously approved by City Council in 2012, the Downtown Master Plan and Action Strategy received a City budget allocation for the first two years to fund priority initiatives. The plan also received an honorary mention at the Canadian Institute of Planners Planning Excellence Awards in 2012.

In Sudbury, it has taken three years to go from a collective visioning process for a Downtown Master Plan through to the project implementation stage. Following City Council's endorsement, many of the 25 Year 1 actions in the Action Strategy have been undertaken. Importantly, a senior planner in the three-member Long-Range Planning Team (part of the Planning Services Division) was tasked with completing a cost-benefit analysis for each of the 10-year recommendations in the Action Strategy. His analysis, combined with community consultation workshops, culminated in a council-approved 10-year Implementation Plan that focuses on a mix of 17 well-defined small-, medium-, and large-scale projects, each with stakeholder backing and at different stages of development. The 17 projects to occur in the next 10 years are:

- The Laurentian School of Architecture
- The Market Square relocation
- The Franklin Carmichael Arts Centre
- The Place des Arts
- The Inno-Tech Park
- The Elgin Greenway
- Residential incentives
- A multi-use conference centre
- The N'Swakamok Native Friendship Centre expansion
- A new central library
- An expanded Cambrian College presence
- A new visitors centre
- A St Anne extension and college underpass
- Civic centre upgrades
- The sharing of Durham Street
- Streetscaping
- The development of retail, public art, and heritage strategies

By the senior planner's calculation, "The Implementation Plan has the potential to generate $208 million in new investment. It is estimated that this new investment would contribute at least $85 million to the city's gross domestic product, generate $53 million in labour income, and generate approximately 845 jobs" (Ferrigan 2012, 5). Thus, the case for City Council support was based on economic development potential and alignment with five priorities (infrastructure, growth and jobs, image, tourism, and healthy communities) outlined in the City of Greater Sudbury Strategic Plan 2012–14. Given the broad scope of the projects, City Council approved the formation of an interdepartmental City Directors Team to resource and steer the enactment of the Implementation Plan. This team is made up of senior managers from six major departments, each with discretionary budget control: Planning, Economic Development, Roads, Water and Wastewater, Leisure, and Assets. It meets regularly and is an important mechanism for efficiently sharing information among project partners while also engaging and empowering a coalition of community members to help run particular projects.

While the Directors Team delivery model has facilitated an aggressive bringing-to-life of the Downtown Master Plan, so too has the CLG. In many respects, the CLG is the grassroots champion of the plan within the wider community as well as the partner responsible for delivering major initiatives (e.g., Laurentian School of Architecture, Franklin Carmichael Arts Centre, Sudbury Central Library, Place des Arts). The directeur général et culturel of the Carrefour Francophone de Sudbury, Stéphane Gauthier, is the co-chair of the CLG. For Gauthier, the CLG has been an important mechanism for uniting different organizations and institutions in the city and for teaching members how to work with each other and with City staff and councillors. Monthly meetings with the Directors Team have ensured that projects are tracked by City staff and remain active and that appropriate budget choices are made with sufficient lead time. Flexible membership in the CLG has allowed new community members to be brought on board to assist with various development projects.

For the CLG and its member arts organizations, the plan has been a valuable document that is "very alive" and "not sitting on a shelf" (Gauthier, personal communication, 4 July 2013)—ownership over it is shared, which creates the potential for community pushback and political accountability. As Gauthier explains, "the plan is fun because it helps you dream but also points you towards potent possibilities." More important, he emphasizes that even as a 63-year-old arts organization with a $3.2 million annual budget, "We couldn't have afforded this type of work and representation of what downtown could be." Thus, in his estimation, the plan has been "useful in the quality of its presentation" for "making projects make sense and seem purposeful" and for opening political doors. As he notes, City staff and politicians are involved with the plan, so when a new development project is presented, "you don't have to convince people, they're already prepared for a new, better, vibrant downtown." The new project that Gauthier is particularly invested in is the Place des Arts, a francophone cultural centre.

The community consultation process for the plan brought eight of Sudbury's francophone arts and culture organizations together to determine what their common priorities were. Le Regroupement des organismes culturels de Sudbury (ROCS) decided that collectively they needed professional space downtown to present, promote, and produce new creative work and that the proposed Place des Arts building needs to represent the dynamism of francophone culture. With the support of City Council and the mayor and $50,000 in funding to support a feasibility study, ROCS has worked to clarify its mandate, develop a business plan, determine its spatial needs, look at opportunity sites, and visit a number of centres for social innovation for inspiration. The plan provided the incentive and the platform for various francophone community organizations to inspire one another and to collectively imagine a new future for downtown Sudbury.

A similar sense of appreciation for the plan is expressed by Karen Tait-Peacock, director of the Art Gallery of Sudbury. As the visioning process gathered momentum through community engagement activities, management of the Art Gallery came to realize the importance of being part of the urban regeneration process, which could expose them to new partners and raise the gallery's profile within the downtown. They were looking to relocate the gallery from its current site in a residential neighbourhood. Tait-Peacock used the plan to explain to her board of directors the importance of a more central location and to clarify the benefits of clustering a new gallery facility with other cultural institutions. The plan became a tool of persuasion. Identification of the Art Gallery as a development priority in years 0–5 of infrastructure investment in the plan has given it political leverage. City planners have been tasked with assisting the Art Gallery in finding a new downtown site for the new Franklin Carmichael Arts Centre, based on cultural facility locations identified in the plan. For Tait-Peacock, the plan has helped the community to focus on the downtown, making it a place that "everyone wants to be"; the plan allows for the "reserving of space for things to happen" and the determining of "who is going to go where and when" (K. Tait-Peacock, personal communication, 4 July 2013). Senior planner Jason Ferrigan agrees: the plan "allowed people to come together and get excited about the downtown.... It allowed stronger relationships between the city and community partners and between partners, creating a new standard of community engagement" (J. Ferrigan, personal communication, 5 July 2013). These sentiments illustrate the strong degree of community and municipal support for the plan and highlight the trust that has been placed in it as a social contract between City officials and local residents.

To date, significant progress has been made to realize the vision outlined in the Downtown Master Plan. The Laurentian School of Architecture on the former Market Square site welcomed its first class of students in fall 2013. The farmers' market has been relocated to the historic CPR station and opened at the end of June 2013 in a temporary model that makes creative use of rail boxcars as vending sites until the station can be renovated for winter use. A local architecture firm has completed renovation plans for an extension to the N'Swakamok Native Friendship Centre. A detailed landscape design has been completed for the Elgin Greenway, a pedestrian and cycle pathway that connects Laurentian University with the downtown core along the waterfront and rail corridor. While funding is uncertain for the Elgin Greenway, planning staff hoped to raise the necessary provincial and federal government support to make it **shovel-ready** in 2014. Other **public realm** improvements include new street furniture, bicycle parking, and an on-street parking pilot project to increase parking close to downtown retailers. Finally, the Greater Sudbury Development Corporation (GSDC) is moving ahead with plans to develop a new Downtown Visitors Centre, along with drafting a new retail strategy and public art policy. Together, these projects demonstrate that the will exists in Sudbury to initiate change. What is holding the bulk of these projects back from realization is lack of capital funding. To assist with revenue generation, priority projects are being considered as part of Sudbury's **development charges** bylaw update. Sudbury voted in favour of extending development charges to include the following general (government, libraries, fire services, police services, parks and recreation, public works, emergency medical services, public transit, and emergency preparedness) and engineering (roads, water, wastewater, and drains) services. But city officials in conjunction with the GSDC will also need to continue to actively lobby other levels of government for financial support and write grant applications to take advantage of matching funding programs.

In realizing the vision outlined in the plan, it is also important that social responsibility remain a primary concern. Sudbury scores poorly on many measures of individual and community well-being (e.g., air quality,

Figure 8.1.6 • Regeneration sites (a) Laurentian School of Architecture and (b) CP Rail Market.

Source: A.L. Bain.

childcare, and diversity). It also has above provincial and federal averages for economic dependency on social programs and poverty rates (Leadbeater 2008). One of the main criticisms of the plan is that the health and quality of life impacts on the poor and homeless are inadequately addressed, potentially making it "easier for the city to evade those responsibilities because they didn't buy into that logic" (Gauthier, personal communication, 4 July 2013). In the early phases of the plan development process, concerns were raised by some local residents about the potential for gentrification in downtown neighbourhoods and the ensuing potential displacement and impoverishment of low-income groups. While the plan is very focused on placemaking and physical infrastructure improvements to generate economic development opportunities (as opposed to non-profit-generating social programming investments), City planners have made sure that a representative of the Social Planning Council is at the decision-making table and that social service agencies and their clients have been consulted with respect to any projects that may affect them. Moving forward, how the most economically marginal and vulnerable social groups in Sudbury will benefit from future development should remain an important consideration in the evaluation of project and funding priorities.

Conclusion

The terms "participation," "empowerment," "partnerships," and "education" have been central to the Downtown Master Plan process in Sudbury. The plan was framed from the outset as a collective visioning process carefully designed to include key stakeholders able to deliver specific agenda items as well as community members who saw opportunities in the downtown. Inevitably, some stakeholders have been under-represented in the consultation process. However, the intent of the consultants and City planners was always to engage as broad a sample of local residents as was willing to participate. Since the completion of the plan, new partners have been brought into the process. The CLG continues to evolve as its delivery priorities shift. Such an engagement strategy has resulted in a widespread sense of ownership and buy-in of the finished product. So involved have individual and community groups become—committed to their individual projects but also to a new vision of the downtown as a destination and an experience—that they have advocated for the CLG to continue to work in tandem with city officials to help deliver elements of the plan.

While "[p]articipation and empowerment are now a convention within urban redevelopment practice" (Pollock and Sharp 2012, 3064), in Sudbury it appears to be more than mere rhetoric. Participation in this case study is not "tyranny" or "tokenism" but rather the mechanism through which some Sudburians have learned to work *with* city staff and councillors on a relatively level playing field (Pollock and Sharp 2012, 3064). Although the City is providing staff resources to aid delivery, it also relies on community partners to build the momentum of the plan, ensure ongoing buy-in, secure additional funding, and prepare business plans and feasibility studies. The Downtown Master Plan process has facilitated relationship-building between community groups; some individuals within these groups have learned how to navigate institutional structures and to operate within the discursive spaces of urban planning decision-making. In Sudbury, the plan process has left a lasting impression on the city, helping to build local "bonds of familiarity and trust" that have "facilitate[d] consensus and collaboration" (Lewis and Donald 2010, 49). Because of the trust developed throughout the project and the genuine relationship forged between the consultant team, the client group, local politicians, and key stakeholders, the plan has garnered significant political and public support. It is these relationships and, perhaps more important, a feeling of ownership over a shared vision and strategic plan for improvement that have supported such a swift uptake of the plan, moving it from vision to lived reality.

KEY TERMS

Development charges (or Development Cost Levies)

Master plan

Public realm

Shovel-ready

REFERENCES

Ardron, R., E. Batty, and I. Cole. 2008. "Devising and delivering masterplanning at neighbourhood level: Some lessons from the New Deal for Communities Programme." Sheffield, UK: Sheffield Hallam University. http://extra.shu.ac.uk/ndc/downloads/general/masterplanning_neighbourhood_level.pdf.

Bain, A.L., and D. McLean. 2011. "Eclectic creativity: Interdisciplinary creative alliances as informal cultural strategy." In A. Lorentzen and B. van Heur, eds, *Cultural Political Economy of Small Cities*, 128–41. London and New York: Routledge.

——. 2013. "From post to poster to post-industrial: Cultural networks and eclectic creative practice in Peterborough and Thunder Bay, Ontario." In M.M. Breitbart, ed., *Creative Economies in Post-industrial Cities: Manufacturing a (Different) Scene*, 97–121. Farnham, UK: Ashgate.

Bell, D., and M. Jayne, eds. 2006. *Small Cities: Urban Experience beyond the Metropolis*. London and New York: Routledge.

Bradley, B. 2010. "Regreening: 3,350 hectares done, but 30,000 hectares to go." *Northern Life* 25 January. http://www.northernlife.ca/news/localNews/2010/01/regreening260110.aspx.

Breitbart, M.M., ed. 2013. *Creative Economies in Post-industrial Cities: Manufacturing a (Different) Scene*. Farnham, UK: Ashgate.

CABE (Commission for Architecture and the Built Environment). 2008. "Creating successful masterplans: A guide for clients." London: CABE. http://webarchive.nationalarchives.gov.uk/20110118095356/http://www.cabe.org.uk/files/creating-successful-masterplans.pdf.

Edinger, E. 2008. "Environmental impacts of nickel mining." In D. Leadbeater, ed., *Mining Town Crisis*, 103–23. Toronto: Fernwood.

Ferrigan, J. 2012. *Report: Downtown Sudbury Master Plan and Action Strategy—10 Year Implementation Plan*. Sudbury: Planning Services.

Garrett-Petts, W. 2005. *The Small Cities Book: On the Cultural Future of Small Cities*. Vancouver: New Star Books.

Leadbeater, D., ed. 2008. *Mining Town Crisis: Globalization, Labour, and Resistance in Sudbury*. Toronto: Fernwood.

Lewis, N.M., and B. Donald. 2010. "A new rubric for 'creative city' potential in Canada's smaller cities." *Urban Studies* 47(1): 29–54.

Nelson, G., ed. 2012. *Beyond the Global City: Understanding and Planning for the Diversity of Ontario*. Montreal and Kingston: McGill-Queen's University Press.

Pollock, L.V., and J. Sharp. 2012. "Real participation or the tyranny of participatory practice? Public art and community involvement in the regeneration of the Raploch, Scotland." *Urban Studies* 49: 3063–79.

Saarinen, O. 1990. "Sudbury: A historical case study of multiple urban economic transformation." *Ontario History* 83(1): 53–81.

——. 2013. *From Meteorite to Constellation City: A Historical Geography of Greater Sudbury*. Waterloo, ON: Wilfred Laurier University Press.

Short, J.R. 1999. "Urban imagineers: Boosterism and the representation of cities." In A. Jonas and D. Wilson, eds, *The Urban Growth Machine: Critical Perspectives Two Decades Later*, 37–54. Albany, NY: State University of New York Press.

Urban Strategies Inc. 2011. *Downtown Sudbury: A Plan for the Future*. Sudbury: City of Greater Sudbury.

van Heur, B. 2010. "Small cities and the geographical bias of creative industries research and policy." *Journal of Policy Research in Tourism, Leisure and Events* 2(2): 189–92.

Wallace, C.M. 2004. *Sudbury: Rail Town to Regional Capital*. Toronto: University of Toronto Press.

INTERVIEWS

J. Ferrigan, 5 July 2013. In-person interview.

G. Gauthier, 4 July 2013. In-person interview.

K. Tait-Peacock, 4 July 2013. In-person interview.

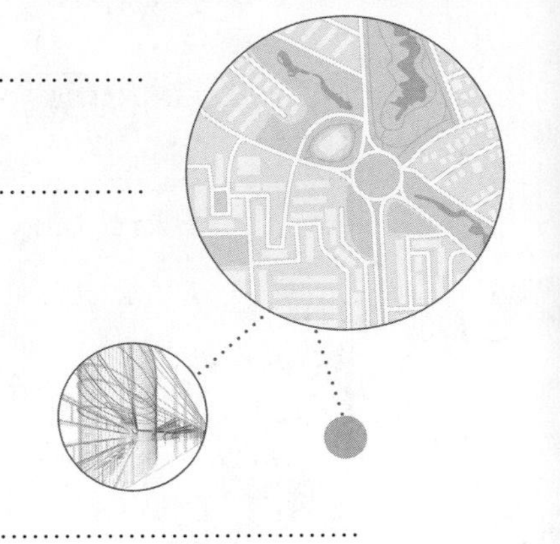

CASE STUDY 8.2

After "Ours": Creating a Sense of Ownership in Winnipeg's Downtown Plan

Gerald H. Couture

SUMMARY

This case study illustrates an innovative planning program designed in support of the creation of a downtown plan for the City of Winnipeg. The intention was to build a sense of "ownership" on the part of downtown stakeholders. The case study explains how the concept of "ownership" can be deliberately pursued as a strategy onto itself through such means as organizational structure, shared decision-making authority, community outreach, consultative processes, and funding commitments. The process was carried out over a two-year period from 1993 to 1994 and resulted in the production of two adopted documents: *CentrePlan: Vision and Strategies* and *CentrePlan: Action Plan 1995–1996*. In 1995, the plan was recognized when it won the Canadian Institute of Planners' Honour Award for Planning Excellence and the International Downtown Association's Award for Outstanding Achievement. The author was project manager for CentrePlan.

OUTLINE

- The CentrePlan process took place in the City of Winnipeg in the early 1990s and was implemented over the following years.
- The CentrePlan process featured a unique organizational structure and decision-making authority that involved a range of stakeholder groups.
- One result of this process was a plan that many citizens saw as "theirs," reflecting their sense of ownership and inclusion. The regeneration of the downtown through the pursuit of shared goals, following decades of decay and disinvestment, can be considered instructive for other mid-sized Canadian cities.

Introduction

It is perfectly natural and reasonable for people to want to have a say in decisions that shape their lives. As planners, we are entrusted with the responsibility to make this happen. But is having a say in decisions enough? How does one not only accommodate public input into a planning program but ensure that participants genuinely feel it is *their* plan in the end, that it accurately reflects *their* aspirations? Can planning structures, processes, and products be devised such that the result is a true community plan, not the planning authority's plan? These were the questions we wrestled with in the mid-1990s as a small team of planners undertook the task of developing a new downtown plan for Winnipeg.

Winnipeg's downtown is expansive and diverse. With the historic intersection of Portage and Main as its heart, the downtown encompasses 780 acres or 1.2 square miles. It is framed along the south and east by the Assiniboine and Red rivers, is crisscrossed by 24 miles of roadway, and is accessed by eight bridges accommodating pedestrian, rail, and vehicular traffic. Home to City Hall and the provincial legislature, the downtown also boasts the city's central business district, two retail malls, a large 20-block historic area called the Exchange District, a modest Chinatown, a cultural district, two long-standing residential enclaves, and the University of Winnipeg.

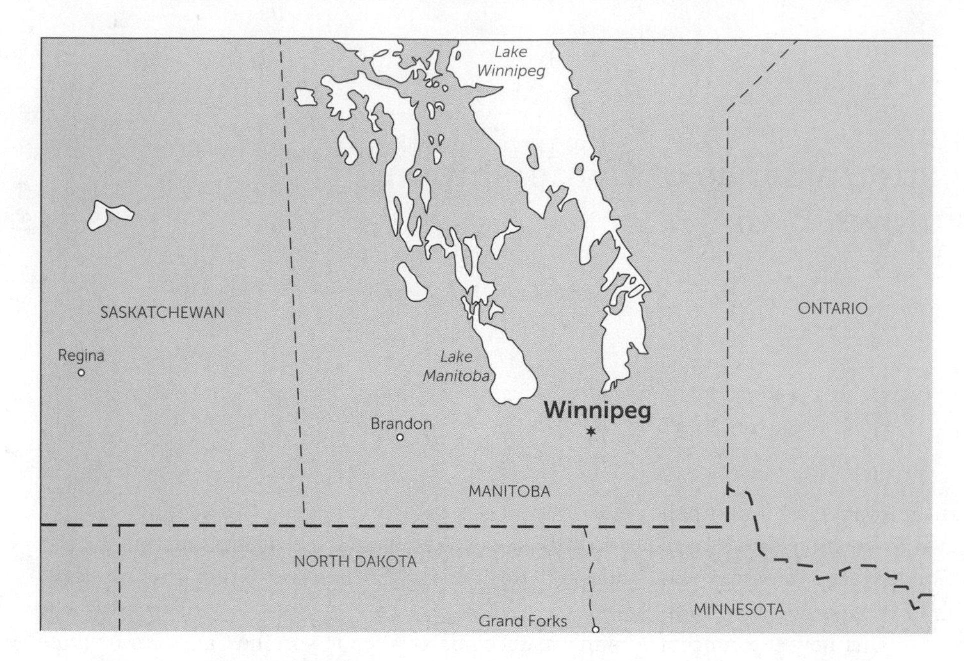

Figure 8.2.1 • Location Map.

In the early 1990s, the downtown had a population of 14,000 in a city of 630,000 and employed 68,000 people, one-quarter of the city's total workforce.

Typical of many North American cities in the 1970s, Winnipeg's downtown was in a state of serious decline following several decades of postwar suburbanization and the decentralization of retail into suburban malls. Decay was evident in the number of boarded-up and vacant buildings, large tracts of surface parking, and the rise of social problems. Significant intervention was needed. To their credit, all three levels of government came together to create the Winnipeg Core Area Initiative, an innovative agreement intended to address the physical, economic, and social challenges through a comprehensive funding commitment broken down into discrete programs.

The initiative allowed diverse public agendas to be pursued in tandem. The federal government could direct its funding toward economic development and employment, the provincial government could fund community development and housing, and the municipal government could emphasize physical, bricks-and-mortar improvements. From 1981 to 1991, nearly $200 million in public funds was spent in the core area (the downtown *plus* its surrounding older neighbourhoods), leveraging more than three times that amount in private-sector investment. Much of this was focused directly on the downtown. Yet in spite of these efforts, the downtown was still suffering.

More needed to be done, and a long-term approach was essential. Long-range planning for the downtown had given way to government-led projects and programs, and while they were recognized as important contributors to revitalization, they did not succeed in building the momentum needed to stimulate ongoing, independent private-sector investment. A plan was deemed necessary—one driven by a compelling vision—and it was seen as self-evident that a comprehensive, inter-jurisdictional approach was the way to go. Government, the private sector, and the not-for-profit sector all had roles to play.

In June 1993, following a year of background research and preparation, CentrePlan was launched. Over the next 18 months, 240 people played active roles on committees, and approximately 900 more participated in various consultative sessions. With the tag-line "Working Together for Winnipeg's Downtown," two documents were produced: *CentrePlan: Vision and Strategies* and *CentrePlan: Action Plan 1995–1996*. In December 1994, both CentrePlan documents were adopted by City Council, and soon after, the boards of key stakeholder organizations were asked to do the same. They garnered unanimous support.

Building Our Structure

While the Planning Department of the City of Winnipeg instigated the planning initiative, there was a deliberate effort to assign leadership responsibilities to the broad community in order to build a sense of ownership. The planning team concluded that three roles needed to be filled: we needed people who could demonstrate community leadership, people who could make things happen

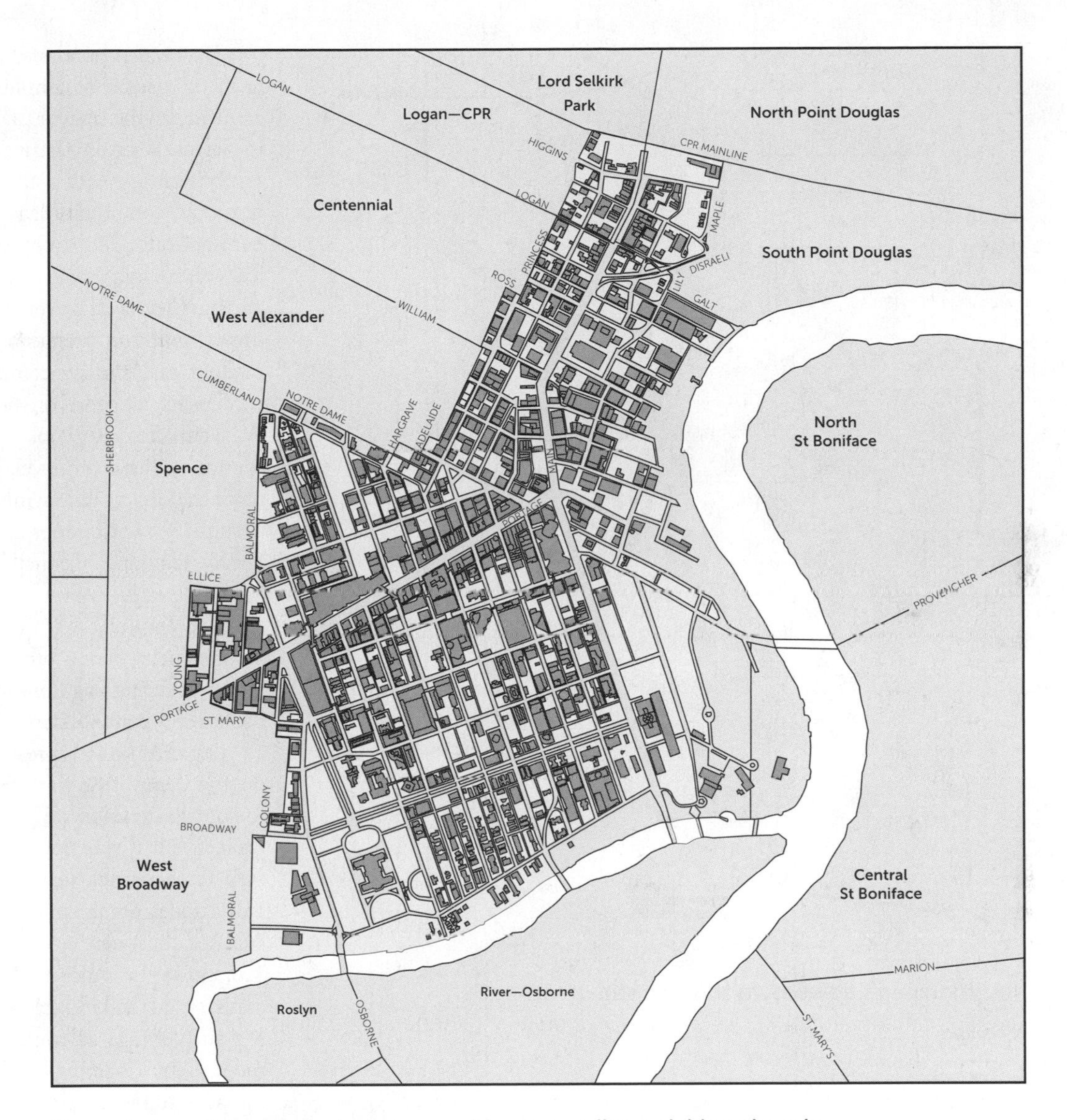

Figure 8.2.2 • Winnipeg's downtown with surrounding neighbourhoods.

Source: Adapted from City of Winnipeg 1994b, inside cover.

through implementation, and people who had strong subject-matter knowledge to point us in the right direction. As a result, we established a Steering Committee, an Advisory Committee, and a series of Strategy Teams. They became the drivers of the plan.

The alternative was to adhere to a more traditional process, one that would be led by the City and driven by municipal planners—albeit with community engagement as a component of the process—leading to the production of a municipal plan. This approach was favoured by many of our colleagues, including our director, and it took some persuasion to gain their support. We were convinced that even a great municipal plan would not be enough; we needed the community to galvanize around its *own* plan.

The Steering Committee

The Steering Committee was established with a wide range of key stakeholders representing the public sector, the private sector, and the not-for-profit sector. The mayor was approached to participate because we felt her

Figure 8.2.3 • Winnipeg's downtown showing sub-areas.

Source: City of Winnipeg.

involvement was necessary to instill a sense of importance and legitimacy to the process. It took some convincing. This was partly due to the demands on her time but, more powerfully, because she was new to office and, having been elected on a "clean-up-city-hall" platform, was wary of being set up to fail. Through a series of informal chats during which we pleaded our case, she became convinced. (She ultimately served two terms, from 1992 to 1998, retiring from office having accomplished much of what she set out to do.)

To build broader political support, the mayor was joined on the Steering Committee by the three City councillors whose wards included portions of downtown, by the provincial minister responsible for municipal affairs, and by the federal member of Parliament representing the downtown riding.

That was a good start, but to instill a sense of community ownership, the mayor agreed to act as co-chair, sharing the responsibility with a well-respected community leader of her choosing. The strong political representation, including all three levels of government, drew significant attention and made it easier to get community leaders to come on board. We went after key people representing, at the board level, such organizations as the Winnipeg Chamber of Commerce, the Social Planning Council, the Winnipeg Convention Centre, and the Manitoba Arts Council, among many, many others. In the end, the Steering Committee included 35 community leaders.

Organizations represented on the committee were asked to contribute financially to the project, but it was not mandatory to their inclusion. About half a dozen organizations contributed between $2000 and $10,000 to the process. Others contributed in-kind by offering such things as conference rooms for meetings. While the additional resources came in handy, the primary intent was to bolster the sense of ownership, putting the "stake" in stakeholder.

The Advisory Committee

The Advisory Committee was configured to mirror the composition of the Steering Committee, but while the Steering Committee was generally made up of elected officials (politicians and board members), the Advisory Committee represented the same organizations from a staff point of view (the CEO, the executive director, or a key staff person). The fact that the chair or another prominent member of the board was represented on the Steering Committee gave us leverage, making it easy to get participation from their administrative counterparts. The Advisory Committee included 45 members.

We appointed two people to co-chair the Advisory Committee, one private-sector, one public-sector. This served two purposes: it created shared responsibility and ownership while ensuring that we had a back-up in case one of them could not make it to a meeting. Also, the two individuals agreed to sit on the Steering Committee as Advisory Committee liaisons, thereby ensuring open and honest communication between committees.

The Strategy Teams

Five Strategy Teams were formed. Early on, we undertook some visioning exercises with the Steering Committee to establish a general sense of direction. Five separate but interconnected statements were produced, and these concepts were used to establish the corresponding strategy teams. The five areas comprised:

- social well-being (a vision of community and belonging);
- sustainable economic development (a vision of prosperity and innovation);
- form and function (a vision of effectiveness and efficiency);
- downtown character (a vision of soul and personality);
- leadership and organization (a vision of direction and commitment).

We sought out experts in these areas to populate the teams. For example, the sustainable economic development team included the director from Economic Development Winnipeg, the general manager of Tourism Winnipeg, and the head of the International Institute for Sustainable Development, among many others. In selecting individuals, we also looked at overall balance to ensure that there would be a good, healthy cross-section of individuals and opinions. The size of the teams ranged from 14 to 49. We appointed two people from the Steering Committee to co-chair each Strategy Team, which helped to mesh the structure.

Support Structure

The organization chart shows how the pieces fit together (Figure 8.2.4). The Steering Committee was the ultimate decision-making authority, with advice coming from the Advisory Committee and from the five Strategy Teams.

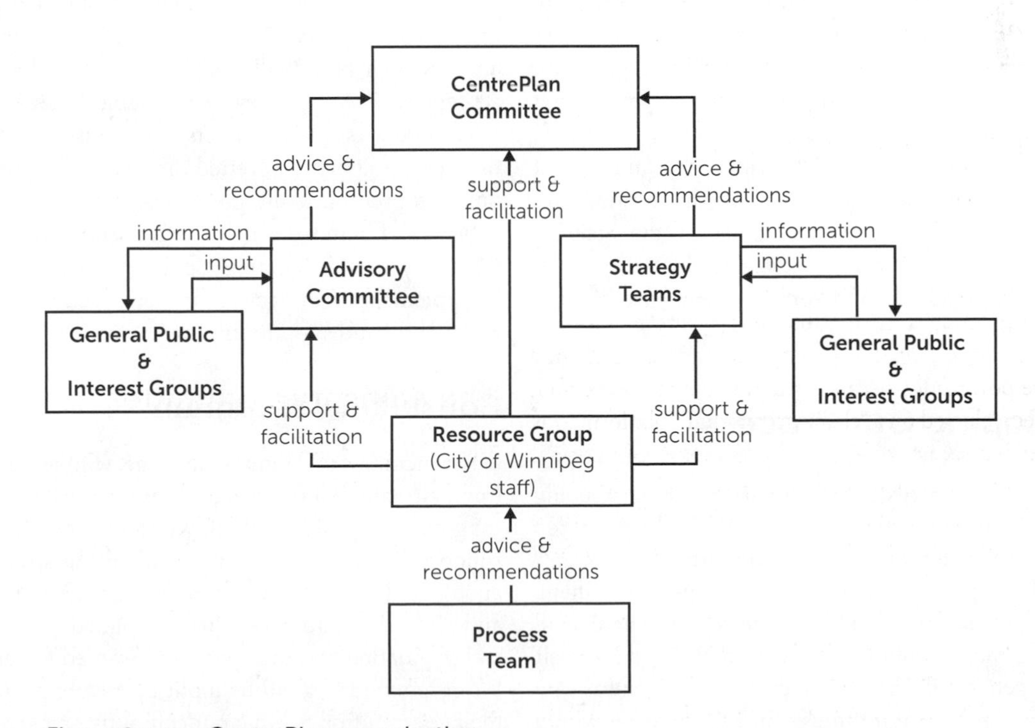

Figure 8.2.4 • CentrePlan organization.

Source: City of Winnipeg 1994a, 58.

As the planning team responsible for the project, we called ourselves the Resource Group to reinforce in the minds of participants that we were not leading the planning process but supporting and facilitating it.

As the Resource Group, our role was to direct the overall project toward success and to bring to bear the resources necessary to make that happen. Perhaps our greatest responsibility was to ensure that the Advisory Committee and the Strategy Teams were well informed. So we ran a parallel consultative program in the broad community while feeding the results to these groups. We also offered our background research and brought in guest speakers for inspiration.

Four Quick Tips on Structure

Build peer-to-peer committees. People relate best to one another in a peer-to-peer context. Populate committees by matching a mayor with a chair of a board, a councillor with a board member, a senior bureaucrat with a senior corporate staffer, and so on. Don't mix apples and oranges.

Use available political clout. Having the mayor take the time to personally call individuals to participate on committees played to their feelings of self-worth and worked well to garner their commitment. There were virtually no refusals.

Be clear on expectations. Many people are willing to lend their time and expertise to planning efforts, but it's critically important to treat them fairly. That begins with a clear and accurate delineation of the time commitment required to fulfill roles and responsibilities.

Make it official. Wanting to entrench commitment while promoting the importance of the posting, we developed letterhead with the names of the Steering Committee members on it. This was used for all project-related correspondence and served as evidence of community ownership.

Developing Our Process

Setting up committees and teams is one thing; keeping them busy and productive is another. We had established terms of reference for each of the committees and teams, so they had a good sense of what was expected of them. However, we had not laid out in detail how the overall planning process would work and how the plan itself would be generated. The reason for this oversight was simple: we hadn't thought it through and assumed we would figure it out as we went along. This was a big mistake.

While the Steering Committee was busy early on with a series of visioning exercises, the Advisory Committee was used as a vetting body for that work, evaluating the feasibility or viability of ideas. It quickly became apparent that the committee members were not comfortable second-guessing the work of their political masters, and they responded by assuming control of the planning process, suggesting what they should be doing instead. In the absence of a defensible process, we were forced to comply, and the initial meetings became consumed with matters of process rather than content.

To address the concerns being raised and wanting to ensure that we continued to build ownership, we decided to form a Process Team, a subcommittee of the Resource Group. We populated the team with those who were particularly vocal about what we should, and should not, be doing. Over the course of several meetings, we managed to produce a comprehensive Process Charter, something that should have been done before the project started. The main fall-out was loss of credibility. As a result, members of the Process Team continued to second-guess the work of the Resource Group to the very end.

The Strategy Teams, however, worked smoothly. These people were problem-solvers and seemed to relish the opportunity to contribute toward solutions. Here, the process was relatively straightforward. We provided each team with the results of our public consultations and research and challenged them to develop relevant strategies. Because they were subject-matter experts (e.g., engineers, social workers, economists), we relied on their best judgment as vetted through a team consensus-building process. Many great ideas were generated, and they were filtered through the Advisory Committee on their way to the Steering Committee—from strategists to implementers to decision-makers. Ideas that survived were sound and supportable.

Consultation Program

Our structure was intentionally large, with representation from virtually every stakeholder organization or interest group we could think of. We were confident that the support of those directly involved in the structure was enough to build a successful plan. Yet we knew that, over and above the 240 people with assigned roles, the general population was interested and wanted to participate. We accommodated public input by having people target their contribution to what our committee most needed to understand about the downtown.

We asked participants three basic questions:

1. What *do you like* about the downtown?
2. What *don't you like* about the downtown?
3. What *changes* would you make to improve the downtown?

For each question, we asked them to establish their top three priorities. Additionally, we asked for three adjectives to describe the ideal downtown. This simple exercise generated words such as "prosperous," "vibrant," "safe," among a host of others, and proved invaluable in helping us build the vision and strategies that formed the core of the plan. Further, it helped to reveal significant consistency in aspirations among stakeholders.

We committed to being fluid. Whoever wanted to participate, individual or group, was accommodated. But this was not enough. If we felt an individual or group needed to be heard and wasn't coming forward on their own, we went after them. One example: we knew the downtown was home to many disenfranchised people, some living on the streets, others living in single-room-occupancy hotels. These people had a right to be heard, but how could we reach them? After some creative brainstorming, we recruited the City's public health nurses and armed them with a short questionnaire outlining our three basic questions. At downtown clinics and on their monthly rounds to visit their clientele, they asked the questions and provided us with the results. The trust relationship they had with their clients facilitated the discussion.

In all, we held 40 workshops with various interest groups. We provided organized groups with two options: we would meet with them at their convenience and facilitate a workshop, or we would provide them with materials and they could do it on their own. Any interested individual could submit an idea as well. Worksheets were available at City Hall, libraries, and other locations in and around the downtown (this exercise predated social media).

We also organized two public forums. They were advertised broadly, and anyone could attend. We recruited members of our Advisory Committee and many of our work colleagues to assist, assembling and instructing 36 facilitators, each assigned to run a series of exercises at a table seating eight to 12 people. Results were recorded on large sheets of paper and pinned to the wall as the evening went along. A local public broadcaster was present and aired the proceedings. It helped to generate some buzz in the community.

To ensure that the forums were well attended, we challenged ourselves to identify barriers to participation and sought solutions to overcome them. For example, we offered rides to seniors and provided on-site childcare with licensed childcare providers to encourage single parents to attend. And, yes, we provided free food, a proven strategy!

Four Quick Tips on Process

Work hard to get participation. It is not enough to provide opportunities for the public to participate. If participation is deemed important to the outcome, then a major effort must be made to ensure attendance. Call. Call again. Meet reasonable demands. Bribe if you have to! But get them out.

Shamelessly seek free professional advice. Professionals are often eager to share their knowledge. For example, engineers, architects, and police officers, if invited to participate in an engaging way, can provide real insight into solutions and do not demand consulting fees.

Validate your results. To be certain that what we were hearing in our workshops and forums was truly indicative of the interests of the broader community, we commissioned a couple of polls at strategic points in time. The polls were scientifically valid and gave us comfort in knowing we were on the right track.

Trust the process. As planners, we often feel we need to shape results toward our own understanding of solutions we believe will work. In this case, though it took some doing, a solid process was established, and we had to trust it would deliver sound results in the end. It did.

Preparing Our Plan

While CentrePlan was largely process driven, it needed to result in a tangible product as evidence of direction and commitment. From the beginning, we understood that one of the greatest challenges we faced was the public's distrust of municipal officials. A few participants insisted that the process was contrived toward predetermined solutions. The content of the plan would have to dispel this impression. We adhered to two principles.

First, we worked hard to make the plan true to what we had heard. We considered ourselves "scribes" as opposed to writers. We scoured all the meeting and workshop notes to ensure that we properly captured what was said. Each discrete idea was reviewed and categorized as a broad goal, a strategic direction, or a specific action. Much of this material reflected a considerable consensus of opinion in that the same idea would emerge from a number of sources. In other cases, an idea would reflect a lower consensus of opinion but at least some shared support. Other ideas were isolated outliers. So we colour-coded the ideas accordingly. Our role was to organize the material, not to edit it. Every idea was classified according to two criteria: the type of idea and the degree of consensus. Further, each idea was traceable to its source.

Second, we sought continual input into the plan itself. Many drafts were produced, and each one was run past the strategy teams and the two lead committees in order to generate an appreciation for the breadth of the material we had to work with and to build ownership over the scrutiny process required to translate it into a comprehensive plan. Because we showed them each draft, stakeholders were able to see the way their previous comments and suggestions had been incorporated into the newer version and how the full range of input was accommodated and balanced. As a result of this iterative process, we were able to pare the results down into a manageable plan.

Because we didn't want to lose any of the interesting and legitimate ideas even if they did not make the cut, we published an accompanying reference document with a comprehensive list. In this way, all participants in the process could see for themselves that a fuller range of ideas was considered.

In the end, the five chapters of the *Vision and Strategies* document (City of Winnipeg 1994a) captured 15 "opportunities" and 65 "strategies." All of this was deemed to represent the community's aspirations for a rejuvenated downtown, explaining *what* needed to be done in pursuit of a comprehensive vision. A second document was created to address the "how"—that is, the actions that would be taken. While *Vision and Strategies* was intended to have some measure of longevity, the *Action Plan* was to be updated on a regular basis. To emphasize the point, a two-year time frame was established initially: *Action Plan 1995–96* (City of Winnipeg 1994b). Thirty-five actions were identified. Each specified the agency responsible, the required resources, the intended benefit, and the first steps needed for getting it done.

What was unique to this plan was that the actions and responsibilities were not assigned solely to City Council (although most were). Some committed other levels of government, some committed arm's-length agencies of government, and others committed private organizations. One action, for example, committed the downtown's two Business Improvement Zones (self-funded by downtown businesses themselves) to implement a downtown safety patrol program. The idea originated with them.

Four Quick Tips on the Plan

Record everything. Especially when a planning initiative is process-heavy, it is critically important to keep a reliable record of all the meetings, consultations, and draft plans. You may need to demonstrate how you got from Point A to Point B, and it's surprising how quickly memories fade.

Build the product together. We hired a graphics/production person early on so that participants could see the document evolve visually. The final draft was not a surprise, because participants helped to shape its look and feel as well as its content.

Use their words. Using quotes verbatim and repeating participant anecdotes at meetings helped to demonstrate that we were listening. We also paid close attention to the specific words used. For example, we changed the word "vibrant" to "animated" when we heard that word used by a member of the Steering Committee. Good word.

Sign-off on commitments. In the end, each organization represented on the Steering Committee was asked to adopt the plan through a formal motion of their board. In cases when an organization was identified in the Action Plan as the implementing agency, it was also asked to endorse the specific action(s), including, in some cases, the commitment to financial expenditures.

After "Ours"

In planning, success often hinges on the creation of a sense of shared ownership and responsibility. The challenge is to ensure buy-in from a wide range of proponents, all of whom can contribute to a plan's success through its

Figure 8.2.5 • Red River College Princess Street Campus, 2003.

Source: Corbett Cibinel Architects.

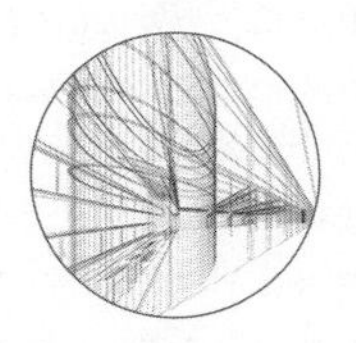

PLANNING THEORY

The Role of Planners in Public Engagement—Degrees of "Ownership"

Communication: sharing information, building understanding
Cooperation: moving separately toward common goals
Co-ordination: moving jointly toward common goals
Collaboration: co-creating, combining multiple interests into one
Partnership: establishing co-ownership, merged interests

Source: Adapted from Arnstein 1969 and Wight 1998.

implementation. People, particularly those in positions of authority, need to feel that the magnitude of the challenges facing their community requires them to be partners in finding and committing to solutions, that the plan cannot succeed without their active participation and ongoing contribution. It is *their* commitment that will ensure the success of the plan. The most valuable lesson that can be passed on to young planners, the one we learned first-hand, is to park your ego. Success lies in taking the Y out of YOUR, turning *your* structure, *your* process, *your* plan into one the community recognizes as *our* structure, *our* process, and *our* plan.

Postscript

At the time that CentrePlan was being prepared, Winnipeg had undergone more than a decade of revitalization efforts through two tri-level government agreements. While the model was seen as innovative and generated an impressive list of projects and investments, the negotiation required to put these agreements in place was formidable. Two years after the conclusion of the second Core Area Initiative and as a result of ongoing discussions, the three orders of governments agreed to a new $75-million agreement called the Winnipeg Development Agreement (WDA). The timing was fortuitous. The WDA was able to use CentrePlan to its advantage (and vice versa), aligning many of its funding programs and initiatives with the plan, knowing it had recently been adopted with broad community support. The community "owned" the plan, and this made political support easily obtainable. The funding agreement lasted six years and helped propel the plan forward.

Just over half of the 35 actions in the *Action Plan 1995–1996* were fully implemented over the two-year period. Others remained works in progress. The few not initiated were carried forward into the subsequent *Action Plan 1997–1999* (City of Winnipeg 1997). The 29 actions in that plan were pursued with varying degrees of success and with the CentrePlan Committee remaining to provide oversight.

Meanwhile, in 1999, as a result of requests for a more physical plan, the *CentrePlan Development Framework* (City of Winnipeg 1999) was prepared, emphasizing urban design and redevelopment improvements. As a parallel initiative, CentreVenture was created as the city's downtown development agency with a mandate to implement the framework. Seed funding was granted by the City for the agency to provide mortgages, gap financing, and

Figure 8.2.6 • Waterfront Drive condos (Exchange District East), 2005.

Source: Wikipedia.

loan guarantees. Over time, these responsibilities grew to include marketing the City's surplus properties for sale or redevelopment.

The CentrePlan Steering Committee evolved into CentreVenture's board of directors, and the Action Plans evolved into CentreVenture's Annual Business Plan. From the outset, the province provided annual operating funds for CentreVenture and continues to do so. The agency remains the driver for downtown revitalization efforts today while periodically refreshing the *CentrePlan Development Framework*. *CentrePlan: Vision and Strategies* has never been updated.

While reference to CentrePlan essentially faded away after seven years or so, it can be argued that the direction it provided, the community support it embodied, and the projects it successfully implemented generated the momentum and commitment that remains evident today.

REFERENCES

Arnstein, S.R. 1969. "A Ladder of Citizen Participation." Journal of the American Institute of Planners 35(4): 216–24.

City of Winnipeg. 1994a. *CentrePlan: Vision and Strategies*. Winnipeg: City of Winnipeg.

———. 1994b. *CentrePlan: Action Plan 1995–1996*. Winnipeg: City of Winnipeg.

———. 1997. *CentrePlan: Action Plan 1997–1999*. Winnipeg: City of Winnipeg.

———. 1999. *CentrePlan Development Framework*. Winnipeg: City of Winnipeg.

Wight, I. 1998. "Building regional community: Collaborative common place-making on a grand scale." Paper presented at the Parkland Institute Annual Conference, 12–14 November, Edmonton.

CASE STUDY 8.3

The Quartier des Spectacles, Montreal

Laurie Loison and Raphaël Fischler

SUMMARY

The Quartier des spectacles (Entertainment District) of Montreal is one of those branded environments that have sprung up in many Western cities as a means to redevelop central areas, stimulate economic development, and project an attractive image to tourists and investors. Initially proposed by local actors as a means of protecting a cluster of artistic and cultural institutions, large and small, from local redevelopment and from suburban competition, the Quartier des spectacles became part and parcel of the efforts of the City of Montreal to develop its tax base through real estate development and to put Montreal on the global map of creative cities. The project underwent a double transition: from bottom-up to top-down planning in terms of process and from functional and social diversity to commercial entertainment and high-end development in terms of substance. The planning of the Quartier des spectacles, which is followed here over the course of about 10 years but is situated in a longer-term historical context, exemplifies the intersection of urban design and economic development policy and highlights the conflict that may arise between metropolitan ambitions and the needs of local stakeholders and residents. It also illustrates the collaboration of various levels of government and the use of alternative modes of governance to implement strategic urban projects.

OUTLINE

- In the early 2000s, the Quartier des spectacles in Montreal transitioned from an ill-defined agglomeration of cultural facilities, entertainment venues, and areas of marginal activity to a high-end, branded neighbourhood open to commercial development.
- The changing goals for the area, the planning actors involved in the process, and the needs of the creative community and local residents have contributed to conflicting visions for the future.
- The case portrays the conflicts between economic development and social equity, between the attraction of private investment and the displacement of low-income or marginal populations.

Introduction

Montreal boasts a very active downtown where hundreds of thousands of people reside or commute to for work, study, or entertainment. It is well served by public transportation (commuter train lines, "metro" lines, and numerous bus routes link it to large parts of the metropolitan area) and has remained the most important shopping destination of the region despite the growth of suburban shopping malls and **power centres**. Like nearly all North American downtowns, it suffered from **modernist** urban renewal in the 1960s and 1970s and was left with many parking lots in the 1980s because of speculative demolitions and the rise of the private automobile as the primary means of transportation.

After the recession of the mid-1990s, when the future of downtown seemed compromised, the City of Montreal launched a series of ambitious plans and projects to boost its vitality and stimulate private development. The main idea behind this policy is well known to those who have studied revitalization efforts in Western cities: an attractive **public realm** and a concentration of similar activities can create a social and economic dynamic of urban (re) development. One of the first projects in which these ideas were implemented, in the early 2000s, was the Quartier international de Montreal (QIM), an area southwest of the central business district where the International Air Transport Association and many other international

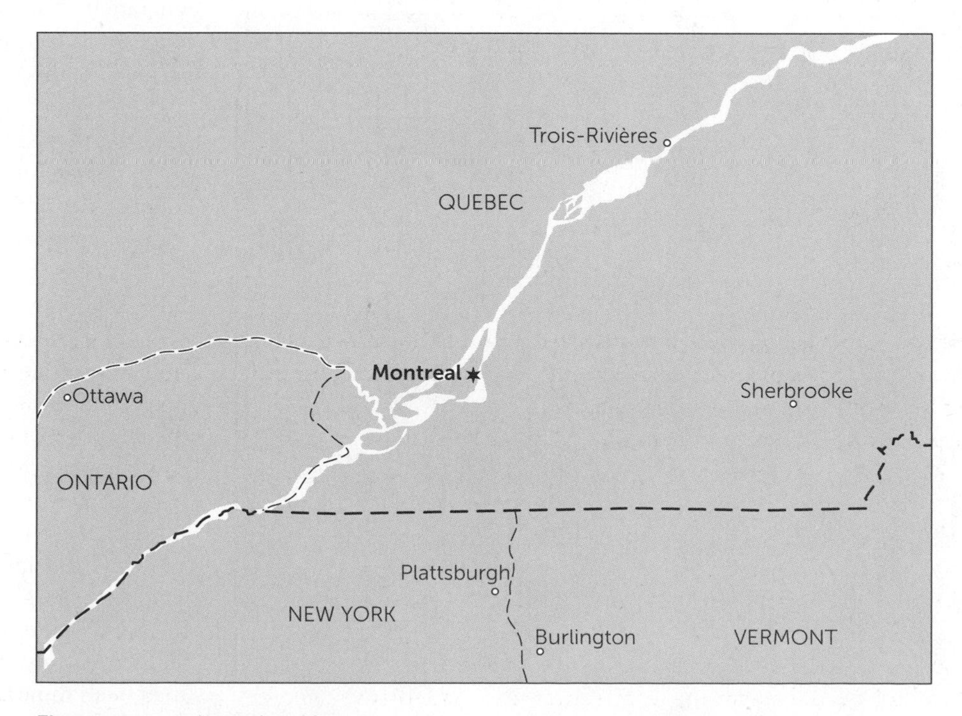

Figure 8.3.1 • Location Map.

organizations are located. Working with a coalition of local property-owners and with higher levels of government, the City used a mix of new public buildings (financed by the Province) and new or redesigned public squares to rebuild a torn urban fabric and create an attractive environment for international agencies, a larger venue for conferences and trade shows, and a new residential district (Aldinger 2007). A hallmark of the QIM was the use of local talent in the design of signature public spaces and street furniture.

The Quartier des spectacles is the latest project that the City has undertaken, again in partnership with local actors and higher levels of government, to redevelop a part of downtown that was functioning well below its potential. When it was initially targeted for change, in the early 2000s, the area was home to a great number of arts and entertainment venues, including Place des Arts (Montreal's equivalent to New York's Lincoln Center), but it also contained many empty lots and was known for its illicit activities related to drugs and sex. Its redevelopment has been part of a campaign to strengthen the city core and of a more general effort to make Montreal a "creative city" with global appeal. This chapter is therefore a contribution to the growing literature on cultural districts and their role in the making of the contemporary city (e.g., Hospers 2003; Markusen and Schrock 2006; Montgomery 2003; Pratt 2008; Zukin and Braslow 2011). Its primary goal, however, is to explain how the planning and implementation of the Quartier des spectacles unfolded and what its creation means for urban policy and planning.

Geographic and Historical Context

The Quartier des spectacles is part of what is historically known as the Faubourg Saint-Laurent, one of the three original suburbs that grew outside the walled city of Montreal. Its main axis was Chemin Saint-Laurent, now known as Boulevard Saint-Laurent. In the second half of the eighteenth century, the faubourg was settled by workers and artisans who could not afford to build in stone, as was required in the walled city (Marsan 1990). The demolition of the fortifications, completed in 1817, united city and suburbs and spurred the latter's development (Burgess 2009). Chemin Saint-Laurent became the dividing line between the eastern and western administrative sections of the city. In the English-speaking population, it became known as "The Main." It also became the axis of immigrant settlement in the city (Anctil 2002). Immigration from the British Isles in the first half of nineteenth century was matched by migration from rural Québec; English speakers generally located to the west of Saint-Laurent, while French speakers generally settled to the east.

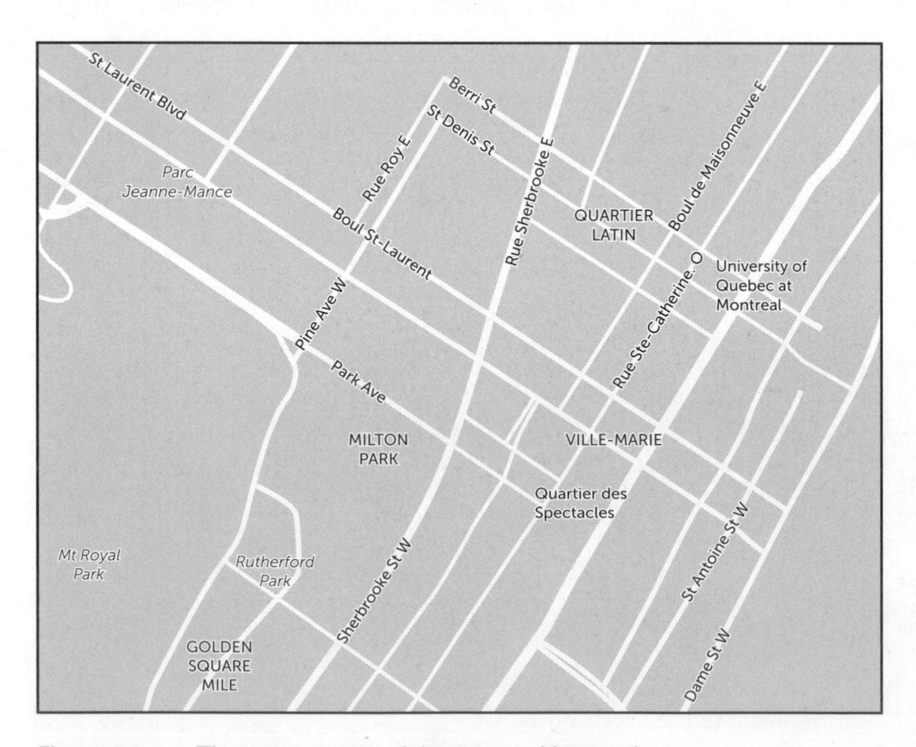

Figure 8.3.2 • The eastern part of downtown Montreal.

With the Industrial Revolution, taller and bulkier structures sprang up, in particular for garment production, and the Faubourg Saint-Laurent became a major employment hub. European immigration and continued rural-to-urban migration provided the necessary labour force. The end of the nineteenth century and the first third of the twentieth century saw a massive influx of eastern European Jews, who created a host of social, cultural, and charitable organizations. On their heels came Chinese workers, in much smaller numbers and mostly men, who came to build the new Canadian railways (Anctil 2002). Today's Chinatown is still located on Boulevard Saint-Laurent, just outside the Old City.

Parts of the Faubourg Saint-Laurent, further east, developed into a bourgeois neighbourhood for the French-Catholic elite. Its leaders built important educational institutions there in the late nineteenth and early twentieth century; these institutions gave part of the faubourg the name "Quartier Latin" after the university district in Paris. On Chemin Saint-Laurent itself, they built the Monument National as a major venue to promote French-Canadian culture. In fact, the building soon became a place of mass entertainment as well as a centre of Jewish cultural and social life (Anctil 2002).

By the early 1920s, the area around the intersection of Boulevard Saint-Laurent (so renamed in 1905) and Rue Sainte-Catherine had become the main entertainment centre of Montreal, with a variety of theatres, cabarets, clubs, bars, and restaurants. Montreal acquired a continent-wide reputation as a place of fun and sin. It became an important destination for Americans in search of liquor during the Prohibition era and an important centre of prostitution for soldiers and civilians alike in the war years. The red-light district along Saint-Laurent of course was a magnet for organized crime as well.

The rise of illicit and criminal activities, together with the growing obsolescence of the building stock, drove the middle class away from the Faubourg Saint-Laurent in the 1940s and 1950s. The French bourgeoisie took its homes and institutions to Outremont and other areas north of Mount Royal (the construction of the Université du Québec à Montreal [UQAM] in the late 1970s restored the Quartier Latin designation of the eastern part of the faubourg). Jews moved north along Saint-Laurent to the Plateau, the Mile End, and Outremont as well. Criminal activities drew the attention of reformist officials, including a young lawyer named Jean Drapeau. In 1954, Drapeau successfully ran for mayor of Montreal on the promise of ridding the city of prostitution and crime and ending collusion among the mob, police, and officials. He wanted to "clean up" the city while at the same time maintaining its standing as a tourist destination, a dual objective that would be echoed half a century later.

Jean Drapeau's time as mayor, from 1954 to 1986 (with a hiatus from 1957 to 1960), coincided with a period of massive urban renewal and modernization. The Faubourg Saint-Laurent was greatly affected by several large-scale transportation and building projects, including the widening of Dorchester Boulevard (now René Lévesque Boulevard), the construction of Place des Arts, Canada's largest performance centre, and the development of Habitations Jeanne-Mance, Montreal's largest public housing complex. Faubourg Saint-Laurent was greatly changed. Physically, the fine-grained, continuous environment of the eighteenth- and nineteenth-century city was cut up into separate parts and partially replaced by megastructures and empty lots. Socially, a vibrant, mixed-use neighbourhood with a gradation of economic classes was transformed into a number of areas with more segregated populations. Over time, the LGBT community settled in the "Village" in the far east section, students concentrated in the Quartier Latin, low-income residents made the Habitations Jeanne-Mance their own, and a marginal population continued to inhabit or frequent the red-light district near Boulevard Saint-Laurent. West of The Main, there was no neighbourhood to speak of anymore; between performances at Place des Arts, summer festivals in

Figure 8.3.3 • Aerial view of part of the Faubourg Saint-Laurent, with Place des Arts on the left and Habitations Jeanne-Mance to the upper right.

Source: Google Earth.

surrounding empty lots, and office work in nearby towers, local residential life had pretty much disappeared.

The transformation of the western part of the faubourg had started with the construction of Place des Arts in the 1960s; it was completed by the creation of the Quartier des spectacles in recent years. Between 1963, when the main concert hall was opened, and 2012, when a new symphony hall was inaugurated, Place des Arts steadily increased the presence of high-brow arts and official events in an area formerly dominated by low-brow culture and marginal activities. The esplanade of Place des Arts and other open spaces around the complex have been used by the very popular Montreal International Jazz Festival (launched in 1980) and the Just for Laughs Festival (begun in 1987). Minor players displaying more marginal cultural forms in smaller, older spaces have given way to major players and public institutions staging mainstream performances in large, modern facilities. The Quartier des spectacles is the latest step in this process.

The Politics of Cultural Development

Like many cities, but perhaps later than some, Montreal has seized on culture as an instrument of economic development. The city has made it one of its priorities, a move owed in no small measure to the work of Helen Foutopoulos, a city councillor who pushed the cultural agenda over two decades. Actors in the cultural industry also provided leadership, in particular Alain Simard, the head of Spectra, the company that produces the International Jazz Festival and the Francofolies festival. Other players in the public and private sectors, most notably in the business community, followed suit in proclaiming the value of culture in economic development.

For Simard and his colleagues, the policy to promote culture in Montreal had to deliver on two specific objectives: to protect the city's entertainment venues from suburban competition and from urban gentrification and to secure the sustainability of festivals downtown. The poor shape of the Faubourg Saint-Laurent, in particular around Place des Arts, and the vulnerability of empty lots to redevelopment, called for local planning to balance the twin needs of redevelopment and protection of venues and open spaces. Given the economic and symbolic weight of Spectra's festivals, Simard's call for action was heard. Less than 10 years later, he would see his vision largely realized, but in a strongly modified form (Table 8.3.1).

The idea of creating a formally recognized entertainment district around Place des Arts was first aired in 2001 by Jacques Primeau, president of ADISQ, the Québec trade group for music, video, and entertainment. He presented the concept at the Montreal Summit of 2002. On January 1 of that year, a new, much larger City of Montreal came into being, the product of a very contentious merger with the other municipalities of the Island of Montreal. The City's first mayor, Gérald Tremblay (who served until 2012), launched a major public consultation process to create consensus on policy goals in a heterogeneous community. The entertainment district was adopted at the summit as one of the projects to be implemented to make Montreal a "creative and innovative metropolis" (Ville de Montreal 2002). Helen Foutopoulos saw this as a vindication of her efforts (H. Foutopoulos, personal communication, 3 July 2013).

Despite its brevity, the story of the Quartier des spectacles is not a simple one: what started as an effort of the arts and culture community to protect its assets and develop its activities became an urban development policy, pushed forward by political actors to create economic and fiscal value. This shift in orientation coincided with a narrowing of focus: the original plan to restore the Faubourg Saint-Laurent into its dual historical roles as mixed-use neighbourhood and as cultural centre of Montreal gave way to an urban design project largely devoted to creating a state-of-art venue for major festivals.

The Initial Vision

One of the six workshops held at the Montreal Summit of 2002 to explore ways of making Montreal into a hub of creation and innovation brought together Jacques Primeau of ADISQ, Dinu Bumbaru of the planning and preservation group Héritage Montreal, and Clément Demers of the QIM, the successful urban redevelopment project southwest of the central business district. They coined the name "Quartier des spectacles" for the cultural district that they thought ought to be created in the western part of the Faubourg Saint-Laurent (C. Demers, personal communication, 2 July 2013). The City hired urban planner Pierre Deschênes to suggest what planning process might be followed to develop and realize the idea. Deschênes recommended the creation of a not-for-profit

Table 8.3.1 • Timeline of the Quartier des Spectacles Project, 2001–13

Year	Milestone
2001	Idea of entertainment district is first aired by leaders in entertainment industry.
2002	Entertainment district becomes official city project; name "Quartier des spectacles" is coined.
2003	Partenariat du Quartier des spectacles is created, made up of private, public, and not-for-profit stakeholders.
2004	First plan for Quartier des spectacles is drafted, with four planning sectors.
2005	Lighting plan is adopted as branding strategy; red dots become signature of former red-light district.
2006	City of Montreal takes planning over from Partenariat; QIM team is put in charge of the project.
2007	Draft PPU (Special Planning Program) for the Place-des-Arts sector is submitted to the City.
2008	Implementation starts; final PPU for Place-des-Arts sector adopted by City Council.
2009	Place des Festivals is inaugurated; other elements of the plan are realized in following years.
2012	Draft PPU (Special Planning Program) for Quartier Latin sector submitted to the City.
2013	City public consultation office issues negative report on PPU re lack of attention to local population and to gentrification.
2013	Final PPU for Quartier Latin adopted by City Council; focus on young professionals, artists, and students remains.

organization with representation from all local stakeholders and with the mandate to draw up a detailed plan for the area. Following the example of the QIM, it was assumed that an independent organization would not fall prey to competition and bureaucracy in the city administration (C. Demers, personal communication, 2 July 2013; P. Deschênes, personal communication, 3 July 2013). In June 2003, City Council formally approved the creation of the Partenariat du Quartier des spectacles.

The 21 members of the Partenariat represented production companies, entertainment venues, community organizations, local institutions, the provincial government, and both the borough of Ville-Marie (downtown) and the City of Montreal, which provided the operating budget. They developed a consensual vision for the area that entailed a "balanced neighbourhood" with a good mix of uses, strong links with other neighbourhoods and a vibrant public sphere; these features would make the Quartier des spectacles into "a hub of creation, innovation, production, exhibition and [diffusion]" and "an international cultural destination and centre for artistic creation," with a distinct identity (Partenariat du Quartier des spectacles 2013).

The architecture firms Nomade Architecture and Brière Gilbert + Associés were given the mandate to draft a physical improvement plan to translate the vision into physical terms. Their 2004 plan, whose subtitle was "Life, Art, Entertainment," reflected the desire to achieve a balance between different uses and activities. Implementation was tailored to the specific character of four historically defined sectors: the area around Place des Arts, the Quartier Latin, the public housing project Habitations Jeanne-Mance, and the commercial corridors of Boulevard Saint-Laurent and Rue Sainte-Catherine.

Although the project represented a consensus among all local stakeholders and enjoyed the support of Mayor Tremblay, it did not please everyone outside the Quartier des spectacles, in particular in the city administration. Planner Deschênes remembers: "Many people [at the City of Montreal] were bothered by it, and especially by the fact that it was a self-organized process led by local stakeholders" (Deschênes, personal communication, 3 July 2013). As we will see later, the City of Montreal would indeed take over planning for the Quartier des spectacles from the Partenariat, with the blessing of certain partners. It would also prepare plans for two sectors rather than four. The smaller, western sector around Place des Arts came to include the intersection of Boulevard Saint-Laurent and Rue Sainte-Catherine. The larger, eastern sector of the Quartier Latin around UQAM came to include the Habitations Jeanne-Mance. The strong presence of educational and research institutions in the Quartier Latin explains the fact that "learning" was added to "living," "creating," and "entertaining" (or "to be entertained") as keywords in the vision statement (Ville de Montréal 2012).

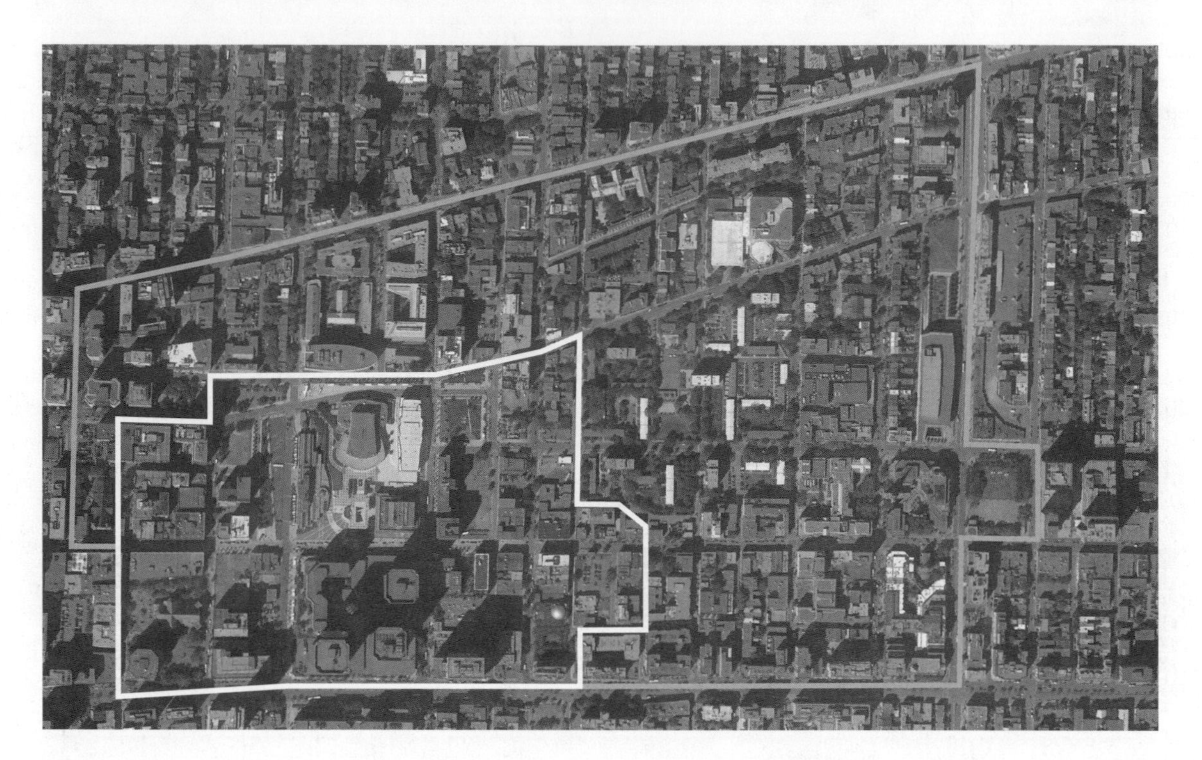

Figure 8.3.4 • Boundaries of the Quartier des spectacles (solid grey line) and of the Place-des-Arts sector (white line).

Source: Ville de Montréal, Arrondissement de Ville-Marie 2008, 2.

In what follows, we will focus on the Place des Arts sector of the Quartier des spectacles, where major interventions in the public domain have been crucial. We will leave aside the Quartier Latin sector, where a much more traditional urban improvement strategy was due to unfold, with a mix of regulatory changes, financial incentives, and street design, and where planning is still ongoing. In the public imagination, the Quartier des spectacles is defined as the area where large festivals occur, i.e., at Place des Arts and surroundings.

From Vision to Implementation

An important factor in the politics of urban development is the creation of a strong identity that can be marketed locally and globally (Ford 1994; Hall 1998). Members of the Partenariat wanted to give their home area a visible, tangible identity that would set it apart in Montreal and in North America. In February 2005, they held an international competition to find a design approach that would rebrand the Quartier des spectacles. The competition was won by the team Integral (Ruedi Baur and Jean Beaudoin), whose proposal was to use light to create local identity. As Anjali Mishra, a former manager for the Partenariat, explains, "[t]he idea of having everyone play with light, a 'Quartier des Lumières,' gave [all venues and festivals] something they could rally to but that didn't compromise their own identity, while at the same time creating a common theme in the area" (A. Mishra, personal communication, 12 June 2013). The concept was worked out with lighting artist Alex Morgenthaler, whose red dots of light, recalling the red-light district, became the signature visual elements of the Quartier. The lighting plan also included video projections on building facades, an art form that has become very popular here as elsewhere for its creative blend of digital and architectural features.

Also critical in the implementation of the vision was the desire to protect the neighbourhood as a living hub of creativity. "Fundamentally," Anjali Mishra states, "the initial project was very much about retaining creative people [and organizations] in the neighbourhood" in the face of development pressure and higher rents (A. Mishra, personal communication, 12 June 2013). What made the Quartier des spectacles "a full project" was the fact that "not only festivals but a lot of the artists and creators made it a cohesive thing" (A. Mishra, personal communication, 12 June 2013). The mix of activities and of actors could be maintained if ongoing development processes were managed or countered. Deschênes argued that local artists had to benefit rather than suffer from rising property prices, and he worked with a local foundation and real estate students at QUÀM to find ways of protecting or building cooperative live–work space for artists (P. Deschênes, personal communication, 3 July 2013). His efforts were not extended in planning work when the City of Montreal took over the leadership of the project from the Partenariat.

Political Power and Planning

Although the Partenariat had developed a common vision for the area and had started to change its image, it did not achieve significant changes in the physical environment around Place des Arts, where major festival organizers had to make do with empty lots and temporary street closures to run their international events. To respond to their urgent demand and also to serve its own interests, the City took over control of the planning process. Arts and culture became one issue among others in the planning process; economic development and urban planning were placed high on the agenda as well (D. Ross, personal communication, 26 June 2013). The Partenariat became an adjunct to the City, a representative of local stakeholders whose new mission was to manage and publicize the activities taking place in the public domain. From this point onward, the Quartier des spectacles would serve the broader goals of selling Montreal to the world as a creative city, strengthening Montreal's downtown, and broadening the city's tax base by means of private real estate development (Ross 2012). It became a strategic project for the City.

The Quartier des spectacles was to be managed in the same way that the Quartier international de Montreal (QIM) had been. The QIM had demonstrated the advantages of using a not-for-profit corporation, controlled and funded by the City but functioning at arm's length from the administration, to realize an ambitious urban project in a limited time. An organization external to the City would therefore be created again to manage an **expedited approvals** and implementation process. The team that had created and was still managing the QIM was brought in to take on this challenge. Because of recent changes in the law, the City could not give it exactly the same autonomy as it had given it to build the Quartier International (for instance, contracts had to be awarded through regular municipal **tendering processes**), but it was still able to "protect" it from the municipal bureaucracy in the planning and design stage (D. Ross, personal communication, 26 June 2013). Once plans were to be presented for official adoption, however, existing rules on public consultation had to be followed.

The resulting organizational structure of the Quartier des spectacles (we will use the acronym QDS to denote the planning team rather than the urban area) was that of a mixed, public–private entity, with the QIM people on the outside and a small group of colleagues inside the administration. In addition to the administrative autonomy of the QDS and the good working relationship that existed between its two parts, the project enjoyed strong political support. It was, one actor said, the "mayor's project" and "an absolute priority" of his administration (M. Maillet, personal communication, 21 June 2013). The usual problems of administrative delay, poor communication, and reluctance to collaborate outside one's "silo" were minimized, and the project moved ahead quickly.

The first decision of the QDS team was to focus its attention on the western sector of the Quartier des spectacles. Clément Demers, head of the QIM, explained that the concentration of investments in a smaller territory—barely 30 per cent of the whole Quartier des spectacles but with 80 per cent of its cultural activities—would make it possible to "reach the critical mass necessary to develop a solid project, capable of generating . . . significant economic leverage" (C. Demers, personal communication, 2 July 2013). Although the vision generated by the Partenariat remained officially in place, the geographic and substantive focus shifted to the physical redevelopment of a specific sector for the sake of major festivals.

Having defined a plan area, the QDS worked to inscribe its objectives into official plans of the City. The instrument it needed is called a Special Planning Program (*Programme particulier d'urbanisme*, or PPU), the adoption of which enables a municipality to augment its master plan with a detailed scheme of public investments, regulatory changes, and fiscal measures for a given territory. The urban design firm Daoust Lestage received the mandate to draft the PPU. After an intense process of consultation with local stakeholders and after receiving City Council approval for a preliminary concept, Daoust Lestage and the QDS submitted their draft PPU in November 2007, barely six months after they had started to work on it. Soon after, the mayor announced that the federal and provincial governments would each contribute $40 million to the project and that the City would commit $67 million, for a total of $147 million.

Most of the budget would go toward the creation of state-of-the-art public spaces around Place des Arts: a principal plaza for festivals (Place des Festivals), secondary festival spaces (Promenade des Artistes, Le Parterre, and Esplanade Clark), and a redesigned Rue Sainte-Catherine across the sector. This work was needed, planners and officials said, to help turn the Quartier des spectacles into "a warm, welcoming and eye-catching district," "a permanent urban venue for culture, cultural enterprises and craftspeople" and "a world-class cultural destination" (Quartier international de Montréal 2007).

Although QDS professed that the plan would "encourage social diversity by retaining existing residents" (Quartier international de Montréal 2007), it was in fact reinforcing trends toward gentrification and the departure of current residents. The network of streets and plazas, designed with flair and built to the highest technical specifications, was clearly meant to provide space for large festivals but also to stimulate private investment in the area and foster economic development. The final version of the PPU, officially adopted in 2008 (after work actually started on the ground), reflects the true intentions of the City better than the preliminary version of 2007. The plan, as adopted, promises that the area would become "a world-class destination" for cultural activities, "a welcoming, balanced and interesting environment," and an exemplary case of "sustainable development" (Ville de Montréal 2008). Protecting and supporting local artists and existing venues was no longer a goal. What was sought, much more modestly, was the inclusion of spaces for cultural activities in new real estate projects.

Figure 8.3.5 • Le Parterre, one of the public plazas where outdoor festivals are held, with renovated loft buildings in the background.

Source: Laurie Tallotte and Laurie Loison.

Anjali Mishra summed up the shift of emphasis over the years as follows:

> As the project evolved, [the objective] became retaining festivals instead of retaining cultural production. And the PPU crystallized that. . . . The tourist mandate really took over. . . . The project shifted. In that sense, a lot of the work that had been done on affordable cultural spaces was passed along to other organizations (A. Mishra, personal communication, 12 June 2013).

The evolution that Mishra describes can be explained in large part by changes in the governance structure of the Quartier des spectacles. The Partenariat of local stakeholders went from being the planner of the project to being a service provider to the City, with "a new mandate: manage public spaces, generate activities and programming for these public spaces" (P. Fortin, personal communication, 3 July 2013). The City and QIM were able to give the project a stronger focus on economic development and its fiscal benefits. Still, the Partenariat is trying to continue its work at the grassroots, so to speak, to aid the local arts community. It is supporting individual creators and cultural organizations by enabling them to showcase their art in the Quartier des spectacles and by promoting the events in which they do so.

Next Steps and Emerging Conflicts

The Quartier des spectacles project has been successfully implemented on a tight schedule. Place des Festivals was inaugurated in 2009, the Promenade des Artistes and the Parterre in 2010, and the partially pedestrianized Rue Sainte-Catherine in 2011–12. Only Esplanade Clark, with its outdoor skating rink, was incomplete at the time of writing. The quality of the work cannot be disputed: towering lighting elements, elegant street furniture, synchronized fountains, and other features (including state-of-the-art technology for shows and other events) give the new public spaces a strong identity, offer their users a very stimulating experience, and provide festival organizers with the platform they need for their events.

Figure 8.3.6 • Place des Festivals: High-tech meets tradition in the amenities of the Quartier des spectacles.

Source: Laurie Tallotte and Laurie Loison.

The Faubourg Saint-Laurent as a whole has been densifying, and its real estate values have been increasing. Condominium conversion or construction projects have multiplied. About 2000 housing units were built between 2004 and 2011 (Bélanger 2013). With unit prices above the city average, projects cater to young urban professionals in search of "buzz and energy" (Louis Bohème 2013). The area where changes are felt most strongly is the sector around Place des Arts, where units are being marketed for the "view on the festivals" that they afford their residents (Vézina-Montplaisir 2012).

It goes without saying that the arrival of a new population in the Faubourg Saint-Laurent is creating tensions. Conflict over the current and future identity of the area is "not just a resident[s] versus workers issue or [an issue of] tourists versus locals, there also is a wealth imbalance in the neighbourhood," in particular between renters and owners (A. Mishra, personal communication, 12 June 2013). For obvious reasons, renters fear the impact of gentrification on their ability to stay in the area. Long-time residents also fear the impact of cultural activities and tourism on their quality of life. While most buyers of new condominium units come to the area with full knowledge of the nuisances they may experience, in particular noise from the festivals, some of them and most of their neighbours who have lived in the area for a while do not easily welcome spill-over effects from the Quartier des spectacles (Cameron and Bélanger 2012; C. Caron, personal communication, 10 July 2013).

The conflict between newcomers and old-timers or, more generally, between the old Faubourg Saint-Laurent and the new Quartier des spectacles is also visible in the reactions elicited by the proposed PPU for the Quartier Latin sector around UQAM. In February 2013, the Office de consultation publique de Montreal (OCPM), the city's public consultation office, issued a report critical of the plan's emphasis on real estate development for a new population of young professionals. The revised plan, the OCPM found, ought "to bring the current residential population back front and centre" and to propose "an integrated housing strategy for the Quartier Latin that would reflect the mixed nature" of that population (Office de consultation publique de Montréal 2013, 2; authors' translation). The lack of attention to local needs has reinforced fears that the Quartier des spectacles will remain alien to its host environment (C. Racine, personal communication, 5 July 2013), that it will be an elite environment in a popular neighbourhood (A. Mishra, personal communication, 12 June 2013), a space designed in isolation from the social dynamics around it (Fischler 2003; Morisset and Noppen 2004).

Another fear, which has also started to become reality, is that the Quartier des spectacles will hurt the artistic and cultural community that it was meant to serve, at least in the original vision put forward by the Partenariat. The poor state of the built environment and the presence of marginal activities in the 1990s, which prompted public action in the 2000s, made the area affordable to artists and available to festival organizers. Since then, many vacant lots have been transformed into beautiful plazas or occupied by private developments. Older structures have been converted or renovated, and those that have not been touched have seen their rents increase. Consequently, many artists have had to leave the area or are under pressure to find work space elsewhere. The main tie between creators and the Quartier des spectacles now consists in the fact that they display their creations in the area rather than living in it (P. Fortin, personal communication, 3 July 2013).

The negative impact of the Quartier des spectacles also is felt by the very performance venues that gave the area its identity and the project its raison d'être. "Festivals," Mishra recalls, "took advantage of vacant lots to organize free outdoor shows that complemented indoor performances" (Mishra 2012, 4). Although the new Quartier des spectacles has clearly added to the dynamism of the area, the strong emphasis on festivals has made it harder for local venues to attract people indoors when so much is to be had outdoors (Boulanger 2013). Instead of helping to diversify venues and activities, as was initially planned, the Quartier des spectacles may well lead to greater homogeneity (P. Deschênes, personal communication, 3 July 2013).

Social homogenization, too, is a likely effect of the project. The marginal population that inhabited the Faubourg Saint-Laurent has been pushed away from the western sector by urban design interventions, by the use of public spaces for cultural events, and by policing and control (P. Fortin, personal communication, 3 July 2013). The displacement is not advertised officially but it is recognized by all and denounced locally by the organizations that cater to the homeless and act on other social issues (Colleu 2011, RAPSIM 2011). It is not the result of a mindless, heavy-handed policy; city officials and representatives of these organizations sat down together to discuss the social impacts of the Quartier des spectacles during the planning stage, and they continue to do so. But the displacement is real. The greater presence of visitors, the installation of a video control system (necessary to manage lights, fountains, video projections, etc.), and the use of private security personnel during festivals have made loitering more difficult. For some, "the Place des Arts sector is a relatively policed sector" today (C. Caron, personal communication, 10 July 2013), and it is seeing the effects of a larger strategy to make downtown Montreal more attractive as a place of residence, commerce, and tourism.

The dispersal of the homeless population has in fact increased conflicts with the residential population in the rest of the faubourg and made it more difficult to deal with the problem. In the easternmost part of the faubourg near Rue Papineau and in Hochelaga-Maisonneuve, the neighbourhood beyond, local residents have noted an increase in the number of homeless people, drug addicts, and prostitutes without a corresponding increase, as yet, of resources to serve their needs or manage their impact on the area (Leduc 2012). A spokesman for the Papineau residents' association wondered aloud whether Montreal "want[s] to create a Quartier des spectacles west and a Quartier de l'itinérance [i.e., a homelessness district] east" (Laurent Vallée, quoted in Russo 2010; authors' translation).

Notwithstanding the negative trends, the homeless are still present in the Quartier des spectacles, and local institutions are not insensitive to their needs. The Place des Arts metro station is an important meeting place for homeless people. There, inside the station itself, the transit authority, a homeless shelter, and a community development organization have created a service centre to cater to their needs (Société de transport de Montréal 2012). The Partenariat du Quartier des spectacles worked with the same community organization to set up a social integration program for the homeless under which a dozen people are trained to work as public information agents during the festivals (Sauves 2013).

Conclusion

The case of the Quartier des spectacles in Montreal can be viewed from a variety of perspectives. It is, first, an example of the large-scale projects that cities are using to improve or even transform older or obsolete parts of their territory. These urban projects serve the goal of enlarging the municipal tax base but also that of improving the city's competitiveness on the national and/or international stage. In this case, the economic niche in which Montreal is trying to gain competitive advantage is that of culture and tourism. The Québec metropolis, which earned the distinction of being a UNESCO City of Design, is positioning itself as a creative city.

Second, the case of the Quartier des spectacles shows the growing importance of urban design in urban policy and planning. Urban design has served multiple purposes here: to create a great scene for festivals and other cultural activities, to make the area more attractive to private developers, and to showcase Québec talent in design and manufacturing. These improvements have of course led to the gentrification of the area.

Third, the case shows, like so many others, that Canadian municipalities depend greatly on property taxes to finance their activities and that cities use real estate development to improve their financial situation. It also exemplifies the increasing reliance of governments on the private or para-public sector—at arm's length from government—to finance, plan, and/or implement its own projects. Montreal's two new "superhospitals," one of which is located just south of the Quartier Latin, are currently being built by means of public–private partnerships. Likewise, the planning of the Quartier international de Montreal and of the Quartier des spectacles was handled by a not-for-profit agency with funding from the City but with a high degree of independence from it.

Finally, the story of the Quartier des spectacles raises important issues in the evolving social geography of our cities. On the one hand, the process of privatization is also felt in the use and management of public space, where security firms exercise greater control over individual and collective behaviour, sometimes with the use of high-tech monitoring systems. On the other hand, in Montreal as in Toronto and elsewhere, poorer and more marginal populations are being pushed out of areas that authorities, developers, and users are finding attractive as new places of residence, work, and entertainment (Catungal, Leslie, and Hii 2009).

In the search for economic development and for global, continental, or national competitiveness, cities will tend to place growth over equity in their list of policy priorities and will try to justify local costs by means of city-wide benefits. Though it did not start the process, the Quartier des spectacles project definitely contributed to gentrification within and around its territory, and it certainly pushed a marginal population of homeless people, prostitutes, and drug addicts from the centre to the periphery of the neighbourhood. New users of the area, be they tourists or residents, probably enjoy the more attractive, cleaner, and seemingly safer environment. But the Quartier des spectacles, a project that has cost taxpayers about $150 million so far and that will represent total investments many times that amount when it is done, has served the public interest in partial ways. In the short term, it may have worsened some of the social problems that existed in the area by sending homeless people and participants in marginal or illegal activities into neighbourhoods that are ill-equipped to deal with them. In the longer term, when tax revenue from new development and increased economic activity accrues to different levels of government, it may help them to enact policies and programs that foster greater social equity. But that remains to be seen.

Authors' Note: The official name of the city is "Ville de Montreal." This name is used in official English-language documents as well. We depart from this practice here; in order to facilitate reading, we use "City of Montreal" in the text to designate the municipality.

KEY TERMS

Expedited approvals process
Modernism
Power centres
Public realm
Tendering process

REFERENCES

Aldinger, F. 2007. "Large scale infrastructure projects as urban redevelopment catalysts: The case of the Quartier international de Montréal." (Supervised research project, School of Urban Planning, McGill University).

Anctil, P. 2002. *Saint-Laurent, la Main de Montréal*. Montréal : Pointe-à-Callière, musée d'archéologie et d'histoire de Montréal.

Bélanger, H. 2013. «Faubourg Saint-Laurent: évolution du foncier résidentiel.» (Unpublished manuscript). Montréal : Table de concertation du Faubourg Saint-Laurent.

Boulanger, L. 2013. «Théâtre d'été . . . ou théâtre endetté?» *La Presse* 21 juillet. http://www.lapresse.ca/arts/spectacles-et-theatre/201307/20/01-4672659-theatre-dete-ou-theatre-endette.php.

Burgess, J. 2009. *Une histoire illustrée du Faubourg Saint-Laurent*. Montréal : Table de concertation du Faubourg Saint-Laurent et Service aux collectivités de L'UQÀM.

Cameron, S., and H. Bélanger. 2012. "Home territories and the atmosphere of spectacle: The experience of residents living in and around Montreal's Quartier des spectacles—A phenomenological inquiry." Paper presented at the 2nd International Congress on Ambiances, Montreal, 19–22 September.

Catungal, J.P., D. Leslie, and Y. Hii. 2009. "Geographies of displacement in the creative city: The case of Liberty Village, Toronto." *Urban Studies* 46(5–6): 1095–114.

Colleu, M. 2011. «Déplacer les sans-abri n'est pas la solution.» Canoe.ca. http://fr.canoe.ca/infos/regional/archives/2011/10/20111020-184503.html.

Fischler, R. 2003. «Quartiers thématiques : enrichissement ou appauvrissement?» *Urbanité* 2(1): 8–9.

Ford, L.R. 1994. *Cities and Buildings: Skyscrapers, Skid Rows, and Suburbs*. Baltimore. MD: Johns Hopkins University Press.

Hall, T. 1998. *Urban Geography*. London and New York: Routledge.

Hospers, G.-J. 2003. "Creative cities: Breeding places in the knowledge economy." *Knowledge, Technology and Policy* 16(3): 143–62.

Leduc, A. 2012. «La zone de tolérance : Québec solidaire plaide plutôt pour la prévention.» http://alexandreleduc.net/2012/07/20/la-zone-de-tolerance-quebec-solidaire-plaide-plutot-pour-la-prevention.

Louis Bohème. 2013. "Condominium project." http://louisbohemecom/Web/pages/index.aspx?disp=lifestyle&lang=EN.

Markusen, A., and G. Schrock. 2006. "The artistic dividend: Urban artistic specialisation and economic development implications." *Urban Studies* 43(10): 1661–86.

Marsan, J.-C. 1990. *Montreal in Evolution: Historical Analysis of the Development of Montreal's Architecture and Urban Environment*. Montreal: McGill-Queen's University Press.

Mishra, A. 2012. "Sustaining emergent culture in Montreal's entertainment district." Paper presented at the 49th International Making Cities Livable Conference, Portland, Oregon.

Montgomery, J. 2003. "Cultural quarters as mechanisms for urban regeneration. Part 1: Conceptualising cultural quarters." *Planning Practice and Research* 18(4): 293–306.

Morisset, L.K., and L. Noppen. 2004. «Le touriste et l'urbaniste (deuxième partie).» *Téoros. Revue de recherche en tourisme* 23(3): 65–9.

Office de consultation publique de Montréal. 2013. «Rapport de consultation publique : PPU du Quartier des spectacles—Pôle du Quartier latin.» http://www.ocpm.qc.ca/sites/default/files/rapport-ppuquartierlatin.pdf.

Partenariat du Quartier des spectacles. 2013. «Vision.» http://www.quartierdesspectacles.com/en/about/vision.

Pratt, A.C. 2008. "Creative cities: The cultural industries and the creative class." *Geografiska Annaler, Series B: Human Geography* 90(2): 107–17.

Quartier international de Montréal. 2007. "Quartier des spectacles: PPU Place des Arts." Official translation taken from the City of Montreal website.

RAPSIM (Réseau d'aide aux personnes seules et itinérantes de Montréal). 2011. *Profilage social et judiciarisation : Portrait de la situation dans l'espace public montréalais*. Montreal : RAPSIM.

Ross, D. 2012. «Le Quartier des spectacles de Montréal.» In J.J. Terrin, ed., *La ville des créateurs/The City of Creators*, 110–33. Marseilles : Parenthèses.

Russo, E. 2010. «Le Refuge des jeunes pense déménager au centre-ville.» *Le Plateau,* 10 mai. http://www.leplateau.com/Vie-de-quartier/Environnement/2010-05-12/article-1512545/Le-Refuge-des-jeunes-pense-demenager-au-centre-ville/1.

Sauves, E. 2013. «Des itinérants retrouvent du travail grâce à la SDSVM.» *Journal de Montréal* 25 juin. http://www.journaldemontreal.com/2013/06/25/des-itinerants-retrouvent-du-travail-grace-a-la-sdsvm.

Société de transport de Montréal. 2012. Communiqué de presse: «Création du 1er pôle de services en itinérance au métro Place des Arts. » http://www.stm.info/fr/presse/communiques/2012/creation-du-1er-pole-de-services-en-itinerance-au-metro-Place des Arts.

Vézina-Montplaisir, G. 2012. «Le Peterson : aux premières loges du Quartier des spectacles.» *Métro* 27 juillet. http://journalmetro.com/plus/immobilier/115218/le-peterson-aux-premieres-loges-du-quartier-des-spectacles.

Ville de Montréal. 2002. «Sommet de Montréal 2002.» http://ville.montreal.qc.ca/portal/page?_pageid=2137,2657439&_dad=portal&_schema=PORTAL.

Ville de Montréal, Arrondissement de Ville-Marie. 2008. «Programme particulier d'urbanisme, Quartier des spectacles—Secteur Place des Arts.» http://ville.montreal.qc.ca/pls/portal/docs/page/plan_urbanisme_fr/media/documents/PPU_QDS-PDA.pdf.

———. 2012. «Programme particulier d'urbanisme, Quartier des spectacles—Pôle du Quartier Latin.» http://ville.montreal.qc.ca/pls/portal/docs/page/arrond_vma_fr/media/documents/vdm_PPU_quartier-latin-mod-ocpm-4b-web.pdf.

Zukin, S., and L. Braslow. 2011. "The life cycle of New York's creative districts: Reflections on the unanticipated consequences of unplanned cultural zones." *City, Culture and Society* 2(3): 131–40.

INTERVIEWS

Personal communications occurred in interviews with officials, planners, and other participants in the project. All interviews were conducted in French, except the interview with Anjali Mishra, which was conducted in English. Translations into English were done by Laurie Loison.

Anonymous, art gallery co-ordinator, Rue Sainte-Catherine, 4 July 2013. Telephone interview.

C. Caron, co-ordinator, Table de concertation du Faubourg Saint-Laurent, 10 July 2013. In-person interview.

C. Demers, director, Quartier international de Montréal, 2 July 2013. In-person interview.

P. Deschênes, Centre Hospitalier Universitaire de Montréal, 3 July 2013. In-person interview.

P. Fortin, director, Partenariat du Quartier des spectacles, 3 July 2013. In-person interview.

H. Foutopulos, former president of the Commission sur la culture, Ville de Montréal, 3 July 2013. Telephone interview.

M. Maillet, former Quartier des spectacles project manager, Ville de Montréal, 21 June 2013. In-person interview.

A. Mishra, former interim director, Partenariat du Quartier des spectacles, 12 June 2013. In-person interview

C. Racine, manager of the Quartier Latin PPU, Arrondissement de Ville-Marie, 5 July 2013. In-person interview.

D. Ross, Quartier des spectacles project manager, Ville de Montréal, 26 June 2013, In-person interview.

N. Trudel, architect, Daoust Lestage, 12 July 2013, Telephone interview.

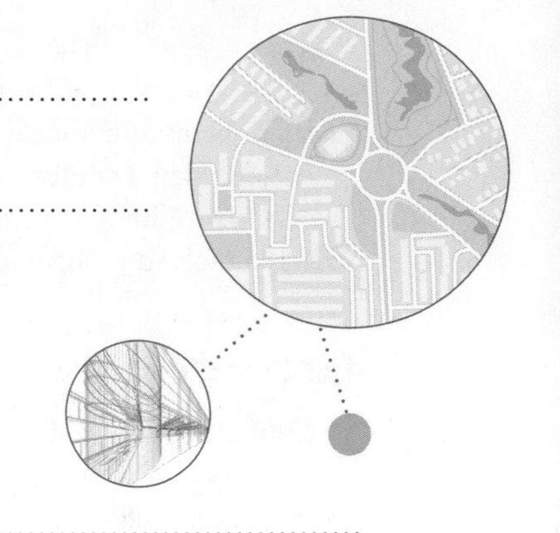

CASE STUDY 8.4

Places to Grow: A Case Study in Regional Planning in a Rapidly Growing Urban Context

Jason Thorne

SUMMARY

This chapter examines Ontario's Growth Plan for the Greater Golden Horseshoe, adopted in 2006, as a model for regional planning in a rapidly growing urban context. With a central goal of creating more compact and complete communities, the Growth Plan put in place a series of new provincial policy directions related to density, intensification, and growth management, which apply to the more than 100 municipalities that comprise the Greater Golden Horseshoe region. The chapter considers the various challenges that led provincial land-use planners to develop this regional approach to managing growth and then discusses the process that was undertaken to develop the Growth Plan. The central features of the Growth Plan are discussed—including its policies related to growth forecasts, density, intensification, urban growth boundaries, and infrastructure investment—as well as the impacts that the plan has had with respect to its original goal of reducing urban sprawl.

OUTLINE

- The provincial Growth Plan for the Greater Golden Horseshoe region of Ontario, which includes 10 of the country's largest cities, is a legally binding plan that aims to direct future development into built-up areas and prevent urban sprawl.
- Key developments of the Places to Grow Act and Growth Plan include density targets, urban growth boundaries, and urban growth centres. The results five years after implementation show some signs of increased density of development.
- The plan marked the Province's significant re-entry into the land-use planning system, with prescriptive growth management policies to which all local land-use planning decisions must conform. Preliminary results suggest that it has had a noticeable impact on how planning is conducted in the region and how development responds to statutory policy directions.

Introduction

In 2006, the Province of Ontario introduced a new land-use plan for the highly urbanized Greater Golden Horseshoe (GGH) region. The Growth Plan for the Greater Golden Horseshoe was a response to growing public concerns in the early 2000s about urban sprawl and associated impacts such as loss of agricultural land, increased traffic congestion, and diminished air quality. Its adoption by the provincial government marked a significant change to Ontario's land-use planning system. With a central goal of creating more compact and complete communities, the Growth Plan has put in place a series of new provincial policy directions related to density, intensification, and growth management, which affect all municipal Official Plans within the GGH. It represents Canada's most concerted effort to date at regional planning in a rapidly growing urban context.

This chapter is based on the author's personal observations as one of the architects of the Growth Plan and, since the plan's adoption, as a professional planning consultant working with municipalities to implement the plan's policies at the local level. The author also reviewed recent published studies related to the implementation of the Growth Plan and interviewed individuals who were closely involved in the plan's development.

Growth in the Greater Golden Horseshoe Region

Ontario's GGH region is the most heavily urbanized area in Canada. Covering more than 30,000 square kilometres, it encompasses 21 upper- or single-tier municipalities and 89 lower-tier municipalities, including 10 of Canada's 30 most populous cities and its largest city—the City of Toronto.

In the early 2000s, this region was one of the fastest-growing urban regions in North America. The 2001 census revealed that from 1995 to 2001, almost half of Canada's total population growth—and more than 90 per cent of Ontario's growth—took place in the GGH (Ontario Growth Secretariat 2003a, 2). With a population of nearly 8 million people in 2001, planners were wrestling with how best to accommodate a forecasted 3.7 million more people and 1.8 million more jobs by the year 2031. Public concerns about the impacts of poorly planned growth were increasing, and calls for stricter controls on growth and development were growing as well.

The potential impacts of continuing historical development patterns in the GGH were documented in a series of reports published by the Neptis Foundation in 2002–3. The report *Implications of Business-as-Usual Development* projects the future impact of adding more than 3 million more people to the region if current trends in infrastructure investment, employment, consumer preferences, and commuting patterns were to continue (Ontario Growth Secretariat 2003a, 4). Among the predicted impacts of business-as-usual development patterns were 45 per cent longer commute times during the morning rush hour and a 42 per cent increase in vehicular carbon dioxide emissions (Ontario Growth Secretariat 2003a, 4). Even though the GGH is heavily urbanized, agriculture is the most significant land use, with more than 44 per cent of the land area classified as farmland by Statistics Canada in 2001 (Walton 2003). Under business-as-usual development patterns, it was estimated that roughly 1069 square kilometres of non-urban land would be consumed (IBI Group 2002, E40).

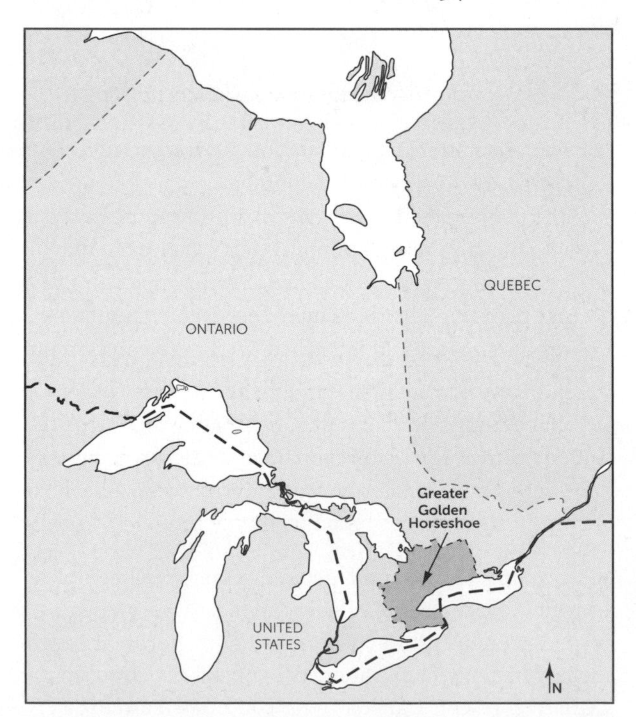

Figure 8.4.1 • The Greater Golden Horseshoe in Ontario.

Source: Adapted from Ontario Ministry of Infrastructure 2006a. Copyright Queen's Printer for Ontario.

The Smart Growth Panels

In February 2002, the provincial government responded to public concerns about growth and development by forming a series of Smart Growth Panels across the province. The Central Ontario Smart Growth Panel, which was responsible for the GGH region, was chaired by Mississauga Mayor Hazel McCallion and included 18 members ranging from local mayors to representatives of development, environmental, and agricultural interest groups. The provincial government also established a Smart Growth Secretariat within the Ministry of Municipal Affairs and Housing to administer and support the panels' work.

After a series of public consultations, the panel released its advice to government in April 2003. The panel (Ontario Growth Secretariat 2003b) set out a broad vision for the region along with 44 recommendations related to:

1. reshaping where and how we live;
2. protecting our environment;
3. unlocking gridlock;
4. rethinking how we manage waste;
5. optimizing other infrastructure.

The Places to Grow Initiative

Shortly after the submission of the Smart Growth Panel's recommendations, the provincial election of 2003 brought a new political party to power. The Dalton McGuinty Liberal Government's pre-election platform had promised to tackle urban sprawl through a series of land-use planning reforms. Within a few months of the election, the previous government's Smart Growth initiative was reinvented and relaunched by the new Liberal government as the Places to Grow initiative. Spearheaded by the newly renamed Ontario Growth Secretariat, now housed in the newly created Ministry of Public Infrastructure Renewal, the Places to Grow initiative was intended to move forward on the Smart Growth Panel's recommendations for the GGH region. The centrepiece of the Places to Grow initiative was to be the Growth Plan for the Greater Golden Horseshoe (the Growth Plan), which was ultimately released in 2006.

The Places to Grow Act

One of the first decision points for the Places to Grow initiative was the status that the new regional plan would have within the broader land-use planning system. Throughout the 1990s, the Province had increasingly withdrawn from direct regulation of development, leaving more responsibility to municipalities. Prior to Places to Grow, the Province relied on a set of broad land-use policy directions set out in the Provincial Policy Statement (PPS) as the basis for provincial direction in land-use planning. The PPS applies province-wide and is therefore limited in its ability to provide specific direction. Each PPS policy must be written such that it is as relevant in large cities as it is in smaller rural communities. In addition to being necessarily broad due to the province-wide context of these policies, the PPS also had a weak legislative standard. The Planning Act (1990) required local planning decisions to "have regard for" these provincial policies. While some supported the flexibility that this planning regime provided, others criticized it for creating too much inconsistency and allowing for too much discretion at the local level.

Whether the new Growth Plan should continue in the vein of the existing PPS or whether it should have a stronger legislative status was decided with the passage of the Places to Grow Act (2005). Under the provisions of the Act, the provincial cabinet was authorized to prepare a Growth Plan for any region of the province. Once adopted, all local planning decisions would be required to "conform with" the provisions of the provincial plan. This conformity provision provides significantly greater weight to Growth Plan policies than that provided by the "have regard for" status of the PPS. The conformity standard also extends to the decisions of the Ontario Municipal Board, the agency that decides on appeals in Ontario's land-use planning system. Furthermore, the Act requires that all local Official Plans be brought into conformity with a Growth Plan within three years of a Growth Plan's adoption.

With the legislative foundation for regional growth planning in place, the Province then set out to adopt the first Growth Plan under the new act, the Growth Plan for the Greater Golden Horseshoe.

The Growth Plan for the Greater Golden Horseshoe

The Growth Plan for the Greater Golden Horseshoe was adopted in 2006. It sets out a series of policy directions intended to guide growth and development in the GGH over a 25-year period. It is structured around four themes:

1. where and how to grow;
2. infrastructure to support growth;
3. protecting what is valuable;
4. implementation.

One of the central challenges facing the architects of the Growth Plan's policies was the level of prescriptiveness appropriate to a provincially adopted regional plan. Because it was a regional plan rather than a provincial plan, the opportunity existed to provide more precise policy direction than the province-wide policies that existed in the PPS. Unlike the PPS, the Growth Plan's policies could be tailored to the unique context of the GGH region. That said, balancing the need for strong, clear, and measurable regional growth management policies with the desire for continued municipal flexibility in pursuing local development objectives was a common point of debate during the three-year Growth Plan consultation process. The final plan includes a mix of broad, higher-level policies alongside a number of very prescriptive policies and targets. While the plan includes dozens of policies across the four themes, the focus of this article is on the policies that, in the author's view, have proved to be the most impactful and transformative in the years following the plan's adoption.

Growth forecasts

One of the most controversial provisions in the Growth Plan is the population and employment forecasts contained within Schedule 3. This schedule contains population and employment forecasts for each upper- and single-tier municipality in the GGH for the years 2011, 2021, and 2031. In 2013, the schedule was amended to add forecasts for the years 2036 and 2041. The forecasts were based on a demographic projection exercise led by the Province (Hemson Consulting 2005). Policy 2.2.1 in the Growth Plan directs that these forecasts "be used for planning and managing growth in the GGH" (Ontario Ministry of Infrastructure 2006a). Each upper-tier municipality is required to allocate the growth forecasts provided in Schedule 3 to their constituent lower-tier municipalities, as outlined in Policy 5.4.2.2(a) (Ontario Ministry of Infrastructure 2006a).

In practice, the Schedule 3 policies have resulted in the Growth Plan's population and employment forecasts being incorporated by municipalities directly into their Official Plans where they become the basis for all local planning decisions, including those related to land budgeting, the need for urban expansions, infrastructure and servicing planning, and so on. This has made the Schedule 3 forecasts one of the most influential policies in the Growth Plan.

Density and intensification

The Growth Plan was in large part born out of the desire to curb what many perceived as an urban-sprawl development pattern. As such, it is not surprising that the plan places a strong emphasis on building more compact communities through higher densities and higher rates of intensification.

Provincial land-use policy had for years encouraged higher densities and intensification of existing urban areas. The PPS, for example, calls for municipalities to establish and implement minimum targets for intensification and for new development to have "densities that allow for the efficient use of land, infrastructure and public service facilities," as outlined in Policies 1.1.3.5 and 1.1.3.7 (Ontario Ministry of Municipal Affairs and Housing 2005).

The Growth Plan has taken a much more prescriptive approach with respect to both intensification and density by establishing specific targets. With respect to intensification, Policy 2.2.3.1 requires that "by the year 2015 and for each year thereafter, a minimum of 40 per cent of all residential development occurring annually within each upper- and single-tier municipality will be within the built-up area" (Ontario Ministry of Infrastructure 2006a). The upper-tier municipalities are required to allocate this intensification requirement to their constituent lower-tier municipalities.

The Growth Plan also establishes a specific target with respect to development densities in greenfield areas. Policy 2.2.7.2 requires that new greenfield areas in each upper- or single-tier municipality "be planned to achieve a minimum density target that is not less than 50 residents and jobs combined per hectare" (Ontario Ministry of Infrastructure 2006a).

For both the intensification and the greenfield density targets, the Growth Plan provides flexibility for smaller and more rural municipalities in the region. In these cases, the Growth Plan permits the minister to prescribe an alternative target, as described in Policies 2.2.3.4 and 2.2.7.5 (Ontario Ministry of Infrastructure 2006a).

To facilitate the implementation of these policies, the Province conducted a mapping exercise and delineated the built-up area for every urban area in the region. The built-up area is defined as the area that was already developed as of the date of the adoption of the Growth Plan. New development within this zone counts toward the achievement of the intensification target. Development outside of this zone is considered greenfield development and must therefore achieve the greenfield density target.

Urban structure

While municipalities are given some degree of latitude to determine how they will achieve the intensification and density targets, the Growth Plan does provide some policy direction with respect to where and how development should occur. Policy 2.2.3.6(e) directs that intensification corridors and major transit station areas be considered as focus areas for intensification (Ontario Ministry of Infrastructure 2006a). The Growth Plan also identifies 25 specific geographic areas across the region as urban growth centres. Urban growth centres are typically the downtowns of the larger municipalities. Policy 2.2.4.4 directs that they accommodate a significant share of population and employment growth and that they be focal areas for infrastructure investment (Ontario Ministry of Infrastructure 2006a). Policy 2.2.4.5 also establishes a minimum density target for urban growth centres: 150 people and jobs combined per gross hectare in mid-sized cities; 200 people and jobs combined per gross hectare in the larger cities; and 400 people and jobs combined per gross hectare in each of the City of Toronto's four urban growth centres (Ontario Ministry of Infrastructure 2006a).

Urban growth boundaries

Establishing urban growth boundaries in municipal Official Plans has long been required by provincial policy.

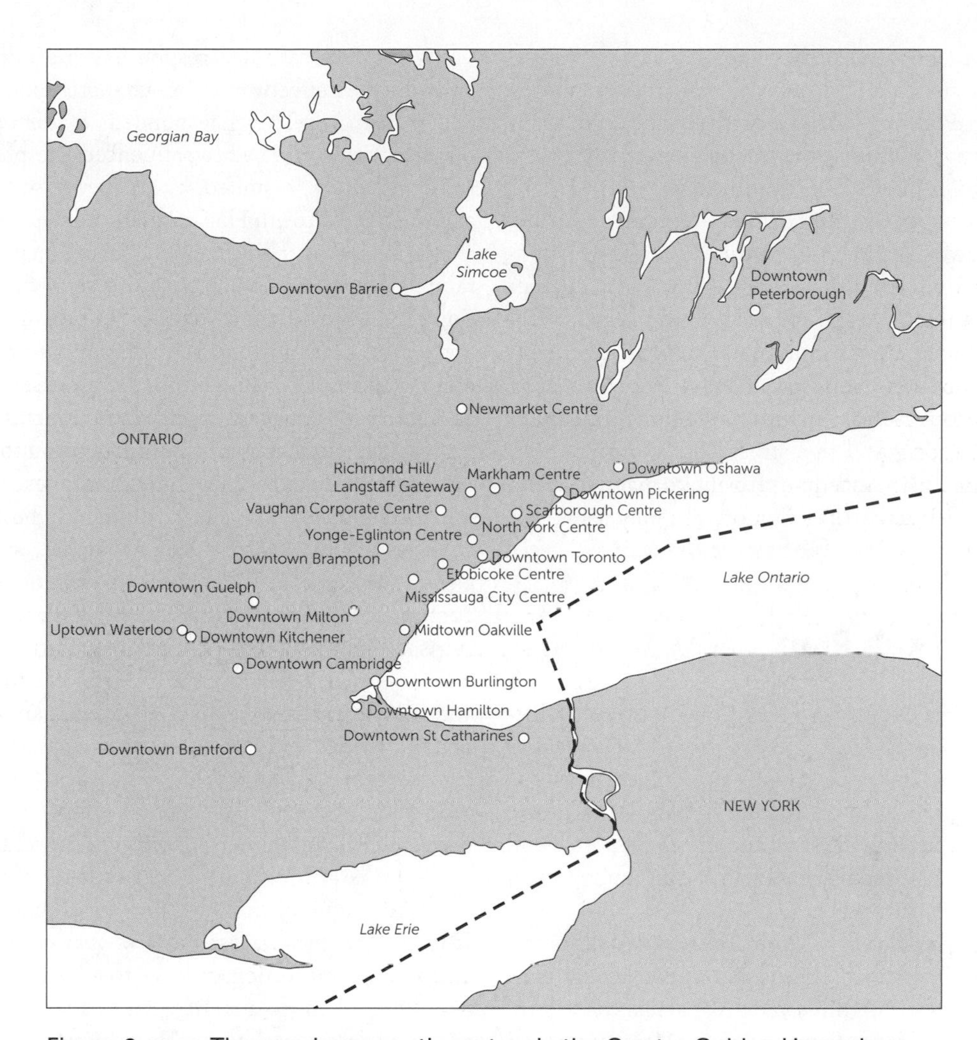

Figure 8.4.2 • The 25 urban growth centres in the Greater Golden Horseshoe.

Source: Adapted from Ontario Ministry of Infrastructure 2006a. Copyright Queen's Printer for Ontario.

The PPS directs municipalities to make available sufficient land to accommodate development for up to 20 years within what is called the settlement area boundary, as described in Policy 1.1.2 (Ontario Ministry of Municipal Affairs and Housing 2005). The PPS then sets out the conditions under which these boundaries can be expanded.

The Growth Plan does not permanently freeze urban boundaries. Urban boundaries can still be expanded. The Growth Plan does, however, provide a more strict set of conditions under which expansions can occur. As a result of the Growth Plan, only municipalities can initiate an expansion to an urban boundary, as outlined in Policy 2.2.8.2 (Ontario Ministry of Infrastructure 2006a). Private landowners are no longer able to initiate an expansion. Policy 2.2.8.2(b) also restates the long-standing requirement that urban boundaries be based on the provision of a maximum 20-year supply of urban land (Ontario Ministry of Infrastructure 2006a). However, the land budgeting that municipalities must undertake to determine their 20-year land needs must now be based on the Growth Plan population and employment allocations, intensification targets, and greenfield density targets. Prior to the Growth Plan, municipalities had the ability to establish their own growth, intensification, and density targets. With the Growth Plan, these key land-budgeting variables are now prescribed, thereby limiting the ability of a municipality to artificially either raise or lower their land-need requirements.

Outside of urban boundaries in the rural areas, development potential is limited by the Growth Plan. For example, Policy 2.2.9.3 states that the creation of more than three units or lots through either plan of **subdivision**, consent, or plan of condominium is no longer permitted outside of urban areas (Ontario Ministry of Infrastructure 2006a).

Infrastructure investment

The Growth Plan explicitly recognizes the significant role that infrastructure investment plays as a growth management tool. This recognition was a key reason for housing the Places to Grow initiative and the Ontario Growth Secretariat within the provincial ministry responsible for infrastructure investment rather than in the ministry responsible for land-use planning. While provincial infrastructure investment decisions are not bound by the Growth Plan, Policies 3.2.1.1 and 3.2.1.2 nonetheless include provisions that infrastructure planning and investment be co-ordinated to implement the plan and that priority be given to infrastructure investments that support the policies of the plan (Ontario Ministry of Infrastructure 2006a).

Impact of the Plan

At the time of the Growth Plan's adoption, it enjoyed widespread support across a number of sectors, including local elected officials, developers, and environmental groups (Ontario Ministry of Infrastructure 2006b). Now, nearly 10 years after the Growth Plan's adoption, it is possible to begin to assess its effectiveness in delivering on its growth management objectives.

Although the Places to Grow Act required all municipal Official Plans to be brought into conformity with the Growth Plan within three years, the actual conformity process in most cases took much longer. Many municipal planners attribute this to delays in the approvals process. In Ontario, land-use decisions by municipalities can be appealed to a provincial tribunal called the Ontario Municipal Board. Many of the decisions by municipalities to amend their Official Plans to conform to the Growth Plan ended up being the subject of such appeals by private landowners. In addition, for some municipalities in the GGH, their Official Plans must be approved by the provincial government, and some municipal planners cited delays in this process as one of the reasons for not meeting the three-year conformity deadline (Thorne 2011, 24). As of October 2012, three years after the original deadline, all 19 upper- and single-tier municipalities with Official Plans had adopted amendments to conform to the Growth Plan, but still less than half of the lower-tier municipalities had done so (Ontario Ministry of Infrastructure 2012).

While the local policy response is one indicator of the Growth Plan's effectiveness, much more important is how development itself has responded. Evidence regarding how the Growth Plan has influenced the physical form of development is limited, but there are some emerging signs that the Growth Plan is having its desired effect.

A significant shift in the housing market has certainly been observed in the years following the adoption of the Growth Plan. According to the Ontario Home Builders' Association, growth patterns across this region and especially in the **inner ring** of the Greater Toronto Area have undergone a fundamental shift from primarily single-family suburban dwellings to more intensified urban dwelling types (Collins-Williams 2012). In 2011, multi-unit housing starts reached their highest level since 1988, while other forms of housing (i.e., single-family) were at near-record lows (Collins-Williams 2012). From 2006 to 2013, the proportion of single detached housing starts declined in both the inner and outer rings of the GGH (see Figure 8.4.3) (Ontario Ministry of Municipal Affairs and Housing 2015b). While these trends are attributable to a multitude of factors, not just the changed regulatory environment, they do indicate a clear shift toward a more diversified housing mix and more compact communities, both of which are central goals of the Growth Plan.

There is also some evidence that the Growth Plan's intensification policies are having an impact. In July 2011, the Province released a "Five-year progress update" on the implementation of the Growth Plan. The update found that of the 63,000 new residential units that were added to the GGH between June 2009 and June 2010, approximately 42,000—or 67 per cent—were located in the existing built-up area. Half of those units were located in the City of Toronto, and many of them—16,000—were in the Growth Plan's identified urban growth centres (Ontario Ministry of Infrastructure 2012).

The "Five-year progress update" also examined alternative growth scenarios, based on computer modelling, with and without the Growth Plan. Among the findings were that with the Growth Plan, more than half of all development over the next 25 years would occur through intensification (Thorne 2011, 24). Without the Growth Plan, future intensification rates were estimated to be 22 per cent. With the Growth Plan, urbanized land in the region was estimated to grow by 14 per cent over the next 25 years compared to 39 per cent without the Growth Plan. And with the Growth Plan, average densities within the total urbanized area of the GGH were expected to increase by

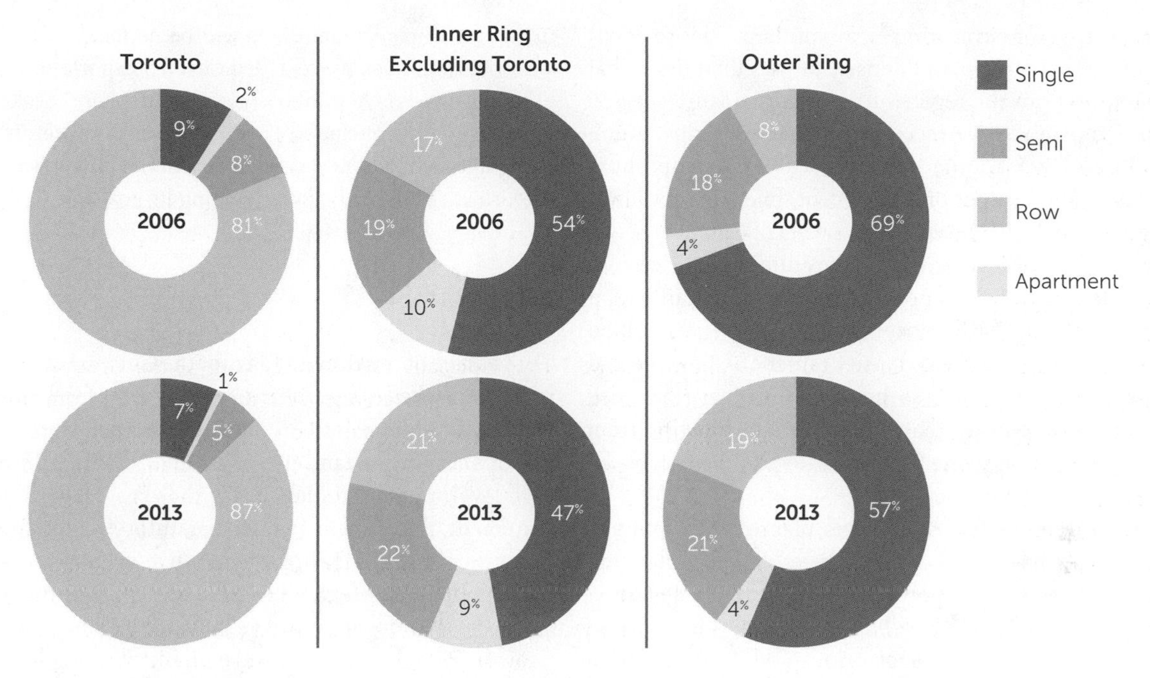

Figure 8.4.3 • Mix of new housing types in Toronto, the inner ring excluding Toronto, and the outer ring.

Source: Adapted from Ontario Ministry of Municipal Affairs and Housing 2015b. Copyright Queen's Printer for Ontario.

as much as 20 per cent by 2031, while without the Growth Plan, a 2 per cent decline in average density was projected.

While the evidence cited above with respect to changes in housing mix, development densities, and intensification rates since the adoption of the Growth Plan are in keeping with the plan's original objectives, some observers have expressed concerns that the plan may be having adverse impacts on land supply and therefore housing prices. The fate of the so-called "whitebelt" lands lying between the current urban boundaries of the inner-ring municipalities and the provincial greenbelt remains a topic of significant debate. These lands are not yet designated for urban development in municipal Official Plans, nor are they identified for permanent protection in either the provincial Growth Plan or Greenbelt Plan. The development industry has stated that development of these lands is critical to the future economic growth of the region. Yet they are concerned that development potential may be negatively affected by "foodbelt," "protected countryside," or similar designations in municipal Official Plans or by expansions to the provincial greenbelt (OHBA and BILDA 2013). Environmental groups, on the other hand, have calculated that there is more than enough whitebelt land available for future development. According to a report by the Friends of the Greenbelt Foundation, there are approximately 58,696 hectares of land available beyond currently designated lands for urban expansion in the whitebelt (Tomalty and Komorowski 2011). Current municipal Official Plans have identified a need for an additional 10,115 hectares to accommodate greenfield population and employment development up to 2031 (Tomalty and Komorowski 2011). This represents just 17.2 per cent of the whitebelt, leaving 82.8 per cent of the whitebelt intact until at least 2031 (Tomalty and Komorowski 2011).

Implementation challenges also remain a concern for some. Some have expressed concerns that implementation of the Growth Plan has been uneven, with some municipalities demonstrating a stronger, more adherent response to the Growth Plan's policies than others (Gombu 2011). This has led these critics to question whether the Growth Plan continues to allow for too much municipal flexibility (Gombu 2011). For example, a 2013 report by the Neptis Foundation criticizes the provincial government for permitting rural municipalities to adopt lower intensification and density targets than those that are required for the region's urban municipalities (Allen and Campsie 2013). According to that study, most municipalities have adopted the Growth Plan's minimum intensification target of 40 per cent, five have set lower targets (ranging from 15 to 32 per cent), and only two have set higher targets than the minimum requirement (Waterloo Region at 45 per cent and the Region of Peel at 50 per cent). Similarly, most municipalities have adopted the minimum greenfield density target of 50 people and jobs combined per gross hectare, nine have adopted density targets lower than the minimum (ranging from 29 to 39), and only one (Waterloo Region) has proposed to exceed the minimum (55).

On the municipal side, some planners and municipal officials have cited a critical need for new planning and fiscal tools to support Growth Plan implementation (Thorne 2011, 24). The need for additional infrastructure investment also remains a top priority. While the adoption by Metrolinx of a 25-year transit plan was an important step, the need for infrastructure investments to support the Growth Plan is a common concern of many municipalities (Thorne 2011, 24).

In 2015–16, the Growth Plan is undergoing a policy review mandated for 10 years after its implementation. This is part of a co-ordinated review of the four regional plans: the Growth Plan, the Greenbelt Plan (approved in 2005), the Oak Ridges Moraine Conservation Plan (approved in 2002), and the Niagara Escarpment Plan (approved in 1985). The goals are to determine the success of the Growth Plan, identify implementation concerns, and co-ordinate the four regional land-use plans (Ontario Ministry of Municipal Affairs and Housing 2015a). Another intention is that the four regional plans will be better co-ordinated with the Metrolinx 25-year Regional Transportation Plan (The Big Move). A public consultation phase began in April 2015 and concluded in fall 2015, and indicators are being updated with new data to determine how successful the Growth Plan has been in meeting its goals.

Conclusions

The passage of the Growth Plan for the Golden Horseshoe in 2006 marked a significant re-entry into the land-use planning system by the Province of Ontario. After a decade of engaging in land-use planning primarily through high-level policy guidance in the Provincial Policy Statement, the Growth Plan is a statutory regional plan that sets out a number of prescriptive growth management policies to which all local land-use planning decisions must conform. In the years since its adoption, the Growth Plan has significantly changed the planning discourse within the GGH and the province. It has introduced a number of new planning concepts that have fundamentally changed how municipal planning is conducted in the region. The plan has not only articulated a vision for more compact urban growth in the region over the next 25 years but instituted a number of statutory policy directions that are intended to make that vision a reality. As with any regional planning exercise, the impact of the Growth Plan will be most properly measured over decades rather than the fewer than 10 years that have passed since its adoption. But preliminary observations suggest that it has had a noticeable impact both on how planning is conducted in the region and how development responds.

KEY TERMS

Inner ring
Smart Growth
Subdivision
Upper-tier, single-tier, and lower-tier municipalities

REFERENCES

Allen, R., and P. Campsie. 2013. *Implementing the Growth Plan for the Greater Golden Horseshoe: Has the Strategic Regional Vision Been Compromised?* Toronto: Neptis Foundation.

Collins-Williams, M. 2012. "The changing face of urban development in the GTA." *Urban Toronto* 7 February.

Gombu, P. 2011. "The region's rise and sprawl." *Toronto Star* 15 January.

Hemson Consulting Ltd. 2005. *The Growth Outlook for the Greater Golden Horseshoe*. Toronto: Hemson Consulting Ltd.

IBI Group. 2002. *Toronto-Related Region Futures Study. Interim Report: Implications of Business-as-Usual Development*. Executive Summary. Toronto: Neptis Foundation.

OHBA (Ontario Home Builders Association) and BILDA (Building Industry and Land Development Association). 2013. *Joint Submission to the Ontario Growth Secretariat on Proposed Amendment 2 to the Growth Plan for the Greater Golden Horseshoe*, 2006. 8 February. OHBA and BILDA.

Ontario Growth Secretariat. 2003a. *Shape the Future: Discussion Paper*. February–March. Toronto: Central Ontario Smart Growth Panel.

——. 2003b. *Shape the Future: Central Ontario Smart Growth Panel Final Report*. Toronto: Central Ontario Smart Growth Panel.

Ontario Ministry of Infrastructure. 2006a. *Growth Plan for the Greater Golden Horseshoe, 2006 (Office Consolidation June 2013)*. Toronto: Ontario Ministry of Infrastructure.

——. 2006b. "Media release: McGuinty government releases landmark plan to shape the future of the Greater Golden Horseshoe." 16 June. Toronto: Ontario Ministry of Infrastructure.

——. 2012. "Five-year progress update." https://www.placestogrow.ca/index.php?option=com_content&task=-view&id=271&Itemid=84.

Ontario Ministry of Municipal Affairs and Housing. 2005. *Provincial Policy Statement*. Toronto: Ontario Ministry of Municipal Affairs and Housing.

——. 2015a. "Our region. community. our home. A discussion document for the 2015 coordinated review." Toronto: Ontario Ministry of Municipal Affairs and Housing. http://www.mah.gov.on.ca/AssetFactory.aspx?did=10759.

——. 2015b. *Performance Indicators for the Growth Plan for the Greater Golden Horseshoe, 2006*. Toronto: Ontario Ministry of Municipal Affairs and Housing.

Places to Grow Act [2005] S.O., c. 13.

Planning Act [1990] R.S.O., c. P. 13.

Thorne, J. 2011. "Four years in review. *Ontario Planning Journal* 26(5): 24–5.

Tomalty, R., and B. Komorowski. 2011. *Inside and Out: Sustaining Ontario's Greenbelt*. Friends of the Greenbelt Foundation Occasional Papers Series. Toronto: Friends of the Greenbelt Foundation.

Walton, M. 2003. *Agriculture in the Central Ontario Zone*. Toronto: Neptis Foundation.

9.0

INTRODUCTION

Transportation and Infrastructure

Transportation and infrastructure planning today includes rail-based systems, road-based systems, and non-motorized systems. All of these systems have been critical to the spatial development and growth of human settlements: thousands of years ago, the size of a town was typically limited by the distance that people could easily walk. The innovation of steam energy to power trains led to a new era of expansion and growth as travelling became faster and easier. Electric streetcars led to compact inner-city neighbourhoods and suburbs on the city's edge. The introduction of motor vehicles in the early twentieth century led to arterials and highways, also contributing to a decline in the use of non-motorized transportation. Today, walking, cycling, and public transit are regaining popularity for environmental and health reasons. Each of these transportation eras has left its mark on Canadian cities and regions, from compact city blocks to suburban street patterns.

Transportation infrastructure played a key role in the development of the Canadian nation: the completion of the railway network shortly after Confederation defined a country spanning from sea to sea. Many prairie and west coast cities owe their beginnings to the Canadian Pacific Railway and, later, to other transcontinental railways, and streetcar lines contributed to dense, walkable grid street layouts in cities like Vancouver, New Westminster, and Montreal.

Transportation issues became particularly contentious in the postwar era when the car was believed to be the next innovation in travel. A rational, scientific approach to transportation decision-making included the adoption of traffic control ordinances and design standards in provincial highway departments and municipal traffic departments (Hess 2009), resulting in remarkably persistent spatial infrastructures such as suburban street layouts and multi-lane highways. The planner, civil engineer, and traffic engineer acted as experts in these processes without consulting the public on decisions that would often turn out to have devastating impacts on communities. However, when massive highway systems proposed in Toronto, Vancouver, and other cities threatened to destroy central city neighbourhoods, citizen protests were successful in ending this destructive practice. Key players in these protests included Jane Jacobs in Toronto and members of Vancouver's Chinese community in Strathcona.

By the late 1980s, environmental concerns prompted many to rethink the decades-old approach to transportation planning (Filion 1999), which relied upon modelling future demand without considering ways to modify demand to decrease infrastructure needs. Municipal budgets needed to stretch further, and existing infrastructure needed to be more efficiently used before new roads, highways, or rail-based infrastructure could be justified. More compact built forms, espoused in the Smart Growth and new urbanist approaches, began to enter theoretical and practical discussions. Public health concerns, such as increasing rates of childhood obesity and diabetes, also began to influence the design of better cycling and walking infrastructure (Frumkin, Frank, and Jackson 2004) and land-use diversity to give people access to goods and services within their own neighbourhoods (Grant 2009). As Meyer and Miller (2013) write, taking a systems approach to transportation acknowledges the social, environmental, and cultural costs of transportation decisions.

Today, transportation remains a key issue in cities and regions. Canadian cities are known for higher public transit ridership and more dense urban centres than comparable cities in the United States, but Toronto, Montreal, and Vancouver account for 67 per cent of the country's public transit ridership (Urban Transit Task Force 2010). There are still only two metro systems in the country (Toronto and Montreal), although other cities (Edmonton, Calgary, Vancouver, and Ottawa) have Light Rapid Transit services; only the three largest cities have regional rail services. Increasingly, transportation solutions try to decrease greenhouse gases, increase physical activity, and prevent sprawl. Transportation and land use have been integrated in approaches such as transit-oriented development, which aims to locate higher-density, mixed-use buildings near transit stops, stations, and interchanges (TransLink 2012).

Transportation governance is complex: under the Constitution Act, 1867, transportation is a municipal responsibility, but all three levels of government are involved in funding transportation infrastructure. Both capital and operating costs have increased over the past decade, and the federal, provincial, and municipal governments have increased their contributions for the construction of transit infrastructure and operating expenses (Urban Transit Task Force 2010). In 2008–9, more than 30 per cent of the transfer of federal gas tax funds to municipalities was allocated to transit (Urban Transit Task Force 2010). In 2015, the federal government announced the creation of a new fund for public transit of $250 million in 2017, increasing to $500 million in 2018 and $1 billion by 2019. Although the amount is minimal, this is the first time the country will have a national funding source for municipal transit infrastructure. In Vancouver and Toronto, regional authorities attempt to co-ordinate transit services across the greater commutershed. Given the power dynamics between provincial, regional, and municipal governments, lack of communication and co-ordination are significant issues in planning across regions (Filion 1999; Hess 2009). In Montreal, transportation governance is fragmented between a large number of different actors, while in Calgary and Edmonton significant efforts are underway to develop regional transit governance models and fare systems.

Transportation planning involves civil engineers, planners, economists, and researchers, all of whom have different communication styles, values, and problem-solving approaches. For example, quantitative modelling using large datasets is common among some groups, while mixed methods approaches are more prevalent among others. Efficiency and cost-effective decisions may be more important to some, while equity issues or characteristics affecting users (such as the design of pedestrian infrastructure) may be more important to others. Transportation and infrastructure planning have fairly poor track records of including public participation in decision-making, contributing to a lack of trust among citizens. The complex governance, actor relationships, and somewhat flawed processes make transportation and infrastructure planning an interesting sub-discipline, as the cases illustrate.

Anna Kramer and Christian Mettke examine a case that has become well known internationally: Toronto's Transit City plan. They are particularly interested in how the plan to extend Light Rail Transit (LRT) to the city's inner suburbs became politicized through political and media interests. Early in the process, this contributed to the development of two opposing forces: those who felt that LRT was the most cost-effective solution to the city's congestion and accessibility issues and those who felt that the plan represented a "war on the car." Issues of public education, informed participation, transparency, and the measurement of efficiency were critical factors in any discussion of the plan. However, this case also illustrates the complex governance involved in addressing regional transportation issues: the merging of six municipalities to form Metro Toronto, the establishment of the first regional transit agency, and the incorporation of the Transit City plan into the larger Big Move regional strategy.

In his exploration of the Gardiner East Environmental Assessment process, Antonio Medeiros discusses a timely issue for municipalities: replacing or repairing infrastructure built during the postwar boom. In Toronto, the Frederick G. Gardiner Expressway has been a contentious structure along the waterfront for decades; as it began to fall apart, Waterfront Toronto began an urgent process to investigate alternatives for its repair or replacement. Using four study lenses, Waterfront Toronto launched an international design process in which teams proposed alternative solutions to Improve, Replace, or Remove the expressway. These alternatives provided an opportunity to broaden public conceptions for possible solutions and engage the public in meaningful dialogue about the broader issues, such as changing transportation patterns, the ability of improved public transit options to meet commuter needs, and the potential of urban design to reintegrate the urban fabric adjacent to the expressway. Waterfront Toronto and the City of Toronto recommended the Remove alternative in March 2014.

Walter Peace, in his case study of Hamilton's Red Hill Valley Parkway, tells a compelling tale that spans half a century. One of the most controversial planning projects in the city's history came to a head in 2007 with completion of the parkway. This was only after decades of conflict between planning goals and public opinion, between the desire to preserve the natural landscape of the Red Hill Valley and the perceived need for economic growth. Lack of public participation in the process seems to have contributed to the polarization of opinions on the topic. With political instability from three levels of government and lack of an overall vision for the future growth of the city, planning goals and the public interest became increasingly muddled. Peace asks whether planners can in fact work in the public interest and whether they could have done a better job of finding common ground between the two opposing groups.

Ugo Lachappelle addresses a transportation demand management approach that is not often discussed: the development of accelerated vehicle retirement programs in the province of Quebec. As a means of improving air quality and encouraging modal shift toward public transit and non-motorized modes, Quebec's *Adieu Bazou / Faites de l'air!* program was first developed in 2002. Offering rebates on new vehicles, free transit passes, car-sharing memberships, and rebates on new bicycles, the program aimed to discourage car use. Between 2009 and 2011, more than 40,000 participants registered in the program and recycled their vehicles, and the vast majority of participants chose transit passes as a way to replace their car use. While such programs represent only one aspect of transportation demand management, they can significantly decrease greenhouse gas emissions and influence travel behaviour.

All of these cases show how complex networks of actors, encompassing all three levels of government as well as regional authorities, politicians, and the public, are involved in transportation planning decisions. Current transportation planning, with its attention to multimodal and sustainable solutions, represents a paradigm shift from the "predict-and-provide" approach common in the 1950s and 1960s. Because Canadian cities grew rapidly in the postwar decades, infrastructure such as highways, overpasses, and arterial roads were the predominant framework used to expand into greenfield areas. They represent, as Grant (2009) and Filion and McSpurren (2007) write, a spatial reality that is difficult to change. Attempts to design roads to include walking, cycling, and public transit, focusing more on accessibility than on mobility approaches represent a major threat to those who believe that car-oriented infrastructure is more efficient and progressive.

However, several significant projects will be completed from 2015 to 2020, including an air–rail link between Pearson International Airport and Union Station in Toronto; rapid transit connecting Waterloo, Kitchener, and Cambridge, Ontario; Alberta's Green Transit Incentives Program aiming to reduce greenhouse gas emissions and provide more sustainable transit options; Vancouver's Evergreen LRT Line; and Ottawa's Confederation LRT line. Longer-term developments include reinvestment in the GO train system in the Toronto region to 15-minute all-day service, bus rapid transit in Hamilton, and expansion of Edmonton's Metro Line LRT. With climate change acknowledged as reality and public health concerns driving more walkable neighbourhoods, Canadian cities are looking for alternatives that decrease congestion, improve air quality, and encourage active transportation. Transportation is one of the last sub-disciplines in planning to incorporate broader public participation and education to stimulate informed decision-making, but this development is critical to the success of sustainable mobility solutions. Politicians, citizens, developers, and municipal staff need to develop informative materials, integrate more transparent decision-making processes, and involve a broader range of actors early on to achieve the shift toward sustainable transportation.

REFERENCES

Filion, P. 1999. "Rupture or continuity? Modern and postmodern planning in Toronto." *International Journal of Urban and Regional Research* 23(3): 421–44.

——. and K. McSpurren. 2007. "Smart Growth and development reality: The difficult coordination of land use and transportation objectives." *Urban Studies* 44(3): 501–23.

Frumkin, H., L.D. Frank, and R. Jackson. 2004. *Urban Sprawl and Public Health: Designing, Planning, and Building for Healthy Communities*. Washington: Island Press.

Grant, J.L. 2009. "Theory and practice in planning the suburbs: Challenges to implementing new urbanism, Smart Growth, and sustainability principles." *Planning Theory and Practice* 10(1): 11–33.

Hess, P.M. 2009. "Avenues or arterials: The struggle to change street building practices in Toronto, Canada." *Journal of Urban Design* 14(1): 1–28.

Meyer, M., and E.J. Miller. 2013. *Transportation Planning: A Decision-Oriented Approach*. M. Meyer.

TransLink. 2012. *Transit-Oriented Communities Guidelines: Creating More Livable Places around Transit in Metro Vancouver*. Vancouver: Translink.

Urban Transit Task Force. 2010. "Recent developments in transit in Canadian cities, 2010." Council of Ministers Responsible for Highway Safety. http://www.comt.ca/english/recent-nov2010.pdf.

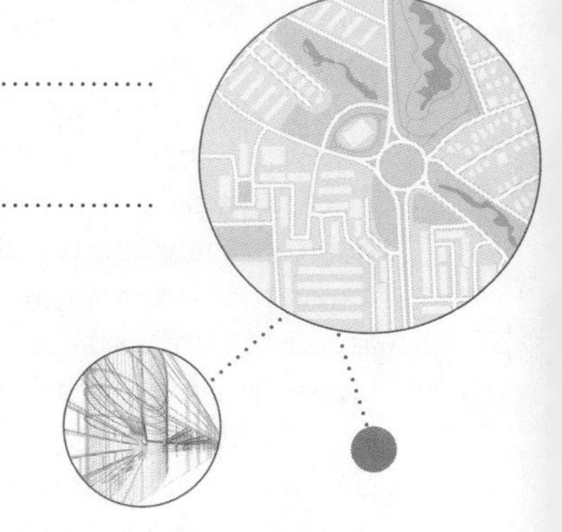

CASE STUDY 9.1

The Death and Life of "Transit City": Searching for Sustainable Transportation in Toronto's Inner Suburbs

Anna Kramer and Christian Mettke

SUMMARY

This is a case study of plans for Light Rail Transit (LRT) expansion in Toronto's inner suburbs. It shows how the original public consultations presented only one option and therefore limited debate on alternatives. Public debate on transit options did not happen until political controversy brought the issue into broader public consciousness. In this case, the presence of political champions both for and against the plan polarized the debate and obscured underlying needs and possibilities for serving those needs. In making the debate so heated, the politicization of these issues makes evidence-based decision-making more difficult.

OUTLINE

- The polarized outcome of Toronto's Transit City plan, including the public consultation during the development of the plan and conflicting political agendas, has been well reported in the media.
- The inclusion of theoretical ideals such as social urbanism, social exclusion, and environmental mismatch in the Transit City plan may not have been evident to the public.
- Transit City seems to have become politicized through political and media voices; are there ways in which such a plan could have been more thoroughly understood and embraced by the public?

Introduction

The Greater Toronto Area (GTA) is increasingly economically and culturally diverse, shaped by socio-spatial transformations. Increasingly, immigrants and the poor live in the inner suburbs of the amalgamated city, while the downtown gentrifies. These changes complicate the traditional understanding of the suburbs as car oriented. Investment in the GTA's public transit infrastructure has not kept up with these changes and recent increases in the regional population. The inner suburbs are underserved by transit.

Drawing from a series of key interviews with politicians and transit planners, a spatial analysis of data from the Toronto Transit Commission (TTC), the Census of Canada, and the Transportation Tomorrow Survey (TTS) and media reports, this case study follows the political and social conversations around extending transit into the inner suburbs. We find that the political rhetoric has obscured the underlying need and shaped the public's understanding more than planning considerations have.

Toronto's Inner Suburbs

In 1998, the six municipalities that had formed Metro Toronto were merged into a single municipality by the Province of Ontario. The Province went ahead with this **amalgamation** despite its rejection by local voters in a referendum. The theory was that a unified megacity would be more efficient at regional governance, a major consideration during an economic downturn.

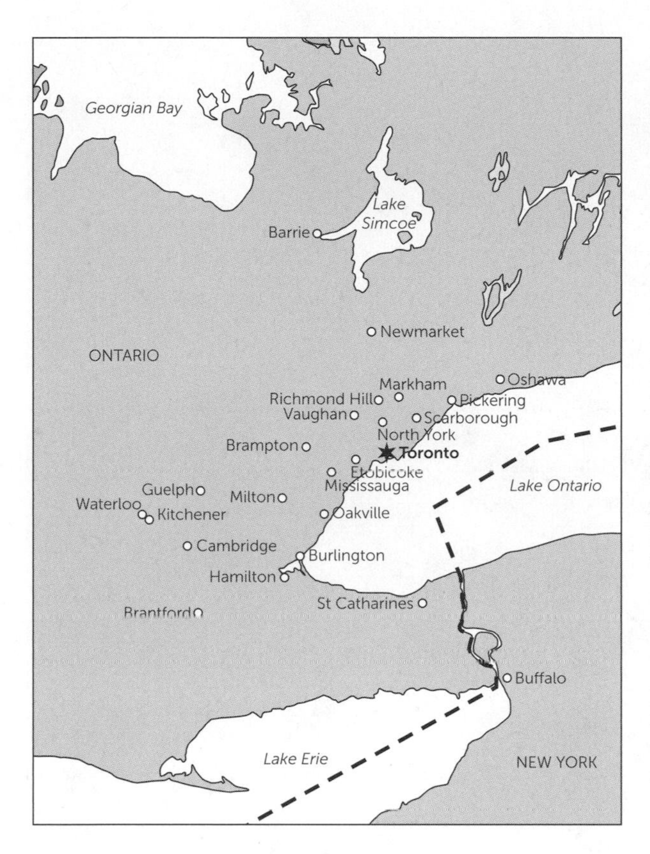

Figure 9.1.1 • Location map.

The amalgamation brought together two very different geographies. The old city of Toronto has a high-density core and a grid of streets that have a mix of land uses, with residential uses on smaller side streets and commercial along main roads. It was mostly built before the postwar era; many neighbourhoods were built around streetcar lines. This resulted in an overall tighter urban form, making destinations more accessible by walking and cycling.

The amalgamated municipalities were the first-ring suburbs built around the old city. These municipalities were built around a wider grid of larger arterial roads with large volumes of faster-moving car traffic in mind. Inside this large grid of **arterials**, uses are separated, and residential land use takes the form of single-family homes. In a later wave of construction throughout the 1960s, 1970s, and early 1980s, high-rise apartment buildings were added along arterials and at major intersections. Although these apartment buildings added considerable density to otherwise low-density neighbourhoods, they are located within an urban form that is not as friendly to alternative modes of transportation; these neighbourhoods were built around the car. Figure 9.1.3 demonstrates the differences in urban form between the old city of Toronto, which has a large high-density core surrounded by mid-density areas of two to five storeys and attached homes, and the inner suburbs, which have a combination of lower-density detached houses within super-blocks, along with high-rise residential buildings at major intersections and along corridors. The difference in street networks can also be seen here.

In the public imagination, suburbs are often conceived as mono-functional areas of single-family homes built around the car as mode of transportation, spreading in rings around cities expanding over time. In many North American cities, the first- or inner-ring suburbs are experiencing declining income and investment. Combined with the **gentrification** of older city areas, these transformations can amount to socio-economic **polarization**, with lower-income households moving to the inner suburbs because of the cost of housing (Walks 2001; Hulchanski 2010). Although there remain low-income residents in the core city, poverty is becoming a "suburban" phenomenon in Toronto (United Way 2004; 2011). In Toronto's inner suburbs, the affordable housing is in the form of high-rise rental housing that was built in between the 1960s and 1980s by private developers with a tax incentive for this type of housing before condos became the more profitable option for high-rise development. These buildings were originally intended for the young urban professional, with some two- and three-bedroom units for families, but now provide an important stock of relatively affordable housing for lower-income households. By 2006, almost half of low-income families in Toronto rented a unit in a high-rise building, and nearly 40 per cent of the families in these buildings were low income (United Way 2011).

Despite being built around the car, the inner suburbs have relatively high transit ridership. Residents in high-rise apartment neighbourhoods have higher-than-average rates of transit use and lower rates of car ownership than surrounding areas (ERA Architects 2009). The transit ridership on the major bus routes in the inner suburbs rivals that of the busiest bus and streetcar routes downtown (Figure 9.1.4). However, interviews with low-income immigrants who live in suburban high-rises found the "lack of adequate transit and convenient access to services, jobs and amenities" a pressing problem (Smith and Ley 2008, 701).

Transportation planners are faced with the challenge of providing transit service to the entire city of Toronto. Although there is a subway running north–south and east–west, the main transit service in the inner suburbs is **local bus** service, which is slower and less frequent than rapid transit. There has not been a significant upgrade to the transit system in Toronto since the Scarborough

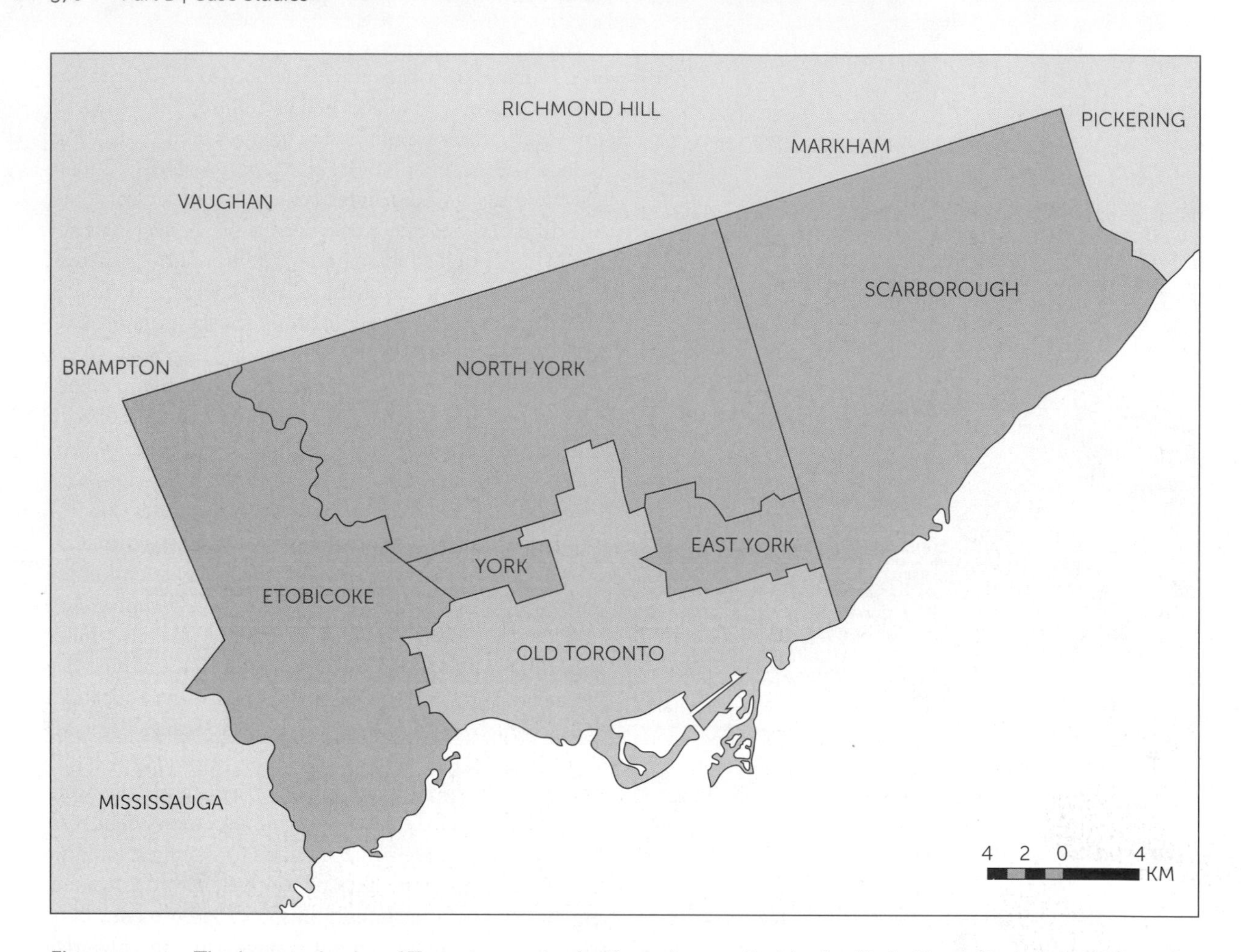

Figure 9.1.2 • The inner suburbs of Toronto are shaded in dark grey: Etobicoke, York, North York, East York, and Scarborough. They were amalgamated with the old City of Toronto in 1998, forming the current City of Toronto.

Source: Anna Kramer.

RT "stubway" extension opened in 1985. This has left the inner suburbs with less than ideal transit service, even as the area continues to absorb population growth along with the rest of the region.

Transit City

In order to improve the public transit connections of these high-density, high-rise, high-ridership pockets in the inner suburbs to the rest of the city, planners at the City of Toronto, the Toronto Transit Commission, and regional transit agency Metrolinx considered possible options. Toronto's existing network consisted of subways, streetcars, and local buses. Subways are a rapid transit option that travel mostly underground, but the cost of digging underground routes is very high. Toronto's existing streetcars and buses run in mixed traffic without dedicated lanes; they travel slowly because they are subject to automobile congestion and because passengers must board and pay at the driver's door. More recent innovations in transit network design offer several more alternatives: **Light Rail Transit (LRT)** and **Bus Rapid Transit (BRT)**. LRT is a longer, faster version of a streetcar that travels on a dedicated lane, separated from car traffic. Bus Rapid Transit is a similar system, using a dedicated lane separate from traffic, and long **articulated buses** can carry about 45 per cent more passengers than regular buses. Both LRT and BRT routes allow passengers to prepay and board the vehicles from all doors instead of lining up at the front

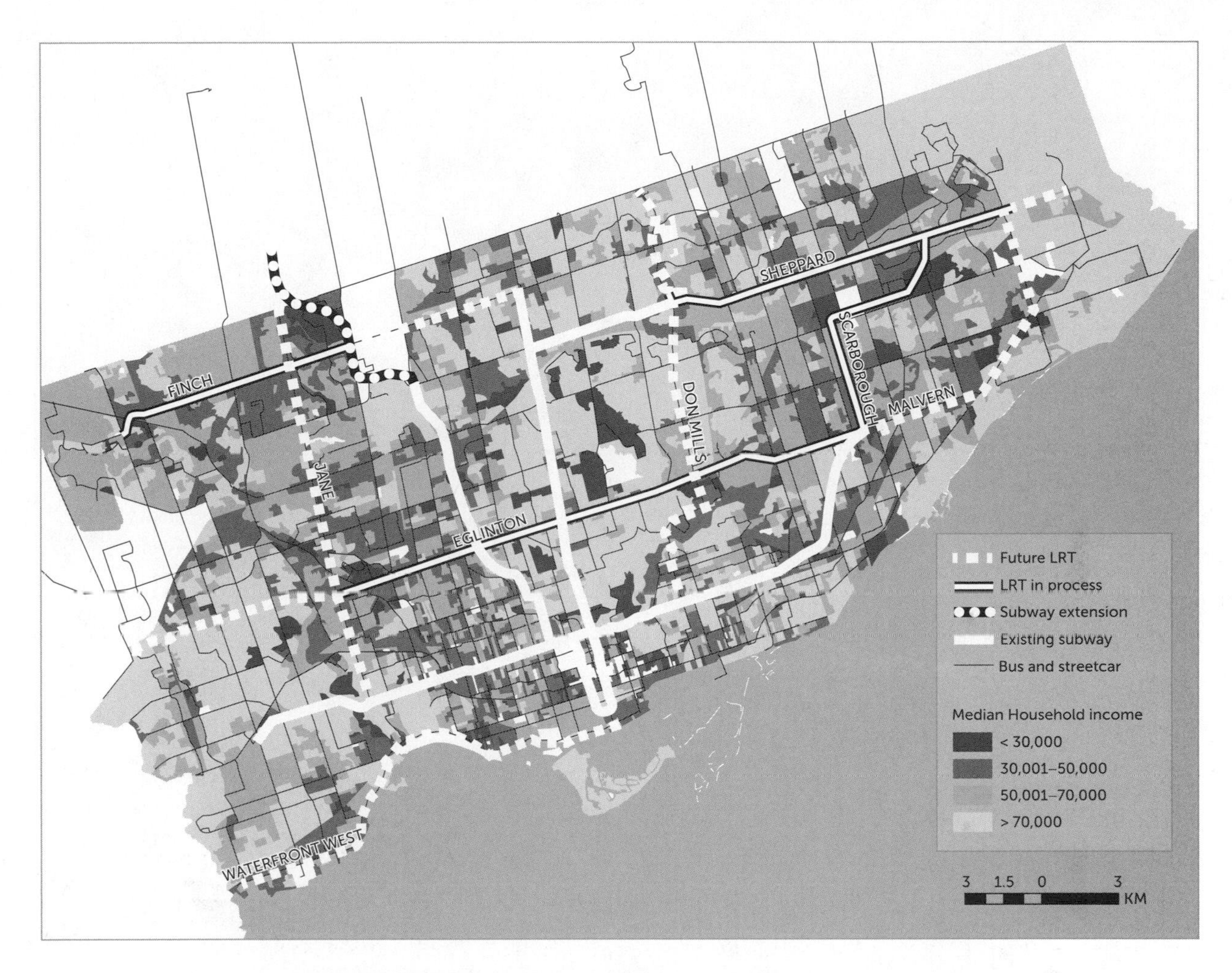

Figure 9.1.3 • Poverty in the inner suburbs (income). This shows the pattern of lower income in Toronto, which follows a "U" shape that widens in the northeast and northwest inner suburbs. The difference in street grids between the older downtown area and the suburban super-blocks can also be seen here.

Source: Anna Kramer with Statistics Canada 2006 Census data.

door to pay the fare to the driver. This, together with the dedicated lane, makes these technologies much faster than streetcars and buses operating in mixed traffic. The stops of these rapid routes are generally spaced farther apart than stops on a local route, allowing the vehicles to pick up speed between stops. Although LRT and BRT generally operate on a surface alignment, at ground level rather than underground, they have more in common with subways than with local routes in terms of volume, speed, and **frequency of service**. They are also much less expensive to build than subways—BRT lines in particular—since they do not need tracks.

Given these considerations, City Council proposed using LRT to connect major east–west and north–south corridors into the existing rapid transit system. Their plan also proposed supplementing these LRT corridors with BRT extensions in some cases. The planners who worked on this plan judged that LRT and BRT would have enough capacity to satisfy projected ridership for some time into the future and that using these technologies would be a cost-effective way to extend rapid transit into these areas. Although faster technologies and alignments were possible, which would have made the trip across the city competitive with the car, the decision was made to recommend at-grade (street-level)

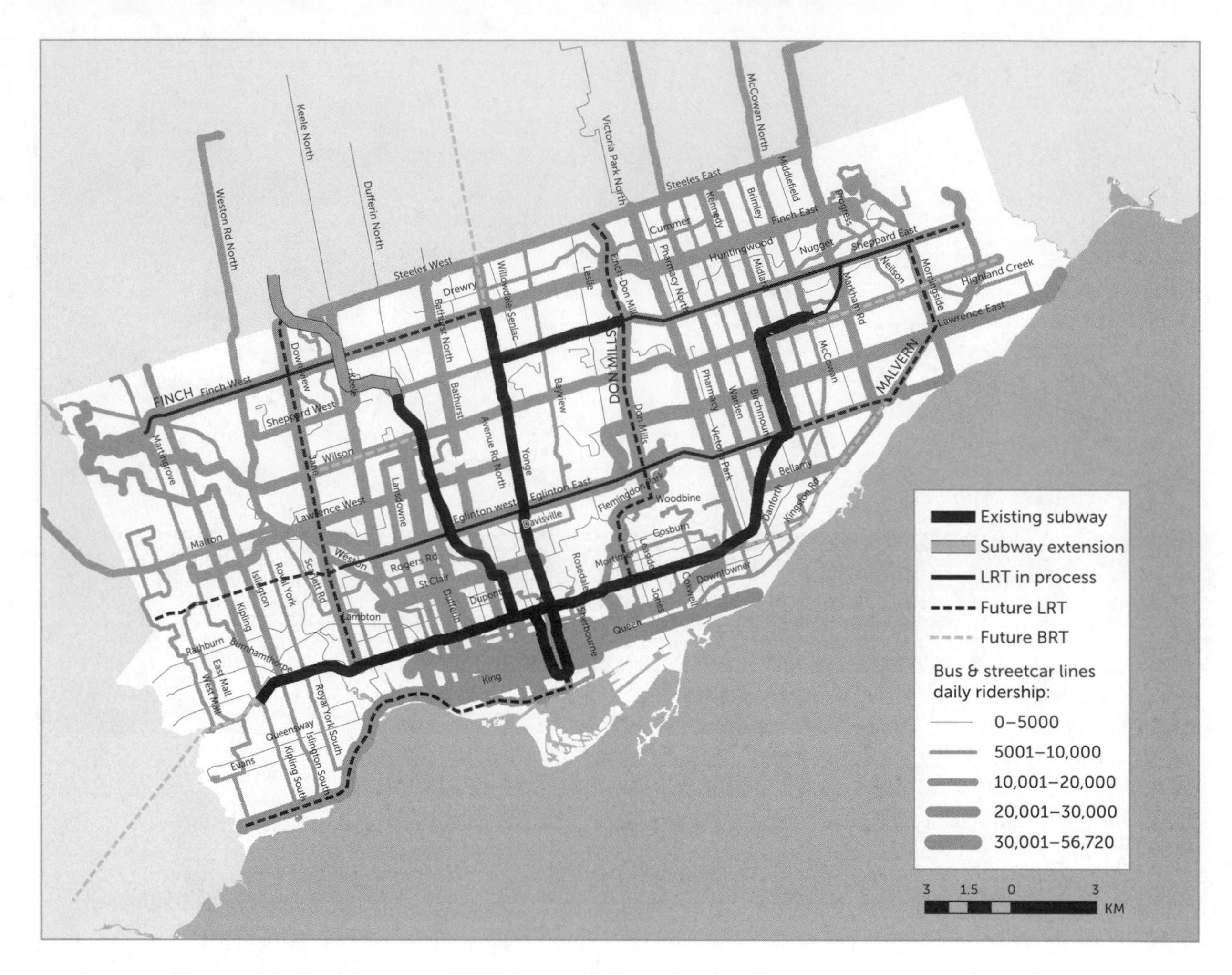

Figure 9.1.4 • Transit ridership in Toronto (data from TTC, 2005–11).

Source: Anna Kramer.

LRT (in all cases except for an urban portion of the Eglinton crosstown, which will be underground) with stops spaced approximately 700 metres apart. A review of the LRT line plans by an independent research organization suggested that stops spaced one kilometre apart in denser areas and even further apart in lower-density, suburban areas would achieve much faster travel speeds (Shabas, in press). An at-grade alignment is less expensive than tunnelling or elevating tracks, although it results in more impacts on traffic, such as the converting of lanes to dedicated tracks and the restriction of left turns for vehicles. Spacing stops closer together means that more residents live near a transit stop, although this also adds considerably to the travel time. In this plan, coverage and cost were weighted as more important than speed and competitiveness with the car.

As required by the Planning Act (1990), the Transit City plan involved multiple public consultations, separately for each proposed line, as part of the environmental assessments. For example, for the Eglinton line alone, the TTC and the City of Toronto held seven open houses across the study corridor in November and December of 2009 during which staff presented the preferred design (City of Toronto 2011). However, it was not until later when the plan became an election issue for candidate Rob Ford, who preferred subways, that this controversy became part of a broader public awareness and debate. This reflects how public consultation, as currently practised, can be problematic and ineffective. Despite planners' efforts to publicize the event, most neighbourhood residents don't attend. Planners may also be more interested in allaying

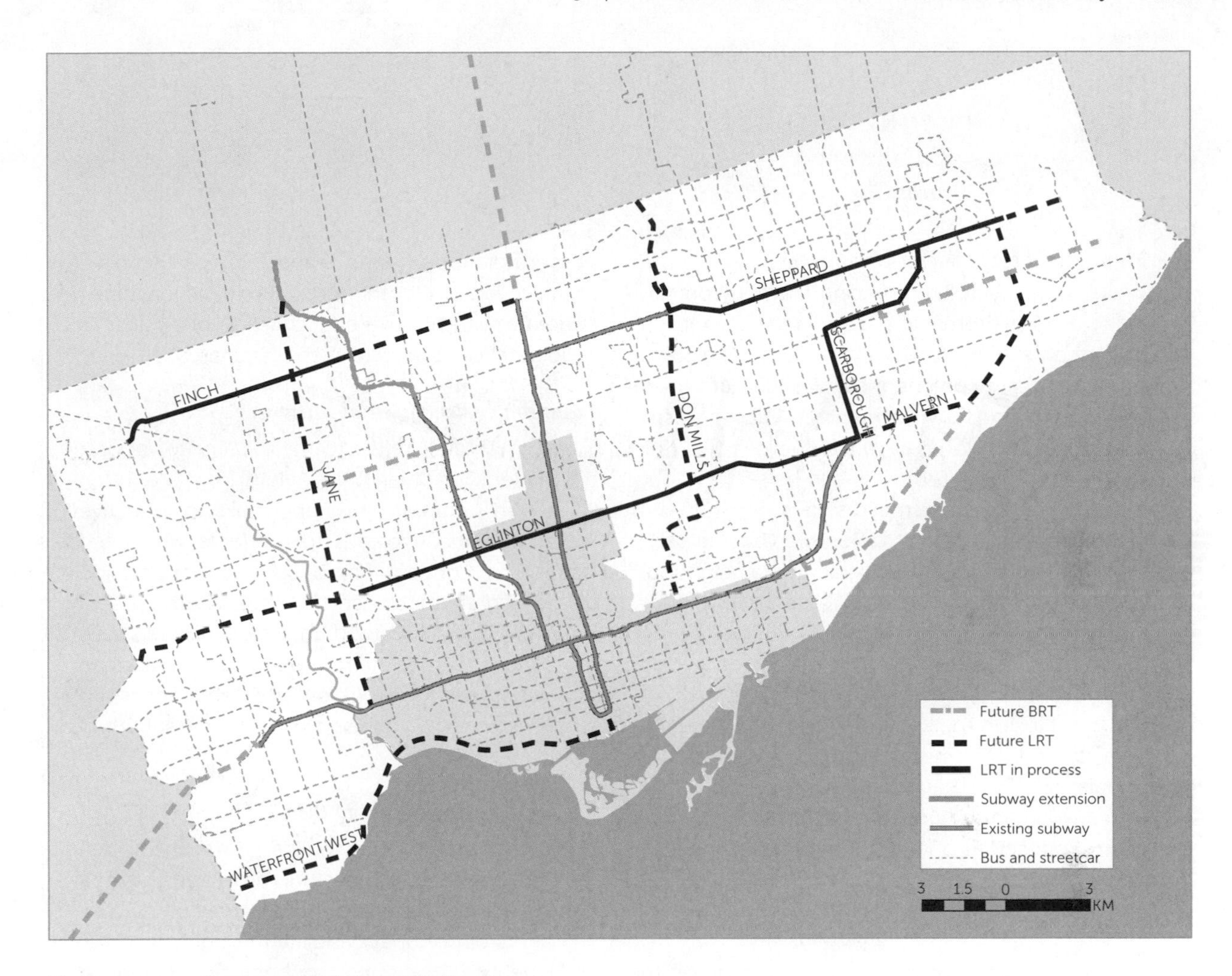

Figure 9.1.5 • Current and future LRT lines.

Source: Anna Kramer.

possible concerns rather than debating options publicly. If only one preferred option is presented rather than several possibilities, is it really possible for the public to have a well-informed debate on the merits of various options?

The plan was originally called "Transit City" and was later pared down and incorporated into a larger regional plan known as The Big Move (Metrolinx 2008).

Innovations: LRT to Connect to High-Rise, High-Density Pockets

In proposing to improve the connection between the affordable rental-housing clusters of high-rise apartment buildings and the rest of the transit network by upgrading crowded local bus routes to Light Rail Transit, the Transit City plan reflected theories of social justice and **transit-oriented development**. Unfortunately, by being implicit rather than explicit, this progressive aspect of the plan was not communicated well to the public and as a result has not been reflected in the debate in the political and public sphere.

The Politics of Transit (Implementation and Challenges)

Transit City was originally championed by Mayor David Miller (mayor from 2003 to 2010). Former Mayor Miller cited "political, environmental, economic and social

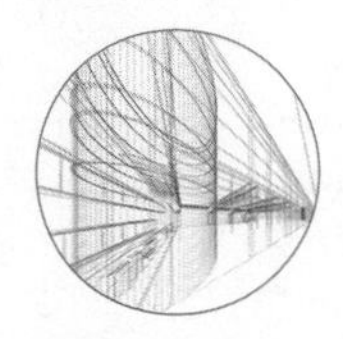

PLANNING THEORY

Transit City

Planning Theory: Social Urbanism

While not explicitly referencing them, by proposing to extend affordable access to the city through investments in improved public transit, the Transit City plan draws on theories of **social urbanism**, **social exclusion**, **environmental justice**, and **spatial mismatch** theories as well as the idea of a "**right to the city**." Social urbanism can be seen in South American cities, where mayors have invested in BRT (Curitiba, Brazil; Bogotá, Colombia), bicycle paths, and metro cable cars to hillside *favela* neighbourhoods (Medellin, Colombia; Rio de Janeiro, Brazil) to improve access for lower-income areas.

Planning Theory: Social Exclusion

Social exclusion is a UK theory describing the barriers faced by marginalized urban and suburban populations to urban opportunity and privileges of many kinds, and improved public transport has been proposed as one way to mitigate some types of exclusion.

Planning Theory: Environmental Justice

The concept of environmental justice arose in the United States in response to the tendency of cities to locate infrastructure with **negative externalities**, like garbage dumps, incinerators, and freeways in poorer neighbourhoods. Transit City can be seen as an inversion to this concept in which infrastructure with positive effects like public transit is deliberately extended to these same areas.

Planning Theory: Spatial Mismatch

Spatial mismatch refers to the location of poorer populations in inner cities while jobs are often located in suburban areas and looks for effective ways of connecting jobs to residents who need them. The concept has a specific connotation in the segregated inner-city neighbourhoods of American cities (Kain 1969; 1976). Again, in this case the concept can be altered to fit a more suburban population that depends on affordable access to jobs located either in the downtown or in suburban areas.

Planning Theory: Right to the City

Finally, Transit City builds on the idea that all residents of an urban region have a right to the city (Lefebvre 1968; Harvey 2008). The city offers opportunity, visibility through presence, and identity; public transit connects populations to the public space of the city. In this way, Transit City could be seen as an innovative practical application of these social justice theories.

justice" reasons for improving public transit (Miller 2011). The Transit City plan secured federal and provincial funding to cover three to five initial lines. Miller called the difficult alignment of three levels of government a "once-in-a-generation" occurrence (Miller 2011). Several key LRT lines (Eglinton crosstown, Finch, and Sheppard) were later adopted as a part of a larger plan, The Big Move, by the regional transportation planning agency Metrolinx (Metrolinx 2008). The Big Move in turn had its context in the provincial Places to Grow Act (2005), which introduced a number of Smart Growth policies in the region, including land-use intensification targets for Urban Growth Centres. The Big Move planned to link these growth centres through improved transportation infrastructure, including rapid public transit.

In 2010, Torontonians elected a new mayor, who garnered 47 per cent of the vote. During his campaign, Rob Ford promised to "stop the gravy train" and eliminate waste at City Hall, running on a platform of "respect for taxpayers" (Lorinc 2011). This message appealed to many people who were tired of garbage and transit strikes and presumably wanted lower taxes. In one of the first statements after he was elected, Mayor Ford declared that "the war on the car is over" and that he would scrap Transit City in favour of building subways (Kalinowski and Rider 2010). His reason for subways rather than surface modes of transportation seemed to be that they do not interfere with car traffic because they are underground. By referencing a "war on the car," Mayor Ford implied that the former administration had been waging battle against drivers by introducing bike lanes and dedicated light rail tracks. This view of driving as a marker of citizenship can also be seen in the rhetoric of postwar planning in the United States (Seiler 2008). The use of the rhetoric of a "war on cars," with drivers as the legitimate road-users, delegitimizes alternative modes of transportation. In a humorous reflection of this politicization of modes of transportation, TV personality Don Cherry wore a pink suit as a shout-out to "all the pinkos out there that ride bicycles" at the mayor's inauguration (Kapral 2010; Cherry 2010). Drawing a contrast between citizen-drivers in the suburbs and left-wing "urban elites" in the city, Rob Ford's rhetoric politicized transportation modes as well as urban forms and lifestyles. This appeal to the car-driving, home-owning "taxpayer" was largely successful; Ford was elected by a majority of (inner) suburban voters, while downtown residents largely voted for his opponent (Topping 2010). The politics of these **inner-ring suburbs** have a history of catering to their "distinctly suburban population [of]... modest homeowners" rather than the high-rise renters (Boudreau et al. 2006, 33).

It may seem surprising that it was largely the inner suburban vote that propelled Mayor Ford's election, given the presence of high-density pockets of vertical poverty. There is a paradox in these areas seeming to vote against their own interests: the Transit City plan was largely aimed at improving the accessibility to rapid transit in these same suburbs, and Ford campaigned against it. This paradox may be partly explained by the discrepancy between those who vote and those who use transit infrastructure, as well as the needs of current residents compared to future populations in these areas.

Another possible explanation is the lack of political representation on the part of many inhabitants of the

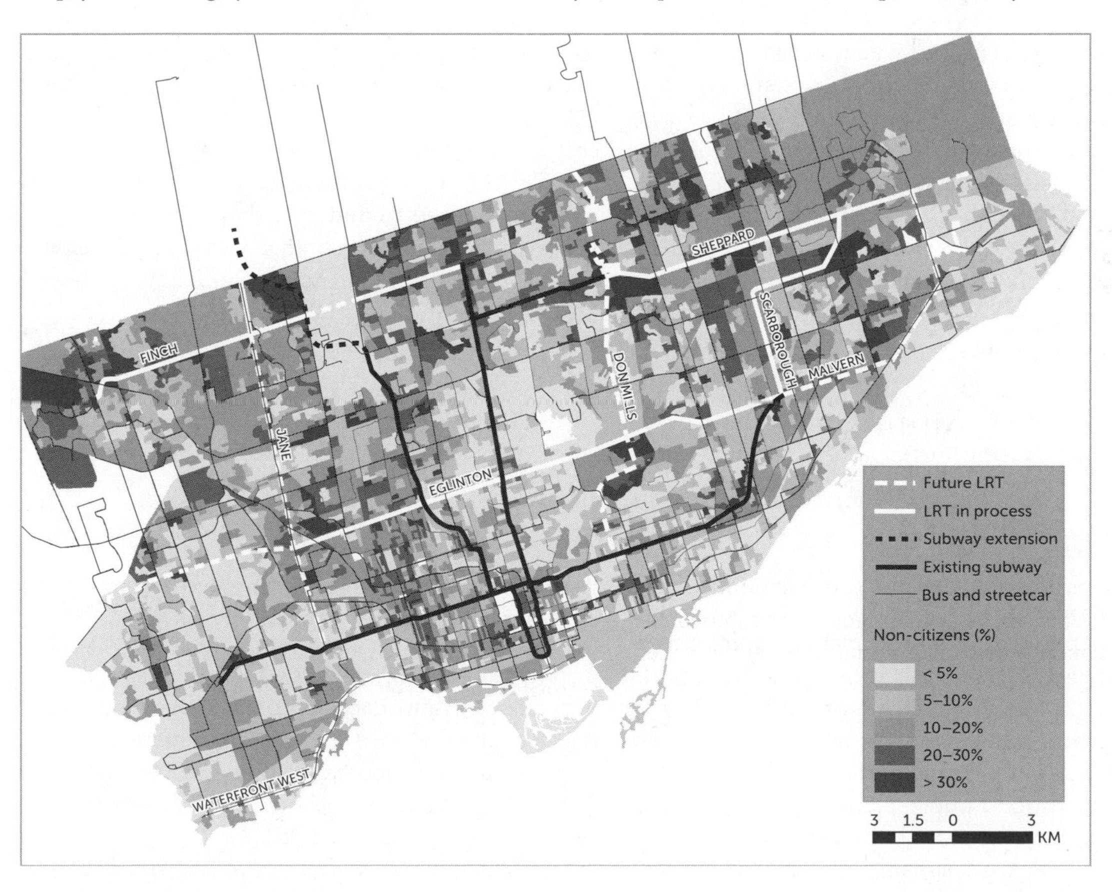

Figure 9.1.6 • Non-citizens living in the metropolitan area.

Source: Anna Kramer with Statistics Canada 2006 Census data.

inner suburbs. One in seven Torontonians is not eligible to vote in municipal elections because they lack Canadian citizenship, even though they pay taxes and reside in the city. Toronto has neighbourhoods in which more than 30 per cent of the residents are non-citizens (Siemiatycki 2010). Many of these neighbourhoods of high concentrations of non-citizen residents are located in the inner suburbs, in areas where Transit City would have improved transit provision.

In fact, urban form can be seen as a predictor of politics in many cases (Walks 2007; 2006). Alan Walks, a researcher who has studied the relationship between politics and urban form in Canadian cities, has found that middle-class households living in the low-rise form tend to vote against the interests of the recent immigrants living in the high-rises the next street over. In this way, he describes polarization as being multi-scalar, with disenfranchisement happening at a neighbourhood as well as a city scale. Therefore, the transformations occurring in the inner suburbs themselves have political implications.

In this way, the Transit City plan became politicized, representing the interests of a diverse, high-rise, transit-dependent population against the interests of a more homogenous, auto-oriented, low-rise population. This latter population has suffered the loss of manufacturing jobs in the suburbs as the **Fordist** economy in which a middle-class population was supported with high-paying, blue-collar jobs (particularly in auto manufacturing in Toronto) is being replaced by a **post-Fordist**, service-based economy in which there is a greater gap between low-paying service jobs and higher-paying "creative-class" jobs requiring a post-secondary education. Instead of embracing transit-oriented intensification, the former may see it as a threat to an auto-oriented, home-owning way of life.

Conclusion: Lessons Learned from Transit City

Transit City was challenged by Mayor Ford, who proposed cancelling it and replacing it with a single subway extension to Scarborough Town Centre. Subways were presented as a mode that was hidden from view underground and therefore did not interfere with car traffic, while LRT was presented as a "second-class" option for the residents of Scarborough. This proposal eventually came to the City Council for a vote, and in preparation for the decision, council asked a panel of experts to present the costs and benefits of each option. This panel demonstrated that the Transit City LRT option would serve more residents for the same cost and be almost as fast as subways and so recommended LRT (City of Toronto 2012; CBC 2012). Citizen's groups mobilized around the debate, creating websites to present arguments on both sides (Save our Subways 2010; Toronto Environmental Alliance 2008).

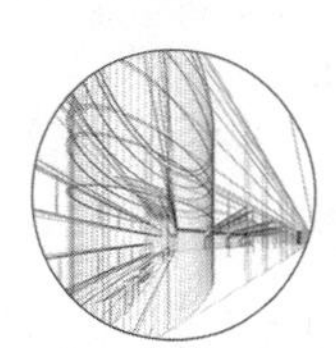

PLANNING THEORIES

Rational Comprehensive, Participatory, Advocacy, and Incremental Approaches

The case study of Transit City demonstrates the importance of clear and evidence-based decision-making (the **rational comprehensive** approach to planning) but also equally the importance of clear communication of these issues in a way that is understandable and relatable to the general public (the **participatory** approach to planning). If no strong public case is made for a plan, no matter how rational or comprehensive it is, it is open to political challenges that can make implementation difficult. This case study also reveals that underlying reasons for plans—in this case for social justice and transit-oriented development—should be argued for explicitly rather than implicitly. If planners explain the reasons for their plans, this may reflect an **advocacy** approach to planning that is justified by the current socio-spatial transformations in the inner suburbs. This case study demonstrates that the successful implementation of plans requires a combination of all three approaches. If that fails, then an **incremental** approach, also known as "muddling through," must be adopted.

In the controversy over the Transit City plan, there was a flurry of press about what Light Rail Transit was, how it was different from streetcars in mixed traffic, and what costs and coverage subways had in comparison to LRT. Issues of land use and transportation were brought into the public debate in a way they had not been when the plan was first introduced. Facts and ideas are contested and sometimes misrepresented, and the debate continues at City Hall, even reaching provincial and federal levels.

The question that arises is whether this debate could have been better informed and less polarized if the original public consultation process had actually considered multiple options and if the Transit City plan had not been so attached to an at-grade LRT technology and alignment from the start. In this case, both Mayor Miller and Mayor Ford became political champions for certain perspectives. It is often said that transit projects benefit from political champions, but in this case it seems that the presence of political champions who were determined to push their project through at any cost seems to have polarized the debate and obscured the underlying needs and alternative possibilities for serving these needs.

KEY TERMS

Advocacy planning
Amalgamation
Arterials
Articulated bus
Bus Rapid Transit (BRT)
Environmental justice
Fordist
Frequency of service
Gentrification
Incremental planning
Inner-ring suburbs
Light Rail Transit (LRT)
Local bus
Negative externalities
Participatory planning
Polarization
Post-Fordist
Rational comprehensive model (RCM)
Right to the city (Droit à la ville)
Social exclusion
Social urbanism
Spatial mismatch
Transit-oriented development (TOD)

REFERENCES

Boudreau, J.-A., P. Hamel, B. Jouve, and R. Keil. 2006. "Comparing metropolitan governance: The cases of Montreal and Toronto." *Progress in Planning* 66(1): 7–59.

CBC. 2012. "Sheppard transit report recommends LRT." 16 March. http://www.cbc.ca/news/canada/toronto/story/2012/03/16/sheppard-avenue-report.html.

Cherry, D. 2010. "What Don Cherry said." *Toronto Star* 7 December.

City of Toronto. 2011. "Eglinton crosstown Light Rail Transit (LRT)." http://www.toronto.ca/involved/projects/eglinton_crosstown_lrt.

——. 2012. "Report of the Expert Advisory Panel regarding transit on Sheppard Avenue East." 15 March. http://www.toronto.ca/legdocs/mmis/2012/cc/bgrd/backgroundfile-45908.pdf.

ERA Architects. 2009. "Tower neighbourhood renewal in the Greater Golden Horseshoe." Centre for Urban Growth and Renewal. http://www.cugr.ca/tnrggh.

Harvey, D. 2008. "Right to the city." *New Left Review* 53: 23–40.

Hulchanski, J.D. 2010. *The Three Cities within Toronto: Income Polarization among Toronto's Neighbourhoods, 1970–2005*. Toronto: Cities Centre, University of Toronto.

Kain, J.F. 1969. *Race and Poverty: The Economics of Discrimination*. Upper Saddle River, NJ: Prentice Hall.

——. 1976. "Race, ethnicity, and residential location." In R. Grieson, ed., *Public and Urban Economics*, 267–92. Lexington, MA: Lexington Books.

Kalinowski, T., and D. Rider. 2010. "'War on the car is over': Ford moves transit underground." *Toronto Star* 2 December.

Kapral, M. 2010. "Don Cherry blasts bike-riding 'pinkos.'" *Canadian Cycling Magazine* 8 December. http://cyclingmagazine.ca/2010/12/news/don-cherry-blasts-bike-riding-pinkos.

Lefebvre, H. 1968. *Le Droit à la ville*. Paris : Anthropos.

Lorinc, J. 2011. "Ford's 'respect for taxpayers' mandate is about to get real." *Globe and Mail* 8 June.

Metrolinx. 2008. *The Big Move: Transforming Transportation in the Greater Toronto and Hamilton Area*. Toronto: Metrolinx.

Places to Grow Act [2005] S.O. 2005, c 13.

Planning Act [1990, R.S.O. c. P. 13].

Save our Subways. 2010. "Move Toronto vs Transit City." http://www.saveoursubways.ca.

Seiler, C. 2008. *Republic of Drivers: A Cultural History of Automobility in America*. Chicago: University of Chicago Press.

Shabas, M. 2013. *Review of Metrolinx's Big Move*. Toronto: Neptis Foundation.

Siemiatycki, M. 2010. "Toronto's lost voters." *The Mark* 12 March.

Smith, H., and D. Ley. 2008. "Even in Canada? The multiscalar construction and experience of concentrated immigrant poverty in gateway cities." *Annals of the Association of American Geographers* 98, 686–713.

Topping, D. 2010. "How Toronto Voted for Mayor." *Torontoist* 28 October. http://torontoist.com/2010/10/which_wards_voted_for_who_for_mayor.

Toronto Environmental Alliance. 2008. "Light Rail Transit FAQ's." http://www.torontoenvironment.org/campaigns/transit/LRTfaq.

United Way. 2004. *Poverty by Postal Code: The Geography of Neighbourhood Poverty, 1981–2001*. Toronto: United Way.

——. 2011. *Vertical Poverty: Declining Income, Housing Quality and Community Life in Toronto's Inner Suburban High-Rise Apartments*. Toronto: United Way.

Walks, A. 2001. "The social ecology of the post-Fordist/global city? Economic restructuring and socio-spatial polarization in the Toronto urban region." *Urban Studies* 38(3): 407–47.

——. 2006. "The causes of city–suburban political polarization? A Canadian case study." *Annals of the Association of American Geographers* 96(2): 390–414.

——. 2007. "The boundaries of suburban discontent? Urban definitions and neighbourhood political effects." *The Canadian Geographer* 51(2): 160–85.

INTERVIEWS

D. Miller, 11 November 2011. In-person interview.

CASE STUDY 9.2

Declining Infrastructure and Its Opportunities: Gardiner East Environmental Assessment

Antonio Medeiros

SUMMARY

As twentieth-century infrastructure built during the post-war boom begins to decline, cities must make tough decisions about their future. Infrastructure such as bridges, water treatment facilities, and highways are in need of significant capital investment and repair. In 2013, the Urban Land Institute estimated that Canada alone was facing a $170-billion infrastructure investment backlog (ULI and Ernst and Young 2013, 53). With major repair bills and constrained fiscal budgets, municipal governments are searching for solutions: should they rehabilitate aging infrastructure for today's needs or build something new?

The Frederick G. Gardiner Expressway and Lake Shore Boulevard corridor in Toronto is 18 kilometres long and was built between 1955 and 1966. Its elevated section has met its life expectancy and faces an uncertain future. After many years of studies and popular debate, the easternmost 2.4-kilometre segment was considered a worthy candidate for a comprehensive re-examination of its role and function. In 2009, the City of Toronto and Waterfront Toronto launched the Gardiner Expressway East and Lake Shore Boulevard Reconfiguration Environmental Assessment and Integrated Urban Design Study (Gardiner East EA) to bring a resolution to the question of its future.

Given the number of opportunities and challenges inherent in the corridor and its potential influence on the future of the waterfront, planners undertook to develop a comprehensive evaluation and decision-making framework through an environmental assessment process. The options under consideration ranged from maintaining the existing elevated expressway to removing it in favour of an at-grade boulevard.

An innovative consultation program was developed to educate, inform, and solicit meaningful feedback on the integrated design and planning approach to developing and assessing the options. Ultimately, the decision framework fostered an informed public discussion about the regional role of the existing transportation corridor and how to integrate it with local emerging waterfront neighbourhoods and needs of a twenty-first-century city.

OUTLINE

- Canadian municipalities must make difficult decisions in terms of aging transportation infrastructure: the cost of maintaining them is high, and other alternatives may be more sustainable and cost-efficient.
- The case of the Frederick G. Gardiner Expressway in Toronto describes the planning process, environmental assessment, and options that were considered in terms of maintaining, replacing, or removing the elevated section. The planning challenges include a polarized public and the politicization of transportation issues in Toronto.
- The option chosen as a result of the environmental assessment best met the needs of the region as well as local neighbourhoods.

Introduction

As infrastructure built during the twentieth-century declines, cities around the world face the daunting task of making tough decisions about their future. The postwar boom brought tremendous infrastructure investment into North American cities, which fuelled their economic growth and expansion. Now, more than a half-century later, infrastructure such as bridges, water treatment facilities, and highways are in need of significant capital investment and repair. In 2013, the Urban Land Institute estimated that Canada alone was facing a $170 billion infrastructure investment backlog (ULI and Ernst and Young 2013, 53). Faced with tremendous repair bills and constrained fiscal budgets, municipal governments have embarked on the necessary search for how to best leverage their scarce investment dollars. What is the decision framework for a democratic twentieth-first-century city to decide between rehabilitating twentieth-century assets for today's needs, or adapting them, or building something else completely?

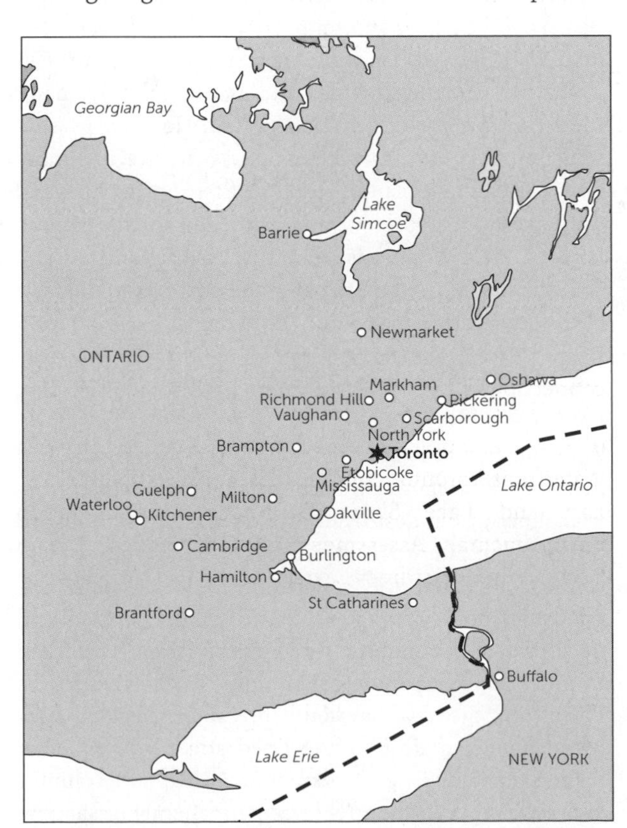

Figure 9.2.1 • Location map.

The Frederick G. Gardiner Expressway and Lake Shore Boulevard corridor in Toronto, Ontario, is one such piece of transportation infrastructure. The 18-kilometre expressway was built between 1955 and 1966, and its elevated section is close to the end of its life expectancy and has an uncertain future. After many years of studies and popular debate, the easternmost 2.4-kilometre segment was considered a worthy candidate for a comprehensive re-examination of its role and function. In 2009, the City of Toronto and Waterfront Toronto launched the Gardiner Expressway East and Lake Shore Boulevard Reconfiguration Environmental Assessment and Integrated Urban Design Study (Gardiner East EA) to bring a resolution to the question of its future and answer the popular query of whether "to keep it up or tear it down."

Figure 9.2.2 • View along Lake Shore Boulevard, just west of the Don River, with the Gardiner Expressway above.

Source: Antonio Medeiros.

Figure 9.2.3 • Aerial view looking west of the study area and eastern segment of the Gardiner Expressway as it connects to the Don Valley Parkway.

Source: Waterfront Toronto 2013.

The answer to the question was urgent. In the spring and summer of 2012, pieces of concrete from the elevated structure rained upon Lake Shore Blvd below. The numerous incidents led to interim emergency repairs of the structure and prompted a comprehensive engineering review. The assessment of the almost 50-year-old expressway concluded that most of the elevated deck structure would need to be demolished and rebuilt on a phased basis over a two-decade period. The rehabilitation cost for the entire expressway, including at-grade and elevated segments, was estimated at $1.3 billion in 2013 net present value dollars (City of Toronto 2014, 39). In 2012, the Gardiner East EA resumed after a pause subsequent to the 2010 municipal election. The urgency of capital repairs thrust the project out of the theoretical realm of study and into the municipal spotlight.

Many questions were asked by the stakeholders, public, project team, and politicians in considering the future of the Gardiner: Why study alternatives or consider removing a link from the larger expressway network? Can the City of Toronto afford to keep it up? How does the expressway fit into Toronto's congestion puzzle? How do planners meaningfully engage the public in making infrastructure investment decisions in Canada's largest city?

Superhighway "A"

The Gardiner Expressway was conceived in the 1940s and 1950s when Toronto's downtown waterfront was still a heavy industrial port with the city's suburbs rapidly expanding. Its alignment originated from a 1943 master plan that outlined a network of limited-access "superhighways" that would later be partially implemented by the expressway named for Frederick Goldwin Gardiner, chairman of the former Metropolitan Toronto Planning Board and Toronto's answer to New York's Robert Moses. Fred Gardiner was as influential as he was controversial and was credited with building a number of major infrastructure projects in that period. He oversaw construction of Toronto's backbone infrastructure, including sewage treatment plants, Toronto's first subway line, and the Lakeshore Expressway later named in his honour. In 1955, after numerous years of planning and design, construction started on the at-grade portions to the west of the city moving eastward. The elevated structure from Dufferin to Leslie, including the Don Valley Parkway connection, was completed from 1958 to 1966 (Hayes 2008, 170).

From the start of construction, the expressway had its critics. The chosen waterfront alignment was controversial because it destroyed prime parkland, demolished the popular Sunnyside Amusement Park, and severed connections from the waterfront to local neighbourhoods. To the west, the new elevated expressway isolated the Parkdale community, which precipitated its decline from a once-prosperous community. To the east in the downtown, the new elevated expressway relegated Lake Shore Boulevard from a tree-lined street to a collector road. Surrounded by the expressway's support columns and deck above it, the street was permanently in shadow. The Gardiner's negative environmental effects galvanized Toronto community organizers, including Jane Jacobs, to oppose the expansion of the "superhighways" network and

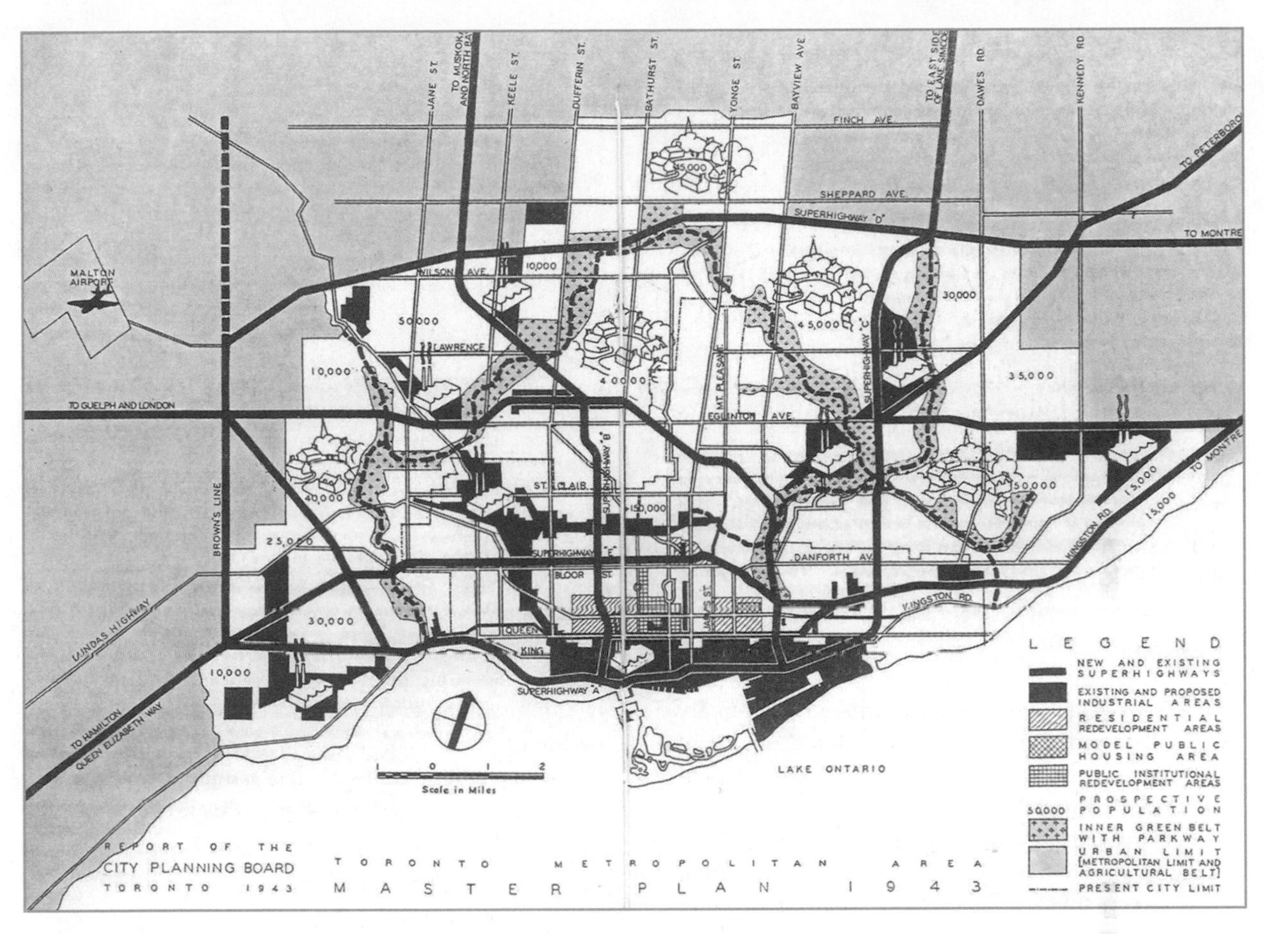

Figure 9.2.4 • Toronto Metropolitan Area Master Plan 1943, which laid out a network of "superhighways" including the Gardiner Expressway, which was noted as superhighway "A" on the map.

Source: Toronto City Planning Board 1943.

were instrumental in stopping the Spadina Expressway plans in the early 1970s.

Since shortly after the expressway was built, there have been calls for its removal, which have been amplified in the face of the structure's decline. As early as 1978, it was in need of repair as "chunks of concrete," some weighing up to 200 pounds, fell from the elevated expressway, prompting emergency repairs and closures (Van Rijn 1978). Most recently, in 2001, a two-kilometre spur between the Don Valley Parkway and Leslie Street was demolished and replaced by a redesigned Lake Shore Boulevard. The decision to do so was a result of the high repair and maintenance costs of the existing structure, which outweighed initial concerns of potential traffic "chaos," which ultimately did not materialize. The Gardiner Expressway/Lake Shore corridor continues to be an important arterial in its current form. However, given Toronto's and others cities' experience with urban highway repairs/removals, the question posed was how to best evaluate transportation alternatives.

Removing Barriers, Making Connections

The future of Gardiner has long been linked to the domain of waterfront revitalization. The expressway and the rail corridor that runs parallel to the north have long been considered physical barriers between the city and its waterfront. Determining their future role and function has always been crucial in the context of Toronto's waterfront revitalization plans.

In 2001, Prime Minister Jean Chrétien, Premier Mike Harris, and Mayor Mel Lastman announced $1.5 billion in seed capital to kick-start the revitalization of Toronto's waterfront. The funding commitment was made in support of the City's bid for the 2008 Olympic Games. Ultimately, Toronto lost the games to Beijing, but the funding was maintained. Building on the recommendations of the Toronto Waterfront Revitalization Task Force Report (2000), also known as the Fung Report, the Toronto Waterfront Revitalization Corporation (operating as Waterfront Toronto) was incorporated as the lead development, planning, and co-ordination agency.

Waterfront Toronto's mandate was to revitalize 2000 acres of waterfront land through public–private partnerships and deliver the infrastructure to support 40,000 new residences, 40,000 new jobs, and hundreds of acres of new parks. The mandate also included determining the future of the elevated Gardiner Expressway, as Fung outlines in his report:

> The redesign of the corridor can be revolutionary in the complexion and future of the waterfront and also emblematic of renewal. The purpose of this effort was to determine whether the challenge could be met cost-effectively, with quantum environmental improvement and without unreasonable disruption, while achieving transport and operational objectives (Toronto Waterfront Revitalization Task Force 2000, 37).

Until recently, Toronto and its waterfront have been mnemonically disconnected, in part because of the linear infrastructure of the Gardiner and rail corridor. For the most part, Torontonians would only go to the waterfront if they had to catch a ferry to the Toronto Islands or get on the Gardiner Expressway. The working port of the past had given way to a patchwork of parking lots, remnant buildings, and scattered light industry. When development arrived, a lack of planning controls led to poor outcomes, such as tall buildings and above-grade parking structures that ran parallel to the water's edge. Torontonians termed the buildings and subsequent development along the waterfront as the "wall of condos." These developments led to a public outcry for the preservation of public waterfront access and views.

In 2003, the City of Toronto adopted the Central Waterfront Secondary Plan, which outlined area-specific planning policies for Toronto's waterfront. The plan set out objectives for public transit, transportation, parks, and designated revitalization areas. The four main principles were:

1. Removing barriers/making connections
2. Building a network of spectacular waterfront parks and public spaces
3. Promoting a clean and green environment
4. Creating dynamic and diverse communities

The Secondary Plan calls for development of precinct plans/ master plans that encode the principles into urban form. Through the precinct planning process, Waterfront Toronto has developed area-specific community plans, built-form zoning, environmental assessments, and municipal infrastructure plans. The Gardiner East EA runs through a number of these communities, including East Bayfront, Keating, Lower Yonge, South Riverdale, and Port Lands.

We Are Not Alone

The public dialogue about the future of the Gardiner Expressway had focused on whether the expressway would be more of a barrier in its elevated form or if it were brought

Figure 9.2.5 • Rendering depicting the East Bayfront precinct, one of the five emerging neighbourhoods that the Gardiner East passes through.

Source: Waterfront Toronto 2013.

down to street level. Public opinion was often split evenly between those who found it difficult to imagine Toronto without a dedicated downtown elevated expressway and those visualizing other options such as a grand boulevard.

Planners had set out to foster an informed public discourse that would ultimately lead to general consensus on the appropriate future for the Gardiner East. Early on, an innovative approach to public engagement was required, in particular to facilitate the understanding of the issues given the understandable emphasis on regional transportation matters and lack of consideration for other important issues. How does one balance regional infrastructure functions with their impacts on local communities?

In developing and consulting on the scope for the environmental assessment (Terms of Reference [TOR]), the City and Waterfront Toronto considered how to best capture this question and create a framework for evaluation and public engagement. Early on in the process, planners held workshops with a Stakeholder Advisory Committee (SAC), which included neighbourhood associations, business improvement associations, community groups, logistics companies, and car associations, among others. Collectively, this group helped the project team define the issues to be studied in the environmental assessment and also provided feedback throughout the study.

While working with the SAC and receiving feedback from public meetings, it was important to convey that Toronto was not alone in making tough decisions about mid-century infrastructure. In fact, many cities in North America and elsewhere were either actively looking or had recently looked at this same problem. These case studies, which included cities such as Buffalo, Seoul, Portland, San Francisco, New York, Seattle, and Montreal, had considered alternatives and retained their elevated expressways. The various approaches were reviewed, and what emerged was a broad categorization for the four alternative solutions that would be studied in the environmental assessment:

- maintain the elevated expressway as it is today;
- improve the urban fabric while maintaining the existing expressway;
- replace it with a new above- or below-grade expressway; and
- remove the elevated expressway and build a new boulevard.

In reviewing the case studies, some members of the public expressed scepticism about the studies' applicability to Toronto. However, although there was no perfect comparators, the case studies assisted planners in moving past the "keep it up or tear it down" debate by introducing a conversation about transportation and non-transportation issues that were common in all the cities studied and referenced throughout the environmental assessment.

Consultation on the particular Gardiner issues and precedents led to creation of four study lenses that grouped the evaluation criteria that would be used in the assessment and selection of a preferred alternative. The four lenses were:

- transportation and infrastructure;
- urban design;
- environment; and
- economics.

The study lenses provided a structured framework for both the debate and the technical evaluation of the comparative benefits and constraints of the alternative solutions. They also proved to be an important organizing principle to solicit meaningful feedback and to develop a "made-in-Toronto solution."

Fostering an Informed Discussion

Often, one of the most challenging aspects of public consultations in an environmental assessment is being able to communicate and foster an informed discussion on often technical ideas, especially if those ideas do not fit the audience's own preconceptions or experiences. For example, for many members of the public, even considering the replacement of an expressway with a boulevard was an obstacle to getting the conversation started. Additionally, the planning process often asks the public early on for their feedback on high-level planning directions that may have unclear ramifications. Directions are adopted, policies defined, strategies developed, implementation tools such as zoning are adopted, and then design starts. However, this linear approach to planning and design can lead to unintended poor built-form responses or less than optimal results.

For the Gardiner East EA, planners decided that pre-visualization of the possible outcomes was key to developing an informed discussion. An international design competition was launched to solicit concepts that would assist in understanding the opportunities and constraints of the expressway corridor. The visual and narrative products of the design competition helped the public to understand what was important to evaluate and the trade-offs involved in each of the alternative solutions being studied (Maintain, Improve, Replace, and Remove).

The international design competition was initiated in 2009, and six teams of architects, planners, landscape architects, engineers, and other professionals were competitively selected to participate. The competition was organized as an eight-week intensive exercise designed to solicit visionary plans based on realizable designs as articulated in a competition design brief. The process started with a kick-off workshop/site visit, and at the mid-point of the competition, teams presented their preliminary findings and received feedback from a group of stakeholders and a technical review team. At the end of the eight-week process, the competition teams made their final submissions, which included materials for public exhibit in the form of presentations, panels, physical models, and a report. For their efforts, each team received an honorarium for complete submissions to cover their production and travel costs.

The six firms were each assigned one of the alternative solutions (Improve, Replace, and Remove): two teams for each alternative solution. No teams were assigned Maintain, because it was the base case, known in the environmental assessment language as the "do nothing" option.

For the "Improve the urban fabric while maintaining the existing expressway" option, the teams of Diller Scofidio + Renfro/architectsAlliance proposed shifting the Lake Shore Boulevard south in order to build public pavilions and recreation space underneath the structure of the expressway. The Kuwabara Payne McKenna Blumberg Architects/Bjarke Ingels Group team took the approach of building a park overtop development parcels that elevated people above the expressway structure to provide views to the lake, thereby reducing the barrier impact of the expressway structure.

For the "Replace with a new expressway" option, two approaches were taken. The team of West 8/DTAH proposed building a new four-lane expressway structure along the existing elevated rail corridor to the north of the existing Gardiner structure. The new expressway would be essentially hidden by new development parcels that lined a new Lake Shore Boulevard where the old expressway had existed. The team of Adrian Smith + Gordon Gill proposed a new four-lane tunnel that eliminated a number of street-level constraints and allowed for a regular set of streets to support the emerging neighbourhoods of East Bayfront and Keating precincts.

Lastly, for "Remove the elevated expressway and build a new boulevard," two radically different proposals were put forth. The first, by the Office of Metropolitan Architecture/AMO, proposed a hyper multimodal hub, which they termed a "Transferium," to the eastern end of the Gardiner at the junction of the Don Valley Parkway. The concept was to have people transfer to other modes of transit at the hub, relieving the auto traffic pressure, and replace the expressway with a boulevard. The James Corner Field Operations team took a more straightforward approach by simply replacing the expressway with a treed boulevard that would have six express lanes, two local lanes and a multi-use path that collectively recreated a grand Lake Shore Boulevard.

Figure 9.2.6 • Artist depiction of a new Lake Shore Boulevard. From the James Corner Field Operations Innovative Design Competition submission for Remove.

Source: Waterfront Toronto 2013.

The competition submissions illustrated the wide variety of solutions and formed a tangible reference for public consultation.

Public Ideas, Design Ideas, Your Ideas

A core objective of the consultation program was to stimulate public discussion and gather meaningful feedback that would inform the development of the alternatives. Presenting the "design ideas" as the competition submissions were presented was key to that strategy, as was including unsolicited "public ideas" as proposed by professionals and members of the public at large.

The public ideas included the Green Ribbon, by Les Klein of Quadrangle Architects, which proposed to add a second deck on top of the existing Gardiner deck for use as public park and maintaining the existing elevated Gardiner. The proposed elevated park would be more than 15 metres high, supported by a new set of columns and accessed by ramps and elevators. Another public idea widely discussed was Jose Gutierrez's Toronto Waterfront Viaduct. The proposal envisioned a new cable-stayed suspension bridge overtop the rail corridor, increasing the number of expressway lanes from six to eight or 10.

At the June 2013 round of consultation, planners asked for feedback on the competition submissions and public ideas while also soliciting those participating to provide "your ideas" for consideration. Feedback was categorized and provided to the project team to assist in the technical development of alternative solutions. In October 2013, the alternative solutions for Maintain, Improve, Replace, and Remove were presented to the public as part of the second round of consultation. Preliminary cost estimates, technical plans, sections, and relative opportunities and constraints were discussed for each of the options and draft evaluation criteria presented. Feedback received during the second round was incorporated into the refinement of the options that would later be presented in the third round of consultation in February 2014 when the project team would present the preliminary evaluation of the alternative solutions (Waterfront Toronto 2014). Approximately 1500 people participated in each round of consultation, which included public meetings, stakeholder meetings, and online outreach.

Transportation Insights

Traffic congestion in the Greater Toronto Area (GTA) was a hot topic during the Gardiner East EA study. A number of factors, including the Places to Grow Act (2005), which sets out regional intensification policies, road construction, and a booming downtown condo and office market, were putting pressure on legacy roadways and transit links. Battling congestion and seeking infrastructure funding was top-of-mind and the focus of various government initiatives, such as Metrolinx's regional transit strategy "The Big Move" and the City of Toronto's "Feeling Congested" consultation.

Given the "congestion" context, many members of the public were sceptical of even considering options that reduced the vehicular capacity of the corridor. The Gardiner Expressway was seen as a critical link in the regional transportation network. By its very definition, an expressway is a limited-access roadway built to move a high number of vehicles over long distances as quickly as possible. However, one of the early findings of the study revealed that only 20 per cent of vehicles used this segment (Yonge to Don Valley Parkway) for its express through-route function. The vast majority (80 per cent) of trips involved entering or exiting the expressway to access downtown and therefore could potentially be served by local streets, although there would be an impact on those making through trips.

The GTA grew from 3.2 million people in 1975 to 6.4 million in 2011. During that time, auto trips to downtown during the peak period have remained relatively flat. Essentially, the expressway has been at capacity for decades, and trip growth has occurred on local and regional transit. Trends indicate that reliance on transit will continue. Projections show that in 2031, 80,000 new trips will be needed to service downtown growth; 95 per cent of that trip growth is expected to be on public transit given existing road constraints. Given the projections, the transit **mode share** will continue to rise, and the vehicular proportion will shrink from approximately 28 per cent in 2011 to 22 per cent in 2031. In 2011, the Gardiner East itself carried a total of 3 per cent of trips (5200), which was roughly the same proportion of trips made by pedestrians and cyclists to the downtown core, granted that the travel distances/options are vastly different for those sets of commuters.

The corridor has a number of other existing and emerging challenges as the city grows around it. In 2013, the Lake Shore Boulevard segment in the study area was in the top 20 per cent of intersections and road segments with the highest rate of collisions in the city. This was due to poor sightlines and free turn on/off ramps that create conflict points with other modes of travel. As growth

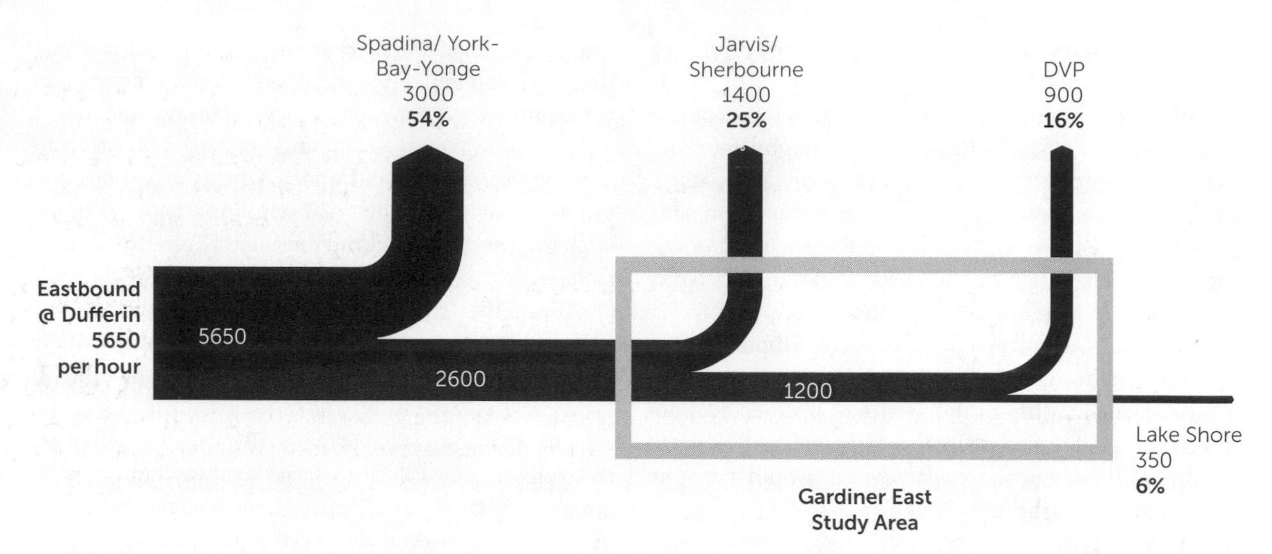

Figure 9.2.7 • Eastbound morning peak-hour trips to downtown.

Source: Waterfront Toronto 2013.

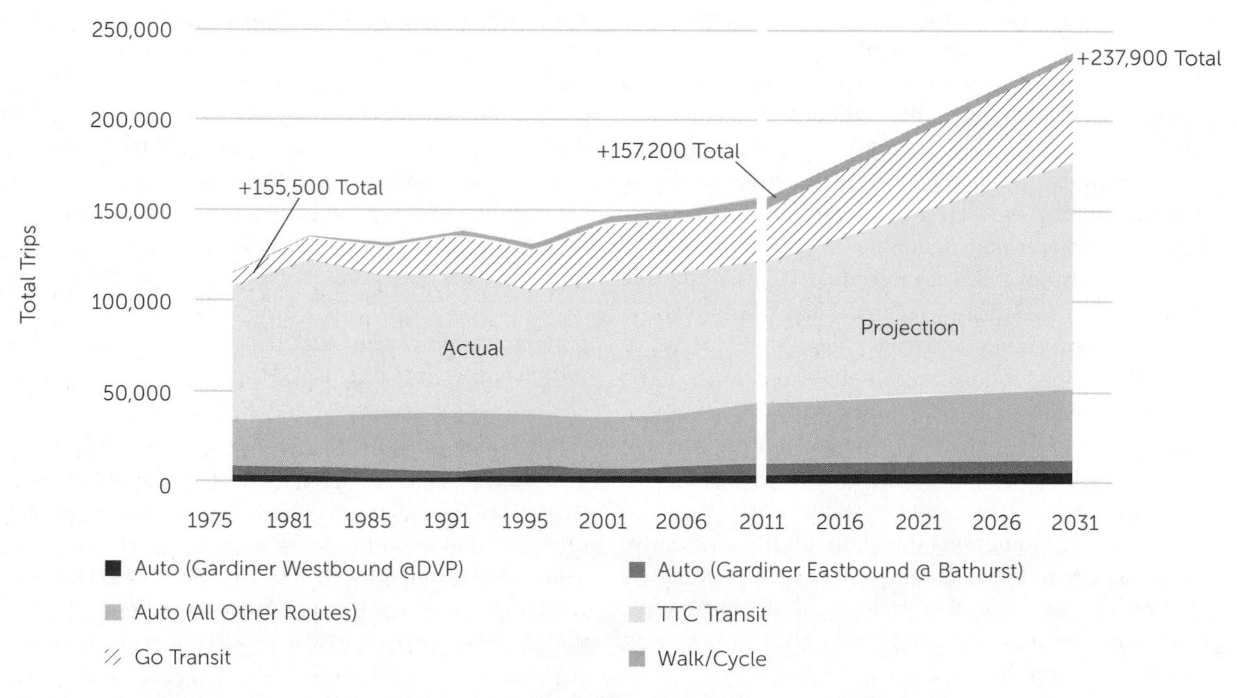

Source: AM Peak Hour Inbound to Downtown: (1) Transportation City Cordon Count (1975–2011); (2) Transportation Model EMME2 Forecast (2011–2031); (3) 2006 Transportation Tomorrow Survey (TTS) For walk/ Cycle Mode and Other data.
Downtown: Defined as Bathurst to Don River and Waterfront to the rail corridor north of Bloor

Figure 9.2.8 • Transportation demand growth: Actual and projected morning peak-hour inbound to downtown trips by mode.

Source: Waterfront Toronto 2013.

continues without redesign and mitigation, conflicts will continue to increase, in particular with pedestrians and bicyclists. Additionally, given the unique geometries of the corridor, there are a number of turning restrictions that make accessing the downtown and emerging waterfront communities difficult.

As part of the study, robust transportation modelling for each alternative solution was undertaken using a 2031 baseline year accounting for land-use, population, and employment growth. Two key findings emerged. As expected, the Remove option reduced the most existing capacity from the corridor and therefore resulted in the highest 2031 projected increase in modelled vehicular travel times of five to 10 minutes as compared to the other options. However, projected travel time increased for all options, which is consistent with existing trends. The second key finding was that for all options, transit was required to meet demand generated by growth in the downtown and waterfront areas. Congestion could not be solved by maintaining the existing expressway function alone; transit expansion was integral to the all four options being studied.

Evaluation through Lenses

Fifty years after the initial construction of the Gardiner Expressway, the City of Toronto and the Waterfront Toronto project team recommended the removal of the Gardiner East section based on the results of the environmental assessment evaluation. The recommendation sought the expressway's replacement by an eight-lane grand boulevard that would serve as both a regional artery and an important local waterfront avenue. The recommendation was made to the City of Toronto Public Works and Infrastructure Committee in March 2014.

A core mandate of Waterfront Toronto has been to develop sustainable communities through integrated planning and design, which has required innovation to find the right solutions to complex problems. In the case of the Gardiner, it meant balancing regional and local needs and ensuring an evaluation process that did not bias one criterion over another. Planners early on in the process understood this challenge given that the study area was primarily understood as a regional transportation corridor and that it would be important to provide a voice for future residents, workers, and visitors to the waterfront.

The evaluation in the environmental assessment consisted of 60 measures that were summarized into 16 criteria groups organized by the four evaluation lenses (transportation and infrastructure, urban design, environment, and economics). Each criteria group received a relative evaluation of least preferred, moderately preferred, or preferred. The evaluation was comparative at the criteria group level, and an evaluation of least preferred did not mean it did not meet a minimum performance threshold. Given that the evaluation consisted of a number of quantitative and qualitative measures, weighting to the lenses on a quantitative basis was not deemed appropriate. A "reasoned argument" approach was adopted as the basis of the evaluation. The evaluation results concluded that Remove best met the goals of the study.

Under the transportation and infrastructure lens, Remove had the highest modelled vehicular delays given its reduction in capacity as compared to the other alternatives. However, it is worth noting that in all cases, the existing capacity of the alternatives would be constrained during lengthy multi-year construction periods. For Maintain, the rehabilitation of the existing structure would mean six years of lane closures on both the Gardiner Expressway and Lake Shore Boulevard as the existing deck was demolished and then rebuilt as compared to three years for Remove. How would six years of lane closures affect commuters' route and mode choice and change behaviours permanently? Given the overall duration of construction and the increased growth, it was expected that drivers would be forced to seek out new routes or change their mode of travel regardless of the option chosen.

Under the environment lens, some of the highlights included reduced emissions for the Remove option given the reduced vehicular demand and increased transit modal split. The reduction in demand and slower speeds in the corridor would also reduce noise levels, and the increase in tree-planting in the corridor would reduce stormwater runoff and increase on-site retention.

Under the urban design lens, the benefits of Remove were numerous. Remove met all the principles of the Central Waterfront Secondary Plan, including "removing barriers/making connections" and "creating dynamic and diverse communities." The creation of a grand boulevard would allow for mixed-use buildings to have a front door along the corridor instead of back-of-house land uses such as parking levels and service entrances, which would be expected under Maintain. The streetscape

Figure 9.2.9a • Maintain

Figure 9.2.9b • Improve

Figure 9.2.9c • Replace

Figure 9.2.9d • Remove

Figure 9.2.9 • Artist's interpretation of the four alternatives in 2031.

Source: Waterfront Toronto 2013.

itself would have long views to the core and improve north–south access to the waterfront. At-grade uses such as cafés and retail would create a pedestrian-friendly environment that would complement the waterfront neighbourhood.

Under the economics lens, Remove also outperformed the other alternatives. At a regional level, research, which included looking at cities that had removed expressways, found that cities thrive with and without through expressways in their core. At the local economic level, Remove was expected to generate more jobs along the corridor given the creation of viable commercial space. The reconstruction of the corridor would allow for a more efficient use of land, and surplus **right-of-way** could be sold to generate revenue for the project in the amount of $220–240 million. Lastly, Remove was the least costly option from a lifecycle costs (capital, maintenance, and operations) perspective. Maintain's lifecycle costs were estimated at $870 million while those of Remove would be $470 million, given that an at-grade street is much less costly to construct and maintain.

Conclusions: Integrated Planning and Design

Developing and fostering a framework for decision-making that is both balanced and objective was key for the Gardiner East project. Transportation projects are inherently difficult, given the large number of variables and stakeholders and the emphasis on auto issues. By adding the economics, environment, and urban design lenses, planners ensured that the discussion among project team members of various disciplines, stakeholders, and the public would inform each other.

Use of the case studies in the project was an important first step. The unveiling of the competition submissions

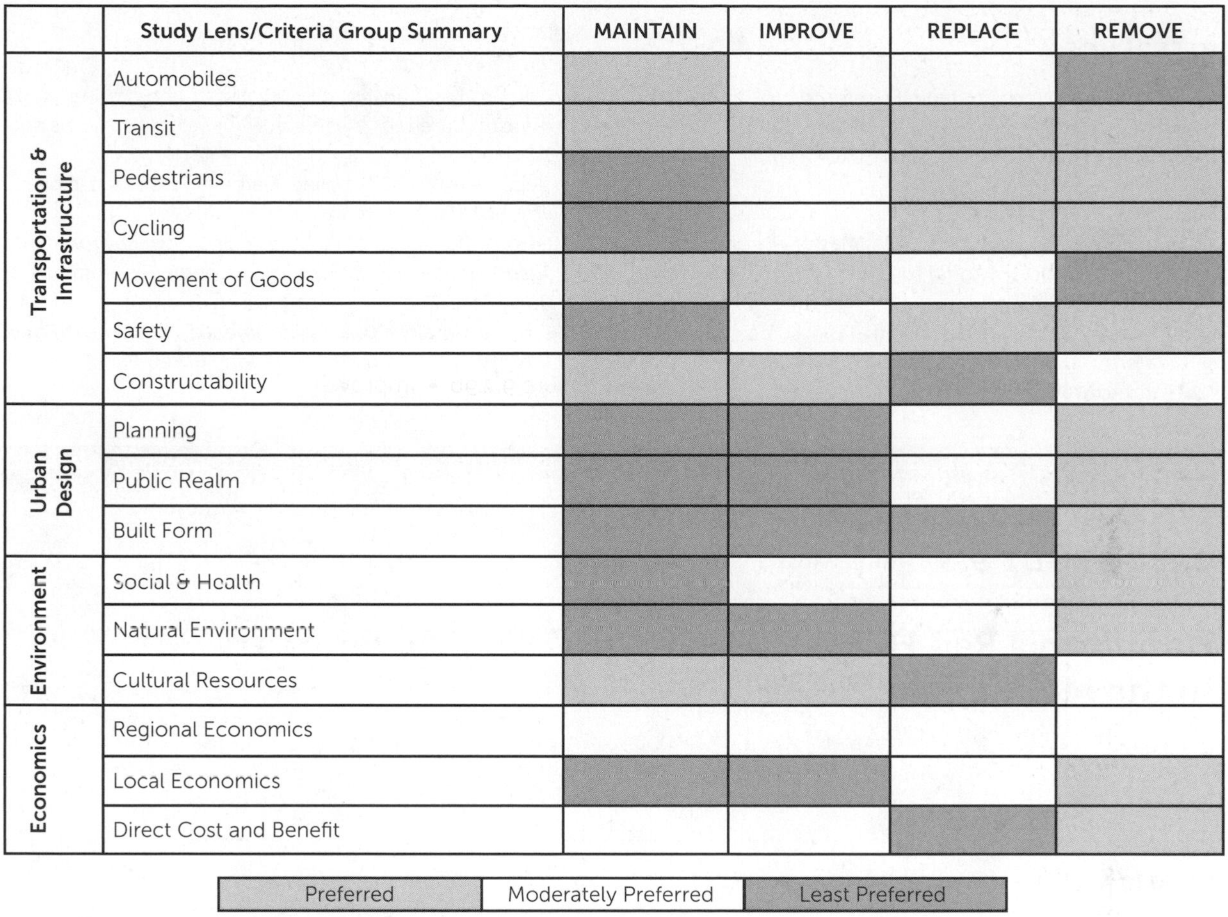

	Study Lens/Criteria Group Summary	MAINTAIN	IMPROVE	REPLACE	REMOVE
Transportation & Infrastructure	Automobiles				
	Transit				
	Pedestrians				
	Cycling				
	Movement of Goods				
	Safety				
	Constructability				
Urban Design	Planning				
	Public Realm				
	Built Form				
Environment	Social & Health				
	Natural Environment				
	Cultural Resources				
Economics	Regional Economics				
	Local Economics				
	Direct Cost and Benefit				

Preferred	Moderately Preferred	Least Preferred

Figure 9.2.10 • Summarized preliminary evaluation results.

Source: Waterfront Toronto 2013.

and public ideas further built upon this learning process in enabling the public to gain insights and better evaluate the trade-offs among the options.

This non-linear approach to planning ensured that potentially abstract goals and objectives were understood, which helped the public to engage in the planning process. The selection of Remove as the preferred option in the environmental assessment is a reflection of planners taking a comprehensive look at all the relevant issues and building a framework that allowed for the opportunities afforded by a declining piece of infrastructure to be understood and evaluated.

Postscript

At the time of writing, Waterfront Toronto and City of Toronto staff, based on the analysis and evaluation of the EA, had recommended the Remove alternative to the City of Toronto Public Works and Infrastructure Committee (PWIC) on 4 March 2014. PWIC requested that staff examine the mitigation of travel time delays associated with the Remove alternative, consider an alternative elevated expressway proposal brought forward to PWIC, and report back in 2015. The project team is preparing to respond to the PWIC direction in early 2015.

KEY TERMS

Mode share
Right-of-way

REFERENCES

City of Toronto. 2014. *Strategic Plan for the Rehabilitation of the F.G. Gardiner Expressway*. Toronto: City of Toronto.

Hayes, D. 2008. *Historical Atlas of Toronto*. Vancouver: Douglas and McIntyre.

Places to Grow Act [2005] S.O., c. 13.

Toronto City Planning Board. 1943. *Master Plan for the City of Toronto and Environs*. Toronto: Toronto City Planning Board.

Toronto Waterfront Revitalization Task Force. 2000. *Our Waterfront: Gateway to the New Canada*. Toronto: Toronto Waterfront Revitalization Task Force.

ULI (Urban Land Institute) and Ernst and Young. 2013. *Infrastructure 2013: Global Priorities, Global Insights*. Washington: ULI.

Van Rijn, N. 1978. "Crumbling Gardiner 'won't be closed.'" *Toronto Star* 3 February.

Waterfront Toronto. 2013. *Gardiner East Environmental Assessment Public Information Centre Presentation*. 13 June. Toronto: Waterfront Toronto.

——. 2014. *Gardiner East Environmental Assessment Public Information Centre Presentation*. 6 February. Toronto: Waterfront Toronto.

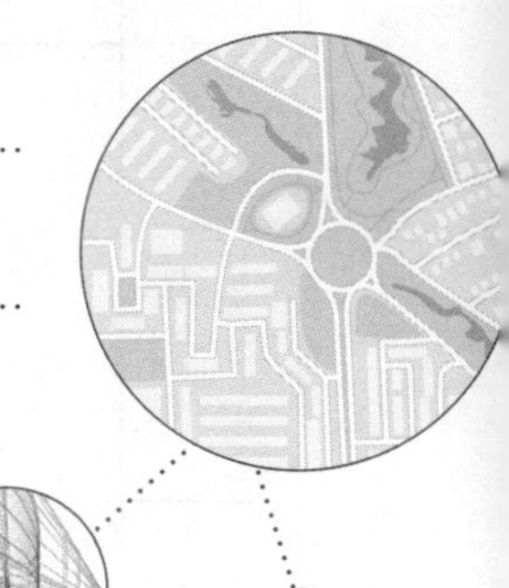

CASE STUDY 9.3

Hamilton's Red Hill Valley Parkway: Fifty-Seven Years in the Making

Walter G. Peace

SUMMARY

Beginning in the 1950s, a number of proposals were put forth to construct an expressway through the Red Hill Creek Valley in the east end of Hamilton, Ontario. This north–south expressway would link up with a major east–west expressway as part of a plan to enhance accessibility both within Hamilton as well as between Hamilton and other urban places in southern Ontario. Over the 50-plus-year history of the project, all three levels of government played critical roles in the final outcome, which was construction of the expressway. In addition, several citizens groups were active participants as both project supporters and opponents in the controversy. Planners played a key role in the survival of the project, especially during the early years of the debate. On the opposing side, the group Friends of Red Hill Valley became one of the most vocal and active citizens group in Hamilton's history. The public's role in the controversy focused mainly on mounting a significant opposition to the project. The project supporters (apart from politicians) were less vocal and less public about their views.

The main challenges encountered at various stages of the project were citizen opposition and conflict between/among the three levels of government over costs, responsibilities, and jurisdiction. Citizen opposition was addressed through legal means (including arrests and the laying of criminal charges), by ignoring the views of opposing citizens, and through attempts to placate the community by means of efforts at environmental remediation during and after the project.

This case study calls into question the way in which planning can and should strive to meet the **public interest**. Among the lessons learned are: (1) citizen participation is an important determinant in planning; (2) this participation must be fully embraced to be meaningful; and (3) failure to incorporate citizen participation in a meaningful way will likely result in an extreme polarization of views, leaving little or no room for the possibility of genuine compromise. In the end, the public interest can best be met through meaningful citizen participation in the planning process.

OUTLINE

- Persistent conflict between planning goals and public opinion is the theme of this case on the construction of Hamilton's Red Hill Valley Parkway.
- Historical documents, media accounts, and current articles show the long-lasting divergence of views on the project, which spanned 50 years from conception to construction.
- How can planning serve the public interest? It can be concluded that the lack of public participation in this planning process was instrumental in the extreme polarization of opinions and the emergence of conflicts that were never resolved.

Introduction

The connection between urban development and transportation-related infrastructure projects is both intuitively obvious and intensely profound. Essentially, neither can occur without the other. A survey of such projects in twentieth-century urban Canada reveals the transformative role played by transportation infrastructure projects. Consider, for example, the significance of Montreal's metro system, the role played by the construction of the Don Valley Parkway and the Gardiner Expressway in Toronto, and, at the regional scale, the construction of Highway 401 linking Quebec City to Windsor. These and countless other projects at the inter- and intra-urban scales have affected the economies, social landscapes, and physical environments of our cities in many significant ways. It is crucial to point out that the majority of these projects have attempted to address concerns related to either accommodating automobiles/vehicular traffic and/or ameliorating the consequences of our reliance on the automobile (in the case of public transit initiatives). For the most part, these projects met little or no opposition from urban residents, who were quite willing to enjoy the benefits and advantages that were to be gained by building more roads, thereby making people, places, and goods more accessible.

This compliant acceptance of more roads (necessary to accommodate the growing number of cars) began to be questioned and ultimately challenged in the 1960s. In the wake of neighbourhood destruction caused by the building of intra-urban expressways in American cities, citizens began to challenge the conventional wisdom of urban transportation policies and planning. In the Canadian context, this challenge came to a head with the successful opposition to the Spadina Expressway in Toronto in the late 1960s. The opposition was based in large part on the perceived detrimental impacts the proposed expressway might have on the social fabric of the inner-city neighbourhoods through which it would cut. As citizen activism became more widespread and better organized, few planning proposals could be expected to go unchallenged, especially those that were perceived to have potentially detrimental impacts on the physical/natural and social environments.

The case of Hamilton's Red Hill Valley Parkway highlights several key aspects of urban planning theory and practice in the twentieth century. The issue evolved over a protracted period of time during which citizen participation became a crucial determinant of both how the planning process would operate and the actual outcomes of the

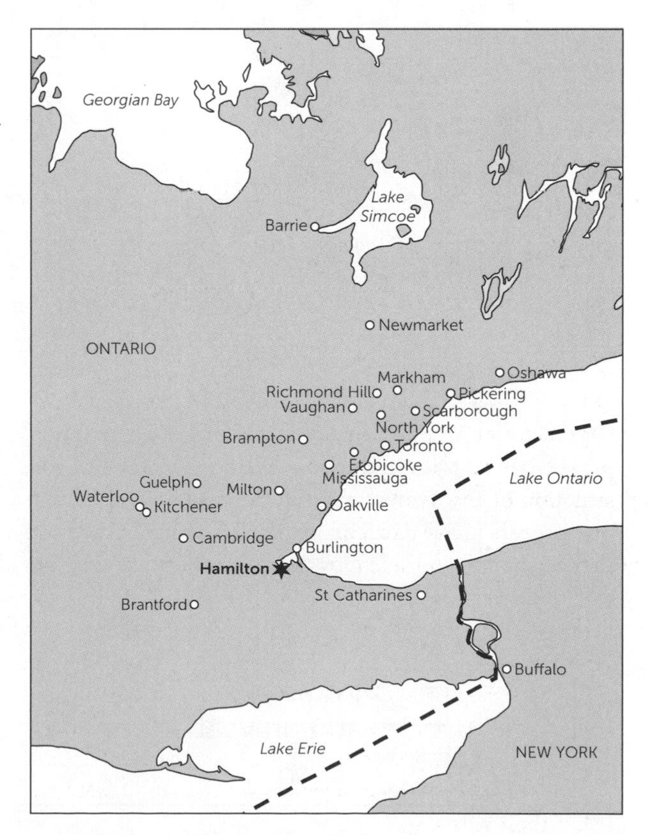

Figure 9.3.1 • Location map.

process itself. The purpose of this chapter is to describe and analyze the Red Hill Valley Parkway issue to determine what lessons might be learned from the controversy, which spanned nearly six decades from the initial proposal to completion of the project. Ultimately, it is the intent of this chapter to consider the extent to which planning can and should act in the public interest. The chapter draws heavily on government documents and media coverage of the project over a period of more than 50 years. In addition, academic sources (e.g., theses and journal articles) provide a basis upon which this critical review rests.

Setting and Context

Site characteristics have played a crucial role in Hamilton's historical evolution. In particular, Burlington Bay (Hamilton Harbour) and the Niagara Escarpment (known to Hamiltonians as "The Mountain") were the key determinants of both the location of the original townsite and the direction of expansion over the city's first century. As seen in Figure 9.3.2, the original townsite was laid out in 1816 by George Hamilton (the city's namesake) at a location some distance removed from the swampy, low-lying bayfront. Following Hamilton's incorporation as a City in 1846, population growth resulted in development of lands between the bay to the north and the escarpment to the south. During the first half of the twentieth century, there was considerable infilling of the harbour to accommodate expansion of the city's industrial sector. Because of the physical constrictions imposed by the harbour (in the north) and the escarpment (in the south), the population began to spread toward the east and to the west. By the middle of the twentieth century, further expansion to the east and west was essentially limited by the city's boundaries with adjacent municipalities. During the 1950s and 1960s, crucial "mountain access" roads were constructed,

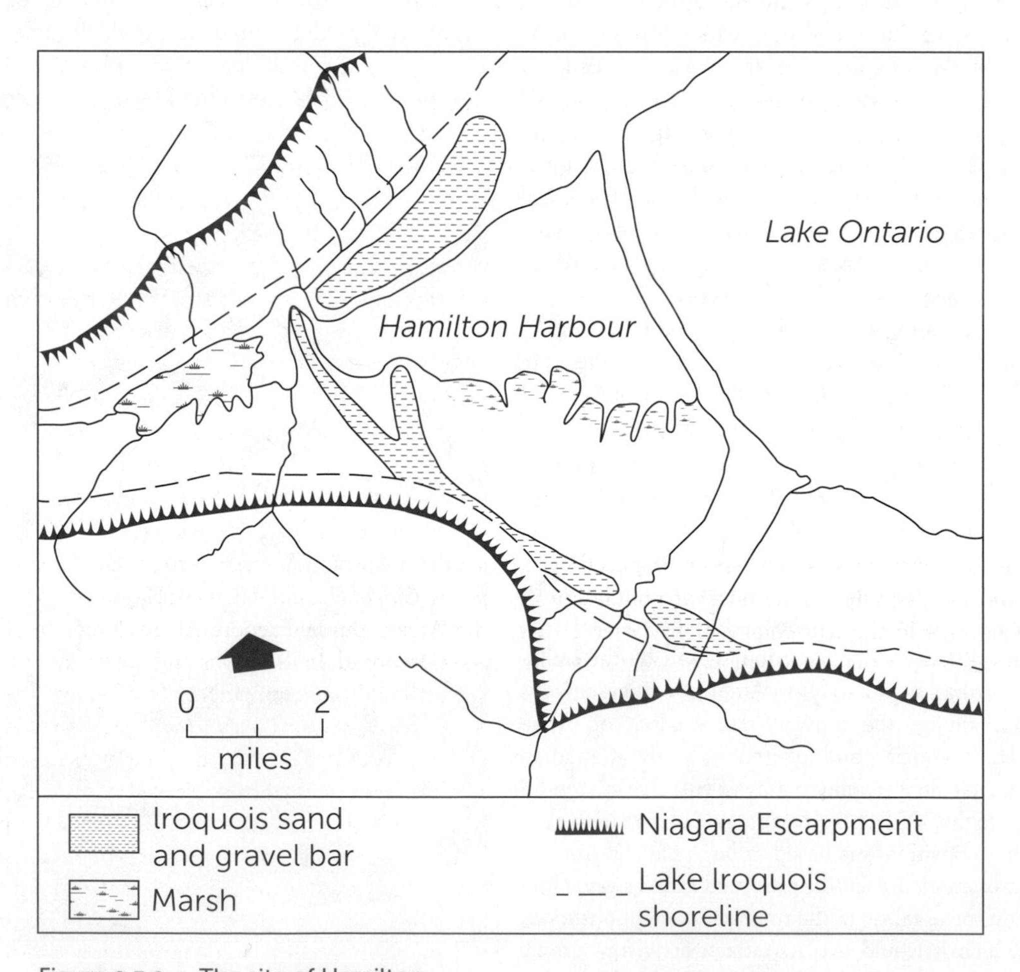

Figure 9.3.2 • The site of Hamilton.

Source: Wood 1961.

thereby providing access to areas above the escarpment. The role played by the city's site characteristics is evidenced by the following figures. In 1951, the city's population was 208,321, of whom about 7000 resided above the escarpment. By 1971, the city's population had risen to 309,173, of whom nearly 100,000 resided above the escarpment (Weaver 1982).

Weaver (1982) notes that Hamilton's postwar years were influenced by the "big-city vision" of Mayor Lloyd D. Jackson. The mountain access roads noted above were a part of this new era in the city's history. And like all other North American cities at this time, Hamilton's form and character was being redefined by the automobile. This provides the link with the setting described above and the context of growth and development in mid-twentieth-century Hamilton. At issue was the fact that the city had grown to a population of just over 200,000 in 1951 and yet there were no major east–west or north–south transportation arteries. In other words, to travel across the city (either from east to west or from north to south), one was forced to drive through the central business district and/or inner-city residential neighbourhoods. This problem was exacerbated by the rapidly expanding population to the south (above the escarpment), thereby intensifying the demand for better accessibility within the city. Thus, transportation planning in the postwar era in Hamilton featured attempts to devise a comprehensive and systematic approach to intra-urban movement that would be structured around a major east–west route across the city along with a major north–south route linking the upper city (above the escarpment) to the lower city (below the escarpment). As it turned out, the location and route of the east–west artery would involve little or no controversy. The north–south component of the system, on the other hand, would become, arguably, the most contentious and bitterly fought planning issue in Hamilton's history.

With these constraints and demands imposed on a growing and increasingly mobile postwar population in mind, let us turn to the chronology of key events that ultimately resulted in the construction of what officially became known as the Red Hill Valley Parkway.

The Red Hill Valley Parkway Process: A Chronology of Events

The Red Hill Creek watershed (see Figure 9.3.3) consisted of approximately 67 square kilometres of land above and below the escarpment in the east end of Hamilton (Peace 1998). The watershed's main tributary, the Red Hill Creek, flows approximately 17 kilometres from its source near Ryckman's Corners on Hamilton Mountain to its mouth at the eastern end of Hamilton Harbour.

In 1929, the City's Board of Parks Management purchased 645 acres of land in the Red Hill Creek Valley. The specific intent of this controversial acquisition was that these lands would constitute a portion of a park belt that would encircle the city. The other components of the park belt were: the Chedoke Creek Valley in the city's west end; the Beach Strip, which separated Hamilton Harbour from Lake Ontario to the north; and the Niagara Escarpment on the south. Initially, the proposed project appears to have been a city initiative. In later years, the road became more highly integrated with provincial plans for highways in southern Ontario. According to Terpstra (1985, 125) the purchase of these properties in the Red Hill Creek Valley made "the Hamilton park system one of the largest in the country." These activities were a product of the City Beautiful ideals (Terpstra 1985, 125). Thomas Baker McQuesten, chair of the Board of Parks Management, was quoted in the *Hamilton Spectator* (*Hamilton Spectator* 1929) as saying, "Large park areas in surroundings of natural beauty fulfill a need which cannot be met in any other way." He further noted that the purchase of these lands would "preserve for all time one of the outstanding spectacular areas in the County of Wentworth" (quoted in Peace 1998, 227). The controversy alluded to above surrounded the amount of money ($198,000) spent on the acquisition of the Red Hill properties (Terpstra 1985, 125). It is highly unlikely that anyone involved at that time could have foreseen these lands becoming much more controversial in the not-too-distant future. This looming controversy would rest on a much different criterion for establishing the value of these lands.

The first mention of the possible use of the Red Hill Creek Valley as the route for Hamilton's major north–south transportation artery is found in *Major Street System for the City of Hamilton* (City of Hamilton 1950), known as the Wilson Bunnell report. At this time, no other routes were proposed. In the 1960s and 1970s, other alternatives were, albeit briefly, considered (see below). The Hamilton Area Transportation Study (HATS), published in 1963, also endorsed the Red Hill Creek Valley route for the north–south transportation artery. Three key aspects of the issue need to be noted here. First, the preference for the Red Hill route expressed by planners was based, in part, on the fact that the city already owned most of the land in the valley, thus eliminating the need for (and cost of) expropriation. Second, from the time of the initial proposal in 1950 until the late 1970s, City Council was opposed to the use of the Red Hill Creek Valley as a route for a major transportation

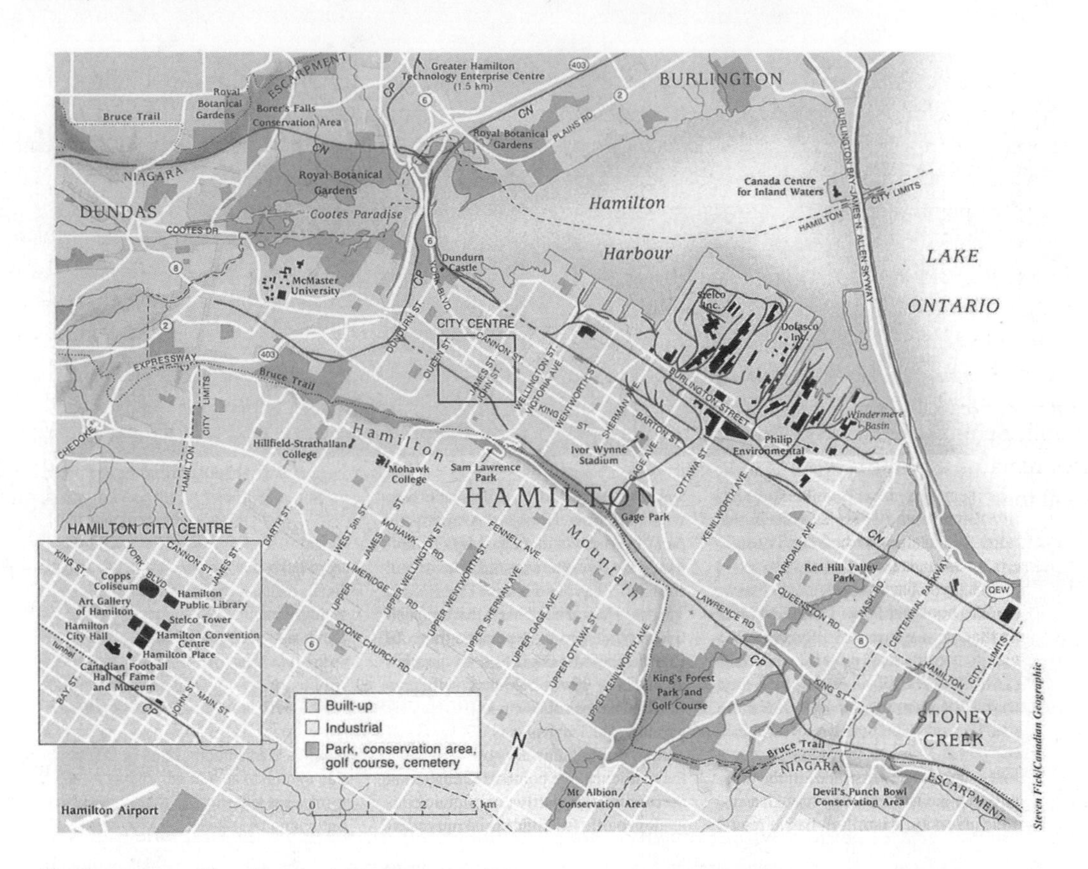

Figure 9.3.3 • Hamilton's urban geography.

Source: Canadian Geographic, July/August 1993.

artery. Finally, as noted above, the original purpose of the land acquisition was related to city beautification and urban aesthetics.

In the mid-1970s, the councils of both the City of Hamilton and the newly formed Regional Municipality of Hamilton-Wentworth had passed resolutions supporting the position that the natural character of the Red Hill Creek Valley be maintained permanently (Peace 1998). Despite the clearly stated position of both councils, a study conducted by the Ontario Ministry of Transportation and Communications (known as the Radbone Study) recommended in early 1975 that the Red Hill Creek Valley route remain in consideration for the north–south portion of the proposed system (Peace 1998). In July 1977, with the threatened withdrawal of provincial government funding (if the Red Hill route was not to be chosen), both the City Council and the Regional Council reversed their long-standing positions and voted in favour of adding the Red Hill Creek Valley to the list of other reasonable route alternatives. In the end, the Red Hill route was justified and chosen based on cost (the City already owned most of the land required for the road) and the minimal amount of disruption to neighbourhoods vis-à-vis the proposed alternative routes. It must be stressed that the role of City staff/planners was crucial in this decision. From the initial proposal of the Red Hill route, city planners continued to support this route despite the fact that both City Council and Regional Council were opposed to it, at least until July 1977. Through a series of reports such as the aforementioned HATS, staff continued to support the Red Hill route. By the mid-1970s, the provincial government began to play a more crucial role in the outcome of the debate by virtue of its economic support, which the city and region desperately needed.

A groundswell of opposition emerged as the possibility of an expressway through the Red Hill Creek Valley became increasingly likely by the late 1970s. Various citizen groups, including Clear Hamilton of Pollution (CHOP), Pollution Probe, and Save the Valley voiced their anti-expressway views. Their opposition to the use of the

Red Hill Creek Valley as a north–south transportation route was based largely on environmental concerns and the fact that a need for the expressway had never been thoroughly justified. In contrast, supporters of the Red Hill route claimed that the expressway would ease traffic problems and bolster the city's economic prospects (Peace 1998). Public participation in the process tended to be confrontational and adversarial throughout the entire project, reaching a peak at the time construction began in 2003. After this, public displays of opposition were less frequent, although opposition remained. The extent to which public opinion was incorporated into the planning processes remains a point of contention for both sides. The extent to which citizens participated in a meaningful way also remains a matter of opinion.

In October 1985, the first-ever Joint Board Hearing (comprised of two members of the Ontario Municipal Board and one member of the Environmental Assessment Board) announced its split (two to one) decision approving the overall transportation project with its east–west and north–south legs; this hearing was necessitated by the Save the Valley Committee's legal challenge to the Hamilton-Wentworth Planning Area **Official Plan**. In March 1987, the Red Hill route received formal approval from the Ontario government. Initial plans called for the two legs of the system to be built in three phases. Phases I and II would comprise the north–south leg of the project (in the Red Hill Creek Valley) while Phase III, the east–west portion, would traverse Hamilton's south mountain (see Figures 9.3.4 and 9.3.5). It was expected that the third phase would be completed by 1999. On 26 June 1990, an official ground-breaking ceremony was held, with *The Hamilton Spectator* reporting: "After years of public opposition the official kick-off to one of Hamilton-Wentworth's most controversial projects was short and smooth" (*Hamilton Spectator* 1990a). For all intents and purposes, it appeared that the controversy had ended as construction on three interchange/overpass projects began in the Red Hill Creek Valley.

On 17 December 1990, the newly elected NDP government in Ontario unexpectedly announced the cancellation of provincial funding for the north–south leg of the project (see Figure 9.3.6). In the announcement, the transportation minister stated that "the Red Hill Creek

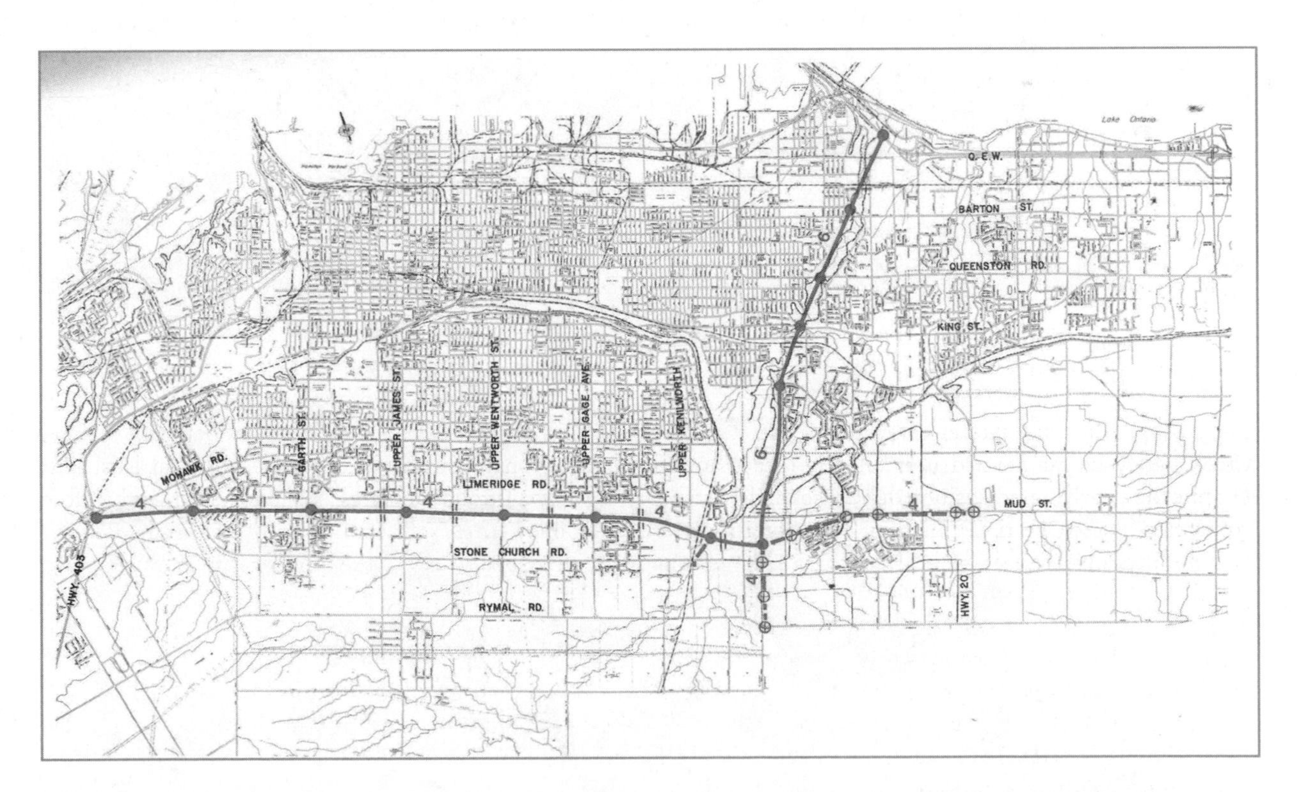

Figure 9.3.4 • The east–west and north–south components of Hamilton's transportation strategy, 1979.

Source: City of Hamilton 1979.

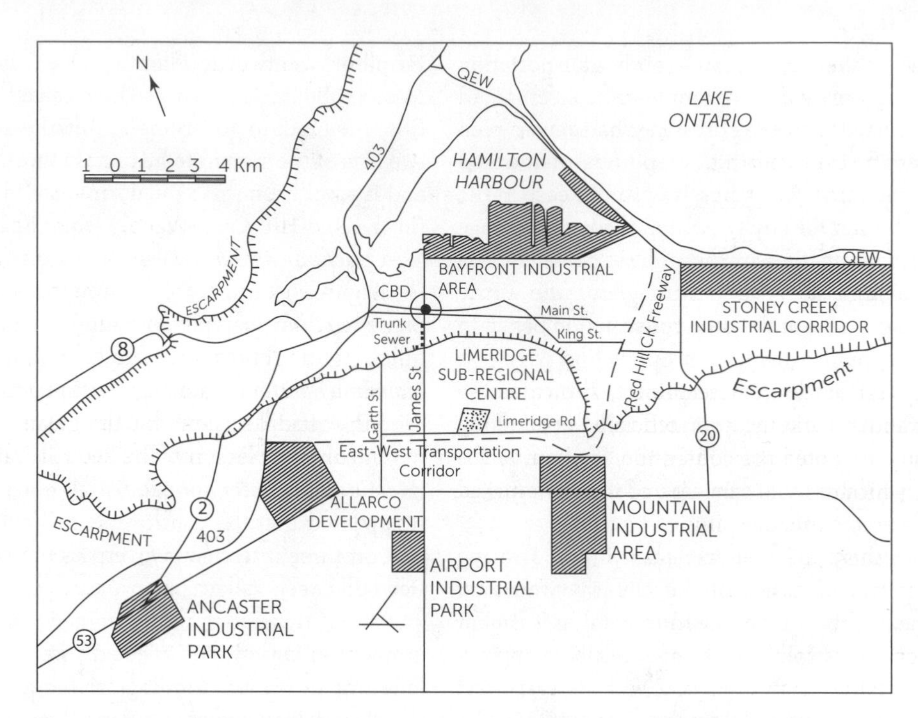

Figure 9.3.5 • Industrial development and transportation infrastructure planning in Hamilton in the 1980s.

Source: Adapted from Dear, Drake, and Reeds 1987. © University of Toronto Press 1987. Reprinted with permission of the publisher.

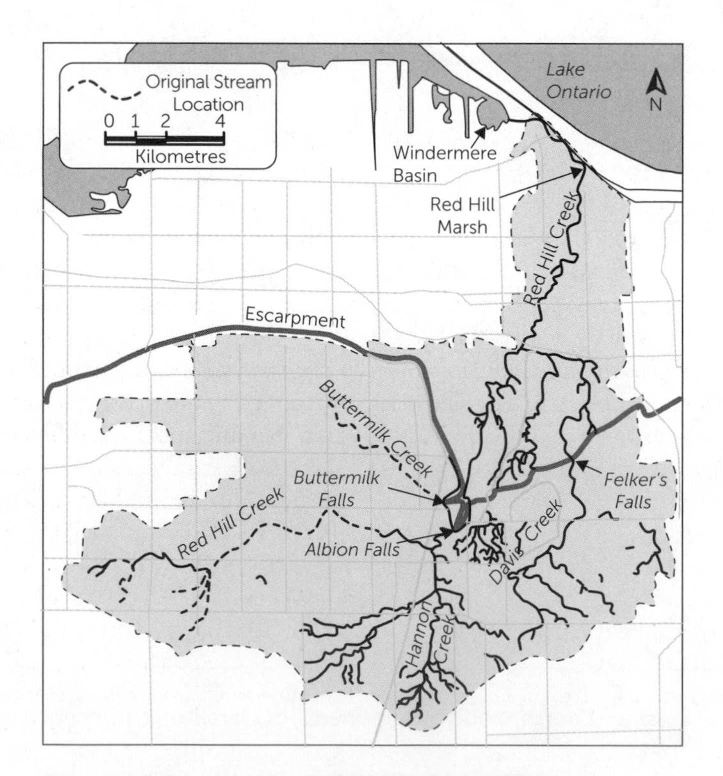

Figure 9.3.6 • Hamilton's Red Hill Creek Valley.

Source: Peace 1998.

Valley is irreplaceable, it is a natural asset that Hamilton must not lose" (*Hamilton Spectator* 1990b). In response to this announcement, the Regional Municipality shifted construction to the east–west phase of the project (which the provincial government was still committed to funding). With the battle lines drawn, the Citizen Expressway Coalition, which supported the Red Hill Expressway, was set to face off against the newly formed Friends of the Red Hill Valley, which, of course, was opposed to the valley route.

The east–west portion of the project, known as the Lincoln Alexander Parkway, was officially opened on 3 October 1997. As noted previously, the location of this route across the south mountain area of the city sparked very little in the way of citizen opposition. In May 1999, the federal environment minister ordered that a federal environmental assessment (FEA) of the Red Hill portion of the project be carried out. In 2000, the Regional Municipality of Hamilton-Wentworth sued the federal government over the need for the FEA. As a result, the FEA, which had already begun public hearings, was put on hold pending the outcome of the region's suit. In 2001, a federal court supported the region's position that an FEA was not necessary. The Ministry of the Environment, in turn, appealed the federal court ruling. Finally, in 2002, the federal government's appeal was rejected, thereby clearing the way for the "new" City of Hamilton to proceed with construction in the valley, pending certain minor approvals.

In preparation for construction in the valley, tree-cutting began in October 2003. The two principal candidates for mayor in the November election were David Christopherson, who was opposed to the expressway, and Larry DiIanni, a staunch expressway supporter. Although the expressway did not become the election issue many had anticipated, DiIanni's victory further bolstered expressway supporters. Despite the impending completion of the project, a group of protesters occupied trees at the location where the expressway would descend the escarpment. This protest lasted 112 days. The City announced its intention to sue the protesters upon their removal from the occupied site. In relative terms, the subsequent two-and-a-half years turned out to be relatively uneventful. On 16 November 2007, the official opening of the Red Hill Valley Parkway took place.

Planning and the Public Interest

The *Hamilton Spectator* was a staunch supporter of the Red Hill route throughout the project's history. On 16 November 2007, the day prior to the official opening of the parkway, the paper's lead editorial was entitled "Red Hill has taught us much." The editorial concluded as follows:

> There are no perfect solutions to the needs of our city, or the needs of the province, which will be home to more than 40 percent of Canada's population by 2021. There is no question highways have negative impacts. But so does the economic stagnation that comes from a lack of an efficient transportation system. Our goal should always be to do our best to mitigate negative impacts, while making sound decisions for the greater good. That's the ultimate lesson of the Red Hill Valley Parkway (*Hamilton Spectator* 2007).

As noted in the previous section, citizen opposition to the project was spearheaded by the group Friends of Red Hill Valley. Speaking to the *Hamilton Spectator* (*Hamilton Spectator* 2007), the group's chair, Don McLean, was asked to reflect on the role played by citizen activism on the eve of the parkway's official opening. McLean's response was succinct and heartfelt: "I'm just a citizen. I'm just doing what I think is best for the city." In the same edition of the paper, Larry DiIanni, Hamilton's mayor, justified his long-time support of the project by stating: "I believed in the rightness and necessity of this direction. But more importantly, I believed that the vast majority of Hamiltonians' intention to see the road constructed could not be denied."

Despite being on opposite sides of the issue, the views of DiIanni and McLean share a crucial commonality: the belief that their actions were in the public interest. The difference between the project's supporters and its opponents, of course, lies in the fact that the two sides had divergent opinions as to what, exactly, was in the public interest. Did the public interest rest on the potential economic benefits the expressway could generate? Or, as the expressway opponents argued, was the public interest better served by preserving the natural character of the valley and spending money in different areas of need?

Herein lies the challenge when one attempts to assess the relative merits of arguments around planning issues like the Red Hill project. Both sides presented largely justifiable arguments. Historically, major investments in transportation infrastructure have resulted in significant (and quantifiable) economic benefits. On the other hand, choices based on aesthetics and environmental considerations—which are typically much more difficult to quantify—also result in benefits to urban residents. The key question is this: how do planners balance these competing interests? In retrospect, it seems that the City supported the Red Hill route on the basis of the "predict-and-provide" model of transportation planning—i.e., predict future traffic levels and build roads that will accommodate these increased levels of demand/automobiles. Indeed, one of the key arguments made by the opponents of the Red Hill route was that the need/demand for the road in the 1990s (before construction began) was based on traffic models and predictions from the 1960s and early 1970s when commuting patterns within the city of Hamilton were substantially different. In the 1960s and 1970s, a significant number of Hamiltonians were employed in the industrial sector on the city's waterfront. By the 1990s, the city had lost much of its industrial base. As a result, commuting patterns in the Hamilton CMA changed to the point where expressway opponents would argue that the project could not be justified.

In retrospect, planners appear to have played a key role in keeping the project "on the back burner" during the 1960s and 1970s when both City Council and Regional Council were opposed to the Red Hill route. Once the Red Hill route was brought back into the picture in 1977, it appears that the project was driven by the political sphere—i.e., the project became more of a political issue than a planning issue, with politicians directing staff/planners to undertake the work necessary to enable completion of the project. This gave rise to other, related concerns (Grant 2005; Hodge and Gordon 2008):

1. Can planners act in the public interest?
2. Is consensus possible?
3. Are planners leaders and visionaries, or are they "process technicians"?

Was the public interest properly and fully served in the case of the Red Hill project? In an interesting and insightful application of ethical principles to the decision, McKay (2000) applied five principles (differing values, equal consideration, equitable participation, distributive justice, and emphasis on quantifiable factors) to the Red Hill project. McKay (2000, 66) concluded that "the [Red Hill Expressway] is a development project which reflects the interests and values of select **stakeholder** groups, ignores the will of a large segment of the stakeholders and overlooks the long-term financial and environmental impacts of the project." Certainly, one of the crucial lessons to be learned from this case is that community concerns must be fully integrated into the planning process. In this instance, it seems that the process became too adversarial to effectively achieve this integration and (ideally) find common ground between two opposing factions. A second lesson from the case centres on the way in which conflicting views from higher levels of government can be navigated. This is especially challenging given such variables as cost-sharing agreements, different political agendas, and so on. With the perspective of hindsight, perhaps the sheer longevity of the project made such cooperation and integration so difficult to achieve. On the positive side, the city has committed itself to the ongoing monitoring of the project's environmental impacts as well as its effects on traffic patterns.

At the time of writing, debate over the relative costs and benefits of the project continues. Cross (2013) notes that "Now, like before the thoroughfare was constructed, there is still no consensus on whether it was a good idea." On the one hand, "The highway has spawned a number of jobs—although so far, short of the 14,000 that were predicted—and it brought in $14,282,714 in taxes and $386 million worth of assessment in 2011. The parkway has also led to Stoney Creek development around it, including big box stores, restaurants, and massive residential growth" (Cross, 2013). In the same article, Cross quotes Don McLean, former chair of Friends of Red Hill Valley: "I think in light of what's happened since, particularly with respect to our understanding of climate change, that we were on the right side when we fought against this." The controversy, it seems, is far from over.

Conclusion

The perceived "success" or "failure" of planning initiatives is dependent on the values underlying an individual's or group's position on the issue in question. Whether or not a particular initiative is seen to be in the public interest is, likewise, a product of a stakeholder's or group's values. In situations where these values are so divergent, it seems unlikely (or impossible) that the opposing sides might agree on which course of action best suits the public interest. In the case of the Red Hill Valley Parkway, conflict over the proposed Red Hill route spanned more than a half-century. The proponents and opponents of the expressway were, understandably, unable to find common ground. Six years after the completion of the project, the two sides seemed unwilling or unable to make any concessions. In the final analysis, there are no perfect solutions to these problems. Perhaps the best that can be hoped for, in an ideal world, is that the two sides in the dispute are able to respect the process and each other such that, regardless of the planning outcome(s), all parties can be satisfied with their efforts to act in the public interest.

KEY TERMS

Official Plan

Public interest

Stakeholder

REFERENCES

City of Hamilton. 1950. *Major Street System for the City of Hamilton*. Prepared by N. Wilson and A.E.K. Bunnell. Hamilton: City of Hamilton.

——. 1979. *Mountain East–West and North–South Corridor Study Information Package, Phase 3—Detailed Evaluation*. Hamilton: City of Hamilton.

Cross, S. 2013. "5 years later, was the Red Hill Parkway worth it?" CBC News, 2 January. http://www.cbc.ca/news/canada/hamilton/news/5-years-later-was-the-red-hill-valley-parkway-worth-it-1.1275560.

Dear, M.J., J.J. Drake, and L.G. Reeds., eds. 1987. *Steel City: Hamilton and Region*. Toronto: University of Toronto Press.

Grant, J. 2005. "Rethinking the public interest as a planning concept." *Plan Canada* 45(2): 48–50.

Hamilton Spectator. 1929. 5 October.

——. 1990a. "Red Hill Expressway project officially rolling." 27 June.

——. 1990b. "Province kills Red Hill Expressway." 17 December.

——. 2007. "Red Hill has taught us much." 16 November.

Hodge, G., and D.L.A. Gordon. 2008. *Planning Canadian Communities*. 5th edn. Toronto: Nelson.

McKay, R.B. 2000. "Applying ethical principles to the decision to build the Red Hill Creek Expressway." *International Journal of Public Sector Management* 13(1): 58–67.

Peace, W.G. 1998. "Farm, forest and freeway: The Red Hill Creek Valley, 1950–1998. In W.G. Peace, ed., *From Mountain to Lake: The Red Hill Creek Valley*, 213–46. Hamilton, ON: W.L. Griffin Printing.

Terpstra, N. 1985. "Local politics and local planning: A case study of Hamilton, Ontario, 1915–1930." *Urban History Review* 14(2): 115–28.

Weaver, J. 1982. *Hamilton: An Illustrated History*. Toronto: James Lorimer.

Wilkins, C. 1993. "Steeltown charts a new course." *Canadian Geographic* 113(4): 42–55.

Wood, H. 1961. *The Site of Hamilton and Its Influence on the Development of the City*. Bulletin no. 7, Education Committee, Canadian Association of Geographers.

CASE STUDY 9.4

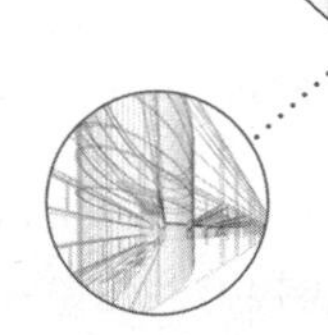

Travel Demand Management and GHG Emission Reductions: Meeting Multiple Objectives through Partnerships and Multi-Level Co-ordination

Ugo Lachapelle

SUMMARY

Quebec's *Adieu Bazou! / Faites de l'air!* **accelerated vehicle retirement program** is designed to reduce vehicle emissions through the replacement of older vehicles with newer, less polluting vehicles. The program also attempts to further reduce emissions by stimulating a **mode shift** among participants and can thus be considered a **Travel Demand Management (TDM)** tool. The program represents a compelling case on the contribution of non-governmental organizations (NGO) and transit planners to supporting senior government efforts to reduce vehicle use and associated impacts. The development of partnerships with providers of alternative travel modes (transit, bicycling, and car cooperative) can contribute to the recycling program's popularity and effectiveness and enhance the attractiveness of existing and improved alternative transportation infrastructure. The chapter explores the evolution of this program, the development of partnerships to provide alternative travel incentives, and the importance of land use and transportation infrastructure in participants' selection of transit passes and other alternative travel incentives. TDM programs based on mutually beneficial partnerships with alternative transportation manufacturers and providers can help city planners attain desired automobile travel reductions. Planners can benefit by co-ordinating efforts with program development at all levels of government because multiple conditions are required to make alternative transportation attractive. The complementarities of planners' and TDM program managers' objectives is thus explored. Interviews with program managers and partners were conducted as well as a review of scientific literature, policy reports, and available documentation on this and other existing programs.

OUTLINE

- Accelerated vehicle retirement programs are typically used to reduce greenhouse gas (GHG) emissions, improve a fleet's fuel efficiency, stimulate the economy, and improve the safety of the vehicle fleet.
- Federal and provincial accelerated vehicle retirement programs in Canada have been designed to stimulate a mode shift to further reduce transportation emissions.
- This case explores the development of partnerships between program managers and alternative transportation manufacturers and providers (such as transit agencies), which enabled the creation of the *Adieu Bazou! / Faites de l'air!* program in Québec.
- Program partnerships can serve to strengthen transit planners' efforts to increase transit use and reduce automobile travel while making the program more attractive.

Introduction

In the context of growing automobile ownership, as well as increases in individual trips and travel distances, improving air quality and contributing to **greenhouse gas (GHG) emission reduction** associated with automobile travel has become a challenging issue. Quebec's *Adieu Bazou! / Faites de l'air!* accelerated vehicle retirement program provides an innovative example of the role of planners in supporting senior government programs aimed at reducing vehicle use and impacts. (*Adieu Bazou* literally means "Goodbye clunker," while *Faites de l'air* is an idiomatic expression meaning "to leave" as well as a pun suggesting "improving air quality"). While planners can enable a mode shift through changes in land uses, urban design, and the provision of transit and bicycling infrastructure, they must work with other TDM program developers to maximize the potential benefits of such infrastructure. Quebec's accelerated vehicle retirement program can be seen as a TDM program that seeks to influence personal travel choices.

This case study is the result of a series of interviews with program managers, funders, and partners as well as a review of scientific literature, policy reports, and available documentation on this and other existing programs. The chapter begins by providing a historical perspective on accelerated vehicle retirement programs around the world. It defines their objectives and functioning and compares them with the particularities of federal and provincial programs—namely, the offer of **alternative travel incentives** such as transit passes, rebates on bicycles, and **car cooperative** memberships. The added benefits of such incentives and their potential influence on travel are discussed. The chapter explores the evolution of the program in the province of Quebec and the development of partnerships to provide travel incentives as well as exemplary vehicle-recycling practices. It also discusses the importance of land use and infrastructure in enabling participants' selection of alternative travel incentives. The case serves to demonstrate that programs based on mutually beneficial partnerships with transit and other alternative transportation providers can help city planners and government officials attain automobile travel reduction goals.

Accelerated Vehicle Retirement Programs: Air Quality, GHG, and Safety

Adieu Bazou! / Faites de l'air! is an accelerated vehicle retirement program, also referred to as vehicle scrappage program. This alternative term is not used in Quebec, because the program requires vehicle-recycling partners to follow criteria for responsible vehicle-recycling practices (AQLPA 2011). Reducing GHG emissions through improved **emission control technologies** and fuel efficiency are central objectives of accelerated vehicle retirement programs. Older vehicles typically emit more pollutants per kilometre driven. The difference between what the retired

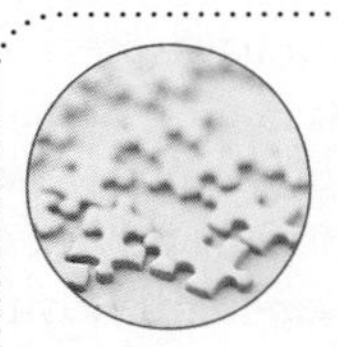

KEY ISSUE:

Changing Travel Behaviour

Reducing automobile travel and changing travel behaviour are objectives supported by all levels of government in Canada (e.g., Transport Canada 2002; MTQ 2009). Such changes in travel behaviour are difficult to enact and require the development of policies, programs, transportation infrastructure, land use, and urban design, as well as changes in travellers' attitudes toward alternative transportation. Co-ordination of efforts between agencies responsible for TDM programs and policies, transit provision, and the integration of transportation and land-use planning is required to maximize the effect of each individual effort.

vehicle would have emitted and emissions from the new vehicle (which are expected to be lower) are calculated as emission reduction benefits and can be used in national GHG accounting. The transportation sector is responsible for an important share of GHG emissions, and governments have had only modest success in reducing emissions from this sector (Ewing et al. 2007). In 2005, the transportation sector accounted for 38.7 per cent of GHG emissions in Québec (MDDEP 2008). By 2009, it accounted for 43.5 per cent (Gouvernement du Québec 2011).

Evolution of Programs around the World

First developed in 1990 by a private company in California, accelerated vehicle retirement programs quickly gained in popularity. Facing new requirements from the Clean Air Act, Unocal Oil Company needed to reduce fixed-source emissions (from industrial smokestacks). In order to comply with new regulations, Unocal developed programs that provided them with "mobile source emission reduction credits" that would compensate for their fixed-source emissions (Dill 2001). Their objective was to develop emissions reduction initiatives at a lower cost than what could have been achieved by replacing filters on their smokestacks (Shaheen, Guensler, and Washington 1994). The South Coast Recycled Auto Project (SCRAP, the first accelerated vehicle retirement program) was considered a theoretical and operational success and was emulated in many jurisdictions around the world. Governments quickly adopted the measure, and most programs became public (some metropolitan regions, provinces, states, and countries adopted the measure) or a mixture of **public–private partnerships (PPPs)** (Dill 2001). Programs around the world have provided either cash-for-scrappage (money in exchange for old vehicles) or cash-for-replacement (rebates on new vehicles) to accelerate vehicle retirement (ECMT 1999; Dill 2001).

The programs varied considerably in nature, objectives, scale, and incentive structure. Following the 2008 economic crisis, many countries developed recycling programs using environmental arguments while adding the objective of stimulating the economy and providing assistance to the ailing car industry (ITF 2011). The US, German, and French programs were recently the subject of a comparative analysis conducted by the International Transportation Forum (ITF 2011). Its predecessor, the European Conference of Ministers of Transport (ECMT 1999) had conducted an evaluation of earlier programs.

Critique of other programs

The ITF report (2011) concludes that programs generate limited emissions reduction and safety benefits and that program design greatly influences outcomes. For example, targeting an eligible fleet by age, ensuring that participating vehicles are actually used by only accepting insured vehicles in good operating condition, and placing restrictions on the types of vehicles that can serve as replacement can all improve program performance.

The ITF study indicated that people often tended to replace their vehicle with a larger one. Even though new vehicles have much stricter emissions standards, a larger-size vehicle will emit more than a newer vehicle of the same class as the recycled vehicle, thereby reducing the discrepancy between emissions from the new and recycled vehicle. This was found to be a program drawback in Germany, where the program did not restrict auto purchase to certain vehicle classes and where many participants used the program to upgrade to larger vehicles (ITF 2011).

Furthermore, media commentators also suggested that while programs act as direct subsidies to car owners, they also provide additional indirect subsidies to automobile manufacturers (Mickey 2009), that they tend to exclude lower-income populations not likely to purchase new vehicles in replacement (Fox 2009), and that they still promote the use of automobiles instead of diverting drivers away from their cars and toward transit or bicycles, a much clearer way of reducing emissions from the transportation sector (Sinclair 2009).

Finally, research suggests that as participants purchase a newer vehicle, they tend to drive more, thereby reducing the benefits achieved through the purchase of a less polluting vehicle (ITF 2011; Dill 2004; Van Wee, Moll, and Dirks 2000). Improvements in fuel efficiency and associated reduced travel costs are likely responsible for the increase in distance travelled (Zolnik 2012). Just like the ITF's study of the experience in Europe, Zolnik's (2012) analysis suggests that the environmental benefits of the US program are tangible but limited.

The Canadian Experience: Additional Mode Shift Objectives

BC SCRAP-IT, the first program in Canada, was developed in 1996. From its inception, the program included partnerships with transit agencies to provide public transit passes to program participants. The program was still in operation in 2015 (SCRAP-IT website). Canadian provincial programs such as BC SCRAP-IT and *Faites de l'air!* (*Faites de l'air!* website), as well as the federal program (Retire your Ride / *Adieu Bazou!*), have the same basic features as other programs developed elsewhere but additionally try to promote a mode shift by incentivizing participants with alternative travel options.

Understanding Behaviour Changes Induced by the Program: The Mode Shift Objective

Reducing emissions through a mode shift

There are reasons for trying to incentivize participants to adopt other travel modes as part of recycling programs instead of promoting the purchase of a new car. By influencing older-car owners to use other modes of transportation and reduce the amount of driving they do, the objective is to reduce vehicle emissions by a greater factor than that achieved through vehicle replacement (AQLPA 2011; Léger Marketing 2010). Potential emission reductions associated with a mode shift are theoretically greater than those accomplished through replacement—for example, the replacement of vehicle travel by bicycle or transit trips would reduce emissions more because these vehicles typically have much lower to nil emission per kilometre driven (Ewing et al. 2007).

Travel choices: Cost, availability, motivation

Changing travel behaviour requires actions on multiple motivators to individual choices. The main factors that influence travel choices are the relative cost of travel and the relative ease with which different modes provide access (Domencich and McFadden 1975; Ben-Akiva and Lerman 1985). Changing travel behaviour requires TDM policies that influence all travel characteristics. Fuel taxes, tolls, fares, insurance premiums, and congestion charges can all be used to alter the cost of travel in favour of alternative modes. Vehicle retirement incentives can thus potentially influence mode shift by altering the relative cost of travel. Personal car use reduction goals can, for example, be stimulated by a free transit incentive that will reduce travel costs (Gärling and Schuitema 2007).

Transition periods in life provide an opportunity to influence travellers to change travel modes (Eriksson, Garvill, and Nordlund 2008). The program's design attempts to take advantage of participants recycling their vehicle to stimulate a shift to other modes. By choosing alternative travel incentives, some participants may decide to permanently drive less and use other modes (Léger Marketing 2010).

Publicity associated with the program is seen as an educational tool. Individuals are offered an opportunity to reduce their use of personal vehicles and to understand both the consequences of vehicle travel and ways to carry on their activities without a vehicle.

Planners' incomplete role in TDM

As cities struggle to reduce automobile travel, economic incentives to shift travel modes can help them meet vehicle congestion, travelled distance, and emission reduction goals. These objectives fall within the broader goals of creating healthy, sustainable cities (Frumkin, Frank, and Jackson 2004). City planners and transportation planners can make transit use more attractive by improving service quality, regional service accessibility, and local urban design through density and intensification of mixed uses (TRB 2001, 2003; Ewing and Cervero 2010). TDM measures work better in combination, especially when linked with land-use changes (Meyer 1999). Combining favourable conditions for mode shift may strengthen the effect of planners' efforts. Planners need to be aware of the other personal factors that influence travel behaviour in order to better influence them or to create partnerships with those who can.

Evolution of the Quebec program *Adieu Bazou! / Faites de l'air!*

In 2002, *Faites de l'air!*, the first pilot program, was developed in the province of Quebec, largely modelled on the BC SCRAP-IT experience. Facing inaction from

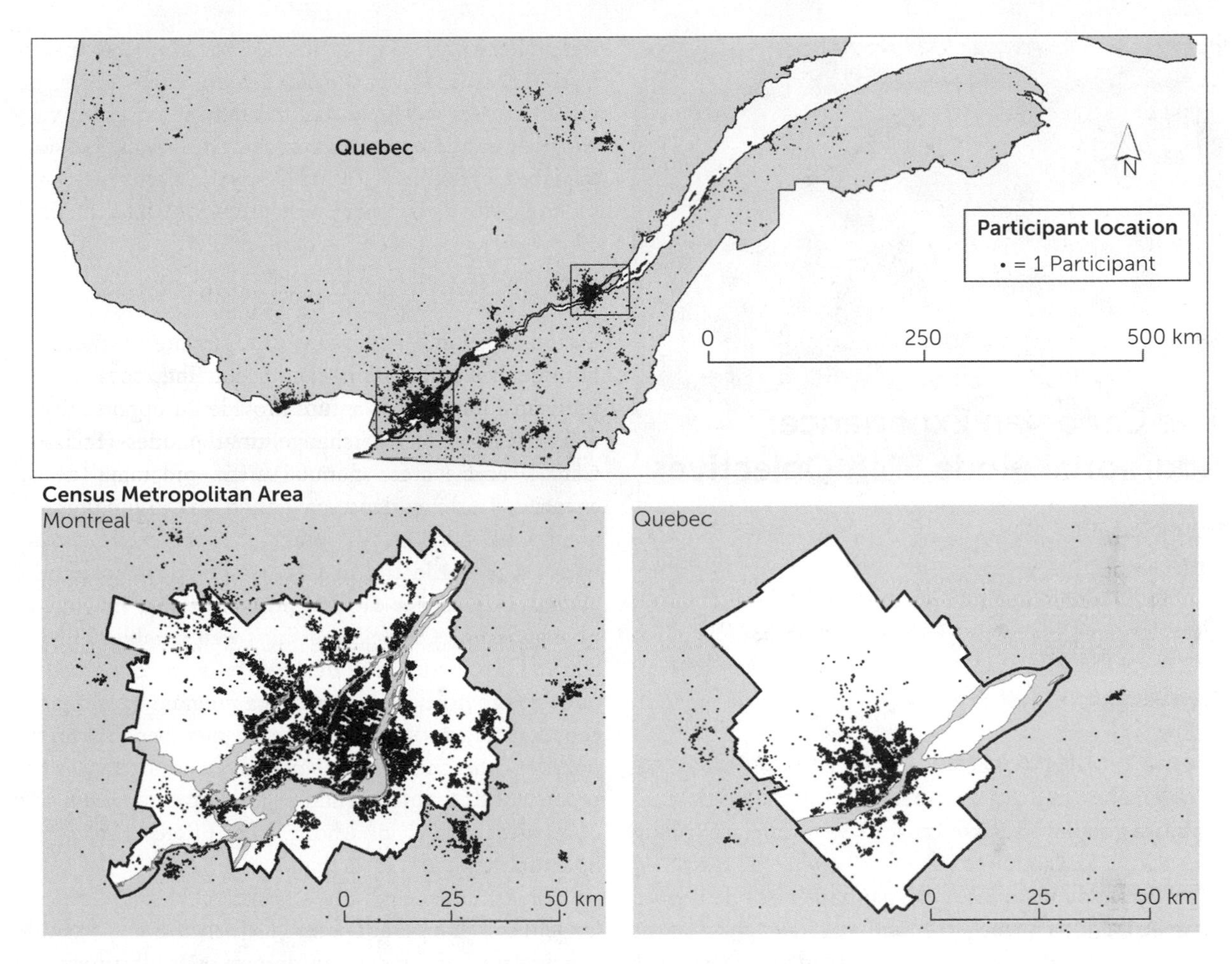

Figure 9.4.1 • Program participants are scattered across the province's populated places and are concentrated in large urban centres. They are, however, unevenly distributed within these centres. Population characteristics, land use, and transit access influence participation rates and incentive selection.

the provincial government, an air quality advocacy group called Association québécoise de lutte contre la pollution atmosphérique (AQLPA) launched a pilot program sponsored by the provincial and federal governments and maintained responsibility for program implementation for six years. The program benefitted from various but unstable sources of funding over the years.

In 2009, following the footsteps of the US government, the Canadian federal government decided to establish Retire your Ride / *Adieu Bazou!*, a federal recycling program. While in BC both provincial and federal programs operated simultaneously, *Adieu Bazou!* (federal) replaced *Faites de l'air!* (provincial) during the two-year period when the federal government fully funded the program. *Adieu Bazou!*, the Quebec component of the Canadian national accelerated vehicle retirement program, was entirely funded by Environment Canada and managed by the Clean Air Foundation (CAF). It allowed owners of vehicles made during or before 1995 (the year when higher emissions restrictions were imposed on vehicles) to recycle their vehicles in exchange for various incentives (Environment Canada 2009) developed through partnerships. Existing incentives established by the AQLPA in Quebec were maintained and expanded, and the organization was mandated by the CAF to implement the program in Quebec because of their existing expertise and experience. In May 2011, federal funding stopped for *Adieu Bazou!*, and the provincial government resumed program support for *Faites de l'air!*. Because of the popularity of *Faites de l'air!*, and the number of vehicles traded, funds were exhausted by the end of 2013. A newly elected government in 2014 did not renew program funding. *Faites de l'air!'s* website states that the program is *temporarily* closed.

Figure 9.4.2 • Older vehicles emit considerably more GHG and air pollutants because of less effective performance emission reduction devices. Newer features such as ABS brakes and airbags also make newer vehicles safer.

Source: R.F.C. de Rijcke.

Program objectives and incentive structure

Between 2009 and 2011, the period when the federal program was deployed, six incentives were in place. Along with $1000 to $3000 rebates on new vehicles, restricted to participating manufacturers, participants could choose between a $300 cheque, six months of free public transit passes (this number was changed to 15 in the second year of the program), a rebate of between $490 and $790 on a new bicycle, a one-year membership in Communauto, Quebec's car-sharing program, and a one-year pass for TRANSDEV, an interregional transit system. Because most alternative travel incentives are admittedly much more accessible and provide a more plausible alternative to car travel within major urban centres, the latter incentive was designed to attract participants interested in mode shift outside of major urban centres.

In the most recent version of the program, after its return to provincial funding in June 2011, *Faites de l'air!* maintained most of the previous incentives save for a few modifications. First and foremost, the $300 cash incentive was abolished. While this incentive provided an opportunity for lower-income owners without the resources to purchase a new vehicle, the incentive was likely to have no to counter-productive effects on GHG. In many cases, the $300 would likely go directly toward the purchase of a slightly more recent used vehicle (Environics Research Group 2010; Lachapelle 2015). Agreements with additional transit agencies were developed, and new participants were offered 12 months of subsidized transit passes instead of six or 15. With respect to automobile rebates, car manufacturers were required to offer discounts on vehicles (new or under four years of age) with lower emissions than the recycled vehicles and to provide higher incentives (up to $1,500) on hybrid electric vehicles (AQLPA 2011).

Co-ordination and Partnerships

Establishing and continuously improving partnerships has been a central task of the AQLPA. They spearheaded program design, sought funding from the government, and developed partnerships. Initial partners included vehicle recyclers, car manufacturers, and major transit agencies. This experience paved the way to the development of partnerships with other alternative travel providers, such as smaller transit agencies, bicycle manufacturer associations, and a car-sharing cooperative (www.communauto.com). Once partnerships were established with associations, individual dealerships were contacted in order to promote awareness of the program and improve communication of its functioning and benefits at the point of sale. The current *Faites de l'air!* program is the product of co-ordination and communication with nearly 400 partners, each of whom have different objectives, potential gains, and costs issues.

For example, for public transit agencies, potential longer-term transit adoption justifies the cost of subsidizing free transit pass incentives. Manufacturers can increase sales and reduce the price of their products to consumers. Dealerships and bicycle stores can also attract more customers through the rebates.

Government funding for the program enabled the creation of an incentive structure that can attract partners, as well as participants, and leverage their investments. The scale, scope, and level of funding of the program can greatly influence partners' interest in program participation. The co-ordination of partnership is particularly daunting in periods of program review and renewal when budgets and higher-level requirements may change the way contributions from each partner are handled.

The Importance of Land Use and Public Transit

The success of *Adieu Bazou! / Faites de l'air!* in promoting a mode shift is highly tied to land use and transportation planning. Choosing a public transit incentive can only be done where transit use is feasible and even competitive with automobile travel (Lachapelle 2013).

Figure 9.4.3 • Membership in Communauto, a car-sharing cooperative in Quebec, is offered as an incentive in the accelerated vehicle retirement program. Unfortunately, car-sharing is still available only in limited areas of the province—namely, larger urban centres.

Source: © *Communauto 2013.*

With **transit-oriented development (TOD)** gaining popularity, urban planners and transportation planners now interact more closely in planning processes. Maintaining a dialogue with policy-makers and program developers is necessary in order to maximize the potential benefits of land-use change and emission reduction programs on mode shift. Partnering with provincial and federal organizations that develop TDM measures outside the hands of planners is one such way.

While city planners were never directly involved, the efforts of transit agencies in collaborating with urban planners have been seen as an important support for the desired modal shift. As is the case with many small- to medium-scale government programs, no direct public consultation was developed. However, the AQLPA is constantly communicating with the public and has sought participant opinion on program improvements through follow-up surveys. Part of the AQLPA's mandate is to educate the public on issues related to atmospheric pollution as well as on actions individuals can take to reduce their household's impacts.

The Program's Success

Between 2009 and 2011, more than 40,000 participants registered in the program in Quebec and recycled their vehicles (AQLPA 2011). The determinants of program participation rates (participants/population) were assessed across the province to better understand where the program was most popular and likely most effective (Lachapelle 2013). Higher participation rates were found in lower-density areas inside metropolitan regions (the suburbs) and in higher-density non-metropolitan areas (mid-sized municipalities).

The same study assessed the characteristics associated with participants' choice of incentives. Transit incentives were more popular in dense, metropolitan centres with higher rates of walking and cycling to work and with more women, younger people, and a wealthier population. Bicycle incentive selection was associated with younger male participants. While these results do not imply a direct, long-term mode shift, it is expected that the incentives were used and likely reduced vehicle travel. Between May 2011 and 2013, program participation decreased because the $300 incentive was no longer available, but effectiveness probably increased because more participants chose transit passes. A follow-up survey of participants for the 2011–12 period revealed that of the 77 per cent of participants who took a transit pass (AQLPA 2011), 69 per cent continued to use it after the incentives expired (Lachapelle 2012). Emission reductions generated by the program were estimated at 30,000 tons per year (AQLPA 2011).

Limitations to Program Effectiveness

While a program designed with mode shift objectives has the potential to produce greater emission reduction benefits than traditional accelerated vehicle retirement programs, challenges have been noted by existing studies, program managers, partnering organizations, and program funders. Two of them are of particular importance. First, most alternative travel incentives are not as frequently available outside of large urban centres. This is the case for public transit passes, car-sharing, and, to a lesser extent, bicycle rebates. Bicycles rebates may be chosen, but the bicycle is more likely to be used for recreational purposes rather than to replace automobile trips, because travel distances are typically greater outside urban centres. As a consequence, program incentives may not attract as many owners of older vehicles outside of major urban centres (Lachapelle 2013) and may not yield emission reduction benefits of the same magnitude.

Second, a survey conducted by Environics Research Group (2009) revealed that owners of pre-1998 vehicles could be broken down into two large categories: wealthier owners of a second, older vehicle that can be more easily foregone and lower-income owners of a single household vehicle who may not be able to purchase a new vehicle (the cost being prohibitive) or make use of a free transit pass (one transit pass or bicycle cannot replace a vehicle used to provide transportation for an entire family). Car-sharing may be the only reasonable option for multiple-person households. This means that the program has difficulty in attracting low-income participants who use their primary and often sole vehicle more intensively.

Lessons and Transferability

This accelerated vehicle retirement program provides an example of the difficulties associated with addressing multiple issues within one program. Whether this version of a vehicle retirement program can successfully address all the issues for which it was designed still deserves attention. Government programs are increasingly being reviewed and compared in terms of cost, efficiency, and effectiveness in addressing issues for which they were designed.

Planners have taken on the task of reducing GHG emissions in cities and enabling a mode shift. Planners, especially transit planners in cities where the program is not in place, can contribute by:

- making the case to senior governments about the added benefits of such programs to city planning objectives;
- networking with TDM policy-makers in order to identify potential win/win solutions that may enhance their efforts to reduce single-occupancy vehicle use and promoting a mode shift;
- increasing funding for transit infrastructure and improving customer satisfaction with public transit to support increased ridership.

Mode shift–oriented accelerated vehicle retirement programs cannot be as effective in the absence of alternative transport provision. The effect of changes in urban design and transportation infrastructure can be improved by the development of programs that educate the population and alter the costs of automobile travel and alternative transportation options. Developing such partnership can be achieved in many jurisdictions and depends on the participation of partners (especially transit agencies) and on the willingness to renounce a portion of fare-box revenue (revenue generated through fares by paying customers).

Conclusion

Mode shifts associated with accelerated vehicle retirement programs are highly dependent on the existence and quality of land use and transportation infrastructure for alternative travel. The importance of the quality of transit service in selecting public transit incentives shows that urban planners' activities can influence the attractiveness and effectiveness of vehicle retirement programs. Senior government incentive programs may not be as effective without planners' active role in changing land use. Inversely, the development of vehicle retirement programs may enhance the attractiveness of existing public transit service and help increase ridership and transit's **mode share** within a region. Together, these factors show that planners can benefit from co-ordinating their efforts with the program development of senior governments. Planners' work and TDM program managers' objectives are complementary and work better in combination.

Acknowledgments

The author thanks AQLPA for sharing their program data and providing insights on the program. The Service aux Collectivités of the Université du Québec à Montréal funded and supported the university–community project upon which this case study is based.

KEY TERMS

Accelerated vehicle retirement program
Alternative travel incentives
Car cooperative
Emission control technologies
Greenhouse gas (GHG) emission reduction
Mode share
Mode shift
Public–private partnerships (PPPs)
Transit-oriented development (TOD)
Travel Demand Management (TDM)

REFERENCES

AQLPA (Association québécoise de lutte contre la pollution atmosphérique). 2011. « Faites de l'air : Le programme québécois de recyclage de véhicules. » Présentation aux membres de l'ATUQ. Montreal : AQLPA.

Ben-Akiva, M.E., and S.R. Lerman. 1985. *Discrete Choice Analysis: Theory and Application to Travel Demand*. MIT Press Series in Transportation Studies. Cambridge, MA: MIT Press.

Dill, J. 2001. "Design and administration of accelerated vehicle retirement programs in North America and abroad." *Transportation Research Record* 1750: 32–9.

———. 2004. "Estimating emissions reductions from accelerated vehicle retirement programs." *Transportation Research Part D: Transport and Environment* 9(2): 87–106.

Domencich, T.A., and D. McFadden. 1975. *Urban Travel Demand: A Behavioral Analysis: A Charles River Associates Research Study*. Amsterdam, NY: North-Holland Publishing; New York: American Elsevier.

ECMT (European Conference of Ministers of Transport). 1999. *Cleaner Cars: Fleet Renewal and Scrappage Schemes*. ECMT Publications.

Environics Research Group. 2009. *National Vehicle Scrappage Program Baseline Public Opinion Survey: Final Report*. Ottawa: Clean Air Foundation.

———. 2010. *Retire Your Ride Program 2010 Participant Survey: Final Report*. Ottawa: Clean Air Foundation.

Environment Canada. 2009. "Government of Canada launches vehicle scrappage program." News release, 11 May. http://www.ec.gc.ca/default.asp?lang=En&n=714D9AAE-1&news=5F2CADD5-0E59-4BFD-9075-54D4288CEC54.

Eriksson, L., J. Garvill, and A.M. Nordlund. 2008. "Interrupting habitual car use: The importance of car habit strength and moral motivation for personal car use reduction." *Transportation Research Part F: Traffic Psychology and Behaviour* 11(1): 10–23.

Ewing, R., K. Bartholomew, S. Winnkelman, J. Walters, D. Chen, B. McCann, and D. Goldberg. 2007. *Growing Cooler: The Evidence on Urban Development and Climate Change*. Washington: Urban Land Institute.

Ewing, R., and R. Cervero. 2010. "Travel and the built environment." *Journal of the American Planning Association* 76(3): 265–94.

Faites de l'air! website. www.faitesdelair.org.

Fox, J. 2009. "The (confused) economics of cash for clunkers." *Time* 4 Aug. http://business.time.com/2009/08/04/the-confused-economics-of-cash-for-clunkers/#ixzz2KRYDNXcD.

Frumkin, H., L. Frank, and R.J. Jackson. 2004. *Urban Sprawl and Public Health: Designing, Planning, and Building for Healthy Communities*. Washington: Island Press.

Gärling, T., and G. Schuitema. 2007. "Travel Demand Management targeting reduced private car use: Effectiveness, public acceptability and political feasibility." *Journal of Social Issues* 63(1): 139–53.

Gouvernement du Québec. 2011. *Le Québec en action VERT 2020 : Plan d'action 2013–2020 sur les changements climatiques*. Québec : Bibliothèque et Archives nationales du Québec.

ITF (International Transport Forum). 2011. *Car Fleet Renewal Schemes: Environmental and Safety Impacts—France, Germany and the United States*. Paris: OECD.

Lachapelle, U. 2012. *Habitudes de déplacement avant et après l'usage des titres de transport en commun offert en incitatif au programme « Faites de l'air! »*. Rapport présenté à l'Association québécoise de lutte contre la pollution atmosphérique, novembre.

———. 2013. "Participation and incentive choice of participants in an early vehicle retirement program in Québec, Canada." *Transportation Research Record* 2375(1): 8–17.

———. 2015, in press. "Using an early vehicle retirement program to support a mode shift: Car purchase and modal intentions following program participation." *Journal of Transportation and Land Use*.

Léger Marketing. 2010. *Habitudes de déplacements des participants au programme « Faites de l'air / Adieu Bazou » ayant choisi les titres de transport en commun comme incitatif*. Montreal : Association québécoise de lutte contre la pollution atmosphérique.

MDDEP (Ministère du développement durable de l'environnement et des Parcs). 2008. *Le Québec et les changements climatiques : un défi pour l'avenir. Plan d'action 2006–2012*. Québec : Gouvernement du Québec.

Meyer, M.D. 1999. "Demand management as an element of transportation policy: Using carrots and sticks to influence travel behavior." *Transportation Research Part A: Policy and Practice* 33(7–8): 575–99.

Mickey, A. 2009. "An outrageous stock market prediction." 22 June. Q1_Publishing. http://www.q1publishing.com/blog/viewblog?contentId=352.

MTQ (Ministère des transports du Québec). 2009. « Plan Stratégique 2008–2012. » http://www.mtq.gouv.qc.ca/portal/page/portal/Librairie/Publications/fr/ministere/plan_strat_2008_2012.pdf.

SCRAP-IT website. www.scrapit.ca.

Shaheen, S., R.L. Guensler, and S.P. Washington. 1994. *Vehicle Scrappage Programs: Are Some Sectors of Society Paying More for Clean Air?* Report UCD-ITS-RP-94-46. Davis: University of California at Davis, Institute of Transportation Studies.

Sinclair, C. 2009. "Another $2B for clunkers? How about bling for bikes?" *Huffington Post* 3 August. http://www.

huffingtonpost.com/cameron-sinclair/another-2b-for-clunkers-h_b_250256.html.

Statistics Canada. 2008. "2006 Census: Ethnic origin, visible minorities, place of work and mode of transportation." *The Daily* 2 April. http://www.statcan.gc.ca/daily-quotidien/080402/dq080402a-eng.htm.

Transport Canada. 2002. « Vision nationale des transports en commun jusqu'en 2020. » http://www.tc.gc.ca/fra/politique/acs-etudestransit-vision-132.htm; modifié 2010.

TRB (Transportation Research Board). 2001. *Making Transit Work: Insights from Western Europe, Canada, and the United States*. Special Report 257. Washington: Transportation Research Board.

——. 2003. "Transit capacity and quality of service manual." 2nd edn. Transit Cooperative Research Program. http://www.trb.org/news/blurb_detail.asp?id=2326.

Van Wee, B., H.C. Moll, and J. Dirks. 2000. "Environmental impact of scrapping old cars." *Transportation Research Part D: Transport and Environment* 5(2): 137–43.

Zolnik, E.J. 2012. "Estimates of statewide and nationwide carbon dioxide emission reductions and their costs from cash for clunkers. *Journal of Transport Geography* 24: 271–81.

Suggestions for Further Reading

Planning Fundamentals

Alfred, T. 2009. "Restitution is the real pathway to justice for Indigenous peoples." In G. Younging, J. Dewar, and M. DeGagne, eds, *Response, Responsibility, and Renewal: Canada's Truth and Reconciliation Journey*, 179–90. Ottawa: Aboriginal Healing Foundation.

Bennet, A., and C. Elman. 2007. "Case study methods in the international relations subfield." *Comparative Political Studies* 40(2): 170–95.

Birch, E. 2008. *The Urban and Regional Planning Reader*. New York: Routledge.

Bourne, L.S., T. Hutton, R. Shearmur, and J. Simmons. 2011. *Canadian Urban Regions: Trajectories of Growth and Change*. Oxford and Toronto: Oxford University Press.

Campbell, H., and R. Marshall. 1999. "Ethical frameworks and planning theory." *International Journal of Urban and Regional Research* 23(3): 464–78.

Dixon-Woods, M., S. Agarwal, D. Jones, B. Young, and A. Sutton. 2005. "Synthesising qualitative and quantitative evidence: A review of possible methods." *Journal of Health Services Research and Policy* 10(1): 45–53.

Dolowitz, D., and D. Marsh. 2000. "Learning from abroad: The role of policy transfer in contemporary policy making." *Governance* 13: 5–24.

Escobar, A. 1992. "Planning." In W. Sachs, ed., *The Development Dictionary: A Guide to Knowledge as Power*, 132–45. London: Zed Books.

Fainstein, S. 2010. *The Just City*. Ithaca, NY: Cornell University Press.

Filion, P., M. Moos, T. Vinodrai, and R. Walker. 2015. *Canadian Cities in Transition: New Directions in the 21st Century*. 5th edn. Toronto: Oxford University Press.

Fischler, R., and J. Wolfe. 2006. "Contemporary planning." In T. Bunting and P. Filion, eds, *Canadian Cities in Transition*, 3rd edn, 338–52. Don Mills, ON: Oxford University Press.

Flyvbjerg, B. 2001. *Making Social Science Matter: Why Social Inquiry Fails and How It Can Succeed Again*. Cambridge: Cambridge University Press.

Friedmann, J. 1986. *Planning in the Public Domain*. Princeton, NJ: Princeton University Press.

Gerring, J. 2004. "What is a case study and what is it good for?" *American Political Science Review* 98(2): 341–54.

Glaser, B. 2014. *Memoing: A Vital Grounded Theory Procedure*. Mill Valley, CA: Sociology Press.

Grant, J. 2006. "Shaped by planning: The Canadian city through time." In T. Bunting and P. Filion, eds, *Canadian Cities in Transition*, 3rd edn, 320–37. Don Mills, ON: Oxford University Press.

———. 2007. "Mixed use in theory and practice: Canadian experience with implementing a planning principle." *Journal of the American Planning Association* 68(1): 71–84.

———. 2008. *A Reader in Canadian Planning: Linking Theory and Practice*. Toronto: Nelson College Indigenous.

Grin, J. 2010. "The governance of transitions." In J. Grin, J. Rotmans, and J. Schot, eds, *Transitions to Sustainable Development: New Directions in the Study of Long Term Transformative Change*, 265–85. New York and London: Routledge.

Hall, P. 2002. "Planning: Millennium retrospect and prospect." *Progress in Planning* 57: 263–84.

Harvey, D. 1973. *Social Justice and the City*. Baltimore, MD: Johns Hopkins University Press.

Hodge, G., and D. Gordon. 2013. *Planning Canadian Communities*. 6th edn. Toronto: Nelson College Indigenous.

Khan, S., and R. VanWynsberghe. 2008. "Cultivating the under-mined: Cross-case analysis as knowledge mobilization." *Forum: Qualitative Social Research* 9(1): article 34. http://www.qualitative-research.net/index.php/fqs/article/view/334/729.

LeGates, R.T., and F. Stout. 2011. *The City Reader*. 5th edn. New York: Routledge.

Little Bear, L. 2000. "Jagged worldviews collide." In M. Battiste, ed., *Reclaiming Indigenous Voice and Vision*, 77–85. Vancouver: University of British Columbia Press.

Lorinc, J. 2006. *The New City: How the Crisis in Canada's Urban Centres is Reshaping the Nation*. Toronto: Penguin Canada.

Maxwell, J.A. 2004. "Using qualitative methods for causal explanation." *Field Methods* 16(3): 243–64.

Miles, M.B., and A.M. Huberman. 1994. *Qualitative Data Analysis: An Expanded Sourcebook*. Thousand Oaks, CA: Sage.

Niewohner, J., and T. Scheffer. 2010. *Thick Comparison: Reviving the Ethnographic Aspiration*. Boston: BRILL.

Rose, R. 2005. *Learning from Comparative Public Policy: A Practical Guide*. London and New York: Routledge.

Sancton, A., and R. Young. 2009. *Foundations of Governance: Municipal Government in Canada's Provinces*. Toronto: University of Toronto Press.

Sandelowski, M., S. Cocherty, and C. Emden. 1997. "Qualitative metasynthesis: Issues and techniques." *Research in Nursing and Health* 20: 365–71.

Sandercock, L. 2000. "When strangers become neighbors: Managing cities of difference." *Planning Theory and Practice* 1(1): 13–30.

———. 2003. *Cosmopolis II: Mongrel Cities*. New York: Continuum.

——. M. Hibbard, M.B. Lane, L. Porter, and J. Beneria-Surkin. 2004. "Interface: Planning and Indigenous communities." *Planning Theory and Practice* 5(1): 95–124.

Saul, J.R. 2008. *A Fair Country: Telling Truths about Canada.* Toronto: Viking.

Schofield, J.W. 2002. "Increasing the generalizability of qualitative research." In A.M. Huberman and M.B. Miles, eds, *The Qualitative Researcher's Companion*, 171–294. Thousand Oaks, CA: Sage.

Sewell, J. 1993. *The Shape of the City: Toronto Struggles with Modern Planning.* Toronto: University of Toronto Press.

Slack, E. 2001. *Municipal Finance and Governance in the Greater Toronto Area.* Toronto: Neptis Foundation.

Spaans, M., and E. Louw. 2009. "Crossing borders with planners and developers and the limits of lesson-drawing." Paper presented to the City Futures 2009 Conference. Madrid: University Rey Juan Carlos of Madrid.

Stern, P.R., and P.V. Hall. 2014. *The Proposal Economy.* Vancouver: University of British Columbia Press.

Stone, D. 1999. "Learning lessons and transferring policy across time, space, and disciplines." *Politics* 19: 51–9.

Thomas, R., and L. Bertolini. 2014. "Beyond the case study dilemma in urban planning: Using a meta-matrix to distil critical success factors in transit-oriented development." *Urban Policy and Research* 32(2): 219–37 DOI: 10.1080/08111146.2014.882256.

Throgmorton, J. 1996. *Planning as Persuasive Storytelling.* Chicago: University of Chicago Press.

Yiftachel, O. 1998. "Planning and social control: Exploring the dark side." *Journal of Planning Literature* 12(4): 395–406.

Yin, R.K. 1981. "The case study crisis: Some answers." *Administrative Science Quarterly* 26(1): 58–68.

——. 2014. *Case Study Research: Design and Methods.* 5th edn. Los Angeles: Sage.

Community Development and Social Planning

Adams, D., and A. Goldbard. 2001. *Creative Community: The Art of Cultural Development.* New York: Rockefeller Foundation.

Alfred, T. 2005. *Wasáse: Indigenous Pathways of Action and Freedom.* Peterborough, ON: Broadview Press.

Amara, A., I. Abu-Saad, and O. Yiftachel. 2012. *Indigenous (In) Justice: Human Rights Law and Bedouin Arabs in the Naqab/ Negev.* Cambridge, MA: Harvard Law School.

Americans for the Arts. 2007. *Arts and Economic Prosperity III: The Economic Impact of Nonprofit Arts and Culture Organizations and Their Audiences in the City of Boulder, Colorado.* Washington: Americans for the Arts.

Angeles, L., O. Shcherbyna, and J. Foster. 2014. "Contemporary social planning: Comparative cases in comprehensive social plan development." *Plan Canada* 54(1): 16–22.

Baeker, G. 2001. *Rediscovering the Wealth of Places: A Municipal Cultural Planning Handbook for Canadian Communities.* St Thomas, ON: Municipal World.

Battiste, M. 2000. *Reclaiming Indigenous Voice and Vision.* Vancouver: University of British Columbia Press.

Borén, T., and C. Young. 2013. "Getting creative with the 'creative city'? Towards new perspectives on creativity in urban policy." *International Journal of Urban and Regional Research* 37(5): 1799–815.

Bradford, Neil. 2002. *Why Cities Matter: Policy Research Perspectives for Canada.* Ottawa: CPRN Discussion Paper F/23.

Brault, S. 2005. "The arts and culture as new engines of economic and social development." *Policy Options* 26(3).

——. 2010. *No Culture, No Future.* tr. J. Kaplansky. Toronto: Cormorant Books. (Original work published 2009).

Bulkens, M., C. Minca, and H. Muzaini. 2014. "Storytelling as method in spatial planning." *European Planning Studies*, 1–17. DOI: 10.1080/09654313.2014.942600.

Carlin, B. 2005. *Location Option Tax Strategies for Supporting the Arts.* Pittsburgh: Carnegie-Mellon University, Heinz School of Public Policy.

Clague, M. 1993. *A Citizen's Guide to Community Social Planning.* Vancouver: Social Planning and Research Council of British Columbia.

Clifton, N., and R. Huggins. 2010. *Competitiveness and Creativity: A Place-Based Perspective.* Toronto: University of Toronto, Martin Prosperity Institute.

Clutterbuck, P., and M. Novick. 2001. *Preserving Our Civic Legacy: Community Consultation on Social Development.* Toronto: Community Social Planning Council of Toronto.

——. 2003. *Building Inclusive Communities: Cross-Canada Perspective and Strategies.* Ottawa: Federation of Canadian Municipalities and Laidlaw Foundation.

Coletta, C. 2008. *Fostering the Creative City.* Chicago: CEOs for Cities.

Cortright, J. 2007. *City Advantage: Variety, Convenience, Discovery, Opportunity.* Chicago: CEOs for Cities.

Di Cicco, P. 2008. "Creativity and the essence of true leadership in municipal government." Keynote address for the City of Hamilton Extended Management Team Meeting.

Gattinger, M. 2008. "Multilevel governance and cultural infrastructure: A review of Canadian and international experiences." In *Under Construction: The State of Cultural Infrastructure in Canada*, Appendix B. Vancouver: Centre of Expertise on Culture and Communities, Simon Fraser University.

Geneau, M., K. Kronstal, G. McGinnis, and B. Sutton. 2010. "Artist live/work space: Best practices and potential models." Ottawa: Canadian Institute of Planners. *Plan Canada* 50(2): 31–3.

Grodach, C., and D. Silver, eds. 2012. *The Politics of Urban Cultural Policy: Global Perspectives.* London and New York: Routledge.

Hallsmith, G. 2003. *The Key to Sustainable Cities: Meeting Human Needs, Transforming Community Systems*. Gabriola Island, BC: New Society Publishers.

Hawkes, J. 2001. *The Fourth Pillar of Sustainability: Culture's Essential Role in Public Planning*. Melbourne: Common Ground and the Cultural Development Network.

——. 2009. "Challenges for local cultural development." Paper prepared for the Barcelona Institute of Culture.

Hill Strategies Research. 2008. *Social Effects of Culture: Detailed Statistical Models*. Hamilton, ON: Hill Strategies Research.

——. 2008. *A Statistical Profile of Artists in Canada Based on the 2006 Census*. Hamilton, ON: Hill Strategies Research.

——. 2010. *New Report Compares Municipal Cultural Investments in Vancouver, Calgary, Toronto, Ottawa and Montréal*. Hamilton, ON: Hill Strategies Research.

Horizons Community Development Associates Inc. 2010. *Nova Scotia Arts and Culture Consultation Summary Report*. Halifax: Culture Division, Nova Scotia Department of Tourism, Culture and Heritage.

Hume, G. 2009. *Cultural Planning for Creative Communities*. St Thomas, ON: Municipal World.

Keenleyside, P. 2007. "The state of cultural infrastructure: Policy and issues dialogue." Regional Roundtable Background Paper, Pacific Region. Vancouver: Simon Fraser University, Centre of Expertise on Culture and Communities.

Kovacs, J. 2011. "Cultural planning in Ontario, Canada: Arts policy or more?" *International Journal of Cultural Policy* 17(3): 321–40.

Landry, C. 2000. *The Creative City: A Toolkit for Urban Innovators*. London: Earthscan.

Legacies Now. 2010. *Cultural Planning Toolkit*. Creative City Network of Canada.

Markusen, A. 2009. *The Economics of Arts, Artists, and Culture: Making a Better Case*. Seattle: Grantmakers in the Arts Reader.

——. 2014. "Creative cities: A 10-year research agenda." *Journal of Urban Affairs* 36(2): 567–89.

Matunga, H. 2013. "Theorizing Indigenous planning." In R. Walker, T. Jojola, and D. Natcher, eds, *Reclaiming Indigenous Planning*, 3–32. Montreal: McGill-Queen's University Press.

Momer, B. 2010. *The Creative Sector in Kelowna, British Columbia: An Economic Impact Assessment*. Kelowna: University of British Columbia Okanagan, Community, Culture and Global Studies.

——. 2011. *Our City, Ourselves: A Cultural Landscape Assessment of Kelowna, British Columbia*. Kelowna: University of British Columbia Okanagan.

Murray, C. 2008. "Cultural infrastructure in BC: Maps, gaps and planning outlook for creative communities." Paper presented at the British Columbia Recreation and Parks Association The Way Forward Symposium, Victoria.

Paris, D., and M.T. Winn, eds. 2014. *Humanizing Research: Decolonizing Qualitative Inquiry with Youth and Communities*. Thousand Oaks, CA: Sage.

Porter, L. 2010. *Unlearning the Colonial Cultures of Planning*. Burlington, VT: Ashgate.

——. 2013. "Coexistence in Cities: The challenge of Indigenous urban planning in the twenty-first century." In R. Walker, T. Jojola, and D. Natcher, eds, *Reclaiming Indigenous Planning*, 283–310. Montreal: McGill-Queen's University Press.

Pstross, M., C.A. Talmage, and R.C. Knopf. 2014. "A story about storytelling: Enhancement of participation through catalytic storytelling." *Community Development* 45(5): 525–38.

Rosenstein, C. 2009. *Cultural Development and City Neighborhoods*. Washington: Urban Institute.

Russo, A., and D. Butler. n.d. *Cultural Planning Toolkit: A Partnership between 2010 Legacies Now and Creative City Network of Canada*. Vancouver: 2010 Legacies Now.

Sandercock, L. 2011. "Out of the closet: The importance of stories and storytelling in planning practice." *Planning Theory and Practice* 4(1): 11–28.

Smith, B. 2012. *Greater Victoria Arts and Culture Sector Economic Activity Study*. Victoria: University of Victoria, Peter Gustavson School of Business.

Smith, L.T. 2010. *Decolonizing Methodologies: Research and Indigenous Peoples*. 2nd edn. London: Zed Books.

Stanger-Ross, J. 2008. "Municipal colonialism in Vancouver: City planning and the conflict over Indian reserves 1928–1950s." *Canadian Historical Review* 89(4): 541–80.

Tofino Arts and Culture Committee. 2012. *Picturing Tofino—A Plan for Arts and Heritage*. Tofino: Tofino Arts and Culture Committee.

United Cities and Local Governments. 2004. "Agenda 21 for culture: An undertaking by cities and local governments for cultural development." http://www.agenda21cultura.net.

Vis-à-vis Management Resources. 2007. *A Case for Investing in Arts, Culture and Heritage Infrastructure*. Duncan, BC: Art and Culture Division, British Columbia Ministry of Tourism, Sport and the Arts.

Wyman, M. 2004. *Why Culture Matters: The Defiant Imagination*. Vancouver: Douglas and McIntyre.

Yiftachel, O. 2012. "Naqab/Negev Bedouins and the (internal) colonial paradigm." In A. Amara, I. Abu-Saad, and O. Yiftachel, eds, *Indigenous (In) justice: Human Rights Law and Bedouin Arabs in the Naqab/Negev*, 289–318. Cambridge, MA: Harvard Law School.

Urban Form and Public Health

Desmarais, E., and A. Miguelez. 2015. "Communities and the streets residents live on: Linking neighbourhood character to zoning." *Plan Canada* 55(1): 25–31.

Dushenko, W., A.D. Terrance, and P.J. Robinson, eds. 2012. *Urban Sustainability: Reconnecting Space and Place*. Toronto: University of Toronto Press.

Frumkin, H., L. Frank, and R. Jackson. 2004. *Urban Sprawl and Public Health: Designing, Planning, and Building for Healthy Communities*. Washington: Island Press.

Glover, T.D. 2003. "The story of the Queen Anne Memorial Garden: Resisting a dominant cultural narrative." *Journal of Leisure Research* 35(2): 190–212.

Godschalk, D. and E. Malizia. 2013. *Sustainable Development Projects: Integrating Design, Development, and Regulation*. Chicago: APA Planners Press.

Isenhour, C., G. McDonogh, and M. Checker, eds. 2015. *Sustainability in the Global City: Myth and Practice*. New York: Cambridge University Press.

Moreno-Leguizamon, C., M. Tovar-Restrepo, C. Irazábal, and C. Locke. 2015. "Learning alliance methodology: Contributions and challenges for multicultural planning in health service provision: A case-study in Kent, UK." *Planning Theory and Practice* 16(1): 79–96. DOI:10.1080/14649357.2014.990403.

Qadeer, M. 1998. "Ethnic malls and plazas: Chinese commercial developments in Scarborough, Ontario." Working Paper. Toronto: Joint Centre of Excellence for Research on Immigration and Settlement (CERIS).

Thompson, D. 2014. *Suburban Sprawl: Exposing Hidden Costs, Identifying Innovations*. Ottawa: University of Ottawa, Sustainable Prosperity.

Wang, S. 1999. "Chinese commercial activity in the Toronto CMA: New development patterns and impacts." *The Canadian Geographer* 43(1): 19–35.

Wheeler, S.M. 2010. "The evolution of urban form in Portland and Toronto: Implications for sustainability planning." *Local Environment: The International Journal of Justice and Sustainability* 8(3): 317–36.

———. 2013. *Planning for Sustainability: Creating Livable, Equitable, and Ecological Communities*. New York: Routledge.

Zala, C.A. 1999. "Operational procedure for automatic true orthophoto generation." *Advances in Computational Mathematics* 11: 211–27.

Natural Resource Management

Columbia Basin Trust. 2012. *Project Concept: Model Climate Resilient Subdivision and Development Servicing Bylaw*. Communities Adaptation to Climate Change Initiative.

David Suzuki Foundation. 2012. *All over the Map: A Comparison of Provincial Climate Change Plans*. Vancouver: David Suzuki Foundation.

———. 2012. *Carbon in the Bank: Ontario's Greenbelt and Its Role in Mitigating Climate Change*. Vancouver: David Suzuki Foundation.

———. 2015. *Making Our Coasts Work: Healthy Oceans, Healthy Economies, Healthy Communities*. Vancouver: David Suzuki Foundation.

Fischhoff, B., and J. Kadvany. 2011. *Risk: A Very Short Introduction*. New York: Oxford University Press.

Foster, J., A. Lowe, and S. Winkelman. 2011. "The value of green infrastructure for urban climate adaptation." Washington: Center for Clean Air Policy. http://www.ccap.org/index.php?component=news&id=402.

Groves, D., E. Bloom, and B. Joyce. 2011. "A decision framework for the 2013 California Water Plan." http://www.waterplan.water.ca.gov/docs/meeting_materials/swan/2011-0513/CWP_2013-RAND_DECISION_FRAMEWORK-2011.05.12.pdf.

Hibbard, M., M. Lane, and K. Rasmussen. 2008. "The split personality of planning: Indigenous peoples and planning for land and resource management." *Journal of Planning Literature* 23: 136–51.

Holling, C.S. 1978. *Adaptive Environmental Assessment and Management*. Chichester, UK: John Wiley.

Kahn, M.E. 2010. *Climatopolis: How Our Cities Will Thrive in the Hotter Future*. New York: Basic Books.

Keeney, R., and T.L. McDaniels. 2001. "A framework to guide thinking and analysis regarding climate change policies." *Risk Analysis* 21(6): 989–1000.

Kellenberg, D.K., and M. Mobarak. 2008. "Does rising income increase or decrease damage risk from natural disasters?" *Journal of Urban Economics* 63: 788–802.

Klain, S.C., T.A. Satterfield, and K.M.A. Chan. 2014. "What matters and why? Ecosystem services and their bundled qualities." *Ecological Economics* 107: 310–20.

Lempert, R.J., and M.T. Collins. 2007. "Managing the risk of uncertain threshold responses: Comparison of robust, optimum, and precautionary approaches." *Risk Analysis* 27(4): 1009–26.

Lempert, R.J., and M.E. Schlesinger. 2000. "Robust strategies for abating climate change." *Climate Change* 45: 387–401.

Loftus, A. 2011. "Adapting urban water systems to climate change: A handbook for decision-makers at the local level." Freiburg, Germany: ICLEI European Secretariat VmbH. http://www.iwahq.org/contentsuite/upload/iwa/all/Water%20climate%20and%20energy/SWITCH_Adaption-Handbook_final_small.pdf.

McBean, G. 2009. "Climate change adaptation and extreme weather: Summary recommendations." Vancouver: Adaptation to Climate Change Team, Simon Fraser University. http://act-adapt.org/wp-content/uploads/2011/03/PDF-WeatherSession_SummaryReport.pdf.

McDaniels, T.L. 2000. "Creating and using objectives for ecological risk assessment and management." *Environmental Science and Policy* 3: 299–304.

———. R. Gregory, and D. Fields. 1999. "Democratizing risk management: Successful public involvement in an electric utility water management decision." *Risk Analysis* 19(3): 491–504.

Mukheibir, P., and G. Ziervogel. 2007. "Developing a Municipal Adaptation Plan (MAP) for climate change: The city of Cape Town." *Environment and Urbanization* 19(1): 143–58.

Ndubisi, F.O., ed. 2014. *The Ecological Design and Planning Reader*. Washington: Island Press.

Picketts, I.M., J. Curry, S.J. Déry, and S.J. Cohen. 2013. "Learning with practitioners: Climate change adaptation priorities in an urban community." *Climatic Change* 118(2): 321–37.

Picketts, I.M., S.J. Déry, and J.A. Curry. 2014. "Incorporating climate change adaptation into local plans." *Journal of Environmental Management* 57(7): 984–1002.

Renn, O. 2004. "The challenge of integrating deliberation and expertise: Participation and discourse in risk management." In T. McDaniels and M.J. Small, eds, *Risk Analysis and Society: An Interdisciplinary Characterization of the Field*, 289–366. Cambridge, MA: Cambridge University Press.

Risbey, J., M. Kandlikar, H. Dowlatabadi, and D. Graetz. 1999. "Scale, context and decision-making in agricultural adaptation to climate variability and change." *Mitigation and Adaptation Strategies for Global Change* 4: 137–65

Speaker, C. 2011. "Community planning through the climate change lens." PICS Conference: Resilience Communities Preparing for Climate Change. http://www.pics.uvic.ca/assets/pdf/resilient_communities/Speaker.pdf.

Wackernagel, M., and W. Rees. 1996. *Our Ecological Footprint: Reducing Human Impact on the Earth*. Gabriola Island, BC: New Society Publishers.

Whitman, E., E. Rapaport, and K. Sherren. 2014. "A conceptual model for balancing management trade-offs between urban forest benefits and wildfire risk." *Plan Canada* 54(4): 17–21.

Housing

August, M., and A. Walks. 2012. "From social mix to political marginalisation? The redevelopment of Toronto's public housing and the dilution of tenant organisational power." In G. Bridge, T. Butler, and L. Lees, eds, *Mixed Communities: Gentrification by Stealth?* 273–97. Bristol, UK: The Policy Press.

Dunn, J.R. 2012. "'Socially mixed' public housing redevelopment as a destigmatization strategy in Toronto's Regent Park: A theoretical approach and a research agenda." *DuBois Review* 9(1): 87–105.

Federation of Canadian Municipalities (FCM). 2009. *Quality of Life in Canadian Communities: Immigration and Diversity in Canadian Cities and Communities*. Ottawa: FCM.

Haan, M. 2005. "The decline of immigrant home-ownership advantage: Life-cycle, declining fortunes and changing housing careers in Montreal, Toronto, and Vancouver." *Urban Studies* 42(12): 2191–212.

——. 2010. "The residential crowding of immigrants to Canada." *Canadian Issues* fall: 16–21. Montreal: Association for Canadian Studies.

Hackworth, J. 2009. "Political marginalisation, misguided nationalism and the destruction of Canada's social housing systems." In S. Glynn, ed., *Where the Other Half Lives: Lower Income Housing in a Neoliberal World*, 257–77. London and New York: Pluto Press.

Hou, F., and G. Picot. 2004. "Visible minority neighbourhoods in Toronto, Montreal and Vancouver." Statistics Canada Catalogue no. 11-008. *Canadian Social Trends* spring: 8–13.

Hulchanski, J.D. 2010. "The three cities within Toronto: Income polarization among Toronto's neighbourhoods, 1970–2005." Update to Research Bulletin 41. Toronto: University of Toronto Cities Centre.

Intervistas. 2012. *Whistler Housing Authority Employer Housing Needs Assessment 2012 Final Report*. Whistler, BC: Whistler Housing Authority.

Macdonald, D. 2010. *Canada's Housing Bubble: An Accident Waiting to Happen*. Ottawa: Canadian Centre for Policy Alternatives.

Oberlander, P.H., and A.L. Fallick. 1992. *Housing a Nation: The Evolution of Canadian Housing Policy*. Prepared by the Centre for Human Settlements, University of British Columbia, for Canada Mortgage and Housing Corporation.

Pembina Institute. 2014. *2014 Home Location Preference Survey*. Toronto: Pembina Institute.

Suttor, G. 2011. "Offset mirrors: Canadian and Australian social housing." *International Journal of Housing Policy* 11(3): 255–83.

Teixeira, C., W. Li, and A. Kobayashi, eds. 2012. *Immigrant Geographies of North American Cities*. Don Mills, ON: Oxford University Press.

Walks, R.A., and L.D. Bourne. 2006. "Ghettos in Canadian cities? Racial segregation, ethnic enclaves and poverty concentration in Canadian urban areas." *The Canadian Geographer* 5(3): 273–97.

Whistler Housing Authority. n.d. *Community Engagement Strategy*. Whistler, BC: Whistler Housing Authority.

——. 2013. *2013 Business and Financial Plan*. Whistler, BC: Whistler Housing Authority.

Participatory Processes

Alberta Ministry of Municipal Affairs. "Resources on digital engagement: Public Input Toolkit, Chapter 3, Using digital engagement for public input." Edmonton: Alberta Ministry of Municipal Affairs. http://www.municipalaffairs.alberta.ca/documents/MDRS/AMA_Public_Input_Toolkit_Sept2014.pdf.

Brabham, D.C. 2013. "Using crowdsourcing in government." Los Angeles: University of Southern California, IBM Centre for Business of Government. http://businessofgovernment.org/sites/default/files/Using%20Crowdsourcing%20In%20Government.pdf.

Cameron, K., and J. Simon. 2014. "Plumbing public values: Greater Vancouver's 40 years' experience with regional public attitude surveys." *Plan Canada* 54(3): 36–41.

David Suzuki Foundation. 2012. *Youth Engagement with Nature and the Outdoors*. Vancouver: David Suzuki Foundation.

Evans-Cowley, J. and J. Hollander. 2010. "The new generation of public participation: Internet-based participation tools." *Planning Practice and Research* 25(3): 397–408.

Federation of Canadian Municipalities (FCM). 2011. *Municipal Youth Engagement Handbook*. Ottawa: FCM.

HB Lanarc Golder. 2012. "Digital sustainability conversations: How local governments can engage residents online." http://www.sustainablecitiesinstitute.org/Documents/SCI/Report_Guide/Guide_USDN_DigitalEngagement1.pdf.

Landry, J., and L. Angeles. 2011. "Institutionalizing participatory governance in municipal planning and policy-making: Initial challenges and lessons from the Plateau-Mont-Royal's comité aviseur model. *Canadian Journal of Urban Research* 20(1): 105–30.

Mandarano, L., M. Mahbubur, and C. Steins. 2010. "Building social capital in the digital age of civic engagement." *Journal of Planning Literature* 25(2): 123–35.

Phillips, S., and M. Orsini. 2002. "Mapping the links: Citizen involvement in policy processes." CPRN Discussion Paper no. F/21. Ottawa: Canadian Policy Research Networks (CPRN).

Urban Design

Alexander, C., S. Ishikawa, and M. Silverstein. 1977. *A Pattern Language: Towns. Buildings. Construction.* New York: Oxford University Press.

Baker, G. 2010. *Rediscovering the Wealth of Places.* Union, ON: Municipal World.

Carmona, M., T. Heath, T. Oc, and S. Tiesdell. 2010. *Public Places Urban Spaces: The Dimensions of Urban Design.* Oxford: Architectural Press.

Conzen, M.R.G. 1960. "Alnwick, Northumberland: A study in town-plan analysis." *Transactions and Papers* (Institute of British Geographers) 27: iii+ixxi+1+3–122.

Cresswell, T. 2004. *Place: A Short Introduction.* Oxford: Blackwell.

Eidelman, G. 2011. "Who's in charge? Jurisdictional gridlock and the genesis of Waterfront Toronto." In G. Desfor and J. Laidley, eds, *Transforming Toronto's Waterfront*, 263–86. Toronto: University of Toronto Press.

Gehl, J. 2010. *Cities for People.* Washington: Island Press.

——. 2011. *Life between Buildings: Using Public Space.* Washington: Island Press.

Grammenos, F., and G.R. Lovegrove. 2015. *Remaking the City Street Grid: A Model for Urban and Suburban Development.* Jefferson, NC: McFarland and Co.

Grant, J. 2002. "Mixed use in theory and practice: Canadian experience with implementing a planning principle." *Journal of the American Planning Association* 68(1): 71–84.

Leahy Laughlin, D., and L.C. Johnson. 2011. "Defining and exploring public space: Perspectives of young people from Regent Park, Toronto." *Children's Geographies* 9(3–4): 439–56.

Lynch, K. 1981. *Good City Form.* Cambridge, MA: MIT Press.

McHarg, I.L. 1992. *Design with Nature.* New York: John Wiley and Sons.

Relph, E. 1987. *The Modern Urban Landscape.* Baltimore, MD: Johns Hopkins University Press.

Sandalack, B.A., and A. Nicolai. 1998. *Urban Structure Halifax: An Urban Design Approach.* Halifax: TUNS Press.

——. 2006. *The Calgary Project: Urban Form/Urban Life.* Calgary: University of Calgary Press.

Schlossberg, M., J. Rowell, A. Amos, and K. Sanford. 2014. *Rethinking Streets: An Evidence-Based Guide to 25 Complete Street Transformations.* Eugene: University of Oregon Sustainable Cities Initiative.

Sorkin, M. 2008. *Indefensible Space: The Architecture of the National Insecurity State.* New York: Routledge.

Tiesdell, S., and D. Adams, eds. 2011. *Urban Design in the Real Estate Development Process.* Chichester, UK: Wiley-Blackwell.

——. 2013. *Shaping Places: Urban Planning, Design and Development.* Abingdon, UK: Routledge.

Urban Regeneration

Neptis Foundation. 2013. *Implementing the Growth Plan for the Greater Golden Horseshoe.* Toronto: Neptis Foundation.

——. 2015. *Understanding the Fundamentals of the Growth Plan.* Toronto: Neptis Foundation.

Ontario. 2006. *Growth Plan for the Greater Golden Horseshoe.* Toronto: Province of Ontario.

Scott, J.C. 1998. *Seeing Like a State: How Certain Schemes to Improve the Human Condition Have Failed.* New Haven, CT: Yale University Press.

Transportation and Infrastructure

Adenot, F. J. 2007. "Ville de Longueil: Place Charles-Le Moyne." Recherche URBATOD étude de cas. Montreal: Département d'études urbaines et touristiques, Université de Québec à Montréal (UQAM).

——. 2007. "Ville Mont-Royal: Les entreprises ferroviaires orientant le développement urbain." Recherche URBATOD étude de cas. Montreal: Département d'études urbaines et touristiques, Université de Québec à Montréal (UQAM).

Bamberg, S., and G. Möser. 2007. "Why are work travel plans effective? Comparing conclusions from narrative and meta-analytical research synthesis." *Transportation* 34: 647–66.

Builing, R.N., R. Mitra, and G. Faulkner. 2009. "Active school transportation in the Greater Toronto Area, Canada: An exploration of trends in space and time (1986–2006)." *Preventive Medicine* 48(6): 507–12.

Butler, G.P, H.M. Orpana, and A.J. Wiens. 2007. "By your own two feet: Factors associated with active transportation in Canada." *Canadian Journal of Public Health* 98(4): 259–64.

California Department of Transportation. 2002. *Statewide Transit-Oriented Development Study: Factors for Success in California.* San Diego: California Department of Transportation.

Cascetta, E., and F. Pagliara. 2008. "Integrated railways-based policies: The Regional Metro System (RMS) project of Naples and Campania." *Transport Policy* 15(2): 81–93.

Cervero, R. 1998. *The Transit Metropolis: A Global Inquiry*. Washington: Island Press.

Curtis, C. 2008. "Evolution of the transit-oriented development model for low-density cities: A case study of Perth's new railway corridor." *Planning, Practice and Research* 23(3): 285–302.

——. and M. Hare. 2014. "Leveraging transit investment to shape the community: Waterloo Region's Central Transit Corridor Community Building Strategy." *Plan Canada* 54(3): 24–9.

Curtis, C., and N. Low. 2012. *Institutional Barriers to Sustainable Transport*. Farnham, UK, and Burlington, VT: Ashgate.

Curtis, C., J.L. Renne, and L. Bertolini, eds. 2009. *Transit Oriented Development: Making It Happen*. Farnham, UK, and Burlington, VT: Ashgate.

Davis, D., and B. Lorenzowski. 1998. "A platform for gender tensions: Women working and riding on Canadian urban public transit in the 1940s." *Canadian Historical Review* 79(3):431–65. DOI: 10.3138/CHR.79.3.431.

Duffhues, J. 2010. "Transportation as a means for densification or the other way around?" Paper presented at Colloquium Vervoersplanologisch Speurwerk, Roermond, The Netherlands.

El-Geneidy, A., L. Kastelberger, and H.T. Abdelhamid. 2011. "Exploring the growth of Montréal's indoor city." *Journal of Transport and Land Use* 4(2): 33–46.

Federation of Canadian Municipalities (FCM). 2012. *Canadian Infrastructure Report Card 2012: Highlights*. Ottawa: FCM.

Filion, P. 2001. "Suburban mixed-use centres and urban dispersion: What difference do they make?" *Environment and Planning A* 33(1): 141–60.

——. K. McSpurren, and B. Appleby. 2006. "Wasted density? The impact of Toronto's residential-density-distribution policies on public-transit use and walking." *Environment and Planning A* 38: 1367–92.

Friman, M., L. Larhult, and T. Gärling. 2013. "An analysis of soft transport policy measures implemented in Sweden to reduce private car use." *Transportation* 40: 109–29.

Gim, T.-H.T. 2012. "A meta-analysis of the relationship between density and travel behavior." *Transportation* 39: 491–519.

Grimsrud, M., and A. El-Geneidy. 2014. "Transit to eternal youth: Lifecycle and generational trends in Greater Montreal public transport mode share." *Transportation* 41: 1–19.

Hanson, S., and G. Giuliano. 2004. *The Geography of Urban Transportation*. 3rd edn. New York: Guildford Press.

Heisz, A., and G. Schellenberg. 2004. "Public transit use among immigrants." *Canadian Journal of Urban Research* 13(1): 170–91.

Knowles, R.D. 2012. "Transit oriented development in Copenhagen, Denmark: From the finger plan to Ørestad." *Journal of Transport Geography* 22: 251–61.

Legacy, C., C. Curtis, and S. Sturup. 2012. "Is there a good governance model for the delivery of contemporary transport policy and practice? An examination of Melbourne and Perth." *Transport Policy* 19: 8–16.

Marzoughi, R. 2011. "Teen travel in the Greater Toronto Area: A descriptive analysis of trends from 1986 to 2006. *Transport Policy* 18(4): 623–30.

Naess, P., A. Strand, T. Naess, and M. Nicolaisen. 2011. "On their road to sustainability? The challenge of sustainable mobility in urban planning and development in two Scandinavian capital regions." *Town Planning Review* 82(3): 285–315.

Pembina Institute. 2014. *Fast Cities: A Comparison of Rapid Transit in Major Canadian Cities*. Toronto: Pembina Institute.

Perl, A., and J. Pucher. 1995. "Transit in trouble? The policy challenge posed by Canada's changing urban mobility." *Canadian Public Policy* 21: 261–83.

Pucher, J., and R. Buehler. 2005. "Why Canadians cycle more than Americans: A comparative analysis of bicycling trends and policies." *Transport Policy* 13(3): 265–79.

Tan, W. G. Z., L. B. Janssen-Jansen, and L. Bertolini. 2014. "Identifying and conceptualising context-specific barriers to transit-oriented development strategies: The case of the Netherlands." *Town Planning Review* 85(5): 639–63. DOI: 10.3828/tpr.2014.38.

——. 2014. "The role of incentives in implementing successful transit-oriented development strategies." *Urban Policy and Research* 32(1): 33–51. DOI: 10.1080/08111146.2013.832668.

Thomas, R., and L. Bertolini. (in press). "Defining critical success factors in TOD implementation using rough set analysis." *Journal of Transport and Land Use Planning*.

——. (in press). "Policy transfer among planners in transit-oriented development." *Town Planning Review*.

TransLink. 2012. *Transit-Oriented Communities Guidelines: Creating More Livable Places around Transit in Metro Vancouver*. Vancouver: TransLink.

Walker, J. 2012. *Human Transit: How Clearer Thinking about Public Transit Can Enrich Our Communities and Our Lives*. Washington: Island Press.

Walter, A.I., and R.W. Scholz. 2007. "Critical success conditions of collaborative methods: A comparative evaluation of transport planning projects." *Transportation* 34: 195–212.

Wheeler, S.M. 2010. "The evolution of urban form in Portland and Toronto: Implications for sustainability planning." *Local Environment: The International Journal of Justice and Sustainability* 8(3): 317–36.

Glossary

KEY TERM	CASE

Aboriginal rights include those described in historical treaties and modern land claims. They range from rights to hunt and trap and engage in traditional practices through to Aboriginal title to land, similar to fee simple. In Canada, Aboriginal rights are recognized and affirmed by Section 35 of the Constitution Act, 1982, and continue to be defined through jurisprudence. See also **Indigenous rights, Indigenous title.** 4.1

Accelerated vehicle retirement program These programs were originally designed to remove older vehicles from the road and replace them with newer vehicles with stricter emissions standards in order to reduce emissions from the transportation sector. 9.4

Action plan An action plan is a sequence of steps that must be taken, or activities that must be performed well, to accomplish a specified goal. An action plan generally includes steps, milestones, measures of progress, responsibilities, assignments, and a time line. 6.3

Active transportation is travel using human-powered means such as walking and cycling. Active commuting refers to regular travel patterns, e.g., between home and school. 6.4

Adaptive capacity is the ability of a human social system to adapt to new threats, trends, and environments. The capacity is determined by how well the system's institutions and networks learn from each other, interact, and store new information to cope with new threats. A system with higher adaptive capacity has a flexible and distributed decision-making framework that allows for stakeholder input. 4.2

Advocacy planning involves the planner taking on the role of advocating on behalf of stakeholders or groups that may not be able to make their own case, usually marginalized populations who are under-represented in the public debate. 9.1, 1.0

Alternative travel incentives Since most trips occur by automobile and most governments attempt to reduce automobile travel, the use of other modes of transportation is often referred to as alternative travel (mainly public transit, walking, bicycling, and car cooperatives). Incentives are typically measures to motivate people to change travel modes through adjustments in costs, ease of access, and use of alternative modes. 9.4

Amalgamation is a process whereby a central municipality is politically merged with adjacent municipalities. 9.1

Appreciative inquiry is a prevalent approach in community planning for inspiring positive social change. It emphasizes what is working well and encourages people to share stories and ask questions. 2.3

Arterials are wide, fast-moving roads designed for automobile traffic above other modes of transportation. 9.1

Articulated bus An articulated bus has two or more segments attached with flexible, accordion-like joints that allow it to turn corners. 9.1

Articulation is the layout or pattern of building elements, including walls, doors, roofs, windows, and decorative elements such as cornices and belt-courses. 7.5

Asian-theme mall A typical Asian theme mall has the following features that distinguish it from a mainstream shopping centre development: concentration of Asian/Chinese-oriented businesses, condominium ownership (vs. leasehold), the absence of conventional anchors (e.g., the Bay, Sears, restaurants, or grocery stores), variable store hours (vs. traditional norms and practices), no control of tenant mix, and smaller store size. In addition, these malls not only act as shopping destinations that offer diversity and choice in the general market but play an important role in immigrants' social lives. With a combination of retailing, dining, entertaining, and business and personal services, these malls serve as community hubs for immigrants and contribute to the retrofit of existing suburban neighbourhoods. 3.4

As-of-right means development that can be achieved under the existing zoning bylaw without any discretionary action needed, e.g., amendment to a bylaw or Official Plan. A proposal for as-of-right development must usually meet a list of specific criteria or conditions set out by the municipality before a development permit will be issued. See also **Zoning bylaws/schedules**. 3.3, 7.5

Blast walls There are many different types of blast walls, but each one acts as protection against a severe explosion. Blast walls will retain their structural integrity against a blast or fire that may result from an explosion. 7.4

Build-out A build-out is an urban planner's estimate of the amount and location of potential development for an area, based on land use or zoning assumptions about density (e.g., residential density, floor area ratio). In the Whistler case (Chapter 5.4), build-out refers to the limit on Whistler's accommodation capacity as part of the municipality's growth management framework. Under its current Official Community Plan, Whistler's accommodation capacity is set at a maximum of 61,750 bed units. 5.4

Bus Rapid Transit (BRT) is a system of dedicated-lane bus routes for which passengers can prepay and board through all doors. 9.1

Bylaw (or by-law) is a "secondary law" established by a local municipality to regulate issues in the spatial area under its jurisdiction. See also **Zoning bylaws/schedules**. 7.2 7.3 7.5 1.0

Canada Mortgage and Housing Corporation (CMHC) is a Crown corporation responsible for federal government housing policy and has a large mortgage and lending role. CMHC was founded in 1946 to implement the National Housing Act (1938) and was responsible for creating most social housing programs in Canada in the 1960s to 1990s. Today it focuses primarily on research and development and mortgage insurance. 5.4 5.3 1.0

Capacity-building refers to strengthening the competencies, skills, and abilities of individuals and organizations so that they can fulfill their developmental goals. For example, as municipal staff develop effective strategies to reach youth and collaborate, youth also learn about their community and about their own capacities for action through participatory processes. 6.4

Car cooperative Car cooperatives provide a fleet of vehicles to members spread out in a number of parking spaces across a city, thereby reducing the need for personal automobile ownership. Different fare structures exist, but members typically pay a registration fee, reserve the vehicle, and pay fees by time and/or kilometres driven. 9.4

Citizen engagement is the process of involving citizens in decision-making about planning-related decisions, especially including those most affected by such decisions. See also **Engagement plan**. 3.2

Climate change adaptation planning Change adaptation refers to actions that reduce negative impacts from climate change effects. Climate change adaptation can be either reactive, in response to climate change impacts, or proactive, in anticipation of potential future impacts. As scientific knowledge about climate change increases, and observed impacts become more severe, planners have been called on to incorporate climate change adaptation into planning practice to help minimize negative outcomes. A proactive approach to climate change adaptation can limit avoidable losses; the initial investment required for anticipatory adaptation generally yields long-term savings. 2.4 1.0

Climatic uncertainty With increasing and more intense climatic change, climate patterns are oscillating and reaching unprecedented levels. These oscillations are making it more difficult for climate models to accurately predict, thus justifying the need to incorporate some uncertainty into the modelling. 4.2

Collaborative planning is based on the premise that planning processes and outcomes can be improved by extending the decision-making process beyond individual planners and planning agencies. It may take various forms, from informal networks that include different agencies, civil society actors, and corporate players to formal consensus-based decision-making forums. 4.3 1.0

Columbia Basin Trust The Columbia Basin Trust (**CBT**) is an organization located in the Columbia Basin that provides support programs to improve social, environmental, and economic well-being for residents of the basin. It aims to create greater self-sufficiency for present and future generations by focusing on local priorities, bringing people together to collaborate, and supporting planning efforts. 4.2

Co-management refers to power-sharing between the state and another level of government, Aboriginal government, or group. The degree to which power is shared can range from advisory to joint decision-making. 4.1 1.0

Community Amenity Contributions CACs are in-kind or cash contributions provided by property developers when a city council grants development rights through rezoning. To lessen the impact of more residents or employees on community services, CACs pay for infrastructure, parks, libraries, childcare facilities, community centres, transportation, and cultural facilities. On large sites, CACs are often provided on-site through the development process. On small sites, CACs are typically a cash contribution, which the city uses to provide community facilities. 4.2 6.1 1.0

Community-based planning research is an approach to research in the planning field that is conducted at the community level, has some interest to that community, and is inclusive of that community in relevant research decisions. 3.2

Community gardens are organized gardening initiatives whereby sections of land are used to produce food or flowers in an urban environment for the personal use or collective benefit of their members who, by participating, share certain resources, such as space, tools, and water. 3.1

Consultation refers to a communication process through which participants receive information and express their views and opinions about determined issues. According to Roger Hart (1997), in the participatory approach, children should be informed about the way their views are taken into account. 6.4 1.0

Contextualization principles In case study methodology, **internal contextualization principles** place the case being studied in a social context that it (the case) operates within, i.e., aspects of certain events and their associated narratives. **External contextualization principles** place a focus on the broad context or setting within which the case is grounded, i.e., provincial or national policies. 3.2

Covenant A covenant is a restriction registered on the title of a property through the land titles office that governs the use and occupancy of the property. For example, in British Columbia they are called Section 219 covenants, and they run with the land, which means that they are transferred from owner to owner. 5.4 1.0

Crowdsourcing is the practice of obtaining needed services, ideas, or content by soliciting contributions from a large group of people and especially from the online community rather than from traditional employees or suppliers. 6.3 6.1

Crown consultation It is the common law duty of the Crown to consult, and where appropriate accommodate, when proposed conduct might adversely affect an established or potential Aboriginal or treaty right. 4.1

Cultural landscapes are defined as "Any geographic area that has been modified, influenced, or given special meaning by people" (Parks Canada 1994, 119). 4.1

Cultural Plan A Cultural Plan is commonly divided into four dimensions: community engagement (working with people), economic development (attracting tourists and creating jobs), land-use planning (altering the spatial layout of the city), and cultural industries (growing industries such as film, media, music, and design). 6.2 2.2

Culture Culture is a broad concept that encompasses the people, places, and things that reflect our community identity and channel creative expression, including cultural heritage, creative cultural workers, creative cultural industries, cultural organizations, festivals and events, natural heritage, stories, values and traditions, and cultural spaces and facilities. 6.2 2.2 1.0

Democratic planning refers to the normative idea that more public involvement in the planning process can lead to better planning outcomes. 7.4

Demonstration (or pilot) project A demonstration project is a one-time project, sometime limited in duration, to test a policy or program. Usually, a pilot project is evaluated to determine its strengths and weaknesses before being improved and adopted on a more permanent basis by planning departments and city councils. In the Dialogues Project context (Chapter 2.3), the term refers to a one-time initiative that aligns with the objectives of the BC Welcoming and Inclusive Communities and Workplaces Program (WICWP) and explores, implements, and evaluates innovative approaches to welcoming and inclusive communities. 2.3 1.0

Design charrette A design charrette is the creative design process that results from an intensive and rapid-paced public participation activity in which groups of experts and ordinary stakeholders work together to propose a solution to a particular issue. The most significant aspect of charrettes is the concept of "consensus." The word "charrette" is French for cart, and the origin of the term "design charrette" goes back to the nineteenth-century l'École des Beaux Arts where a cart was used to collect the students' assignment submissions from around the studio at the end of the day. In order to produce their drawings in the limited time allowed, the students not only continued to work until the last minute but also worked frantically and in a surge of energy before their work was carted away for final review by their professors. 3.1 1.0

Design guidelines A recent instrument of development control in Canadian towns, design guidelines are statements of intent as to the preferred use, density, and general appearance of new development. Typically, design guidelines are recommendations for developers rather than legally binding requirements such as performance standards or zoning bylaws. See also **Zoning bylaws/schedules**. 3.3 7.3

Development charges (or Development Cost Levies) New development increases demands on city services. Development charges (known as **Development Cost Levies** or DCLs in British Columbia) are a way for a municipality to fund the cost of growth related to residential, commercial, and industrial capital projects. DCLs are a fee paid by property developers as a condition of development, usually payable on the date that a building permit is issued. Fees are usually based on the square footage of new construction. DCLs help to pay for new services such as parks and childcare facilities. 6.1 8.1 1.0

Dissemination area Dissemination areas are the smallest scale at which census data are available. They each have a population of approximately 400 to 700 people but may vary significantly in geographic area depending on population density (Statistics Canada 2012). 2.4

Downloading refers to the process of decreasing funding from a higher-tier to a lower-tier government while increasing responsibilities for the lower-tier government. For example, the federal government reduced its transfers to the provinces and territories by $5 billion ($29.8 billion in 1995–6 to $24.6 in 1996–7), and the provinces cut their transfers to municipalities by $2 billion (from $10 to $8 billion) (McMillan 2003). 2.1 3.3 1.0

Emergency management planning Emergency management plans outline actions to prevent, prepare for, respond to, and recover from emergencies such as severe weather, natural disasters, industrial accidents, infectious diseases, or intentional acts. Considerations in emergency management plans include public safety, health and health services, critical infrastructure, and the environment. 2.4

Emission control technologies comprise a series of devices added to a vehicle to enhance its fuel efficiency and reduce GHG emissions per kilometre driven (e.g., catalytic converters). See also **Greenhouse gas emission reduction**. 9.4

Engagement plan An engagement plan details, for a given project, the public consultation objectives, audience, timeline, responsibilities, techniques, and measures of effectiveness. See also **Citizen engagement**. 6.3

Environmental analysis There is an inherent logic in the evolution of any city or town, and often this has something to do with landscape, topography, and hydrography. Environmental analysis allows better understanding of natural form and process and provides information that can contribute to the development of cities and towns with more environmental responsiveness, authentic identity, and sense of place. It is essential to the development of sustainable places and can also provide design determinants. 7.3

Environmental impact assessment The International Association for Impact Assessment (IAIA) defines an environmental impact assessment as "the process of identifying, predicting, evaluating and mitigating the biophysical, social, and other relevant effects of development proposals prior to major decisions being taken and commitments made." 4.1

Environmental justice means balancing the environmental risks and benefits of infrastructure or development, ensuring a fair and equitable distribution of benefits and burdens rather than assuming that some communities or members will absorb all of the negative externalities. The term has also been adapted to issues such as climate change, where communities in coastal or low-lying areas may be more affected by the effects of rising sea levels ("climate justice"). See also **Negative externalities**. 9.1

Ethnoburbs Ethnoburbs can be recognized as suburban clusters of residential areas and business districts in large metropolitan areas, especially in "global cities." Ethnoburbs are multi-ethnic communities in which one ethnic minority group has a significant concentration but does not necessarily constitute a majority. The establishment of the ethnoburb overlays an ethnic economy onto a local mainstream economy, which is then transformed into a global economic outpost and direct target for international capital investment, making the ethnoburb a vital area of economic activities and job opportunities (Li 1998, 504). 3.4

Exclusionary zoning is a form of zoning associated with the explicit exclusion of various socio-economic groups or certain types of development from a particular development or neighbourhood. For example, security measures that delimit urban spaces and prescribe how space can be used, or housing policies that do not allow low-income housing development, are two types of exclusionary zoning. 7.4

Expedited approvals process Development applications go through a process of approval during which a planner with experience in zoning and bylaws examines all elements of the proposed project to determine whether it is in agreement with existing regulations. This process can be lengthy depending on the number of applications the municipality receives during a given period of time. A common tool that municipalities use to signal their planning and development priorities is to deal with certain types of development proposals first, minimizing the time they spend waiting for approval. For example, the City of Toronto prioritized development applications for mid-rise buildings on the Avenues (Chapter 7.5). 7.5 8.3

Facade A facade is the exterior wall of a building. 7.5

Festival marketplace refers to a leading development form in the 1970s and 1980s, particularly in the revitalization strategies of urban waterfronts, and typically included retail and souvenir shops, a food court offering international cuisine, restaurants, and live entertainment. The focus was typically on drawing tourists and visitors to the area. 7.1

First Nations First Nations is an umbrella term used to designate all the original inhabitants of the territories now known as Canada who are neither Inuit nor Métis. In the Dialogues Project context (Chapter 2.3), the term refers to the Musqueam, Squamish, and Tsleil-Waututh Nations. 2.3 1.0

First-ring suburb (or inner-ring suburb) This type of suburb is usually located very close to the downtown and is characterized by a relatively fine-grained urban fabric, with a grid pattern of residential development—single-family detached and semi-detached homes—centred on a commercial main street. Many first-ring suburbs, built until the early 1950s, are well served by public transit and other services. Second-ring suburbs, typically built in the 1960s and later, were usually located farther out from the city centre, with much more dispersed, single-family homes on large lots and few land uses other than residential, and were designed around car accessibility. In the case of a region such as the Greater Golden Horseshoe in Ontario, the inner ring refers to those municipalities sharing a boundary with the largest central municipality. See also **Upper-tier, single-tier, and lower-tier municipalities**. 3.3 9.1 8.4

Fordist refers to a political and economic arrangement based on a large number of high-paying manufacturing jobs, a strong middle class, and suburban home- and auto-ownership, which flourished in North America in the second half of the twentieth century. The term comes from Henry Ford, whose assembly-line manufacturing of cars revolutionized the industry. Many Western countries began to make the transition to post-Fordist economies during the 1970s. See also **Post-Fordist**. 9.1 1.0

Free entry system refers to the system through which the Crown allocates mineral rights. It is characterized by: a right of free access to lands in which the minerals are in public ownership, a right to take possession of them and acquire title by one's own act of staking a claim, and a right to proceed to develop and mine the minerals discovered. 4.1

Frequency of service Known in transit planning as headways, this is the amount of time between vehicles travelling along a transit route; in other words, shorter headways mean more frequent service and shorter waiting times. 9.1

Gentrification refers to the gradual displacement of lower-income populations to higher-income populations because of the rising costs of housing, land, goods, and services in urban neighbourhoods. Gentrification was described as long ago as ancient Rome and Britain, and sociologist Ruth Glass coined the term in the 1960s to describe middle-class people buying and renovating modest homes in working-class districts. 9.1 1.0

Geographic Information System A Geographic Information System (GIS) is a system designed to capture and analyze geographic data. A GIS allows multiple layers of information to be displayed on a single map and can be a valuable tool for planners in analyzing complex information and informing planning decisions. 2.4

Greenhouse gas (GHG) emission reduction As part of the Kyoto Protocol, countries are required to design policies, programs, and strategies to reduce the greenhouse gas emissions that are responsible for climate change. All levels of government can contribute to enacting and carrying out such strategies. 9.4

High-growth areas are communities experiencing rapid population growth. 5.1

Housing continuum A housing continuum is the spectrum of housing in a community, involving income, economic and social independence, and level of support needs. It ranges from no house, to emergency shelters/supportive and transitional housing, to government-supported housing, to non-profit/cooperative/community-sponsored, and to private-market rental and ownership. 5.1

Housing First Housing First is an approach to homelessness based on the concept that the primary need of the homeless is stable housing and that other issues facing the household should be addressed after housing is obtained. 2.5

Housing Plan A Housing Plan is intended to provide information to better understand housing needs across the entire continuum within a community or region and to provide a model for residential development that is effective, efficient, and sustainable. 5.1 1.0

Ideation tool Ideation is the creative process of generating, developing, and communicating new ideas. Ideation tools support and facilitate the creative brainstorming and idea-creation process. 6.3

Immigrants Generally, people who were not born in Canada and were not Canadian citizens by birth (first-generation); those born in Canada to immigrant parents are known as second-generation immigrants. In some cases, such as the Dialogues Project context (Chapter 2.3), the term refers to all people who are not First Nations, Inuit, or Métis, regardless of when they or their ancestors arrived in the area. 2.3 1.0

Impact pathway refers to a conceptual diagrammatic approach to illustrate the linkages between a climate change hazard (e.g., flooding) and how the hazard may affect the community's social, economic, environmental, and cultural fabric. It is used to help communities visually understand climate risks that complicated climate models would not be capable of doing. 4.2

Income and asset limits are the upper limits on income and assets that determine the ability to pay for housing. Individuals and families above the set income limit are expected to be able to afford private-market housing; as well, those with total assets above the asset limit are expected to liquidate their assets to fund their housing costs. See also **Income testing**. 5.1

Income testing means restricting eligibility for housing to those with incomes below a certain level. For example, applicants for social housing units are required to provide evidence of their income to determine their eligibility for social housing. 5.4

Incremental planning is a planning process whereby plans are implemented in small steps, incrementally, in response to the political and economic feasibility at any given time. Plans are constantly being modified and delayed in this process. 9.1

Indigenous rights are the practices, customs, and traditions that are integral to the distinct culture and society of an Indigenous group and that were exercised at the time of European contact; the courts have stated that these rights are not absolute and may be infringed upon in certain circumstances. See also **Aboriginal rights**. 4.3

Indigenous title is a particular form of Indigenous right that refers to the right to the land itself; it is a communally held property right, and any infringements on that right must not disrupt Indigeĺnous peoples' unique relationship to their traditional territory. Indigenous title is often interpreted as giving rise to some form of decision-making right over how traditional territories are used and managed. See also **Indigenous rights**, **Aboriginal rights**. 4.3

Infill development Infill is the use of land within a built-up area for further construction. Infill buildings are constructed on vacant or underutilized property or between existing buildings. Infill development can make more efficient use of existing services and provide a choice of housing types. 6.1

Inner-ring suburbs or **Inner ring**. See **First-ring suburb**. 9.1

Interprovincial migration is the movement of people from one province or territory to another, involving a change in the usual place of residence. 5.1

Jersey barriers The Jersey barrier takes its name from the US state where it was designed, New Jersey. Designed in the 1950s, the concrete barrier was initially developed as a divider for multi-lane highways. It has since been repurposed to act as protection against any vehicle suicide bomb attack. 7.4

Kitchen table meetings A kitchen table meeting or discussion is a small, informal group that meets in someone's home to talk, share, listen, and build on subjects of mutual interest (State Government of Victoria 2013). Because kitchen table discussions are less formal than a typical public participation process, participants are usually more willing to share their ideas. These discussions are led by a facilitator who fosters an environment in which people's contributions are valued and participants are encouraged to listen. 4.2

Land claim agreements These agreements are negotiated in areas of the country where Aboriginal rights and title have not been addressed by treaty or through other legal means. They are modern treaties between Aboriginal claimant groups, Canada, and the relevant province or territory. 4.1

Light Rail Transit (LRT) refers to rail-based transit running on a dedicated track, usually at grade (street level), which allows passengers to prepay and board at all doors. 9.1

Local bus refers to bus routes that have stops approximately every 500 metres and run in mixed traffic. 9.1

Low Income Cut-off There is no official poverty rate measure or definition of what constitutes "poverty" in Canada. Statistics Canada's Low Income Cut-off (LICO) rate is not a measure of poverty per se; it is a relative measure, calculated by household type, and fluctuates slightly. Its data, however, provide a useful indicator of individuals and families in the community who spend a disproportionate share of their income on basic necessities such as food, clothing, and shelter (e.g., City of Richmond website). For example, about one-fifth of the Canadian population lives below the LICO, and lone parents in BC are at approximately three-fifths of the LICO. 2.1, 2.5

Master plan A master plan is a strategic document (or series of documents) that set(s) out a co-ordinated planning and design framework for a defined spatial area. Master plans typically contain a combination of two- and three-dimensional visualizations, design and development principles, and strategies for implementation. 7.2, 8.1, 1.0

Microblogging is a broadcast medium that differs from a traditional blog in the length of its entries, which are shorter and allow users to exchange small elements of content. Examples include Twitter, Tumblr, and the status update feature on Facebook. 6.5

Mid-rise building generally refers to a building that is five to 12 storeys or up to a height that is no taller than the right-of-way width of the street on which it is located. 7.5

Mixed methods research Mixed methods research combines quantitative and qualitative analysis. 7.5

Mixed-use building refers to a building or set of buildings used for multiple purposes. It may include a combination of residential, employment, retail, or institutional uses and often includes retail on the ground floor with office or residential uses above. 2.4

Mobile apps are applications designed for cellphones, tablets, laptops, and other mobile devices to allow the creation and exchange of user-generated content. Some apps are location- and time-sensitive (e.g., Foursquare) while others are not (e.g., YouTube). 6.5

Modernism was an architectural movement beginning at the turn of the twentieth century and becoming popular after the end of World War II. The movement was characterized by its bold, simple style and use of futuristic materials such as concrete, glass, and steel. One major criticism of modernist architecture, appearing in works such as Jane Jacobs's *The Death and Life of Great American Cities* (1961), was its concentration on non-traditional forms at the expense of users, e.g., large, monolithic buildings that took up entire city blocks and presented a blank facade to pedestrians on the street. These buildings, such as high-rise apartment blocks and office towers, were often built after the demolition of the smaller, fine-grained buildings characteristic of historical cities. Many of the modernist housing projects, such as Regent Park in Toronto, were poorly designed, lacked through streets, amenities, and open spaces, and were thought to concentrate poverty. By the 1990s, they began to be demolished, redesigned, and rebuilt. 5.2, 8.3, 1.0

Mode share refers to the proportion of users of each transportation mode (e.g., cycling, public transit, car) in a population. It can be calculated for specific trip types such as the commute or for overall travel. Statistics Canada's (2008) 2006 national census reported that 80 per cent of commuters (journey to work) were drivers or passengers of automobiles, 11 per cent used public transit, and the remainder mostly walked or cycled. However, mode share for public transit is much higher in Canada's large municipalities. 9.4, 9.2

Mode shift refers to the act of changing travel mode for an individual or the objective of policies and infrastructures that attempt to change the proportion of users of each travel mode in a population. Many municipalities have adopted policies or programs to encourage a mode shift away from driving. See also **Mode share, Travel Demand Management** (TDM). 9.4

Multi-stakeholder planning is a specific type of collaborative planning in which representatives from various interest or stakeholder groups directly participate in the development of the plan, often through a series of face-to-face meetings. Stakeholders usually receive policy direction and support from a planning professional, who may also facilitate stakeholder meetings. See also **Stakeholder**. 4.3

Municipal Development Plan See **Official Community Plan** (OCP). 7.3

Natural growth Natural growth includes the births and deaths in a population and does not take into account migration (e.g., interprovincial or across provinces) or immigration (from other countries). 5.1

Needs assessment A needs assessment is a systematic process for determining and addressing needs between current conditions and desired conditions. 5.1

Negative externalities are the negative impacts of a project or industry that are generally not taken into account when planning or running the project or industry, e.g., various forms of pollution. 9.1

New urbanism is an urban design movement that arose in the US in the 1980s and was formalized in the Congress for New Urbanism (1993). The movement is characterized by architectural forms common in American communities before the 1950s, including front porches, rowhouses, and garage apartments. New urbanism is based on the larger ideas of regional planning for open space, mixed-use development, and pedestrian-oriented design such as the use of the grid street pattern, narrow streets, and well-screened parking. New urbanism includes principles such as transit-oriented development (TOD) and traditional neighbourhood development (TND). Some Canadian examples of new urbanism include the Cornell development in Markham, Ontario, and Simon Fraser University's UniverCity neighbourhood. See also **Traditional main-street zoning**. 5.2, 1.0

Nimbyism NIMBY (an acronym for "not in my backyard") describes the views of citizens who object to new developments in their neighbourhoods. Nimbyism describes the phenomenon of citizen protest, particularly to developments that may affect their property values but may have no other negative effects. 3.3, 5.4, 6.1, 1.0

Official Community Plan (OCP) An Official Community Plan is a comprehensive plan created by a community or region that dictates public policy on a broad range of topics over a long term (25 to 30 years), typically reviewed every five years. This type of plan is created in accordance with provincial Planning Acts, has slightly different names in each province: in British Columbia and Saskatchewan, it is called an Official Community Plan; in Ontario, it is known as an Official Plan, and in Alberta as a Municipal Development Plan. For example, in British Columbia local governments and districts are permitted to adopt an OCP through Part 26 of the Local Government Act (LGA). These plans are used to assist local government or district municipalities in addressing planning issues such as housing, transportation, infrastructure, parks, economic development, and, more recently, climate change adaptation. Under the BC Local Government Act Section 875, an OCP is "a statement of objectives and policies to guide decisions on planning and land use management, within the area covered by the plan, respecting the purposes of local government." This provincial legislation provides local governments and districts with the flexibility to respond to the different needs and changing circumstances of their communities. 2.1, 3.3, 4.2, 5.1, 5.4, 6.1, 7.2, 9.3, 1.0

Official Plan See **Official Community Plan (OCP)**.

Open data The term "open data" generally refers to the concept of making data free to anyone to use, re-use, and redistribute. Recently, there has been a movement to make government data open on the basis of transparency and accountability and to allow third parties to develop applications and services that address public and private demands. The Canadian Government Open Data website (www.data.gc.ca) was established in 2011 and contains a list of municipal and provincial open data websites. 6.5

Orthomaps Orthomaps are mosaics of aerial photographs that are produced through a complex and costly process involving the acquisition of aerial photographs and ground control data, aerial triangulation, and sophisticated processing of the raw images in association with a digital elevation model. As such, orthomaps are geometrically corrected to yield a uniform and undistorted scale akin to a map and thus represent true distances and measures. 3.1

Overlook An overlook is the view to adjacent properties from higher residential units or communal outdoor spaces (e.g., patios). 7.5

Pacific Climate Impacts Consortium (PCIC) The Pacific Climate Impacts Consortium, based at the University of Victoria, is a regional climate service centre that provides information about short- and long-term physical impacts of climate change in various regions of British Columbia and Yukon. PCIC works closely with regional stakeholders to develop practical information and tools that support long-term planning for climate change. See also **Stakeholder**. 4.2

Participatory planning is planning that takes the opinions and input of the public into account. Generally, this involves public meetings but increasingly other forms of media like the Web. In Canada, public participation in many municipal planning processes is required by provincial Planning Acts. 9.1 1.0

Pedestrian-oriented refers to public-realm design that makes pedestrian movement safe, attractive, and comfortable for all ages and abilities. See also **Public realm**, **Walkability**. 7.5 1.0

Placemaking is a term used to describe a multi-faceted approach to land-use planning that employs community engagement, visioning, and spatial masterplanning to engender design sensitivity in the development/redevelopment process. See also **Public realm**, **Sense of place**. 7.2

Podcasts Based on the word "broadcast" and the popularity of the iPod in the early 2000s, a podcast is a program available in digital format (usually audio or video) for immediate Internet downloading. Users can subscribe to specific shows online and receive regular podcast updates or stream them online. 6.5

Polarization is the phenomenon in which the middle of a spectrum migrates to the extremes; for example, income polarization is when the middle class is increasingly separated (or polarized) into lower and upper classes. Spatial polarization in cities can be seen where increasingly concentrated groups are located in particular areas. Spatial polarization can occur by income, by race, by dwelling type, and by other factors. 9.1

Post-Fordist refers to the political and economic system that has emerged since Fordism was weakened by globalization and other structural changes. Post-Fordism is associated with the rise of the service economy, income polarization, reurbanization, and the loss of manufacturing jobs. 9.1

Power centres are unenclosed shopping centres of between 250,000 square feet (23,000 m^2) and 750,000 square feet (70,000 m^2), usually containing three or more big-box retailers and various smaller retailers located in strip plazas, with a common parking area shared among the retailers. It is likely that more money is spent on features and architecture than is the case for a traditional big-box shopping centre. Power centres are primarily designed around car accessibility and are often on the outskirts of a municipality. 8.3

Precedents Precedent study is a method used in planning and design in which past examples are studied in order to develop new ideas and concepts. Precedents should be selected based on their relevance and appropriateness to the scale, context, and scope of the project under consideration. 7.3

Precinct A precinct is a place developed around the activities and identity of a specific geographic or demographic group. 7.1

Preliminary screening is an initial examination of a development proposal to determine whether the project might have significant adverse environmental, social, and cultural impacts or be the cause of public concern. The purpose is to determine whether further, more in-depth assessment is required. If not, the application proceeds to the permitting phase. 4.1

Privately owned public space refers to an amenity provided and maintained by the property-owner for public use, such as public plazas, atriums, and the common areas inside shopping malls. See also **Public space**. 7.1

Private spaces are places controlled by individuals, organizations, or companies to offer users privacy, exclusive use, and protection from the intrusion of others. 7.1

Public interest There are at least two schools of thought when defining the public interest. One interpretation suggests that it is impossible to define the public interest, since there are many publics, and that a planner must balance competing interests to determine the most significant set of concerns. A second and more conventional way of understanding the public interest suggests that it is determined through elections. In other words, elected officials are given a mandate by the citizenry to carry out plans and policies that are within the electorate's public interest. Planning in Canada focuses on the public interest, while in the United States planning focuses on protecting individual property rights. 7.4 9.3 1.0

Public life Public life is the political and social pursuits outside home and domestic (private) life. 7.1

Public markets are markets in privately or publicly owned spaces where independent merchants sell food items and handcrafted goods. 7.1

Public–private partnerships (PPPs) are government services or private business ventures that are funded and operated through a partnership of government and one or more private-sector companies. In projects that are aimed at creating public goods, as in the infrastructure sector, the government may provide a capital subsidy in the form of a one-time grant so as to make it more attractive to the private investors. In other cases, the government may suppor the project by providing revenue subsidies, including tax breaks, or by removing guaranteed annual revenues for a fixed time period. In theory, PPPs enable the public sector to harness the expertise and efficiencies that the private sector can bring to the delivery of certain facilities and services traditionally procured and delivered by the public sector. PPPs are structured so that the public-sector body seeking to make a capital investment does not have to do any borrowing. In practice, PPPs have had mixed success in reducing public costs. 9.4 1.0

Public realm The public realm includes spaces that are perceived as being publicly accessible, including publicly owned streets, sidewalks, pathways, rights-of-way, parks, publicly accessible open spaces, public and civic buildings and facilities, and building forecourts. 7.5 8.3 7.3 8.1

Public spaces are places accessible to everyone with little or no restrictions on users, who go there for individual or group activities. See also **Privately owned public space**. 7.1

Rational comprehensive model (RCM) RCM is a model in municipal planning decision-making that takes a scientific approach to problem-solving; a full analysis is employed to assess all possible variables affecting a given set of circumstances and all possible alternatives to resolving the problem under consideration. This model leaves little room for meaningful public involvement in decision-making and tends to rely on quantitative, scientific evidence rather than other types of information, such as interview data or community preferences. 4.2 9.1 1.0

Rent Geared to Income (RGI) Rent Geared to Income housing is subsidized housing. The rent is based directly on the tenant's income, usually 30 per cent of the gross monthly household income. If a tenant receives social assistance, the rent charges are based on the rent benefit set by the provincial government rather than 30 per cent of the gross monthly income. RGI housing subsidies are most often available in publicly owned social housing but are also available in cooperative, non-profit, and private housing. 5.2 5.3

Resilience In the case of a neighbourhood, town, or city, resilience refers to the ability to maintain its identity and integrity in the face of fluctuating environmental and human use and to recover from change or misfortune. Resilience can refer to physical or social factors. 7.3 1.0

Right-of-way refers to the part of the street space that is publicly owned and lies between the property lines, including the roadway and the sidewalk area. 7.5 9.2 7.2

Right to the city (Droit à la ville) is the idea that the public space of the city is important both symbolically and practically and belongs to all people. 9.1

Risk communication In the context of climate change, risk communication is a detailed portrayal of the historical, current, and, if possible, future risks surrounding a climate hazard such as a flood, heat wave, or drought. Risk communication acknowledges uncertainty and conveys this using means such as probability estimates and visualizations to an audience of non-scientists. 4.2

Rural health service planning refers to planning processes related to any type of health service that has its greatest relevance in a rural context. 3.2

Secondary frontage is the side of a retail or commercial building that is considered secondary to the primary facade used for entry and display of goods, typically on corner lots. See also **Facade**. 7.3

Security is the condition of being free from danger. In terms of planning, it can be understood as the reactionary and precautionary measures taken to address a crisis or emergency or to prevent the possibility of a disruptive incident such as a riot or a terrorist act. 7.4

Sense of place refers to how people perceive or feel about a place, space, or neighbourhood. Often, sense of place is understood as the collective character generated by a combination of the physical and social characteristics of the public realm. See also **Public realm**, **Placemaking**. 7.2

Setback refers to the distance between a property line and the front, side, or rear of a building. 7.5 7.2

Settler The term generally refers to people of European descent whose ancestors were among the early colonizers in Canada. 2.3

Shovel-ready is a term used by government officials to describe a construction project (usually larger-scale infrastructure) when project planning, engineering, and funding have advanced to the stage where labourers may immediately be employed to begin work. The term originated in the United States at the time of the economic crisis beginning in 2008. 5.3 8.1

Smart Growth is a type of urban development inspired by a movement of urbanists, urban planners, and architects that promotes compact urban growth following neo-traditional concepts, e.g., small-scale, mid-rise, mixed-use buildings on main streets and development of transportation corridors where new growth will be concentrated. See also **New urbanism**. 3.3 8.4 1.0

Social Determinants of Health The Social Determinants of Health are a set of social conditions that contribute to marginalization and exclusion, which in turn cause negative health outcomes. A strong body of Canadian health and social sciences research demonstrates that these social determinants, rather than lifestyle choices or medical treatments, have the most profound impact on health. 2.4

Social exclusion refers to the exclusion of various segments of the population from urban and lifestyle amenities and opportunities, whether because of economic, social, or institutional barriers or other forms of lack of access. 9.1

Social housing is housing developed under federal or provincial government programs that provide housing affordable for low- and in some cases moderate-income households through the provision of an ongoing operating subsidy. Some or all residents may pay rent based on their income, usually 30 per cent of income. These programs were terminated in 1993, and no new units are being created in this way. Existing social housing projects continue to operate, and many will stop receiving operating subsidies in the next 20 years as the agreements expire. 5.4 5.2 7.2 1.0

Social media are a group of Internet applications that allow the creation and exchange of user-generated content, built upon the ideas and technology of Web 2.0. They depend on mobile and Web based technologies to create interactive platforms allowing individuals and groups to share, co-create, and discuss and modify content. Examples include Facebook and Twitter. 6.5 6.1 6.3

Social mix In the area of housing, social mix refers to combining social and market housing to produce a community that is economically, socially, and culturally heterogeneous. 5.2

Social planning Social planning addresses community needs and interests and seeks solutions to problems by building on community assets. Activities may include research, education, and advocacy related to social, environmental, economic, and health issues. Values inherent in social planning include social justice, equity, and inclusion. 2.4 2.5 1.0

Social urbanism means extending services and infrastructure to excluded and marginalized neighbourhoods with the goal of improving access to opportunity and equity. 9.1

Social vulnerability Social, cultural, and political forces accord different levels of access to resources to different groups in society. Socially vulnerable people are those who have less access to social, economic, and political resources and may experience marginalization. Social vulnerability is not an inherent trait but rather the result of social, cultural, political, and historical conditions. Therefore, the factors that contribute to social vulnerability may vary from place to place. 2.4

Social vulnerability to climate change and natural hazards Social vulnerability to climate change and natural hazards is the interaction between climate change/natural hazard impacts and pre-existing social vulnerability. The factors that make people socially vulnerable make them socially vulnerable to climate change and natural hazards by reducing their access to the resources they need to prepare for, withstand, and recover from adverse impacts 2.4

Spatial mismatch refers to the location of jobs far away and inaccessible to the populations that need them. Researcher J.F. Kain theorized in the 1960s that low-income African-American populations living in segregated inner-city neighbourhoods in the United States would not have access to higher-paying jobs in the rapidly growing suburbs. 9.1

Stakeholder A stakeholder is an individual or group of individuals with an interest or "stake" in the process and outcomes of planning; unlike government, a stakeholder has no direct political authority or jurisdiction over the decision. See also **Multi-stakeholder planning**. 4.3 6.2 9.3 1.0

Standard deviation Standard deviation is a term used in statistics and probability theory that refers to the measurement of the dispersion of a set of data from its mean. It is a mathematical formula for the average distance from the average of a set of numbers. 2.4

Stepback refers to the setting back of the upper storeys of a building. 7.5

Strategic natural resource planning is an integrated approach to the management of Crown or public lands that considers a range of natural resource uses (e.g., forestry, recreation) and seeks to address potential conflicts through land-use zoning; it often involves the designation of new protected areas. 4.3

Subdivision is a planning process that divides large areas of land into separate parcels in readiness for real estate development. The subdivision process typically leads to a "plan of subdivision" that denotes each individual parcel as well as the organization of roads, sidewalks, sewage, and other services. 7.2 7.3 1.0

Tendering process is the process by which bids are invited from interested contractors to carry out specific packages of construction work or refurbishment or maintenance of an existing facility. The most commonly used method is single-stage selective tendering based on the invitation of firms from a pre-approved list, chosen because they meet certain minimum standards with respect to financial standing, experience, capability, and competence. The competition element of the tender is provided on the basis of price and quality. In Canada, anyone who wants to bid on work for a public agency must be able to do so, provided that the goods, services, or construction being proposed are above a certain value (e.g., $100,000 for services). The work must be advertised broadly, typically through a request for proposal (RFP), request for information (RFI), request for tender (RFT) on a public website, and the procurement process must be transparent and fair to all bidders. Legislation to regulate public procurement in Canada is in place in only 8.3 5.3

a few Canadian jurisdictions, and there is no Canada-wide statute that applies uniformly.

Tokenism means making a minimal or merely symbolic effort to include the public as participants in decision-making processes. Unlike authentic participation, tokenism leads to frustration and represents a form of manipulation (Arnstein 1969). 6.4 1.0

Traditional knowledge refers to knowledge and values that have been acquired through experience, observation, from the land or from spiritual teachings, and handed down from one generation to another. 4.1

Traditional mainstreet zoning is a form of zoning inspired by the new urbanism and Smart Growth movements to ensure that main streets in a city conform to design concepts such as small-scale buildings, adequate provision for pedestrians, and decreased space for cars. For example, in Ottawa (Chapter 3.3) this particular form of zoning applies to a number of streets that run through neighbourhoods developed primarily prior to 1945 and intends to foster a "traditional" urban form characterized by the prevalence of small-scale, mixed-use buildings set close to the street (City of Ottawa 2006). See also **New urbanism**. 3.3

Traditional territory refers to a geographic area that has been identified by an Indigenous group as being the lands that were occupied and/or used by their ancestors. 4.3

Transit-oriented development (TOD) means developing and organizing land use around transit hubs, stations, and stops to increase the potential of public transit to attract users and provide work, commercial, and recreational destinations within walking/cycling distance. TOD is strongly linked to **New urbanism** and **Smart Growth**. 9.4 9.1 1.0

Travel Demand Management (TDM), or Transportation Demand Management, involves measures used to reduce automobile ownership and facilitate travel reductions (distance and time travelled), emission reductions, and mode shift. TDM measures can influence the quality and quantity of transportation infrastructure, adapt land use to surrounding transit systems, influence attitudes toward driving, or modify the taxation policies and fee structure of travel options. Many municipalities and regions now employ TDM measures as well as building new infrastructure to manage growing transportation demand. See also **Mode shift**. 9.4

Upper-tier, single-tier, and lower-tier municipalities Ontario has a two-tier system of local government in place for much of the province. Upper-tier municipalities, typically referred to as regions or counties, include multiple lower-tier municipalities (e.g., the Region of Peel includes the City of Mississauga, the City of Brampton, and the Town of Caledon). The upper-tier municipality is typically governed by councillors either directly elected or appointed from among the elected representatives of each lower-tier municipality. The upper-tier municipalities deliver certain services for the lower-tier municipalities within its geographical boundaries. In some cases, a single tier exists in which the functions of both the upper and lower tiers are vested in one municipal government. These municipalities, such as the City of Toronto and the City of Hamilton, are referred to as single-tier municipalities. 8.4

Urban Aboriginals Urban Aboriginals is a term imposed by the Canadian government to refer to First Nations, Inuit, and Métis peoples residing in urban areas. 2.3

Urban revitalization is a process that aims to reverse an area's decline through social and economic policy measures and restoration of the built environment. 7.1

Vehicle inspection and maintenance program These programs were developed to reduce the negative impacts of older vehicles by ensuring that pollution control technologies are in good working condition and by making their upgrade mandatory. 9.4

Walkability is the degree to which an environment allows and encourages the activity of walking, the most simple form of circulation for humans. Factors that positively affect the size of the "walkshed," which is the area that a person can reach conveniently and within a reasonable time on foot, include a permeable block pattern (such as the grid), continuous sidewalks or paths, the presence and quality of streetscape elements such as street trees, lighting, and signage, and the location of destinations such as stores, services, and amenities. See also **Pedestrian-oriented**. 7.3 1.0

Windfall profits are unexpected profits obtained through unforeseen circumstances and generally as a result of activities not controlled by the recipient. In the Whistler case (Chapter 5.4), this would refer to appreciation or difference between the cost of purchasing and the sale price of a home. 5.4

Youth The United Nations defines youth as persons between the ages of 15 and 24. 6.4

Zoning bylaws/schedules are tools used by local governments to manage the use of land. Zoning schedules specify allowable land uses, height, setbacks, and floor space. To create a zoning schedule, a city council is required to hold a formal public hearing before approval. Once a zoning schedule is approved, city officials are permitted to approve developments that comply with the schedule. Developments that do not comply with existing zoning must be approved through a public hearing before they can proceed. See also **As-of-right**, **Exclusionary zoning**. 6.1 7.2 7.3 3.3 1.0

Index

Aboriginal communities: Kwakwaka'wakw Nations, 172; Nanwakolas Council, 169–70; 172–4, pre-contact, 6; Tsawwassen First Nation, 24; Wanapitei First Nation, 329; Whitefish Lake First Nation, 329; Yellowknives Dene First Nation, 145–54
Aboriginal governments, 17, 24, 29, 169; government-to-government approach, 32, 168–74; land claims, 143; sovereignty, 152; *see also* treaties
Aboriginal people: intercultural understanding, 75–82; lone-parent families; 96, 98; on reserve, 329; oral traditions, 78; participation in planning, 24, 78, 144, 169, 332; Social Determinants of Health, 88; traditional territories, 75, 169; urban, 329
Aboriginal rights, 32, 145–54, 168–74
accelerated vehicle retirement program, 405–12
active transportation, 40, 254–5, 273, 371; for children, 104
activism, 17, 65, 111, 397, 402
activity economy, 276
Adams, Thomas, 10–11
adaptive capacity, 157–8, 164
Adieu Bazou!, 405–12
Adrian Smith + Gordon Gill, 390
advocacy approach to planning, 23, 382
affordable housing crisis, 22, 177
age-friendly practices, 55
Agricultural and Rural Development Act, 16
Agricultural Rehabilitation and Development Act, 16
Alberta, 9, 11, 16, 20, 116–24; Green Transit Incentives Program, 373
amalgamation, 131, 330, 374–5
American pragmatism, 29
Ames, Herbert Brown, 9
AMO, 390
anti-terrorism security measures, 304–14
appreciative inquiry, 76, 80
archaeological sites, 146, 149–50
architectsAlliance, 390
Arnstein, Sherry, 31
Art Gallery of Sudbury, 336
Artibise, Alan, 6
arts: public funding, 68, 73; *see also* culture
Asian-theme malls, 134–42
as-of-right development, 129, 318
aspirational branding, 245
Assiniboine Park, 9
Atlantic Climate Adaptation Solutions Association (ACAS), 50, 86, 93
Atlantic Development Board, 16
Avon River, 9

Baker House, 201–6
Banff, AB, 9, 210, 213
Bell, Carole, 138
Bellefair Kew Beach Residences, 322
Benalto, AB, 290–303
best practices, 4, 52, 59, 204
bicycle lanes, 40, 380
Bjarke Ingels Group, 390
blogs, 262
Booth, Charles, 32
Brabham, Daren C., 246
Brandon, MB, 7
Brière Gilbert + Associés, 353
British Columbia; and Aboriginal governance, 168–74; central coast, 168–74; extreme weather events, 156; Housing Renovation Partnership, 199; SCRAP-IT program, 408; treaty process, 172; Welcoming and Inclusive Communities and Workplaces Program, 76
British North America Act, *see* Constitution Act
British Planning Act, 11
Broadbent, Marianne, 119
Brook McIlroy Planning + Urban Design, 315, 318
brownfield redevelopment, 130, 224
Brundtland Report, 21, 37, 132
Bumbaru, Dinu, 352
Bunce, Susannah, 127
Bureau d'aménagement de l'est du Québec, 16
Burgess, E.W., 32
Bus Rapid Transit, 376–7

Calgary, AB, 7, 14; social housing retrofits, 200–6
Calthorpe, Peter, 37
Cambrian College, 330; First Nations College Experiences Program, 329
Cambridge, ON; community gardens, 108–13; rapid transit, 373
Campbell, Gordon, 222, 224
Canada Health Act, 123
Canada Land Inventory, 295
Canada Mortgage and Housing Corporation (CMHC), 13–14, 36, 176–7, 182; Neighbourhood Improvement Program, 19; social housing programs, 21, 198–200, 203, 209; statistics, 181–2; Residential Rehabilitation Assistance Program, 19
Canada Water Act, 18
Canada's Economic Action Plan, 178, 196
Canadian Charter of Rights and Freedoms, 283
Canadian Federation of Municipalities, 97
Canadian Institute of Planners (CIP), 4, 14, 25; Planning Excellence Awards, 335, 339
Canadian Pacific Railway, 6–7, 273, 330, 370
Canadian Public Health Association, 21
Canadian Urban Institute, 70
car culture, 21, 23, 40, 254–5; shift away from, 35, 255, 405–12; youth, 254
car ownership, 16, 90, 375
Carbon Tax Act, 156
car-oriented development, 25, 27, 269, 270, 275
Carrefour action municipale et famille, 252–3
car-share programs, 406, 410
case study; definition of, 118; as method, 4–5, 116–24; types of, 117
census data, use in planning, 88–9, 94
Central Coast Land and Resource Management Plan, 168–74
Central Park, 9
Charlottetown, PEI, 6
charrettes, 24, 38, 105, 107–8; limitations of, 38; student participation, 111–12
Cherry, Don, 381
Child-Friendly Municipalities, 251–3
childcare, 56, 98, 100–1; as employment, 25; to enable participation in planning, 345

Chinatown: Montreal, QC, 351; Toronto, ON, 140; Winnipeg, MB, 339
Christopherson, David, 401
citadelization, 308
citizen engagement: controlling variables, 81; degrees of "ownership," 347; in health planning, 105, 120–2; non-traditional strategies, 60; recruiting participants, 81; *see also* collaborative planning, participatory planning
citizen opposition, 396–404
City Beautiful ideals, 399
City Spaces Consulting Ltd, 211
Clean Air Act, 407
Clean Air Foundation, 409
climate change, 54, 143–4, 155; adaptation planning, 84, 94, 144, 155–66; mitigation, 156, 164; social vulnerability, 84–94
Climate Change Action Plan, 86
Climate Change Conference, 156
climatic uncertainty, 157–8
Coaffee, Jon, 308
collaborative planning, 28, 30–1, 123, 144; and digital participation, 265; as area of research, 162–3, 217; natural resources, 169, 172; *see also* participatory planning, citizen engagement
Collège Boréal, 330; Métis and Aboriginal Centre, 329
colonization, 6, 79
Columbia Basin Trust, 158
Commission for Architecture and the Built Environment, 331
Communauto, 409
communicative planning, 29–31, 162, 217
Community Amenity Contributions, 22, 227
community development, 5, 21, 48–50, 189, 278; and culture, 65, 233; assets-based, 80; funding for, 340
community dinners, 279
community gardens, 106–13, 279
community-based research, 88–9, 119–20
commuting, 221, 362, 371–2, 402; for school, 254
concentric model of cities, 32
condominiums; conversion, 22, 182, 357; development, 35, 177, 193, 195, 326, 357, 388; model for retail complex, 137, 139; legislation, 19, 177
Congress for New Urbanism, 37
Conservation Authorities Act, 13
Consolidated Goldwin Ventures, 150–1
Constitution Act, 1867, 8, 12, 143
conversation, 78–9
cooperative housing, 19, 22; energy-efficient retrofits, 204–5
coordinative planning, 197
Corner Brook, NL, 11
corporate suburbs, 14
Council of Women, 8–9, 49
covenants, 12, 178, 213, 279
creative class, 65, 233
creative economy, 64, 73
cross-case study, as research method, 4, 123
crowdsourcing, 223, 240–51, 263, 266
Crown consultation, 147, 149, 152–4
Crown land, 20, 143, 209; mineral rights, 147
Cube Lofts, 323
cultural exchange visits, 79–80
cultural landscape, 146
cultural mapping, 234
cultural planning, 49, 72, 64–73, 231–9; data collection, 72; and economic growth, 65, 352; and tourism, 65, 70; and quality of life, 65, 67
cultural policy, 71, 233
Culture Days, 69
CultureLink, 109
cynicism, addressing, 230

Daniels Corporation, 189–90, 194
Daoust Lestage, 355
Darke, Peta, 119
data privacy, 246
Davidoff, Paul, 28, 36
Debogorski Diamond, 146
Debogorski, Alex, 151
Delgamuukw v. British Columbia, 72
Demers, Clément, 352, 355
demographic projection, 362, 364
demolition replacements, 132
demonstration project, *see* pilot project
density targets, 128
Department of Regional Economic Expansion, 16
Deschênes, Pierre, 352–3
design guidelines, 129, 292
development levies, 4, 210–11, 227, 336
Dewey, John, 29
Dialogues Project, 49, 75–82
diamond mining, 146–7, 151
Diggable Communities Collaborative, 107
digital participation, 246, 249, 218, 229, 265; *see also* social media
DiIanni, Larry, 402
Diller Scofido + Renfro, 390
Dillon Consulting, 310
dissemination area, 86, 88–9
Dominion Housing Act, 12, 176–7
Don Mills, ON, 14
downloading of responsibilities, 54, 62, 126
Drapeau, Jean, 351
Drybones Bay, NWT, 145–54
DTAH, 390
Duany, Andres, 21, 37

Economist, The, 190
Edmonton, AB, 7, 14
Elford, BC, 155–66
Ellis, Cliff, 26, 38
emergency management planning, 84, 91, 94
emission reduction credits, 407
energy consumption, 19, 130, 199, 206
energy-efficient retrofits, 196–206
Environics Research Group, 411
Environment Canada, 409
environmental activism, 400–2
environmental analysis, 291, 294
environmental impact assessments, 19, 27, 145–54, 389, 401; legislation, 18
environmental justice, 31–2, 379–80
environmental management, 8, 23, 143
environmental planning, 117, 144
environmental protection, 18, 21, 27, 127, 241–51
environmental renewal, 330
ERA Architects, 315, 318
Esri ArcMap, 88
ethnic shopping malls, 134–42
ethnoburbs, 136
exclusionary zoning, 308
expedited approvals process, 319, 355
experience economy, 276
experiential learning, 76, 79–80
extreme weather events, 143, 156

Facebook, 68, 248, 256, 262–4, 319
Fainstein, Susan, 27, 29–31, 101, 217
Falconbridge Ltd, 330
family-friendly practices, 56
farmers' markets, 272, 336
feminism, 26, 28
Ferrigan, Jason, 336
festival marketplaces, 272–6
festivals, 67, 69, 352, 355, 357
Filion, Pierre, 26
First Capital City Development Corporation, 273

First Nations, *see* Aboriginal people
first responders, 92
First World War, 11; *see also* World War I
first-ring suburbs, 125, 375
flanerie, 275
Flickr, 248
flooding, 156, 161–3; storm surge, 88, 91, 94
Florida, Richard, 39, 64–5, 70, 233
Flyvbjerg, Bent, 4, 26, 31, 39, 44, 124, 217, 219, 232, 240
food security, 105; and gardens, 107–8
food waste collection, 279
Ford, Rob, 378, 380–3
Fordist era, 33
Forester, John, 30, 217
Fort Albert, BC, 6
Foutopoulos, Helen, 352
Francofolies festival, 352
Franco-Ontarian culture, 329
Friedan, Betty, 35, 104
Friedmann, John, 26, 30
Friends of Red Hill Valley, 396, 402, 404

Garden City principles, 11, 283
gardens, 105–13, 268, 278–9
Gardiner Expressway, 325, 384–95; public consultation, 391
Gardiner, Frederick Goldwin, 386
Gaspé, QC, 16
gated communities, 308
Gauthier, Stéphane, 335, 337
generalizability, 123
gentrification, 38, 337, 356–7, 375
Geographic Information System, 88, 94
global climate model, 158
globalization, 33, 54
GO train system, 140, 373
Google Maps, 262
governance, 8, 40; Aboriginal government-to-government approach, 32, 168–74; allocation of powers, 19–20, 51, 53, 225; and natural resources, 32, 143; downloading, 54, 62, 126; local, 29, 53, 225, 232, 235, 243; network, 197; processes, 34; transportation, 371
Grant McNeil Place, 201–6
Grant, Jill, 6, 26, 29, 37
Great Depression, 12
Greater Toronto Area (GTA), 134–5; transportation issues, 374–83
Gréber, Jacques, 128
Greek agora, 274–5
Green Climate Fund, 156
greenbelt plan, 367–8
Greenbelt, National Capital, 128
Greenbrook, 200
greenfield development, 39, 126–8, 177, 224, 364–8
greenhouse gas emission reduction, 405–12
ground lease, 210, 214
groundwater absorption coefficient, 299
growth management, 23, 127, 180, 361–8, 391
Gutierrez, Jose, 391

Habermas, Jürgen, 29
habitat preservation, 143
Habitations Jeanne-Mance, 351, 353
Haida Nation v. British Columbia, 172–4
"halal flats," 190
Halifax, NS, 9, 24; Africville, 32, 36
Hamilton Spectator, 402
Hamilton, George, 398
Hamilton, ON, 233, 397–9; cultural plan, 231–9; Red Hill Valley Parkway, 396–404
Harris, Mike, 131
Harvey, David, 31
Haussmann, Georges-Eugène, 268, 307
Healey, Patsy, 29–30, 34,162, 217, 237
health planning, 117–24
Healthy Cities Movement, 21
Héritage Montréal, 352
heritage properties, 69, 284, 310; demolition of, 330
heritage studies, 150
homeless population, 52, 55, 101, 198, 358
homeownership, 12, 36, 176–7
hospice care, 120–3
housing, 8–9, 31, 101, 176–86; Aboriginal population, 17; continuum, 179, 183; inclusionary, 40, 214; industry, 13; market, 32, 185, 215, 366; suburban ideals, 13, 21; temporary, 191; for workers, 11, 48, 208–16; *see also* cooperative housing, public housing, rental housing, social housing
housing career model, 36
Housing First, 101
Housing Plans, 22, 183
housing policy, 12–13, 17, 19, 21, 36
Hoyt, Homer, 32
Hudson's Bay Company, 6
Hugh, Michael, 292

Ideas Fair, 223
Ideas Slam, 248
ideation tools, 241, 244–5
LGBT community, 351
immigrants; income inequality, 25; integration, 32, 55, 109; intercultural understanding, 75–82; language barrier, 113; participation in planning process, 32, 58, 72; Social Determinants of Health, 88; suburbanization, 56, 134–42
immigration, 14, 135–6; as source of growth, 24, 40; Canada compared to US, 326; geographic distribution, 136, 350–1
Immigration Act, 21
impact pathways, 161–2, 164
inclusionary housing, 40, 214
inclusionary zoning, 210–11
Inco Ltd, 330
incremental approach to planning, 382
Indigenous peoples, *see* Aboriginal people
Indigenous rights, *see* Aboriginal rights,
industrial parks, 33
Industrial Revolution, 350
industrialization, 7, 35, 104, 176
infill development, 129–30, 226, 315
intercultural understanding, 75–82
intergenerational issues, 77
International Transportation Forum, 407
interprovincial migration, 180–1
Inuit, *see* Aboriginal people

Jackson, Lloyd D., 398
Jacobs, Jane, 36–7, 64–5, 371, 387
James Bay, QC, 18
James Corner Field Operations, 390
Jersey barriers, 308, 312–13
Just for Laughs Festival, 352

Kapuskasing, ON, 11
Kelbaugh, Doug, 37
Kelowna, BC, 67–9
Kingston, ON, 69–70
kitchen design, cultural significance, 190
kitchen table meetings, 163
Kitchener, ON, 11; community gardens, 108–13; rapid transit, 373
Klein, Les, 391
Koetter Kim and Associates, 287–8
Koetter, Fred, 287
Kuwabara Payne McKenna Blumberg Architects, 390
Kwakwaka'wakw Nations, 172

Lachine Canal, Montreal, 21
Ladder of Citizen Participation, 31
Lagendijk, Arnoud, 122
Lakehouse Beach Residences, 322–3
land claims, 148, 150, 152, 154
landscape ecology, 294
laneway houses, 226
language; accommodating multiculturalism, 223–4; as barrier, 58, 249; bilingual communities, 72, 332, 334; legislation, 56; Social Determinants of Health, 88
Laurentian University, 330; Native Studies Program, 329; School of Architecture, 330, 335–6
Leadership in Energy and Environmental Design (LEED), 299
League for Social Reconstruction, 12
Le Corbusier, 36
levies on development, 4, 210–11, 227, 336
light rail, 374–83
London, UK; securitization, 308
lone-parent families, 96–102
Low Income Cut-off, 51, 98
Lunenberg, NS, 6
Lynch, Kevin, 28, 37, 252

McCallion, Hazel, 362
McGill University, 14
McGuinty, Dalton, 363
McHarg, Ian, 294
Mackenzie Valley Environmental Impact Review Board, 146, 148
Mackenzie Valley Land and Water Board, 149
McLean, Don, 402, 404
McQuesten, Thomas Baker, 399
Magnetic Hill, 71
Mahmoudi, Dillon, 246
Manuel, Patricia, 86
Marcuse, Peter, 308
Maritime Marshland Rehabilitation Administration, 12
Markham, ON, 136–42
Marsden, Blair, 293
masterplanning, 36, 282–9, 331
Medicine Hat, AB, 7
Meehan-Prins, Kristina, 293
methodology, *see* research methods
Métis, *see* Aboriginal people
Metro Toronto Housing Authority, 189
Metrolinx, 141, 368, 376, 379–80
Metro Vancouver, 54, 273
Mikisew decision, 154
Miller, David, 380, 383
Mill Woods, AB, 14
mineral rights, 145–54; free entry system, 147
mining, 146–7, 151
Ministry of Natural Resources, 143
Ministry of Urban Affairs, 17
Mishra, Anjali, 354, 356
mobile apps, 262–3
modernism, 18, 27–8, 36, 177, 349
Moncton, NB, 70–2
Montreal International Jazz Festival, 352
Montreal Summit, 352
Montreal, QC, 6, 9, 12, 14; Faubourg Saint-Laurent, 350–1; Lachine Canal, 20; metro system, 17; Office de consultation publique, 357; Quartier des spectacles, 348–59; Quartier international, 349, 355; redevelopment, 325
Moose Jaw, SK, 7
Morgenthaler, Alex, 354
Multiculturalism Act, 21
multiple-case study, 122–3
multi-stakeholder planning, 171
Murray, Glen, 64–5

Nanwakolas Council, 169–70, 172–4
National Building Code, 13
National Capital Commission (NCC), 309, 311–13
National Capital Greenbelt, 128
National Housing Act, 13, 17, 36, 176–7
national parks, 9, 18, 213
National Planning Act, 14
Native Friendship Centres, 329, 332, 335–6
natural capital, 296
natural resource management, 32, 48, 143–5, 169; co-management, 24, 148
Natural Resources Canada, 86
needs assessment, 183, 216
Neighbourhood Improvement Program, 18–19
neo-liberal welfare reforms, 98
Neptis Foundation, 361, 367
network governance, 197
new urbanism, 27, 37–9, 189
New Westminster, BC, 7, 272–81
New York, NY, 9; Central Park, 106; community gardens, 110
Newcomers Guide to First Nations and Aboriginal Communities, 50, 78
nimbyism, 22, 129, 213, 227
Nomade Architecture, 353
Northwest Territories, resource management, 148

Obama, Barack, 305
obesity, 38, 105, 255, 371
Occupy Wall Street, 268
Office of Metropolitan Architecture, 390
Official Community Plans, 4, 8, 54, 225, 184; Whistler, BC, 209
Official Plans, 361–2; amendments, 366; Ottawa, ON, 128, 313; Hamilton-Wentworth, ON, 400; secondary plans, 285; Sudbury, ON, 330; Toronto, ON, 285, 316, 320
Okanagan Valley, BC, 67
Olds, AB, 121–2
Olmstead, Frederick Law, 9, 106
Olympic Games, Vancouver 2010, 55
online tools: citizen engagement, 218, 229, 240–51; ideation tool, 241, 244–5; *see also* digital participation, social media
Ontario Building Code Act, 129
Ontario Home Builders' Association, 366
Ontario Municipal Board, 128–9, 225, 363
Ontario regional planning, 361–8
Ontario Trillium Foundation, 113
open data, 219, 262–3
orthomaps, 111
Ottawa Citizen, 130
Ottawa, ON, 9; Byward Market, 273, 310; light rail, 373; Major's Hill Park, 310; Regional Municipality of Ottawa-Carleton, 127; urban intensification, 125–32; US Embassy, 304–14; US presidential visit, 305

Pacific Climate Impacts Consortium, 158
Pacific Mall, 136–9
Pallium educational initiative, 123
Paris, France, 268, 307
parks: industrial, 33; national, 9, 18, 213; public, 9, 11, 399; skate, 251, 255–7
Parks and Playgrounds movement, 9
Parks Canada, 143
participatory planning, 22, 24, 107, 217–19, 382; as area of research, 217; barriers to, 218; CentrePlan example, 339–47; CityPlan example, 220–30; engagement plan, 244; exclusion, 249; follow-up, 248; reaching diverse groups, 223, 242, 344–5; and sense of ownership, 339–47; spectrum of

options, 221–2; student involvement, 107, 111–12, 258; youth engagement, 77, 251–9, 296; *see also* collaborative planning, citizen engagement
Pathways to Education, 190
Pearson International Airport, 373
Pecha Kucha, 247
pedestrian infrastructure, 14, 287, 372
pedestrian-oriented planning, 21, 35–8, 132, 140, 315
Pei, I.M., 19
performance standards, 315–16, 318–23
Petitcodiac River, 71
Phillips, Art, 221
pilot project, 3, 75–7
placelessness, 282, 284
placemaking, 284, 289
Places to Grow Act, 37, 128, 361–8; and transportation planning, 380
planning: Acts, 11, 128; as formal discipline, 2–3, 5–6, 14, 17, 25, 117; history of, 2–3, 6–25, 104; process, 37, 217, 221; theory, 26–39
Planning Institute of British Columbia, 156
Plater-Zyberg, Elizabeth, 21, 37
policy transfer, 4–5
politicization; of development, 130, 402; of transportation, 374–83, 385
Portage la Prairie, MB, 6
positivism, 26–9
postmodernism, 18–23, 27–8
poverty, 51, 96–102, 337, 375
Prairie Farm Rehabilitation Act, 12
precinct plans, 285–7
preliminary screenings, 149
Prince Albert Mission, SK, 6
Prince Charles, 37
privacy: as middle-class privilege, 99–100; of data, 246
private spaces, 271–81
privatization of public space, 269, 275–6, 308, 358
Project for Public Space, 279
property rights, 8, 12, 40, 283, 289
property tax, 21, 24, 54, 227, 358
prostitution, 101, 351, 358
public engagement, *see* citizen engagement,
public funding: accountability, 73; arts and culture, 68, 73
public health, 8, 13, 39, 104–6
public housing, 3, 14, 21, 105, 177; design, 36; Habitations Jeanne-Mance, 351, 353; in Aboriginal communities, 17; Regent Park redevelopment, 187–95; replacing slums, 14, 17, 27; right of return, 188–9, 191–3; *see also* social housing
public markets, 271–81
public meeting, legal requirements, 29, 264
public parks, 9, 11, 399
public spaces, 271–81; as extension of domestic space, 99; degrees of publicness, 276–7; privatization of, 269, 275–6, 308, 358; securitization of, 305–6; surveillance of, 101, 305–8
public transportation, 21, 370–2; funding for, 371; incentives to use, 99, 410–11; mobile apps, 263; ridership rates, 22, 40, 140, 375; and social justice, 379–80, 382; Taxi 12–17, 251–2, 254–7; transit-oriented development, 37–8, 132, 371; and youth, 219, 251–2, 254–7; *see also* transit, transportation
public–private partnerships, 21, 40, 189, 269
Punter, John, 4
Putnam, Robert, 39

Quadrangle Architects Ltd, 315, 318, 391
Quebec City, QC, 6

rail, 9–11, 286, 330, 370; abandoned railways, 290–1, 297; air-rail link, 373; Canadian Pacific Railway, 6–7, 273, 330, 370; light rail, 372, 374, 376, 379, 383; regional, 371
Rapaport, Eric, 86
Rapid Transit, 128, 274, 371, 373, 376–7
rational comprehensive model, 26, 162, 382
rationality, 26–9, 152; strategic, 197
reconciliation, 78, 145, 154
Red Deer County, AB, 291–2
redistributive era, 34
red-light district, 351
Regent Park, Toronto, 187–95
Regina, SK, 7
regional planning, 2, 13, 16–17, 21, 132, 361–8
regional science, 19
regulatory era, 34
relationship-building, 34, 50, 79–81, 337
Relative Happiness Index, 253
Renewable Energy Initiative, 178, 198
rent banks, 101
Rent Geared to Income (RGI), 193, 200
rental housing, 19, 177, 182–3; as part of new development, 288; preservation of, 22, 177; price of, 99, 182, 185; vacancy rate, 182
resale restrictions, 213
research ethics board, 111
research methods, 18, 28; case study as, 4–5, 116–24; cross-case study, 4, 123; multiple-case study, 122–3; townscape analysis, 291, 293–4
Residential Rehabilitation Assistance Program, 19
resort municipalities, 208–16
resource-based economy, 170
Retire your Ride program, 408–9
Richardson, Bill, 248
Richmond, BC, 51–62
right of return, 188–9, 191–3
"right to the city," 379–80
rights-of-way, 288, 319–23, 394
risk; assessment, 158, 160, 163; communication, 157–8, 161–5; discourse of, 98; management, 157; tolerance, 73, 163, 261
River Market, 276–80
Robertson, Gregor, 242
Rockwood Park, 9
Rothert, Gene, 112
Rouse, James, 275
Royal Canadian Mounted Police (RCMP), 309, 311
rural redevelopment, 326

Sainte-Julie, QC, 251–9
Saint John, NB, 6, 9
St John's, NL, 19, 323
San Francisco, CA: Ghirardelli Square, 272–3
Sandalack, Beverly A., 293
Sandercock, Leonie, 104, 268
Saskatchewan: affordable housing, 179–86; economy, 180; population growth, 180; social housing, 182
Saskatoon, SK, 7
Scarborough, ON, 140
Seattle, WA, 110
Second World War, 12; *see also* World War II
secondary suites, 226
securitization of urban space, 305–6
security planning, 313
segregation, 21, 32, 326
Seltzer, Ethan, 246
seniors, 55, 72, 77; care for, 105, 121; gardening, 109–10; housing needs,

190, 213; Social Determinants of Health, 88; social vulnerability, 84, 91
seniors centres, 120–3
Sennett, Richard, 104
sense of place, 109, 283, 292, 294, 297
service-dependent neighbourhoods, 99
Sewell, John, 189
sex work,. *see* prostitution
Shanks, Graeme, 119
shopping malls: condominium ownership model, 137, 139; ethnic, 134–42; parking, 138; as public–private spaces, 269, 276, 280; replacing main streets, 273, 275; residential component, 140; as security risk, 311
sidewalks: age-friendly, 55; lack of, 296; sun penetration, 318–9, 321; width, 319, 321–2
Simard, Alain, 352
skate parks, 251, 255–7
skating rinks, 140
skyscrapers, 19
SkyTrain, 274
slums, 8–9, 176; clearance of, 14, 17, 177–8, 188; and urban renewal, 27, 97
Smart Growth, 37–40, 126–31, 271, 327, 362–3
Smith, Dorothy, 119
social capital, 39, 58, 109, 232, 238–9
Social Determinants of Health, 87–9
social exclusion, 79, 374, 379–80
social homogenization, 358
social housing, 12, 21, 185; as part of new development, 287; Baker House, 201–6; energy-efficient retrofits, 196–206; for resort workers, 209–10; Grant McNeil Place, 201–6; Greenbrook development, 200; resident participation, 190–1; Toronto, 190; Villa Otthon, 201–6; *see also* public housing
social justice, 39, 96–9, 255; and public transportation, 379–80, 382; and spatial planning, 101
social media, 241, 260–6; and anonymity, 265; and public consultation, 58, 60, 68, 248, 250; for information dissemination, 264, 319; for information gathering, 264; *see also* digital participation, online tools
social mix, 189, 194
social planning, 23, 48–50, 56, 96–102
social urbanism, 379–80
social vulnerability, 84–94
social welfare delivery model, 96
Société d'habitation du Québec, 21
socio-economic polarization, 375
soil classification, 295
Soja, Edward, 31
solar panels, 199
Sørensen, Eva, 197
Sorkin, Michael, 308
sovereignty, 152
Spadina Expressway, opposition to, 16, 387, 397
Sparrow decision, 172
spatial justice, 31–2
spatial mismatch, 379–80
Special Planning Program, 355
Splendid China Tower, 139–40
sprawl, 38, 127, 283, 330; combating, 22–3, 361, 363–4
Stake, Robert, 123–4
standard deviation, 88
Stelter, Gilbert, 6
Stone, Edward Durrell, 19
storm surge, 86, 88, 91, 94
storytelling, 76–9
Stratford, ON, 9
street trees, 296, 299, 303, 321
streetcar suburbs, 9, 35, 41
streetcars, 9, 125, 370
strengths-based approach, 80
suburbs, 32–9, 375; car dependency, 375; commercial space in , 25, 134–42, 272, 275; corporate, 14; first-ring, 125, 375; growth of, 14, 23–4; housing ideals, 13, 21; and immigrants, 56, 134–42; and isolation, 37; as lifestyle, 25, 177; political polarization, 381; and poverty, 375; streetcar, 9, 35, 41; transportation issues, 255, 374–83
Success by Six, 97
Sudbury, ON, 328–38
Summit Action Fund, 184
sunlight penetration; and gardens, 110; and sidewalk, 318–9, 321
surveillance, 101, 305–8
sustainability: and culture, 69, 73; and economic development, 169, 343; environmental, 196–206, 221–2, 241–51, 294, 299; and food production, 107, 109; as goal of planning, 25, 37, 40, 126–7, 218, 269; and growth, 330; principles of, 39, 57, 32, 226; social, 56–7; and transportation, 35, 40, 373–83; and urban intensification, 130–2
Suzuki, David, 241
Swift Current, SK, 7
SWOT analysis, 58

Tait-Peacock, Karen, 336
Talen, Emily, 38
Taxi 12–17, 251–2, 254–7
technical optimism, 33
Témiskaming, QC, 11
tendering process, 200, 355
terrorism, 304–5, 308
tidal bore, 71
Together 4 Health, 110
tokenism, 31, 235, 237, 259, 337
Torfing, Jacob, 197
Torgan Group, 138
Toronto Community Housing Corporation (TCHC), 188–93, 198–200
Toronto Waterfront Revitalization Corporation, 284–5
Toronto, ON, 7, 11, 19; amalgamation, 374–5; community garden, 109; Community Housing Corporation, 188; East Bayfront, 286–7; Gardiner Expressway, 325, 384–95; light rail system, 374–83; mid-rise buildings, 315–23; Parkdale, 325; Regent Park, 14, 187–95; social housing, 198, 200–6; Tall Buildings Guidelines, 320; urban intensification, 315–23; waterfront development, 282–9, 388
tourism, 71, 178, 269, 325, 357; cultural, 65, 69–70, 358
Town Planning Institute of Canada, 6, 11, 283
townscape analysis, 291, 293–4
traditional main street, 130
transit, 368; *see also* public transportation
Transit City plan, 374–83
transition theory, 41
transit-oriented development, 37, 379, 411
transportation, 370–3; issues in suburbs, 255, 374–83; mode share, 40, 391; mode shift, 405–12; modelling, 392; politicization of, 374–83, 385; sustainable, 35, 40, 373–83; youth, 251, 254–5; *see also* public transportation,
Transportation Association of Canada, 13
transportation planning, 38, 140–1, 371
travel demand management, 405–12
treaties: historical, 152; modern, 152, 172
Tremblay, Gérald, 352–3
Trois-Rivières, QC, 6

Tsawwassen First Nation, 24
Twitter, 248, 250, 262–3

United Nations, 22, 177, 330; Convention on the Rights of the Child, 252; Population Fund, 54
United States Embassy in Ottawa, 304–14
University of British Columbia, 4, 13, 160, 247
University of Calgary, Urban Lab, 290, 293, 303
University of Manitoba, 14
University of Waterloo, 107
Unocal Oil Company, 407
urban design, 130, 268–71
urban ecology, 294
urban fortress, 308
urban growth boundaries, 361, 365
urban intensification, 38, 364, 270; as-of-right, 129; and environmental protection, 127; opposition to, 129–31; Ottawa, ON, 125–32; Toronto, ON, 315–23
urbanization, 7–8, 54
Urban Land Institute, 384
Urban Marketing Collaborative, 315, 318
urban morphology, 294–5
urban redevelopment, 22, 325–7, 337
urban renewal, 14, 17, 22, 36; and culture, 65; failures of, 97; and markets, 272; Montreal, QC, 14, 349, 351–2; opposition to, 16, 27; Sudbury, ON, 328–38
Urban Strategies, 330
Urban Task Force, 331
Uribe, Francisco Alaniz, 270, 293

value-based judgments, 153
Vancouver, BC, 9, 16, 19, 220–1; 2010 Olympic Games, 55; bicycle lanes, 40; CityPlan, 220–30; Expo 1986, 22, 274; Granville Island, 273; intercultural understanding, 75–82; light rail, 373; lone-parent families, 96–102; online citizen engagement, 241–51; social housing, 199–206; sustainability, 241–51; Year of Reconciliation, 78
van der Rohe, Mies, 19
Varro, Krisztina, 122
Vaux, Calvert, 9
vehicle recycling, 406–7
Victorian Order of Nurses, 92–3
video projections, 354
Villa Otthon, 201–6
visually impaired people, 112–13
Vulnerable Persons Registry, 92–3

walkability, 38, 299
Walks, Alan, 382
Wanapitei First Nation, 329
"war on cars," 380–1
Waterfront Toronto, 284, 372, 384, 388–9, 392
Waterloo, ON; community gardens, 106–13; rapid transit, 373
water supply, 9, 163
Web 2.0, 260–6; *see also* social media
welfare reforms, 98
West 8, 390
Westminster Quay Public Market, 272–81
Whistler Housing Authority, 211–2
Whistler Valley Housing Society, 210, 211
Whistler, BC, 208–16
whitebelt, 366–7
Whitefish Lake First Nation, 329
windfall profits, 213
Winnipeg, MB, 6–7, 9, 13, 339–47
World War I, before, 10, 11; *see also* First World War
World War II, after, 14, 25, 96, 97, 330; *see also* Second World War
Woodhouse, Nathalie, 293
Woodsworth, J.S., 8
workfare, 98
Wright, Frank Lloyd, 36

Yarmouth, NS, 84–94
Yellowknives Dene First Nation, 145–54
Yiftachel, Oren, 27, 268
Yin, Robert K., 4
Young, Iris Marion, 31
youth engagement, 72, 77, 223, 251–9, 296, 332; CAMF guidelines, 258
YouTube, 248, 262–3

Zoldak, Natalia, 293
zoning, 11–13, 26, 282–4, 292; amendments and variances, 129–30, 191, 195, 264; as-of-right, 319; bylaws, 11–13, 104, 177; changes in , 221, 226, 318; design issues, 288–9; exclusionary, 270, 308; inclusionary, 210; and natural resources, 172; studies, 130; and urban intensification, 125, 128–9, 131, 226–9
Zumundo Consultants, 160